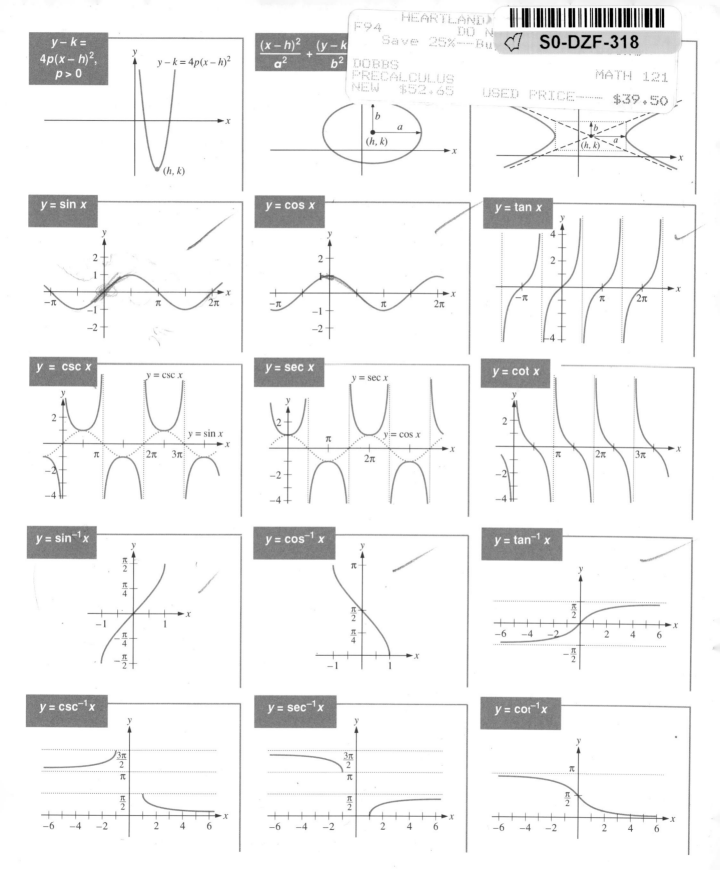

WARNING!

THIS BOOK SOCKS AT EXPLAINING THE PRECEDURES FOR THE SOLUTION

YOU SHOULD STUDY YOUR BUTT OFF TO GET AT LEAST A "B" IN MATH 096 I GOT A B+ FOR THE CLASS AND AM HAVING PROBLEMS RIGHT NOW — THATS HOW HARD IT IS!

PRECALCULUS

David E. Dobbs

*University of Tennessee—
at Knoxville*

John C. Peterson

*Chattanooga State Technical
Community College*

TOM LYON
Sales Representative

District Office
1419 124th Avenue
Hopkins, MI 49328
616-793-7674

WM. C. BROWN PUBLISHERS
—— A Division of Wm. C. Brown Communications, Inc. ——

 Wm. C. Brown Publishers

Dubuque, Iowa • Melbourne, Australia • Oxford, England

Book Team

Editor *Dwala A. Canon*
Developmental Editor *Theresa Grutz*
Publishing Services Coordinator (Production) *Julie A. Kennedy*
Permissions Editor *Vicki Krug*
Publishing Services Coordinator (Design) *Barbara J. Hodgson*

Wm. C. Brown Publishers
A Division of Wm. C. Brown Communications, Inc.

Vice President and General Manager *George Bergquist*
National Sales Manager *Vincent R. Di Blasi*
Assistant Vice President, Editor-in-Chief *Edward G. Jaffe*
Marketing Manager *Elizabeth Robbins*
Advertising Manager *Amy Schmitz*

Publishing Services Manager *Karen J. Slaght*
Permissions/Records Manager *Connie Allendorf*
Managing Editor, Production *Colleen A. Yonda*
Manager of Visuals and Design *Faye M. Schilling*

Wm. C. Brown Communications, Inc.

Chairman Emeritus *Wm. C. Brown*
Chairman and Chief Executive Officer *Mark C. Falb*
President and Chief Operating Officer *G. Franklin Lewis*
Corporate Vice President, Operations *Beverly Kolz*
Corporate Vice President, President of WCB Manufacturing *Roger Meyer*

Cover photo by Michael Stuckey/Comstock Inc.
Copyedited by Lisa Jacobson
Production by Lifland et al. Bookmakers
Interior and Cover Design by Karen Mason
Illustrations by LM Graphics

Copyright © 1993 by David E. Dobbs and John C. Peterson

Library of Congress Catalog Card Number: 92–72534

ISBN 0–697–16235–4

No part of this publication may be reproduced, stored in a retrieval system, or transmitted, in any form or by any means, electronic, mechanical, photocopying, recording, or otherwise, without the prior written permission of the publisher.

Printed in the United States of America by Wm. C. Brown Communications, Inc., 2460 Kerper Boulevard, Dubuque, IA 52001

10 9 8 7 6 5 4 3 2 1

Our deepest thanks go to our families, especially our wives, Elaine Dobbs and Marla Peterson, to whom this book is dedicated.

Contents

To the Student viii
To the Instructor x

CHAPTER 0   *Review Material*   1

CHAPTER 1   *Fundamental Algebraic and Geometric Concepts*   21

1.1 Algebraic Inequalities 21
1.2 Rectangular Coordinates, Graphs, and Slope 31
1.3 Equations of Lines 41
1.4 Circles, Parabolas, and Tangents 54
1.5 Complex Numbers 65
1.6 Chapter Summary and Review Exercises 75

CHAPTER 2   *Functions and Their Graphs*   78

2.1 Functions 78
2.2 Symmetry and Transformations 97
2.3 Linear and Quadratic Functions 110
2.4 New Functions from Old 120
2.5 Chapter Summary and Review Exercises 132

CHAPTER 3   *Polynomial and Rational Functions*   135

3.1 Polynomial Functions 135
3.2 Synthetic Division and Nested Multiplication 143
3.3 Graphing Polynomial Functions 149
3.4 Roots of a Complex Polynomial 163
3.5 Counting and Bounding Real Roots; High-Degree Polynomial Inequalities 175
3.6 Rational Functions and Their Graphs 185

Contents

*3.7 Partial Fractions   197
3.8 Chapter Summary and Review Exercises   205

CHAPTER 4 Exponential and Logarithmic Functions   208

4.1 Exponential Functions   208
4.2 Logarithmic Functions   218
4.3 Properties of Logarithms   225
4.4 The Exponential Function e^x; Applications Involving Exponential Growth and Decay   230
4.5 Exponential and Logarithmic Equations and Inequalities   236
4.6 Chapter Summary and Review Exercises   246

CHAPTER 5 Trigonometric Functions and Their Graphs   248

5.1 Angles and Their Measures   248
5.2 Defining the Trigonometric Functions   258
5.3 Circular Functions   267
5.4 Values of the Trigonometric Functions   273
5.5 Graphs of the Sine and Cosine Functions   282
5.6 Graphs of the Other Trigonometric Functions   294
5.7 The Inverse Trigonometric Functions   302
5.8 Trigonometric Equations and Inequalities   311
5.9 Chapter Summary and Review Exercises   319

CHAPTER 6 Analytic Trigonometry   323

6.1 Basic Identities   323
6.2 The Addition and Subtraction Identities   330
6.3 The Double- and Half-Angle Identities   338
6.4 Graphs of Combinations of Sine and Cosine Curves   347
*6.5 Sum and Product Identities   358
6.6 Polar Form of a Complex Number   363
6.7 Powers and Roots of Complex Numbers   371
6.8 Chapter Summary and Review Exercises   379

*Optional section

CHAPTER 7 — Applications of Trigonometry 381

- 7.1 Right Triangles 381
- 7.2 The Law of Sines and the Ambiguous Case 387
- 7.3 The Law of Cosines 397
- 7.4 Vectors in the Plane 403
- 7.5 Dot Product 416
- *7.6 Harmonic Motion 428
- 7.7 Chapter Summary and Review Exercises 436

CHAPTER 8 — Conic Sections and the General Second-Degree Equation in Two Variables 441

- 8.1 Parabolas 442
- 8.2 Ellipses 454
- 8.3 Hyperbolas 465
- 8.4 Rotation of Axes and the General Quadratic Equation in x and y 476
- 8.5 Chapter Summary and Review Exercises 488

CHAPTER 9 — Systems of Linear Equations and Linear Inequalities 491

- 9.1 Systems of Linear Equations 491
- *9.2 Matrices and Their Use in Solving Systems of Linear Equations 507
- *9.3 Determinants and Cramer's Rule 522
- 9.4 Linear Inequalities and Linear Programming 538
- 9.5 Chapter Summary and Review Exercises 553

CHAPTER 10 — Additional Topics in Geometry and Algebra 558

- 10.1 Polar Coordinates 558
- 10.2 Curves in Polar Coordinates 573
- *10.3 Intersection of Polar Graphs 584
- 10.4 Parametric Equations 588
- 10.5 Mathematical Induction, Summation Notation, and Sequences 595
- 10.6 Binomial Theorem 609
- 10.7 Chapter Summary and Review Exercises 620

*Optional section

APPENDICES

Tables 623

I Powers of e 623
II Common Logarithms   624
III Natural Logarithms   626
IV Values of Trigonometric Functions—Degree Measure   627
V Values of Trigonometric Functions—Radian Measure   639

Answers to All Exercises in Chapter 0   643
Answers to Odd-Numbered Exercises in Chapters 1–10   645
Index   694

To the Student

This precalculus textbook has been written for you. As you study it, you will see certain topics being emphasized. These include functions, graphing, inequalities, trigonometry, and analytic geometry. The material in this book will extend what you already know and will also give a taste of some of the topics that await you in calculus. After completing this book, you should have the background to succeed in the calculus sequence.

In writing this book, we assumed that you have completed two years of algebra and have a good background in geometry. We understand that your memory of this background material may need to be refreshed. Chapters 0 and 1 are for this purpose.

Chapter 0 gives a very condensed, quick overview of algebra. Most of the material in it should be review. Your instructor may cover chapter 0 in class or assign it as reading. In the latter case, you may want to begin by working the exercises at the end of chapter 0, checking your work with the answers in the back of the book and studying sections corresponding to exercises where you had difficulty.

We do not expect you to be as familiar with chapter 1. While much of this chapter's material is also of a review nature, it also lays the foundation for the topics mentioned above. Many instructors will cover much of chapter 1 in class.

Chapter 2 introduces the notion of a "function." This unifying concept is fundamental for the rest of the book. If you are interested now in the specific contents of chapters 2–10, turn to "To the Instructor."

This book has been written with an experience-oriented approach to new ideas. Often, a concept is introduced through examples. Then the new idea is defined, it is related to earlier material, and examples are given to further clarify the new idea. The text contains 487 worked examples. As you read these, be sure to keep pencil and paper handy to check the calculations: mathematics is not a spectator sport! The exercises give you an opportunity to practice and extend what you've learned. Working through the examples and exercises is of primary importance if you want to get the most out of this book.

The first exercises in each section should help you determine if you have understood that section's concepts. If you have difficulty with these initial problems, read the worked examples in the text, which proceed progressively from the routine to the challenging. You can become proficient at all the routine procedures in any non-review section by working either all the odd-numbered exercises or all the even-numbered exercises for that section.

The exercise sets are designed not only to complement the text material but also to supplement it. While many exercises provide an in-depth study of concepts already introduced, other exercises introduce and analyze related concepts. Each exercise that is referred to (in the text or in another exercise) is

To the Student

"flagged" with the symbol ◆. You should at least read any flagged exercise because it is associated with an important theme. In fact, you may wish to skim all the exercises, but be warned: *each* exercise set contains problems to challenge the best of students.

Each chapter ends with a review section, containing a checklist and an exercise set. The checklist gives an alphabetic list of the chapter's major concepts. This should help you check your recall of these topics without the help that would automatically result if topics were listed in the order in which they were introduced in the text. Similarly, each review exercise set gives you a chance to confront a chapter's topics without specific reference to the sections in which they were introduced. Some review exercises also give you an opportunity to integrate a chapter's topics with those from earlier chapters. Be sure to consider *all* the exercises in each review section (not just the odd-numbered ones). By working all these exercises, you will confront all the skills in the chapter.

Answers to the odd-numbered exercises for chapters 1–10 may be found in the back of the book. You may wish to purchase the *Student's Solution Guide*. This contains detailed solutions for all the odd-numbered exercises and answers to all the even-numbered exercises.

We assume that most of you have access to a calculator. No exercises have been marked as "calculator exercises" since we feel that a calculator can be used whenever it seems appropriate. Instructions on the use of calculators are included whenever we introduce a new concept whose study involves a new calculator key. It is suggested that you have a calculator that uses either Reverse Polish Notation (better known as RPN) logic or algebraic logic.

We have used a Texas Instruments TI-55 III calculator to illustrate algebraic logic, with an occasional reference to a Radio Shack EC-4014. A Hewlett-Packard HP-15C was used to illustrate RPN logic. A Texas Instruments *TI-81* and a Casio *fx-7700G* were used to illustrate graphing with a calculator. You should always consult the owner's manual for your particular calculator. Further information about graphing calculators appears in "To the Instructor."

Finally, we mention that the text includes some instructions for using a computer to evaluate trigonometric and inverse trigonometric functions. This was done so that you can derive certain defined BASIC functions not included in the BASIC computer programming language.

This preface was intended to let you know how to use this book. We would be glad to hear from you after you have done so. Good luck!

D.E.D.
J.C.P.

To the Instructor

In this preface, we discuss some of our writing methods and some other noteworthy features of this text, a chapter-by-chapter summary of the book, possible syllabi, and available supplemental materials. For brevity, we assume that you have already read "To the Student."

Writing Methods and Noteworthy Features

We use three general methods to help give students a firm foundation for calculus. First, we use a "cyclic approach," introducing many important ideas early and re-examining them later in relevant situations. Second, the exercise sets are unusually rich, not only providing drill on the relevant skills but also providing applications to earlier material, glimpses of later material, and abundant opportunities for syllabus enrichment and discovery activities. Third, we make use of the numerous examples to develop skills and introduce connections between topics. After discussing these three general methods, we will comment on our approach toward graphing and functions, graphing calculators, trigonometry, precalculus teaching problems, and preparation for calculus.

■ **The Cyclic Approach**—One particular example of the cyclic approach is the *recurring theme* of inequalities. Inequalities are reviewed in chapter 0 and solved using sign charts (section 1.1). Sign charts are used repeatedly (cycled) to provide a uniform method for solving inequalities. Their use extends into absolute value and radical inequalities (section 1.1), higher-degree polynomial inequalities (section 3.5), rational function inequalities (section 3.6), exponential and logarithmic inequalities (section 4.5), trigonometric inequalities (section 5.8), and systems of inequalities (section 9.4).

With this background in inequalities, your students will be well prepared to apply differential calculus in situations where they must determine the intervals over which a first or second derivative has a given sign. The use of sign charts is just one intuitive way in which we prepare the student for his or her study of "continuity" in calculus. As this example suggests, our cyclic approach emphasizes the interplay between algebra and geometry.

Another feature of the cyclic approach is the use of *recurring figures,* such as figure 1.14 which reappears in chapter 7 during the proof of the Cosine Law.

Recurring results provide another means of cycling ideas. For example, the negative reciprocal result for perpendicular lines is proved via Pythagoras' Theorem in section 1.2, exercise 58, and is proved again via trigonometry in chapter 6 and via vectors in chapter 7. Another recurring result is the geometric triangle inequality. This is first proved in section 1.2; another proof of it using a trigonometric identity and properties of the cosine function is sketched in section 7.3, exercise 40. A third example of a recurring result concerns concurrent altitudes of a triangle. Not only is this in section 1.2, exercise 60, but we prove it again using vectors in section 7.5, example 7.38.

We believe the emphasis on recurring themes, recurring figures, and recurring results gives the book a direction and provides internal motivation for some of the later, deeper topics.

■ **Exercises**—*The 3469 exercises in the text's exercise sets are a major vehicle of the cyclic approach.* Each exercise set tests all the routine procedures *and* contains problems to challenge the strongest of your students. Many exercises lay a foundation for ideas in subsequent sections; others introduce students to connections between the material of that section and previous concepts. To help you locate these exercises, we have "flagged" them with the symbol ◆. The *Instructor's Solutions Guide* includes a complete list of flagged exercises and a rationale for flagging them. You may use this information to help decide which of the flagged exercises to include in your syllabus.

Each exercise set consists of several clusters of exercises. Each cluster revolves around one topic. The level of difficulty runs the entire spectrum from the easiest computational exercises to fundamental discovery-type exercises often integrating several previous topics and/or exploring more advanced ideas. Access to the "flagged" exercises and the enrichment material in the exercises should significantly enhance the scope and flexibility of courses based on this text, as the instructor is free to choose how challenging or non-traditional his or her course will be.

■ **Examples**—Our experience-oriented approach develops the text's material progressively through 487 worked examples which are supported by 620 figures. Examples fall into four categories: (1) clarification of definitions with examples which either do or do not satisfy that definition; (2) computations illustrating the book's techniques; (3) proofs of theorems or identities; (4) word problems. The level of examples generally gets progressively more difficult throughout each section. Each example is supported by a number of odd and even exercises.

■ **Graphing and Functions**—Chapter 1 includes linear and simple quadratic graphs. Thus, once functional concepts have been introduced in chapter 2, they can be immediately illustrated with the help of familiar substantial data.

Algebraic and geometric approaches are integrated at every opportunity, beginning with the graphing of solution sets of inequalities (section 1.1). We try to help students see that a function can be understood either graphically or algebraically. We motivate the definition of even functions in section 2.2 by algebraically analyzing the geometric criteria for symmetry and by using the midpoint formula. In section 2.2, we initiate stretching and shrinking of graphs, an idea that continues throughout the book. Students are encouraged to use graphs to estimate ranges of complicated functions in section 2.4. This fosters experimentation and implicitly introduces continuity.

We use graphing as a vehicle for teaching trigonometric, exponential, and logarithmic functions. For example, in section 4.1, exponential functions are defined and then graphed. The calculator is used as a tool to build tables for graphing, and graphs are then studied to determine properties of exponentials. This approach begins when functions are introduced in chapter 2 and continues with the introduction of each new type of function.

In addition, instructors are encouraged to use graphing software whenever they feel it is appropriate.

■ **Graphing Calculators**—Any precalculus text in this closing decade of the twentieth century must come to grips with the emerging technology of calculators that generate graphs. We have done so in a number of ways that are compatible with the philosophy of this text. We believe that our treatment of graphing calculators reinforces the text's geometric and graphical motivations, together with an intuition about approximations that is linked to the study of inequalities. At eleven places in the text, sections 2.1, 3.3, 3.4, 4.5, 5.3, 6.4, 8.1 (twice), 9.4 (twice), and 10.1, there are extensive discussions indicating the strengths (and occasionally the foibles) of graphing calculators. The text also includes a number of open-ended exercises where students are encouraged to use a graphing calculator to discover facts and formulate conjectures. At several other points in the exercises, we remind students to use whatever is available, such as graphing calculators, in exploring possible generalizations. In truth, such reminders could be placed in each section, and we encourage both teachers and students to become familiar with this exciting technology.

■ **Trigonometry**—Trigonometry is organized as follows: (1) concepts and basic ideas of trigonometry (chapter 5); (2) algebraic manipulations and beginning applications (chapter 6); and (3) an in-depth coverage of applications (chapter 7). We approach trigonometric functions by first considering angles in standard position and then continue with an approach using the unit circle. Graphing and the knowledge of inverse relations (i.e., the horizontal line test) are used to introduce inverse trigonometric functions as a basic concept which, therefore, belongs in the first trigonometry chapter. Additional information about the trigonometry chapters is given below.

■ **Precalculus Teaching Problems**—Precalculus students vary a great deal in their background and their ability. To accommodate this diversity, we have adopted a "user-friendly" writing style. We hope students find this book accessible and stimulating.

It seems that by the time many precalculus students need to apply fundamental facts from algebra and geometry, they have forgotten these topics. We have several topics in each section to keep students' attention and to prevent them from thinking prematurely that they know the material. For example, let's consider section 1.1. In this section the sign chart method is applied to algebraically more complicated functions than are found in many texts. In section 1.1, students also confront a topic that many of them find difficult—namely, continued inequalities. Here, a sign chart method is preferable because it does not focus attention on the logical problems with "and" and "or" statements. Finally, section 1.1, exercise 69 is a word problem where students have to solve a quadratic inequality using modeling skills of the level needed by calculus students.

■ **Preparation for Calculus**—We mention the following twenty items:
1. Functions defined by more than one equation—In addition to the usual definitions of the absolute value and the greatest integer functions in the text (section 2.1), we include other examples of such functions in the exercises.

2. Preparation for the Chain Rule—The exercises in section 2.4 help students identify given functions as compositions $f \circ g$ for suitable functions f and g.
3. Increasing and decreasing functions—Students are asked to determine whether a function is increasing or decreasing in section 2.1, exercises 68–72. These exercises support the earlier example 2.9.
4. Different quotients—Introduced in section 2.1, example 2.7, and supported in section 2.1, exercises 57–67.
5. Continuity—See our earlier comments on graphing and sign charts. Continuity is explicitly introduced in section 3.3. There, the Intermediate Value Theorem and sign charts are used to help students understand continuity.

    The work with asymptotes uses continuity. We define all three major types of linear asymptotes—vertical, horizontal, and slant—as well as the general concept in section 3.6. Asymptotes are considered for rational functions (section 3.6), exponential and logarithmic functions (sections 4.1, 4.2), trigonometric functions (section 5.6), and hyperbolas (section 8.3).
6. Tangents—Tangents to quadratic graphs are in section 1.4 and chapter 8. We do not take a limit approach but rather use the classical notion that reinforces the quadratic formula.
7. Limits—*Every* type of limit that students study in first-year calculus is met at least once in an *intuitive* way. We do not use limit notation. In particular, we consider asymptotes via limits at infinity and via one-sided limits.
8. Max-min for quadratic functions—See the word problem exercises in section 2.3, exercises 40–46.
9. Max-min context for calculus—For example, see section 2.3, exercises 36–37 and 48; section 2.4, exercise 32; and section 3.8, exercise 54.
10. Related rates—Anticipated in section 2.1, exercise 73; presented in the context for calculus in section 2.4, exercise 31.
11. Area formulas for triangles—Heron's formula (section 1.4, exercise 60) and related formulas in section 7.3.
12. Triangle inequality—Both an algebraic version (chapter 0) and a geometric version (section 1.1).
13. Partial fractions—Section 3.7 gives an unusually thorough treatment.
14. Mathematical induction—Section 10.5 contains applications including summation formulas.
15. Polar coordinates—Extensive discussion in sections 10.2–10.4, including some graphs not in other texts.
16–20. To accommodate institutions which have relocated "noncalculus" material that had traditionally been in calculus courses, we include extensive discussions of (16) conic sections, (17) planar vectors and dot product, (18) parametric equations, (19) systems of linear equations and inequalities, and (20) matrices and determinants. These topics can easily be avoided by those following a more traditional syllabus, as

can the early introduction of complex numbers in section 1.5. In keeping with our "cyclic" approach, we hope that your course syllabus will include at least some of this material that traditionally (re)appears in calculus courses.

Chapter-by-Chapter Summary

Chapter 0 provides a condensed overview of algebra and reviews the definition of Cartesian coordinates. This material is included for the sake of completeness and for reference purposes. You may well decide to skip most of chapter 0, treating it as assigned reading.

Chapter 1 begins the work in analytic geometry, synthesizing algebra and geometry. There is an extensive treatment of graphing, and a foundation is laid for the discussion of functions in chapter 2 and the more thorough discussion of analytic geometry in chapter 8. To permit a complete discussion of the quadratic formula, chapter 1 also includes section 1.5 concerning complex numbers. This early introduction of complex numbers allows for their use throughout the book. In particular, a discussion of nonreal complex roots of polynomials is included in chapter 3, and the Argand diagram for complex numbers is in chapter 6.

Chapter 2 introduces functions. Heavy emphasis is placed on this concept. We show how to graph a function, exploiting various types of symmetry and translations of standard graphs. Operations with functions, composition of functions, and inverse functions are also covered in chapter 2.

Chapters 3, 4, and 5 each focus on a particular type of function. Chapter 3 addresses the polynomial and rational functions; chapter 4, the exponential and logarithmic functions; and chapter 5, the trigonometric functions. The sections on polynomial functions include an extensive analysis of procedures for finding the roots of a polynomial. Together with the material on solving inequalities, this will be of great help when your students study differential calculus.

Chapter 4 begins by introducing the exponential functions and the number e. The logarithmic functions are then studied by emphasizing their inverse-function relationship with the exponential functions. It is our recent experience that, with the increased use of calculators and computers, students often do not get a sufficient hands-on understanding of the computational properties of logarithms. Thus, our coverage of these properties of logarithms focuses on those aspects which must be mastered before the operations of calculus can be performed on these transcendental functions. Chapter 4 also contains numerous illustrations of exponential growth or decay.

Chapters 5, 6, and 7 constitute a unit on trigonometry. This material has been organized according to the following principles. Chapter 5 concerns the concepts of trigonometry, chapter 6 deals with trigonometric identities and their applications, and chapter 7 treats a wide variety of physical and geometric applications of trigonometry.

In chapter 5, an angular approach is initially emphasized because students will need skills involving right triangles in science courses and when they study integration techniques in calculus. However, section 5.3 also contains a full discussion of the circular functions approach to the trigonometric functions;

circular functions are revisited in the exercises of later sections. Chapter 5 emphasizes all the ways in which the trigonometric functions can be viewed as functions of a real variable, and should develop in students the function-theoretic attitude to trigonometry that is needed for calculus.

Chapter 6 continues the study of trigonometry, emphasizing the identities needed by calculus students. Applications include the polar form of complex numbers and De Moivre's Theorem. Chapter 7 looks at many additional applications of trigonometry. These include the sine and cosine laws, vectors in the plane, simple harmonic motion, wave motions, and alternating electrical currents. Chapter 7 also contains an unusually thorough coverage of the ambiguous case of the Sine Law.

Chapter 8 returns to analytic geometry, which students met in chapters 0 and 1. This chapter includes a thorough discussion of each type of conic section and the conics' graphs in rectangular coordinates. It includes translation and rotation of axes, which permit a complete analysis of the general quadratic polynomial equation in two variables.

Systems of linear equations are covered in chapter 9. These are included and solved using algebraic methods. Next, matrices and determinants are introduced, and enough of their properties are developed so that students can become proficient at using them to solve any system of linear equations. As with exercise sets throughout the book, the exercises chosen for section 9.3 (covering determinants) emphasize the connections between that section's topic and earlier material. Section 9.4 merges two themes of the text by studying systems of inequalities.

Chapter 10 considers additional topics in geometry and algebra. The former includes polar coordinates, intersections of polar graphs, and parametric equations. These ideas are explored in a function-theoretic way and are illustrated with examples that both reinforce and extend the earlier material. The algebraic topics included in this chapter are the binomial theorem and mathematical induction. These will be of help in calculus, as will the optional material on limits of sequences in the exercises of section 10.5. Optional material on combinatorics, of help for later study of discrete mathematics, appears in the exercises of section 10.6.

Possible Syllabi

A class that covers most of chapters 0 and 1 can expect to complete chapters 2–8 in one semester (meeting four times per week), possibly with some omissions from chapters 3 and 7. An exceptionally well-prepared class that omits most of chapters 0 and 1 can aim to complete the remainder of the book in one semester. In any event, we expect that most semester courses will be structured to include portions of chapters 9 and 10, perhaps at the expense of portions of chapters 3 and 7. As with the exercise sets, such rearrangement and choice is generally both feasible and desirable.

A one-quarter course (meetings four times per week) can expect to complete most of chapters 0–7, possibly by covering sections 1.4 and 3.5 lightly and omitting sections 1.5, 3.6, 7.4, 7.5, and 7.6.

Supplemental Materials

The student's preface describes the *Student's Solution Guide*. You may obtain a copy of the *Instructor's Solution Guide,* which contains solutions for all the text's exercises, authors' comments on each chapter and section, and authors' advice on syllabi. Also available to institutions adopting this text is the *Test Bank.*

This preface was intended to indicate our text's organization and the features in it that prepare your students for calculus. After you've used the text, please let us know your reactions to it.

Acknowledgments

We would like to thank the following reviewers, who offered many helpful suggestions and comments:

Debra Adler, Michigan Technological University; John E. Alberghini, Manchester Community College; Karen E. Barker, Indiana University at South Bend; Dr. William Beyer, The University of Akron; Paul R. Boltz, Harrisburg Area Community College; Charles Brennan, Bloomsburg University; Paul W. Britt, Louisiana State University; Janet Brougher, Olympic College; Warren J. Burch, Brevard Community College; Tim Carroll, Eastern Michigan University; Rebecca Conti, SUNY at Fredonia; Cherlyn Converse, California State University Fullerton; William Conway, University of Arizona; Margaret Dolga, University of Delaware; C. J. Gardiner, Eastern Michigan University; Stephen Gendler, Clarion University; Pat Gilbert, Diablo Valley College; Stuart Goldenberg, California Polytechnic State University, S.L.O.; Frank Goulard, Portland Community College; Arjun Gupta, Bowling Green State University; Louis Hoelzle, Bucks County Community College; Ed Huffman, Southwest Missouri State University; Lynne Ipina, University of Wyoming; Annamarie Langlie, North Hennepin Community College; Gerald Leibowitz, University of Connecticut; Sam Lesseig, Northeast Missouri State University; Donald E. Marshall, University of Washington; Donna McCracken, University of Florida; Michael McDonald, Duke University; Ann Megaw, U. of Texas at Austin; Deborah Mirdamadi, Penn State University; Gary A. Phillips, Oakton Community College; Paul Porch, Mt. Hood Community College; Mahmood Pournazari, Adelphi University; Antonio R. Quesada, The University of Akron; Howard L. Rolf, Baylor University; John Rose, Miami-Dade Community College/South Campus; John Scott, Montgomery College; Frank O. Smith, Kent State University; Rita L. Stout, Texas A & M University; E. H. Tjoelker, California State University, Sacramento; Fredric Tufte, University of Wisconsin–Platteville; John L. Van Iwaarden, Hope College; Dr. Mahbobeh Vezvaei, Kent State University; Sandy Wager, Univ. of North Carolina at Wilmington; David S. Weinstein, Moorpark College; Mary Winter, Michigan State University; C. T. Wolf, Millersville University; Sandra Wray-McAfee, University of Michigan–Dearborn.

Special thanks for their technical and professional support go to Joe Calimaneri at Casio* and Tom Ferrio at Texas Instruments.

D.E.D.
J.C.P.

---

*Casio® is a registered trademark of the Casio Computer Co., Ltd.

Chapter

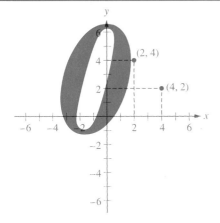

Review Material

As the title indicates, you should already be familiar with most of the material in this chapter. We suggest that you begin by working the exercises at the end of the chapter. Check your work with the answers at the back of the book. If you find an exercise that you cannot solve, review the material in this chapter and then rework the exercise.

SETS

A **set** is a collection of objects called its **elements**. **Equal sets** contain exactly the same elements. If S and T are sets and each element of S is an element of T, then S is a **subset** of T. If S and T are sets, the **union** of S and T, written $S \cup T$, is the set of all elements that belong to S or T (or both S and T). The **intersection** of S and T, written $S \cap T$, is the set of all elements that belong to both S and T. The set with no elements is called the **empty set**. Two sets are **disjoint** if their intersection is the empty set.

Some useful sets of numbers, and associated notation, are

integers, $Z = \{\ldots, -3, -2, -1, 0, 1, 2, 3, \ldots\}$. The set of **positive integers** is $Z^+ = \{1, 2, 3, 4, \ldots\}$. The set of **negative integers** is $\{\ldots, -4, -3, -2, -1\}$. The integer 0 is neither positive nor negative. The set of **nonnegative integers** is $Z^+ \cup \{0\}$.

rational numbers, $Q = \left\{\frac{p}{q} : p \text{ and } q \text{ are integers and } q \neq 0\right\}$. Each rational number can be expressed in a decimal form that either terminates or

1

repeats. Examples of rational numbers that terminate are $\frac{1}{4} = 0.25$, $-\frac{15}{8} = -1.875$, and $\frac{295}{1} = 295$. Examples of rational numbers that can be expressed as repeating decimals include $-\frac{5}{3} = -1.6666\ldots = -1.\overline{6}$, $\frac{71}{22} = 3.2272727\ldots = 3.2\overline{27}$, and $-\frac{3800}{11} = -345.454545\ldots = -345.\overline{45}$. The repeating series of digits is indicated by placing a bar over it.

irrational numbers. These are numbers whose decimal representations neither terminate nor repeat. For example, 3.10110111011110…, 4.123112233111222333…, $\pi \approx 3.14159265359$, $\sqrt{2} \approx 1.4142136$, and $-\sqrt{18} \approx -4.2426407$ are each irrational. (The symbol $\approx$ means "is approximately equal to.")

real numbers, R. This set is the union of Q and the set of irrational numbers. Each real number is either rational or irrational, but not both.

The symbol $\in$ may be used to indicate that a certain element belongs to a set. For instance, $-2 \in Z$, $4 \in Z^+$, and $\sqrt{17} \in R$. To indicate that an element does not belong to a given set, we use the symbol $\notin$, as in $\frac{1}{2} \notin Z$ or $\pi \notin Q$.

THE REAL NUMBER SYSTEM

The **real number system** consists of the set R of real numbers, together with the familiar operations of **addition** and **multiplication**. **Subtraction** is defined by $a - b = a + (-b)$. If $b \neq 0$, the multiplicative inverse of b is denoted by b^{-1} and satisfies $b \cdot b^{-1} = 1$. If $b \neq 0$, **division** is defined by $a \div b = \frac{a}{b} = a \cdot b^{-1}$. If $b \neq 0$, then $\frac{0}{b} = 0$. But $\frac{a}{0}$ is undefined for all values of a.

R can be visually represented by a **real number line**. There is a **one-to-one correspondence** between R and the points on a real number line. The point corresponding to a particular real number is called the **graph** of that number, and the number associated with a given point is the **coordinate** of that point. The **origin**, O, is the graph of 0. A horizontal real number line is shown in figure 0.1. The arrow at its right indicates that the line's **positive** or **increasing direction** is the direction from **0** to **1**; the line's other direction is its **negative** or **decreasing direction**.

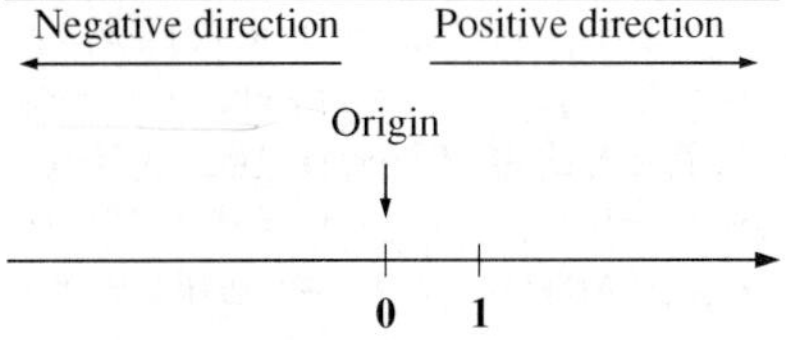

A real number is positive if its graph is to the right of the origin, and is negative if its graph is to the left of the origin. Figure 0.2 shows the points with coordinates -5.2, $-\sqrt{7}$, $\frac{3}{2}$, and $3\sqrt{5}$.

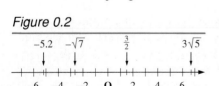

Figure 0.1

Figure 0.2

Arithmetic Properties of Real Numbers

Properties of addition and multiplication on R are listed in table 0.1 for real numbers a, b, c.

Table 0.1

Property	Explanation/Example
Closure	The sum or product of two real numbers is a real number.
Commutative	$a + b = b + a \qquad ab = ba$
Associative	$a + (b + c) = (a + b) + c \qquad a(bc) = (ab)c$
Identity elements	0 is the **additive identity** because $0 + a = a = a + 0$.
	1 is the **multiplicative identity** because $1 \cdot a = a = a \cdot 1$.
Inverse elements	Each real number a has an **additive inverse**, $-a$, such that $a + (-a) = 0 = (-a) + a$.
	Each nonzero real number a has a **multiplicative inverse**, a^{-1} or $\frac{1}{a}$, such that $a\left(\frac{1}{a}\right) = \left(\frac{1}{a}\right)a = 1$.
Distributive	$a(b + c) = ab + ac$

The **generalized associative property** allows terms to be grouped in any convenient manner when a sum or product is computed. The **generalized distributive property** states that $a(b_1 + b_2 + b_3 + \cdots + b_n) = ab_1 + ab_2 + ab_3 + \cdots + ab_n$. The **zero product property of real numbers** states that if $ab = 0$, then either $a = 0$ or $b = 0$ (or both).

INEQUALITIES

If a and b are real numbers such that $a - b$ is positive, then we say that *a* **is greater than** *b* and write $a > b$. It is equivalent to say that *b* **is less than** *a*, written $b < a$. On a real number line, if A and B are points with coordinates a and b respectively, then $a > b$ (or $b < a$) means that A lies to the right of B. Inequalities of the form $a < b$ or $b > a$ are called **strict inequalities**. If *a* **is less than or equal to** *b*, we write $a \leq b$. If *a* **is greater than or equal to** *b*, we write $a \geq b$.

There is a compact notation to assert that two inequalities which "point the same way" both hold. For example, "$a < b$ and $b < c$," can be written as $a < b < c$; this guarantees that *b* **is between** *a* **and** *c*. Similarly, $a \leq b \leq c$ means that $a \leq b$ and $b \leq c$; $a \leq b < c$ means that $a \leq b$ and $b < c$; and $a < b \leq c$ means that $a < b$ and $b \leq c$. For instance, $-5 < 2 < 3$ and $-\frac{2}{3} \leq -\frac{1}{3} < 0$.

Consider real numbers a and b such that $a < b$. The **open interval** (a, b) is the set of all real numbers between a and b, or $(a, b) = \{x: a < x < b\}$. The **closed interval** $[a, b]$ is the union of (a, b) and $\{a, b\}$; thus, $[a, b] = \{x: a \leq x \leq b\}$. **Half-open intervals** (or **half-closed intervals**) are defined as $(a, b] = \{x: a < x \leq b\}$ and $[a, b) = \{x: a \leq x < b\}$. In interval notation, parentheses are used to indicate that an endpoint is not included in the interval, and brackets are

4 Chapter 0 / Review Material

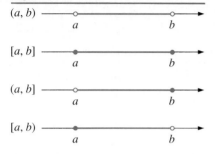

Figure 0.3

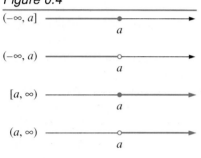

Figure 0.4

used to indicate that an endpoint is included in the interval. The four types of intervals are shown graphically in figure 0.3. In each figure, a is the left endpoint and b is the right endpoint. In figure 0.3, an open circle ○ is used to denote an endpoint if it is not included in the interval, and a filled circle ● is used for an endpoint that is included in the interval.

The **trichotomy property** states that for any real numbers a and b, exactly one of the following is true:

$$a < b \quad a = b \quad a > b.$$

If $b = 0$, then this says that any real number a is exactly one of the following: negative, zero, or positive.

Other important subsets of $\mathbf{R}$ include

$$(-\infty, a] = \{x : x \leq a\};$$
$$(-\infty, a) = \{x : x < a\};$$
$$[a, \infty) = \{x : x \geq a\};$$
$$(a, \infty) = \{x : x > a\};$$

and

$$(-\infty, \infty) = \mathbf{R}.$$

These five types of intervals are shown in figure 0.4.

ABSOLUTE VALUE

Let a real number a be the coordinate of a point A on a real number line. The distance from A to the origin is called the **absolute value** of a, and is denoted $|a|$. We have

$$|a| = \begin{cases} a & \text{if } a \geq 0 \\ -a & \text{if } a < 0 \end{cases}$$

Properties of Absolute Value

Absolute value has the following properties for all real numbers a and b:

(1) $|a| \geq 0$
(2) $-|a| \leq a \leq |a|$
(3) $|ab| = |a| \cdot |b|$
(4) $|a + b| \leq |a| + |b|$.   Triangle inequality

If points A and B on a real number line have coordinates a and b, respectively, then the **distance between A and B**, denoted $d(A, B)$, is $d(A, B) = |b - a|$. The number $d(A, B)$ is sometimes called the **length of the line segment**

Equations

AB. Since $d(B, A) = |a - b| = |b - a| = d(A, B)$, we conclude $d(B, A) = d(A, B)$. This is illustrated by the three examples in figures 0.5(a), (b), and (c).

Figure 0.5

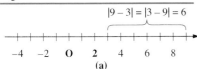
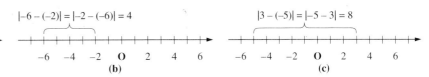

EQUATIONS

A **conditional equation** is an equation in one variable whose truth or falsity depends on which number is substituted for that variable. A number r is a **(particular) solution** or **root of an equation** if a true statement results when r is substituted for the variable in the equation. If r is a solution of an equation, we say that r **satisfies** that equation. The **solution set** of an equation is the set of all particular solutions of the equation; to **solve an equation** means to find its solution set. If a conditional equation is true regardless of which value is substituted for its variable, the equation is called an **identity**. If no number satisfies an equation, its solution set is empty and we say the equation is **inconsistent**. Two equations are **equivalent** if they have exactly the same solution sets.

A **linear equation in one variable** is equivalent to an equation of the form $ax + b = 0$, where $a \neq 0$. The solution set of the linear equation $ax + b = 0$ is $\left\{-\dfrac{b}{a}\right\}$.

A **quadratic equation in one variable** is equivalent to an equation of the form $ax^2 + bx + c = 0$, where $a \neq 0$ (and a, b, c are constants). One way to solve a quadratic equation uses the technique of **completing the square**:

$$x^2 + kx = \left(x + \frac{k}{2}\right)^2 - \left(\frac{k}{2}\right)^2.$$

Another way to solve a quadratic equation is to use the **quadratic formula**:

$$x = \frac{-b \pm \sqrt{b^2 - 4ac}}{2a}.$$

The expression $b^2 - 4ac$ is called the **discriminant** of the quadratic equation $ax^2 + bx + c = 0$. If the discriminant equals 0, then the quadratic formula is interpreted to mean that the only solution is $x = -\dfrac{b}{2a}$.

▶ **EXAMPLE 0.1** Solve the equation $3x^2 - 4x = 7$ by (a) completing the square and (b) using the quadratic formula.

Solutions (a) The given equation is equivalent to $x^2 - \frac{4}{3}x = \frac{7}{3}$. Completing the square on the left-hand side with $k = -\frac{4}{3}$, we find

$$x^2 - \tfrac{4}{3}x = (x - \tfrac{2}{3})^2 - (-\tfrac{2}{3})^2$$
$$= (x - \tfrac{2}{3})^2 - \tfrac{4}{9}.$$

Thus, the given equation is equivalent to
$$(x - \tfrac{2}{3})^2 - \tfrac{4}{9} = \tfrac{7}{3}$$
$$(x - \tfrac{2}{3})^2 = \tfrac{7}{3} + \tfrac{4}{9} = \tfrac{25}{9}$$
$$x - \tfrac{2}{3} = \pm\sqrt{\tfrac{25}{9}}.$$

This equation's solution set is $\{\tfrac{2}{3} + \sqrt{\tfrac{25}{9}}, \tfrac{2}{3} - \sqrt{\tfrac{25}{9}}\} = \{\tfrac{7}{3}, -1\}$.

(b) Before the quadratic formula can be applied, the equation $3x^2 - 4x = 7$ must be rewritten in the equivalent form $ax^2 + bx + c = 0$, namely $3x^2 - 4x - 7 = 0$. With $a = 3$, $b = -4$, and $c = -7$, the quadratic formula leads to
$$x = \frac{-(-4) \pm \sqrt{(-4)^2 - 4(3)(-7)}}{2(3)}$$
$$= \frac{4 \pm \sqrt{100}}{6}$$
$$= \frac{4 \pm 10}{6}.$$

Now, $\dfrac{4+10}{6} = \dfrac{7}{3}$ and $\dfrac{4-10}{6} = -1$, and so the required solution set is $\{\tfrac{7}{3}, -1\}$. ◀
This is the same solution set we obtained in (a).

Let r_1 and r_2 be the roots of a quadratic equation, $ax^2 + bx + c = 0$, as determined by the quadratic formula. Then

(1) $\quad r_1 + r_2 = -\dfrac{b}{a}$

(2) $\quad r_1 \cdot r_2 = \dfrac{c}{a}$

(3) $\quad ax^2 + bx + c = a(x - r_1)(x - r_2)$.

The factorization in (3) helps us rewrite the equation in example 0.1 as $(3x - 7)(x + 1) = 0$. According to the zero product property of real numbers, solutions of this equation satisfy either $3x - 7 = 0$ or $x + 1 = 0$. In other words, the equation's solution set is $\{x: 3x - 7 = 0\} \cup \{x: x + 1 = 0\} = \{\tfrac{7}{3}\} \cup \{-1\} = \{\tfrac{7}{3}, -1\}$, as above.

RECTANGULAR COORDINATES

Two perpendicular real number lines that intersect at their origins lead to the **rectangular** or **Cartesian coordinate system** for a plane. The point of intersection of the axes is called the **origin, O**. The horizontal axis is the **x-axis** and

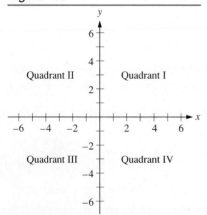

Figure 0.6

7
Exponents

Figure 0.7

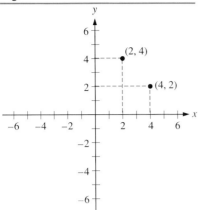

the vertical axis is the **y-axis**. The axes separate the plane into four regions called **quadrants**.

There is a one-to-one correspondence between the set of points in the plane and the set of ordered pairs of real numbers. If P is a point in the plane, the vertical line through P intersects the x-axis at a point whose coordinate is the **x-coordinate** or **abscissa** of P. The second coordinate of the ordered pair corresponding to P is the **y-coordinate** or **ordinate** of P; it is the coordinate of the point where the horizontal line through P intersects the y-axis. Since coordinates are given in a specific order, the coordinates of a point form an **ordered pair**. You can see in figure 0.7 that the point $(4, 2)$ is different from the point $(2, 4)$.

EXPONENTS

If n is a positive integer, then a^n denotes the product of n factors of a:

$$a^n = \underbrace{a \cdot a \cdot a \cdots a}_{n \; a\text{'s}}.$$

The expression a^n is read "a to the nth power" or "a to the n." We call a the **base** and n the **exponent**. You should be aware that $(-a)^n$ and $-a^n$ do not have the same meaning and may actually be unequal. For instance, $(-3)^4 = (-3)(-3)(-3)(-3) = 81$, but $-3^4 = -(3^4) = -81$. The expression a^{-n} means $\frac{1}{a^n}$.

Laws of Exponents

Let a and b be nonzero real numbers and let m and n be integers. We have the properties in table 0.2, with examples of each. (These properties hold whenever all the expressions appearing in them are meaningful real numbers.)

Table 0.2

Property	Example (assuming x, y nonzero)
1. $a^m a^n = a^{m+n}$	$2^3 2^5 = 2^{3+5} = 2^8$
2. $\dfrac{a^m}{a^n} = a^{m-n}$	$\dfrac{5^6}{5^2} = 5^{6-2} = 5^4$; $\dfrac{x^9}{x^5} = x^{9-5} = x^4$
3. $\dfrac{1}{a^n} = a^{-n}$; $\dfrac{1}{a^{-n}} = a^n$	$\dfrac{1}{3^4} = 3^{-4}$; $\dfrac{1}{y^5} = y^{-5}$; $\dfrac{1}{x^{-3}} = x^3$
4. $a^0 = 1$	$x^0 = 1$; $5^0 = 1$; $(x^3 + \frac{7}{3})^0 = 1$
5. $(ab)^n = a^n b^n$	$(3x)^4 = 3^4 x^4 = 81x^4$
6. $(a^m)^n = a^{mn}$	$(x^2)^4 = x^{2 \cdot 4} = x^8$
7. $\left(\dfrac{a}{b}\right)^n = \dfrac{a^n}{b^n}$	$\left(\dfrac{2}{y}\right)^4 = \dfrac{2^4}{y^4} = \dfrac{16}{y^4}$; $\left(\dfrac{x+y}{3}\right)^2 = \dfrac{(x+y)^2}{9}$

These properties are not used in isolation. In order to simplify an expression, you will often need to use several of the above properties. For instance, combining properties 5 and 6 produces $(2x^3)^4 = 2^4(x^3)^4 = 16x^{3 \cdot 4} = 16x^{12}$. This is illustrated further in example 0.2.

▶ **EXAMPLE 0.2**

Rewrite each of the following using positive exponents, and simplify:
(a) $\dfrac{(-a^{-3}b)^2}{(a^{-2}b)^3}$ and (b) $\left(\dfrac{y^{-2}}{5z^{-1}}\right)\left(\dfrac{xz^2}{5y}\right)^{-2}$. Assume that a, b, x, y, and z are nonzero.

Solutions

(a) $\dfrac{(-a^{-3}b)^2}{(a^{-2}b)^3} = \dfrac{(-a^{-3})^2 b^2}{(a^{-2})^3 b^3} = \dfrac{(-1)^2(a^{-3})^2 b^2}{(a^{-2})^3 b^3} = \dfrac{a^{-6} b^2}{a^{-6} b^3} = a^{-6-(-6)} b^{2-3}$

$= a^0 b^{-1} = \dfrac{1}{b^1} = \dfrac{1}{b}.$

(b) $\left(\dfrac{y^{-2}}{5z^{-1}}\right)\left(\dfrac{xz^2}{5y}\right)^{-2} = \left(\dfrac{z}{5y^2}\right)\left(\dfrac{5y}{xz^2}\right)^2 = \left(\dfrac{z}{5y^2}\right)\left(\dfrac{25y^2}{x^2z^4}\right) = \dfrac{5}{x^2 z^3}.$ ◀

POLYNOMIALS

A **polynomial (of degree n) in x** is an expression of the form

$$a_n x^n + a_{n-1} x^{n-1} + \cdots + a_2 x^2 + a_1 x + a_0,$$

where n is a nonnegative integer, each **coefficient** a_k is a number, and the **leading coefficient** a_n is nonzero. Each expression $a_k x^k$ is a **term** of the polynomial. A polynomial with only one nonzero term is called a **monomial.** Two polynomials are equal if and only if they have the same degree and their coefficients of corresponding terms are equal. A **constant polynomial** is a polynomial of degree 0. The **zero polynomial** is denoted by 0 and does not have a degree.

Polynomials are added (or subtracted) by adding (or subtracting) the coefficients of terms with the same degree. Polynomials are multiplied by using the generalized distributive property as follows. Each term of one polynomial is multiplied by each term of the other polynomial, these products of individual terms are added together, and like terms are combined to simplify the resulting answer.

Products and Factors of Polynomials

Some common products of polynomials are shown in table 0.3. An example is given of each.

Polynomials

Table 0.3

Property	Example
1. $(x-y)(x+y) = x^2 - y^2$	$(5x^2 - 3)(5x^2 + 3) = (5x^2)^2 - 3^2$ $= 25x^4 - 9$
2. $(x+a)(x+b) = x^2 + (b+a)x + ab$	$(x+5)(x-7) = x^2 + (-7+5)x + 5(-7)$ $= x^2 - 2x - 35$
3. $(x+y)^2 = x^2 + 2xy + y^2$	$(3a+5)^2 = (3a)^2 + 2(3a)(5) + 5^2$ $= 9a^2 + 30a + 25$
4. $(x-y)^2 = x^2 - 2xy + y^2$	$(x-4z)^2 = x^2 - 2x(4z) + (4z)^2$ $= x^2 - 8xz + 16z^2$
5. $(x+y)^3 = x^3 + 3x^2y + 3xy^2 + y^3$	$(a+2b)^3 = a^3 + 3a^2(2b) + 3a(2b)^2 + (2b)^3$ $= a^3 + 6a^2b + 12ab^2 + 8b^3$
6. $(x-y)^3 = x^3 - 3x^2y + 3xy^2 - y^3$	$(2x^2 - 5y)^3$ $= (2x^2)^3 - 3(2x^2)^2(5y) + 3(2x^2)(5y)^2 - (5y)^3$ $= 8x^6 - 60x^4y + 150x^2y^2 - 125y^3$

The **method of repeated multiplication** can be used to rewrite high powers of polynomials. In general, the method uses the generalized distributive property to expand

$$(a+b)^n = (a+b)^{n-1}(a+b).$$

For instance,

$$(a+b)^4 = (a+b)^3(a+b) = (a^3 + 3a^2b + 3ab^2 + b^3)(a+b)$$
$$= a^4 + 4a^3b + 6a^2b^2 + 4ab^3 + b^4.$$

When a polynomial is written as the product of other polynomials, each of the polynomials being multiplied together is called a **factor** of the original polynomial. The process of expressing a polynomial as a product of factors is called **factoring.** The most common methods for factoring are shown in table 0.4. An example is given of each method.

Table 0.4

Factoring Method	Example
1. Factoring by grouping $ax + ay = a(x+y)$	$2x^4 + xy^2 - 2x^3y - y^3$ $= x(2x^3 + y^2) - y(2x^3 + y^2)$ $= (x-y)(2x^3 + y^2)$
2. $x^2 + 2xy + y^2 = (x+y)^2$	$9a^2 + 30a + 25 = (3a+5)^2$

(continued)

Table 0.4 (continued)

Factoring Method	Example
3. $x^2 - 2xy + y^2 = (x - y)^2$	$x^2 - 8xz + 16z^2 = (x - 4z)^2$
4. Difference of squares $x^2 - y^2 = (x - y)(x + y)$	$25x^4 - 9 = (5x^2 - 3)(5x^2 + 3)$
5. Sum of cubes $x^3 + y^3 = (x + y)(x^2 - xy + y^2)$	$8 + x^3 = (2 + x)(4 - 2x + x^2)$
6. Difference of cubes $x^3 - y^3 = (x - y)(x^2 + xy + y^2)$	$8y^3 - z^6 = (2y - z^2)(4y^2 + 2yz^2 + z^4)$

Irreducible Polynomials

A nonconstant polynomial with real number coefficients is **irreducible over R** if it cannot be expressed as a product of polynomials with real coefficients and positive degree. Any quadratic with real number coefficients and a negative discriminant is irreducible over R. For instance, $x^2 + 2$ is irreducible over R. A quadratic with real number coefficients and a positive discriminant is not irreducible over R. For instance, $x^2 - 2 = (x - \sqrt{2})(x + \sqrt{2})$ is not irreducible over R, but it is irreducible over Q.

Division of Polynomials

Division of polynomials, just like ordinary division of whole numbers, uses the terms **quotient, remainder, divisor,** and **dividend.** The relationship connecting these four parts of a division problem can be expressed as

$$\text{dividend} = (\text{quotient})(\text{divisor}) + \text{remainder}$$

Long division for polynomials follows the same three-step cycle used in long division of whole numbers: divide, multiply, and subtract. The cycling procedure is continued until you reach a remainder that either is 0 or has degree less than that of the divisor. You will find it convenient to write both the dividend and the divisor in decreasing powers of x. Do not omit any terms in the dividend whose coefficient is 0.

▶ **EXAMPLE 0.3** Divide $7x^4 - 5x^2 + 6x + 9$ by $2x + 3x^2$.

Solution First check to see whether the divisor and the dividend are written with descending powers of x. For this example, rewrite the divisor as $3x^2 + 2x$.

Fractional Expressions

$$\begin{array}{r}
\frac{7}{3}x^2 - \frac{14}{9}x - \frac{17}{27} \\
3x^2 + 2x\overline{\smash{)}7x^4 + 0x^3 - 5x^2 + 6x + 9} \\
\underline{7x^4 + \frac{14}{3}x^3} \\
-\frac{14}{3}x^3 - 5x^2 + 6x + 9 \\
\underline{-\frac{14}{3}x^3 - \frac{28}{9}x^2} \\
-\frac{17}{9}x^2 + 6x + 9 \\
\underline{-\frac{17}{9}x^2 - \frac{34}{27}x} \\
\frac{196}{27}x + 9
\end{array}$$

Since the degree of $\frac{196}{27}x + 9$ is less than the degree of $3x^2 + 2x$, the problem is finished. Thus, when $7x^4 - 5x^2 + 6x + 9$ is divided by $2x + 3x^2$, the quotient is $\frac{7}{3}x^2 - \frac{14}{9}x - \frac{17}{27}$ and the remainder is $\frac{196}{27}x + 9$. In other words,

$$7x^4 - 5x^2 + 6x + 9 = (\tfrac{7}{3}x^2 - \tfrac{14}{9}x - \tfrac{17}{27})(2x + 3x^2) + (\tfrac{196}{27}x + 9).$$ ◀

FRACTIONAL EXPRESSIONS

Quotients of algebraic expressions are called **fractional expressions**. A quotient of two polynomials is a **rational expression**. **Simplifying** a rational expression amounts to "canceling" all the common factors (having positive degree) of the numerator and the denominator. In other words, both the numerator and the denominator are to be divided by the product of their common factors. The resulting expression is said to be **in lowest terms**.

To add or subtract rational expressions, it is helpful to first rewrite them using a common denominator, preferably the **least common multiple (LCM)** of their denominators. Then the actual addition or subtraction can be performed in much the same way as for rational numbers.

▶ **EXAMPLE 0.4**   Perform the indicated operation and simplify $\dfrac{15ax^2}{x^2 - 16} \div \dfrac{21a^2x}{x^2 + x - 12}$.

Solution
$$\frac{15ax^2}{x^2 - 16} \div \frac{21a^2x}{x^2 + x - 12} = \frac{15ax^2}{x^2 - 16} \cdot \frac{x^2 + x - 12}{21a^2x}$$

$$= \frac{15ax^2}{(x - 4)(x + 4)} \cdot \frac{(x + 4)(x - 3)}{21a^2x}$$

$$= \frac{5x(x - 3)}{7a(x - 4)}.$$ ◀

▶ **EXAMPLE 0.5**   Perform the indicated operations and simplify

$$\frac{2x}{2x^2 + 9x - 5} + \frac{x}{x^2 + 2x + 1} - \frac{4x - 3}{2x^2 + x - 1}.$$

Chapter 0 / Review Material

Solution

Begin by factoring the denominators: $2x^2 + 9x - 5 = (2x - 1)(x + 5)$, $x^2 + 2x + 1 = (x + 1)^2$, and $2x^2 + x - 1 = (2x - 1)(x + 1)$. The LCM of these denominators is $(2x - 1)(x + 5)(x + 1)^2$. The terms are rewritten using this common denominator:

$$\frac{2x}{2x^2 + 9x - 5} = \frac{2x(x + 1)^2}{(2x - 1)(x + 5)(x + 1)^2},$$

$$\frac{x}{x^2 + 2x + 1} = \frac{x(2x - 1)(x + 5)}{(2x - 1)(x + 5)(x + 1)^2},$$

and

$$\frac{4x - 3}{2x^2 + x - 1} = \frac{(4x - 3)(x + 5)(x + 1)}{(2x - 1)(x + 5)(x + 1)^2}.$$

Thus, the given expression simplifies as

$$\frac{2x(x + 1)^2 + x(2x - 1)(x + 5) - (4x - 3)(x + 5)(x + 1)}{(2x - 1)(x + 5)(x + 1)^2}$$

$$= \frac{2x(x^2 + 2x + 1) + x(2x^2 + 9x - 5) - (4x - 3)(x^2 + 6x + 5)}{(2x - 1)(x + 5)(x + 1)^2}$$

$$= \frac{-8x^2 - 5x + 15}{(2x - 1)(x + 5)(x + 1)^2}. \blacktriangleleft$$

An effective way to solve a rational equation is to instead solve a simpler polynomial equation. You can obtain a simpler polynomial equation by multiplying both sides of the given equation by the LCM of the denominators. This multiplication may not produce an equivalent equation. So, you must check each particular solution of the polynomial equation by substituting it into the given rational equation.

▶ **EXAMPLE 0.6**

Solve $\dfrac{x - 1}{x} - \dfrac{x}{x + 3} = \dfrac{2x + 33}{x^3 + 3x^2}$.

Solution

The denominator of the right-hand side factors as $x^2(x + 3)$, and so you can see that the LCM of the three denominators is $x^2(x + 3)$. Multiplying both sides of the given equation by this LCM produces

$$x^2(x + 3)\frac{x - 1}{x} - x^2(x + 3)\frac{x}{x + 3} = x^2(x + 3)\frac{2x + 33}{x^3 + 3x^2}.$$

This leads to the polynomial equation

$$x(x + 3)(x - 1) - x^3 = 2x + 33.$$

Expanding its left-hand side, you can rewrite this equation as

$$x^3 + 2x^2 - 3x - x^3 = 2x + 33$$

$$2x^2 - 5x - 33 = 0$$

$$(2x - 11)(x + 3) = 0.$$

This polynomial equation has the particular solutions $\frac{11}{2}$ and -3.

Since the rational expression $\frac{x}{x+3}$ is not defined when $x = -3$, we see that -3 is *not* a particular solution of the given rational equation. But you can check that this equation *is* satisfied when x is replaced by $\frac{11}{2}$. Thus, the required solution set is $\{\frac{11}{2}\}$. ◀

RADICALS

Let n be an integer greater than 1, and let y, b be real numbers. Then y is the **principal *n*th root of *b*,** and we write $y = \sqrt[n]{b}$, in case

(i) n is odd and $y^n = b$

or   (ii) n is even, $b \geq 0$, $y^n = b$ and $y \geq 0$.

Here is a compact way to remember the above definition. The *n*th power of $\sqrt[n]{b}$ is b, and if more than one real number has its *n*th power equal to b, then $\sqrt[n]{b}$ is chosen to be the positive real number with this property.

In the radical expression $\sqrt[n]{b}$, $\sqrt{}$ is the **radical sign**, n the **index** and b the **radicand.** We define $b^{1/n} = \sqrt[n]{b}$; this symbol does not define a real number if n is even and b is negative.

A rational number $\frac{m}{n}$ is in **lowest terms** if 1 is the only positive integral common factor of m and n. If the rational number $\frac{m}{n}$ is in lowest terms, then

$$b^{m/n} = \sqrt[n]{b^m}$$

(provided that this radical is a real number). It can be shown that $b^{m/n} = (b^m)^{1/n} = (b^{1/n})^m = (\sqrt[n]{b})^m$, provided $b^{1/n}$ is defined.

The laws of exponents, as stated above, can be shown to hold for rational number exponents, indeed for real number exponents provided that all the relevant expressions are real numbers.

In dealing with radicals, it is often easier to change to fractional exponents (changing $\sqrt[n]{b}$ to $b^{1/n}$), rewrite using laws of exponents, and then change back to radicals (using $b^{m/n} = \sqrt[n]{b^m} = (\sqrt[n]{b})^m$). For instance,

$$\sqrt[n]{ab} = (ab)^{1/n} = a^{1/n}b^{1/n} = \sqrt[n]{a}\,\sqrt[n]{b}.$$

Properties of Radicals

Let a, b be real numbers and m, n positive integers greater than 1. Then, provided that all the relevant radical expressions are real numbers, we have the following properties of radicals listed in table 0.5. An example is given of each property.

Table 0.5

Property	Example
1. $(\sqrt[n]{a})^n = a$	$(\sqrt[5]{4})^5 = 4$
2. $\sqrt[n]{a^n} = \begin{cases} |a| & \text{if } n \text{ is even} \\ a & \text{if } n \text{ is odd} \end{cases}$	$\sqrt[8]{(-5)^8} = |-5| = 5$ $\sqrt[7]{(-5)^7} = -5$
3. $\sqrt[n]{ab} = \sqrt[n]{a}\sqrt[n]{b}$	$\sqrt[3]{64x^2} = \sqrt[3]{64}\sqrt[3]{x^2} = 4\sqrt[3]{x^2}$
4. $\sqrt[n]{\dfrac{a}{b}} = \dfrac{\sqrt[n]{a}}{\sqrt[n]{b}}$	$\sqrt{\dfrac{15}{16}} = \dfrac{\sqrt{15}}{\sqrt{16}} = \dfrac{\sqrt{15}}{4}$
5. $\sqrt[m]{\sqrt[n]{a}} = \sqrt[mn]{a}$	$\sqrt[3]{\sqrt{12}} = \sqrt[3]{\sqrt[2]{12}} = \sqrt[6]{12}$
6. $\sqrt[n]{a^m} = (\sqrt[n]{a})^m$	$\sqrt[4]{(5x)^3} = (\sqrt[4]{5x})^3$

RATIONALIZING FRACTIONAL EXPRESSIONS

To **rationalize the denominator** (or **numerator**) of a fractional expression means to rewrite the expression without a radical in the denominator (or numerator). The following two methods are often used to rationalize denominators (or numerators).

(1) If the denominator (or numerator) is of the form $a\sqrt[n]{b^m}$ with $m < n$, multiply both numerator and denominator by $\sqrt[n]{b^{n-m}}$.

(2) If the denominator (or numerator) is of the form $a\sqrt{b} + c\sqrt{d}$, multiply both numerator and denominator by $a\sqrt{b} - c\sqrt{d}$.

▶ **EXAMPLE 0.7**   Rationalize (a) the denominator of $\dfrac{\sqrt[5]{3x^4}}{\sqrt[5]{8x^2}}$ and (b) the numerator of $\dfrac{2-\sqrt{3}}{7}$.

Solutions

(a) $\dfrac{\sqrt[5]{3x^4}}{\sqrt[5]{8x^2}} = \dfrac{\sqrt[5]{3x^4}}{\sqrt[5]{2^3}\sqrt[5]{x^2}}$

$= \dfrac{\sqrt[5]{3x^4}}{\sqrt[5]{2^3}\sqrt[5]{x^2}} \cdot \dfrac{\sqrt[5]{2^2}\sqrt[5]{x^3}}{\sqrt[5]{2^2}\sqrt[5]{x^3}}$

$= \dfrac{\sqrt[5]{3 \cdot 2^2 x^7}}{\sqrt[5]{2^5}\sqrt[5]{x^5}} = \dfrac{x\sqrt[5]{12x^2}}{2x}$

$= \dfrac{\sqrt[5]{12x^2}}{2}$

(b) $\dfrac{2-\sqrt{3}}{7} = \dfrac{2-\sqrt{3}}{7} \cdot \dfrac{2+\sqrt{3}}{2+\sqrt{3}}$

$= \dfrac{4-3}{14+7\sqrt{3}} = \dfrac{1}{14+7\sqrt{3}}$

◀

MISCELLANEOUS SIMPLIFICATIONS

In differential calculus, it is often desirable to simplify expressions obtained by "differentiating" products and quotients. This may be done by adapting the above methods. It is often helpful to first rewrite the given expression by eliminating radical notation and negative exponents, and then add or subtract by using a common denominator.

▶ **EXAMPLE 0.8** Perform the indicated operations and simplify, without using radical notation or negative exponents:

(a) $2x^{3/2}(x^2 - 1)^{-1/2} + x^{-1/2}\sqrt{x^2 - 1}$ and (b) $\dfrac{2x\sqrt{x^2 + 1} - x^3(x^2 + 1)^{-1/2}}{x^2 + 1}$.

Solutions (a) $2x^{3/2}(x^2 - 1)^{-1/2} + x^{-1/2}\sqrt{x^2 - 1}$

$$= \frac{2x^{3/2}}{(x^2 - 1)^{1/2}} + \frac{1}{x^{1/2}} \cdot (x^2 - 1)^{1/2}$$

$$= \frac{2x^{3/2}x^{1/2}}{x^{1/2}(x^2 - 1)^{1/2}} + \frac{(x^2 - 1)^{1/2}(x^2 - 1)^{1/2}}{x^{1/2}(x^2 - 1)^{1/2}}$$

$$= \frac{2x^2 + (x^2 - 1)}{x^{1/2}(x^2 - 1)^{1/2}}$$

$$= \frac{3x^2 - 1}{x^{1/2}(x^2 - 1)^{1/2}}$$

(b) $\dfrac{2x\sqrt{x^2 + 1} - x^3(x^2 + 1)^{-1/2}}{x^2 + 1}$

$$= \frac{2x(x^2 + 1)^{1/2} - \dfrac{x^3}{(x^2 + 1)^{1/2}}}{x^2 + 1}$$

$$= \frac{\left[\dfrac{2x(x^2 + 1)^{1/2}(x^2 + 1)^{1/2} - x^3}{(x^2 + 1)^{1/2}}\right]}{x^2 + 1}$$

$$= \frac{2x(x^2 + 1) - x^3}{(x^2 + 1)^{1/2}(x^2 + 1)}$$

$$= \frac{x[2(x^2 + 1) - x^2]}{(x^2 + 1)^{3/2}}$$

$$= \frac{x[x^2 + 2]}{(x^2 + 1)^{3/2}}. \quad \blacktriangleleft$$

SOLVING RADICAL EQUATIONS

A **radical equation** is an equation in which at least one radical expression contains a variable. To solve a radical equation, use the **principle of powers**:

$$\text{if } a = b, \text{ then } a^n = b^n.$$

The "isolate-power-check" procedure is often used for solving a radical equation:

(1) If necessary, find an equivalent equation with a single radical term on one side.
(2) Raise both sides of this equation to the same power as the index of the isolated radical.
(3) If a radical still appears in the new equation, repeat steps 1 and 2.
(4) Solve the resulting equation.
(5) Check each possible solution in the original equation.

▶ **EXAMPLE 0.9**

Solution

Solve $x + \sqrt{1+x} = 5$.

The given equation is equivalent to

$$\sqrt{1+x} = 5 - x.$$

Using the principle of powers, square both sides:

$$1 + x = 25 - 10x + x^2$$

$$x^2 - 11x + 24 = 0$$

$$(x-8)(x-3) = 0$$

$$x = 8 \quad \text{or} \quad x = 3.$$

Substituting $x = 8$ into the *original* equation produces

$$8 + \sqrt{1+8} = 5$$

$$8 + 3 = 5$$

or

$$11 = 5,$$

which is false. Thus, $x = 8$ is not a particular solution of the given equation.
Substituting $x = 3$ into the original equation produces $3 + \sqrt{1+3} = 5$ or $5 = 5$, which is a true statement. Hence, the required solution set is $\{3\}$. ◀

▶ **EXAMPLE 0.10**

Solution

Solve $\sqrt{5x+6} - \sqrt{x+3} = 3$.

Begin by rewriting the original equation so that one of its radical terms is isolated.

$$\sqrt{5x+6} = 3 + \sqrt{x+3}$$

Squaring both sides produces

$$5x + 6 = 9 + 6\sqrt{x+3} + (x+3)$$

or

$$4x - 6 = 6\sqrt{x+3},$$

Solving Absolute Value Equations

which is equivalent to

$$2x - 3 = 3\sqrt{x + 3}.$$

Squaring both sides of this new equation, you find

$$4x^2 - 12x + 9 = 9(x + 3)$$

or

$$4x^2 - 21x - 18 = 0$$

$$(4x + 3)(x - 6) = 0.$$

This equation has the particular solutions $-\frac{3}{4}$ and 6.

You should verify that $x = -\frac{3}{4}$ does not satisfy the original equation and that $x = 6$ does satisfy it. Thus, the solution set of the given equation is $\{6\}$. ◄

SOLVING ABSOLUTE VALUE EQUATIONS

Solving an equation involving absolute value notation requires you to consider cases. You will often need to use the Properties of Absolute Value listed on page 4.

▷ **EXAMPLE 0.11**

Solve $|2x - 5| = 17$.

Solution

The only numbers with absolute value of 17 are 17 and -17. Thus, the given equation is equivalent to

either $\quad 2x - 5 = 17 \quad$ or $\quad 2x - 5 = -17.$

The solution set of the linear equation $2x - 5 = 17$ is $\{11\}$; and the solution set of $2x - 5 = -17$ is $\{-6\}$. Thus, the solution set of $|2x - 5| = 17$ is $\{11, -6\}$. ◄

▷ **EXAMPLE 0.12**

Solve $\left|\dfrac{4x + 3}{5}\right| = |x + 5|$.

Solution

Two numbers have the same absolute value if and only if the numbers are either equal or additive inverses of each other. Thus, the given equation is equivalent to

either $\quad \dfrac{4x + 3}{5} = x + 5 \quad$ or $\quad \dfrac{4x + 3}{5} = -(x + 5).$

Solving $\dfrac{4x + 3}{5} = x + 5$, you obtain

$$4x + 3 = 5x + 25$$

or $\quad\quad\quad\quad\quad\quad\quad\quad\quad x = -22.$

Solving $\dfrac{4x + 3}{5} = -(x + 5)$ produces

$$4x + 3 = -5x - 25$$

or $\quad\quad\quad\quad\quad\quad\quad\quad\quad x = -\dfrac{28}{9}.$

Chapter 0 / Review Material

Thus, the solution set of $\left|\dfrac{4x+3}{5}\right| = |x+5|$ is $\left\{-22, -\dfrac{28}{9}\right\}$.

EXAMPLE 0.13 Solve $\left|\dfrac{4x+3}{5}\right| = x + 5$.

Solution The given equation is equivalent to

$$\left|\dfrac{4x+3}{5}\right| = |x+5| \quad \text{and} \quad x + 5 \geq 0.$$

According to example 0.12, the solution set of $\left|\dfrac{4x+3}{5}\right| = |x+5|$ is $\{-22, -\dfrac{28}{9}\}$; and the solution set of $x + 5 \geq 0$ is $[-5, \infty)$. Therefore, the solution set of the given equation is

$$\{-22, -\tfrac{28}{9}\} \cap [-5, \infty) = \{-\tfrac{28}{9}\}.$$

Review Exercises

Sets

1. Determine which of the following sets is/are a subset of Q:
   (a) Z (b) R
   (c) the set of irrational numbers (d) Z^+
2. Determine which of the following sets is/are equal to $\{1, 2\}$:
   (a) $\{3\}$ (b) $\{1 + 1, 1\}$ (c) $\{3, 4\}$
   (d) $\{x: (x-1)(x-2) = 0\}$
3. Determine which of the following real numbers is/are rational:
   (a) $-\tfrac{3}{2}$ (b) $\sqrt{16}$ (c) π
   (d) $-\sqrt{2}$ (e) $-12.0\overline{47}$ (f) 12.04

The Real Number System

4. On a real number line, draw the points with coordinates $0, 1, -2, 3, \tfrac{1}{3},$ and $-\tfrac{3}{5}$.
5. Determine which arithmetic properties of R are illustrated in the following:
   (a) $\pi\sqrt{2}$ is a real number
   (b) $3a = a \cdot 3$ for each real number a
   (c) $(3 + 5) + 4 = 3 + (4 + 5)$
   (d) $(2 + 5) \cdot 7 = 14 + 35$
6. For each of the following real numbers, find the additive inverse:
   (a) 2 (b) $-\tfrac{2}{3}$ (c) π (d) 0
7. For each of the following nonzero real numbers, find the multiplicative inverse:
   (a) 2 (b) $-\tfrac{2}{3}$ (c) $\sqrt{2}$ (d) -1
8. Simplify, using arithmetic properties of R:
   (a) $(a + b)(c + d)$ (b) $(x + 3)^2$
   (c) $(x + y + 1)^2$ (d) $(a + b + c)(x + y + z + 1)$

Inequalities

9. Determine which of the four inequality symbols ($<, >, \leq, \geq$) can replace $\square$ in each of the following:
   (a) $-3 \square -9$ (b) $3 \square 9$ (c) $-3 \square -3$
   (d) $3 \square 0$ (e) $0 \square -\tfrac{3}{2}$ (f) $3 \square \pi$
10. Graph and use interval notation to describe the following:
    (a) $\{x: -2 < x < 1\}$ (b) $\{x: x \geq 3\}$ (c) $\{x: 0 \leq x \leq 2\}$
    (d) $\{x: x < 5\}$ (e) $(-\infty, 2] \cup (1, 3)$
    (f) $[-1, 3) \cap [2, 5]$
11. Show that if a and b are unequal real numbers, then either $a < b$ or $b < a$. (Hint: can $b - a$ and $a - b$ both be negative?)

Absolute Value

12. Express without using absolute value notation:
    (a) $|4| - |9|$ (b) $|4| + |-9|$ (c) $|4 - 9|$
    (d) $||4| - |9||$ (e) $|\pi - 3|$ (f) $|4 - 3\sqrt{2}|$

Properties of Absolute Value

13. The points $A, B, C, D, E,$ and O on a real number line have coordinates $-14, -2, 6, 2, \tfrac{1}{3},$ and 0, respectively. Find the following distances:

Review Exercises

(a) $d(O, B)$ (b) $d(D, E)$ (c) $d(C, A)$
(d) $d(B, E)$ (e) $d(A, D)$ (f) $d(A, O)$

◆ 14. Show that $|-a| = |a|$ for each real number a. (Hint: consider $a > 0$, $a = 0$, and $a < 0$ as three separate cases. Notice that $a > 0$ if and only if $-a < 0$.)

Equations

15. Determine whether the given equation is an identity, inconsistent, or neither:
    (a) $2(x - 12) = 2x - 12$ (b) $2(x - 12) = 2x - 24$
    (c) $(x + 3)^2 = x^2 + 5x - 9$ (d) $x^2 - 25 = (x - 5)(x + 5)$

16. Solve the following linear equations:
    (a) $\frac{2}{5}x + 3 = \frac{17}{5}x - 5$ (b) $12t + 3 = 5t - 11$

17. Rewrite, using the method of completing the square:
    (a) $x^2 - 25x$ (b) $t^2 + \frac{3}{7}t$ (c) $5x^2 + 30x$

18. Use completing the square to solve the following quadratic equations:
    (a) $2x^2 + 5x = 12$ (b) $x^2 + 2\sqrt{2}x - 6 = 0$

19. Use the quadratic formula to solve the following quadratic equations:
    (a) $4x^2 + 4x - 3 = 0$ (b) $4x^2 + 4x - 11 = 0$

20. Use factoring to solve the given quadratic equations:
    (a) $x^2 - x - 6 = 0$ (b) $3x^2 - 2x - 1 = 0$

◆ 21. Find the dimensions of a rectangle having area 64 ft² and perimeter $16\sqrt{10}$ ft.

Rectangular Coordinates

In exercises 22–23, plot the indicated points relative to a pair of coordinate axes.

22. $A(2, 3)$ and $B(-2, 3)$
23. $C(-2, -3)$, $D(2, -3)$, and $E(0, 0)$

Exponents and Laws of Exponents I

24. Rewrite the given expressions as simply as possible:
    (a) 2^0 (b) $2^4 - 3^{-2}$
    (c) $\left(-\frac{3x}{2}\right)^5$ (d) $\frac{x^4}{x^7} - 7x^{-2}x^3x^{-4}$

25. Rewrite the given expressions using positive exponents, and simplify:
    (a) $\frac{(-xy^{-2})^5}{(3x^4y)^{-2}}$ (b) $(\frac{1}{4}u^5v^{-6})(8u^{-2}v^4)$
    (c) $\left(\frac{3u^2v^{-1}}{u^5v^{-2}}\right)^{-3}$ (d) $(-2x^{-3})^3\left(\frac{-2xy^{-1}}{x^2y^2}\right)^2 x^{16}$

Polynomials

26. Find (i) the degree and (ii) the leading coefficient (if they exist) for the given polynomial:
    (a) $3x^5 + 5x^3 - 3x - 5$
    (b) $12x^2 - x^3 + 1$ (c) 12 (d) 0

27. Perform the indicated operations and simplify the resulting polynomial:
    (a) $(\frac{7}{3}x^4 - 2x^3 + x^2 - 4) + (\frac{2}{3}x^4 - x^3 + 5x)$
    (b) $(2x^5 - 2x^4 + 2x^3 + 3x - 5) - (x^5 - 2x^4 - 3x^3 + 4x^2 - 7)$
    (c) $(5x^2 - x + 2)(-3x^3 + 3x^2 - 1)$

Products and Factors of Polynomials

28. Simplify the following products:
    (a) $(3x - \frac{2}{3})(6x + 9)$ (b) $(3x^2 + y)^2$
    (c) $(4u^3 - 2v)^3$ (d) $(u + 3v)^4$

29. Use the method of repeated multiplication to simplify:
    (a) $(2x - 3y)^4$ (b) $(a + b)^5$

30. Factor the following expressions:
    (a) $3x^2y + 6xy^2 - 5x - 10y$ (b) $\frac{9}{16}x^2 - 3x + 4$
    (c) $36x^2y^6 - 1$ (d) $v^3 + 64$
    (e) $4x^2 - 12x - 40$ (f) $2x^2 + 5x - 3$

Irreducible Polynomials

31. Which of the following polynomials is/are irreducible over $\mathbf{R}$?
    (a) $\sqrt{2}x$ (b) $x^2 - 4x - 3$
    (c) $x^2 - 4x + 5$ (d) $\pi x + \frac{2}{3}$

Division of Polynomials

32. Divide:
    (a) $8x^2 - 13x + 11$ by $x - 3$ (b) $8x^2 - 4x - 61$ by $4x + 1$
    (c) $x^3 - 6x^2 - \frac{x}{3} + 4$ by $3x^2 + 1$

Fractional Expressions

33. Simplify the following rational expressions:
    (a) $\frac{x^3 - 49x}{x^2 - x - 42}$ (b) $\frac{x^2 - 4y^2}{3x^2y - 6xy^2 + 5x - 10y}$
    (c) $\frac{x^3 - 64}{x^2 - 8x + 16}$

34. Perform the indicated operations and simplify:
    (a) $\frac{x^2 - 2x - 15}{6xy^2} \cdot \frac{3x^2y}{x^2 - 25}$ (b) $\frac{1 - 8x^3}{x - 1} \div \frac{4x^2 - 1}{x + 1}$
    (c) $\frac{1}{x + 4} - \frac{x - 6}{x^2 - 16}$ (d) $2 - \frac{1}{x - 1} + \frac{2}{x} + \frac{1}{x + 1}$

Chapter 0 / Review Material

35. Simplify $\dfrac{\dfrac{1}{(x+h)^3} - \dfrac{1}{x^3}}{h}$.

36. Determine which of the following rational expressions is/are in lowest terms. (Any "cancellation" is to be done with polynomials having real coefficients.)
    (a) $\dfrac{x^3+1}{x^2-1}$ (b) $\dfrac{x^3-1}{x^2+1}$
    (c) $\dfrac{x^2+3x-18}{x^2-6x+9}$ (d) $\dfrac{5x}{x^2-7x^3}$

Solving Rational Equations

37. Solve the following rational equations:
    (a) $\dfrac{2}{x+2} + \dfrac{1}{x-2} = \dfrac{3x-10}{x^2-4}$
    (b) $\dfrac{1}{x} + \dfrac{2}{x^2} = 1$ (c) $5 - \dfrac{2}{x} = 8 - \dfrac{7}{x}$

Radicals

38. Simplify the following radical expressions:
    (a) $\sqrt{16}$ (b) $\sqrt[3]{\dfrac{-8}{27}}$
    (c) $\sqrt[4]{\dfrac{256}{81}}$ (d) $\sqrt[5]{\dfrac{1}{32}}$ (e) $\sqrt{0}$

39. Rewrite using fractional exponents:
    (a) $\sqrt{x^2+y^2}$ (b) $\sqrt[3]{4}$
    (c) $\sqrt[3]{-z^2}$ (d) $\left(\sqrt[5]{x^{2/3}}\right)^4$

Laws of Exponents II

40. Simplify the following expressions, using laws of exponents if necessary:
    (a) $27^{-1/3}$ (b) $-4^{1/2} + 8^{4/3}$
    (c) $(-8x^9)^{2/3} + x^{-2/3}x^{1/2}$ (d) $\dfrac{x^{5/7}}{x^{1/3}} + \left(\dfrac{x^5y^6}{x^3y^5}\right)^{-3/2} - 3^0$

Properties of Radicals

41. Use properties of radicals to simplify the following expressions:
    (a) $\sqrt[3]{(-\pi)^3}$ (b) $(\sqrt{\pi})^2$ (c) $\sqrt{\dfrac{7}{9}}$
    (d) $\sqrt[3]{8x}$ (e) $\dfrac{\sqrt[5]{128x^4y^7}}{\sqrt[5]{x^6y^2}}$ (f) $\sqrt{50} - \sqrt{18} + \sqrt{32x^2}$

Rationalizing Fractional Expressions

42. Rationalize each denominator:
    (a) $\dfrac{\sqrt{7}}{\sqrt{3}}$ (b) $\dfrac{5}{2\sqrt{3}-\sqrt{6}}$ (c) $\dfrac{\sqrt[3]{-80x^2y^2}}{\sqrt[3]{-10xy^4}}$

43. Rationalize each numerator:
    (a) $\dfrac{2\sqrt{3}-\sqrt{6}}{5}$ (b) $\dfrac{\sqrt{3x}}{\sqrt{2y}}$ (c) $\dfrac{\sqrt{-7(x+h)}-\sqrt{-7x}}{h}$

Miscellaneous Simplifications

44. Perform the indicated operations and simplify, without using radical notation or negative exponents:
    (a) $6x(x^3-1) - x^{-2}(x^3-1)^2$ (b) $\dfrac{x-2}{2\sqrt{x+1}} + \sqrt{x+1}$
    (c) $\dfrac{2x^2(1-x^2)^{-1/2} + \sqrt{1-x^2}}{x^2}$ (d) $\dfrac{\dfrac{x^3+1}{2\sqrt{x}} - 3x^2\sqrt{x}}{(x^3+1)^2}$

Solving Radical Equations

In exercises 45–47, solve the given radical equation.

45. $\sqrt[3]{1-2x} = 5$ 46. $\sqrt{9-x} = 7-x$
47. $\sqrt{x+1} + \sqrt{x-1} = 2$

Properties of Absolute Value

48. Use properties of absolute value to simplify:
    (a) $|-6x^2 - 17|$ (b) $|2x^5 + 10|$
    (c) $|-5xy^2|$ (d) $\left|\dfrac{x^2}{9y}\right|$

Solving Absolute Value Equations

In exercises 49–53, solve the given absolute value equation.

49. $|7-3x| = 5$ 50. $|2x+5| = |3x-4|$
51. $x^2 = |x|$ 52. $\left|\dfrac{x-2}{x+2}\right| = \left|\dfrac{x+2}{x-2}\right|$
53. $|2x+5| = 3x-4$

Chapter 1

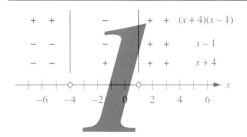

Fundamental Algebraic and Geometric Concepts

In this chapter we discuss material that is basic for the rest of this book. First, we shall show how the "sign chart" method of graphing on a number line can help determine the solution of an inequality. Next, the graphing idea will use the fact that each point in a plane is represented by an ordered pair of real numbers. This permits algebraic ideas to be pictured graphically in the plane, with different kinds of equations having distinctive graphs. The relationship between each new concept and its graphical representation will be explored throughout the text. Chapter 1 also contains an algebraic and geometric study of systems of linear or quadratic equations, as well as an introduction to complex numbers.

1.1 ALGEBRAIC INEQUALITIES

In chapter 0, you saw how to solve many different types of equations. However, many situations can be mathematically described only by an inequality. In fact, in the study of differential calculus, you are expected to be able to solve inequalities. In this section, we show how to solve inequalities by building on your equation-solving skills. Inequality- and equation-solving techniques will be applied throughout the book.

Properties of Inequalities

Solving an inequality often requires you to use the properties stated below. These properties have been stated in terms of the "is less than" or "<" symbol, but are also true for $\leq$, $>$, and $\geq$.

Chapter 1 / Fundamental Algebraic and Geometric Concepts

Properties of Inequalities

For real numbers a, b and c, the following are true:

1. If $a < b$ and $b < c$, then $a < c$.   Transitive property
2. If $a < b$, then $a + c < b + c$.   Addition property
3. If $a < b$ and $c > 0$, then $ac < bc$.   Positive multiplication property
4. If $a < b$ and $c < 0$, then $ac > bc$.   Negative multiplication property

Most of the above properties are similar to those you use to rewrite equations. Perhaps only property 4 is not intuitively obvious. It asserts that if both sides of an inequality are multiplied (or divided) by a negative number, the direction of the inequality sign reverses. An example will show you that this is reasonable. For instance, if you multiply the inequality $-3 < 2$ through by -4, you obtain $12 > -8$, which is true (not $12 < -8$, which is false).

However, an example is not the same as a proof! We can prove the above properties of inequalities by using two facts. First, as noted in chapter 0, the inequality $x < y$ is equivalent to the statement that $y - x$ is positive. Second, any sum or product of positive numbers is positive.

Let's see how to use the above facts to prove property 1. By the first fact, the assumption that $a < b$ and $b < c$ amounts to stating that $b - a$ and $c - b$ are both positive. By the second fact, $(b - a) + (c - b)$ is positive. But $(b - a) + (c - b) = c - a$, and so $c - a$ is positive. This means $a < c$, which is the asserted conclusion.

Exercises 4–6 will ask you to give similar proofs of properties 2–4.

What does it mean to solve an inequality? It means simply to find all the real numbers for which the inequality holds. The set of all these **particular solutions** is the **solution set** of the inequality. Two inequalities are **equivalent** if they have the same solution sets. For instance, $2x > 6$ is equivalent to $x > 3$, and so the solution set of $2x > 6$ is the interval $(3, \infty)$. To **graph an inequality** means to graph the solution set of the inequality.

Solving Linear Inequalities

To solve a linear inequality, such as $ax + b > cx + d$, you use property 2 to obtain an equivalent inequality of the form $ex > f$ or $f > ex$. Then you apply property 3 or property 4. The next two examples illustrate the procedure.

▷ **EXAMPLE 1.1**

Solve the inequality $7x + 24 > 3x - 6$. Describe the solution set using interval notation and graph the inequality.

Solution

$$7x + 24 > 3x - 6$$
$$7x + 24 + (-3x - 24) > 3x - 6 + (-3x - 24)$$
$$7x - 3x > -6 - 24$$

Algebraic Inequalities

$$4x > -30$$
$$x > -\frac{30}{4}$$
$$x > -\frac{15}{2}.$$

The solution set is the interval $(-\frac{15}{2}, \infty)$, as shown in figure 1.1.

Figure 1.1

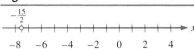

▶ **EXAMPLE 1.2** Solve the inequality $3(\frac{14}{3} - 2x) \leq \frac{2}{5}(x - 5)$. Describe its solution set using interval notation and graph the inequality.

Solution

$$3(\tfrac{14}{3} - 2x) \leq \tfrac{2}{5}(x - 5)$$
$$14 - 6x \leq \tfrac{2}{5}x - 2$$
$$-\tfrac{32}{5}x \leq -16$$
$$x \geq -16(-\tfrac{5}{32})$$
$$x \geq \tfrac{5}{2}.$$

Notice that the direction of the inequality sign was reversed when we divided by $-\frac{32}{5}$.

The solution set is the interval $[\frac{5}{2}, \infty)$, as shown in figure 1.2.

Figure 1.2

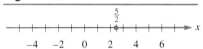

▶ **EXAMPLE 1.3** Solve the inequality $-4 \leq 5 - 3x < 23$. Describe its solution set using interval notation and graph the inequality.

Solution The given inequality is equivalent to

$$-4 \leq 5 - 3x \quad \text{and} \quad 5 - 3x < 23.$$

We shall first solve each of *these* inequalities. The intersection of their solution sets will be the required solution set.

$$-4 \leq 5 - 3x \quad \text{and} \quad 5 - 3x < 23$$
$$-9 \leq -3x \quad \text{and} \quad -3x < 18$$
$$3 \geq x \quad \text{and} \quad x > -6.$$

Thus, the solution set of $-4 \leq 5 - 3x < 23$ is the intersection of the intervals $(-\infty, 3]$ and $(-6, \infty)$. This is the interval $(-6, 3]$, namely the set of all real numbers x such that $-6 < x \leq 3$, and is shown in figure 1.3.

Figure 1.3

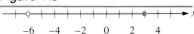

A shorter procedure for solving problems like example 1.3 is to deal with both inequalities simultaneously. For instance, example 1.3 could have been worked as follows:

$$-4 \leq 5 - 3x < 23$$
$$-9 \leq -3x < 18$$
$$3 \geq x > -6$$

or

$$-6 < x \leq 3.$$

Chapter 1 / Fundamental Algebraic and Geometric Concepts

EXAMPLE 1.4

Solve $7 < \dfrac{\pi x - 4}{2} < 12$.

Solution

$$7 < \frac{\pi x - 4}{2} < 12$$
$$14 < \pi x - 4 < 24$$
$$18 < \pi x < 28$$
$$\frac{18}{\pi} < x < \frac{28}{\pi}.$$

The solution set is the interval $\left(\dfrac{18}{\pi}, \dfrac{28}{\pi}\right)$. ◀

You should notice that if example 1.4 were changed to have a negative denominator in its middle term, as in $7 < \dfrac{\pi x - 4}{-2} < 12$, then your first step in the solution would be to apply the negative multiplication property. This would lead to $-14 > \pi x - 4 > -24$ and, ultimately, to the solution set $\left(-\dfrac{20}{\pi}, -\dfrac{10}{\pi}\right)$.

As noted above, **when the statement of an inequality problem involves the word "and," the solution involves an intersection of solution sets**. We shall see next how to solve inequality problems that involve the word "or."

EXAMPLE 1.5

Find all x such that either $4x + 3 \leq -5$ or $4x + 3 > 5$. Describe the solution set using interval notation and graph this solution set.

Solution

$$4x + 3 \leq -5 \quad \text{or} \quad 4x + 3 > 5$$
$$4x \leq -8 \quad \text{or} \quad 4x > 2$$
$$x \leq -2 \quad \text{or} \quad x > \tfrac{1}{2}.$$

Thus, the solution consists of a union of intervals $(-\infty, -2] \cup (\tfrac{1}{2}, \infty)$, as shown in figure 1.4. ◀

Figure 1.4

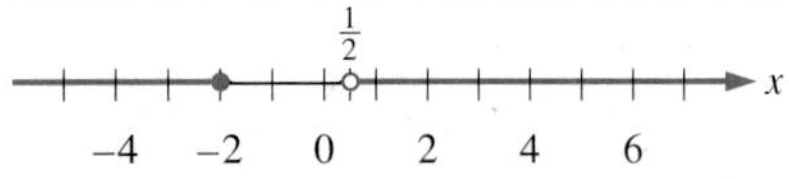

As example 1.5 illustrated, **when the statement of an inequality problem involves the word "or," the solution involves a union of solution sets.**

Before moving to another type of inequality, we must give an important warning about notation. The solution set in example 1.5, which was described by "$x \leq -2$ or $x > \tfrac{1}{2}$," *cannot* also be described by "$\tfrac{1}{2} < x \leq -2$." In fact, the latter describes the empty set, for no real number can satisfy both $\tfrac{1}{2} < x$ and $x \leq -2$.

Inequalities Involving Products or Quotients

Suppose that an inequality can be rewritten equivalently so as to have only a 0 on one side of the inequality symbol. If the other side of the inequality symbol can be factored or is a quotient of expressions that can be factored, we can then use a "sign chart" to determine the solution set of the original inequality. The next example illustrates this useful method.

Algebraic Inequalities

▶ **EXAMPLE 1.6** Solve the inequality $x^2 + 3x > 4$. Describe its solution set using interval notation and graph the inequality.

Solution We begin by rewriting the inequality so as to get a 0 on the right-hand side; next, we factor the left-hand side.

$$x^2 + 3x - 4 > 0$$

$$(x + 4)(x - 1) > 0.$$

We next "graph" each factor on a sign chart using a display of signs. For example, the sign chart of the factor $x + 4$ is shown in figure 1.5(a). An open circle has been placed on the number line at -4 to indicate two things. First, this circle means that the factor $x + 4$ is 0 when $x = -4$. (A root of any factor is called a **key point**. So, -4 is the key point for $x + 4$.) Second, the circle is open because the given inequality is strict and hence does not contain -4 in its solution set.

A vertical line has been placed at -4. This line acts as a line of demarcation. The factor $x + 4$ is positive for every real number x greater than -4. This is indicated by the row of + signs to the right of the demarcation line. The row of − signs to the left of the demarcation line indicates that $x + 4$ is negative for all real numbers x less than -4. At one end of the row of + and − signs, the expression $x + 4$ has been written to remind us that this is the sign chart for the factor $x + 4$.

Next, a sign chart is created for the factor $x - 1$ above the sign graph for $x + 4$. Since $x - 1$ has the root 1, the factor $x - 1$ has the key point 1, where an open circle is placed in figure 1.5(b).

A third row is next added to the sign chart. This row, shown in figure 1.5(c), represents the signs of the product $(x + 4)(x - 1)$. Remember, a product or quotient of two nonzero real numbers is positive if and only if these two numbers are either both positive or both negative.

From the top, or product, row of the sign chart in figure 1.5(c), you can see that $(x + 4)(x - 1) > 0$ for x in $(-\infty, -4) \cup (1, \infty)$. But $(x + 4)(x - 1) > 0$ is equivalent to $x^2 + 3x > 4$. Thus, $(-\infty, -4) \cup (1, \infty)$ is the solution set for the given inequality, and is graphed in figure 1.5(d). ◀

Figure 1.5

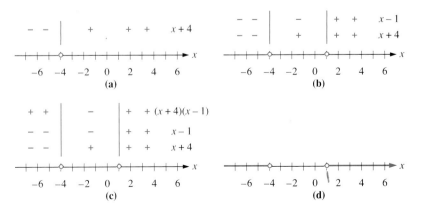

An important advantage of using sign charts is that you can immediately read off the solution set of any "associated" inequality from the top row. For example, the solution set of $x^2 + 3x < 4$ is $(-4, 1)$.

As you will see in example 1.7, **the sign chart method can be used to solve inequalities involving any number of factors**. As you move through this book, you will use this method time and again, often with different types of factors, not just linear ones.

▶ **EXAMPLE 1.7**

Solve and graph the inequality $-2x(x + 3)^2(2 - x) \leq 0$.

Solution

The sign chart in figure 1.6(a) has a row for the signs of each factor. Because the inequality is nonstrict, each key point is a particular solution. The key points are filled in to indicate that $0, -3$, and 2 are solutions for this inequality. Since the square of any real number is nonnegative, the row corresponding to $(x + 3)^2$ consists of $+$ signs (except over its key point, -3). Notice that in the row for the factor $2 - x$, the $+$ signs are to the left of the demarcation line at 2. At the very top of figure 1.6(a) is a row of $-$ signs for the factor -2. The product row has been added to the sign chart in figure 1.6(b). From this row, you can see that the solution set is $\{-3\} \cup [0, 2]$. This solution set is graphed in figure 1.6(c).

Figure 1.6

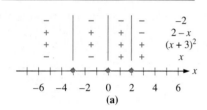 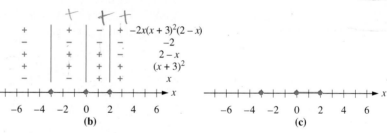

(a)   (b)   (c)

▶ **EXAMPLE 1.8**

Solve $\dfrac{2x + 3}{x - 1} \geq 1$.

Solution

First, notice that $\dfrac{2x + 3}{x - 1}$ is undefined at $x = 1$. Thus, even though the given inequality is not strict, an open circle should be placed at 1 to indicate that it cannot be in the solution set.

The given inequality is equivalent to

$$\frac{2x + 3}{x - 1} - 1 \geq 0$$

or

$$\frac{2x + 3}{x - 1} - \frac{x - 1}{x - 1} \geq 0$$

which simplifies to

$$\frac{x + 4}{x - 1} \geq 0.$$

Algebraic Inequalities

The sign chart for each factor is shown in figure 1.7(a). Remember, we began with an open circle at the key point 1. The circle at the key point −4 *is* filled in because −4 is in the inequality's solution set. The top, or quotient, row of figure 1.7(b) leads to the solution set, which is $(-\infty, -4] \cup (1, \infty)$.

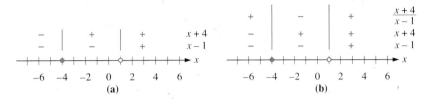

Figure 1.7

It is important to stress how to handle inequalities involving fractional expressions, as in example 1.8. Values of x for which the numerator becomes zero *may* belong to the solution set of the inequality. Values of x for which the denominator becomes zero *cannot* belong to the solution set.

Be careful! It is tempting to try to solve an inequality such as the one in example 1.8 by multiplying through by the denominator, just as you would solve an equation. Many students would have attempted to solve example 1.8 by multiplying through by $x - 1$ to obtain

$$2x + 3 \geq x - 1$$

or

$$x \geq -4.$$

This is not the same solution set we obtained in example 1.8 by using the sign chart. So, this second "method" is wrong. Why? Because, when we multiplied by $x - 1$ in the second "solution," we assumed that $x - 1$ was positive. Thus, our attention was restricted to $x > 1$. To find the correct solution set, we also need to consider the case when $x - 1 < 0$. As you can see, the sign chart method is not only easier because it organizes the cases automatically, it is also a surer method to solve many inequalities.

Absolute Value Inequalities

In chapter 0, you saw how four properties involving absolute value could be used to solve certain equations. (See examples 0.9 and 0.10.) To solve inequalities involving absolute value, we need to point out two additional properties. As usual, these properties will be stated for the strict inequality symbols, but their analogues for ≤ and ≥ are also valid. You will be asked to prove these properties in exercise 51.

Additional Properties of Absolute Value

For any positive real number a,

1. $|x| < a$ if and only if $-a < x < a$;
2. $|x| > a$ if and only if either $x > a$ or $x < -a$.

The type of inequality in example 1.9 appears often in calculus courses in the study of limits.

▶ **EXAMPLE 1.9** Solve the inequality $|3x + 5| \leq 2$.

Solution By the analogue of property 1 for the $\leq$ symbol, this inequality is equivalent to

$$-2 \leq 3x + 5 \leq 2$$

or

$$-7 \leq 3x \leq -3$$

$$-\tfrac{7}{3} \leq x \leq -1.$$

Thus, the solution set is the interval $[-\tfrac{7}{3}, -1]$. ◀

The work in the next example is reminiscent of example 1.5.

▶ **EXAMPLE 1.10** Solve the inequality $|\tfrac{3}{4} - 5x| > 3$.

Solution By property 2, this inequality is equivalent to

$$\tfrac{3}{4} - 5x > 3 \quad \text{or} \quad \tfrac{3}{4} - 5x < -3$$

$$-5x > \tfrac{9}{4} \quad \text{or} \quad -5x < -\tfrac{15}{4}$$

$$x < -\tfrac{9}{20} \quad \text{or} \quad x > \tfrac{3}{4}.$$

Thus, the solution is a union of intervals, $(-\infty, -\tfrac{9}{20}) \cup (\tfrac{3}{4}, \infty)$. ◀

Radical Inequalities

The principle of powers is often used to solve radical equations. It was reviewed in chapter 0. (See examples 0.9 and 0.10.) An analogous technique is used for solving radical inequalities.

Properties Used in Solving Radical Inequalities

Solving radical inequalities depends on the following two facts, where n is a positive integer and a and b are real numbers:

1. If n is odd, then $a < b$ is equivalent to $a^n < b^n$.
2. If n is even, then $0 \leq a < b$ is equivalent to $0 \leq a^n < b^n$.

To illustrate these properties, consider, for instance, $-4 < 2$. The first property, with $n = 3$, produces $(-4)^3 < 2^3$, or $-64 < 8$. Similarly, with $n = 5$, we have $(-4)^5 < 2^5$, or $-1024 < 32$. On the other hand, consider $3 < 5$. Since both 3 and 5 are positive, $3^n < 5^n$ for any n, even or odd. So, using $n = 4$, we have $3^4 < 5^4$, or $81 < 625$. Now, let's use these techniques to solve some inequalities.

▶ **EXAMPLE 1.11** Solve the inequality $\sqrt[3]{x^2 + 3} \geq 2$.

Solution Here, the index is $n = 3$, which is odd. Thus, by (1) above, the given inequality is equivalent to

Algebraic Inequalities

Figure 1.8

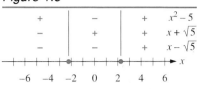

or
$$(\sqrt[3]{x^2+3})^3 \geq 2^3$$
$$x^2+3 \geq 8$$
$$x^2-5 \geq 0.$$

Factoring, we rewrite this as $(x-\sqrt{5})(x+\sqrt{5}) \geq 0$. From the sign chart in figure 1.8, the solution set is the union of intervals $(-\infty, -\sqrt{5}] \cup [\sqrt{5}, \infty)$. ◀

▶ **EXAMPLE 1.12**   Solve the inequality $\sqrt{x-5} < 2$.

Solution   As principal square roots are nonnegative, the given inequality is equivalent to
$$0 \leq \sqrt{x-5} < 2.$$

The index here is $n = 2$, which is even. Squaring, we obtain the equivalent condition
$$0 \leq x-5 < 4$$
or
$$5 \leq x < 9.$$

Thus, the solution set is the interval $[5, 9)$. ◀

Until now, all the fractional expressions appearing in our inequalities have been rational expressions. The next example involves a denominator which is not a polynomial. It illustrates how to use a sign chart in order to solve an inequality involving any fractional expression.

▶ **EXAMPLE 1.13**   Solve $\dfrac{5-x}{\sqrt{x^2-4}} \geq 0$.

Solution   Begin by considering the denominator. As $x^2 - 4$ factors as $(x-2)(x+2)$, we may build a sign chart as in figure 1.9(a).

In figure 1.9(b), a row for $\sqrt{x^2-4}$ is added. A horizontal line is drawn under this row since it is the only row thus far that will be used to determine the quotient row.

Figure 1.9

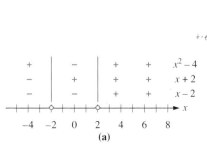
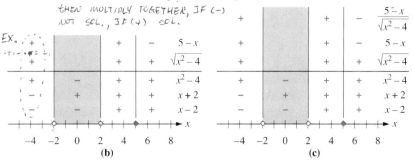

Chapter 1 / Fundamental Algebraic and Geometric Concepts

The radical $\sqrt{x^2-4}$ is defined only if $x^2 - 4 \geq 0$. Since a denominator cannot be 0, $x^2 - 4 > 0$ describes the numbers for which $\dfrac{5-x}{\sqrt{x^2-4}}$ is meaningful.

But $x^2 - 4 > 0$ rules out the numbers in the interval $[-2, 2]$, and so the section of the sign chart corresponding to this interval has been crossed out.

The sign graph for the numerator, $5 - x$, has been placed in figure 1.9(b) above the sign graph for $\sqrt{x^2 - 4}$.

Figure 1.9(c) contains the quotient row giving the sign of $\dfrac{5-x}{\sqrt{x^2-4}}$. From this, you can see that the solution set is $(-\infty, -2) \cup (2, 5]$.

Exercises 1.1

1. Given $2 < 4$ and $c = -3$, what can you conclude by applying property 4 of inequalities?
2. Given $-4 \leq -2$ and $c = 3$, what can you conclude by applying property 3 of inequalities?
3. Explain why the pair of facts $2 < 3, -8 > -12$ does not contradict property 3 of inequalities.
♦ 4. Prove property 2 of inequalities.
♦ 5. Prove property 3 of inequalities.
♦ 6. Prove property 4 of inequalities.

In exercises 7–50, solve the given inequality, describe its solution set using interval notation, and graph the inequality.

7. $9x - 2 \leq 3x + 1$
8. $7x + \sqrt{2} < 2x - \sqrt{3}$
9. $\frac{2}{3}x + 2 > \frac{2}{7}x - \pi$
10. $2x - 7 \geq 5x - \frac{1}{2}$
11. $2(x + \frac{1}{5}) < \frac{5}{2}(2 - x)$
12. $-3(2x - 1) \leq -3x$
13. $7x \geq -4(6 - x)$
14. $\frac{2}{3}(8x - 3) > -\frac{4}{3}(3 - x)$
15. $-14 \leq 3x - 5 < 7$
16. $-2 \leq \dfrac{x+4}{3} \leq -1$
17. $-3 < \dfrac{4-x}{2} < 2$
18. $-\sqrt{2} < \dfrac{\pi x}{-\sqrt{2}} \leq 3\sqrt{2}$
19. Either $3x - 4 < -10$ or $3x - 4 > 10$
20. Either $4 - 3x \geq 10$ or $4 - 3x \leq -10$
21. $(x - \sqrt{2})(x + 3) \geq 0$
22. $x^2 + 3x < 4x + 6$
23. $-2(x - 3)(x + 4) > 0$
24. $-3(x + \pi)x \leq 0$
25. $5x^2 + \frac{3}{2} \leq 3x^2 + \frac{11}{2}$
26. $-4(3 - x)(2 - x) \geq 0$
27. $3x^2 + 4x + 5 < x^2 + 2x + 7$
28. $5x^2 + 4 > 2x^2 - 4$
29. $(x - 1)(x - 2)(x - 3)(x - 4) \leq 0$
30. $-6x(x + 1)(x - 2) \geq 0$
31. $6x^3 (x - \frac{3}{2})^2 (4 - x) > 0$
32. $(x - 3)^6 (x - \frac{3}{2})(x - 4) < 0$
33. $-2(x - 3)^5 (x - \frac{3}{2})(x - 4) < 0$
34. $6x^3(x - \frac{3}{2})^5 (4 - x) > 0$
35. $\dfrac{3x-2}{2x+6} \geq 1$
36. $\dfrac{2-3x}{2x+6} \leq 2$
37. $\dfrac{3-2x}{3x+6} < -3$
38. $\dfrac{2x}{3x+6} > \dfrac{-4x-9}{x+2}$
39. $|2x - 3| < 5$
40. $|x - 5| \leq 0.01$
41. $|5 - 2x| \leq 0.4$
42. $|\sqrt{2}x + 1| < \sqrt{2}$
43. $|x^2 + 5| < 5$
44. $|x^3 - 8| < 0$
45. $|-3x| \geq 4$
46. $|3x - 2| > 4$
47. $|x - \pi| \geq 1$
48. $|2 - x| > 0.5$
49. $\left|\dfrac{3}{x-1}\right| \geq 2$. (Hint: $\left|\dfrac{a}{b}\right| = \dfrac{|a|}{|b|}$ if $b \neq 0$.)
50. $|x^2 - 7x - 10| > 0$

♦ 51. (a) Prove that if a is positive, then $|x| < a$ if and only if $-a < x < a$. (Hint: it follows from chapter 0 that the distance from x to the origin is $|x|$.)
    (b) Prove that if a is positive, then $|x| > a$ if and only if either $x > a$ or $x < -a$.
52. Prove that if a and b are real numbers, then $||a| - |b|| \leq |a - b|$. (Hint: use the triangle inequality and exercise 14 of chapter 0.)
♦ 53. Prove that if a and b are real numbers which are not both 0, then $a^2 + ab + b^2 > 0$. (Hint: consider cases corresponding to the possible signs of a and b. If $a > 0$ and $b < 0$, consider $(a + b)^2 - ab$.)
54. (There is no "variable analogue" of properties 3 and 4 of inequalities.) Show that the inequality $x^2 < 2x$ is equivalent to neither $x < 2$ nor $x > 2$.

In exercises 55–64, solve the given radical inequality and describe its solution set using interval notation.

55. $\sqrt[3]{x^2 - 17} < 2$
56. $\sqrt{x^2 - 17} \leq 2$
57. $\sqrt[4]{x - 17} \leq 2$
58. $\sqrt[5]{x - \pi} > -1$
59. $\sqrt[4]{x - \pi} < -1$
60. $\sqrt{2 - 3x} \leq \sqrt{3 + 2x}$
61. $\dfrac{x-5}{\sqrt{x^2-9}} \geq 0$
62. $\dfrac{2-x}{\sqrt{x^2-9}} \leq 0$
63. $\dfrac{2-x}{\sqrt{9-x^2}} > 0$
64. $\dfrac{5-x}{\sqrt{9+x^2}} < 0$

Rectangular Coordinates, Graphs, and Slope

65. Fahrenheit (F) and Celsius (C) degrees are related by the equation $F = \frac{9}{5}C + 32$. On a day when the temperature varied from 68°F to 84°F, over what interval did the Celsius degree readings vary?
66. According to Hooke's law, if a certain spring is stretched or compressed x inches from its natural length L, a force of $F = 4x$ pounds is needed to maintain this compression. During an experiment, the spring was stretched from a length of $L + 8$ inches to a length of $L + 11$ inches. Over what interval did the applied force F vary?
67. Smartz-Rent-a-Car charges $40 per day plus 35¢ per mile. If John and David rent a car from Smartz for one day and wish to spend at most $61.40 (excluding tax), what is the furthest they should drive? (Hint: what is the cost if they drive x miles?)
68. The product of two real numbers is at least 312. If one of these numbers is −26, what can you conclude about the other number?
69. A child throws a ball into the air. After t seconds, the ball is $-16t^2 + 32t + 4$ feet above the ground. During which time interval is the ball at least 8 feet above the ground?
70. A rectangular garden is four times as long as it is wide. The area of the garden is A square meters and the perimeter is p meters. What can you conclude if $A > p$? (Hint: let the width and length of the garden be x meters and y meters, respectively.)
71. A parallel electrical circuit has more than one possible path for current to flow. The current in the figure below has two paths that it can use to get from A to B. Then R, the resistance (in ohms, Ω) in such a circuit, may be calculated by using the formula

$$\frac{1}{R} = \frac{1}{R_1} + \frac{1}{R_2}$$

where R_1 and R_2 measure the resistances in each of the two paths. What can you say about R_2 if $R_1 = 12$ and $R < 3$?

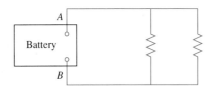

72. If an object is located at a distance p units from a converging lens of focal length f units, then the object's image will be located at a distance q units from the lens according to the relationship

$$\frac{1}{f} = \frac{1}{p} + \frac{1}{q}.$$

What can you say about the value of f if $p = 15$ and $q < 10$?

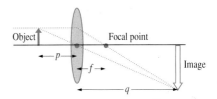

1.2 RECTANGULAR COORDINATES, GRAPHS, AND SLOPE

There is a one-to-one correspondence between the set of ordered pairs of real numbers and the set of points in the plane. This correspondence was reviewed in chapter 0. We shall now use it to study planar graphs.

The ability to plot points permits a geometric interpretation of equations or inequalities involving two variables, say x and y. For example, only the points with coordinates $(x, y) = (x, 1)$ satisfy the equation $y = 1$. By plotting such points, you can see that they are all on the horizontal line situated 1 unit above the x-axis. We say that this line is the graph of the equation $y = 1$. More generally, the **graph** of an algebraic statement in two variables is obtained by plotting all the points whose coordinates (x, y) satisfy the given algebraic condition. Often the difficulty is determining how many points are "enough" to produce a reasonably accurate picture of the graph. Precalculus will give you some answers as to how many points are enough; calculus will give more answers as to how many points are enough and will help you locate them.

Chapter 1 / Fundamental Algebraic and Geometric Concepts

▶ **EXAMPLE 1.14**   Graph each of the following: (a) $y = 1$, (b) $y = -x$, (c) $x \leq -1$, and (d) $|x| \leq 2$ and $y = 1$.

Solutions   The graphs are in figures 1.10(a)–(d).

Figure 1.10

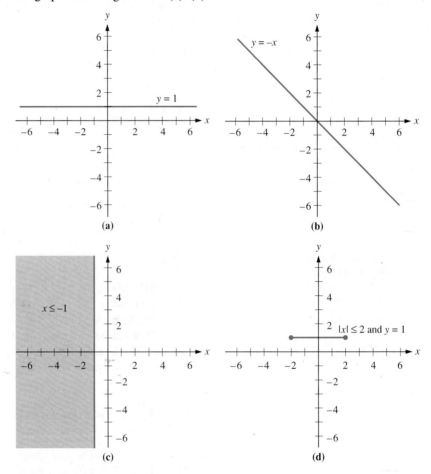

The Distance Formula

Figure 1.11

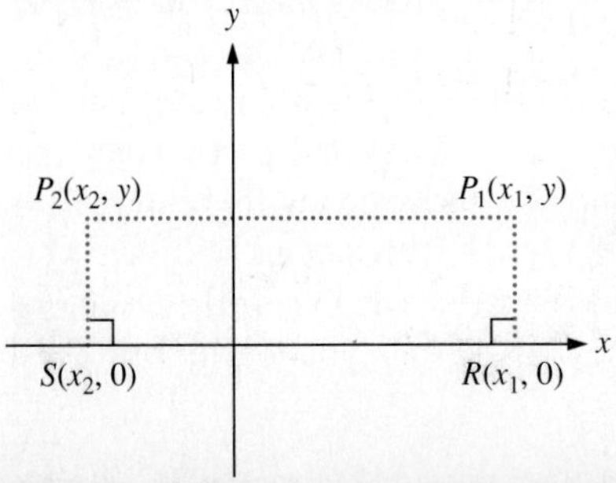

We often need to determine the distance between two points. Given their coordinates, we can do this by a useful formula that amounts to a version of Pythagoras' Theorem. Before developing this formula in general, let's first consider the distance between two points $P_1(x_1, y)$ and $P_2(x_2, y)$ on a horizontal line, as shown in figure 1.11. If perpendiculars from P_1 and P_2 meet the x-axis at R and S respectively, these points have coordinates $R(x_1, 0)$ and $S(x_2, 0)$. Thus, the distance between R and S, denoted $d(R, S)$, is $d(R, S) = |x_2 - x_1|$. The figure $P_1 R S P_2$ is a rectangle, and so the opposite sides $\overline{P_1 P_2}$ and $\overline{RS}$ are congruent. Hence, $d(P_1, P_2) = d(R, S) = |x_2 - x_1|$. We have just shown that the distance between two points on a horizontal line is determined by taking the absolute value of the difference of their x-coordinates.

Rectangular Coordinates, Graphs, and Slope

In a similar way, you can show that the distance between two points on a vertical line is determined by taking the absolute value of the difference of their y-coordinates.

EXAMPLE 1.15

Find the distance between $(-2, 5)$ and $(-2, -4)$.

Solution

As shown in figure 1.12, these two points are on a vertical line. The distance between them is $|5 - (-4)| = 9$.

Figure 1.12

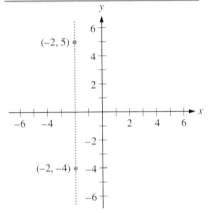

Let $P_1(x_1, y_1)$ and $P_2(x_2, y_2)$ be any two points which do not lie on the same horizontal or vertical line. Draw a vertical line through P_2 and a horizontal line through P_1. As in figure 1.13, these two lines meet at the point Q with coordinates (x_2, y_1). Notice that $\triangle P_1QP_2$ is a right triangle, with hypotenuse $\overline{P_1P_2}$. Because $\overline{P_2Q}$ is parallel to the y-axis, the above case gives $d(P_2, Q) = |y_2 - y_1|$. Similarly, $d(P_1, Q) = |x_2 - x_1|$. Hence, by Pythagoras' Theorem,

$$[d(P_1, P_2)]^2 = [d(P_1, Q)]^2 + [d(P_2, Q)]^2$$
$$= (x_2 - x_1)^2 + (y_2 - y_1)^2.$$

We have just developed the formula for finding the distance between two points in a coordinate plane.

Distance Formula

The distance between $P_1(x_1, y_1)$ and $P_2(x_2, y_2)$ is

$$d(P_1, P_2) = \sqrt{(x_2 - x_1)^2 + (y_2 - y_1)^2}$$

which is a nonnegative number.

Figure 1.13

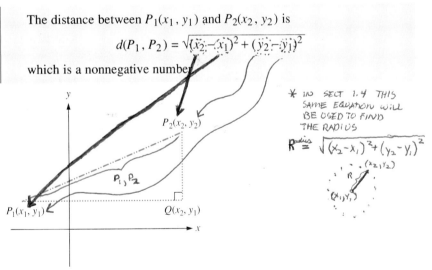

▶ **EXAMPLE 1.16**

Determine the distance between the points $(3, -4)$ and $(-5, 2)$.

Solution

It does not matter which point we call P_1. If we call the first point P_1, then $x_1 = 3$, $y_1 = -4$, $x_2 = -5$, and $y_2 = 2$. Substituting these values into the distance formula, we find $d(P_1, P_2)$ is

$$\sqrt{(-5 - 3)^2 + [2 - (-4)]^2} = \sqrt{(-8)^2 + 6^2} = \sqrt{64 + 36} = \sqrt{100} = 10.$$

At this point, we return to the triangle inequality. It states that if a and b are any real numbers, then $|a + b| \leq |a| + |b|$. You can show, moreover, that if

a and b are nonzero, then $|a + b| = |a| + |b|$ if and only if a and b have the same sign. The name for the triangle inequality is explained by the following geometric analogue.

Let P_1, P_2, and P_3 be three distinct points. If these points are collinear (that is, all lie on one line), one of them, say P_2, is between the other two. In this case, you can see that $d(P_1, P_2) + d(P_2, P_3) = d(P_1, P_3)$. Conversely, what we are now going to show is that if the points are *not* collinear, then this equation does *not* hold.

Geometric Triangle Inequality

The **geometric triangle inequality** asserts that if P_1, P_2, and P_3 are noncollinear points, then

$$d(P_1, P_2) + d(P_2, P_3) > d(P_1, P_3).$$

In other words, the sum of the lengths of any two sides of a triangle is greater than the length of the third side.

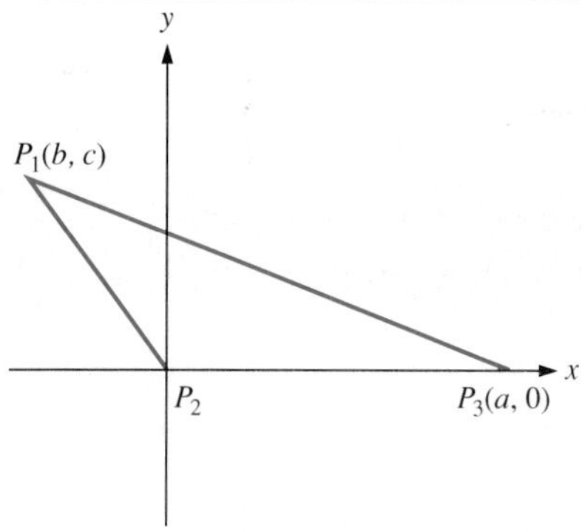

To prove the geometric triangle inequality, we orient the axes as in figure 1.14, so that P_1 has coordinates (b, c) with $c \neq 0$, P_2 is the origin $(0, 0)$, and P_3 has coordinates $(a, 0)$ with $a > 0$. By the distance formula, what we are trying to prove amounts to

$$\sqrt{(b-0)^2 + (c-0)^2} + \sqrt{(a-0)^2 + (0-0)^2} > \sqrt{(b-a)^2 + (c-0)^2}$$

or

$$\sqrt{b^2 + c^2} + \sqrt{a^2} > \sqrt{(b-a)^2 + c^2}.$$

Figure 1.14

Applying various principles from section 1.1, we see that the required assertion is equivalent to each of the following assertions:

$$(\sqrt{b^2 + c^2} + \sqrt{a^2})^2 > (\sqrt{(b-a)^2 + c^2})^2$$

$$b^2 + c^2 + 2a\sqrt{b^2 + c^2} + a^2 > b^2 - 2ab + a^2 + c^2$$

$$2a\sqrt{b^2 + c^2} > -2ab$$

$$\sqrt{b^2 + c^2} > -b.$$

This inequality is true, and hence so is the desired one. In fact, since $c \neq 0$, we have

$$\sqrt{b^2 + c^2} > \sqrt{b^2} = |b| \geq -b.$$

This completes the proof.

The Midpoint Formula

Let $P(x, y)$ be the midpoint of the segment joining two points $P_1(x_1, y_1)$ and $P_2(x_2, y_2)$. Draw lines parallel to the axes, as in figure 1.15. Then $\triangle P_1PQ$ and $\triangle PP_2R$ are congruent right triangles. As a result, the corresponding sides $\overline{P_1Q}$ and $\overline{PR}$ have the same length. Thus,

Rectangular Coordinates, Graphs, and Slope

$$x - x_1 = x_2 - x$$

which means that

$$x = \frac{x_1 + x_2}{2}.$$

This proof assumed $x_1 < x_2$, but you can check that similar reasoning leads to the same formula for x if $x_1 > x_2$. Also note that if $x_1 = x_2$, then $x_1 = x = x_2$, and so, once again, $x = \frac{x_1 + x_2}{2}$.

A similar argument shows that

$$y - y_1 = y_2 - y$$

or

$$y = \frac{y_1 + y_2}{2}.$$

From this, we conclude that the coordinates of the midpoint of $\overline{P_1P_2}$ are given by the following **midpoint formula.**

Midpoint Formula

The midpoint of the line segment containing $P_1(x_1, y_1)$ and $P_2(x_2, y_2)$ is

$$\left(\frac{x_1 + x_2}{2}, \frac{y_1 + y_2}{2}\right).$$

Figure 1.15

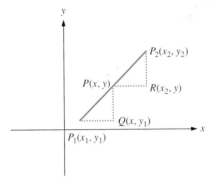

▶ **EXAMPLE 1.17**   Determine the coordinates of the midpoint of the segment joining $(2, -4)$ and $(-5, 2)$.

Solution   Let's decide to call $(2, 4)$ the point P_1. Then $x_1 = 2$, $y_1 = -4$, $x_2 = -5$, and $y_2 = 2$. Applying the above formula, we find that the midpoint is

$$\left(\frac{2 + -5}{2}, \frac{-4 + 2}{2}\right) = \left(-\frac{3}{2}, -1\right).$$
◀

When P is the midpoint of the segment $\overline{P_1P_2}$, then $d(P_1, P) = \frac{1}{2}d(P_1, P_2)$. We could just as easily find a formula for a point Q on $\overline{P_1P_2}$ which is one-third as far from P_1 as is P_2; in this case $d(P_1, Q) = \frac{1}{3}d(P_1, P_2)$. A generalization of the midpoint formula to handle situations like this will appear in exercise 32.

Slope

Given the three points $P(4, 2)$, $Q(1, 1)$ and $R(3, 5)$ as shown in figure 1.16, draw the lines $\overleftrightarrow{PQ}$ and $\overleftrightarrow{RQ}$. As you can see, $\overleftrightarrow{RQ}$ is "steeper" than $\overleftrightarrow{PQ}$. The concept of slope is a number which indicates how steep a line is and in which direction it tilts.

Figure 1.16

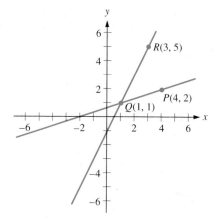

Definition of Slope

Let $P_1(x_1, y_1)$ and $P_2(x_2, y_2)$ be points such that $x_1 \neq x_2$. Then the **slope** of the line $\overleftrightarrow{P_1P_2}$ is denoted by **m** and defined by

$$m = \frac{y_2 - y_1}{x_2 - x_1}.$$

The slope of a vertical line is undefined.

Exercise 34 will ask you to prove that a line is horizontal if and only if its slope equals 0.

▶ **EXAMPLE 1.18**

Determine the slopes of the lines $\overleftrightarrow{PQ}$ and $\overleftrightarrow{RQ}$ in figure 1.16.

Solution

We shall let m_{PQ} denote the slope of $\overleftrightarrow{PQ}$, and m_{RQ} the slope of $\overleftrightarrow{RQ}$.

$$m_{PQ} = \frac{1-2}{1-4} = \frac{1}{3}$$

$$m_{RQ} = \frac{1-5}{1-3} = 2.$$

Notice that the steeper line has the larger slope. ◀

▶ **EXAMPLE 1.19**

Given the points $A(-2, 4)$, $B(4, -5)$, and $C(2, -2)$, determine the slopes of $\overleftrightarrow{AB}$ and $\overleftrightarrow{AC}$.

Solution

Figure 1.17

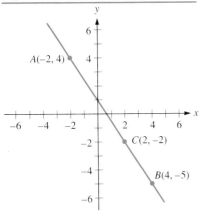

The indicated points and lines appear in figure 1.17.

$$m_{AB} = \frac{-5-4}{4-(-2)} = \frac{-9}{6} = -\frac{3}{2}$$

$$m_{AC} = \frac{-2-4}{2-(-2)} = \frac{-6}{4} = -\frac{3}{2}.$$

The lines in figure 1.16 tilt upward to the right and have positive slopes. The line in figure 1.17 tilts downward to the right and has a negative slope. As indicated in figure 1.18, these are general phenomena. You can see this as follows. Suppose $P_2(x_2, y_2)$ is to the right of $P_1(x_1, y_1)$, in the sense that $x_1 < x_2$. Then $x_2 - x_1$ is positive, and so $m = \dfrac{y_2 - y_1}{x_2 - x_1}$ has the same sign as $y_2 - y_1$. In particular, P_2 is above P_1 if m is positive, and P_2 is below P_1 if m is negative.

Figure 1.18 also indicates the following general results.

Figure 1.18

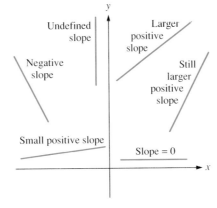

Proposition

The closer a line's slope is to zero, the closer the line is to being horizontal. The larger the absolute value of the slope, the closer the line is to being vertical.

These assertions may be proved by noticing how to use the definition of slope to sketch the graph of a nonvertical line. In going from $P_1(x_1, y_1)$ to $P_2(x_2, y_2)$, we would "run" $x_2 - x_1$ units horizontally and "rise" $y_2 - y_1$ units vertically. For instance, if $P_1(-3, 6)$ is on a line with slope $\frac{2}{5}$, we could find another point $P_2(x_2, y_2)$ on this line by starting at P_1 and moving 5 units to the right and 2 units up. This leads to $x_2 = 2$ and $y_2 = 8$. By connecting P_1 to $(2, 8)$, we obtain a sketch of the desired line. As expected since its slope is positive, it tilts upward to the right.

Let's return to figure 1.17. It appears that A, B, and C are collinear. In fact, this is so. You can verify it by showing that $d(A, C) + d(C, B) = d(A, B)$. (Doing so involves showing that $\sqrt{52} + \sqrt{13} = \sqrt{117}$.) A quicker way is to use the calculations in example 1.19. These indicate the following key fact.

Figure 1.19

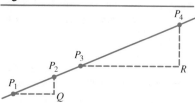

The slope of a nonvertical line ℓ does not depend on which two points on ℓ are used to calculate m.

This fact may be proved as follows. Consider points $P_1(x_1, y_1)$, $P_2(x_2, y_2)$, $P_3(x_3, y_3)$, and $P_4(x_4, y_4)$ on a nonvertical line ℓ, as in figure 1.19.

Draw dashed vertical and horizontal lines as in the figure. The result is a pair of similar (not necessarily congruent) right triangles $\triangle P_1 P_2 Q$ and $\triangle P_3 P_4 R$. You know from geometry that corresponding sides of similar triangles are proportional. In particular,

$$\frac{d(P_2, Q)}{d(P_1, Q)} = \frac{d(P_4, R)}{d(P_3, R)}.$$

When the points are arranged as in figure 1.19, this equation becomes

$$\frac{y_2 - y_1}{x_2 - x_1} = \frac{y_4 - y_3}{x_4 - x_3}.$$

You can check that an equivalent equation results for any other arrangement of the given four points on ℓ. Thus,

$$m_{P_1 P_2} = m_{P_3 P_4}$$

and the proof is complete.

By once again using the fact that corresponding sides of similar triangles are proportional, you may similarly prove the following useful result. A proof is requested in exercise 59.

Nonvertical lines are parallel if and only if their slopes are equal.

Through a given point, there is exactly one line parallel to any given line. Thus, by the above two boldfaced results, we obtain the following general conclusion that example 1.19 suggested.

Using Slopes to Detect Collinearity

Three points A, B, and C with different x-coordinates are collinear if and only if $m_{AB} = m_{AC}$.

Slope allows us to determine if lines are parallel. It can also be used to determine whether two lines are perpendicular, as in the following numerical criterion.

Slopes of Perpendicular Lines

Let ℓ_1 and ℓ_2 be nonvertical lines, with slope m_1 and m_2, respectively. Then ℓ_1 and ℓ_2 are perpendicular if and only if $m_1 m_2 = -1$.

Rectangular Coordinates, Graphs, and Slope

Some people remember this assertion by calling it the "negative reciprocals result," since the condition $m_1 m_2 = -1$ is equivalent to $m_2 = \dfrac{-1}{m_1}$. For instance, if a line ℓ has slope $\tfrac{2}{7}$, then any line perpendicular to ℓ has slope $\dfrac{-1}{2/7} = -\dfrac{7}{2}$.

You will frequently need the negative reciprocals result. As one indication of its importance, this book contains four proofs of it. Below, we give a proof using similar triangles. Exercise 58 asks you to prove it by using Pythagoras' Theorem. Additional proofs, using trigonometry and vectors, are in chapters 6 and 7.

Our first proof of the negative reciprocals result will depend on figure 1.20. In this figure, we have located the axes so that ℓ_1 and ℓ_2 intersect at the origin.

Because ℓ_1 is perpendicular to ℓ_2, the right triangles $\triangle P_1 Q O$ and $\triangle O R P_2$ are similar. Hence, corresponding sides are proportional:

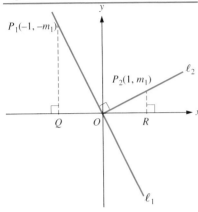

Figure 1.20

$$\frac{d(P_1, Q)}{d(Q, O)} = \frac{d(O, R)}{d(R, P_2)}$$

or

$$\frac{\sqrt{m_1^2}}{1} = \frac{1}{\sqrt{m_2^2}}$$

$$\sqrt{m_1^2}\,\sqrt{m_2^2} = 1$$

$$|m_1||m_2| = 1.$$

The previous comments show that, since ℓ_1 and ℓ_2 tilt opposite ways, m_1 and m_2 have opposite signs. Thus, $m_1 m_2$ is negative, and so it follows that $m_1 m_2 = -1$.

Next, we prove the converse. Suppose $m_1 m_2 = -1$. Replace ℓ_2 in figure 1.20 with the line ℓ_3 (with slope m_3) passing through O and perpendicular to ℓ_1. The previous argument gives $m_1 m_3 = -1$. Thus, $m_3 = \dfrac{-1}{m_1} = m_2$. It follows that ℓ_3 is parallel to ℓ_2. Since these two lines pass through O, they coincide. In particular, ℓ_3 is perpendicular to ℓ_1, and the proof of the negative reciprocals result is complete.

▶ **EXAMPLE 1.20**

Given $P_1(2, -3)$, $P_2(4, -8)$, $P_3(-6, -8)$, $P_4(-10, 2)$, $P_5(\tfrac{5}{2}, \tfrac{3}{2})$, and $P_6(\tfrac{5}{8}, \tfrac{3}{4})$, determine if (a) $\overleftrightarrow{P_1 P_2} \parallel \overleftrightarrow{P_3 P_4}$, (b) $\overleftrightarrow{P_1 P_2} \parallel \overleftrightarrow{P_5 P_6}$, (c) $\overleftrightarrow{P_1 P_2} \perp \overleftrightarrow{P_3 P_5}$ and (d) $\overleftrightarrow{P_1 P_2} \perp \overleftrightarrow{P_5 P_6}$.

Solutions

(a) By the definition of slope, we have

$$m_{P_1 P_2} = \frac{-8 - (-3)}{4 - 2} = -\frac{5}{2} \quad \text{and} \quad m_{P_3 P_4} = \frac{2 - (-8)}{-10 - (-6)} = \frac{10}{-4} = -\frac{5}{2}.$$

Hence, since their slopes are equal, $\overleftrightarrow{P_1 P_2} \parallel \overleftrightarrow{P_3 P_4}$.

(b) Here, we have

$$m_{P_5P_6} = \frac{\frac{3}{4} - \frac{3}{2}}{\frac{5}{8} - \frac{5}{2}} = \frac{-\frac{3}{4}}{-\frac{15}{8}} = \frac{2}{5}.$$

Since $m_{P_1P_2} \neq m_{P_5P_6}$, $\overleftrightarrow{P_1P_2} \not\parallel \overleftrightarrow{P_5P_6}$.

(c) Calculating, we see that

$$m_{P_3P_5} = \frac{\frac{3}{2} - (-8)}{\frac{5}{2} - (-6)} = \frac{\frac{19}{2}}{\frac{17}{2}} = \frac{19}{17}.$$

Since $m_{P_1P_2} \cdot m_{P_3P_5} = (-\frac{5}{2})(\frac{19}{17}) = -\frac{95}{34} \neq -1$, we see that $\overleftrightarrow{P_1P_2}$ is not perpendicular to $\overleftrightarrow{P_3P_5}$.

(d) We have seen that $m_{P_1P_2} = -\frac{5}{2}$ and $m_{P_5P_6} = \frac{2}{5}$. Since these are negative reciprocals, we conclude that $\overleftrightarrow{P_1P_2} \perp \overleftrightarrow{P_5P_6}$.

Exercises 1.2

In exercises 1–14, sketch the graph of the given algebraic condition. (It may help first to build a table of some ordered pairs (x, y) satisfying the given condition. Next, plot the corresponding points and look for a pattern.)

1. $y = 2$
2. $x = 4$
3. $x = -4$
4. $y = -3$
5. $x = 0$
6. $|x| = 0$
7. $|y| = 0$
8. $|y| = 2$
9. $y = x$
10. $y = |x|$
11. $x = -y$
12. $x = \sqrt{y}$
13. $y = \sqrt{x}$
14. $1 \leq x \leq 2$ and $y = 4$

In exercises 15–20, find (a) the distance from P_1 to P_2 and (b) the coordinates of the midpoint of the line segment $\overline{P_1P_2}$.

15. $P_1(-2, -3), P_2(-2, -6)$
16. $P_1(2, 8), P_2(-4, 8)$
17. $P_1(2, 8), P_2(10, 14)$
18. $P_1(-2, -3), P_2(1, -1)$
19. $P_1(-\frac{1}{2}, 0), P_2(-\frac{7}{2}, \sqrt{2})$
20. $P_1(\pi, -2), P_2(0, 5)$

21. Determine the possible value(s) of x if the distance from $(8, 3)$ to $(x, 3)$ is 6.
22. Determine the coordinates of P_2 if P_1 has coordinates $(-2, 4)$ and the midpoint of $\overline{P_1P_2}$ is $(3, -\frac{5}{2})$.
23. Prove that $A(-2, 3), B(5, 4)$, and $C\left(5 + \frac{1}{\sqrt{2}}, 4 - \frac{7}{\sqrt{2}}\right)$ are the vertices of a right triangle. (Hint: verify that $d(A, B)^2 + d(B, C)^2 = d(A, C)^2$.)
24. Prove that $A(0, 2), B(4, 10), C(12, 6)$, and $D(8, -2)$ are the vertices of a square. (Hint: show that $ABCD$ is equilateral and has a right angle.)

In exercises 25–28, determine whether the given three points are collinear. (Hint: can they be named A, B, C in such a way that $d(A, B) + d(B, C) = d(A, C)$?)

25. $(2, 3), (4, 7), (5, 9)$
26. $(0, 0), (3, 4), (11, 12)$
27. $(0, 0), (3, 4), (-4, -3)$
28. $(2, 3), (4, 1), (-2, 7)$

In exercises 29–31, find and simplify an algebraic condition on the coordinates (x, y) of the points P satisfying the given property.

29. The distance from the origin to P is 5.
30. P is on the perpendicular bisector of $\overline{AB}$, given $A(-1, 2)$ and $B(2, 14)$. (Hint: equivalently, P is equidistant from A and B.)
31. $d(A, P) + d(B, P) = 4$, given $A(-1, 0)$ and $B(1, 0)$.
◆ 32. (This generalizes the midpoint formula.) Consider distinct points $P_1(x_1, y_1)$ and $P_2(x_2, y_2)$ and a real number r such that $0 < r < 1$. Show that the point P on $\overline{P_1P_2}$ such that $\frac{d(P_1, P)}{d(P_1, P_2)} = r$ has coordinates $(rx_2 + (1 - r)x_1, ry_2 + (1 - r)y_1)$.
33. Consider the points $P_1(-\frac{1}{2}, \frac{2}{3})$ and $P_2(\frac{1}{4}, -\frac{1}{6})$. Use the result in exercise 32 to find the coordinates of the point on $\overline{P_1P_2}$ which is three-fourths of the way from P_1 to P_2.
◆ 34. Prove the assertion in the text that a line is horizontal if and only if its slope is 0.

In exercises 35–38, determine the slope of the line through P and Q.

35. $P(2, -3), Q(-1, -4)$
36. $P(-5, 4), Q(\frac{1}{2}, \frac{3}{2})$
37. $P(\sqrt{2}, 4\sqrt{2}), Q(2\sqrt{2}, 3\sqrt{2})$
38. $P(\pi, -1), Q(3, -1)$

In exercises 39–40, determine which line has (a) positive slope, (b) negative slope, (c) slope equal to 0, (d) undefined slope.

39.

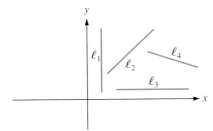

40.

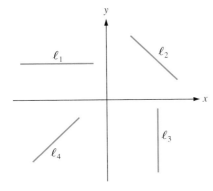

In exercises 41–42, determine which line's slope is greater.

41.

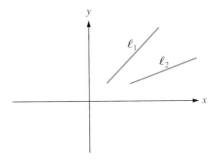

42.

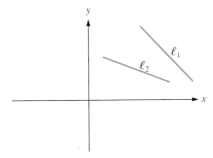

Equations of Lines

◆ In exercises 43–46, find another point on the line, with slope m, passing through P_1. Hence sketch the graph of this line.

43. $P_1(-1, 2)$, $m = -3$
44. $P_1(1, -2)$, $m = 4$
45. $P_1(2, -\frac{2}{3})$, $m = \frac{5}{3}$
46. $P_1(-\frac{9}{7}, 1)$, $m = -\frac{2}{7}$

In exercises 47–48, determine whether the lines $\overleftrightarrow{P_1P_2}$ and $\overleftrightarrow{P_3P_4}$ are parallel.

47. $P_1(1, 12)$, $P_2(3, 4)$, $P_3(5, 6)$, $P_4(4, 10)$
48. $P_1(\frac{1}{2}, \frac{1}{3})$, $P_2(\frac{5}{2}, \frac{5}{3})$, $P_3(4, 8)$, $P_4(7, 6)$

In exercises 49–50, determine whether $P_1P_2P_3P_4$ is a parallelogram.

49. $P_1(0, 0)$, $P_2(3, 4)$, $P_3(10, 9)$, $P_4(4, 1)$
50. $P_1(-1, 1)$, $P_2(2, 7)$, $P_3(5, 7)$, $P_4(2, 1)$

In exercises 51–52, use calculations involving slope in order to determine whether A, B, and C are collinear.

51. $A(\frac{1}{2}, \frac{1}{4})$, $B(\frac{13}{2}, \frac{15}{4})$, $C(\frac{25}{12}, \frac{25}{4})$
52. $A(1, 2)$, $B(-3, 6)$, $C(5, -2)$

In exercises 53–56, determine whether the lines $\overleftrightarrow{P_1P_2}$ and $\overleftrightarrow{P_3P_4}$ are perpendicular.

53. $P_1(0, -2)$, $P_2(3, 5)$, $P_3(-1, 3)$, $P_4(6, 0)$
54. $P_1(1, 2)$, $P_2(3, 10)$, $P_3(-1, -2)$, $P_4(-2, 2)$
55. $P_1(\frac{1}{2}, \frac{1}{4})$, $P_2(\frac{3}{2}, \frac{7}{4})$, $P_3(1, -1)$, $P_4(3, 2)$
56. $P_1(-5, 1)$, $P_2(2, 0)$, $P_3(-4, 1)$, $P_4(-3, 8)$
57. Use calculations involving slope to rework exercise 23.
◆ **58.** Use Pythagoras' Theorem and its converse to give another proof of the negative reciprocals result. (Hint: in figure 1.20, ℓ_1 and ℓ_2 are perpendicular if and only if $d(O, P_1)^2 + d(O, P_2)^2 = d(P_1, P_2)^2$.)
◆ **59.** Let ℓ_1 and ℓ_2 be nonvertical lines, with slopes m_1 and m_2, respectively. Show that ℓ_1 and ℓ_2 are parallel if and only if $m_1 = m_2$.
◆ **60.** Prove that the altitudes of any triangle are concurrent; that is, given any triangle, some point lies on all three of its altitudes. (Hint: you can suppose that the vertices of the triangle are $A(b, c)$, $B(0, 0)$, and $C(a, 0)$, as in figure 1.14. Show that the altitudes from A and B intersect at $P\left(b, \frac{(a-b)b}{c}\right)$. Next, show that $\overleftrightarrow{PC}$ is perpendicular to $\overleftrightarrow{AB}$.)

1.3 EQUATIONS OF LINES

Roughly speaking, much of calculus and its applications consists of approximating pieces of graphs with suitable line segments. It is therefore important for you to understand lines and relations having linear graphs.

Suppose, for example, that you wish to describe a type of financial opportunity in which the resulting profit is twice as great as the initial investment. If x denotes the initial investment and y denotes the resulting profit, the underlying algebraic relation is the equation $y = 2x$. Its graph is a line, with slope 2, passing through the origin. In general, $y = mx$ describes a relation of direct proportionality between y and x, with constant of proportionality m. Graphically, as you will see in this section, $y = mx$ corresponds to a line, with slope m, passing through the origin. This section emphasizes equations whose graphs are lines. Linear relations are treated lightly here, primarily as applications in example 1.26 and exercises 61–70, since the more general subject of functions is the topic of chapter 2.

Which equations have lines as their graphs? Corresponding to **vertical** lines, we have equations of the form $x = $ constant. Figure 1.21 illustrates the following.

Equations of Vertical Lines

The graph of $x = c$ is a vertical line. It is c units to the right of the y-axis if $c > 0$, and is $|c|$ units to the left of the y-axis if $c < 0$.

The graph of $x = 0$ is the y-axis.

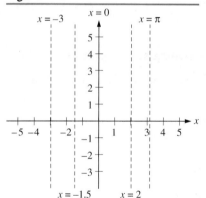

Figure 1.21

What about **nonvertical** lines? As you saw in section 1.2, any such line has a slope. Let's suppose that ℓ is a nonvertical line, with slope m, passing through a given point $P_1(x_1, y_1)$. By the discussion in section 1.2, you can see that if $P(x, y)$ is another point, then P is on ℓ if and only if the slope of $\overleftrightarrow{P_1P}$ is m. By the definition of m, this condition may be expressed as $\dfrac{y - y_1}{x - x_1} = m$, or equivalently, $y - y_1 = m(x - x_1)$. The coordinates of P_1 also satisfy this equation. So a point is on ℓ if and only if its coordinates satisfy this equation. Thus, we have seen how P_1 and m determine the **point-slope form** of an equation of ℓ.

Point-Slope Form

$$y - y_1 = m(x - x_1)$$

is an equation for the line passing through $P_1(x_1, y_1)$ and having slope m.

▶ **EXAMPLE 1.21**

Find, and simplify, an equation of the line which passes through the point $(-2, 5)$ and has slope $-\frac{2}{3}$.

Solution

We shall use the point-slope form, with $x_1 = -2$, $y_1 = 5$, and $m = -\frac{2}{3}$. The resulting equation is

$$y - 5 = -\tfrac{2}{3}[x - (-2)]$$
$$= -\tfrac{2}{3}(x + 2).$$

This simplifies to

Equations of Lines

$$3(y - 5) = -2(x + 2)$$
$$3y - 15 = -2x - 4$$
$$3y + 2x = 11.$$ ◀

Given an equation, we often need to plot several points before seeing the shape of its graph. Let's consider the equation $3y + 2x = 5$, for example. This is similar to the equation found in example 1.21. In order to plot points on the graph, we begin by solving $3y + 2x = 5$ for y in terms of x. We obtain $y = -\frac{2}{3}x + \frac{5}{3}$. If we set $x = 1$, then $y = -\frac{2}{3}(1) + \frac{5}{3} = 1$. If $x = 4$, then $y = -\frac{2}{3}(4) + \frac{5}{3} = -1$. If $x = 0$, then $y = \frac{5}{3}$. A table showing these, and several other, values is given below.

x	-5	-2	-1	0	1	4	7
y	5	3	$\frac{7}{3}$	$\frac{5}{3}$	1	-1	-3

If the corresponding points (x, y) are plotted and connected, we seem to get the line shown in figure 1.22. Later in this section, you will learn how to recognize when an equation has a linear graph. Then you would only need to plot two points before completing the sketch.

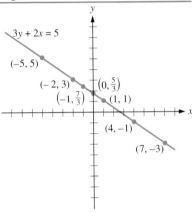

Figure 1.22

EXAMPLE 1.22

Find an equation for the line with slope 3, passing through the point $(1, -5)$.

Solution

Here, we are given $m = 3$, $x_1 = 1$, and $y_1 = -5$. The point-slope form of the equation is

$$y - (-5) = 3(x - 1).$$

This simplifies to

$$y + 5 = 3x - 3$$
$$y = 3x - 8.$$

The given point and the line are graphed in figure 1.23. ◀

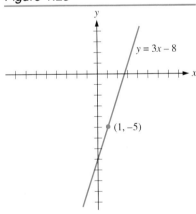

Figure 1.23

Consider the form in which we left the equation in example 1.22. The coefficient of x was the given slope, 3. Also, you can see from figure 1.23 that the constant term, -8, is the coordinate of the point where the line crosses the y-axis. This is a general phenomenon. Indeed, suppose that a nonvertical line ℓ with slope m crosses the y-axis at the point $(0, b)$. By the point-slope form, an equation for ℓ is

$$y - b = m(x - 0)$$

or

$$y = mx + b.$$

Slope-Intercept Form

The equation $y = mx + b$ is called the **slope-intercept form** for an equation of the line with slope m that intersects the y-axis at $(0, b)$.

Chapter 1 / Fundamental Algebraic and Geometric Concepts

EXAMPLE 1.23   Find the slope and the y-intercept of the line which is the graph of $5x - 2y + 6 = 0$.

Solution   To answer the question, we shall determine the slope-intercept form for this equation. This is done by solving the given equation for y. We obtain

$$2y = 5x + 6$$
$$y = \tfrac{5}{2}x + 3.$$

This equation is now in slope-intercept form, with $mx + b = \tfrac{5}{2}x + 3$. Thus, we see that the slope is $m = \tfrac{5}{2}$ and the y-intercept is $(0, b) = (0, 3)$. ◀

One consequence of the slope-intercept form is that the graph of $y = mx$ is a line, with slope m, passing through the origin.

Another consequence is analogous to the earlier discussion of vertical lines.

Equations of Horizontal Lines

> If b is constant, the graph of $y = b$ is a **horizontal** line. If $b > 0$, it is b units above the x-axis; and if $b < 0$, it is $|b|$ units below the x-axis. The graph of $y = 0$ is the x-axis.

We next recall some important facts that were proved in section 1.2.

Review

> Let ℓ_1 and ℓ_2 be two nonvertical lines with slopes m_1 and m_2, respectively. Then
>
> ℓ_1 and ℓ_2 are parallel if and only if $m_1 = m_2$;
>
> and ℓ_1 and ℓ_2 are perpendicular if and only if $m_1 m_2 = -1$.

EXAMPLE 1.24   Find an equation of the line containing the point $(-2, 4)$ and parallel to the line $2x + 4y - 5 = 0$.

Solution   If we rewrite the given equation in the form $y = mx + b$, we have $y = -\tfrac{1}{2}x + \tfrac{5}{4}$. Thus, the slope of the given line, and any line parallel to it, is $m = -\tfrac{1}{2}$. Using the point-slope form to find an equation for the line through the point $(-2, 4)$, and with slope $-\tfrac{1}{2}$, we obtain

$$y - 4 = -\tfrac{1}{2}(x + 2)$$

or $2y + x = 6$. ◀

EXAMPLE 1.25   Find an equation of the line containing $(-2, 4)$ and perpendicular to the line $2x + 4y - 5 = 0$.

Solution   We are given the same point and line as in example 1.24. We have seen that the slope of the given line is $m_1 = -\tfrac{1}{2}$. If m_2 is the slope of a line perpendicular to the given line, then $m_2 = \dfrac{-1}{m_1} = 2$. Thus, the slope of the desired line is 2. Using the point-slope form, with the point $(-2, 4)$ and slope 2, we find

Figure 1.24

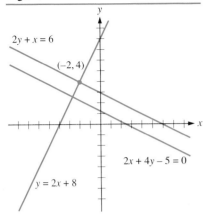

$$y - 4 = 2(x + 2)$$
or
$$y = 2x + 8.$$

The graphs of the data and solutions of examples 1.24 and 1.25 are in figure 1.24.

The General Linear Equation

We have seen that any vertical line has an equation of the form $x = c$, and that any nonvertical line has an equation $y = mx + b$. Rewriting these, we see that any line has an equation of the form $Ax + By + C = 0$, where A, B, and C are real numbers and A and B are not both zero. Next, we shall show that *any* equation of this type has a line as its graph. For this reason, we have the following.

General Linear Equation

$Ax + By + C = 0$ (with not both A and B being zero) is called the **general linear equation**.

Consider any such equation, $Ax + By + C = 0$. If $B \neq 0$, we can proceed as in example 1.24, solving for y as $y = -\frac{A}{B}x - \frac{C}{B}$. Viewing this as slope-intercept form, $y = mx + b$, we recognize the graph as a nonvertical line with slope $m = -\frac{A}{B}$ and y-intercept $(0, b) = \left(0, -\frac{C}{B}\right)$.

On the other hand, if $B = 0$, then $A \neq 0$ and so $Ax + By + C = 0$ is equivalent to $x = -\frac{C}{A}$. Its graph is a vertical line, with x-intercept $\left(-\frac{C}{A}, 0\right)$.

You don't need to memorize the above formulas. For any linear equation $Ax + By + C = 0$, you can sketch its graph as follows. If $B \neq 0$, solve for $y = -\frac{A}{B}x - \frac{C}{B}$ and then, as in section 1.2, sketch the line, with slope $-\frac{A}{B}$, passing through $\left(0, -\frac{C}{B}\right)$. If $B = 0$, solve for $x = -\frac{C}{A}$, and then sketch this vertical line.

The next example is a typical "word problem application" of linear equations.

▶ **EXAMPLE 1.26**

The net income of a small business grows "linearly." If the business's income was $50,000 in 1988 and $55,000 in 1989, what income would you expect it to have in 2000?

Solution

Let's summarize the given information by using points whose first coordinate is the year and whose second coordinate is the net dollar income for that year. In this way, we obtain the points (1988, 50,000), (1989, 55,000), and (2000, y_3). The question is to find y_3, the net dollar income in the year 2000. Since the net income grows linearly, one way to proceed is to find an equation for the line, ℓ, that fits the given data.

Using the first two points, we find that the slope of ℓ is

$$m = \frac{55{,}000 - 50{,}000}{1989 - 1988} = 5000.$$

Next, we apply the point-slope form, with $x_1 = 1988$, $y_1 = 50{,}000$, and $m = 5000$. The resulting equation for ℓ is

$$y - 50{,}000 = 5000(x - 1988).$$

This simplifies to

$$y = 5000x - 5000(1988) + 50{,}000$$
$$= 5000x - 9{,}890{,}000.$$

To find the net income for the year 2000, we let $x = 2000$ and determine that

$$y = 5000(2000) - 9{,}890{,}000$$
$$= 110{,}000.$$

Thus, we expect the business to have a net income of \$110,000 in the year 2000. ◀

Distance from a Point to a Line

We shall next state a formula for determining the distance from a point to a line.

Distance from a Point to a Line

If ℓ is a line described by a given linear equation $Ax + By + C = 0$ (in which not both A and B are 0) and if $P_1(x_1, y_1)$ is a point, then the (perpendicular) distance from P_1 to ℓ is

$$\frac{|Ax_1 + By_1 + C|}{\sqrt{A^2 + B^2}}.$$

You will be asked to prove this result in exercise 40. Additional proofs of it will be developed in exercise 29 of section 5.2 and in section 7.6. The next example illustrates how to use it.

▶ **EXAMPLE 1.27** Find the distance from (a) the point $(2, -1)$ to the line $-3x + 4y - 8 = 0$; (b) the point $(4, 2)$ to the line $y = 2x - 5$; and (c) the point $(-1, 2)$ to the line $12x + 5y = 4$.

Solutions (a) Here, $x_1 = 2$, $y_1 = -1$, $A = -3$, $B = 4$, and $C = -8$. Substituting into the above formula, we find that the distance is

$$\frac{|-3(2) + 4(-1) - 8|}{\sqrt{(-3)^2 + 4^2}} = \frac{|-18|}{\sqrt{25}} = \frac{18}{5}.$$

(b) First, we must rewrite the given equation in general linear form. The result is $2x - y - 5 = 0$. Thus, the required distance is

$$\frac{|2(4) + (-1)(2) - 5|}{\sqrt{2^2 + (-1)^2}} = \frac{1}{\sqrt{5}}.$$

(c) Rewriting the given equation in general linear form as $12x + 5y - 4 = 0$, we find the distance as

$$\frac{|12(-1) + 5(2) - 4|}{\sqrt{12^2 + 5^2}} = \frac{|-6|}{\sqrt{169}} = \frac{6}{13}.$$ ◀

Systems of Linear Equations

Consider the two lines given by

$$3x + y = 5$$
$$2x - 3y = -4.$$

From the graphs in figure 1.25(a), we see that these two lines intersect at a point, which appears to be (1, 2). You can verify this guess by checking that (1, 2) satisfies both of the given equations.

Figure 1.25(b) shows the graphs of the lines given by

$$3x + y = 5$$
$$2x - y = 1.$$

These two lines also intersect at a point. But it is not easy to determine the coordinates of that point by looking at the graph. The following material is an introduction to methods for determining the coordinates of any points of intersection.

In each of the above two examples, the two equations were considered as a pair and, as such, give an example of a **system** of equations. The first system

Figure 1.25

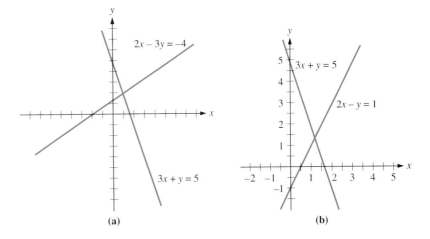

(a)   (b)

has the particular solution (1, 2) and the **solution set** {(1, 2)}. You can verify that the second system has the particular solution (1.2, 1.4) and the solution set {(1.2, 1.4)}. In general, an ordered pair of numbers is a **particular solution** of a system if the ordered pair satisfies each equation in the system. (It is common to refer to a particular solution of a system as a solution of the system.) **Solving** a system means finding its **solution set**, namely the set of all its particular solutions. A system that has one or more particular solutions is said to be **consistent.** If its solution set is empty, a system is **inconsistent.**

Solving systems of linear equations is a skill you will need often. This section will introduce the topic and provide you with the tools to solve systems with two variables.

The next example shows how to determine whether a given ordered pair is a solution of a given system.

▶ **EXAMPLE 1.28**

Consider the system

$$\begin{cases} 3x + 2y = 7 \\ x - 3y = 6. \end{cases}$$

Determine which of the following is/are a particular solution of this system: (a) (1, 2), (b) (6, 0), (c) (3, −1).

Solutions

(a) You can check that (1, 2) is a solution of the first equation. However, it does not satisfy the second equation, since $1 - 3(2) \neq 6$. Thus, (1, 2) is not a particular solution of the system.

(b) Although the ordered pair (6, 0) satisfies the second equation, it does not satisfy the first. Hence, (6, 0) is not a solution of the system.

(c) (3, −1) satisfies both equations and, hence, is a solution of the system. ◀

Figures 1.26(a)–(c) show the three possible relationships between two lines in a plane. Hence, there are three possible types of solution sets for a system of two linear equations in two variables. Figure 1.26(a) illustrates the case where the equations' graphs are two parallel lines. This is an example of an inconsis-

Figure 1.26

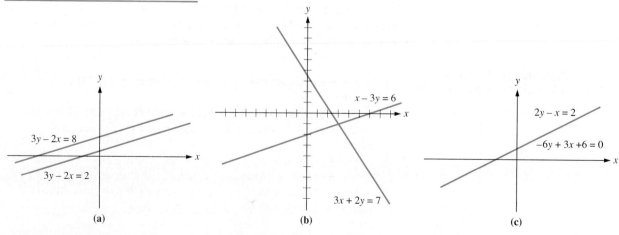

Equations of Lines

tent system. Figure 1.26(b) shows the graphs of the equations from example 1.28. The result is two lines intersecting in exactly one point. Thus, the system in example 1.28 is an example of a consistent system with exactly one particular solution. Figure 1.26(c) illustrates the situation of two linear equations with identical graphs. The corresponding system of equations has infinitely many solutions. A system of linear equations with infinitely many solutions is said to be **dependent.** It can be shown that a consistent system of linear equations either has one unique particular solution or is dependent.

One may consider the solution sets for systems of three (or more) linear equations in two (or more) variables. For example, the system

$$\begin{cases} x + y = \frac{3}{2} \\ x + 2y = 5 \\ 2x - y = -\frac{15}{2} \end{cases}$$

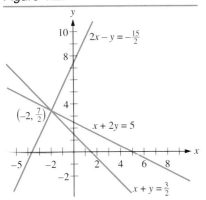

Figure 1.27

is graphed in figure 1.27. It has the solution set $\{(-2, \frac{7}{2})\}$.

While it is fairly easy to pick two lines that intersect in a point, it is much harder to do so with three or more lines. For instance, consider the following system of three linear equations in two variables:

$$\begin{cases} x + y = 5 \\ \frac{1}{2}x - y = -2 \\ 5x - 4y = 16. \end{cases}$$

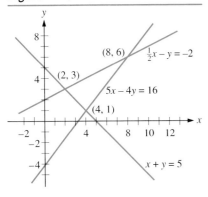

Figure 1.28

The associated graphs are in figure 1.28. As you can see, no point lies on all three lines. Thus, the system has no particular solution, and is therefore inconsistent.

Until now, we have solved systems by graphing. This method works well only if the graphs are accurate. Hence, we next move on to an algebraic method for solving systems of linear equations.

Solution by Substitution

The following method is probably familiar to you in the case of two linear equations in two variables, but it applies more generally. After its statement, several examples will be given to show how to use it.

Substitution Method for Solving Systems of Equations

1. Solve one of the equations for one variable in terms of the other variables.
2. Substitute the result of step 1 into the other equation(s). This will produce a system with one less equation and at least one less variable.
3. Repeat steps 1 and 2 until you reach one equation in one variable.
4. Solve the equation obtained in step 3.
5. Substitute what you found in step 4 to rewrite the system. Repeat the process to find all the solutions of the system.

Chapter 1 / Fundamental Algebraic and Geometric Concepts

▶ **EXAMPLE 1.29**   Use the substitution method to solve the system

$$\begin{cases} 4x + 5y = 7 & (1) \\ 2x - y = -1. & (2) \end{cases}$$

Solution   Let's solve equation (2) for y:

$$y = 2x + 1.$$

Substituting this expression into equation (1) produces

$$4x + 5(2x + 1) = 7$$

$$14x = 2$$

$$x = \tfrac{1}{7}.$$

Using this result, we rewrite the other equation, (2), getting

$$2(\tfrac{1}{7}) - y = -1$$

$$y = \tfrac{9}{7}.$$

The only particular solution is $(x, y) = (\tfrac{1}{7}, \tfrac{9}{7})$.   ◀

Example 1.30 illustrates the fact that if a system is dependent, then it is possible to use the substitution method on it and eventually produce an equation of the form $b = b$. On the other hand, producing an equation of the form $b = b$ by applying the substitution method to a system does *not*, by itself, tell you whether that system is dependent or even consistent.

▶ **EXAMPLE 1.30**   Use the substitution method to solve the system

$$\begin{cases} 6x - 4y = -2 & (1) \\ -9x + 6y = 3. & (2) \end{cases}$$

Solution   Solving equation (1) for x produces

$$6x = 4y - 2$$

$$x = \tfrac{2}{3}y - \tfrac{1}{3}. \qquad (3)$$

Using equation (3) allows us to rewrite equation (2) as

$$-9(\tfrac{2}{3}y - \tfrac{1}{3}) + 6y = 3$$

$$-6y + 3 + 6y = 3$$

$$3 = 3.$$

The existence of this last equation, which is of the form $b = b$, *suggests* (but does not yet prove) to us that the given system is dependent. In fact, equation (2) can be obtained by multiplying equation (1) through by $-\tfrac{3}{2}$. Thus, (1) and (2) have the same graph, which is a certain line. The ordered pairs corresponding to the points on this line are the particular solutions of the system.   ◀

Equations of Lines

Example 1.31 illustrates how to use the substitution method to determine if a system is inconsistent.

Using Substitution to Detect Inconsistent Systems

Using the substitution method on an inconsistent system will lead to a numerical contradiction.

▶ **EXAMPLE 1.31**

Use the substitution method to solve the system

$$\begin{cases} 3x - y = 5 & (1) \\ 6x - 2y = 12. & (2) \end{cases}$$

Solution

We begin by solving equation (1) for y:

$$y = 3x - 5. \qquad (3)$$

Figure 1.29

Next, we substitute (3) into (2):

$$6x - 2(3x - 5) = 12$$
$$6x - 6x + 10 = 12$$
$$10 = 12. \qquad (4)$$

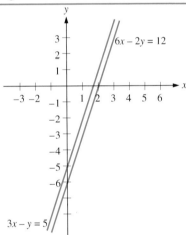

Equation (4) is false. Since this contradiction followed logically from (1) and (2), we see that no ordered pair (x, y) can satisfy both (1) and (2). In other words, the given system of equations is inconsistent. This is indicated geometrically in figure 1.29, which shows that the graphs of the given equations are distinct parallel lines. ◀

Our final example, example 1.32, illustrates how to use the substitution method for systems of linear equations with more than two variables. In a calculus course, you will learn how to graph a linear equation with three variables. Without such graphing ability, one must rely on algebra to solve example 1.32.

▶ **EXAMPLE 1.32**

Solve the system of linear equations

$$\begin{cases} 2x + 8y - 3z = 2 & (1) \\ -4x + 4y + z = 6 & (2) \\ 6x - 4y + 5z = -1. & (3) \end{cases}$$

Solution

Solving equation (2) for z, we find

$$z = 4x - 4y + 6.$$

Substituting this into equation (1) produces

$$2x + 8y - 3(4x - 4y + 6) = 2$$

or

$$-10x + 20y = 20. \qquad (4)$$

Substituting the above expression for z into (3) gives

Chapter 1 / Fundamental Algebraic and Geometric Concepts

$$6x - 4y + 5(4x - 4y + 6) = -1$$

or

$$26x - 24y = -31. \qquad (5)$$

Thus, the original system of three equations in three variables is "reduced" to a system of two equations in two variables:

$$\begin{cases} -10x + 20y = 20 & (4) \\ 26x - 24y = -31. & (5) \end{cases}$$

We return to step 1 and solve equation (4) for x, with the result

$$x = 2y - 2.$$

Substituting this into (5) produces

$$26(2y - 2) - 24y = -31$$
$$28y = 21$$
$$y = \tfrac{3}{4}.$$

We have reached one equation in one variable, thus completing step 3. Substituting $y = \tfrac{3}{4}$ into equation (4), we have

$$-10x + 20(\tfrac{3}{4}) = 20$$
$$x = -\tfrac{1}{2}.$$

Finally, we substitute $x = -\tfrac{1}{2}$ and $y = \tfrac{3}{4}$ into equation (2), to solve for z.

$$-4(-\tfrac{1}{2}) + 4(\tfrac{3}{4}) + z = 6$$
$$5 + z = 6$$
$$z = 1.$$

Thus, the only particular solution is $x = -\tfrac{1}{2}$, $y = \tfrac{3}{4}$, $z = 1$; more compactly, we may write $(x, y, z) = (-\tfrac{1}{2}, \tfrac{3}{4}, 1)$. ◀

Exercises 1.3

1. Write equations whose graphs are the vertical lines (a) ℓ_1, (b) ℓ_2, (c) ℓ_3, and (d) ℓ_4 shown in the figure.

2. Graph the following: (a) $x = -1$, (b) $x = 4$, (c) $x - 1 = 0$, and (d) $2x - 3 = 0$.

In exercises 3–8, write and simplify an equation of the line which passes through P_1 and has slope m.

3. $P_1(-\tfrac{7}{2}, 1)$, $m = -2$
4. $P_1(8, -4)$, $m = \tfrac{1}{2}$
5. $P_1(6, -2)$, $m = \tfrac{1}{3}$
6. $P_1(-1, -2)$, $m = 0$
7. $P_1(2, 1)$, $m = 0$
8. $P_1(1, 2)$, $m = -3$

In exercises 9–14, write and simplify an equation of the line with y-intercept $(0, b)$ and slope m.

9. $b = -8$, $m = 2$
10. $b = 2$, $m = -\tfrac{1}{3}$
11. $b = \tfrac{1}{3}$, $m = 0$
12. $b = 0$, $m = 0$
13. $b = \sqrt{14}$, $m = -\tfrac{1}{2}$
14. $b = -2$, $m = \pi$

Equations of Lines

In exercises 15–18, find (a) the slope and (b) the y-intercept of the line which is the graph of the given equation.

15. $y - 3x + 4 = 0$
16. $y = -4$
17. $2y - 7 = 0$
18. $14x + 7y = 28$
19. Graph the following: (a) $y = 4$, (b) $y = -1$, (c) $y + 2 = 0$, and (d) $2y - 7 = 0$.
20. Write equations whose graphs are the horizontal lines (a) ℓ_1, (b) ℓ_2, (c) ℓ_3, and (d) ℓ_4 shown in the figure.

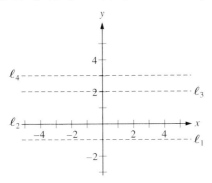

In exercises 21–24, write and simplify equations for the lines through P_1 and either (a) parallel or (b) perpendicular to the given line.

21. $P_1(-7, \frac{1}{2})$; $4x + 5y + 6 = 0$
22. $P_1(\sqrt{2}, \pi)$; $2x - 9 = 0$ **23.** $P_1(\sqrt{2}, \pi)$; $2y + 7 = 0$
24. $P_1(3, 2)$; $2x - 3y + 4 = 0$
25. Find an equation of the line passing through $(-2, 1)$ and $(3, 16)$. (Hint: first find its slope.)
26. Find an equation of the line with x-intercept $(a, 0)$ and y-intercept $(0, b)$. (Comment: your answer should involve the nonzero constant "parameters" a and b.)

In exercises 27–31, sketch the graph of the given linear equation.

27. $3y - 2x + 6 = 0$ **28.** $3x - 2y + 6 = 0$
29. $3x + 6 = 0$ **30.** $2y - 6 = 0$
31. $x + y + 6 = 0$
32. Determine which of the following is/are a linear equation:
(a) $x + \pi y - \sqrt{2} = 0$ (b) $x^{1/3} - 2 = 0$
(c) $x^2 - y = 0$ (d) $5 - y = 0$

In exercises 33–38, find the distance from P_1 to the given line.

33. $P_1(2, 1)$; $-3x + 4y + 7 = 0$
34. $P_1(1, 2)$; $5x - 12y + 4 = 0$
35. $P_1(3, 10)$; $y = 2x + 3$ **36.** $P_1(-2, 2)$; $x + 2y = 1$
37. $P_1(-4, 2)$; $3x + 5y + 2 = 0$
38. $P_1(-2, 4)$; $y = -3x - 2$

39. Let S be the set of points P such that the distance from P to the line $-3x + 4y + 7 = 0$ is 2. Find an equation whose graph is S. (Hint: find an algebraic condition on the coordinates of P.)

◆ **40.** Prove the formula given in the text for the distance from a point to a line. (Hint: if P_1 is on ℓ, the assertion reduces to the truism $0 = 0$. Suppose P_1 is not on ℓ. Let P_2 be the intersection of ℓ with the line $A(y - y_1) = B(x - x_1)$. Why is $d(P_1, P_2)$ relevant? Calculate it.)

In exercises 41–46, graph the given equations and hence determine whether the given system is dependent, consistent with a unique particular solution, or inconsistent.

41. $\begin{cases} 2x + 3y = 9 \\ -3x + 3y = 4 \end{cases}$ **42.** $\begin{cases} -3x + 3y = 4 \\ 12x - 12y + 16 = 0 \end{cases}$

43. $\begin{cases} \frac{x}{2} - \frac{y}{3} - 1 = 0 \\ -3x + 2y + 6 = 0 \end{cases}$ **44.** $\begin{cases} x + y = 2 \\ y - x = 0 \\ 2x - y = 1 \end{cases}$

45. $\begin{cases} x + y = 0 \\ -x + 2y = 3 \\ 4x - 2y = -9 \end{cases}$ **46.** $\begin{cases} \frac{x}{2} - \frac{y}{3} - 1 = 0 \\ -3x + 2y + 12 = 0 \end{cases}$

In exercises 47–50, determine which, if any, value(s) of k make the given system (a) dependent, (b) consistent with a unique particular solution, or (c) inconsistent.

47. $\begin{cases} -2x + 3y = 4 \\ 6x - 9y = k \end{cases}$ **48.** $\begin{cases} -2x + ky = 0 \\ x + y = 1 \end{cases}$

49. $\begin{cases} x + 2y = k \\ 2x - y = 2k \end{cases}$ **50.** $\begin{cases} -2x + 3y = 4 \\ 3x - y = 8 \\ kx + y = 12 \end{cases}$

51. Determine which of the ordered pairs (a) $(35, -21)$, (b) $(-52, 37)$, (c) $(37, -52)$, and (d) $(4, -3)$ is/are a particular solution of the system

$$\begin{cases} 2x + 3y = 7 \\ -5x - 7y = 1. \end{cases}$$

52. Use the substitution method to **indicate** that the system

$$\begin{cases} 10x + 20y - 15 = 0 \\ -6x - 12y + 9 = 0 \end{cases}$$

is dependent. Then **prove** this geometrically.

In exercises 53–54, use the substitution method to show that the given system is inconsistent.

53. $\begin{cases} 10x + 20y = 15 \\ 6x + 12y = -10 \end{cases}$ **54.** $\begin{cases} 3x + 2y + 5z = 4 \\ 2x - 3y + z = 2 \\ 13y + 7z = 0 \end{cases}$

In exercises 55–60, use the substitution method to solve the given system.

Chapter 1 / Fundamental Algebraic and Geometric Concepts

55. $\begin{cases} 3x - y = 7 \\ 4x + 5y = 3 \end{cases}$

56. $\begin{cases} 4x + 3y = 7 \\ 2x + y = 2 \end{cases}$

57. $\begin{cases} 3x - y = 7 \\ 30x - 5y = 5 \\ x - 2y = 24 \end{cases}$

58. $\begin{cases} x + 2y = 0 \\ 4x - 4y = 6 \\ 17x - 26y = 30 \end{cases}$

59. $\begin{cases} x + 2y - 2z = -6 \\ 2x - 3y - z = -1 \\ 3x + y + 2z = 25 \end{cases}$

60. $\begin{cases} x + 2y - 2z = -2 \\ 3y - z = 0 \\ 4z = -12 \end{cases}$

◆ 61. Suppose that it costs $0.45 per mile to drive a car. Let x denote the number of miles that a car is driven and let y denote the associated cost (in dollars). Write an equation expressing a "linear relationship" between x and y.

◆ 62. John and David agree to pay a typist $5 per hour plus $0.80 for each page that is typed. If the typist works for 6 hours, types x pages, and receives a wage of y dollars, write an equation expressing a "linear relationship" between x and y.

◆ 63. John rents a car from a company that charges $0.29 per mile plus a flat rate of $54. David rents a car from another company that charges $0.33 per mile plus a flat rate of $28. John and David drive the same distance (meeting halfway between Tennessee and Maine) and discover that their car rentals cost the same. (a) How far did each man drive? (b) How much did each man pay to rent his car?

◆ 64. A rectangular playground's length is twice its width. If the playground's perimeter is 360 yards, determine the playground's dimensions.

◆ 65. The population of a small town grows "linearly." If the town's population was 2000 in 1980 and 2480 in 1988, what population would you expect it to have in 1998?

◆ 66. Tickets for a charity event cost $5 each for adults and $3 each for children. If a total sale of 75 tickets produces gross revenue of $315, how many adult tickets were sold?

◆ 67. As a fund-raising project, schoolchildren sell candy bars for $1 apiece. Each candy bar only costs the school $0.48, but the school also has a fixed expense of $85 associated with the project. If sales of x candy bars lead to a net profit of y dollars, write an equation expressing a "linear relationship" between x and y.

◆ 68. A radiator contains 6 quarts of fluid that is 40 percent antifreeze (by volume) and 60 percent water. How much of this fluid should be drained off so that, if one then adds pure antifreeze, the radiator will contain 6 quarts of fluid that is 50 percent antifreeze?

◆ 69. Find the real numbers a and b such that the line $ax + by + 1 = 0$ has slope $-\frac{2}{3}$ and passes through the point $(-7, 11)$.

◆ 70. Working together, a husband and wife require 2 hours to paint a certain room. If the wife could have painted this room in 5 hours working by herself, how long would it have taken the husband to paint the room by himself?

1.4 CIRCLES, PARABOLAS, AND TANGENTS

In order to develop the material in chapter 2, we shall need several geometric examples. The last two sections have provided examples of graphs determined by linear expressions. We shall also need graphs which are described by quadratic expressions. Examples of graphs described by quadratic expressions are the circles and parabolas studied in this section. Later in the book, you will study parabolas in greater detail. Also, just as section 1.3 considered geometric and algebraic methods for solving systems of linear equations, this section will examine methods for solving systems of quadratic equations in two variables.

Just what is a circle? When you draw a circle with a compass, the pencil marks are all the same distance from the compass point. This observation leads us to make the following definition.

Definition of Circle

Let C be a point and let $r > 0$. The **circle** with **center** C and **radius** r is the set of points each of which is at a distance of r units from C.

Consider the circle with center $C(h, k)$ and radius r, as in figure 1.30. According to the distance formula from section 1.2, a point $P(x, y)$ is on this circle if and only if

Circles, Parabolas, and Tangents

Figure 1.30

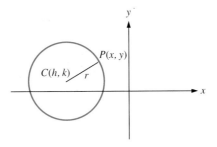

$$\sqrt{(x-h)^2 + (y-k)^2} = r.$$

Squaring both sides, we obtain the standard equation for a circle with center (h, k) and radius r.

Equation of a Circle

The **standard equation for the circle with center (h, k) and radius r** is

$$(x-h)^2 + (y-k)^2 = r^2.$$

▶ **EXAMPLE 1.33**

Find and simplify an equation for the circle with center $(-\frac{1}{2}, 3)$ and radius 4.

Solution

Here, $h = -\frac{1}{2}$, $k = 3$, and $r = 4$. Substituting into the standard equation for a circle, we have

$$(x + \tfrac{1}{2})^2 + (y - 3)^2 = 4^2.$$

Simplifying, we rewrite this equation as

$$x^2 + y^2 + x - 6y - \tfrac{27}{4} = 0$$

or

$$4x^2 + 4y^2 + 4x - 24y - 27 = 0. \qquad \blacktriangleleft$$

▶ **EXAMPLE 1.34**

Determine an equation for the circle which passes through the point $(3, -6)$ and has center $(-2, 6)$.

Solution

The data are shown in figure 1.31, with the circle indicated by a dotted curve.

Figure 1.31

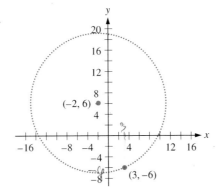

Chapter 1 / Fundamental Algebraic and Geometric Concepts

Since the center is (–2, 6), we know that the equation has the form $(x + 2)^2 + (y - 6)^2 = r^2$. We need to find the value of r^2.

The radius r is the distance between the two given points. So, by the distance formula,

$$r^2 = (3 + 2)^2 + (-6 - 6)^2 = 25 + 144 = 169.$$

Thus, a suitable equation is

$$(x + 2)^2 + (y - 6)^2 = 13^2$$

or

$$x^2 + y^2 + 4x - 12y - 129 = 0.$$

▶ **EXAMPLE 1.35**

Determine the center and radius of the circle given by $9x^2 + 9y^2 - 18x + 6y - 215 = 0$.

Solution

To find the center $C(h, k)$ and the radius r, we shall rewrite the equation in standard form, $(x - h)^2 + (y - k)^2 = r^2$. To do this, we complete the square:

$$9x^2 + 9y^2 - 18x + 6y - 215 = 0$$
$$9(x^2 - 2x) + 9(y^2 + \tfrac{2}{3}y) = 215$$
$$9(x^2 - 2x + 1) + 9(y^2 + \tfrac{2}{3}y + \tfrac{1}{9}) = 215 + 9 + 1$$
$$9(x - 1)^2 + 9(y + \tfrac{1}{3})^2 = 225$$
$$(x - 1)^2 + (y + \tfrac{1}{3})^2 = 25 = 5^2.$$

Viewing this as standard form, we see that the center is $(1, -\tfrac{1}{3})$ and the radius is 5.

▶ **EXAMPLE 1.36**

Determine the graph of $2x^2 + 2y^2 + 12x - 16y + 50 = 0$.

Solution

Since this appears to be the equation of a circle, we shall proceed as in example 1.35 and complete the square:

$$2(x^2 + 6x + 9) + 2(y^2 - 8y + 16) = -50 + 18 + 32$$
$$2(x + 3)^2 + 2(y - 4)^2 = 0$$
$$(x + 3)^2 + (y - 4)^2 = 0 = 0^2.$$

This is a "degenerate" circle consisting of the point(s) whose distance from (–3, 4) is 0. In other words, the graph of this equation is the set consisting of the point (–3, 4).

▶ **EXAMPLE 1.37**

Determine the graph of $x^2 + y^2 - 5x - 10y + 35 = 0$.

Solution

As in the preceding two examples, we complete the square:

Circles, Parabolas, and Tangents

$$(x^2 - 5x + \tfrac{25}{4}) + (y^2 - 10y + 25) = -35 + \tfrac{25}{4} + 25$$
$$(x - \tfrac{5}{2})^2 + (y - 5)^2 = -\tfrac{15}{4}.$$

The two terms on the left-hand side of this equation are nonnegative. Hence, so is their sum. As a result, there are no real numbers that can be substituted for x and y to satisfy this equation. Thus, the (real) graph contains no points and is the empty set. ◀

The last three examples indicate how one would establish the following result.

Proposition

Let A, B, C, and D be real numbers such that $A \neq 0$. Then the (real) graph of

$$Ax^2 + Ay^2 + Bx + Cy + D = 0$$

is either a circle, a set containing only one point, or the empty set.

Parabolas

Just as lines are graphs of linear equations, parabolas are graphs of certain quadratic equations in two variables. Chapter 2 will continue this discussion, and parabolas will be thoroughly studied in section 8.1.

Basic Types of Parabolas

The parabolas studied in this section arise as graphs of equations of the form

$$y = kx^2 \quad \text{or} \quad x = ky^2$$

where k is a nonzero real number.

Example 1.38 will show how to graph one such parabola.

EXAMPLE 1.38

Sketch the graph of $y = x^2$.

Solution

Here, $y = kx^2$ with $k = 1$. We begin with a table of values.

x	-3	-2	-1	0	1	2	3
y	9	4	1	0	1	4	9

The corresponding points (x, y) are plotted in figure 1.32(a). You will see later that these points can be smoothly connected, resulting in the graph in figure 1.32(b). The lowest point of this graph is at the origin and is called the **vertex** of the parabola. You can see why we say that this graph "opens upward." ◀

Figure 1.32

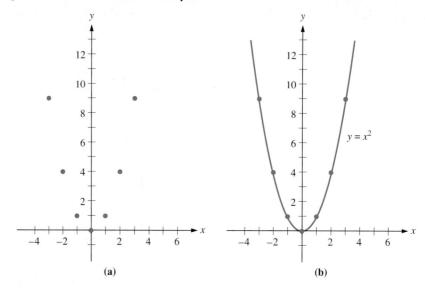

Figure 1.33

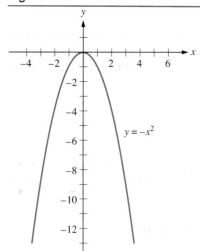

Notice that if we had instead used $k = -1$, we would have found the graph of $y = -x^2$. This graph, which you will find in figure 1.33, is a parabola with nearly the same shape as the parabola in figure 1.32(b). The difference is that the graph of $y = -x^2$ "opens downward." Its vertex is also at the origin, but on this parabola, the vertex is the highest point of the graph.

Similarly, the graphs of $x = y^2$ and $x = -y^2$ are shown in figures 1.34(a) and (b). Each of these parabolas has its vertex at the origin. The graph of $x = y^2$ "opens to the right," while the graph of $x = -y^2$ "opens to the left."

What do the graphs of parabolas look like if $k \neq 1$? Consider the graph of $y = 3x^2$. The following table contains some values for x, x^2 and y.

x	-3	-2	-1	0	1	2	3
x^2	9	4	1	0	1	4	9
$y = 3x^2$	27	12	3	0	3	12	27

Figure 1.34

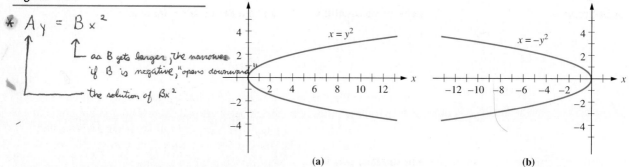

Figure 1.35

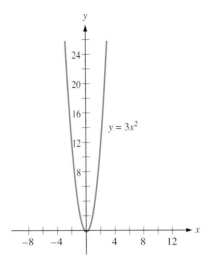

The graph of $y = 3x^2$ is shown in figure 1.35.

Intersections of Lines and Circles

In section 1.3, we studied systems of linear equations. Not all systems are of this type. A **nonlinear system of equations** is a system at least one of whose equations is not linear. One example of a nonlinear system is

$$\begin{cases} x - y = 1 \\ x^2 + y^2 = 25 \end{cases}$$

Its solution set consists of the coordinates of points of intersection of the line $x - y = 1$ and the circle $x^2 + y^2 = 25$.

A line and a circle can intersect in at most two points. As the next example illustrates, the method of substitution can often be used to find the points of intersection.

EXAMPLE 1.39

Use the substitution method to solve the nonlinear system

$$\begin{cases} x - y = 1 & (1) \\ x^2 + y^2 = 25. & (2) \end{cases}$$

Solution

Let's solve equation (1) for y and then substitute that value into equation (2).

$$y = x - 1$$

$$x^2 + (x - 1)^2 = 25$$

$$x^2 + x^2 - 2x + 1 = 25$$

$$x^2 - x - 12 = 0$$

$$(x - 4)(x + 3) = 0$$

and so either $x = 4$ or $x = -3$.

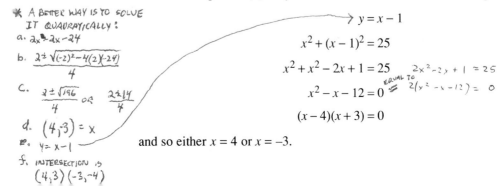

Chapter 1 / Fundamental Algebraic and Geometric Concepts

Figure 1.36

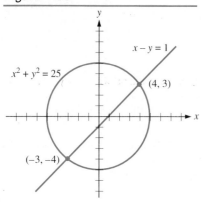

If $x = 4$, then from equation (1), we see that $y = 3$; and if $x = -3$, then $y = -4$. Thus, the solution set consists of the two points $(4, 3)$ and $(-3, -4)$, as is shown by the graphs in figure 1.36. ◀

Nonlinear systems arise in a number of ways. For instance, in example 1.56 of section 1.5, we will use the substitution method to solve the system

$$\begin{cases} a^2 - b^2 = 1 \\ 2ab = -1. \end{cases}$$

This may be considered a "quadratic system in the variables a and b," but the quadratic equations in it do not have circles or parabolas as their graphs. In chapter 8, you will study such hyperbolas and related graphs. The next example illustrates the various types of solution sets that quadratic systems may have.

▶ **EXAMPLE 1.40**

Determine the solution sets of the following systems:

(a) $\begin{cases} (x - 4)^2 + (y - 2)^2 = 1 \\ xy = 0 \end{cases}$ (b) $\begin{cases} (x - 2)^2 + (y + 1)^2 = 1 \\ xy = 0 \end{cases}$

(c) $\begin{cases} x^2 + (y - 2)^2 = 1 \\ xy = 0 \end{cases}$ (d) $\begin{cases} x^2 + y^2 - x - y = 0 \\ xy = 0 \end{cases}$

(e) $\begin{cases} x^2 + y^2 = 4 \\ xy = 0 \end{cases}$

Solutions

The solution set of $xy = 0$ is the union of the x-axis and the y-axis. Using this fact, we can graph the equations in these systems. Figure 1.37(a) relates to systems (a), (b), and (c); and figure 1.37(b) relates to systems (d) and (e).

Figure 1.37

*How the HELL THEY CAME UP W/ ANY OF THESE SOLUTIONS I DON'T KNOW

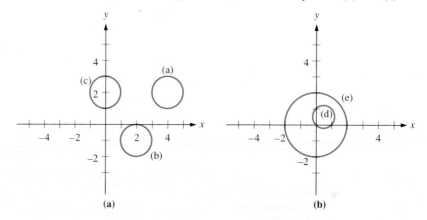

(a) (b)

Using the figures, we know how many particular solutions each system has. You can then determine that the solution set

of system (a) is empty;

of system (b) is $\{(2, 0)\}$;

of system (c) is $\{(0, 1), (0, 3)\}$;

Circles, Parabolas, and Tangents

of system (d) is $\{(0, 0), (1, 0), (0, 1)\}$;

of system (e) is $\{(2, 0), (-2, 0), (0, 2), (0, -2)\}$. ◀

Tangents to Circles

Recall from plane geometry that if a line ℓ intersects a circle at only one point, then ℓ is said to be **tangent** to the circle at that point. Also, if a line ℓ intersects a circle at a point P, then ℓ is tangent to this circle if and only if ℓ is perpendicular to the "radius vector" at P.

The earlier material on solving systems can be used to study tangents. You will be asked to show in exercise 28 that if Q is a point outside a given circle, then exactly two tangent lines to this circle pass through Q.

▷ **EXAMPLE 1.41** Find equations of the tangent lines to the circle $(x - 2)^2 + (y + 3)^2 = 4$ that pass through the point $(8, 4)$.

Solution The data are shown in figure 1.38(a).

You can see from the figure that $(8, 4)$ is outside the circle and that the vertical line $x = 8$ is not tangent to the circle. Thus, the required tangent lines have slopes. Now, according to the point-slope form, the line that passes through $(8, 4)$ and has slope m is $y - 4 = m(x - 8)$. To find where this line intersects the circle, we should solve the system

$$\begin{cases} y - 4 = m(x - 8) & (1) \\ (x - 2)^2 + (y + 3)^2 = 4. & (2) \end{cases}$$

From (1), we have $y = mx + 4 - 8m$. Substituting this into (2) produces

$$(x - 2)^2 + (mx + 7 - 8m)^2 = 4$$
$$x^2 - 4x + 4 + m^2x^2 + 49 + 64m^2 + 14mx - 16m^2x - 112m = 4$$
$$(1 + m^2)x^2 + (-4 + 14m - 16m^2)x + (49 - 112m + 64m^2) = 0. \quad (3)$$

Figure 1.38

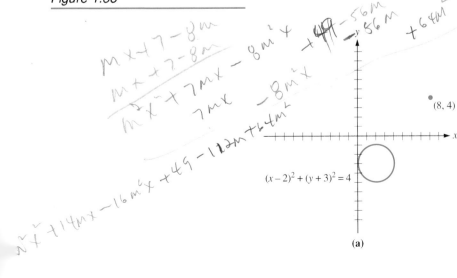
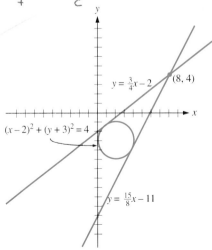

(a)   (b)

Now, (3) is a quadratic equation in the variable x. Because a tangent intersects a circle in exactly one point, (1) is tangent to the given circle if and only if exactly one real number is a root of (3). According to the quadratic formula, this is equivalent to the condition that the discriminant of (3) is 0, namely $(B^2) - 4(a)(c) = 0$

*put in $0 = B^2 - 4ac$ soon

$$(-4 + 14m - 16m^2)^2 - 4(1 + m^2)(49 - 112m + 64m^2) = 0.$$

You can check that this equation simplifies to

$$-128m^2 + 336m - 180 = 0$$

or

$$32m^2 - 84m + 45 = 0.$$

By factoring or the quadratic formula, the solutions for m are $\frac{15}{8}$ and $\frac{3}{4}$.

SIMPLIFIED FURTHER TO
$Y = \frac{15}{8}x - 11$

Corresponding to $m = \frac{15}{8}$, we have the tangent line $y - 4 = \frac{15}{8}(x - 8)$, or $15x - 8y - 88 = 0$; and corresponding to $m = \frac{3}{4}$, we have the tangent line $y - 4 = \frac{3}{4}(x - 8)$, or $3x - 4y - 8 = 0$. These tangent lines are shown in figure 1.38(b).

In the previous example, setting the discriminant equal to 0 allowed us to determine equations of the desired tangent lines. The technique of setting a discriminant equal to 0 is also useful in the following type of problem.

▶ **EXAMPLE 1.42**

Solution

Determine which tangents to the circle $x^2 + y^2 = 9$ have slope -5.

By the slope-intercept form, lines with slope -5 have equations of the form $y = -5x + b$. We are thus interested in the values b such that the system

$$\begin{cases} y = -5x + b & (1) \\ x^2 + y^2 = 9 & (2) \end{cases}$$

has only one solution for x. Substituting (1) into (2) produces

$$x^2 + (-5x + b)^2 = 9$$

or

$$26x^2 - 10bx + (b^2 - 9) = 0.$$

This quadratic has only one root if and only if its discriminant is 0. This condition is equivalent to

$$(-10b)^2 - 4(26)(b^2 - 9) = 0$$

or

$$-4b^2 + 936 = 0$$

$$b^2 = 234.$$

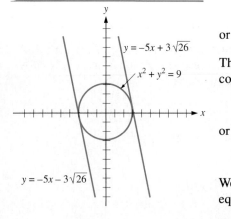

Figure 1.39

We find $b = \pm\sqrt{234} = \pm 3\sqrt{26}$. Thus, the required tangent lines are given by the equations $y = -5x + 3\sqrt{26}$ and $y = -5x - 3\sqrt{26}$, as shown in figure 1.39. ◀

In exercise 38, you will be asked to use the above method to derive equations for the two tangent lines, with given slope, to a given circle.

Circles, Parabolas, and Tangents

Intersections of Circles

Two distinct circles can intersect in at most two points. In order to determine the coordinates of these points, you ordinarily begin by eliminating the second-degree terms and then apply the substitution method. This is illustrated in the following example.

▶ **EXAMPLE 1.43**

Solve and interpret geometrically:

$$\begin{cases} (x-4)^2 + y^2 = 25 & (1) \\ (x+1)^2 + (y-3)^2 = 9. & (2) \end{cases}$$

Solution

We begin by expanding both sides and rewriting so that like terms are written above one another.

$$x^2 - 8x + y^2 \qquad - 9 = 0 \qquad (3)$$
$$x^2 + 2x + y^2 - 6y + 1 = 0. \qquad (4)$$

We can eliminate the second-degree terms by subtracting. The result is

$$10x - 6y + 10 = 0. \qquad (5)$$

Next, we shall use the substitution method to solve the system consisting of equation (5) and one of the given equations. Let's solve (5) for y, getting $y = \frac{5}{3}x + \frac{5}{3}$. Substituting this into equation (1) produces

$$(x-4)^2 + (\tfrac{5}{3}x + \tfrac{5}{3})^2 = 25.$$

You can check that the roots of this quadratic equation are -1 and $\frac{28}{17}$. Substituting these values of x into equation (5), we may solve for the corresponding values of y. The solution set consists of $(-1, 0)$ and $(\frac{28}{17}, \frac{75}{17})$. As shown in figure 1.40, these are the coordinates of the points of intersection of the circles determined by equations (1) and (2). ◀

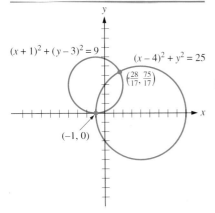

Figure 1.40

Intersections of Parabolas with Lines or Circles

You have seen the substitution method at work in examples 1.38 and 1.40–1.42. As example 1.44 shows, the substitution method can also be used to find the points where a parabola intersects another curve.

▶ **EXAMPLE 1.44**

Solve and interpret geometrically:

$$\begin{cases} x^2 + (y-1)^2 = 13 & (1) \\ y = x^2. & (2) \end{cases}$$

Solution

Substituting x^2 for y in equation (1), we obtain

$$x^2 + (x^2 - 1)^2 = 13$$

or

$$x^4 - x^2 - 12 = 0.$$

Factoring leads to $(x^2 - 4)(x^2 + 3) = 0$. Thus, either $x^2 = 4$ or $x^2 = -3$. The equation $x^2 = -3$ has no solution in the set of real numbers. Hence $x^2 = 4$, and so x is either 2 or -2. We can see from equation (2) that $x = 2$ and $x = -2$ each give $y = 4$. Thus, the solution set consists of $(2, 4)$ and $(-2, 4)$. As shown in figure 1.41, these are the coordinates of the points of intersection of the circle $x^2 + (y-1)^2 = 13$ and the parabola $y = x^2$.

Figure 1.41

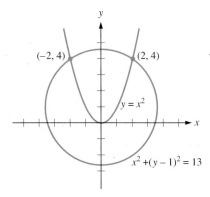

Exercises 1.4

In exercises 1–8, find and simplify an equation for the circle with the given properties.

1. Center $(-2, 4)$, radius 3
2. Center $(\frac{1}{2}, -1)$, radius 2
3. Center $(0, \pi)$, radius $\sqrt{2}$
4. Center $(-\frac{2}{3}, 0)$, radius $\sqrt{3}$
5. Center $(0, 0)$; passes through $(-12, 5)$
6. Center $(0, 0)$; passes through $(2, -2)$
7. Center $(\frac{1}{2}, -1)$; passes through $(-\frac{3}{2}, 1)$
8. Center $(\pi, 2)$; passes through $(\pi + 3, 6)$

In exercises 9–12, determine the center and radius of the circle with the given equation.

9. $x^2 + y^2 + 6x - 8y + 9 = 0$
10. $2x^2 + 2y^2 - 2x + 6y - 3 = 0$
11. $9x^2 + 9y^2 + 12x - 6y - 76 = 0$
12. $x^2 + y^2 + x - y = 0$

In exercises 13–16, determine the (real) graph of the given equation.

13. $x^2 + y^2 + 6x - 8y + 25 = 0$
14. $x^2 + y^2 + 2x + 1 = 0$
15. $9x^2 + 9y^2 + 12x - 6y + 23 = 0$
16. $2x^2 + 2y^2 + 2x - 2y + 7 = 0$

In exercises 17–24, graph the given parabola and determine its vertex.

17. $y = 2x^2$
18. $y = \frac{1}{2}x^2$
19. $x = -2y^2$
20. $x = -\frac{1}{2}y^2$
21. $y + 1 = x^2$
22. $x - 1 = y^2$
23. $x = (y + 1)^2$
24. $y = (x - 1)^2$

In exercises 25–27, find and simplify an equation for the circle with the given properties.

25. Center $(-1, 2)$; tangent to the line $12x - 5y - 4 = 0$. (Hint: use the formula in section 1.3 for the perpendicular distance from a point to a line.)
26. Center on the line $y - x + 1 = 0$; tangent to the line $3x - 4y + 7 = 0$; passes through $\left(6 - \frac{1}{\sqrt{2}}, 5 + \frac{1}{\sqrt{2}}\right)$. (Hint: show that the x-coordinate, h, of the center satisfies $49h^2 - 578h + 1704 = 0$. Note that there are two circles with the given properties.)
27. Passing through the points $(-1, 0)$, $(1, 0)$, and $(0, 1)$
♦ 28. Let b and r be real numbers such that $0 < r < b$.
   (a) Show that exactly two of the lines that pass through $P(0, b)$ are tangent to the circle $x^2 + y^2 = r^2$.
   (b) Let ℓ be either of the tangent lines discussed in (a). Let Q be the point at which ℓ intersects the circle $x^2 + y^2 = r^2$. Show that $d(P, Q) = \sqrt{b^2 - r^2}$.

In exercises 29–32, find an equation for each of the tangent lines to the given circle that pass through the given point.

29. $x^2 + y^2 = 25$; (0, 7)
30. $(x+3)^2 + (y-2)^2 = 4$; (5, 5)
31. $(x+3)^2 + (y-2)^2 = 4$; (4, 4)
32. $(x+3)^2 + (y-2)^2 = 4$; (−1, 2)

In exercises 33–34, you may assume that the Earth's radius is 3960 miles.

33. On a clear day, John and David climb to the top of a skyscraper that is 984 feet high. From that vantage point, they see a boat, carrying a shipment of Wm. C. Brown Publishers texts, appear on the horizon. To the nearest half-mile, how far are John and David from the boat?

34. Looking at the ocean, a surfer who is 6 feet tall sees a swimmer appear on the horizon. To the nearest mile, how far is the swimmer from the surfer?

In exercises 35–37, determine equations for the tangents to the given circle which have the given slope m.

35. $x^2 + y^2 = 4$; $m = -3$ **36.** $x^2 + y^2 = 25$; $m = 6$
37. $(x-2)^2 + (y-1)^2 = 9$; $m = -2$

◆ 38. Show that $y - k = m(x - h) \pm r\sqrt{m^2 + 1}$ are equations for the tangents to the circle $(x - h)^2 + (y - k)^2 = r^2$ which have slope m. (Hint: reason as in example 1.42.)

In exercises 39–58, solve the given system. Interpret your answer geometrically in exercises 39–54.

39. $\begin{cases} y - x = 1 \\ x^2 + y^2 = 25 \end{cases}$
40. $\begin{cases} y = -2x - 1 \\ x^2 + y^2 = 9 \end{cases}$
41. $\begin{cases} x - 2y + 8 = 0 \\ x^2 + y^2 = 9 \end{cases}$
42. $\begin{cases} x - y = 2 \\ x^2 + y^2 = 100 \end{cases}$
43. $\begin{cases} x^2 + y^2 = 25 \\ xy = 0 \end{cases}$
44. $\begin{cases} (x-2)^2 + y^2 = 1 \\ xy = 0 \end{cases}$
45. $\begin{cases} (x+2)^2 + (y-1)^2 = 1 \\ xy = 0 \end{cases}$
46. $\begin{cases} (x+2)^2 + (y-4)^2 = 1 \\ xy = 0 \end{cases}$
47. $\begin{cases} x^2 + (y-1)^2 = 20 \\ (x-1)^2 + y^2 = 26 \end{cases}$
48. $\begin{cases} (x+2)^2 + (y+1)^2 = 1 \\ (x+2)^2 + (y+3)^2 = 4 \end{cases}$
49. $\begin{cases} (x+2)^2 + y^2 = 4 \\ (x-2)^2 + y^2 = 4 \end{cases}$
50. $\begin{cases} (x+2)^2 + y^2 = 2 \\ (x-2)^2 + y^2 = 2 \end{cases}$
51. $\begin{cases} x^2 + y^2 = 6 \\ y = -x^2 \end{cases}$
52. $\begin{cases} x^2 + (y+\sqrt{6})^2 = 6 \\ y = 2x^2 \end{cases}$
53. $\begin{cases} x + 2 = 0 \\ x = 3y^2 \end{cases}$
54. $\begin{cases} y - x = 12 \\ x = -y^2 \end{cases}$
55. $\begin{cases} x^2 + y^2 = 13 \\ xy = 6 \end{cases}$
56. $\begin{cases} x^2 + y^2 = 10 \\ xy = 3 \end{cases}$
57. $\begin{cases} 2x + 3y = 14 \\ xy + y^2 = 20 \end{cases}$
58. $\begin{cases} x^2 + y^2 = 25 \\ 3x^2 - 4y^2 = 12 \end{cases}$

59. Consider the altitude in the figure.

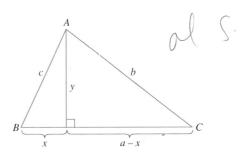

(a) Show that x and y satisfy the system
$$\begin{cases} x^2 + y^2 = c^2 \\ (a-x)^2 + y^2 = b^2. \end{cases}$$
(b) Solve the system in (a).

◆ 60. (A proof of Hero's formula for the area of a triangle; you will meet this formula again in section 7.3.) Let a, b, and c be the lengths of the sides of a given triangle. Let s denote $\dfrac{a+b+c}{2}$. Show that the area of the triangle is $\sqrt{s(s-a)(s-b)(s-c)}$. (Hint: you may label the triangle as in the above diagram. Take the positive solution for y in exercise 59(b) and use a factoring method from chapter 0 to simplify $\dfrac{ay}{2}$.)

1.5 COMPLEX NUMBERS

Equations such as $x^2 + 1 = 0$ have no real number solutions. More generally, the quadratic formula shows that a quadratic equation with a negative discriminant has no real number roots. Nearly two hundred years ago, the need to solve this type of quadratic equation led to a new type of number and a new number system, the complex number system. This section will serve as a formal algebraic introduction to complex numbers. This will help us present a more complete discussion of the quadratic formula than was possible in chapter 0. It will also be of help in chapter 3 when we study roots of real polynomials. Later, in sections 6.6–6.7, we shall study the underlying geometry and algebra more deeply.

We begin the study of complex numbers with the following definitions.

Imaginary Numbers

> The **imaginary unit**, denoted i, is defined to be a number such that $i^2 = -1$. A **pure imaginary number** z is a product bi, where b is any nonzero real number and i is the imaginary unit. As a special case, $1i = i$. (In many areas of engineering, the symbol j is used rather than i, because the symbol i also denotes electrical current.)

Principal Square Root

> The **principal square root of a negative real number** is defined as follows. If b is a positive real number, then $\sqrt{-b} = \sqrt{b}\,i$.

The preceding definition is motivated by the calculation

$$(\sqrt{b}\,i)^2 = (\sqrt{b})^2 i^2 = (b)(-1) = -b.$$

Notice also that, according to the above definition, $\sqrt{-1} = \sqrt{1}\,i = 1i$, or more compactly, we have

$$\sqrt{-1} = i.$$

Pure imaginary numbers (and, in particular, i) are not real numbers. They are a special type of "complex number." Before we give a formal definition of complex numbers, we practice working with the pure imaginaries.

▶ **EXAMPLE 1.45**   Express $\sqrt{-7}$ and $\sqrt{-9}$ in terms of the imaginary unit i.

Solution
$$\sqrt{-7} = \sqrt{7}\,i \qquad \text{commonly known as } i\sqrt{7}$$
$$\sqrt{-9} = \sqrt{9}\,i = 3i. \qquad ◀$$

Do not assume that *all* the earlier laws of algebra carry over to complex numbers. For instance, the properties of radicals do not. In fact, *the following calculation is wrong*:

$$\sqrt{-9}\,\sqrt{-16} \neq \sqrt{(-9)(-16)} \neq \sqrt{144} \neq 12.$$

However, *the following calculation is correct*:

$$\sqrt{-9}\,\sqrt{-16} = (3i)(4i) = 12i^2 = (12)(-1) = -12.$$

Similarly, $\sqrt{-5}\,\sqrt{-3} = (\sqrt{5}\,i)(\sqrt{3}\,i) = \sqrt{5}\,\sqrt{3}\,i^2 = \sqrt{15}(-1) = -\sqrt{15}$.

The next example shows how to add, subtract, and divide pure imaginary numbers. Formal definitions for these operations will not be given at this time.

Complex Numbers

They will be incorporated into the definitions for the larger set of complex numbers after the next example.

EXAMPLE 1.46 Simplify (a) $\sqrt{-4} + \sqrt{-25}$, (b) $\sqrt{-5} - \sqrt{-45}$, and (c) $\dfrac{-\sqrt{15}}{\sqrt{-3}}$.

Solutions

(a) $\sqrt{-4} + \sqrt{-25} = 2i + 5i = (2+5)i = 7i$.

(b) $\sqrt{-5} - \sqrt{-45} = \sqrt{5}\,i - 3\sqrt{5}\,i = (\sqrt{5} - 3\sqrt{5})i = -2\sqrt{5}\,i$.

(c) $\dfrac{-\sqrt{15}}{\sqrt{-3}} = \dfrac{-\sqrt{15}}{\sqrt{3}\,i} = \dfrac{-\sqrt{15}}{\sqrt{3}\,i} \cdot \dfrac{i}{i} = \dfrac{-\sqrt{15}\,i}{\sqrt{3}\,i^2} = \dfrac{-\sqrt{15}\,i}{-\sqrt{3}} = \sqrt{\dfrac{15}{3}}\,i = \sqrt{5}\,i$. ◀

Complex Numbers

A **complex number** has the form $a + bi$, where a and b are real numbers and i is the imaginary unit. We call a the **real part** of $a + bi$ and b the **imaginary part** of $a + bi$. If $a = 0$ and $b \neq 0$, then $a + bi = 0 + bi = bi$, which is a pure imaginary number. If $b = 0$, then $a + bi = a + 0i = a$, which is a real number.

Examples of complex numbers include $\sqrt{5} + 3i$, $\tfrac{1}{2}i$, $-2 + \tfrac{1}{2}\sqrt{3}\,i$, and 7.

We generally try to simplify complex numbers into the $a + bi$ form. For such simplification, we need the following precise definitions. These definitions generalize what you already know in the special cases of real numbers and pure imaginaries.

Basic Operations on Complex Numbers

Let $a + bi$ and $c + di$ be complex numbers, where i denotes the imaginary unit and a, b, c, and d are real numbers. Then we have the following definitions:

(i) $a + bi = c + di$ if and only if $a = c$ and $b = d$ — Equality
(ii) $(a + bi) + (c + di) = (a + c) + (b + d)i$ — Addition
(iii) $(a + bi) - (c + di) = (a - c) + (b - d)i$ — Subtraction
(iv) $(a + bi)(c + di) = (ac - bd) + (ad + bc)i$ — Multiplication
(v) $\dfrac{a + bi}{c + di} = \dfrac{ac + bd}{c^2 + d^2} + \dfrac{-ad + bc}{c^2 + d^2}i$, if $c + di \neq 0$ — Division

You do *not* need to memorize the above formulas (i)–(v). For (i)–(iv), all you need to remember is that arithmetic on complex numbers is performed almost as if you were working with polynomials. However, since i is not a variable, you must remember to replace i^2 with -1 after doing the "polynomial" part of a calculation. As for formula (v), it can be worked out as needed in any given case, by "rationalizing the denominator," in the sense of chapter 0, as follows:

$$\frac{a+bi}{c+di} = \frac{a+bi}{c+di} \cdot \frac{c-di}{c-di} = \frac{(a+bi)(c-di)}{(c+di)(c-di)} = \frac{(ac+bd)+(-ad+bc)i}{c^2+d^2}$$

$$= \frac{ac+bd}{c^2+d^2} + \frac{-ad+bc}{c^2+d^2}i.$$

The next example will illustrate complex number arithmetic.

▶ **EXAMPLE 1.47** Let $z_1 = 3 + 2i$ and $z_2 = 5 - 4i$. Simplify $z_1 + z_2$, $z_1 - z_2$, $z_1 z_2$, $\frac{z_1}{z_2}$, and $\frac{1}{z_1}$.

Solution

$z_1 + z_2 = (3 + 2i) + (5 - 4i) = (3 + 5) + (2 - 4)i = 8 - 2i$

$z_1 - z_2 = (3 + 2i) - (5 - 4i) = (3 - 5) + [2 - (-4)]i = -2 + 6i$

$z_1 z_2 = (3 + 2i)(5 - 4i) = 3 \cdot 5 - 3 \cdot 4i + 2i \cdot 5 + 2i(-4i)$
$\qquad = 15 - 12i + 10i - 8i^2 = 23 - 2i$

$\frac{z_1}{z_2} = \frac{3+2i}{5-4i} = \frac{3+2i}{5-4i} \cdot \frac{5+4i}{5+4i} = \frac{(3+2i)(5+4i)}{(5-4i)(5+4i)} = \frac{7+22i}{41} = \frac{7}{41} + \frac{22}{41}i$

$\frac{1}{z_1} = \frac{1}{3+2i} = \frac{1}{3+2i} \cdot \frac{3-2i}{3-2i} = \frac{3-2i}{(3+2i)(3-2i)} = \frac{3}{13} - \frac{2}{13}i.$ ◀

The integral powers of i form an easily remembered pattern. You know that $i^0 = 1$, $i^1 = i$, and $i^2 = -1$. From this, you can see that

$i^3 = i^2 \cdot i = (-1)i = -i;$

$i^4 = i^3 \cdot i = -i(i) = -i^2 = -(-1) = 1;$

$i^5 = i^4 \cdot i = (1)i = i;$

and so on.

The block of answers $1, i, -1,$ and $-i$ repeats.

Powers of i

If n is a positive integer, we may calculate i^n as i^r, where r is the remainder upon dividing n by 4.

▶ **EXAMPLE 1.48** Calculate i^{79}.

Solution Since $79 = 4 \cdot 19 + 3$, we have $i^{79} = i^{4 \cdot 19 + 3} = i^{4 \cdot 19} i^3 = (i^4)^{19} i^3 = 1^{19} i^3 = i^3 = -i.$ ◀

The Conjugate and the Absolute Value of a Complex Number

The **conjugate of the complex number** $z = a + bi$ is defined to be the complex number $\bar{z} = a - bi$.

Complex Numbers

▶ **EXAMPLE 1.49**   (a) The conjugate of $z = \frac{1}{2} + i\sqrt{3}$ is $\overline{z} = \frac{1}{2} - i\sqrt{3}$.
(b) The conjugate of $z = 3\sqrt{5} - 2.7i$ is $\overline{z} = 3\sqrt{5} + 2.7i$.
(c) The conjugate of $z = 14i$ is $\overline{z} = -14i$.
(d) The conjugate of $z = \frac{1}{2}$ is $\overline{z} = \frac{1}{2}$. *SHOULDN'T IT BE $\overline{z} = -\frac{1}{2}$? No, BECAUSE $\overline{z} = \frac{1}{2} + 0i$. YOU ONLY CHANGE $+bi \to -bi$ OR VICE VERSA, $\frac{1}{2}$ REMAINS THE SAME*

Another way to write the preceding definition is as follows.

Conjugate of a + bi

$$\overline{a + bi} = a - bi.$$

Thus, the "rationalize the denominator" step in a division problem just amounts to

$$\frac{z_1}{z_2} = \frac{z_1}{z_2} \cdot \frac{\overline{z_2}}{\overline{z_2}}.$$

The product of the denominators is $z_2 \cdot \overline{z_2}$, which is the real number $c^2 + d^2$, assuming that $z_2 = c + di$. Also, for any complex number $z = a + bi$, we have $z + \overline{z} = (a + bi) + (a - bi) = 2a$, which is a real number. On the other hand, $z - \overline{z} = (a + bi) - (a - bi) = 2bi$, which is either 0 or a pure imaginary number. We summarize this discussion as follows.

Proposition

Let $z = a + bi$ be a complex number, with conjugate $\overline{z} = a - bi$. Then

(i) $z + \overline{z} = 2a$, which is a real number. $(a+bi) + (a-bi) = 2a$
(ii) $z \cdot \overline{z} = a^2 + b^2$, which is a real number. $(a+bi)(a+(-bi)) = a^2+b^2$
(iii) $z - \overline{z} = 2bi$, which is either 0 or a pure imaginary number. $(a+bi) + (-a+bi) = 2bi$
(iv) $\overline{z} = z$ if and only if $b = 0$, that is, if and only if z is a real number. $a - (0)i = a + (0)i \Rightarrow a = a$

We shall next explore further the interplay between conjugation and complex arithmetic. Consider complex numbers $z = a + bi$ and $w = c + di$. Then

Goal: want to become more simple in other words complex $\to$ simple

$$\overline{z + w} = \overline{(a + bi) + (c + di)} = \overline{(a + c) + (b + d)i} = (a + c) - (b + d)i$$
$$= (a - bi) + (c - di) = \overline{z} + \overline{w}$$
$$\overline{zw} = \overline{(a + bi)(c + di)} = \overline{(ac - bd) + (ad + bc)i} = (ac - bd) - (ad + bc)i$$
$$= (a - bi)(c - di) = \overline{z} \cdot \overline{w}.$$

These calculations establish facts (i) and (ii) below. In exercise 17, you will be asked to prove (iii).

Properties of Conjugates

For any complex numbers z and w,

(i) $\overline{z+w} = \overline{z} + \overline{w}$.
(ii) $\overline{zw} = \overline{z} \cdot \overline{w}$.
(iii) The conjugate of $\dfrac{z_1}{z_2}$ is $\dfrac{\overline{z_1}}{\overline{z_2}}$.
(iv) $\overline{z^n} = (\overline{z})^n$, where n is any positive integer.

The above properties of conjugates will be used in section 3.4 when we study complex roots of polynomials.

The absolute value of a complex number is defined as follows:

Absolute Value

The **absolute value** of $z = a + bi$ is the nonnegative real number given by $|z| = \sqrt{a^2 + b^2}$.

If z is a real number, then $b = 0$ and this definition agrees with the earlier definition of the absolute value of a real number.

▶ **EXAMPLE 1.50** Determine the absolute value of (a) $2 - 5i$, (b) $-\frac{3}{5} + \frac{4}{5}i$, and (c) i.

Solution

(a) $|2 - 5i| = \sqrt{2^2 + (-5)^2} = \sqrt{29}$.

(b) $\left|-\frac{3}{5} + \frac{4}{5}i\right| = \sqrt{(-\frac{3}{5})^2 + (\frac{4}{5})^2} = \sqrt{\frac{9}{25} + \frac{16}{25}} = \sqrt{1} = 1$.

(c) $|i| = |0 + 1i| = \sqrt{0^2 + 1^2} = \sqrt{1} = 1$. ◀

Linear or Quadratic Equations Involving Complex Numbers

We solve linear equations with complex coefficients just as we solve linear equations with real coefficients.

▶ **EXAMPLE 1.51** Solve $3ix - 4i + 5 = (3 + i)x - 7i$.

Solution

$$3ix - 4i + 5 = (3 + i)x - 7i$$
$$3ix - (3 + i)x = -7i + 4i - 5$$
$$[3i - (3 + i)]x = -7i + 4i - 5$$
$$(-3 + 2i)x = -5 - 3i$$
$$x = \frac{-5 - 3i}{-3 + 2i}$$
$$= \frac{-5 - 3i}{-3 + 2i} \cdot \frac{-3 - 2i}{-3 - 2i} = \frac{9 + 19i}{13} = \frac{9}{13} + \frac{19}{13}i.$$

◀

Complex Numbers

If a linear equation involves conjugation, the solution is often easier to find if we write $x = a + bi$ (and $\bar{x} = a - bi$) and proceed to solve for a and b. Example 1.52 illustrates the procedure.

▶ **EXAMPLE 1.52**

Solve $5x + \frac{5}{4}\bar{x} = \frac{9}{5} - 6i$.

Solution

Let's substitute $x = a + bi$ (and $\bar{x} = a - bi$). We obtain

$$5(a + bi) + \frac{5}{4}(a - bi) = \frac{9}{5} - 6i$$

$$5a + 5bi + \frac{5}{4}a - \frac{5}{4}bi = \frac{9}{5} - 6i$$

$$(5a + \frac{5}{4}a) + (5b - \frac{5}{4}b)i = \frac{9}{5} - 6i$$

$$\frac{25}{4}a + \frac{15}{4}bi = \frac{9}{5} - 6i.$$

The definition of equality of complex numbers allows us to equate corresponding real and imaginary parts. This gives us two simple equations for the variables a and b. Thus, $\frac{25}{4}a = \frac{9}{5}$ and $\frac{15}{4}b = -6$. Hence, we see that $a = \frac{36}{125}$ and $b = -\frac{24}{15} = -\frac{8}{5}$. Thus, the only particular solution to the given equation is $x = \frac{36}{125} - \frac{8}{5}i$. ◀

Until now, all our quadratic equations have had real coefficients and nonnegative discriminants. With the availability of complex numbers, we can remove these two restrictions. You will see, in particular, that if a real quadratic has a negative discriminant, then it has two nonreal complex number roots which are conjugates of one another. The key task is to interpret the quadratic formula when complex numbers are involved.

▶ **EXAMPLE 1.53**

Solve $3x^2 - 2x + 1 = 0$.

Solution

By the quadratic formula, the roots are

$$x = \frac{2 \pm \sqrt{4 - 12}}{6} = \frac{2 \pm \sqrt{-8}}{6} = \frac{2 \pm 2\sqrt{2}i}{6} = \frac{1}{3} \pm \frac{\sqrt{2}}{3}i.$$

Thus, the particular solutions, or roots, are $\frac{1}{3} + \frac{\sqrt{2}}{3}i$ and $\frac{1}{3} - \frac{\sqrt{2}}{3}i$. ◀

▶ **EXAMPLE 1.54**

Solve $4x^2 - 12x + 9 = 0$.

Solution

By the quadratic formula, the roots are

$$x = \frac{12 \pm \sqrt{144 - 144}}{8} = \frac{12 \pm \sqrt{0}}{8} = \frac{12 \pm 0}{8} = \frac{4}{3}.$$

The solution set in this example is $\{\frac{4}{3}\}$, with only one element. Since both the $+$ sign and the $-$ sign in the quadratic formula led to $\frac{4}{3}$, we refer to $\frac{4}{3}$ as a **double root** of $4x^2 - 12x + 9$. Notice that examples 1.41 and 1.42 of section 1.4 gave some geometric situations in which the "double root" concept arises naturally. In section 3.4, we shall study "multiple roots" of high-degree polynomials more deeply. ◀

Chapter 1 / Fundamental Algebraic and Geometric Concepts

The introduction of complex numbers and the preceding examples allow us to completely describe the impact of the discriminant of a quadratic equation.

Theorem

> Let $ax^2 + bx + c = 0$ be a quadratic with real coefficients a ($\neq 0$), b, and c. Then
>
> - if $b^2 - 4ac > 0$, the quadratic has two unequal real roots;
> - if $b^2 - 4ac = 0$, the quadratic has only one root, and it is real;
> - if $b^2 - 4ac < 0$, the quadratic has two nonreal complex roots, and they are conjugates of one another.
>
> The polynomial $ax^2 + bx + c$ is irreducible over $\mathbf{R}$ if and only if $b^2 - 4ac < 0$.

The next example indicates how to combine various skills developed thus far.

▶ **EXAMPLE 1.55**

Solve $x^3 + 1 = 0$.

Solution

Factoring the sum of cubes as in chapter 0, we have $(x + 1)(x^2 - x + 1) = 0$. If $x + 1 = 0$, then $x = -1$. If $x^2 - x + 1 = 0$, then by the quadratic formula,

$$x = \frac{1 \pm \sqrt{3}i}{2} \quad \Rightarrow \quad x \pm \frac{\sqrt{1-4(1)(1)}}{2}$$

Thus, the solution set is $\left\{-1, \frac{1}{2} + \frac{\sqrt{3}}{2}i, \frac{1}{2} - \frac{\sqrt{3}}{2}i\right\}$. ◀

What is the solution set of the equation $x^2 - 4 = 0$? By factoring or the quadratic formula, you arrive quickly at $\{2, -2\}$. Just as quickly, you could solve $x^2 - b = 0$ for any positive number b, getting $\{\sqrt{b}, -\sqrt{b}\}$.

How about an equation like $x^2 + 4 = 0$? For this, we may apply the quadratic formula, getting $x = \pm\sqrt{-4}$. The solution set is thus $\{2i, -2i\}$.

What about $x^2 - (1 - i) = 0$, a quadratic equation with nonreal coefficients? The next example shows how to proceed.

▶ **EXAMPLE 1.56**

Solve $x^2 - (1 - i) = 0$.

Solution

The quadratic formula would lead us to $x = \pm\sqrt{1 - i}$, but we have not seen how to interpret such (misuse of) radical notation. (We understand the notation $\sqrt{B}$ only for real numbers B, and $1 - i$ is not a real number.) So, we shall proceed differently. With $x = a + bi$ (for a and b real, as usual), we may rewrite $x^2 - (1 - i) = 0$ as

$$a^2 - b^2 + 2abi - (1 - i) = 0$$

$$a^2 - b^2 + 2abi = 1 - i.$$

Equating the corresponding real parts and the corresponding imaginary parts leads to the equivalent quadratic system

Complex Numbers

$$a^2 - b^2 = 1 \quad \text{and} \quad 2ab = -1.$$

The system's second equation gives $b = -\dfrac{1}{2a}$. Substituting this into the first equation, we have

$$a^2 - \left(\dfrac{-1}{2a}\right)^2 = 1$$

or

$$a^2 - \dfrac{1}{4a^2} = 1$$

$$4a^4 - 1 = 4a^2$$

$$4a^4 - 4a^2 - 1 = 0.$$

Using a new variable $u = a^2$, we may rewrite the preceding equation as $4u^2 - 4u - 1 = 0$. By the quadratic formula, you can check that its solutions are $u = \dfrac{1 \pm \sqrt{2}}{2}$. Since $u = a^2$, we must reject the negative solution for u. We now find suitable values of a, namely $a = \pm\sqrt{\dfrac{1+\sqrt{2}}{2}}$, with the corresponding $b = \dfrac{-1}{2a} = \mp\dfrac{1}{2}\sqrt{\dfrac{2}{1+\sqrt{2}}}$.

To sum up, the required solution set is $\{x, -x\}$, where

$$x = \sqrt{\dfrac{1+\sqrt{2}}{2}} - \dfrac{1}{2}\sqrt{\dfrac{2}{1+\sqrt{2}}}\, i.$$

◀

In section 1.4, you saw geometric interpretations for some systems of quadratic equations, though not necessarily of the kind met while solving example 1.56. In example 1.58, you will see another example "of quadratic type," like the fourth-degree polynomial (in the variable a) in the previous solution. Incidentally, to find nth roots of complex numbers, you must await an application of trigonometry in section 6.7. Next, however, we show how to use the skill developed in example 1.56.

▶ **EXAMPLE 1.57**

Solution

Solve $x^2 - 2x + i = 0$.

The quadratic formula leads to $x = \dfrac{2 \pm \sqrt{4-4i}}{2} = \dfrac{2 \pm 2\sqrt{1-i}}{2} = 1 \pm \sqrt{1-i}$.

Thanks to example 1.56, we know how to interpret such notation. Thus, the roots are given by

$$x = 1 \pm \left(\sqrt{\dfrac{1+\sqrt{2}}{2}} - \dfrac{1}{2}\sqrt{\dfrac{2}{1+\sqrt{2}}}\, i\right).$$

◀

▶ **EXAMPLE 1.58**

Solve $(x-1)^{2/3} - (x-1)^{1/3} - 12 = 0$.

Solution   Using a new variable $u = (x - 1)^{1/3}$, we may rewrite the given equation as $u^2 - u - 12 = 0$. By factoring or the quadratic formula, we find the solutions to *this* quadratic equation, namely $u = 4, -3$. Now, since $u = (x - 1)^{1/3}$, we can solve for x as in chapter 0, getting $x = u^3 + 1$. With $u = 4$, this gives $x = 4^3 + 1 = 65$; while for $u = -3$, we have $(-3)^3 + 1 = -26$. Thus, the solution set of the given equation is $\{65, -26\}$. ◀

Exercises 1.5

In exercises 1–2, express the given quantities in terms of the imaginary unit i.

1. (a) $\sqrt{-16}$ (b) $\sqrt{-15}$ (c) $\sqrt{-\pi}$ (d) $\sqrt{\frac{-4}{9}}$

2. (a) $\sqrt{-17}$ (b) $\sqrt{-25}$ (c) $\sqrt{-\frac{9}{4}}$ (d) $\sqrt{-2}$

In exercises 3–4, perform the indicated operations and simplify.

3. (a) $\sqrt{-5} + \sqrt{-20}$ (b) $\sqrt{-4} - \sqrt{-49}$
   (c) $\sqrt{-9}\sqrt{-49}$ (d) $\frac{\sqrt{-15}}{\sqrt{-5}}$

4. (a) $\sqrt{-3} - \sqrt{-75}$ (b) $\sqrt{-64} + \sqrt{-16}$
   (c) $\frac{\sqrt{-7}}{\sqrt{-63}}$ (d) $\sqrt{-\frac{9}{4}}\sqrt{\frac{-100}{121}}$

5. Find (i) the real part and (ii) the imaginary part for each of the following complex numbers:
   (a) $\frac{3}{2} - 5i$ (b) $-\sqrt{2}$ (c) $7i$
   (d) $-\frac{3}{2} + 5i$ (e) 7 (f) 0

6. Determine which of the following complex numbers is (i) pure imaginary, (ii) real:
   (a) $-2i$ (b) $-i$ (c) -2
   (d) $2 - i$ (e) $2 + i$ (f) 7

In exercises 7–14, perform the indicated operations and simplify answers to the form $a + bi$ (with a and b real).

7. (a) $(2 - 6i) + (\frac{3}{4} + 4i)$
   (b) $(-2 + i) + (\sqrt{2} - \frac{2}{3}i)$
   (c) $(-\frac{3}{2} + \frac{1}{3}i) - (-\frac{5}{2} + \frac{7}{3}i)$ (d) $6i - (6 - 4i)$

8. (a) $(-2 + \frac{1}{6}i) + (-1 + \frac{2}{3}i)$ (b) $(2 - 3i) - (4 - 5i)$
   (c) $3 - (5 - \sqrt{2}i)$ (d) $\left(\frac{1}{2} + \frac{\sqrt{3}}{2}i\right) + \left(\frac{1}{2} - \frac{\sqrt{3}}{2}i\right)$

9. (a) $(2 + 4i)(3 + 5i)$ (b) $(-2 + 4i)(3 - 5i)$
   (c) $(\frac{1}{2} - \frac{2}{3}i)(6 + \frac{2}{3}i)$ (d) $-7i(i - 1)$

10. (a) $(2 - 4i)(5 + 3i)$ (b) $(\frac{1}{2} - i)(\sqrt{2} - 4i)$
    (c) $3(-\sqrt{2} + 2i)$ (d) $(-5 + \frac{2}{3}i)(-6 + 9i)$

11. (a) $\frac{2+i}{2-i}$ (b) $\frac{\sqrt{2}+3i}{-3i}$ (c) $\frac{4}{-5+4i}$ (d) $\frac{-8+6i}{2}$

12. (a) $\frac{-5+3i}{4+2i}$ (b) $\frac{4}{-2+3i}$
    (c) $\frac{12-6i}{-3}$ (d) $\frac{\pi^2+7\pi i}{\pi i}$

13. (a) $-(4 - 2i)$ (b) $(2 + 3i)^3$
    (c) $2 - 3i - 4i(5 - 6i)$ (d) i^{71}

14. (a) $\frac{1}{4-2i}$ (b) i^{86}
    (c) $-2 + 3i - 4(5i - 6)$ (d) $(5 - 3i)^2$

In exercises 15–16, determine the given conjugates.

15. (a) $\overline{\sqrt{3} + \frac{1}{2}i}$ (b) $\overline{2 - i}$ (c) $\overline{-2 + i}$
    (d) $\overline{3i}$ (e) $\overline{-5}$ (f) $\overline{0}$

16. (a) $\overline{-2 - 3i}$ (b) $\overline{\sqrt{3}}$ (c) $\overline{5}$
    (d) $\overline{-1 + 2i}$ (e) $\overline{-3i}$ (f) $\overline{i}$

◆ 17. If z and w are complex numbers and $w \neq 0$, prove that the conjugate of $\frac{z}{w}$ is $\frac{\overline{z}}{\overline{w}}$. (Hint: write $z = a + bi$ and $w = c + di$.)

18. Simplify:
    (a) $\overline{6 + 7i - 3i(4 - 5i)}$ (b) $\overline{(\frac{3}{2} - \frac{4}{5}i)(\frac{3}{2} + \frac{4}{5}i)}$

19. Determine the absolute value of:
    (a) $\frac{3}{2} - 2i$ (b) $\sqrt{3}i$ (c) $-\frac{11}{13} + \frac{12}{13}i$ (d) -7

20. Determine the following:
    (a) $|-2 + 3i|$ (b) $|-3i| - |4 - 3i|$

In exercises 21–24, let z and w be complex numbers and prove the given statements.

21. (a) $|z| = \sqrt{z\overline{z}}$ (b) $|\overline{z}| = |z|$
22. $|zw| = |z||w|$. (Hint: see the hint for exercise 17. Chapter 0 is relevant too.)
23. If $w \neq 0$, then:
    (a) $\left|\frac{1}{w}\right| = \frac{1}{|w|}$. (Hint: use exercise 22.)
    (b) $\left|\frac{z}{w}\right| = \frac{|z|}{|w|}$. (Hint: use (a) and exercise 22.)
24. (a) The conjugate of $\overline{z}$ is z.   (b) If $\overline{z} = \overline{w}$, then $z = w$.

Chapter Summary and Review Exercises

In exercises 25–32, solve the given equation (for x).

25. $2ix + 3 - 4i = 5ix - 6 + 7i$
26. $(2 + i)x + 3i = -4ix + 5$
27. $(-1 + 4i)x - 5 = 3x + 5i$
28. $2x - \pi = \pi i - 3x$
29. $\frac{5}{4}x - 5\bar{x} = 9 + \frac{6}{5}i$
30. $-5x + 6\bar{x} = 2 - 3i$
31. $(1 - i)x + 4i = \frac{2}{3}(-6i + 9) + x$
32. $2 - 3i = (1 + i) + xi$

In exercises 33–38, calculate the discriminant and hence determine if the given quadratic equation has two unequal real roots, a real double root, or two nonreal complex roots.

33. $2x^2 - 3x + 4 = 0$
34. $\sqrt{2}x^2 + \frac{3}{2}x - 1 = 0$
35. $4x^2 - 4\sqrt{3}x + 3 = 0$
36. $2x^2 - 3x - 4 = 0$
37. $-3x^2 + 2x + 1 = 0$
38. $x^2 - 12x + 36 = 0$

In exercises 39–58, solve the given polynomial equation.

39. $3x^2 - 2x - 4 = 0$
40. $\frac{1}{4}x^2 + x + 3 = 0$
41. $\frac{1}{4}x^2 + \frac{2}{3} = 0$
42. $-3x^2 - 18 - 0$
43. $9x^2 - 12x + 4 = 0$
44. $4x^2 + \sqrt{2}x - 1 = 0$
45. $x^2 - 2x + 1 - \frac{\pi^2}{4} = 0$
46. $9x^2 + 6x + 1 = 0$
47. $x^3 + 8 = 0$
48. $x^3 - 64 = 0$
49. $x^2 - i = 0$
50. $x^2 + 1 - i = 0$
51. $x^2 - 1 - i = 0$
52. $x^2 - 2x + 1 - i = 0$
53. $(1 - i)x^2 + 2x + 1 = 0$
54. $ix^2 - 2x + 1 = 0$
55. $x^4 - 1 = 0$. (Hint: there are four roots. Let $u = x^2$.)
56. $x^4 - 16 = 0$
57. $x^4 - 5x^2 - 36 = 0$
58. $(2x + 1)^{2/5} - (2x + 1)^{1/5} - 20 = 0$
59. (This exercise shows why "positive" and "negative" only refer to real numbers.) Show that there is no way to write the set of nonzero complex numbers as a union $P \cup N$, where P and N are sets such that (i) $z^2 \in P$ for each nonzero complex number z, (ii) the sum of any two elements of P is in P, and (iii) the sum of any two elements of N is in N. (Hint: show that either i or $-i$ would be in P. Notice $1 \in P$.)
60. Factor $x^4 + x^2 + 1$ as a product of polynomials which are irreducible over R. (Hint: add and subtract x^2.)

1.6 CHAPTER SUMMARY AND REVIEW EXERCISES

Major Concepts

- Circle
- Complex numbers
  Absolute value
  Conjugate
  Imaginary part
  Pure imaginary
  Real part
- Distance formula
- Distance from a point to a line
- Double root
- Equations of a line
  General linear equation
  Point-slope form
  Slope-intercept form
- Geometric triangle inequality
- Graph of an algebraic condition
- Imaginary unit
- Inequalities
  Properties of
- Key point
- Midpoint formula
- Parabolas
  Vertex
- Particular solution
  Of an equation
  Of an inequality
- Principal square root of a negative real number
- Properties of
  Absolute value
  Inequalities
  Radicals
- Pure imaginary number
- Quadratic formula
- Sign chart
- Slope, m
- Solution set
  Of an equation
  Of an inequality
- Substitution method for solving equations
- System of equations
  Consistent
  Dependent
  Inconsistent
  Solving by the substitution method

Review Exercises

In exercises 1–11, solve the given inequality, describe its solution set using interval notation, and graph the inequality.

1. $18(\frac{4}{3}x - 2) \leq \frac{5}{2}(6 - 2x)$
2. $\frac{-7}{3} < \frac{5 - 4x}{3} < 3$
3. Either $3 - 5x < -12$ or $3 - 5x > 12$
4. $(x - \sqrt{3})(x + 2) \geq 0$
5. $5x^2 + 3x + 4 \leq 4x^2 + 2x + 10$
6. $-5x^5(x - \frac{3}{5})^2(x - 4) < 0$
7. $\frac{2 - 3x}{3 - 2x} > 1$
8. $|3x - 2| \geq 4$

Chapter 1 /

9. $|x + 6| < 0.01$
10. $\sqrt[3]{x + 4} > -2$
11. $\dfrac{3 - x}{\sqrt{x^2 - 16}} \leq 0$
12. A juggler throws a ball straight up into the air. After t seconds, the ball is $-16t^2 + 64t + 6$ feet above the ground. During which period of time is the ball at most 6 feet above the ground?
13. Sketch the graph of the given algebraic condition:
    (a) $y = 0$ (b) $y = -x$
    (c) $x \geq 1$ and $y \leq 3$ (d) $x = -2$
14. Find (i) the distance from P_1 to P_2 and (ii) the coordinates of the midpoint of the line segment $\overline{P_1 P_2}$:
    (a) $P_1(-4, 5), P_2(-6, 9)$ (b) $P_1(2, -3), P_2(-1, 1)$
15. Determine whether the given three points are collinear:
    (a) $(0, 0), (5, -12), (10, -24)$
    (b) $(0, 0), (-5, -12), (3, 4)$
16. Find and simplify an algebraic condition on the coordinates (x, y) of the points on the perpendicular bisector of $\overline{AB}$, given $A(3, -1)$ and $B(7, 7)$.
17. Determine the slope of the line through P and Q:
    (a) $P(-1, -4), Q(-3, 2)$ (b) $P(2, 5), Q(\sqrt{2}, 5)$
    (c) $P(1, 2), Q(3, 7)$
18. Determine which line has (a) positive slope, (b) negative slope, (c) slope equal to 0, (d) undefined slope.

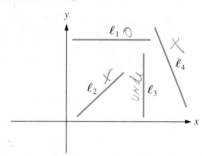

19. Find another point on the line, with slope -2, passing through $(-2, 4)$. Hence sketch the graph of this line.
20. Determine whether the lines $\overleftrightarrow{P_1 P_2}$ and $\overleftrightarrow{P_3 P_4}$ are parallel, perpendicular, or neither:
    (a) $P_1(-2, 3), P_2(7, -5), P_3(-5, 0), P_4(4, -8)$
    (b) $P_1(-1, 1), P_2(0, 2), P_3(3, -1), P_4(2, 1)$
    (c) $P_1(8, 7), P_2(4, 17), P_3(-1, 3), P_4(4, 5)$
21. Graph the following:
    (a) $x = 0$ (b) $2x + 1 = 0$ (c) $x = 3$
22. Write and simplify an equation of the line with the given properties:
    (a) passes through the point $(-\frac{1}{3}, 4)$; slope 3
    (b) passes through the point $(0, -5)$; slope -2
    (c) passes through the points $(4, -6)$ and $(-6, 14)$
    (d) passes through the point $(-2, 3)$; parallel to the line $2x - 3y - 6 = 0$
    (e) passes through the point $(5, 7)$; perpendicular to the line $2x - 3y - 6 = 0$
23. Find (a) the slope and (b) the y-intercept of the line which is the graph of $7x + 14y + 28 = 0$. Hence, (c) sketch this line.
24. Find the distance from P_1 to the given line:
    (a) $P_1(1, 4), 12x - 5y + 3 = 0$
    (b) $P_1(1, -1), y = -3x + 2$ (c) $P_1(-1, 1), x + 2y = 0$
25. By graphing the given equations, determine whether the given system is dependent, consistent with a unique particular solution, or inconsistent:
    (a) $\begin{cases} 5x - 6y = 7 \\ -5x + 7y = 6 \end{cases}$ (b) $\begin{cases} \frac{2}{3}x + \frac{12}{7}y - \frac{9}{7} = 0 \\ 14x + 36y - 29 = 0 \end{cases}$
    (c) $\begin{cases} 14x + 36y - 27 = 0 \\ \frac{2}{3}x + \frac{12}{7}y - \frac{9}{7} = 0 \end{cases}$
26. Determine which, if any, values of k make the system
    $$\begin{cases} 5x - 6y = 7 \\ -10x + 12y = k \end{cases}$$
    (a) dependent, (b) consistent with a unique particular solution, or (c) inconsistent.
27. Use the substitution method to **indicate** that the system
    $$\begin{cases} 3x - y = 14 \\ -15x + 5y = -70 \end{cases}$$
    is dependent. Then **prove** this geometrically.
28. Use the substitution method to show that the system
    $$\begin{cases} 12x - 8y = 13 \\ x + 2y = 4 \\ 16x - 10y = 15 \end{cases}$$
    is inconsistent.

In exercises 29–30, use the substitution method to solve the given system.

29. $\begin{cases} 2x + y = 4 \\ 7x + 2y = 2 \end{cases}$ 30. $\begin{cases} x + y + z = -1 \\ 2x - y - 3z = -4 \\ x - y + z = 11 \end{cases}$

31. Rachel sells 85 tickets to a show at her school. They cost $4 each for adults and $1.50 each for children. Her expenses amount to $6. If x of the 85 tickets were for adults and the net profit from Rachel's sales is y dollars, write an equation expressing a "linear relationship" between x and y.
32. A radiator contains 6 quarts of fluid that is 35 percent antifreeze (by volume) and 65 percent water. How much of this fluid should be drained so that, if one adds

a solution that is 80 percent antifreeze (and 20 percent water), the radiator will contain 6 quarts of fluid that is 55 percent antifreeze?

33. Find and simplify an equation for the circle with the given properties:
   (a) center $(2, -4)$, radius $\sqrt{3}$
   (b) passes through the point $(2, 1)$; center $(-1, 5)$
   (c) center $(6, 2)$; tangent to the line $3x - 4y = 9$
   (d) passes through the points $(0, 0)$, $(1, 0)$, and $(2, 3)$

34. Determine the center and radius of the circle with the given equation:
   (a) $x^2 + y^2 - 10x + 4y - 13 = 0$
   (b) $16x^2 + 16y^2 + 16x - 8y - 31 = 0$

In exercises 35–36, determine the (real) graph of the given equation.

35. $x^2 + y^2 - 4x + 5 = 0$
36. $x^2 + y^2 - 10x + 4y + 29 = 0$
37. Find equations of all the tangent lines to the circle $(x - 3)^2 + y^2 = 9$ that pass through the point $(-2, 1)$.
38. Determine equations for the tangents to the circle $x^2 + (y - 1)^2 = 16$ which have slope -4.

In exercises 39–43, solve the given system. Interpret your answer geometrically in exercises 39–42.

39. $\begin{cases} y - x = 1 \\ x^2 + y^2 = 1 \end{cases}$

40. $\begin{cases} (x - 1)^2 + (y + 4)^2 = 9 \\ (x - 1)^2 + (y - 2)^2 = 9 \end{cases}$

41. $\begin{cases} (x - 2)^2 + y^2 = 9 \\ (x + 7)^2 + y^2 = 25 \end{cases}$

42. $\begin{cases} x^2 + (y - 3)^2 = 1 \\ xy = 0 \end{cases}$

43. $\begin{cases} x^2 + y^2 = 20 \\ xy = 8 \end{cases}$

44. Graph each of the following parabolas, identifying its vertex:
   (a) $y = -2x^2$ (b) $x = \frac{1}{2}y^2$
   (c) $x + 1 = y^2$ (d) $y = -(x + 3)^2$

45. Perform the indicated operations and simplify:
   (a) $\sqrt{-6} + \sqrt{-54}$ (b) $\sqrt{-16} - \sqrt{-64}$
   (c) $\sqrt{\frac{-4}{9}}\sqrt{\frac{-2}{25}}$ (d) $\frac{\sqrt{-80}}{\sqrt{-5}}$

46. Find (i) the real part and (ii) the imaginary part for each of the following complex numbers:
   (a) $-\frac{3}{2} + 7i$ (b) $2 - 7i$ (c) $\sqrt{2}$ (d) i

In exercises 47–48, perform the indicated operations and simplify answers to the form $a + bi$ (with a and b real).

47. (a) $(-\frac{2}{3} + 4i) + (\frac{8}{3} - 6i)$ (b) $(4 - \frac{2}{3}i) - (6 - \frac{11}{3}i)$
   (c) $(-4 + 2i)(6 - 3i)$ (d) $-5i(2 - i)$
   (e) $-2 + 3i - 5i(6i - 4)$

48. (a) $\frac{2 - i}{2 + i}$ (b) $\frac{4}{4 - 5i}$
   (c) $\frac{12 + 3i}{3}$ (d) $(3 - 2i)^3$ (e) i^{67}

49. Determine (i) the conjugate and (ii) the absolute value of each of the following complex numbers:
   (a) $\frac{1}{\sqrt{2}} + \frac{1}{\sqrt{2}}i$ (b) $3 - 4i$ (c) $-\frac{\pi}{2}$ (d) $\sqrt{2}i$

50. Solve the following equations for x:
   (a) $(1 - i)x + 2 - 3i = 5x - i$ (b) $5x + 7\overline{x} = 36 + 4i$

In exercises 51–53, calculate the discriminant and hence determine whether the given quadratic equation has two unequal real roots, a real double root, or two nonreal complex roots. Next, use the quadratic formula to solve the quadratic equation.

51. $3x^2 - 2x + 4 = 0$ 52. $4x^2 + 12x + 9 = 0$
53. $3x^2 + 2x - 4 = 0$

In exercises 54–56, find all the complex roots of the given polynomial equation.

54. $x^3 - 27 = 0$ 55. $x^2 + 2i = 0$
56. $x^4 + 5x^2 - 36 = 0$

Chapter 2

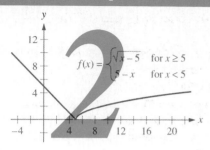

Functions and Their Graphs

In the last chapter, you looked at equations and graphs of lines, circles, and parabolas, and ways in which some of these interact. Each of these equations involved a relationship between two variables. In this chapter, you will study a special type of relationship between two variables, called a function. The notion of function is perhaps the most important idea in all of mathematics. In particular, this concept forms the basis for the remainder of this text.

Topics in this chapter include various types of symmetry that may be exhibited by the graph of a function; numerous examples of functions, including applications of quadratic functions; and ways of constructing functions, such as functional composition. In section 2.1, you will also find this text's first discussion of graphing calculators.

2.1 FUNCTIONS

Consider the circle which is the graph of the equation $x^2 + y^2 = 100$. Some of the ordered pairs that satisfy this equation include (0, 10), (0, –10), (6, 8), (6, –8), (–6, 8), (–6, –8), (8, 6), (8, –6), (–8, 6), (–8, –6), (10, 0), and (–10, 0). Figure 2.1 shows the graph of this circle and these twelve points. This set of twelve ordered pairs, and the set of all ordered pairs that satisfy the equation $x^2 + y^2 = 100$, are each examples of a relation. A **relation** is any set of ordered pairs of elements. When the elements are real numbers, this means that a relation is any subset of the Cartesian plane.

We are interested in a certain type of relation, called a function.

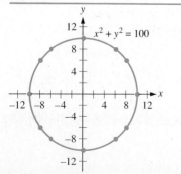

Figure 2.1

Functions

Fundamental Definitions

A **function** is a set of ordered pairs such that for each first element there is a unique second element. Because the second element is determined by the first element, the second element is referred to as the **dependent variable** and the first element as the **independent variable**.

The set of all values assumed by the independent variable is called the **domain,** denoted D, and the set of all values assumed by the dependent variable is the **range**, denoted R.

The term **natural domain** refers to the set of all real numbers that are *sensible* replacements for the independent variable. They are "sensible" in that the resulting dependent variable values are well-defined real numbers.

You may assume that a given function's domain is its natural domain unless we say otherwise.

First, you will be studying real-number functions; that is, we will ignore functions that have nonreal complex numbers as members of their domains or ranges.

Example 2.1

Some examples of functions are:

(a) $\{(0, 10), (6, 8), (-6, 8), (10, 0)\}$. The domain is $D = \{-6, 0, 6, 10\}$ and the range is $R = \{0, 8, 10\}$. (While the elements of the domain and range were each written in increasing order, this is not essential. For instance, you could have written $D = \{0, 6, -6, 10\}$.)

(b) $y = 4x - 3$. (Here we have used an equation to describe, or determine, the ordered pairs of this function. In ordered pair notation, this function would be defined as $\{(x, y): y = 4x - 3\}$. This clearly shows that x is the independent variable and y is the dependent variable.) Its natural domain and range are both the set of real numbers, that is, $D = R = \mathbf{R}$. This is seen by inspecting the "extent" of the (linear) graph of the given equation.

(c) $y = \sqrt{x}$. The domain and range are both the set of nonnegative real numbers, so $D = \{x: x \geq 0\}$ and $R = \{y: y \geq 0\}$. This is seen by recognizing the graph as the top half of the parabola $x = y^2$.

(d) $y = \sqrt{x^2 - 9}$, where $D = \{x: |x| \geq 3\} = \{x: x \leq -3 \text{ or } x \geq 3\}$ and $R = \{y: y \geq 0\}$. Here, in order to find D, we had to solve the inequality $x^2 - 9 \geq 0$.

(e) $y = \dfrac{5}{x - 7}$. If $x = 7$, then $\dfrac{5}{x - 7}$ is not defined, so 7 is not in D. In fact, the domain consists of all real numbers except 7, and so we can write $D = \{x: x \neq 7\}$. Since the numerator cannot be zero, it is not possible that $y = 0$. Indeed, $R = \{y: y \neq 0\}$.

(f) $y = |x + 5|$ has $D = \mathbf{R}$ and $R = \{y: y \geq 0\}$. The identification of D and R follows by graphing $y = |x + 5|$. Section 2.2 shows a quick way to obtain this graph on the basis of knowing how to graph $y = |x|$. To prepare for this, we shall soon give examples of graphs of some useful "standard" functions, including $y = |x|$.

(g) $y^3 = x^2 + 5$. This can be written as $y = (x^2 + 5)^{1/3} = \sqrt[3]{x^2 + 5}$. Here $D = \mathbf{R}$. Because x^2 is nonnegative, $x^2 + 5 \geq 5$, and the range is $[\sqrt[3]{5}, \infty)$.

EXAMPLE 2.2

Some examples of relations that are not functions are:

(a) $\{(0, 0), (1, 1), (1, -1), (2, 4), (2, -4), (2, 5)\}$. The first element 1 has two second elements, 1 and -1; and the first element 2 has three second elements, 4, -4, and 5.

(b) $x^2 + y^2 = 100$. The associated relation, $\{(x, y): x^2 + y^2 = 100\}$, was graphed in figure 2.1. Some of these points have the same first element but different second elements. For example, consider $(6, 8)$ and $(6, -8)$; or $(0, 10)$ and $(0, -10)$. For any first element x in the open interval $(-10, 10)$, there are two second elements y such that $x^2 + y^2 = 100$. That is why this set of ordered pairs is not a function.

(c) $y^2 = x + 2$. Some of the ordered pairs that satisfy this equation are $(2, 2)$, $(2, -2)$, $(7, 3)$, $(7, -3)$, $(10, \sqrt{12})$, and $(10, -\sqrt{12})$. Thus, some values of x have more than one second element. (Even though the first element $x = -2$ corresponds to only one second element, $y = 0$, this relation is not a function.)

(d) $y^4 = \sqrt{x} - 1$. Some ordered pairs that satisfy this relation are $(1, 0)$, $(4, 1)$, $(4, -1)$, $(5, \sqrt[4]{\sqrt{5} - 1})$, and $(5, -\sqrt[4]{\sqrt{5} - 1})$.

A function is often best understood by its graph. In a preliminary way, you can do this via a table of values. As parts (b)–(g) of example 2.1 indicate, it is often possible to view a function by means of an underlying equation. However, as indicated by the next four paragraphs, you must be careful.

Although the functions $y = \dfrac{x^2 - 1}{x - 1}$ and $y = x + 1$ agree at all values of $x \neq 1$, they are not equal sets of points. The reason is that they have different domains. The first function's natural domain is $\{x: x \neq 1\}$ and the second's is $\mathbf{R}$. Thus, the functions are not equal as sets since only the second one contains $(1, 2)$.

Another way to think of this is to ask the question "Do $y = \dfrac{x^2 - 1}{x - 1}$ and $y = x + 1$ define exactly the same set of points?" In particular, is the point $(1, 2)$ a point on both graphs? Since this point's coordinates satisfy $y = x + 1$ but do not satisfy $y = \dfrac{x^2 - 1}{x - 1}$, these equations do not define the same function.

Functions

To view the difference between $y = \dfrac{x^2 - 1}{x - 1}$ and $y = x + 1$ graphically, notice that the graph of the first function is obtained from that of the second (which is a line) by removing the point $(1, 2)$, as indicated in figures 2.2(a) and (b).

We can expand on the above ideas. For instance, the two functions given by $y = \dfrac{(x^2 - 1)(x + 2)}{(x - 1)(x + 2)}$ and $y = \dfrac{(x^2 - 1)(x + 2)(x - 3)}{(x - 1)(x + 2)(x - 3)}$ are other examples of functions that are not equal to the function given by $y = x + 1$. The graphs of these two functions are lines with two and three points removed, respectively, as shown in figures 2.2(c) and (d).

Figure 2.2

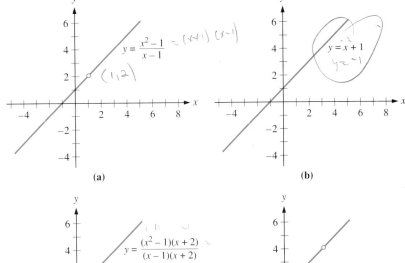

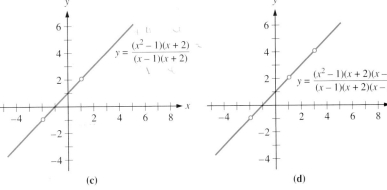

Functional Notation

A function is often named by a letter or a group of letters. Examples of letter designations are f, g, h, j, F, G, H, K, and P. Examples of groups of letters are sin, cos, tan, ABS, SGN, ln, and log. If x is the independent variable, then the dependent variable is designated by $f(x)$, $g(x)$, $h(x)$, $F(x)$, ABS (x), tan (x), and so on. The notation $f(x)$ is read "f of x." We say that y is a function of x and often write $y = f(x)$. Thus, the functions in parts (b)–(g) of example 2.1 could be described by $f(x) = 4x - 3$, $g(x) = \sqrt{x}$, $h(x) = \sqrt{x^2 - 9}$, $F(x) = \dfrac{5}{x - 7}$, $T(x) = |x + 5|$, and $C(x) = (x^2 + 5)^{1/3}$; in each of these cases, for lack of

Chapter 2 / Functions and Their Graphs

additional information, the natural domain is supposed. The function in part (a) of example 2.1 could be defined by $F(x) = \sqrt{100 - x^2}$ with domain $\{0, 6, -6, 10\}$.

Functional notation can be used to make more precise the above discussion of equality of functions:

Equal Functions

Two functions f and g are **equal functions** if (1) they have the same domains and (2) $f(x) = g(x)$ for each x in the domain.

Functional notation is especially useful for denoting the **value of a function at a specific point.** For example, consider the function f given by $f(x) = 4x - 3$. This function's value at the point 7 is denoted $f(7)$. This is obtained by substituting 7 for x in the expression for $f(x)$. Here, $f(7) = 4 \cdot 7 - 3 = 25$. Thus, the function f contains the ordered pair $(7, f(7)) = (7, 25)$.

▶ **EXAMPLE 2.3**

The value of the function h given by $h(x) = \sqrt{x^2 - 9}$ at the point -12 is denoted by $h(-12)$, and $h(-12) = \sqrt{(-12)^2 - 9} = \sqrt{144 - 9} = \sqrt{135}$. In particular, $(-12, h(-12)) = (-12, \sqrt{135})$ is in the function h. ◀

▶ **EXAMPLE 2.4**

If $s(t) = 8t^2 - 2t - 21$, find $s(3.6)$ and $s(-1.5)$.

Solution

$$s(3.6) = 8(3.6)^2 - 2(3.6) - 21$$
$$= 8(12.96) - 2(3.6) - 21$$
$$= 75.48.$$

$$s(-1.5) = 8(-1.5)^2 - 2(-1.5) - 21$$
$$= 8(2.25) - 2(-1.5) - 21$$
$$= 0.$$ ◀

Zero or Root of a Function

If r is a number such that $f(r) = 0$, then r is called a **zero** or **root** of the function f.

In Example 2.4, -1.5 is a zero of s; you can check that $\frac{7}{4}$ is also a zero of s.

▶ **EXAMPLE 2.5**

If $h(x) = 3x^3 + 5x^2 - 3x + 2$, find $h(-x)$.

Solution

Each x in $h(x)$ is replaced with $-x$, and so

$$h(-x) = 3(-x)^3 + 5(-x)^2 - 3(-x) + 2$$
$$= -3x^3 + 5x^2 + 3x + 2.$$ ◀

▶ **EXAMPLE 2.6**

If $g(x) = 5x^2 + 3x - 10$, find $g(4 + h)$.

Solution

Here each x in $g(x)$ is replaced with $4 + h$, producing

Functions

$$g(4 + h) = 5(4 + h)^2 + 3(4 + h) - 10$$
$$= 5(16 + 8h + h^2) + 3(4 + h) - 10$$
$$= 80 + 40h + 5h^2 + 12 + 3h - 10$$
$$= 82 + 43h + 5h^2.$$

EXAMPLE 2.7 If $j(x) = 2x^2 - 3$ and $h \neq 0$, find $\dfrac{j(x_0 + h) - j(x_0)}{h}$.

Solution We have

$$j(x_0 + h) = 2(x_0 + h)^2 - 3$$
$$= 2(x_0^2 + 2x_0h + h^2) - 3$$
$$= 2x_0^2 + 4x_0h + 2h^2 - 3$$

and $j(x_0) = 2x_0^2 - 3$.
As a result,

$$\frac{j(x_0 + h) - j(x_0)}{h} = \frac{(2x_0^2 + 4x_0h + 2h^2 - 3) - (2x_0^2 - 3)}{h}$$
$$= \frac{4x_0h + 2h^2}{h}$$
$$= 4x_0 + 2h.$$

Functions Defined by More Than One Equation

Nothing requires that a function be defined by a single equation such as $f(x) = x^3 + 4x^2 - 7$. For example, consider the function described by

$$f(x) = \begin{cases} \sqrt{x - 5} & \text{if } x \geq 5 \\ 5 - x & \text{if } x < 5 \end{cases}$$

The domain of this function is the entire set of real numbers, and the range is the set of nonnegative real numbers. The graph of this function is in figure 2.3.

A practical situation that uses such a bifurcated functional rule is a laboratory experiment, with x representing the time t and $f(t)$ a measured quantity at time t. The two cases (bifurcation) indicate that at time $t = 5$, one begins to observe phenomena following a new pattern.

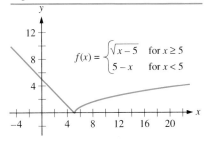

Figure 2.3

$$f(x) = \begin{cases} \sqrt{x - 5} & \text{for } x \geq 5 \\ 5 - x & \text{for } x < 5 \end{cases}$$

A familiar example of a function defined by more than one equation is the **absolute value function,** given by

$$f(x) = |x| = \begin{cases} x & \text{if } x \geq 0 \\ -x & \text{if } x < 0 \end{cases}$$

Another example of a function defined by more than one equation is the **greatest integer function.** This function is denoted by the symbol [ ]. Specifically, $[x]$ means the greatest integer less than or equal to x.

On both calculators, the ALPHA key is pressed in order to input an alphabetic character; the Range key is used to view or restrict the domain and range of the display; and the (−) key is pressed prior to a numeric value if one wishes to make that value negative. In that way the (−) key is equivalent to the +/− key on some other calculators. Do not confuse the (−) key with the −, or subtraction, key. On the *TI-81*, the ENTER key is pressed to obtain the result of a computation, draw a graph, advance to the next execution after a computation result, or advance to the next data input. On the Casio, the EXE, or EXEcute, key acts the same as the *TI-81's* ENTER key.

▶ **EXAMPLE 2.11**

Use a graphing calculator to sketch the graph of $f(x) = x^3 - 2.3x^2 + 0.2625x + 0.675$. Use the sketch to approximate the roots of the function f.

Solution   *TI-81* **Version**

First, press the MODE key to show the mode settings. A blinking block will indicate the location of the cursor. The selected mode has a dark background and white letters. If you do not have the settings shown in figure 2.9(a), use the ▲ and ▼ keys to move the cursor to the row that contains the setting you want changed. Then, press the ◀ or ▶ keys to move the cursor to the desired setting. Pressing the ENTER key will select the setting.

To leave the MODE setting, you can press either the Y=, GRAPH, RANGE, or 2nd QUIT keys. Press the RANGE key; the screen should show something like that in figure 2.9(b).

The domain to be shown on the screen is indicated by Xmin (the smallest value in the domain of interest), Xmax (the corresponding largest *x*-value), and Xscl (the *x*-axis's scale or, more precisely, the distance between "tick marks" on the *x*-axis). The range is similarly restricted by Ymin, Ymax, and Yscl. At the very bottom of the screen is Xres (the plotting resolution). If you do not have the values shown in figure 2.9(b), use the ▲ and ▼ keys to move the cursor to the character that you want to change. Press CLEAR and then type in the new value. Remember, if you want a negative number, you should first use the (−) key.

Now, press the Y= key. You should see the display in figure 2.9(c). If anything is written to the right of any of the four equal signs, use the ▲ and ▼ keys to move the cursor to the appropriate line, and then press the CLEAR key. When you have finished clearing, move the cursor to the top line, the line that begins ":Y1 = ".

Figure 2.9

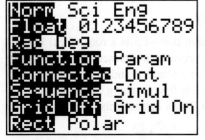

(a)   (b)   (c)

Functions

You are now ready to input a formula describing the function you want graphed. To input $f(x) = x^3 - 2.3x^2 + 0.2625x + 0.675$, press the sequence of keys

$\boxed{\text{X|T}}$ $\boxed{\wedge}$ 3 $\boxed{-}$ 2.3 $\boxed{\text{X|T}}$ $\boxed{x^2}$ $\boxed{+}$ 0.2625 $\boxed{\text{X|T}}$ $\boxed{+}$ 0.675

When you have finished, the result should look like the screen pictured in figure 2.9(d). If you have made a mistake, use the $\boxed{\blacktriangleleft}$ and $\boxed{\blacktriangleright}$ keys to move the cursor to the location of the error. Press $\boxed{\text{DEL}}$ to delete an unwanted character. If you have omitted a character, press $\boxed{\text{INS}}$ and type the missing character. As you have probably noticed, the $\boxed{\text{DEL}}$ key is used to DELete a keystroke and the $\boxed{\text{INS}}$ key is used to INSert one.

When you have the equation as you want it, press $\boxed{\text{GRAPH}}$. The result is the graph shown in figure 2.9(e).

In order to approximate the coordinates of a particular point, such as an x-intercept, on the graph, you need to use the trace function and the cursor keys $\boxed{\blacktriangleleft}$ and $\boxed{\blacktriangleright}$. After a graph has been sketched, press $\boxed{\text{TRACE}}$ and a blinking point will appear at the center of the graph. The x- and y-coordinates of this point are approximated on the bottom line of the display screen. Figure 2.9(f) shows the result of pressing $\boxed{\text{TRACE}}$ after the graph in figure 2.9(e) was displayed. The blinking point is indicated by the ⋈ on the graph.

To move the blinking point to the right, press the $\boxed{\blacktriangleright}$ key; to move it to the left, press $\boxed{\blacktriangleleft}$. By repeatedly pressing or holding down the $\boxed{\blacktriangleleft}$ key, we can move the blinking point to, for instance, what appears to be the first place where the graph crosses the x-axis. Actually, you may notice that the blinking point jumps from just above the x-axis to just below it. As shown on the calculator's screen, the values of the x-coordinates at these two places are $x = -0.3157895$ and $x = -0.5263158$. The corresponding y-coordinates are $y = 0.33125091$ and -0.2460709, respectively. Neither of these y-values is 0, and so neither -0.3157895 nor -0.5263158 is a root. However, the graph strongly suggests that there is a root of f between -0.3157895 and $x = -0.5263158$. By pressing the $\boxed{\blacktriangleright}$ key repeatedly, we see similarly that there appears to be another root between $x = 0.73684211$ and $x = 0.94736842$; and a third root at $x = 2$.

In order to get more accurate approximations of these three roots, we use one of the procedures that allow us to magnify or reduce the part of a graph

Figure 2.9 (continued)

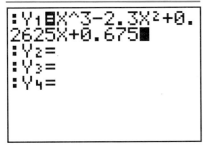

(d)

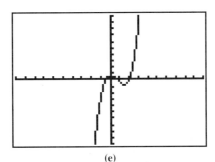

(e)

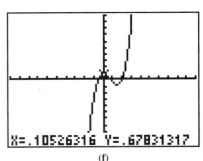

(f)

Figure 2.9 (continued)

(g)

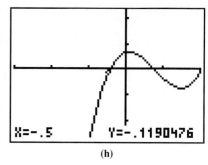
(h)

that is displayed on the screen. Let's consider the root between $x = -0.3157895$ and $x = -0.5263158$. Locate the cursor near the point $(-0.5, 0)$ because this procedure magnifies around the location of the cursor. Now press the ZOOM key. You will see the display in figure 2.9(g). Because we want to enlarge our view of the graph, press 2 ENTER to "zoom in." The screen will clear and then the graph shown in figure 2.9(h) will be drawn. By pressing TRACE and using the ◀ and ▶ keys as above, we see that the first root of f is approximately $x = -0.4473684$.

By pressing the ◀ and ▶ keys repeatedly on either the original graph (figure 2.9(e)) or the graph in figure 2.9(h), you can approximate the roots of f one at a time. You should determine that $x = -0.4473684$, $x = 0.71052632$, and $x = 2$ appear to be the roots. In fact, the actual roots are $x = -0.45$, $x = 0.75$, and $x = 2$. It is not at all unusual that you did not find the exact decimal expansions of the roots. This is caused in part by the standard screen settings for this calculator and by the size of the dots on the screen. However, the ability to magnify or zoom can sometimes give you sufficiently close approximations. In more complicated examples, possibly involving irrational roots, you cannot hope to do more than find good approximations of the decimal expansions of the roots. You will often need more advanced sections of precalculus and even calculus to determine very precise information about the roots of a function.

Casio fx-7700G Version

First, enter Shift Cls EXE to clear the graph screen.

We begin by setting the domain and range we want displayed on the screen. Press the Range key. The screen should look something like the one in figure 2.9(i).

The domain to be shown on the screen is indicated by Xmin (the smallest value in the domain of interest), Xmax (the corresponding largest x-value), and Xscl (the x-axis's scale or, more precisely, the distance between "tick marks" on the x-axis). The range is similarly restricted by Ymin, Ymax, and Yscl. These values can be changed, and we will do that in later graphing calculator activities. However, for this example, we will use the "standard" preset values. If you do not have the values shown in figure 2.9(i), press F1 and you should get them.

Functions

You are now ready to input a formula describing the function you want graphed. First, press the [Range] key twice to exit the screen settings mode. To input $f(x) = x^3 - 2.3x^2 + 0.2625x + 0.675$, press the sequence of keys

[GRAPH] [X,θ,T] [x^y] 3 [−] 2.3 [X,θ,T] [Shift] [x²] [+] 0.2625 [X,θ,T] [+] 0.675 [EXE]

When you have finished, the result should look like the screen pictured in figure 2.9(j). If you have made a mistake, use the [◄] and [►] keys to move the cursor to the location of the error. Press [DEL] to delete an unwanted character. If you have omitted a character, press [Shift] [INS] and type the missing character. As you have probably noticed, the [DEL] key is used to DELete a keystroke and the [Shift] [INS] keys are used to INSert one.

In order to approximate the coordinates of a particular point, such as an x-intercept, on the graph, you need to use the trace function and the cursor keys [◄] and [►]. After a graph has been sketched, press [Shift] [Trace] and a blinking point will appear at the extreme left part of the graph. The blinking point is indicated by the ✖ on the graph. The coordinates of this point will be approximated on the bottom line of the display screen. Figure 2.9(k) shows the result of pressing [Shift] [Trace] after the graph in figure 2.9(j) was displayed.

To move the blinking point to the right, press the [►] key; to move it to the left, press [◄]. By repeatedly pressing or holding down the [►] key, we can move the blinking point to, for instance, what appears to be the first place where the graph crosses the x-axis. Actually, you will notice that the blinking point jumps from just below the x-axis to just above it. As shown on the calculator's screen, the values of the x-coordinates at these two places are $x = -0.5$ and $x = -0.4$. The corresponding y-coordinates are $y = -0.15625$ and 0.138, respectively. Neither of these y-values is 0, and so neither -0.5 nor -0.4 is a root. However, the graph strongly suggests that there is a root of f between -0.5 and -0.4. By continuing to press the [►] key, we see that there appears to be another root between $x = 0.7$ and $x = 0.8$; and a third root at $x = 2$.

In order to get more accurate approximations of these three roots, we use one of the procedures that allow us to magnify or reduce the part of a graph that is displayed on the screen. Let's consider the root between -0.5 and -0.4. Locate the cursor near the point $(-0.5, 0)$ because this procedure magnifies around the location of the cursor. Now press the [Shift] [Zoom] keys. The screen should look something like the one displayed in figure 2.9(l).

Figure 2.9 (continued)

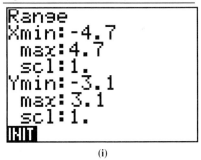

(i)

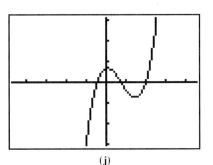

(j)

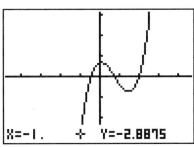

(k)

(l)

Chapter 2 / Functions and Their Graphs

Figure 2.9 (continued)

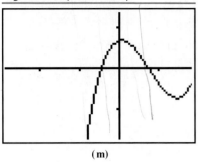

(m)

Press the F3 key to "zoom in." The screen will clear and then the graph shown in figure 2.9(m) will be drawn. By pressing Shift Trace and using the ◄ and ► keys, we see that the first root of f is approximately $x = -0.45$.

Apply the Shift and Shift Trace features repeatedly on the original graph (figure 2.9(j)) to approximate the roots of f one at a time. You should determine that $x = -0.45$, $x = 0.75$, and $x = 2$ appear to be the roots. In fact, they are! In more complicated examples, possibly involving irrational roots, you cannot hope to be so lucky as to find the exact decimal expansions of the roots. However, as you will see later, the ability to magnify or zoom can sometimes give you sufficiently close approximations. You will often need more advanced sections of precalculus and even calculus to determine very precise information about the roots of a function.

Exercises 2.1

In exercises 1–12, use either the definition of a function or the Vertical Line Test to determine whether the relation being described is a function.

1. $\{(0, 1), (4, 7), (-4, 2)\}$
2. $\{(-1, 1), (-1, -1), (0, 2)\}$
3. $\{(-2, 4), (-2, 9), (4, 8)\}$
4. $\{(0, 0), (-1, 1), (1, 1)\}$
5. $y = \sqrt{4 - x^2}$
6. $x^2 + y^2 = 4$

7.

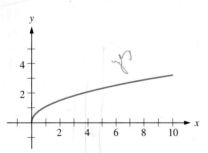

8.

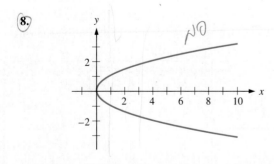

9.

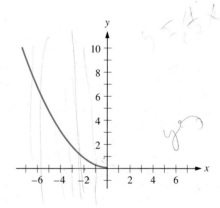

10.

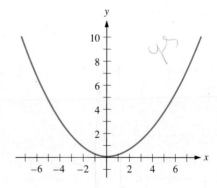

11.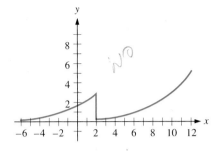

12.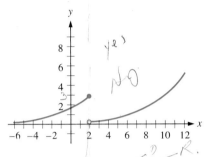

13. Consider the relation given by $\{(x, y): y$ is the husband of x as of noon, Eastern Standard Time, January 1, 1991$\}$.
  (a) Explain why this is a function.
  (b) What is its domain?
  (c) What is its range?
  (d) If "husband" is changed to "brother," is the resulting relation a function?

14. (a) Explain why the relation $\{(x, y):$ at some time, y has been the husband of $x\}$ is not a function.
  (b) Find at least three subsets of this relation which are functions.

15. Is each subset of a function necessarily a function? Explain.

In Exercises 16–32, find the natural domain of the given function.

16. $\{(0, -1), (1, -1), (2, 4)\}$
17. $f(x) = \sqrt{4 - x}$
18. $g(x) = -\sqrt{5 + x}$
19. $h(x) = \pi\sqrt{4 - x^2}$
20. $k(x) = 2\sqrt{5 - x^2}$
21. $m(x) = 2|x + 5|$
22. $n(x) = 2|x| - 5$
23. $p(x) = 8\sqrt{5 + x^2}$
24. $q(t) = \sqrt[3]{4 - t^2}$
25. $r(t) = \frac{2}{7}t^3 - t + 5$
26. $s(t) = -6$
27. $u(t) = \frac{t^9 - 7t^5 + 6}{2t + 3}$
28. $v(t) = \frac{t^9 - 7t^5 + 6}{2t^4 + 3}$
29. $w(t) = \frac{t^3 - 1}{t - 1}$
30. $F(t) = \frac{t^3 + 1}{t - 1}$
31. $G(t) = \frac{t^2 + t - 12}{\sqrt{t - 2}}$
32. $H(t) = \frac{t + 4}{\sqrt{t^2 - t - 2}}$

33–40. Graph and find the range of each of the functions given in exercises 17–24.

41. Find the domain and range of each of these functions.

(a)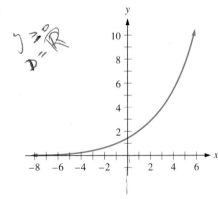

(b)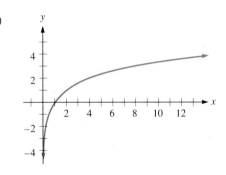

(c)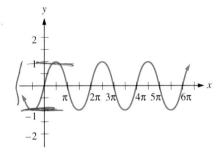

(d)

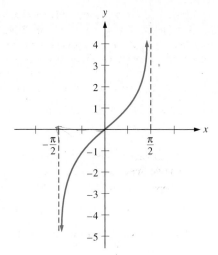

(e)

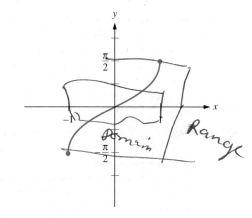

42. In the country of Ecalpon, the cost of mailing an envelope weighing less than 1 gram is 7 spendies, of mailing an envelope weighing at least 1 gram but less than 2 grams is 14 spendies, of mailing an envelope weighing at least 2 grams but less than 3 grams is 21 spendies, and so on. Let $c(g)$ denote Ecalpon's cost in spendies of mailing an envelope weighing g grams.
   (a) Find the domain and the range of the function c.
   (b) Graph c.
   (c) Prove that $c(g) = 7[g+1]$ for all $g > 0$.
   (d) Are the functions given by $c(g)$ and $7[g+1]$ equal? Explain.

43. (See the preceding exercise.) Ecalpon's neighbor, Lavir, has a postage function given by $d(g) = 9 + 6[g]$ for all $g > 0$.
   (a) Describe its underlying rule in words. (b) Graph d.
   (c) Find all values of g such that $c(g) = d(g)$.

In exercises 44–56, find $f(0), f(1), f(-1), f(2.7), f(3+h)$, $f(x+h), f(-x)$, and $f\left(\frac{1}{x}\right)$, if they exist, for the given choice of $f(x)$.

44. $f(x) = \sqrt{4-x}$
45. $f(x) = \pi\sqrt[3]{5-x}$
46. $f(x) = \sqrt{4-x^2}$
47. $f(x) = -4x^2 + 2$
48. $f(x) = \frac{2}{7}x^3 - x + 5$
49. $f(x) = -6$
50. $f(x) = \frac{2}{x}$
51. $f(x) = \frac{x^2 - 3x + 5}{x+2}$
52. $f(x) = \frac{x^3 - 8}{x-2}$
53. $f(x) = 2|x-3|$
54. $f(x) = 2|x| - 3$
55. $f(x) = x + |x|$
56. $f(x) = x + 2|x|$

The *average rate of change* of a function f between x_0 and $x_0 + h$ is $\frac{f(x_0+h) - f(x_0)}{h}$. In exercises 57–67, compute this average for the given data.

57. $f(x) = 5x + 7$, $x_0 = x$. Generalize to $f(x) = 5x + b$.
58. $f(x) = -7x + 5$, $x_0 = x$. Generalize to $f(x) = mx + b$.
59. $f(x) = [x]$, $x_0 = \frac{2}{3}$, $-\frac{1}{6} < h < \frac{1}{6}$. Generalize to x_0 not an integer, $|h| < |x_0 - $ nearest integer$|$.
60. $f(x) = [x]$, $x_0 = 4$, $h = \frac{1}{2}$. Generalize to x_0 an integer, $0 < h < 1$.
61. $f(x) = [x]$, $x_0 = 4$, $h = -\frac{1}{2}$. Generalize to x_0 an integer, $-1 < h < 0$.
62. $f(x) = |x|$, $x_0 = 9.7$, $h = -0.2$. Generalize to x_0 positive and not an integer, $|h| < |x_0 - $ nearest integer$|$.
63. $f(x) = |x|$, $x_0 = -6.5$, $h = 0.07$. Generalize to x_0 negative and not an integer, $|h| < |x_0 - $ nearest integer$|$.
64. $f(x) = |x|$, $x_0 = 0$, $h > 0$.   65. $f(x) = |x|$, $x_0 = 0$, $h < 0$.
66. $f(x) = \frac{3}{2}x^2 - 7x + 3$, $x_0 = x$.
67. $f(x) = -7x^2 + 3x + 5$, $x_0 = x$.
68. Prove that each of the functions given by the following rules is increasing: $f(x) = x$, $g(x) = x + 3$, $h(x) = 7x - 4$, and $k(x) = x^3 + 2$.
69. Prove that each of the functions given by the following rules is decreasing: $f(x) = -x$, $g(x) = -x - 3$, $h(x) = -7x + 4$, and $k(x) = -x^3 - 2$.
70. Prove that the function $m(x) = x^2$ is increasing on the interval $(0, \infty)$ and decreasing on the interval $(-\infty, 0)$.
71. Prove that the function $f(x) = x^3 - 3x$ is increasing on $(-\infty, -1)$, decreasing on $(-1, 1)$, and increasing on $(1, \infty)$.
◆ 72. Let m and b be real numbers and consider the function $f(x) = mx + b$. Prove that f is increasing if $m > 0$; f is decreasing if $m < 0$; and f is constant if $m = 0$. Interpret geometrically.

Functions

♦ 73. Any circle with radius r has area A that can be expressed as a function of r according to the rule $A = A(r) = \pi r^2$. Write the functional rule which expresses A as a function of the diameter d.

♦ 74. Any square with side of length s has area A given by $A = s^2$. Let d be the length of a diagonal of the square. Express s in terms of d. (Hint: apply Pythagoras' Theorem.) Hence find the rule by which A is a function of d.

♦ 75. After nearly colliding, two ships steam off at right angles, with speeds of 15 mph and 20 mph. Express the distance the ships are apart (in miles) as a function of time (t, measured in hours).

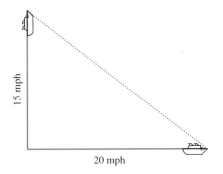

♦ 76. A wire is to be cut into two pieces each x units long. One piece will be bent into a circle and the other will be bent into a square. Express the sum of the areas of the circle and the square as a function of x.

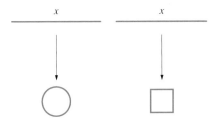

In exercises 77–78, (a) sketch the graph of the given function f; (b) calculate $f(4)$; (c) determine the intervals on which f is increasing; (d) determine the intervals on which f is decreasing.

77. $f(x) = \begin{cases} \sqrt{x} & \text{if } 0 \leq x < 4 \\ x - 1 & \text{if } 4 \leq x \end{cases}$

78. $f(x) = \begin{cases} x & \text{if } x \leq 4 \\ 4 & \text{if } 4 < x < 5 \\ -2x + 14 & \text{if } 5 \leq x \end{cases}$

♦ 79. At Tohoot University, an enterprising student rents out electric fans for $20 apiece per month. Beyond the first five fans, each additional dozen fans that he rents costs him $6 per month in maintenance expenses.
(a) Show that his monthly net income in dollars resulting from renting n fans is given by
$$f(n) = \begin{cases} 20n & \text{if } n = 0, 1, 2, 3, 4, 5 \\ 20n - 6\left[\dfrac{n-5}{12}\right] & \text{if } n = 6, 7, 8, \ldots \end{cases}$$
(b) Prove that $f(n_1) < f(n_2)$ whenever n_1 and n_2 are positive integers such that $n_1 < n_2$.
(c) Consider the function g, with domain $[6, \infty)$, defined by the rule $g(x) = 20x - 6\left[\dfrac{x-5}{12}\right]$. Prove that g is *not* an increasing function. (Hint: compute $g(16.951)$.)

Note: The following two exercises are designed to be used with a graphing calculator or graphing software. Two versions of the instructions are given in the exercises. One set of instructions is for a Texas Instruments *TI-81* graphics calculator and the other is for a Casio *fx-7700G* graphics calculator. If you are not using either of these calculators, you may have to adapt the directions.

80. (a) Graph the following functions using your graphing calculator: $f(x) = |x|$, $g(x) = |3x|$, $h(x) = -0.25|3x|$, and $k(x) = -0.25|3x| - 1.5$.
TI-81 **Version**—You will enter your commands so that the functions f, g, h, and k are assigned to the functions Y1, Y2, Y3, and Y4, respectively, on the Y= menu. The graphs will be drawn sequentially and the final screen will show all four graphs. The ENTER key is used to tell the calculator to move to a new function. Begin by pressing Y= . Now enter the four functions by pressing the following sequence of keys. (If a function is already entered on one of these lines, you should begin by pressing the CLEAR key. This erases the displayed formula.) Each line includes the necessary keystrokes for one of the functions f, g, h, and k.

2nd ABS X|T ENTER
2nd ABS 3 X|T ENTER
(−) .25 2nd ABS 3 X|T ENTER
(−) .25 2nd ABS 3 X|T − 1.5 ENTER

Next, press ZOOM 6. This will clear the calculator's display screen, set the domain and range to the calculator's standard settings, and then

Chapter 2 / Functions and Their Graphs

graph the four functions. The results on your screen should look sequentially like the figures (a)–(d) below.

Casio fx-7700G Version—You can enter your commands on one line so that the graphs will be drawn sequentially and the final screen will show all four graphs. The Shift ↵ keys are used to tell the calculator to begin graphing a new function. Begin by clearing the calculator's display screen and setting the domain and range to the standard settings: press Shift Cls Range F1 Range Range EXE. Now enter the four functions by pressing the following sequence of keys. Each line that begins with GRAPH starts the necessary keystrokes for one of the functions f, g, h, and k.

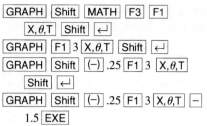

The results on your screen should look sequentially like the figures (e)–(h) below.

(b) Graph $f(x) = |x|$, $g(x) = |-2x|$, $h(x) = 0.4|-2x|$, and $k(x) = 0.4|-2x| - 1.75$.

(c) Graph $f(x) = |x|$, $g(x) = |0.5x|$, $h(x) = -3|0.5x|$, and $k(x) = -3|0.5x| + 2$.

(d) Each of the above parts considered functions of the form $f(x) = |x|$, $g(x) = |bx|$, $h(x) = a|bx|$, and $k(x) = a|bx| + d$. What effect do the numbers a, b, and d seem to have on the basic graph of f?

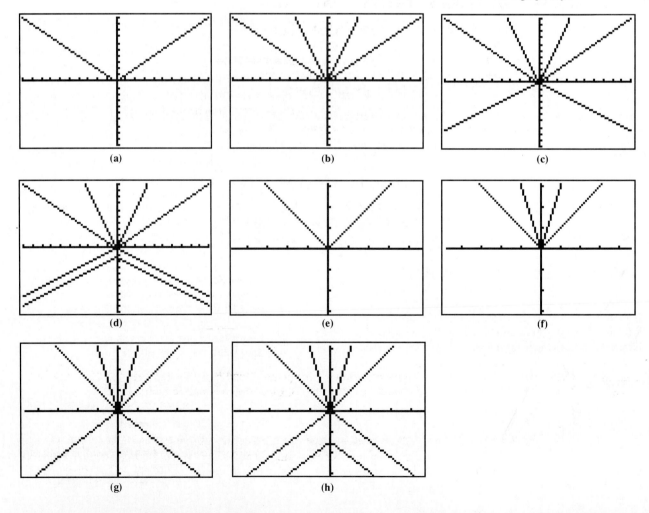

81. Use your graphing calculator and sketch the graphs of each of the following functions. (See exercise 80 above.)
    (a) Graph $f(x) = |x|$, $g(x) = |2x|$, $h(x) = |2x + 3|$, and $k(x) = |2x + 3| - 1$. (Hint: to graph h and k, you will need to put parentheses around $2x + 3$. For instance, on a TI-81, h is keyed in by pressing [2nd] [ABS] [(] 2 [X|T] [+] 3 [)]. On a Casio fx-7700G, h is keyed in by pressing [GRAPH] [Shift] [MATH] [F3] [F1] [(] 2 [X,θ,T] [+] 3 [)] .)
    (b) Graph $f(x) = |x|$, $g(x) = |-0.3x|$, $h(x) = |-0.3x - 2|$, and $k(x) = |-0.3x - 2| + 1.5$.
    (c) Both (a) and (b) considered functions of the form $f(x) = |x|$, $g(x) = |bx|$, $h(x) = |bx + c|$, and $k(x) = |bx + c| + d$. What effect do the numbers b, c, and d seem to have on the basic graph of f?

2.2 SYMMETRY AND TRANSFORMATIONS

In the last section, we defined the general notion of "function" and then looked at examples, such as constant functions and the greatest integer function. The vertical line test provided a way to determine geometrically whether a given relation was a function. We also used geometry to find intervals on which functions were increasing or decreasing. In this section, we shall look at some other geometric properties of functions.

Line Symmetry

Consider two points $A(x_1, y_1)$, $B(x_2, y_2)$, and a line ℓ.

Line Symmetry

If ℓ is the perpendicular bisector of the segment $\overline{AB}$, then we say that the points A and B are **symmetric with respect to the line ℓ**, and ℓ is called a **line of symmetry** of A and B. If A is a point on ℓ, we say that A is symmetric to itself with respect to ℓ.

> **EXAMPLE 2.12**

What is the line of symmetry for the points $A(-4, -1)$ and $B(2, 3)$?

Solution

The midpoint of $\overline{AB}$ is $(-1, 1)$. The slope of the line through $\overline{AB}$ is $\frac{2}{3}$, so a line perpendicular to $\overline{AB}$ will have slope $-\frac{3}{2}$. Thus the line of symmetry has the slope $-\frac{3}{2}$ and passes through the point $(-1, 1)$. Using the point-slope form, we get

$$y - 1 = -\frac{3}{2}[x - (-1)]$$
$$y = -\frac{3}{2}(x + 1) + 1$$
$$y = -\frac{3}{2}x - \frac{1}{2}$$
$$2y = -3x - 1.$$

The line of symmetry for points A and B, as shown in figure 2.10, is the line $2y = -3x - 1$. Notice that the line $\overleftrightarrow{AB}$ is perpendicular to this line of symmetry. ◀

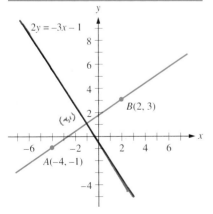

Figure 2.10

We are seldom interested in the line of symmetry for just two points. More often, we want a line of symmetry for an entire set of points—usually a curve

Figure 2.11

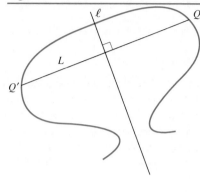

of some kind. We shall use the idea of the line of symmetry for two points to determine lines of symmetry for a figure.

To say that a line ℓ is a line of symmetry for a set of points means that for each point P in the set, there is a point P' in the set such that P and P' are symmetric with respect to ℓ. Because a line of symmetry acts like a mirror, a line of symmetry is also referred to as a **line of reflection.**

Look at the curve in figure 2.11. Pick any point, Q, on the curve and draw a line, L, through Q so that L is perpendicular to ℓ. Suppose line L intersects the curve in a second point, Q', and ℓ is the perpendicular bisector of segment QQ'. (As a degenerate case, we permit $Q = Q'$ to be a point on ℓ.) If this is true for any point Q you select on the curve, then ℓ is a line of symmetry for the curve.

One reason to want to recognize instances of symmetry is suggested by the mirror effect discussed above. If a set S is symmetric about a line ℓ, then in order to graph S, you graph the "half" of S that lies on one side of ℓ, and then reflect these points through ℓ. The images of this reflection, together with the half that had been plotted first, form the entire graph of S.

Axes as Lines of Symmetry

A coordinate axis is often a line of symmetry, as we show in example 2.13.

▶ **EXAMPLE 2.13**

The curve $y = x^2$, shown in figure 2.12(a), is symmetric with respect to the y-axis. The curve $x = y^2$ in figure 2.12(b) is symmetric with respect to the x-axis.

It is possible for a curve to have more than one axis of symmetry. The curve $9x^2 + 16y^2 = 144$, shown in figure 2.12(c), is symmetric with respect to both the x- and the y-axes. ◀

Figure 2.12

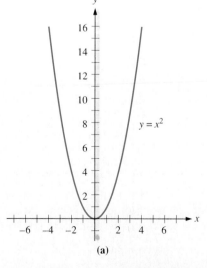
(a)

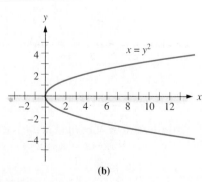
(b)

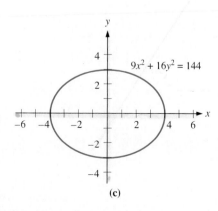
(c)

Symmetry and Transformations

After we look at some other types of symmetry, we shall determine algebraic criteria for the axes to be lines of symmetry.

Point Symmetry

Symmetry about a point is another type of symmetry. As you just saw, two distinct points are symmetric about a line if that line does two things: (1) passes through the midpoint of the segment formed by the two points and (2) is perpendicular to that segment. Point symmetry considers only the first condition. Suppose that one has two distinct points, A and B, and that the point P is the midpoint of $\overline{AB}$. We say that P is a point of symmetry for A and B, or that A and B are symmetric about the point P.

Point Symmetry

> A set of points (such as a curve) is **symmetric about a point P** if for each point A in the set of points, there is a corresponding point B in the set of points such that P is the midpoint of $\overline{AB}$. (Possibly, P may be in the given set of points. In that case, if we consider $A = P$, we allow $B = P$ also, and agree that P is the "midpoint" of the "degenerate" segment $\overline{PP}$.)

▶ **EXAMPLE 2.14**

One curve that has point symmetry about the origin is $y = x^3$, shown in figure 2.13(a). Another example, also with point symmetry about the origin, is the graph in figure 2.12(c). ◀

In exercise 63, you will be asked to show that the graph of $y = x^3 - 3x^2 - x + 5$, shown in figure 2.13(b), is symmetric about the point (1, 2).

Figure 2.13

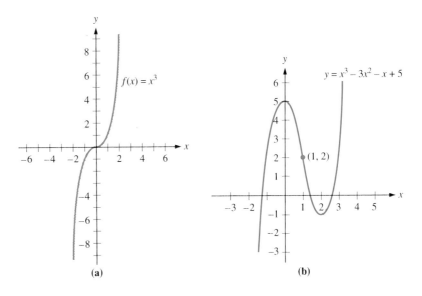

(a)   (b)

Even and Odd Functions

The next two properties provide ways to quickly determine if the graph of a function is symmetric with respect to either an axis or a point.

Consider a function f whose graph is symmetric about the y-axis, and let $P(a, f(a))$ be any point on this graph. Suppose that $a \neq 0$; that is, P is not on the y-axis. Since the y-axis is a line of symmetry, there is a point $P'(b, f(b))$ on this graph such that the y-axis is the perpendicular bisector of $\overline{PP'}$. Since the y-axis is a vertical line, $\overline{PP'}$ must be horizontal and, hence, has a slope of 0. So $f(b) = f(a)$. By the midpoint formula, the midpoint of $\overline{PP'}$ is $\left(\frac{b+a}{2}, \frac{f(b)+f(a)}{2}\right) = \left(\frac{b+a}{2}, f(a)\right)$. Since $(0, f(a))$ is the midpoint of $\overline{PP'}$, $0 = \frac{b+a}{2}$ and $b = -a$. Combining this with the fact that $f(b) = f(a)$, we see that $f(a) = f(-a)$. We have discovered the following test to determine if a function's graph is symmetric with respect to the y-axis.

Even Function

A function is **even** if $f(-x) = f(x)$ for all values of x in the domain of f. The graph of a function f is symmetric about the y-axis if and only if f is even.

▶ **EXAMPLE 2.15**

Determine whether the following functions are even:
(a) $f(x) = \frac{3}{5}x^4 - 6x^2 + \pi$ and (b) $g(x) = 5x^3$.

Solutions

(a) To determine if a function f is even, we check to see if $f(-x) = f(x)$. For $f(x) = \frac{3}{5}x^4 - 6x^2 + \pi$, we see that $f(-x) = \frac{3}{5}(-x)^4 - 6(-x)^2 + \pi = \frac{3}{5}x^4 - 6x^2 + \pi = f(x)$. Since $f(x) = f(-x)$, the function f is even. Its graph is symmetric with respect to the y-axis. This is illustrated in figure 2.14(a).

(b) We next determine if the function given by $g(x) = 5x^3$ is even. Evaluating $g(-x)$, we see that $g(-x) = 5(-x)^3 = -5x^3$. If g were even, then $g(x) = g(-x)$ for every point x in the domain of g. It is true that $g(0) = g(-0) = 0$. But $g(2) = 40$ and $g(-2) = -40$. This one case, where $g(2) \neq g(-2)$, is sufficient for us to say that $g(x) \neq g(-x)$ as functions. Thus, the function g is not even. As can also be seen from figure 2.14(b), the graph of the function g is not symmetric with respect to the y-axis. ◀

There is a related result for graphs of relations.

Symmetry and Transformations

Figure 2.14

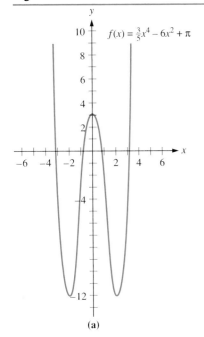
(a)

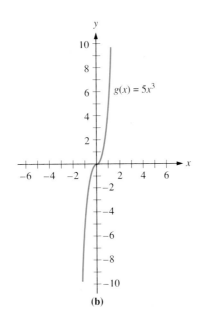
(b)

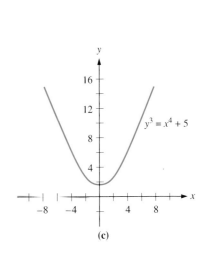
(c)

Symmetry about y-axis

The graph of an equation is **symmetric with respect to the y-axis** if replacing x by $-x$ produces an equivalent equation.

▶ **EXAMPLE 2.16**

The graph of $y^3 = x^4 + 5$, shown in figure 2.14(c), is symmetric about the y-axis since $y^3 = (-x)^4 + 5$ is equivalent to $y^3 = x^4 + 5$. ◀

A similar test can be used to determine if a curve is symmetric with respect to the x-axis.

Symmetry about x-axis

The graph of an equation is **symmetric with respect to the x-axis** if replacing y by $-y$ produces an equivalent equation.

▶ **EXAMPLE 2.17**

Let's reconsider the graph of $x = y^2$. This was shown in figure 2.12(b) to be symmetric about the x-axis. Replacing y with $-y$ in this equation produces $x = (-y)^2$, which is equivalent to the original equation since $(-y)^2 = y^2$. Hence, we have an algebraic proof that the graph of $x = y^2$ is symmetric with respect to the x-axis. ◀

Chapter 2 / Functions and Their Graphs

One of the most useful points of symmetry is the origin.

Symmetry about the Origin

> The graph of a function is **symmetric about the origin** if, for any point $(a, f(a))$ on the graph, there is a point $(-a, f(-a)) = (-a, -f(a))$ that is also on the graph.

Such a function is called an odd function. Specifically, we have the following.

Odd Function

> A function is **odd** if $f(-x) = -f(x)$ for all values of x in the domain of f. The graph of a function f is symmetric about the origin if and only if f is odd. More generally, the graph of an equation is symmetric with respect to the origin if replacing x by $-x$ and y by $-y$ produces an equivalent equation.

▷ **EXAMPLE 2.18**

Determine whether the following functions are odd:
(a) $f(x) = x^3$ and (b) $h(x) = \frac{3}{4}x$.

Solutions

(a) To determine if a function f is odd, we check to see if $f(-x) = -f(x)$. For the function given by $f(x) = x^3$, we see that $f(-x) = (-x)^3 = -x^3$ and that $-f(x) = -(x^3) = -x^3$. Since $f(-x) = -f(x)$, this function is odd, and hence its graph is symmetric about the origin. This graph was given in figure 2.13(a).

(b) We next check to see if the function given by $h(x) = \frac{3}{4}x$ is odd. $h(-x) = \frac{3}{4}(-x) = -\frac{3}{4}x = -h(x)$; and since $h(-x) = -h(x)$, the function h is an odd function. By looking at the graph of h in figure 2.15(a), you can also see directly that it is symmetric with respect to the origin. ◁

▷ **EXAMPLE 2.19**

Determine whether the graphs of the following equations are symmetric with respect to the origin: (a) $4y = 3x$ and (b) $y^2 = x^4 + 5$.

Solutions

(a) Replacing y with $-y$ and x with $-x$ in the equation $4y = 3x$ produces $4(-y) = 3(-x)$; equivalently, $-4y = -3x$ or $4y = 3x$. Since this is the same as the original equation, the graph of this equation is symmetric about the origin. In fact, $4y = 3x$ amounts to the function $h(x) = \frac{3}{4}x$ in example 2.18(b); its graph was in figure 2.15(a).

(b) In order to determine if the curve $y^2 = x^4 + 5$ is symmetric about the origin, we replace y with $-y$ and x with $-x$ to get $(-y)^2 = (-x)^4 + 5$, or $y^2 = x^4 + 5$. Again, we get an equation equivalent to the original. Thus its graph, as shown in figure 2.15(b), is symmetric about the origin. ◁

Symmetry and Transformations

Figure 2.15

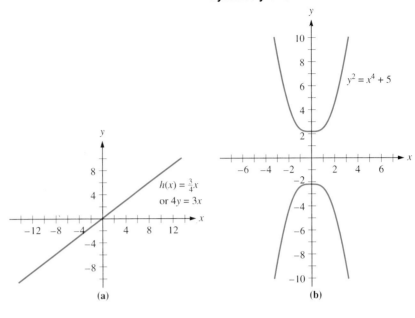

Figure 2.16

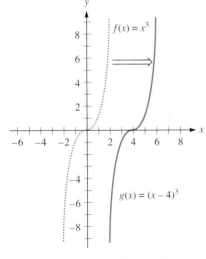

Horizontal Translations

Look at the graphs of the functions $f(x) = x^3$ and $g(x) = (x - 4)^3$ shown in figure 2.16. The two graphs are copies of each other. The only difference is that the graph of g is shifted, or translated, 4 units to the right of the graph of f. For example, the point $(4, 0)$ on g corresponds to the point $(0, 0)$ on f. This is easier to see if you examine a table of values for the two functions.

x	−2	−1	0	1	2	3	4	5	6	7	8
$f(x) = x^3$	−8	−1	0	1	8	27	64	125	216	343	512
$g(x) = (x - 4)^3$	−216	−125	−64	−27	−8	−1	0	1	8	27	64

What you have just seen illustrates the following **horizontal translation theorem**.

How to Graph
$y = f(x - a)$

If f is a function and a is a constant, then the graph of $y = f(x - a)$ translates the graph of $y = f(x)$ horizontally a distance of $|a|$ units. This translation is to the right if $a > 0$ and to the left if $a < 0$.

Proof: Let (x_1, y_1) be a point on the graph of $y = f(x - a)$. This implies that $y_1 = f(x_1 - a)$, and so the point $(x_1 - a, y_1)$ is on the graph of $y = f(x)$. The point $(x_1 - a, y_1)$ is $|a|$ units from (x_1, y_1). If $a > 0$, then $x_1 > x_1 - a$ and so (x_1, y_1) is $|a|$ units to the right of $(x_1 - a, y_1)$. If $a < 0$, then $x_1 < x_1 - a$ and (x_1, y_1) is $|a|$ units to the left of $(x_1 - a, y_1)$.

Chapter 2 / Functions and Their Graphs

▶ **EXAMPLE 2.20**

Solution

Figure 2.17

Draw the graph of $g(x) = (x + 3)^2$.

The graph of $y = x^2$ is a parabola through the origin and opening upward, as shown in figure 2.17. Since $(x + 3)^2 = \{x - (-3)\}^2$, we have $a = -3$ and $a < 0$. The graph of $g(x) = (x + 3)^2$ is obtained by translating the graph of $f(x) = x^2$ three units to the left. ◀

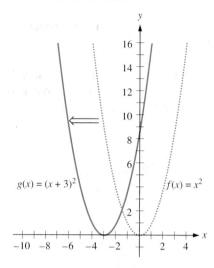

Vertical Translations

We just saw that replacing x by $x - a$ in an equation produces a horizontal translation of the graph. In the same way, replacing y in an equation by $y - b$, where b is a constant, produces a **vertical translation** of the graph.

How to Graph
$y - b = f(x)$

> If f is a function and b is a constant, then the graph of $y - b = f(x)$ translates the graph of $y = f(x)$ vertically a distance of $|b|$ units. This translation is upward if $b > 0$ and downward if $b < 0$.

▶ **EXAMPLE 2.21**

Solution

Figure 2.18

Draw the graph of $y - 3 = |x|$.

The graph of $y = |x|$ is shown by the dashed curve in figure 2.18. In the equation $y - 3 = |x|$, we see that $b = 3$. So to get its graph, we know the dashed graph should be translated 3 units upward, yielding the solid curve in the figure. ◀

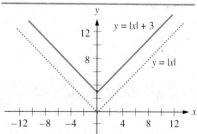

Interesting and useful graphs often result from a succession of translations. For instance, the graph of $y - 3 = |x + 4|$ results by translating the graph of $y = |x|$ to the left 4 units and 3 units upward. In general, we have the following.

Symmetry and Transformations

How to Graph
y − b = f(x − a)

The graph of $y - b = f(x - a)$ is obtained by positioning the graph of $y = f(x)$ as though (a, b) were the origin of a new coordinate system.

Vertical Stretchings and Shrinkings

Horizontal and vertical translations are two types of transformations. Stretchings and shrinkings are other types. We consider vertical stretching and shrinking next.

Once again, look at the graph of the function given by $f(x) = x^2$. Compare this to the graphs of the functions $g(x) = 2x^2$ and $h(x) = \frac{1}{2}x^2$ shown in figure 2.19(a). Now compare the graphs of $f(x) = x^2$ and $g(x) = 2x^2$ with those of

Figure 2.19

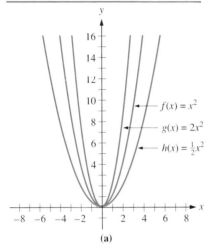
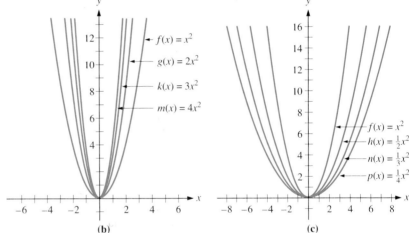

Figure 2.20

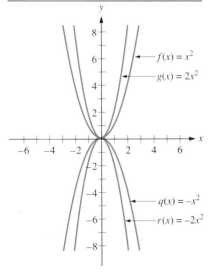

$k(x) = 3x^2$ and $m(x) = 4x^2$ shown in figure 2.19(b). Similarly, compare the graphs of $f(x) = x^2$ and $h(x) = \frac{1}{2}x^2$ with those of $n(x) = \frac{1}{3}x^2$ and $p(x) = \frac{1}{4}x^2$ in figure 2.19(c).

A pattern should begin to appear. All these graphs have the same general appearance. They each look somewhat like the graph of $f(x) = x^2$ and come from an equation of the form $y = cf(x)$, where c is a positive constant. If $c > 1$, the graphs are obtained by vertically stretching the graph of $y = f(x)$. If $0 < c < 1$, the graphs are obtained by vertically shrinking the graph of $y = f(x)$.

What happens if $c < 0$? Figure 2.20 shows the graphs of $f(x) = x^2$, $g(x) = 2x^2$, $q(x) = -x^2$, and $r(x) = -2x^2$. Here again, the functions are all of the form $y = cf(x)$, where c is a constant. If $c < 0$, the graph of $y = |c|f(x)$ has been reflected across the x-axis; the stretching-shrinking effect is the same as when c is positive.

The information about vertical stretching and shrinking is summarized below.

Chapter 2 / Functions and Their Graphs

How to Graph y = cf(x)

The graph of the function $y = cf(x)$, where c is a constant such that $c \neq 0$ and $c \neq 1$, is obtained by changing the graph of $y = f(x)$ as follows:

(i) if $|c| > 1$, stretch vertically;
(ii) if $|c| < 1$, shrink vertically;
(iii) if $c < 0$, also reflect across the x-axis.

Just as there were horizontal and vertical translations, there are horizontal and vertical stretchings and shrinkings. We consider the horizontal ones next. Consider the graph of the function $y = f(x)$ shown in figure 2.21(a). In figure 2.21(b), you find the graphs of $y = f(x)$ and $y = f(2x)$; and in figure 2.21(c), the graphs of $y = f(x)$ and $y = f(4x)$. In a similar way, the graphs of $y = f(x)$, $y = f(\frac{1}{2}x)$, and $y = f(\frac{1}{3}x)$ are shown in figure 2.21(d).

Figure 2.21

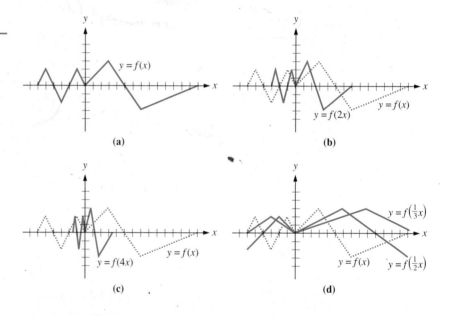

Once again you should notice that the graphs are stretched or shrunk, only this time horizontally instead of vertically. Each of these functions is of the form $y = f(dx)$, where d is a nonzero positive constant.

What happens if $d < 0$? Figure 2.22 shows the graph of $y = f(x)$ and $y = f(-x)$. In general, one can discuss functions of the form $y = f(dx)$, where d is a constant. If $d < 0$, the graph of $y = f(|d|x)$ is reflected across the y-axis; apart from this, the stretching-shrinking effect is the same as it was when d was positive.

Summarizing horizontal stretchings and shrinkings, we have the following.

Figure 2.22

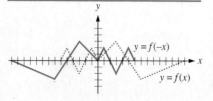

Symmetry and Transformations   107

How to Graph y = f(dx)

The graph of the function $y = f(dx)$, where d is a constant such that $d \neq 0$ and $d \neq 1$, is obtained by changing the graph of $y = f(x)$ as follows:

(i)   if $|d| < 1$, stretch horizontally;
(ii)   if $|d| > 1$, shrink horizontally;
(iii) if $d < 0$, also reflect across the y-axis.

Exercises 2.2

In exercises 1–4, find an equation for the line of symmetry for the given points.

1. $(-2, 3)$ and $(7, 6)$ 2. $(\pi, 2)$ and $(4, 0)$
3. $(-3, 5)$ and $(2, 5)$ 4. $(\frac{1}{2}, -6)$ and $(\frac{1}{2}, 2)$
5. If the line $x = 7$ is the line of symmetry for the points $(4, 9)$ and P, find the coordinates of P.
6. If the line $y = 2x + 5$ is the line of symmetry for the points $(0, 0)$ and P, find the coordinates of P. (Hint: find the intersection of the given line with the perpendicular to the given line from the origin.)

In exercises 7–10, determine how many lines of symmetry the given curve has.

7.

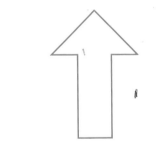

8.

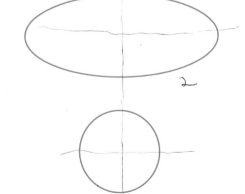

9.

10.

11. Consider a point $P(a, b)$. Show that P and $(-a, b)$ are symmetric about the y-axis; that P and $(a, -b)$ are symmetric about the x-axis; and that P and $(-a, -b)$ are symmetric about the origin.

12. (a) Using exercise 11, prove that if a set S of points is symmetric about both the x-axis and the y-axis, then S is also symmetric about the origin.
    (b) What may you conclude if S is symmetric with respect to the x-axis and the origin?

13. (a) Show that $f(x) = \frac{3}{7}x^4 - 9x^2 + \pi$ describes an even function.
    (b) Show that $g(x) = \frac{3}{7}x^5 - \frac{2}{3}x^3 + 4x$ describes an odd function.

14. Show that the function defined by $h(x) = 2x^2 - x$ is neither even nor odd.

In exercises 15–41, test the graph of the given equation or function for symmetry with respect to the y-axis, the x-axis, and the origin.

15. $y = \frac{2}{3}x$ 16. $x = \frac{4}{7}y$
17. $f(x) = 6x - 1$ 18. $y = 3x^2$
19. $y + 1 = 3x^2$ 20. $y = 3(x - 1)^2$
21. $x = -2y^2$ 22. $x - 3 + 2y^2 = 0$
23. $x = -2(y - 5)^2$ 24. $g(x) = \dfrac{5x^2 + 3}{9x^4 - 6x^2 + \pi}$
25. $2x^2 - 2y^2 = 3$ 26. $xy = \sqrt{5}$
27. $(x - 4)y = 5$ 28. $24x^2 + 16y^2 = 9$
29. $24(x + 2)^2 + 16y^2 = 9$ 30. $2y^3 = 3x^2$
31. $-2y^2 = 3x^3$ 32. $-2(y + 6)^2 = 3x^3$
33. $-2(y + 6)^2 = 3(x - 1)^3$ 34. $h(x) = |x|$
35. $y = |x| + 2$ 36. $y = |x + 2|$
37. $y = \sqrt{x}$ 38. $y^2 = x$

Chapter 2 / Functions and Their Graphs

39. $x = \dfrac{1}{(y^4 + 3)^5}$ **40.** $x = y^3$

41. $y = x^4$

In exercises 42–56, use the transformation theory from this section to draw the graph of the given equation.

42. $x - 2 = y^2$ **43.** $x + 2 = (y - 3)^2$
44. $x + 1 = y^2 - 6y + 8$ **45.** $x = (y - 3)^2$
46. $x - 2 = \sqrt{y + 4}$ **47.** $y = 2|x|$
48. $y = -\tfrac{1}{2}|x|$ **49.** $y = \tfrac{2}{3}x^3$
50. $y + 2 = 2(x - 1)^4$ **51.** $y = \left|\dfrac{x}{2}\right|$
52. $y = |2x|$ **53.** $y = |2x - 3|$
54. $y - 1 = 4|2x - 3|$ **55.** $y - \tfrac{1}{2} = [x + 1]$
56. $y = -\tfrac{1}{2}[2x + 3]$

57. Let f be a function with domain $[-7, 5) = \{x : -7 \le x < 5\}$. Define functions g, h, and k by the rules $g(x) = f(x - 4)$, $h(x) = -4f(x)$, and $k(x) = f(4x)$. For each of these new functions, determine its domain and range. (Describe the ranges in terms of the range of f.)

58. (a) Let a, b, c, and d be positive numbers and f a given function with domain $\mathbf{R}$. Suppose that the graph of $y = f(x)$ is repositioned as though $\left(\dfrac{a}{d}, b\right)$ were the origin of a new coordinate system. Next, suppose that the repositioned graph is subjected to a vertical stretch/shrink by a factor of c and a horizontal stretch/shrink by a factor of d. Show that, relative to the original coordinate system, the resulting figure is the graph of the equation $y - b = cf(dx - a)$.
(b) Using (a), graph $y - 4 = 3|2x + 5|$.
(c) Modify the analysis in (a) in case the positivity of a, b, c, and d is not assumed.
(d) Using (c), graph $y + 4 = -3|2x + 5|$.

59. The function defined by $f(x) = \sqrt{x + 2}$ has the graph shown below. For each of the graphs in (a) through (e), find constants a, b, c, and d such that the function given by $y - b = cf(dx - a)$ has the given graph. (Hint: apply exercise 58(a).)

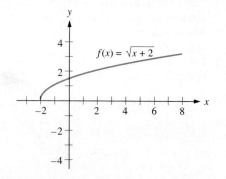

(a)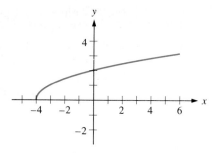

(b)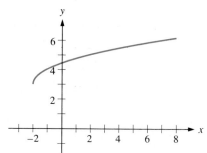

(c)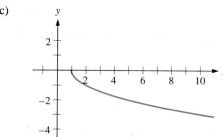

(d)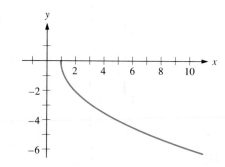

(e)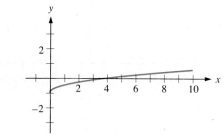

Symmetry and Transformations

60. The function defined by $g(x) = \dfrac{1}{x}$ has the graph shown below. For each of the graphs in (a) through (e), find constants a, b, c, and d such that the function given by $y - b = cg(dx - a)$ has the given graph.

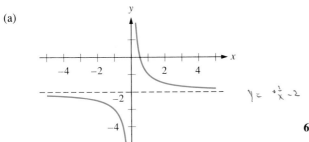

(a)

(b)

(c)

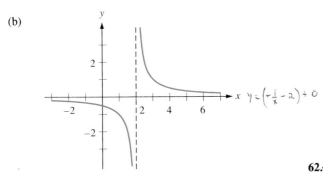

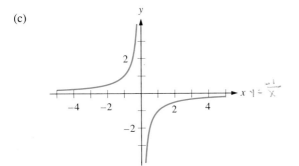

(d)

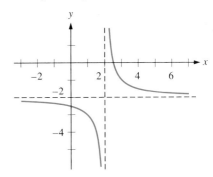

(e)

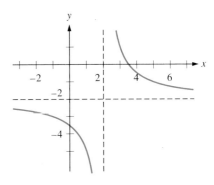

61.♦(a) Prove that (a, b) and (b, a) are symmetric with respect to the line $y = x$. (Hint: modify the hint for exercise 6.)

(b) Using (a), show that, if interchanging x and y in a given equation produces an equivalent equation, then the graph of the given equation is symmetric about the line $y = x$.

(c) Using (b), show that the graphs of $4x + 4y - 5 = 0$ and $\sqrt{x} + \sqrt{y} = 9$ are each symmetric about the line $y = x$.

(d) Show that the graph of $2x^2 + 3y^2 = 4$ is not symmetric about the line $y = x$. (A specific numerical counterexample to symmetry is needed here.)

62.♦(a) Prove that (a, b) and $(-a + 2h, -b + 2k)$ are symmetric about the point (h, k). (Hint: imagine a new coordinate system with origin at (h, k), and then apply exercise 11.)

(b) Using (a), show that $(-\tfrac{2}{3}, \tfrac{1}{7})$ and $(\tfrac{32}{3}, \tfrac{13}{7})$ are symmetric about $(5, 1)$.

(c) Are $(\tfrac{1}{7}, -\tfrac{2}{3})$ and $(\tfrac{13}{7}, \tfrac{32}{3})$ symmetric about $(1, 5)$? Explain your answer. Generalize!

♦ **63.** Show that the graph of $y = x^3 - 3x^2 - x + 5$ is symmetric about the point $(1, 2)$. (Hint: using exercise 62, you only need to verify that if $y = x^3 - 3x^2 - x + 5$, then $-y + 4 = (-x + 2)^3 - 3(-x + 2)^2 - (-x + 2) + 5$.)

2.3 LINEAR AND QUADRATIC FUNCTIONS

Among the graphs of equations studied in chapter 1, you saw lines and parabolas. In this section, we consider whether such curves are graphs of functions. We then study the kinds of functions and "real-world" applications that have lines or parabolas as their graphs.

Linear Functions

Since a vertical line cannot pass the vertical line test, no vertical line can be the graph of a function. However, this is not true of nonvertical lines. Each nonvertical line is the graph of a function determined by $y = mx + b$.

Linear Function

> Each function expressed by $f(x) = mx + b$, with m and b constants, is called a **linear function.** Its graph is a nonvertical line with slope m and y-intercept $(0, b)$.

Consider any linear function, $f(x) = mx + b$. If $m = 0$, then $f(x) = b$ describes a constant function. Hence, a constant function is a special type of linear function. The graph of the constant function given by $f(x) = 4$ is in figure 2.23(a).

If $m > 0$, then $f(x) = mx + b$ describes an increasing function; and if $m < 0$, then $f(x) = mx + b$ gives a decreasing function. (Recall exercise 72 in section 2.1.) The graph of the increasing linear function $g(x) = 3x - 2$ is in figure 2.23(b); that of the decreasing linear function $h(x) = -4x - 2$ is in figure 2.23(c).

Figure 2.23

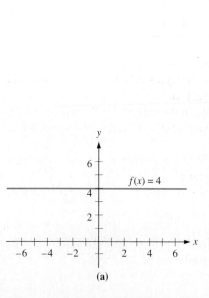
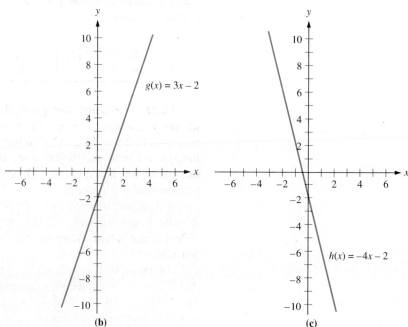

Linear and Quadratic Functions

The domain of any linear function $f(x) = mx + b$ is $(-\infty, \infty)$. If $m \neq 0$, its range is also $(-\infty, \infty)$; but if $m = 0$, the range of f is just $\{b\}$.

Quadratic Functions

Quadratic Function

> Any function f given by $f(x) = ax^2 + bx + c$, with $a \neq 0$, is called a **quadratic function**.

Although $a \neq 0$, it is possible that b or c or both may be zero. No quadratic function is a linear function.

▶ **EXAMPLE 2.22**

The following give examples of quadratic functions:

$$f(x) = 3x^2 + 5x - 4$$
$$g(x) = \sqrt{5}x^2 - 7x + \tfrac{3}{2}$$
$$h(x) = \pi x^2 + \sqrt{17}$$
$$j(x) = -4\pi\sqrt[3]{7}x^2$$
$$k(x) = 2x^2.$$

$\dfrac{-b \pm \sqrt{b^2 - 4ac}}{2a}$

◀

Graphs of Quadratic Functions

The natural domain of any quadratic function is $(-\infty, \infty)$. However, its range is definitely not $(-\infty, \infty)$! In order to identify the range, let's take time to examine the graph of the quadratic function given by $f(x) = x^2$. We begin with a table of values.

x	−4	−3	−2	−1	0	1	2	3	4
$f(x) = x^2$	16	9	4	1	0	1	4	9	16

These nine points are graphed in figure 2.24(a), and when they are smoothly connected, we get the curve in figure 2.24(b). As we know from section 1.4, this is the graph of a parabola. In fact, the graph of any quadratic function is a parabola. Parabolas are used in the design of reflecting telescopes, satellite dishes, and spotlights. These, and other, applications of parabolas will be examined in chapter 8.

A quadratic function has a graph with the general shape shown in figure 2.24(b). Every parabola has this general shape even though it may have a different orientation. However, not every parabola is the graph of a quadratic function.

For example, figure 2.24(c) is the graph of the parabola $x = y^2 + 4$. Note that this is not the graph of a function. Similarly, the graph of $16x^2 - 24xy + 9y^2 + 100x - 200y + 100 = 0$, shown in figure 2.24(d), is also a parabola. It, too, is not the graph of a function.

Chapter 2 / Functions and Their Graphs
Figure 2.24

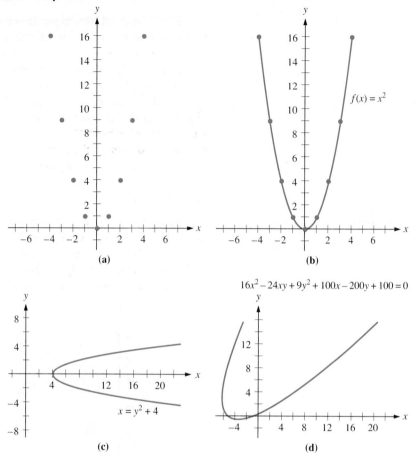

Examine the parabolas in figure 2.24. In addition to the general shape of the parabola, notice that every parabola has symmetry about a line. For the function in figure 2.24(b), the line of symmetry is the y-axis. The parabola in figure 2.24(c) has the x-axis as a line of symmetry, and the parabola in figure 2.24(d) has the line $y = \frac{4}{3}x + \frac{20}{3}$ as its line of symmetry. This line of symmetry is called the **axis** of the parabola. The point where the axis intersects the parabola is called the **vertex** of the parabola.

Starting with the basic parabola $f(x) = x^2$, we can apply the transformations from the last section to obtain several variations. We shall list, and then examine, each of these variations.

$F(x) = ax^2$	Multiply by a constant
$g(x) = x^2 + k$	Add a constant, k
$j(x) = (x - h)^2$	Replace x with $x - h$, with h a constant
$m(x) = a(x - h)^2 + k$	All of the above, successively

In the last section we learned the effects that these first three variations produce. For example, we know that if the function is multiplied by a constant, its graph is stretched, shrunk, or flipped over the x-axis. The effects these had on the function $f(x) = x^2$ were shown in figures 2.19 and 2.20.

Now, let's look at the second variation. You may not recognize this since it is not quite in the same form examined in the last section. You may remember that replacing y by $y - b$ produces a vertical translation. Also, the function given by $y = x^2$ has a parabola as its graph. Thus, each equation of the form $y - k = x^2$ gives a parabola (with a vertical translation). Solving this equation for y leads to $y = x^2 + k$, or $f(x) = x^2 + k$. Thus, the second variation produces a vertical translation of the graph of the function $f(x) = x^2$ by $|k|$ units. This translation is upward if k is positive and downward if k is negative.

▶ **EXAMPLE 2.23**   Graph $g(x) = x^2 + 5$.

Solution   The graph of $f(x) = x^2$ is shown by the dashed curve in figure 2.25. For the function $g(x) = x^2 + 5$, we see that the constant $k = 5$, and so we know the graph of g will be obtained by translating that of f upward 5 units, as shown by the solid curve in the figure. The vertex of this parabola is the point $(0, 5)$, and the y-axis (the line $x = 0$) is the parabola's axis.   ◀

Figure 2.25

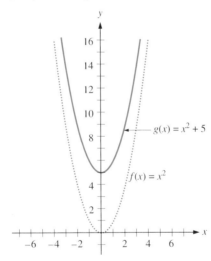

In the third variation, x is replaced by $x - h$. Once again, we may appeal to the work in the last section. There we learned that replacing x by $x - a$ produces a horizontal translation. Thus, the third variation effectively causes the graph of $y = x^2$ to undergo a horizontal translation of $|h|$ units. The translation will be to the right if h is positive and to the left if h is negative.

▶ **EXAMPLE 2.24**   Graph $j(x) = (x - 3)^2$.

Solution   As usual, the graph of $f(x) = x^2$ is shown by the dashed curve in figure 2.26. For the function $j(x) = (x - 3)^2$, we see that the constant $h = 3$, and so we know the graph of f should be translated 3 units to the right, as shown by the solid curve in the figure. The result is the graph of j. The vertex of this parabola is the point $(3, 0)$, and its axis is the line $x = 3$.   ◀

Chapter 2 / Functions and Their Graphs

Figure 2.26

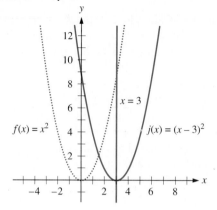

The final variation, replacing x^2 with $a(x-h)^2 + k$, is a combination of the previous three variations and, thus, produces all three effects. The next example probably explains this best.

▶ **EXAMPLE 2.25**

Graph $m(x) = -2(x-3)^2 + 5$.

Solution

Once again, the graph of $f(x) = x^2$ is shown by the dashed curve in figure 2.27(a). To graph $m(x) = -2(x-3)^2 + 5$, note first that a stretching effect is produced by the 2, as shown in figure 2.27(b). The negative sign at the beginning has the effect of flipping the graph over so that it opens downward (see figure 2.27(c)). Next, the parabola is translated 3 units to the right (figure 2.27(d)) and 5 units up to produce the solid curve in figure 2.27(e). The vertex of this parabola is the point (3, 5), and the axis of the parabola is the line $x = 3$. Notice that when a parabola opens downward, as this one does, the vertex is the highest point on the graph. When a parabola opens upward, its vertex is the lowest point. ◀

Figure 2.27

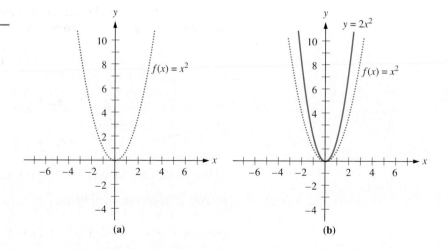

Figure 2.27 (continued)

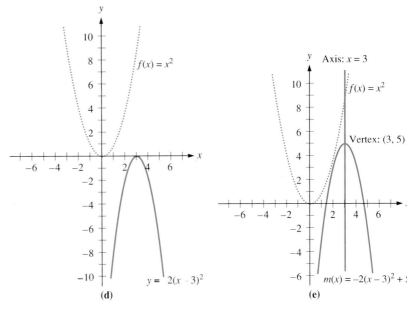

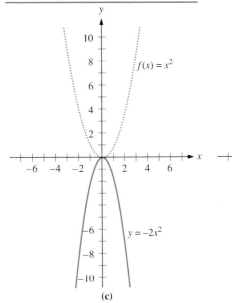

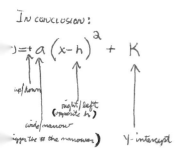

Analyzing the four variations leads to the following summary.

Graph of a Quadratic Function

The graph of a quadratric function defined by

$$f(x) = a(x - h)^2 + k \text{ (where } a \neq 0)$$

(i) is a parabola with vertex (h, k) and has the vertical line $x = h$ as its axis;
(ii) opens upward if $a > 0$, and downward if $a < 0$;
(iii) compared to the graph of $f(x) = x^2$, is stretched if $|a| > 1$ and shrunk if $0 < |a| < 1$.

We can now determine the range of a quadratic function f. We have $f(x) = ax^2 + bx + c$, with $a \neq 0$. By completing the square,

$$f(x) = a\left(x^2 + \frac{b}{a}x\right) + c$$

$$= a\left(x + \frac{b}{2a}\right)^2 + c - \frac{b^2}{4a}$$

$$= a\left(x + \frac{b}{2a}\right)^2 + \frac{4ac - b^2}{4a}.$$

This is of the form $a(x - h)^2 + k$. The graph of $f(x) = ax^2 + bx + c$ (with $a \neq 0$) is thus a parabola with vertex at the point $(h, k) = \left(\frac{-b}{2a}, \frac{4ac - b^2}{4a}\right)$. This parabola's axis is the line $x = h$, that is, $x = -\frac{b}{2a}$. Hence, the range of f is

$\{y: y \geq \frac{4ac - b^2}{4a}\}$ if $a > 0$ and $\{y: y \leq \frac{4ac - b^2}{4a}\}$ if $a < 0$. Rather than memorize this description of the range, you should work it out in any given example by completing the square as above or by remembering that the **x-coordinate of the vertex is $\frac{-b}{2a}$.**

▶ EXAMPLE 2.26

Describe and graph the function $f(x) = 2x^2 - 8x + 4$.

Solution

By completing the square, $f(x) = 2(x - 2)^2 - 4$. Its graph is a parabola, with vertex at $(h, k) = (2, -4)$ and axis the line $x = h = 2$. This parabola opens upward (since $a = 2 > 0$). Its graph appears in figure 2.28. Note that, compared to the graph of x^2, it is stretched vertically. ◀

Figure 2.28

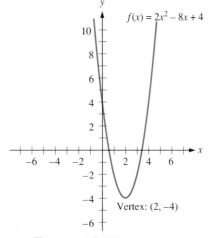

Word Problem Applications of Linear and Quadratic Functions

To close this section, we will look at three word problems. These are typical of the many ways in which linear or quadratic functions arise naturally in applications. The first of these, example 2.27, is a simple problem involving two linear functions. We include it here in order to review the algebraic approach to problem solving. After reading it, you may wish to review the "linear" word problems in example 1.26 and exercises 61–70 of section 1.3.

▶ EXAMPLE 2.27

An absent-minded David leaves John's house and drives 50 miles per hour towards Wm. C. Brown Publishers. Twelve minutes after David's departure, John notices that David forgot to take the galley proofs of *Precalculus* with him. John immediately leaves, driving 60 miles per hour, in pursuit of David. When will John overtake David?

Solution

The "unknown" in this word problem is the amount of time that John must drive before he overtakes David. Let the variable x be this amount, measured in hours. Since

$$\text{distance} = \text{rate} \cdot \text{time}$$

John must drive $60x$ miles. At the moment they meet, how far has David driven? Since David has traveled 12 minutes longer than John, David has been driving for $x + \frac{12}{60} = x + \frac{1}{5}$ hours, covering $50(x + \frac{1}{5})$ miles. Since John and David have traveled the same distance, we have

$$60x = 50(x + \tfrac{1}{5})$$
$$60x = 50x + 10$$
$$10x = 10$$
$$x = 1.$$

Linear and Quadratic Functions

This means that John drives for one hour in order to overtake David. ◀

In exercise 21 of chapter 0, you saw a word problem involving a quadratic function. The next two word problems involve more substantial, calculus-type applications of quadratic functions. You can find additional applications in the last thirteen exercises of this section.

▶ **EXAMPLE 2.28**   A tire company buys tires at $30 apiece. The company's manager knows that she can sell 20 tires daily if she charges $90 apiece. For each $2 that she reduces the price of a tire, she can sell one extra tire per day. How many tires should she hope to sell in order to maximize her profit?

Solution   The "unknown" here is the number of $2 reductions in the price of a tire. Let this number be x. The result is a daily sale of $20 + x$ tires, each selling for $(90 - 2x)$. This leads to daily **revenue** of $(20 + x)(90 - 2x)$, or $(1800 + 50x - 2x^2)$. The daily **cost** for the company to buy tires is $(20 + x)30 = (600 + 30x)$. Since

$$\text{profit} = \text{revenue} - \text{cost}$$

the manager seeks to maximize the profit, $P(x)$, as follows:

$$P(x) = (1800 + 50x - 2x^2) - (600 + 30x)$$
$$= 1200 + 20x - 2x^2$$
$$= -2(x - 5)^2 + 1250.$$

The graph of $y = P(x)$ is a parabola, opening downward, with its highest point at the vertex $(5, 1250)$. This means that daily profit is maximized (as $1250) when $x = 5$, that is, when $20 + x = 20 + 5 = 25$ tires are sold daily at $(90 - 2x) = (90 - 10) = 80$ per tire. ◀

▶ **EXAMPLE 2.29**   A commercial artist needs to make a rectangular poster whose perimeter will be 80 inches. How can he do this so as to maximize the area of the poster?

Solution   It will definitely help to sketch a picture of a rectangular poster, using variables to label the relevant quantities. This is done in figure 2.29, with the rectangle's dimensions being x inches by y inches.

Figure 2.29

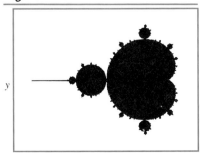

We know that the perimeter is 80 inches, and so

$$80 = 2x + 2y = 2(x + y)$$

or

$$40 = x + y$$
$$y = 40 - x.$$

The artist must maximize the rectangle's area. To do this, the area is expressed as a function, $A(x)$, of a single variable, x:

$$\text{Rectangle's area} = \text{length} \cdot \text{width}$$

$$A(x) = yx$$
$$= (40 - x)x$$
$$= 40x - x^2$$
$$= -(x - 20)^2 + 400.$$

The graph of this quadratic function is a parabola opening downward, with its highest point at the vertex (20, 400). This means that the optimal poster has width $x = 20$ inches and length $y = 40 - x = 40 - 20 = 20$ inches. In other words, the poster must be a square measuring 20 inches per side. ◀

Sketches like figure 2.29 often help you solve maximization/minimization problems. You will see this again in the exercises and very often in calculus.

Exercises 2.3

In exercises 1–6, graph the given linear function; state whether it is increasing, decreasing, or neither; and find the graph's y-intercept.

1. $f(x) = 3x - 4$
2. $g(x) = -4x + 3$
3. $h(t) = -\pi t$
4. $j(t) = \frac{3}{4}t$
5. $k(s) = -\sqrt{7}$
6. $m(s) = 0$

7. Suppose that the population of a slowly growing town at time t is given by a linear function, $f(t) = mt + b$. If the population in 1980 was 4112 and the population in 1987 was 4259, what will be the population in 1999? (Hint: you should find m. Do you need to find b? Review example 1.26.)

8. What is a line of symmetry of the graph of $f(x) = 2$? Of $g(x) = -2$? Of $h(x) = 0$? Of any constant function $k(x) = b$?

9. Consider the graph of a nonconstant linear function, $f(x) = mx + b$.
   (a) Show that the graph of f is symmetric about the line $y = x$ in only the following cases: $f(x) = x$ and $f(x) = -x + b$.
   (b) Show that the graph of f is symmetric about the origin if and only if $b = 0$.
   (c) Show that the graph of f is never symmetric about the x-axis or the y-axis.
   (Hints: For (a), review exercise 61 in section 2.2. For (b) and (c), review exercise 11 in section 2.2.)

10. Let $f(x) = mx + b$, a linear function.
    (a) Under what conditions on m and b is f an even function?
    (b) When is f an odd function?
    (c) Under what conditions is f both even and odd?

In exercises 11–18, explain how the graph of the given function may be obtained from that of $f(x) = x^2$. Also identify the domain and the range of the given function.

11. $g(x) = \sqrt{2}x^2$
12. $h(x) = -\frac{2}{3}x^2$
13. $j(x) = (x + \sqrt{3})^2 - 5$
14. $k(x) = (x - 7)^2 - \pi$
15. $m(x) = -2(x + \sqrt{3})^2$
16. $n(x) = 2(x - 7)^2$
17. $p(x) = 2(x - 7)^2 - 2\pi$
18. $q(x) = -2(x + \sqrt{3})^2 - 5$

19. For each of the parabolas which are the graphs of the functions in exercises 11–18, find the vertex and an equation for the axis.

20. (a) Find all the values of a such that $(-2, 25)$ lies on the graph of $f(x) = ax^2 - 7x + 3$.
    (b) Find all the values of a such that $(a, 25)$ lies on the graph of $g(x) = -2x^2 - 7x + 30$.
    (c) Find all the values of a such that $(a, 25)$ lies on the graph of $h(x) = -2x^2 - 7x + 3$.

21. Explain why the graph of a quadratic function is neither increasing nor decreasing.

22. Explain why no graph of a quadratic function is symmetric about the origin.

23. (a) Determine necessary and sufficient conditions on a, b, and c so that a given quadratic function $f(x) = ax^2 + bx + c$ is an even function.
    (b) Redo (a), replacing "even" with "odd."
    (c) Using (a) and (b), determine which quadratic functions have graphs symmetric to the y-axis, and which have graphs symmetric to the origin.

In exercises 24–25, write each of the given functional expressions in the form $f(x) = ax^2 + bx + c$.

Linear and Quadratic Functions

24. $-2(x + 3)^2 + 8$
25. $-\frac{3}{8}(x - 6)^2 - 5$

In exercises 26–31, use the completion of squares to write each of the given functional expressions in the form $f(x) = a(x - h)^2 + k$; state whether the function's graph opens upward or downward; and identify the highest or lowest point on that graph.

26. $f(x) = 2x^2 + 8x - \sqrt{5}$
27. $f(x) = 2x^2 - 7x - 5$
28. $f(x) = -2x^2 + 8x - \sqrt{5}$
29. $f(x) = -2x^2 - 7x + 5$
30. $f(x) = \pi x^2 + 7$
31. $f(x) = -\pi x^2 + \pi$

32. Prove that the y-intercept of the graph of a quadratic function $ax^2 + bx + c$ is $(0, c)$; and that this graph has an (at least one) x-intercept precisely in case $b^2 \geq 4ac$.

33. Using the preceding exercise and the quadratic formula, find the y-intercept and x-intercepts (if they exist) for the graphs of each of the following quadratic functions:
    (a) $f(x) = 2x^2 - 3x - 4$
    (b) $g(x) = 2x^2 - 3x + 4$
    (c) $h(x) = \pi x^2 + 2\sqrt{\pi} x + 1$
    (d) $j(x) = -3x^2 + 2$
    (e) $k(x) = -3x^2$

34. Perform the following operations successively on the graph of the function given by $y = x^2$: expand vertically by a factor of 2; reflect through the x-axis; and translate 3 units to the right and 4 units downward. Show that what results is the graph of the function described by $y = -2(x - 3)^2 - 4$.

35. Perform the operations described in exercise 34 in the following order: the expansion, the translations, then the reflection. Show that the result is the graph of the function described by $y = -2(x - 3)^2 + 4$.

36. At time $t = 0$, a juggler on a roof 60 feet above the ground throws a ball nearly straight up, but fails to catch it. The ball's height above the ground at time t (measured in seconds) is $f(t) = -16t^2 + \sqrt{385}\, t + 60$ feet.
    (a) What is the ball's height above the ground at $t = 1$?
    (b) When does the juggler have the best chance to catch the ball? (Hint: find $t > 0$ so that $f(t) = 60$.)
    (c) When does the ball strike the ground?

37. For the function in exercise 36, observe that $f(t_1) = 0$ for $t_1 = \dfrac{-65 + \sqrt{385}}{32}$. But $t_1 < 0$, and so t_1 does not correspond to a time after the juggler's toss. Let t_2 denote the answer to exercise 36(c). Describe an enriched scenario, involving an accomplice of the juggler at ground level, for which it is physically meaningful to say that $f(t)$ describes the height of the ball above the ground for all t so that $t_1 \leq t \leq t_2$.

38. When point masses of sizes m and M are situated r units apart, the resulting gravitational attraction is $F = \dfrac{GmM}{r^2}$, where G is a certain nonzero constant. (This is the "inverse square law" of universal gravitation.)

    (a) For fixed m and r, show that F is a linear function of M.
    (b) Let $x = \dfrac{1}{r}$. Then, for fixed nonzero m and M, show that F is a quadratic function of x.

39. Let F be as in exercise 38.
    (a) What is the effect on F if m and r are fixed but M is doubled? Generalize (to the situation of fixed m and r, with M magnified by a factor of k).
    (b) What is the effect on F if m and M are fixed and r is doubled? Generalize!
    (c) Give physical interpretations for your answers in (a) and (b).

40. Find two real numbers whose difference is 4 and whose product is as small as possible. (Hint: find the vertex or axis of the graph of $f(x) = (x + 4)x$.)

41. Find two real numbers whose sum is 4 and whose product is as large as possible.

◆ 42. A farmer wishes to use 200 meters of fencing to enclose a rectangular region in a large field. Show that the area enclosed is maximized if the rectangular region is square. (Hint: find the highest point on the graph of $(100 - x)x$.) Generalize!

◆ 43. Another farmer wishes to use a total of 200 meters of fencing along three sides of a rectangular region, with the fourth side running along a straight river bank (and hence needing no fence). Show that the area enclosed is maximized if the dimensions of the rectangle are 100 meters by 50 meters. Which of those dimensions is the optimal length of the side parallel to the river bank? Generalize!

◆ 44. Find the point on the graph of $y = -2x$ which is closest to $(2, 3)$. (Hint: why is it enough to find the lowest point on the graph of $(x - 2)^2 + (-2x - 3)^2$?)

◆ 45. Find the points on the graph of $y = -2x^2 + 1$ which are closest to the origin. (Hint: minimize a certain function of x^2, finding the square of the desired values of x.)

◆ 46. Wilma acquires widgets wholesale at \$2 apiece. She knows that she can sell 30 widgets if she charges \$6 apiece. For each dime that she cuts the price of widgets, she can sell an extra widget.
    (a) Show that Wilma's profit, based on a sale of $30 + x$ widgets, is $P(x) = \dfrac{-x^2}{10} + x + 120$ dollars.
    (b) How many widgets should Wilma hope to sell in order to maximize her profit?

◆ 47. Humble Hamburger Hut (HHH) hires two competing architects to draw blueprints for its proposed trademark, a single parabolic arch, opening downward, with base 20 units and maximum height 48 units. The first architect produces the graph of the function given by

Chapter 2 / Functions and Their Graphs

$f(x) = 48 - \frac{12}{25}x^2$. The second architect deals with the function $g(x) = \frac{48}{5}x - \frac{12}{25}x^2$. Show that both architects have solved HHH's problem correctly. Which horizontal translation of the graph of f would produce the graph of g?

◆ 48. An airline offers a charter flight for a student excursion during Spring Break. The fare will be $300 per student if at least 50 but fewer than 100 students sign up. Each sign-up beyond the hundredth causes the fare for each passenger to be reduced by $0.60. What is the maximum total revenue that the airline can obtain if the plane has 269 seats? [Hint: consider the revenue function given by

$$R(x) = \begin{cases} 300x & \text{for } 50 \leq x \leq 100 \\ (300 - 0.60x)x & \text{for } 100 < x \leq 269 \end{cases}].$$

2.4 NEW FUNCTIONS FROM OLD

In this section, we look at ways in which new functions can be formed from given ones. First, we look at the formation of a new function by adding, subtracting, multiplying, or dividing "old" functions. Next, we look at a different technique, called composition, for forming functions. Lastly, we examine the inverse function of a function, when it exists, and how it can be determined.

Operations on Functions

Functional Operations

Suppose that f and g are two functions. Then their sum, $f + g$; their difference, $f - g$; their product, $f \cdot g$; and their quotient, f/g, are functions defined by

$$(f + g)(x) = f(x) + g(x)$$

$$(f - g)(x) = f(x) - g(x)$$

$$(f \cdot g)(x) = f(x) \cdot g(x)$$

$$(f/g)(x) = \left(\frac{f}{g}\right)(x) = \frac{f(x)}{g(x)}, g(x) \neq 0.$$

In each of these cases, the domain of the new function is the common domain of f and g, that is, the intersection of the domains of f and g, with the extra stipulation that the domain of f/g excludes any values of x for which $g(x) = 0$.

▶ **EXAMPLE 2.30**

Find $(f + g)(x)$, $(f - g)(x)$, $(f \cdot g)(x)$, and $(f/g)(x)$, and their respective domains and ranges, if $f(x) = x^2 - 16$ and $g(x) = 2x - 8$. Graph each of the new functions.

Solution

The domain of both f and g is $\mathbf{R}$, the set of real numbers. The range of f is $\{y: y \geq -16\}$ and the range of g is $\mathbf{R}$.

$$(f + g)(x) = f(x) + g(x) = (x^2 - 16) + (2x - 8) = x^2 + 2x - 24 = (x + 1)^2 - 25.$$

This is a parabola that opens upward with vertex $(-1, -25)$ and axis $x = -1$. The domain is $D = \mathbf{R}$ and the range is $R = \{y: y \geq -25\}$. The graphs of f, g, and $f + g$ are in figure 2.30(a).

New Functions from Old

$(f - g)(x) = f(x) - g(x) = (x^2 - 16) - (2x - 8) = x^2 - 2x - 8 = (x - 1)^2 - 9.$

This is also a parabola that opens upward. Its vertex is (1, –9) and it has the axis $x = 1$. The domain is $D = \mathbf{R}$ and the range is $R = \{y: y \geq -9\}$. The graphs of f, g, and $f - g$ are in figure 2.30(b).

$(f \cdot g)(x) = f(x) \cdot g(x) = (x^2 - 16)(2x - 8) = 2x^3 - 8x^2 - 32x + 128.$

The domain and the range are both the set of real numbers and so $D = R = \mathbf{R}$. (At present you do not have the background to determine the range of this function. That will be discussed in the next chapter. So, for now, we ask you to "believe" that the range is $\mathbf{R}$. However, you can intuitively see that this is the case from the graph of $f \cdot g$ in figure 2.30(c).)

$(f/g)(x) = \dfrac{f(x)}{g(x)} = \dfrac{x^2 - 16}{2x - 8} = \dfrac{(x + 4)(x - 4)}{2(x - 4)} = \dfrac{x + 4}{2} = \dfrac{1}{2}x + 2.$

The domain is $D = \{x: x \neq 4\}$ and the range is $R = \{y: y \neq 4\}$. You can see that D does not include $x = 4$ because this substitution will make the denominator zero. Even though the expression $\frac{1}{2}x + 2$ looks like its graph will be a straight line, the fact that 4 is not in the domain of f/g means that there is no value for $(f/g)(4)$ (although $\frac{1}{2}(4) + 2 = 4$). The graph of the function f/g is a straight line with a "hole" at the point (4, 4), as shown in figure 2.30(d).

Figure 2.30

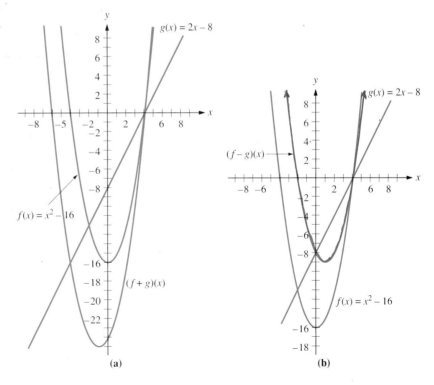

Chapter 2 / Functions and Their Graphs

Figure 2.30 (continued)

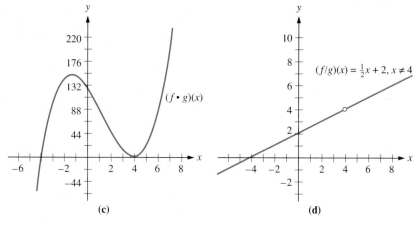

(c) (d)

▶ **EXAMPLE 2.31**

Solution

If $f(x) = \sqrt{x+5}$ and $g(x) = \sqrt{x-4}$, then find $(f+g)(x)$, $(f-g)(x)$, $(f \cdot g)(x)$, and $(f/g)(x)$, and their respective domains. Graph each function and use the graphs to determine their respective ranges.

The domain of f is $\{x: x \geq -5\}$ and the domain of g is $\{x: x \geq 4\}$. Both f and g have the same range: $\{y: y \geq 0\}$. Each of the four new functions, their domains and ranges are given below. Their graphs are in figure 2.31.

$(f+g)(x) = \sqrt{x+5} + \sqrt{x-4}$.

$D = \{x: x \geq 4\}$ and $R = \{y: y \geq 3\}$. (See figure 2.31(a).)

$(f-g)(x) = \sqrt{x+5} - \sqrt{x-4}$.

$D = \{x: x \geq 4\}$ and $R = \{y: 0 < y \leq 3\}$. The graph is in figure 2.31(b).

$(f \cdot g)(x) = (\sqrt{x+5})(\sqrt{x-4})$ simplifies as

$$\sqrt{(x+5)(x-4)} = \sqrt{x^2 + x - 20} = \sqrt{(x+\tfrac{1}{2})^2 - \tfrac{81}{4}}.$$

The graph of $f \cdot g$ is in figure 2.31(c). $D = \{x: x \geq 4\}$ and $R = \{y: y \geq 0\}$. (Note that $\sqrt{x^2 + x - 20}$ is defined for $x \leq -5$. But these values of x are not in the domain of g and therefore are not in the domain of $f \cdot g$.)

$(f/g)(x) = \dfrac{\sqrt{x+5}}{\sqrt{x-4}}$ simplifies as $\sqrt{\dfrac{x+5}{x-4}}$.

Since $x - 4$ cannot be zero, the domain cannot contain $x = 4$. In fact, $D = \{x: x > 4\}$. (Again, even though $\sqrt{\dfrac{x+5}{x-4}}$ is defined for $x \leq -5$, such values of x are not in the domain of g and hence they are not in the domain of f/g.) One can see from the graph of f/g in figure 2.31(d) that $R = \{y: y > 1\}$. One

New Functions from Old

can arrive at this set algebraically. Let $y = \sqrt{\dfrac{x+5}{x-4}}$. Hence, $y^2 = \dfrac{x+5}{x-4} = 1 + \dfrac{9}{x-4}$, and $y^2 - 1 = \dfrac{9}{x-4}$. Since $x > 4$, $\dfrac{9}{x-4} > 0$, so $y^2 - 1 > 0$; that is, the conditions on y are $y \geq 0$ and $y^2 > 1$. Thus, the possible values of y are in the interval $(1, \infty)$, and this is the range indicated in figure 2.31(d).

Figure 2.31

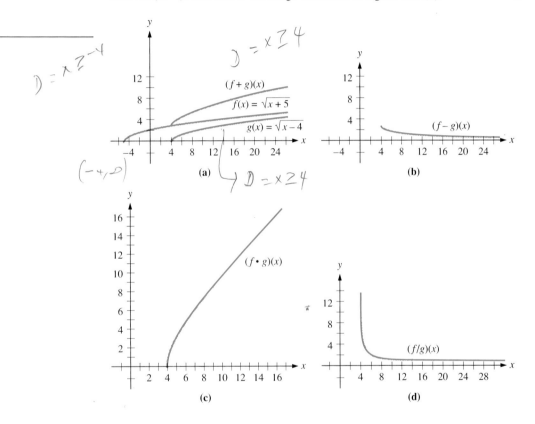

Composite Functions

There is another method for combining two functions to form a new function. This new function is called a composite function.

Composite Function

If f and g are functions, the **composite function**, or **composition**, of f and g, written $f \circ g$, is defined by

$$(f \circ g)(x) = f(g(x)).$$

The domain of $f \circ g$ is all the values of x in the domain of g such that $g(x)$ is in the domain of f.

The rule for $f \circ g$ is obtained by first applying the rule for g to x, noting the result, and then applying the rule for f to the noted result.

▶ **EXAMPLE 2.32** If $f(x) = \sqrt{x-5}$ and $g(x) = x^2 + 4$, then find $(f \circ g)(x)$ and $(g \circ f)(x)$ and their respective domains and ranges. Also find $(f \circ g)(6)$ and $(g \circ f)(6)$.

Solution The domain of f is $\{x: x \geq 5\}$ and the domain of g is $\mathbf{R}$. The range of f is $\{y: y \geq 0\}$ and the range of g is $[4, \infty)$.

$$(f \circ g)(x) = f(g(x)) = f(x^2 + 4) = \sqrt{(x^2 + 4) - 5} = \sqrt{x^2 - 1}.$$

The domain is $\{x: x \leq -1 \text{ or } x \geq 1\}$ and the range is $\{y: y \geq 0\}$.

$$(g \circ f)(x) = g(f(x)) = g(\sqrt{x-5}) = (\sqrt{x-5})^2 + 4 = (x - 5) + 4 = x - 1.$$

According to the above definition, the domain of $g \circ f$ is the set of x such that $f(x)$ is in the domain of g. Thus, the domain of $g \circ f$ is $\{x: x \geq 5\}$. As $(g \circ f)(x) = x - 1$ for all x in the domain of $g \circ f$, we see that the range is $\{y: y \geq 4\}$. (Warning: do not use the formula "rule" for a composite function to determine its range. In this case, $(g \circ f)(x) = x - 1$ and one might be tempted to say the range of $g \circ f$ is $\mathbf{R}$. But the range of a function depends not only on its rule but also on its domain. Since the domain of $g \circ f$ is the interval $[5, \infty)$, the range of $g \circ f$ is accordingly restricted.)

$$(f \circ g)(6) = \sqrt{6^2 - 1} = \sqrt{35}.$$

$$(g \circ f)(6) = 6 - 1 = 5. \quad \blacktriangleleft$$

Notice that in example 2.32, $(f \circ g)(6) \neq (g \circ f)(6)$. It is "usually" true, as in this example, that $(f \circ g)(x) \neq (g \circ f)(x)$. However, as will be seen next when we study inverse functions, $f \circ g$ and $g \circ f$ are equal in certain important special cases.

Inverse Functions

Many operations in mathematics have the effect of "undoing," or reversing, each other. For example, subtraction reverses the process of addition, in the sense that $(a + b) - b = a$; and, similarly, division reverses the process of multiplication. When two operations are related in this way, we call them inverse operations. Similarly, two functions f and g that each "undo" what the other function has done are called **inverse functions.**

In order to understand how to determine the inverse of a function, we return to our earlier work with relations. A relation is a set of ordered pairs. **To find the inverse of any relation R, simply interchange the first and second elements.** For example, if $R = \{(1, 2), (3, 5), (4, 6), (4, -2)\}$, then its inverse relation would be the set $R^{-1} = \{(2, 1), (5, 3), (6, 4), (-2, 4)\}$.

If the given relation had been a function, the process of finding its inverse would have been similar. In general, the inverse of a function is only a relation, not necessarily a function. We are mostly interested in the inverses of functions that are themselves functions.

New Functions from Old

To determine the *inverse relation* of $y = f(x)$, you first solve for x. If this solution is also a function, say g, then g is the **inverse function** of f. We use the notation f^{-1} to designate the inverse function of f. You should not think of the -1 in this notation as an exponent; f^{-1} does not mean $\dfrac{1}{f}$!

A function and its inverse interchange their domain and range. Thus, if f has an inverse function, then the domain of f is the range of f^{-1}, and the range of f is the domain of f^{-1}.

▷ **EXAMPLE 2.33** Find the inverses of the functions given by (a) $f(x) = 9x + 2$, (b) $g(x) = x^3 - 4$, (c) $h(x) = x^2$, (d) $j(x) = \dfrac{2x - 3}{-5x + 6}$, and (e) $k(x) = \dfrac{-6x + 24}{4x - 16}$. Determine if the inverse relation is a function. Whenever possible, find the domain and the range of the inverse function.

Solutions

(a) Let $y = f(x) = 9x + 2$ and solve for x. Thus, $x = \dfrac{y - 2}{9}$. This is a function, and so $x = f^{-1}(y)$, or $f^{-1}(y) = \dfrac{y - 2}{9}$. By tradition, we normally let x represent the independent variable. Thus, the formula for the inverse function of f is $f^{-1}(x) = \dfrac{x - 2}{9}$. Notice that both f and f^{-1} have domain and range each equal to $\mathbf{R}$.

(b) Let $y = g(x) = x^3 - 4$. Solving for x, we get $x = \sqrt[3]{y + 4}$. This x is a function of y, and so $g^{-1}(x) = \sqrt[3]{x + 4}$. Again, both g and g^{-1} have domain and range each equal to $\mathbf{R}$.

(c) Let $y = h(x) = x^2$. Solving for x, we get $x = \sqrt{y}$ or $x = -\sqrt{y}$, which is often written as $x = \pm\sqrt{y}$. This is not a function; that is, the inverse relation h^{-1} is not a function.

(d) Let $y = j(x) = \dfrac{2x - 3}{-5x + 6}$. Let's solve for x. First, we find that $y(-5x + 6) = 2x - 3$, or $-5yx + 6y = 2x - 3$. Hence $-5yx - 2x = -3 - 6y$, or $x(-5y - 2) = -3 - 6y$. Thus $x = \dfrac{-3 - 6y}{-5y - 2}$. This x is a function of y, and so $j^{-1}(x) = \dfrac{-3 - 6x}{-5x - 2}$. The domain of j^{-1} is $\{x: -5x - 2 \neq 0\} = \{x: x \neq -\tfrac{2}{5}\}$. The range of j^{-1} is the domain of j, namely $\{y: -5y + 6 \neq 0\} = \{y: y \neq \tfrac{6}{5}\}$.

(e) Let's put $y = k(x)$ and try to solve for x. Proceeding as in (d), we find eventually that $x(4y + 6) = 24 + 16y$, and so it seems that $x = \dfrac{24 + 16y}{4y + 6} = 4$. This is *not* reasonable because 4 is not in the domain of k! Hence division by $4y + 6$ is impossible. It follows that $4y + 6 = 0$, or $y = -\tfrac{6}{4} = -\tfrac{3}{2}$, a fact that you can also see by simplifying $\dfrac{-6x + 24}{4x - 16} = \dfrac{-6(x - 4)}{4(x - 4)} = -\dfrac{6}{4} = -\dfrac{3}{2}$. In particular, $k = \{(x, y): x \neq 4, y = -\tfrac{3}{2}\}$ and $k^{-1} = \{(x, y): x = -\tfrac{3}{2}, y \neq 4\}$. Thus you can see that k^{-1} is not a function. ◀

Chapter 2 / Functions and Their Graphs

Look again at example 2.33(a). The natural domain and range were used as the example was solved. There will be later applications when the domain needs to be restricted (and as a result, the range will be correspondingly restricted).

▶ **EXAMPLE 2.34**

Suppose that the function in example 2.33(a), given by $f(x) = 9x + 2$, had a domain of $[-1, 2)$. Its range would then be $[-7, 20)$. The inverse of f would be given by the same functional rule as it was in example 2.33(a), namely $f^{-1}(x) = \frac{x-2}{9}$. However, this new f^{-1} would have domain $[-7, 20)$ and range $[-1, 2)$. ◀

▶ **EXAMPLE 2.35**

If $f(x) = 9x + 2$ and $g(x) = x^3 - 4$ as in example 2.33, determine each of the following: (a) $f \circ f^{-1}$, (b) $f^{-1} \circ f$, (c) $g \circ g^{-1}$, and (d) $g^{-1} \circ g$.

Solutions

From example 2.33, we know $f^{-1}(x) = \frac{x-2}{9}$ and $g^{-1}(x) = \sqrt[3]{x+4}$.

(a) $(f \circ f^{-1})(x) = f(f^{-1}(x)) = f\left(\frac{x-2}{9}\right) = 9\left(\frac{x-2}{9}\right) + 2 = x$. So, $f \circ f^{-1}$ is the "identity function" with domain **R**.

(b) $(f^{-1} \circ f)(x) = f^{-1}(f(x)) = f^{-1}(9x + 2) = \frac{(9x+2)-2}{9} = \frac{9x}{9} = x$. So, $f^{-1} \circ f$ is the "identity function" with domain **R**.

(c) $(g \circ g^{-1})(x) = g(g^{-1}(x)) = g(\sqrt[3]{x+4}) = (\sqrt[3]{x+4})^3 - 4 = (x + 4) - 4 = x$. So, $g \circ g^{-1}$ is the "identity function" with domain **R**.

(d) $(g^{-1} \circ g)(x) = g^{-1}(g(x)) = g^{-1}(x^3 - 4) = \sqrt[3]{(x^3 - 4) + 4} = \sqrt[3]{x^3} = x$. So, $g^{-1} \circ g$ is the "identity function" with domain **R**. ◀

Example 2.35 illustrates that, in general, we have the following.

> If f is a function with inverse function f^{-1}, then
> $$(f \circ f^{-1})(x) = x \text{ for each } x \text{ in the range of } f,$$
> and
> $$(f^{-1} \circ f)(x) = x \text{ for each } x \text{ in the domain of } f.$$

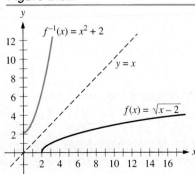

Figure 2.32

Graphing an Inverse Relation

Consider the graph of $y = f(x) = \sqrt{x-2}$ shown in figure 2.32. Pick any point on the graph of f, say $P(x_1, y_1) = (x_1, \sqrt{x_1 - 2})$. The corresponding point on the graph of its inverse relation is $P'(\sqrt{x_1 - 2}, x_1)$. The line of symmetry for the points P and P' is the line $y = x$. Since P was any point on the graph of f, this

New Functions from Old

means that $y = x$ is the line of symmetry for the union of the graphs of f and its inverse relation. The line $y = x$ is the dashed line in figure 2.32. Reflecting the graph of $f(x) = \sqrt{x - 2}$ in $y = x$, we get the colored curve in the figure. This is the graph of $f^{-1}(x) = x^2 + 2$, for $x \geq 0$.

In view of the preceding discussion and exercise 61(a) in section 2.2, we may summarize as follows.

Graph of Inverse Function

If a function f has an inverse function, the graph of f^{-1} may be obtained from the graph of f by reflecting through the line $y = x$.

Theoretically, this can be accomplished by folding the graph paper along the line $y = x$ and tracing the new position of the graph of f. One does not always find that this paper-folding technique is easily accomplished. By double-reflection (or by folding, tracing the image, and then tracing *its* image), we see that $(f^{-1})^{-1} = f$.

▶ **EXAMPLE 2.36** Graph the inverse function of $f(x) = x^3$.

Solution The graph of $y = x^3$ is shown as a solid black curve in figure 2.33. The line $y = x$ is shown by a dashed line. Reflecting $y = x^3$ in the line $y = x$ produces the graph of $f^{-1}(x) = \sqrt[3]{x}$, as shown by the colored curve. ◀

Figure 2.33

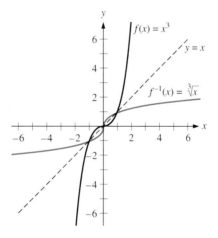

Horizontal Line Test

As the parabola in example 2.33(c) indicates, it is not always obvious whether a given function has an inverse function. However, the graph of a function f provides one of the easiest ways to determine if f has an inverse function. Recall that we used the Vertical Line Test to determine if a given graph was the graph of a function. A similar test, called the Horizontal Line Test, is used to determine if a function has an inverse function.

Horizontal Line Test

The **Horizontal Line Test** states that a function f has an inverse function if no horizontal straight line intersects the graph of f in more than one point.

The function in figure 2.34(a) has an inverse function because it is not possible to draw a horizontal line that will intersect the graph more than once. The function in figure 2.34(b) is $h(x) = x^2$. As you can see, each horizontal line above the vertex intersects the graph in two points. Thus, h does not have an inverse function. The Horizontal Line Test has confirmed what we noticed in example 2.33(c).

Figure 2.34

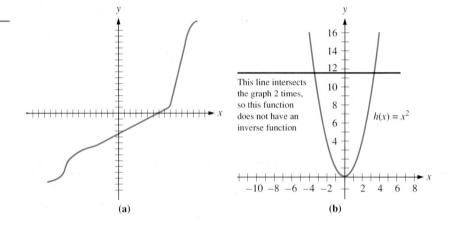

In order for a function f to have an inverse function, it is essential that different numbers in the domain always give different values of f. We refer to such a function as a one-to-one function. A formal definition is given next.

One-to-One Function

A function f with the property

$$x_1 \neq x_2 \quad \text{implies that} \quad f(x_1) \neq f(x_2)$$

is called a **one-to-one function.**

Any increasing function is one-to-one; so is any decreasing function. In particular, $f(x) = x^3$ is one-to-one. The function $g(x) = x^2$ is not one-to-one since $2 \neq -2$ and $g(2) = 4 = g(-2)$.

By the Horizontal Line Test, we have the following conclusion:

Theorem

A function h has an inverse function if and only if h is one-to-one.

Exercises 2.4

In exercises 1–6, find the rules or underlying sets for the functions $f + g$, $f - g$, $f \cdot g$, and f/g. Find the domains of each of these functions and also (estimate, by graphing calculator, if necessary) their ranges.

1. $f(x) = x + 3$, $g(x) = 9 - x^2$
2. $f(x) = x^2 + 3$, $g(x) = 9 - x^2$
3. $f(x) = \sqrt{x - 3}$, $g(x) = 2\sqrt{x}$
4. $f(x) = \dfrac{2}{x}$, $g(x) = x - 1$
5. $f = \{(1, 2), (-1, 2), (0, 2), (3, 4)\}$, $g = \{(-1, -2), (3, -4)\}$
6. $f = \left\{\left(\dfrac{2}{3}, 1\right), (\pi, \pi), \left(\dfrac{3}{4}, 0\right)\right\}$,
   $g = \left\{\left(\dfrac{2}{3}, 0\right), \left(\pi, \dfrac{\pi}{2}\right), \left(0, \dfrac{3}{4}\right)\right\}$

7. (a) If f, g, and h are functions, prove that $(f + g) + h = f + (g + h)$ and $(f \cdot g) \cdot h = f \cdot (g \cdot h)$.
   (b) Find examples of functions f, g, and h such that $(f - g) - h \neq f - (g - h)$ and $(f/g)/h \neq f/(g/h)$.

8. (a) Let f and g be even functions. Prove that $f + g$, $f - g$, $f \cdot g$, and f/g are each even functions.
   (b) Let f and g be odd functions. Prove that $f + g$ and $f - g$ are each odd functions and that $f \cdot g$ and f/g are each even functions.
   (c) What conclusions concerning symmetry about the y-axis or the origin may be inferred from (a) and (b)?

9. Let f and g be increasing functions.
   (a) Prove that $f + g$ is an increasing function.
   (b) Show, by choosing appropriate increasing f and g, that none of $f - g$, $f \cdot g$, and f/g need be increasing. (Hint: there are linear examples.)

10. Find an increasing function f and a decreasing function g such that $f + g$ is a constant function. Hence show that a sum of one-to-one functions need not be one-to-one.

In exercises 11–24, find the rules or underlying sets for the functions $f \circ g$ and $g \circ f$. Find the domains of both these functions and also (graphically, if necessary) their ranges.

11. $f(x) = x^4$, $g(x) = 2x - 1$
12. $f(x) = x^3 + 1$, $g(x) = x^2$
13. $f(x) = x + 2$, $g(x) = \sqrt{x} - 1$ (Hint: for the range of $g \circ f$, review section 2.2.)
14. $f(x) = \dfrac{1}{3x + 5}$, $g(x) = \sqrt{5x - 3}$
15. $f(x) = x^2 + 2$, $g(x) = \dfrac{1}{x}$
16. $f(x) = x^2 + 3$, $g(x) = \dfrac{1}{2x - 1}$
17. $f(x) = |x + 2|$, $g(x) = x - 2$
18. $f(x) = [x]$, $g(x) = x + \dfrac{1}{2}$
19. $f(x) = 2x^{4/5}$, $g(x) = x - 3$
20. $f(x) = x^{1/7}$, $g(x) = x^7$
21. $f(x) = x^{1/n}$, $g(x) = x^n$, where n is any positive integer.
22. $f(x) = 5x - 3$, $g(x) = \dfrac{x + 3}{5}$
23. $f = \{(0, 1), (-1, 1), (2, 3)\}$,
    $g = \{(-1, 2), (3, -1), (\pi, 3), (\sqrt{2}, \pi)\}$
24. $f = \{(\sqrt{2}, \pi), (-7, 6), (0, 1)\}$,
    $g = \{(6, -7), (1, 0), (\pi, \sqrt{2})\}$

In exercises 25–30, express each of the given functions as $f \circ g$, where neither f nor g is the given function. (There may be more than one correct answer for certain exercises.)

25. $h(x) = 2(x - 3)^{4/5}$ (Hint: let $g(x) = x - 3$.)
26. $j(x) = 2(x - 3)^{4/5} + 6$
27. $k(x) = -3(x + 4)^2 - \pi$
28. $m(x) = \dfrac{x^2}{x + 1}$. $\left(\text{Hint: consider } x - 1 + \dfrac{1}{x + 1} = x - 1 + \dfrac{1}{(x - 1) + 2}.\right)$
29. $n = \{(0, 2), (1, -1), (\pi, \sqrt{2})\}$
30. $p(x) = x$

31. At time $t = 0$, a stone is dropped into a calm pool of water at point P. At time $t > 0$, the rippling effect has produced a circle with center P, radius $r(t)$, and area $A = \pi r(t)^2$. If $r(t) = 3t^{2/5}$, express A explicitly as a function of t. How may A be regarded as a composite function?

32. A right circular cylindrical tube, with radius r and height h, has a closed bottom and open top. If $h = f(r) = \dfrac{r + 1}{3}$, express the tube's surface area A as an explicit quadratic function of r. Find a quadratic function $g(r)$ so that $A = g \circ f$.

33. Let f be any function with domain equal to $\mathbf{R}$ and let g be any nonconstant linear function. Show that the domain of $f \circ g$ is $\mathbf{R}$. What is its range?

34. State and prove an analogue of exercise 7(a) regarding the possible associativity of composition of functions. Interpret the result in terms of applying three machines successively. Generalize!

35. Let f and g be functions.
    (a) If g is even, prove that $f \circ g$ is even.
    (b) If f and g are each odd, prove that $f \circ g$ is odd.
    (c) What conclusions about symmetry follow from (a) and (b)?

36. (a) If f and g are each increasing functions, show that $f \circ g$ is also an increasing function.

(b) Does the assertion in (a) remain valid if "increasing" is changed throughout to "decreasing"? (Hint: consider $f(x) = -x = g(x)$.)

(c) Let $f(x) = \begin{cases} x^2 & \text{if } 0 \leq x \leq 1 \\ x^2 + 3 & \text{if } x > 1 \end{cases}$

and $g(x) = \begin{cases} x & \text{if } 0 \leq x \leq 1 \\ x - \frac{1}{2} & \text{if } x > 1. \end{cases}$

Show that f is increasing, g is neither increasing nor decreasing, $f \circ g$ is neither increasing nor decreasing, and $g \circ f$ is increasing.

37. (a) Let $h(x) = x$. Show, for all functions f and g, that $f \circ h = f$ and $h \circ g = g$.
   (b) Find functions f and k, each having domain $\mathbf{R}$, such that $f \circ k = f$ and $k(1) = k(-1)$.
   (c) Prove that if f and k are functions, each having domain $\mathbf{R}$, such that $f \circ k = f$ and f is one-to-one, then $k(x) = x$.

38. For each real number a, consider the function f_a given by $f_a(x) = \dfrac{x}{x-a}$, with its natural domain, $D = \{x: x \neq a\}$.

   (a) Prove that $(f_a \circ f_a)(x) = \dfrac{x}{(1-a)x + a^2}$ for all x in the domain of $f_a \circ f_a$.
   (b) Show that the domain of $f_a \circ f_a$ is D if $a = 1$, and is $\left\{x: x \neq a, \dfrac{a^2}{a-1}\right\}$ if $a \neq 1$.
   (c) Using (a) and (b), show that $f_a \circ f_a = f_a$ for exactly one value of a, namely 0.

39. Suppose that f and g are functions, each having an inverse function. Show that $f \circ g$ also has an inverse function, given by $(f \circ g)^{-1} = g^{-1} \circ f^{-1}$.

40. (a) Explain why no constant function has an inverse function.
    (b) Explain why no quadratic function has an inverse function.

41. Let $f(x) = mx + b$ be a nonconstant linear function. Show that f has an inverse function, given by $f^{-1}(x) = \dfrac{x-b}{m}$.

42. Using exercise 41, find the inverse functions of
    (a) $-\frac{2}{3}x + 7$ and (b) $\frac{3}{2}x - \pi$.

43. If a function f has domain $\mathbf{R}$ and its graph has a vertical line of symmetry, explain why f does not have an inverse function.

44. Prove that each even function whose domain contains a nonzero real number does not have an inverse function.

45. (a) Explain why (the odd function) x^3 is one-to-one (and, hence, has an inverse function).
    (b) Show that (the odd function) $x^3 - x$ is not one-to-one (and, hence, does not have an inverse function).

46. Give an example of a function f which is neither increasing nor decreasing, such that f has an inverse function.

In exercises 47–56, use the Horizontal Line Test to determine whether the function with the given graph has an inverse function.

47.

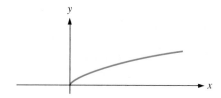

48.

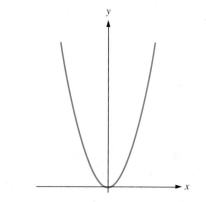

49.

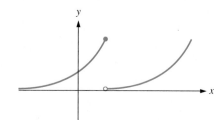

50.

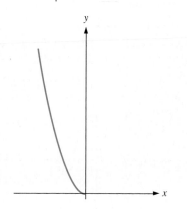

New Functions from Old

51.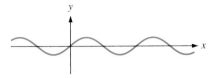

52.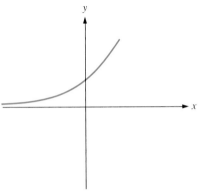

53.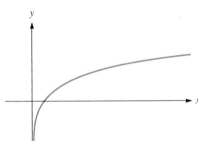

54.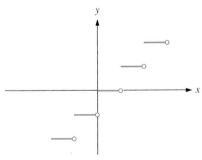

55.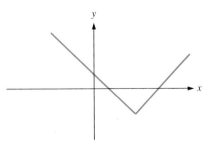

56.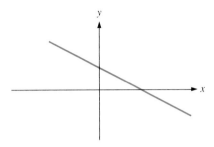

In exercises 57–64, describe R^{-1}, where R denotes the given relation.

57. $\{(0, 0), (2, 7), (\pi, \frac{7\pi}{2}), (-1, -\frac{7}{2})\}$

58. $\{(0, 1), (0, 2), (0, 3), (1, 3), (2, -4)\}$

59. $xy^2 = 7$ **60.** $x^2y = -7$

61. $(x + 1)(y - 2) = 3$ **62.** $3\sqrt{x - 1} = \sqrt{y}$

63. $3\sqrt{x - 1} = y$ **64.** $xy^7 = 1$

65. (a) Let $f(x) = \frac{x^2 + 1}{2}$, with its natural domain, R.

Explain why the function f does not have an inverse function.

(b) Let $g(x) = \frac{x^2 + 1}{2}$, with domain $\{x : x \leq 0\}$. Show that g^{-1} is a function, given by $g^{-1}(x) = -\sqrt{2x - 1}$, with domain $\{x : x \geq \frac{1}{2}\}$.

(c) Let $h(x) = \frac{x^2 + 1}{2}$, with domain $\{x : x \geq 0\}$. Prove that h has an inverse function and find its domain.

66. (a) Prove that $(R^{-1})^{-1} = R$ for each relation R.

(b) Give an example of a relation R such that R^{-1} is a function but R is not a function. (Hint: consider $x = y^2$.)

In exercises 67–82, determine whether the given function has an inverse function. If it does, determine the formula (or rule or underlying set) for the inverse function, graph both functions on the same coordinate system, and identify all domains and ranges.

67. $\{(0, 0), (2, 7), (\pi, \frac{7\pi}{2}), (-1, -\frac{7}{2})\}$

68. $\{(0, 0), (1, \pi), (-2, 0)\}$ **69.** $f(x) = x^2 + 3$

70. $g(x) = x^2 + 3$, with domain $\{x : x \leq 0\}$

71. $h(x) = -\frac{3}{2}x + \frac{4}{3}$, with domain $\{x : x \leq -8\}$

72. $j(x) = -\frac{3}{2}x + \frac{4}{3}$ **73.** $k(x) = x^3 - 7$

74. $m(x) = x^3 - x$ **75.** $n(x) = |x|$

76. $p(x) = |x|$, with domain $\{x : x < -2\}$

77. $q(x) = -\frac{2}{x}$, with domain $\{x : x > 0\}$

78. $r(x) = -\dfrac{2}{x}$
79. $s(x) = (2x - 7)^{1/3}$
80. $u(x) = x^{3/4}$
81. $v(x) = \sqrt{x - 3}$
82. $w(x) = \dfrac{4x^2 + 5}{x^2 - 1}$, with domain $\{x: x > 1\}$

In exercises 83–86, determine whether the given function has an inverse function. If it does, determine the formula for the inverse function, its domain, and its range.

83. $F(x) = \dfrac{-6x + 24}{-5x + 10}$
84. $G(x) = \dfrac{2x - 3}{-6x + 9}$
85. $H(x) = \dfrac{-6x + 24}{-3x + 12}$
86. $K(x) = \dfrac{2x - 3}{3x + 2}$

2.5 CHAPTER SUMMARY AND REVIEW EXERCISES

Major Concepts

- Composition
- Dependent variable
- Domain
- Function
  - Composite
  - Constant
  - Decreasing
  - Difference
  - Even
  - Greatest integer
  - Increasing
  - Inverse
  - Linear
  - Odd
  - One-to-one
- Product
  - Quadratic
  - Quotient
  - Sum
- Horizontal Line Test
- Independent variable
- Line of reflection
- Line of symmetry
- Natural domain
- Range
- Reflection, line of
- Relation
  - Inverse
- Shrink
- Stretch
- Symmetry
  - About a point
  - About the origin
  - About the x-axis
  - About the y-axis
  - Line of
  - Point of
  - With respect to a line
  - With respect to the line $y = x$
- Translation
  - Horizontal
  - Vertical
- Vertical Line Test

Review Exercises

In exercises 1–4, determine whether the relation being described is a function.

1. $\{(-1, 0), (1, 2), (\pi, 0), (5, 2)\}$
2. $x = \sqrt{4 - y^2}$
3.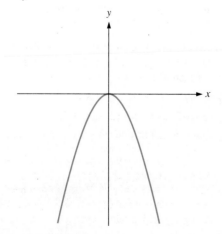
4.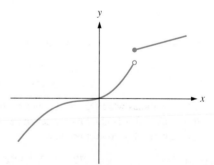

In exercises 5–10, find the natural domain of the given function, graph the function, and find the function's range.

5. $\{(-1, 0), (1, 0), (\pi, 1)\}$
6. $f(x) = -\dfrac{2}{3}x - 7$
7. $g(x) = \sqrt{2}$
8. $h(x) = x^3 - x$
9. $j(x) = -3\sqrt{9 - x^2}$
10. $k(x) = \dfrac{x^2 - 9}{x - 3}$

Chapter Summary and Review Exercises

11. Are the functions $f(x) = x$ and $g(x) = \sqrt{x^2}$ equal? Explain.

In exercises 12–16, find $f(0), f(-1), f(3.5), f\left(\dfrac{2}{x}\right)$, and $\dfrac{f(x+h) - f(x)}{h}$ for the given choices of $f(x)$ and h.

12. $f(x) = -3x + 5$, $h = 0.13$
13. $f(x) = 4x^2 - \pi x + \frac{1}{2}$, $h > 0$
14. $f(x) = \dfrac{x+6}{x^2 - 35}$, h an arbitrary real number
15. $f(x) = \sqrt{5-x}$, $h < 0$
16. $f(x) = x + |x|$, h arbitrary. (Hint: several cases arise for the last calculation.)
17. Classify the following functions as increasing, decreasing, or neither:
    (a) $-7x - 4$ (b) $x^3 + 1$
    (c) $-2x^2 + \sqrt{2}x - \pi$ (d) $\sqrt{2}$
18. Find an equation for the line of symmetry for the points $(\frac{2}{3}, -\frac{1}{5})$ and $(\frac{8}{3}, \frac{7}{5})$.
19. If the line $y = -7$ is the line of symmetry for the points $(-9, -4)$ and P, find the coordinates of P.
20. Classify the following functions as even, odd, or neither:
    (a) $-\frac{7}{3}x^5 - \frac{2}{9}x + \sqrt{2}$ (b) $x^2 - 2x$ (c) $x^8 - 8x^2 + 7$

In exercises 21–26, test the graph of the given equation or function for symmetry with respect to the y-axis, the x-axis, the origin, and the line $y = x$.

21. $xy = -2$
22. $2y^3 = 3x^5$
23. $x + 1 = 3y^2$
24. $f(x) = -\sqrt{x}$
25. $x^2 + y^2 = 9$
26. $g(x) = |x| - 2$

In exercises 27–30, draw the graphs of the given equations.

27. $y - 2 = 3(x+1)^2$
28. $y = 3x^2 + 6x + 6$
29. $y + 2 = 3\sqrt{x-4}$
30. $y + \frac{1}{2} = 2|3x + 4|$

For each of the parabolas which are the graphs of the functions described in exercises 31–34, find the vertex, an equation for the axis, and the highest or lowest point.

31. $f(x) = -7x^2$
32. $y = 2x^2 + 8x + 2$
33. $g(x) = \pi\sqrt{2}(x-3)^2 - 7$
34. $h(x) = -2x^2 + 8x + 5$
35. Find all values of a such that $(a, 5)$ lies on the graph of $f(x) = 2x^2 - 7x - 3$.
36. Find the y-intercept and the x-intercepts (if any) for the graphs of each of the following quadratic functions:
    (a) $f(x) = 2x^2 - 4x + 1$ (b) $g(x) = 2x^2 - 4x + 2$
    (c) $t(x) = 2x^2 - 4x + 3$
37. What sequence of expansion, reflection, and translations may be applied to the graph of the function x^2 in order to produce the graph of the function $k(x) = -24x^2 + 96x + 24$?
38. Find two real numbers whose sum is 5 and whose product is as large as possible.
39. Find the point on the graph of $y = \frac{2}{3}x$ that is closest to $(6, -9)$.
40. A parabolic arch, opening downward, has a base of 20 feet and maximum height (above the ground) of 60 feet. Tabby breaks free of his leash and begins to climb up the arch. When Tabby is directly overhead a point 4 feet from the midpoint of the arch's base, the rescue squad urges Tabby to jump. Tabby knows that he would hit the squad's net at $8\sqrt{h}$ feet per second, where h is Tabby's height above the ground. The net will break if Tabby's speed on impact exceeds 56 feet per second. Should Tabby jump? (Hint: Tabby has only one life and wants to live to climb another day. So, determine whether Tabby's jump would break the net.)
41. The current is flowing in a 120-volt circuit having a resistance of 10 ohms. The power W in watts is given by the formula $W = 120I - 10I^2$, where I measures the current in amperes. What is the maximum power that can be delivered in this circuit?
42. A salesperson's income is often a combination of two salaries, namely a base salary and a commission. The base salary is a guaranteed minimum and the commission is a percentage of that salesperson's sales. Thus, the total salary S is a linear function of the amount of sales, x. For one salesperson, the salary is $75 when the total sales are $300, and the salary is $800 when sales amount to $9000.
    (a) Write a linear function that describes this salesperson's salary.
    (b) What is the base salary?
    (c) What percentage of sales is commission?
    (d) What is the total salary when sales are $8400?
    (e) How much must this salesperson sell in order to have a total salary of $1000?

In exercises 43–46, find the rules or underlying sets for the functions $f + g, f - g, f \cdot g, f/g$, and $f \circ g$. Find the domains and ranges for each of these functions.

43. $f = \{(1, \pi), (2, 0), (3, \sqrt{2})\}$,
    $g = \{(3, 2), (1, \pi), (2, 1), (-3, 2)\}$
44. $f(x) = 2x + 1$, $g(x) = -\dfrac{3}{x}$ (Hint: you will need to approximate several of the ranges "intuitively," possibly using a graphing calculator.)
45. $f(x) = \sqrt{x}$, $g(x) = x^2 + 4$ (Hint: you will need to approximate several of the ranges "intuitively," possibly using a graphing calculator.)

46. $f(x) = \sqrt{x}$, $g(x) = \sqrt{2+x}$

In exercises 47–50, find the rule for $f \circ g$, and then find the domain and the range for $f \circ g$.

47. $f(x) = 3x - 4$, $g(x) = \sqrt{x-2}$

48. $f(x) = \dfrac{2}{x}$, $g(x) = 3\sqrt{x+2}$

49. $f(x) = 3x^{2/7}$, $g(x) = 2x - 5$

50. $f(x) = g(x) = -\dfrac{1}{x}$

51. Express each of the following as a composite function in a nontrivial way:

(a) $3(5x+2)^{7/2} + 6$ (b) x (c) $\dfrac{x^2}{x-1}$

52. (a) If f and g are any linear functions corresponding to lines with slope 1, show that $(f \circ g)^{-1} = f^{-1} \circ g^{-1}$.
(b) Find functions f and g, each having an inverse function, such that $(f \circ g)^{-1} \neq f^{-1} \circ g^{-1}$. (Hint: try $f(x) = x^2$ with a suitable domain and $g(x) = x+1$.)

53. Use the Horizontal Line Test to determine whether the function with the given graph has an inverse function.

(a)
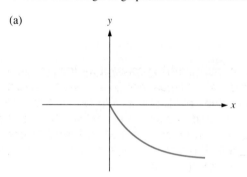

(b)
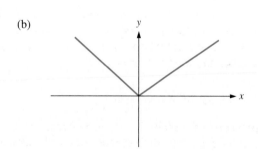

(c)
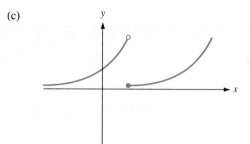

(d)
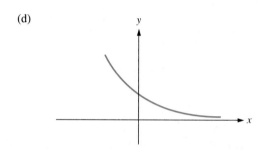

54. Describe R^{-1}, where R denotes the given relation:
(a) $\{(0, 1), (1, 2), (3, 0)\}$ (b) $3\sqrt{y-1} = x$
(c) $y^7 x = -1$

In exercises 55–59, determine whether the given function has an inverse function. If it does, determine the formula (or rule or underlying set) for the inverse function, graph both functions on the same coordinate system, and identify all domains and ranges.

55. $f(x) = \tfrac{2}{3}x - 1$, with domain $\{x: x > 2\}$

56. $g(x) = (7x - 2)^{1/5}$ **57.** $h(x) = 2x^3 + 7$

58. $k(x) = [x]$, with domain $\{0, 2, \pi, 5\}$

59. $m(x) = \sqrt{\dfrac{x-4}{x+5}}$

60. During the celebration pictured on the front cover, a skyrocket is shot directly upward from ground level at time $t = 0$ seconds. Six seconds later, the skyrocket explodes and continues directly upward in a beautiful stream of light. The height in feet of the skyrocket above the ground at time t seconds is given by

$$s(t) = \begin{cases} 400t - 16t^2 & \text{if } 0 \leq t \leq 6 \\ 648 + 292t - 16t^2 & \text{if } 6 \leq t \leq T \end{cases}$$

where T is the time at which the skyrocket strikes the ground. Find the total distance traveled by the skyrocket.

Chapter 3

Polynomial and Rational Functions

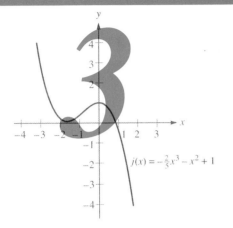

$j(x) = -\frac{2}{5}x^3 - x^2 + 1$

Functions are an important unifying concept for the remainder of this course and for the study of calculus. For that reason, chapter 2 was devoted to an intensive study of functions. You learned about their domains and ranges; how to add, subtract, multiply, divide, and compose functions; what it means for a function to be increasing or decreasing; how to determine if a function has an inverse function; and, perhaps most important of all, how to graph functions. You also learned how to determine whether the graph of a function is symmetric with respect to an axis or the origin.

In this chapter, we begin to examine some special types or classes of functions by focusing on polynomial functions and rational functions. In the next four chapters, 4–7, our attention will be on the classes known as exponential, logarithmic, and trigonometric functions.

3.1 POLYNOMIAL FUNCTIONS

Chapter 0 introduced polynomials. Section 2.3 discussed constant functions, linear functions, and quadratic functions. Each of these kinds of functions is an example of a polynomial function.

Polynomial Definitions

Let n be a nonnegative integer. A **polynomial function of degree n,** or an ***n*th-degree polynomial,** is a function P defined by an expression of the form

Chapter 3 / Polynomial and Rational Functions

$$P(x) = a_n x^n + a_{n-1} x^{n-1} + \cdots + a_2 x^2 + a_1 x + a_0$$

where $a_n \neq 0$. The numbers $a_n, a_{n-1}, \ldots, a_2, a_1, a_0$ are called the **coefficients** of the polynomial function P, and a_n is the **leading coefficient** of P.

If $a_n, a_{n-1}, \ldots, a_2, a_1, a_0$ are all real numbers, $P(x)$ is a **real polynomial of degree n**; and if $a_n, a_{n-1}, \ldots, a_2, a_1, a_0$ are all complex numbers, $P(x)$ is a **complex polynomial of degree n**.

The **zero polynomial** is the function Q that is identically zero: $Q(x) = 0$ for all x. The zero polynomial function is not assigned a degree.

▶ **EXAMPLE 3.1**

If c is a nonzero number, then the constant function f defined by $f(x) = c$ is a zero-degree polynomial. For instance, $g(x) = 5$, $h(x) = -\frac{7}{3}$, $k(x) = 2 + \sqrt{7}$, and $m(z) = 4 + 2i$ are all zero-degree polynomials. All four of these are complex zero-degree polynomial functions; in addition, the functions g, h, and k are real polynomials of degree 0. ◀

▶ **EXAMPLE 3.2**

A linear function f, defined by $f(x) = mx + b$, is a polynomial of degree 1 if $m \neq 0$. Thus, $g(x) = 2x$, $h(x) = -x + 5$, $k(x) = \frac{7}{2}x - 3$, and $j(x) = (3 - 2i)x + 7$ are all first-degree complex polynomials. Moreover, the functions g, h, and k are also real polynomials. ◀

▶ **EXAMPLE 3.3**

A quadratic function f, defined by $f(x) = ax^2 + bx + c$ where $a \neq 0$, is a second-degree polynomial. Hence, $g(x) = 3ix^2 - 5x + 2 - 7i$, $h(x) = 4x^2$, $j(t) = \frac{7}{3}t^2 - 1$, and $k(m) = -3m^2 + \frac{\sqrt{7}}{3}m - 4\sqrt{5}$ are all second-degree complex polynomials. Moreover, the functions h, j, and k are also real polynomials. ◀

▶ **EXAMPLE 3.4**

(a) $f(x) = \sqrt{9}x^7 + 3$ is a real (and complex) polynomial of degree 7.

(b) $g(t) = (4 + 2i)t^9 + 5t^3$ is a complex ninth-degree polynomial.

(c) $h(u) = 3u^{2/3} + 5$ does not define a polynomial function. (It does not *appear* to because the exponent of u is not a nonnegative integer. In exercise 46(b), we'll indicate why no polynomial expression can define a function such as h.)

(d) Similarly, $j(z) = 3z^4 + z^{-3} + 5z + 1$ does not define a polynomial function because the z^{-3} term has a negative exponent. ◀

One special type of polynomial function is a power function.

Power Function of Degree n

A **power function of degree n** is a polynomial of the form $f(x) = ax^n$, where $a \neq 0$ and n is a positive integer.

Thus, a first-degree power function is a linear function whose graph is a straight line, with nonzero slope a, that passes through the origin. The graph of a

Polynomial Functions

second-degree power function $f(x) = ax^2$ is a parabola with vertex at the origin; this parabola opens upward if $a > 0$, and opens downward if $a < 0$.

In section 2.2, we discussed even functions and odd functions. Recall that a function f is even if and only if the graph of $y = f(x)$ is symmetric with respect to the y-axis; and odd if and only if the graph of $y = f(x)$ is symmetric with respect to the origin. Are power functions even or odd (or neither)? Four power functions are graphed in figure 3.1(a), namely $f(x) = x$, $g(x) = x^2$, $h(x) = x^3$, and $j(x) = x^4$. Two more, $k(x) = x^5$ and $m(x) = x^6$, are graphed in figure 3.1(b). Observe that all six graphs pass through the origin and through $(1, 1)$. It also looks as if three of the graphs are symmetric with respect to the y-axis and, hence, come from even functions. These are $g(x) = x^2$, $j(x) = x^4$, and $m(x) = x^6$. The other three graphs all appear to be symmetric with respect to the origin, and so we conjecture that $f(x) = x$, $h(x) = x^3$, and $k(x) = x^5$ are odd functions. This conjecture is easy to verify algebraically because $(-x)^n = x^n$ if n is even and $(-x)^n = -x^n$ if n is odd. This type of reasoning leads directly to the following theorem.

> **EXAMPLE 3.9**
>
> **Solution**

Figure 3.1

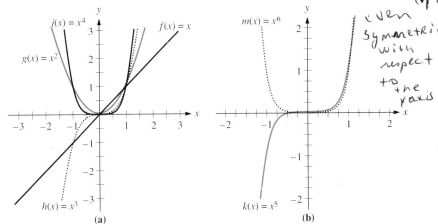

> **Division Algor**

Theorem

A power function of degree n, $f(x) = ax^n$ with $a \neq 0$, is an even function if n is even and is an odd function if n is odd. Its graph contains the origin and the point $(1, a)$. This graph also contains $(-1, a)$ if n is even, and contains $(-1, -a)$ if n is odd.

> **EXAMPLE 3.5**

Sketch the graphs of $f(x) = \frac{1}{3}x^2$, $g(x) = \frac{1}{3}x^4$, and $h(x) = \frac{1}{3}x^6$ relative to the same coordinate system.

> **Solution**

Since 2, 4, and 6 are even, the functions f, g, and h are even. Therefore, we can sketch each curve by first considering points corresponding to $x \geq 0$ and then reflecting such points in the y-axis to get the portion of the graph corresponding to $x \leq 0$. To begin building tables of values, note that the graphs of f, g, and h each contain the points $(-1, \frac{1}{3})$, $(0, 0)$, and $(1, \frac{1}{3})$; f also contains $(2, \frac{4}{3})$ and $(\frac{1}{2}, \frac{1}{12})$; g also contains $(2, \frac{16}{3})$ and $(\frac{1}{2}, \frac{1}{48})$; and h also contains $(2, \frac{64}{3})$ and $(\frac{1}{2}, \frac{1}{196})$. The completed graphs are in figure 3.2.

> **EXAMPLE 3.10**

Chapter 3 / Polynomial and Rational Functions

Figure 3.2

EXAMPLE 3.6

Solution

Figure 3.3

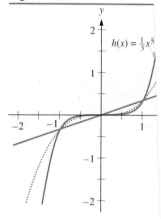

EXAMPLE 3.7

Solution

EXAMPLE 3.8

Solution

Chapter 3 / Polynomial and Rational Functions

EXAMPLE 3.11

Solution

$$\begin{array}{r} 4x^2 + 6x + 15 \\ x-2\overline{)4x^3 - 2x^2 + 3x - 5} \\ \underline{4x^3 - 8x^2} \\ 6x^2 + 3x - 5 \\ \underline{6x^2 - 12x} \\ 15x - 5 \\ \underline{15x - 30} \\ 25 \end{array}$$

Thus, $q(x) = 4x^2 + 6x + 15$ and $r(x) = 25$. By the Division Algorithm, $4x^3 - 2x^2 + 3x - 5 = (4x^2 + 6x + 15)(x - 2) + 25$.

If $P(x) = 4x^3 - 2x^2 + 3x - 5$ and $D(x) = x - 1$, what are $q(x)$ and $r(x)$?

$$\begin{array}{r} 4x^2 + 2x + 5 \\ x-1\overline{)4x^3 - 2x^2 + 3x - 5} \\ \underline{4x^3 - 4x^2} \\ 2x^2 + 3x - 5 \\ \underline{2x^2 - 2x} \\ 5x - 5 \\ \underline{5x - 5} \\ 0 \end{array}$$

Thus, $q(x) = 4x^2 + 2x + 5$ and $r(x) = 0$.

EXAMPLE 3.12

Solution

If $P(x) = 4x^3 - 2x^2 + \frac{3}{2}x - 5i$ and $D(x) = x - \sqrt{7}$, what are $q(x)$ and $r(x)$?

$$\begin{array}{r} 4x^2 + (4\sqrt{7} - 2)x + \left(\frac{59}{2} - 2\sqrt{7}\right) \\ x-\sqrt{7}\overline{)4x^3 - 2x^2 + \frac{3}{2}x - 5i} \\ \underline{4x^3 - 4\sqrt{7}x^2} \\ (4\sqrt{7} - 2)x^2 + \frac{3}{2}x - 5i \\ \underline{(4\sqrt{7} - 2)x^2 - (28 - 2\sqrt{7})x} \\ \left(\frac{59}{2} - 2\sqrt{7}\right)x - 5i \\ \underline{\left(\frac{59}{2} - 2\sqrt{7}\right)x + \left(14 - \frac{59}{2}\sqrt{7}\right)} \\ \frac{59}{2}\sqrt{7} - 14 - 5i \end{array}$$

Thus, $q(x) = 4x^2 + (4\sqrt{7} - 2)x + \left(\frac{59}{2} - 2\sqrt{7}\right)$ and $r(x) = \frac{59}{2}\sqrt{7} - 14 - 5i$.

In examples 3.7 and 3.8 we evaluated $P(x) = 4x^3 - 2x^2 + 3x - 5$ at $x = 2$ and $x = 1$, finding $P(2) = 25$ and $P(1) = 0$. In example 3.10, when this same $P(x)$ was divided by $x - 2$, the remainder was 25; and in example 3.11, when $P(x)$

was divided by $x - 1$, the remainder was 0. This coincidence leads one to conjecture the final two theorems of this section, the Remainder Theorem and the Factor Theorem. You will see that the Remainder Theorem is a corollary of the Division Algorithm, and the Factor Theorem is a corollary of the Remainder Theorem. These results will have profound implications in section 3.4 when we examine roots of a complex polynomial.

Remainder Theorem

If a polynomial $P(x)$ is divided by a first-degree polynomial of the form $D(x) = x - k$, then the remainder is $P(k)$, the value of the polynomial function at k.

Proof: From the Division Algorithm, we know that $P(x) = (x - k)q(x) + r(x)$. Since r is either 0 or has degree less than the degree of $x - k$, it follows that $r(x)$ describes a constant function; i.e., $r(x) = r$.

On the other hand, if we evaluate $P(k)$ we get

$$P(k) = (k - k)q(k) + r$$
$$= 0 \cdot q(k) + r$$
$$= r.$$

Factor Theorem

A polynomial $P(x)$ has a linear factor of the form $x - k$ if and only if $P(k) = 0$.

Proof: If $x - k$ is a factor of $P(x)$, then $P(x) = q(x) \cdot (x - k)$ for some polynomial q. By the uniqueness assertion in the Division Algorithm, the remainder when P is divided by $x - k$ is 0. Hence, by the Remainder Theorem, $P(k) = 0$.

On the other hand, if $P(k) = 0$, the Remainder Theorem guarantees that $P(x) = (x - k) \cdot q(x) + 0$ for some polynomial q, in which case we see that $x - k$ is a factor of $P(x)$.

▶ **EXAMPLE 3.13**

Use the Factor Theorem to show that both $x - 2$ and $x + 3$ are factors of $P(x) = 2x^5 + 2x^4 - 12x^3 - 9x^2 - 9x + 54$.

Solution

Observe that $P(2) = 2(2^5) + 2(2^4) - 12(2^3) - 9(2^2) - 9(2) + 54 = 64 + 32 - 96 - 36 - 18 + 54 = 0$. Since $P(2) = 0$, the Factor Theorem allows us to conclude that $x - 2$ is a factor of $P(x)$.

Next, we evaluate $P(-3)$: $P(-3) = 2(-3)^5 + 2(-3)^4 - 12(-3)^3 - 9(-3)^2 - 9(-3) + 54 = -486 + 162 + 324 - 81 + 27 + 54 = 0$. Since $P(-3) = 0$, the Factor Theorem yields that $x - (-3) = x + 3$ is a factor of $P(x)$. ◀

Exercises 3.1

1. Consider the polynomial functions f, g, h, k, and m, given by $f(x) = 2x - 3$, $g(x) = \sqrt{2}x^3 - 7x + \frac{1}{2}$, $h(x) = \frac{3}{2}x^7 + 4x^2$, $k(x) = -x^9 + ix^8 + 5$, and $m(x) = 0$.
   (a) Which of these five functions are real polynomials?
   (b) Which of these five functions are complex polynomials?

2. (a) Define what ought to be meant by a "rational polynomial."

(b) Using (a), determine which of the five polynomials listed in exercise 1 are rational polynomials.

3. (a) Define what ought to be meant by an "integral polynomial."
   (b) Using (a), determine which of the five polynomials listed in exercise 1 are integral polynomials.

4. For each of the five polynomials listed in exercise 1, determine whether the polynomial has a degree, and if it does, find that degree.

5. For the polynomial functions g and h given in exercise 1, simplify the functional expressions for the functions $g + h$, $g - h$, $g \cdot h$, and $3g - 2h$.

6. For the polynomial functions f and g in exercise 1, simplify the functional expressions for the functions $f \circ g$ and $g \circ f$.

7. (a) Based on your calculations in exercise 5, conjecture whether a sum, difference, or product of polynomial functions must be a polynomial function.
   (b) Give an (intuitive, if necessary) argument supporting the conjecture that you made in (a).

8. (a) Based on your calculations in exercise 6, conjecture whether a composite of polynomial functions must be a polynomial function.
   (b) Give an (intuitive, if necessary) argument supporting the conjecture that you made in (a).

9. Explain why the natural domain of any real polynomial function is **R**.

10. Prove the assertion in example 3.4 that the function given by $j(z) = 3z^4 + z^{-3} + 5z + 1$ is not a polynomial function. (Hint: reasoning as in exercise 7, reduce matters to considering whether z^{-3} defines a polynomial function. Then apply exercise 9.)

11. Give an example of a real polynomial function that is neither increasing nor decreasing. (Hint: section 2.3.)

12. Give an example of a real polynomial function that is neither even nor odd. (Hint: section 2.3.)

In exercises 13–16, determine whether the given functions are even, odd, or neither; and graph them relative to one coordinate system. Establish all valid symmetries with respect to the axes or the origin.

13. (a) $f(x) = 3x^2$ (b) $g(x) = 2x^3$
    (c) $h(x) = -\frac{3}{2}x^5$ (d) $k(x) = x^3 + x^2$

14. (a) $f(x) = 4x^6$ (b) $g(x) = -7x^5$
    (c) $h(x) = \frac{2}{3}x^4$ (d) $k(x) = 3x^3 - 2x^2$

15. (a) $f(x) = x^3$ (b) $g(x) = (x - 2)^3$
    (c) $h(x) = (x - 2)^3 - 5$ (d) $k(x) = (2x - 2)^3 + \frac{1}{2}$

16. (a) $f(x) = x^4$ (b) $g(x) = \frac{1}{2}x^4$
    (c) $h(x) = -2x^4$ (d) $k(x) = 2(\frac{1}{2}x + 1)^4 - 1$

17. Using the graphs in exercise 15, determine the range for each of the four given functions.

18. Using the graphs in exercise 16, determine the range for each of the four given functions.

19. For each real number r, construct the functional expression for a specific real polynomial function having range $[r, \infty)$. (Hint: section 2.3.)

20. (a) For each real number r, construct the functional expression for a specific real polynomial function having range $(-\infty, r]$.
    (b) Find a real polynomial function with range **R**.

In exercises 21–24, determine which of the values $\frac{2}{3}, 0, -1, \sqrt{2}$, and i are roots of the given polynomial function P.

21. $P(x) = x^5 - x^3 - 2x$
22. $P(x) = 3x^4 - 14x^3 - 7x^2 + 10x$
23. $P(x) = 3x^2 - 3x + 18$ 24. $P(x) = 0$

In exercises 25–30, find the quotient $q(x)$ and remainder $r(x)$ arising from the given dividend $P(x)$ and divisor $D(x)$. In exercises 25 and 26, use your calculations and the Remainder Theorem to determine the value of k such that $P(k) = r(x)$.

25. $P(x) = -2x^3 - 3x + 7$, $D(x) = x - 9$
26. $P(x) = 2x^3 - 3x + 7$, $D(x) = 6x - 9$
27. $P(x) = 2x^5 - ix^3 + 5$, $D(x) = 3ix^2 - 4$
28. $P(x) = -\frac{2}{3}x^3 - \sqrt{2}x$, $D(x) = 2x^3 - 3x + 5$
29. $P(x) = -\frac{2}{3}x^3 - \sqrt{2}x$, $D(x) = ix^4 + 7\sqrt{2}x - 1989$
30. $P(x) = -2x^5 - ix^4 - 2ix^2 + x$, $D(x) = ix^4 - x$

31. Use the Factor Theorem to show that both $x + 2$ and $x - 3$ are factors of $P(x) = 2x^4 - 2x^3 - 13x^2 + x + 6$.

32. Use the Factor Theorem to show that $x - 4$ is *not* a factor of $4x^3 - x^2 - 4x + 1$.

33. Use the Factor Theorem to show that $7i$ is a root of $P(x) = 3x^3 - 21ix^2$.

34. Use the Remainder and Factor Theorems to show that -2 is *not* a root of $P(x) = 2x^3 + 4x^2 - x + 2$. (Hint: divide $P(x)$ by $x + 2$. Is the remainder the zero polynomial?)

35. Without substituting the value $\frac{2}{5}$ into $P(x) = \frac{1}{2}x^3 - x + 2$, use the Remainder Theorem to find $P(\frac{2}{5})$. (Hint: perform a "long division.")

◆ In exercises 36–41, use the Factor Theorem to determine the roots of the given polynomial. (Hint: you may need to "factor further" to find "all" the linear factors.)

36. $P(x) = (x - 1)^4$
37. $P(x) = -2(x - 4)(3x + 2)(x^2 - 2)$
38. $P(x) = (2x - 1)(3x + 1)^2(x^2 + 1)$
39. $P(x) = (3x^2 - 7x + 6)x(x - 5)$

40. $P(x) = (x^3 + 8)(3x - 1)x^2$ **41.** $P(x) = 0$

42. (a) Verify that

$$4x^5 - 2x + 1$$
$$= (4x^4 + 4x^3 + 4x^2 + 4x + 2)(x - 1) + 3$$
$$= (4x^4 + 4x^3 + 4x^2)(x - 1) + 4x^2 - 2x + 1.$$

Conclude that the assertion of uniqueness in the statement of the Division Algorithm would fail if we did not restrict the degree of the remainder.

(b) Find polynomials Q and R, other than those in (a), such that $4x^5 - 2x + 1 = Q(x)(x - 1) + R(x)$.

◆ **43.** (Well-definedness of degree) In section 3.4, it will be shown that a polynomial of degree d cannot have more than d roots.

(a) Using this, prove that if polynomial functions

$$P(x) = a_n x^n + a_{n-1} x^{n-1} + \cdots + a_0 \text{ (with } a_n \neq 0\text{)}$$

and $Q(x) = b_m x^m + b_{m-1} x^{m-1} + \cdots + b_0$ (with $b_m \neq 0$)

are equal (as functions), then $n = m$, $a_n = b_m$, $a_{n-1} = b_{m-1}$, ..., $a_1 = b_1$, and $a_0 = b_0$. (Hint: what are the roots of the polynomial function $P - Q$?)

(b) Use (a) to show that each nonzero polynomial function has a uniquely determined degree.

◆ **44.** Let P and Q be nonzero polynomial functions of degree m and n, respectively.

(a) Explain why PQ is a polynomial function of degree $m + n$.

(b) Explain why $P + Q$ is either the zero polynomial or a polynomial whose degree is at most the larger of m and n.

◆ **45.** Use exercise 44 to prove the uniqueness assertion in the statement of the Division Algorithm. (Hint: suppose $P(x) = D(x)q(x) + r(x) = D(x)q_1(x) + r_1(x)$, with $q \neq q_1$.

Use exercise 44 to compute the degree of $D(q - q_1) = r_1 - r$ in two different ways. How does this degree compare with the degree of D?)

◆ **46.** It will follow from the theory developed in section 3.4 that if $P(x)$ is an nth degree polynomial whose only root is 0, then $P(x) = cx^n$ for some constant c.

(a) Using this, prove that the function f defined by $f(x) = x^{2/3}$ is not a polynomial function. (Hint: find all values k such that $f(k) = 0$. Notice that $x^{2/3} = cx^n$ implies $x^2 = c^3 x^{3n}$. Apply exercise 43. Is 2 an integral multiple of 3?)

(b) Use (a) to prove the assertion in example 3.4 that the function given by $h(u) = 3u^{2/3} + 5$ is not a polynomial function. (Hint: you'll also need exercise 7.)

◆ **47.** Construct a third-degree polynomial P with roots $\frac{1}{2}, -3,$ and 0. (Hint: express P as a product of three suitable linear factors.)

◆ **48.** Construct a fourth-degree polynomial whose only roots are $\frac{1}{2}, -3,$ and 0.

49. If k is a nonzero constant and n is a positive integer, prove the following four assertions.

(a) $x - k$ is a factor of $x^n - k^n$.

(b) $x - k$ is not a factor of $x^n + k^n$.

(c) $x + k$ is a factor of $x^n + k^n$ if and only if n is odd.

(d) $x + k$ is a factor of $x^n - k^n$ if and only if n is even.

◆ **50.** Explain the connection between the real roots of a real polynomial function P and the x-intercepts of the graph of $y = P(x)$.

51. The dimensions (length, width, and height) of a certain rectangular box are three consecutive integers, and the volume of the box is 336 cubic units. Find the dimensions of the box. (Hint: you will need to find a root of a third-degree polynomial.)

3.2 SYNTHETIC DIVISION AND NESTED MULTIPLICATION

Dividing one polynomial by another is time-consuming and requires a lot of space. In this section, you will learn more rapid procedures that can be used whenever the divisor is a linear polynomial, including one procedure that is especially useful when using a scientific calculator or a computer.

Synthetic Division

Synthetic division is a procedure that streamlines the work in any polynomial division problem involving a divisor of the form $x - k$. Let's reexamine example

3.10. There, we divided the dividend $P(x) = 4x^3 - 2x^2 + 3x - 5$ by the divisor $D(x) = x - 2$, getting the following result.

$$\begin{array}{r}
4x^2 + 6x + 15 \\
x - 2 \overline{\smash{)}4x^3 - 2x^2 + 3x - 5} \\
\underline{4x^3 - 8x^2 } \\
6x^2 + 3x - 5 \\
\underline{6x^2 - 12x } \\
15x - 5 \\
\underline{15x - 30} \\
25
\end{array}$$

When we set up and worked this problem, we were careful to follow the basic procedure where the divisor and the dividend were each written in descending powers of x. If a term had been "missing," we would have written 0 for its coefficient, as we did in example 3.9. We can simplify the basic procedure by eliminating the variables in the dividend. Since the coefficient of the x term in the divisor is 1, it can also be omitted. With these changes, the worked problem now looks like this:

$$\begin{array}{r}
4 + 6 + 15 \\
-2 \overline{\smash{)}4 - 2 + 3 - 5} \\
\underline{4 - 8 } \\
6 + 3 - 5 \\
\underline{6 - 12 } \\
15 - 5 \\
\underline{15 - 30} \\
25
\end{array}$$

You can see that, in the space including and below the dividend, some numbers are repeated. For example, there are two 4's, two 3's, three −5's, and so on. If we remove all but the first occurrence of such numbers, then the work appears as follows.

$$\begin{array}{r}
4 + 6 + 15 \\
-2 \overline{\smash{)}4 - 2 + 3 - 5} \\
\underline{- 8 } \\
6 \\
\underline{- 12 } \\
15 \\
\underline{- 30} \\
25
\end{array}$$

Synthetic Division and Nested Multiplication

If the work beneath the dividend is condensed vertically, the problem is presented as follows.

$$\begin{array}{r} 4 + 6 + 15 \\ -2\overline{)4 - 2 + 3 - 5} \\ \underline{-8 - 12 - 30} \\ 4 \quad 6 \quad 15 \quad 25 \end{array}$$

Notice that in order to get the bottom row, -8 was subtracted from -2, -12 was subtracted from 3, and -30 was subtracted from -5. Most people find it easier and faster to add than to subtract. So, if we were to change the signs of each of the numbers in the middle row, we could produce the same effect by adding instead of subtracting. In order to accomplish this, we also change the sign of the constant term in the divisor. The final result looks like this.

$$\begin{array}{r} 2\overline{)4 - 2 + 3 - 5} \\ \underline{8 \quad 12 \quad 30} \\ 4 \quad 6 \quad 15 \quad 25 \end{array}$$

We began this discussion with a problem that we had previously considered: $P(x) = 4x^3 - 2x^2 + 3x - 5$ divided by $D(x) = x - 2$. When this was worked in example 3.10, we found that the quotient was $q(x) = 4x^2 + 6x + 15$ and the remainder was $r(x) = 25$. Now, look at the last row of the above synthetic division. The first three numbers in it are the coefficients of the quotient, and the row's last number is the remainder. It can be proved in general that this compact, rapid organization for **synthetic division always displays the coefficients of the quotient and the remainder when a polynomial is divided by a divisor of the form $x - k$**. Perhaps you can see why synthetic division is sometimes referred to as "the method of detached coefficients."

▶ **EXAMPLE 3.14**   If $P(x) = 5x^6 - 42x^4 - 9x^3 - 7x^2 + 30x - 9$ and $D(x) = x - 3$, use synthetic division to find $q(x)$ and $r(x)$ when $P(x)$ is divided by $D(x)$.

Solution   Since there is "no" x^5 term in $P(x)$, we must write a 0 for the coefficient of x^5. We begin by writing the coefficients of $P(x)$ as a top row. Immediately to its left, we write the additive inverse of the constant term in the divisor. Subsequent work leads to the following display.

$$\underline{3)} \quad 5 \qquad 0 \qquad -42 \qquad -9 \qquad -7 \qquad 30 \qquad -9$$

$$\begin{array}{ccccccc} \downarrow & \xrightarrow{\times 3} & 15 & \xrightarrow{\times 3} & 45 & \xrightarrow{\times 3} & 9 & \xrightarrow{\times 3} & 0 & \xrightarrow{\times 3} & -21 & \xrightarrow{\times 3} & 27 \\ 5 & & \downarrow + 15 & & \downarrow + 3 & & \downarrow + 0 & & \downarrow + -7 & & \downarrow + 9 & & \downarrow + 18 \end{array}$$

The leading coefficient, 5, was "brought down." It was then multiplied by 3 to get 15. Next, we added to get the second number in the bottom row, 15. This was multiplied by 3 to get 45, which was added to -42 to get the 3 in the bottom row, and so on. The last number in the bottom row is the remainder, and the other numbers are the coefficients of the quotient. Since the degree of the quotient is 1 less than the degree of the dividend, we can

Chapter 3 / Polynomial and Rational Functions

conclude that $q(x) = 5x^5 + 15x^4 + 3x^3 - 7x + 9$ and $r(x) = 18$. Notice that the term in x^2 in the quotient was not written because its coefficient is 0. ◀

EXAMPLE 3.15 If $P(x) = 4x^4 + 3x^3 - 5x^2 + 2x + 1$ and $D(x) = x + 2$, what are $q(x)$ and $r(x)$ when $P(x)$ is divided by $D(x)$?

Solution Synthetic division produces the following display:

$$\begin{array}{r|rrrrr} -2) & 4 & 3 & -5 & 2 & 1 \\ & & -8 & 10 & -10 & 16 \\ \hline & 4 & -5 & 5 & -8 & 17 \end{array}$$

The quotient is $4x^3 - 5x^2 + 5x - 8$, and the remainder is 17. ◀

EXAMPLE 3.16 What are the remainder and the quotient when $P(x) = 2x^5 + x^3 - \frac{3}{2}x + \frac{3}{2}$ is divided by $D(x) = x + 1$?

Solution We shall again use synthetic division. We need to be careful here because two terms of $P(x)$ have coefficients of 0.

$$\begin{array}{r|rrrrrr} -1) & 2 & 0 & 1 & 0 & -\frac{3}{2} & \frac{3}{2} \\ & & -2 & 2 & -3 & 3 & -\frac{3}{2} \\ \hline & 2 & -2 & 3 & -3 & \frac{3}{2} & 0 \end{array}$$

The quotient is $2x^4 - 2x^3 + 3x^2 - 3x + \frac{3}{2}$, and the remainder is 0. Thus, $x + 1$ is a factor of $P(x)$ and, by the Remainder Theorem in section 3.1, -1 is a root of $P(x)$. ◀

EXAMPLE 3.17 What are the remainder and the quotient when $P(x) = x^4 - 3ix^3 + 2x^2 - 6ix - x + 3i$ is divided by $D(x) = x - 3i$?

Solution As usual, we shall use synthetic division. To do so, we must first identify the (complex) coefficients of the dividend and the divisor. By collecting terms, we may rewrite the dividend as $P(x) = x^4 - 3ix^3 + 2x^2 + (-1 - 6i)x + 3i$. Then the synthetic division can proceed:

$$\begin{array}{r|rrrrr} 3i) & 1 & -3i & 2 & (-1-6i) & 3i \\ & & 3i & 0 & 6i & -3i \\ \hline & 1 & 0 & 2 & (-1+0i) & 0 \end{array}$$

The quotient is $x^3 + 2x - 1$, and the remainder is 0. Notice, by the Remainder Theorem, that $3i$ is a root of $P(x)$. ◀

EXAMPLE 3.18 What are the remainder and the quotient when $P(x) = 4x^3 - 5x^2 + \frac{2}{3}x - \frac{5}{2}$ is divided by $2x - 3$?

Synthetic Division and Nested Multiplication

Solution

We need to be careful in applying synthetic division because the given divisor is not in the form $x - k$. However, we can factor a 2 out of $2x - 3$, getting $2x - 3 = 2(x - \frac{3}{2})$. Notice that $x - \frac{3}{2}$ is in the form $x - k$, with $k = \frac{3}{2}$.

Let's divide $P(x)$ by $x - \frac{3}{2}$. Synthetic division leads to

$$\begin{array}{r|rrrr} \frac{3}{2}) & 4 & -5 & \frac{2}{3} & -\frac{5}{2} \\ & & 6 & \frac{3}{2} & \frac{13}{4} \\ \hline & 4 & 1 & \frac{13}{6} & \frac{3}{4} \end{array}$$

Thus, when $4x^3 - 5x^2 + \frac{2}{3}x - \frac{5}{2}$ is divided by $x - \frac{3}{2}$, the quotient is $4x^2 + x + \frac{13}{6}$ and the remainder is $\frac{3}{4}$. In other words,

$$4x^3 - 5x^2 + \tfrac{2}{3}x - \tfrac{5}{2} = (4x^2 + x + \tfrac{13}{6})(x - \tfrac{3}{2}) + \tfrac{3}{4}.$$

Notice that the first term of the right-hand side can be rewritten as

$$(4x^2 + x + \tfrac{13}{6})(x - \tfrac{3}{2}) = (2x^2 + \tfrac{1}{2}x + \tfrac{13}{12})(2x - 3).$$

Thus, $\quad 4x^3 - 5x^2 + \tfrac{2}{3}x - \tfrac{5}{2} = (2x^2 + \tfrac{1}{2}x + \tfrac{13}{12})(2x - 3) + \tfrac{3}{4}.$

This means that when $4x^3 - 5x^2 + \frac{2}{3}x - \frac{5}{2}$ is divided by $2x - 3$, the quotient is $q(x) = 2x^2 + \frac{1}{2}x + \frac{13}{12}$ and the remainder is $\frac{3}{4}$. ◀

In general, when you have to divide a polynomial $P(x)$ by a linear polynomial $ax - b$, you can proceed as above. Notice that the quotient obtained from synthetic division must be divided by a in order to get the desired quotient.

Nested Multiplication

Synthetic division gives a fast method for evaluating real polynomials. Nested multiplication is a *substitute* for synthetic division (and the Remainder Theorem), and allows us to more quickly determine $P(k)$ with the aid of a calculator. Consider the polynomial in example 3.15: $P(x) = 4x^4 + 3x^3 - 5x^2 + 2x + 1$. Each term except the constant term contains a factor of x. If we factor x out of each of these terms, we can rewrite the polynomial as

$$P(x) = (4x^3 + 3x^2 - 5x + 2)x + 1.$$

Now, x appears in each of the first three terms inside the parentheses. If x is factored out of each of these terms, the polynomial is rewritten as

$$P(x) = ([4x^2 + 3x - 5]x + 2)x + 1.$$

Continuing, we see that x can be factored from the first two terms inside the brackets, thus leading to

$$P(x) = ([(4x + 3)x - 5]x + 2)x + 1.$$

The above procedure that was used to rewrite $P(x)$ is known as the **nested multiplication** format for writing a polynomial. It is applied to any polynomial by successively factoring x out of each nonconstant term. The particularly nice aspect about nested multiplication is that it provides a method to more easily evaluate a real polynomial with the use of a calculator or computer. For instance, if we wanted to evaluate the above polynomial $P(x)$ at $x = k$, then we would enter the following calculator steps:

$$4 \;\boxed{\times}\; k \;\boxed{+}\; 3 \;\boxed{=}\; \boxed{\times}\; k \;\boxed{-}\; 5 \;\boxed{=}\; \boxed{\times}\; k \;\boxed{+}\; 2 \;\boxed{=}\; \boxed{\times}\; k \;\boxed{+}\; 1.$$

Additional time can be saved, and errors reduced, by storing k and then using the $\boxed{\text{RCL}}$ key each time k is needed.

▶ **EXAMPLE 3.19** Evaluate $P(x) = 4x^4 + 3x^3 - 5x^2 + 2x + 1$ at $x = -2$ by using a calculator.

Solution This is the same polynomial considered in example 3.15. We shall begin by storing -2 in the calculator's memory, and then proceed as outlined above.

Enter	Display
2 $\boxed{+/-}$ $\boxed{\text{STO}}$	-2
4 $\boxed{\times}$ $\boxed{\text{RCL}}$ $\boxed{+}$ 3 $\boxed{=}$	-5
$\boxed{\times}$ $\boxed{\text{RCL}}$ $\boxed{-}$ 5 $\boxed{=}$	5
$\boxed{\times}$ $\boxed{\text{RCL}}$ $\boxed{+}$ 2 $\boxed{=}$	-8
$\boxed{\times}$ $\boxed{\text{RCL}}$ $\boxed{+}$ 1	17

The answer, 17, is the same as the remainder we got in example 3.15, when $P(x)$ was divided by $x + 2$. Of course, this is no coincidence, for the Remainder Theorem guarantees, for this polynomial, that the remainder is $P(-2)$. ◀

Nested multiplication, with its associated display, also describes the quotient when $P(x)$ is divided by $x - k$. The leading coefficient for the quotient is the same as the leading coefficient of $P(x)$. Notice that the nonleading coefficients of the quotient (-5, 5, and -8, as in example 3.15) are displayed in turn after the $\boxed{=}$ key is pressed.

While we have indicated how nested multiplication works with real polynomials, you should know that this method also applies to complex polynomials. For implementation, you need a scientific calculator that can handle arithmetic operations with complex numbers.

Exercises 3.2

In exercises 1–26, use synthetic division wherever possible. In exercises 1–10, find the quotient $q(x)$ and the remainder r arising when the given dividend $P(x)$ is divided by the given divisor $D(x)$.

1. $P(x) = 2x^3 - 3x - 7$, $D(x) = x - 9$
2. $P(x) = -2x^4 + x^2 + 7$, $D(x) = x - 2$
3. $P(x) = -\frac{3}{2}x^4 - x^2 + 25$, $D(x) = x + 2$
4. $P(x) = \frac{1}{6}x^4 + 3x^2 - \frac{7}{3}$, $D(x) = x + 4$
5. $P(x) = 3x^5 + x^4 + \sqrt{2}x^3 - x + 5$, $D(x) = 2x + \sqrt{2}$
6. $P(x) = 2x^3 - 3x^2 + x - \sqrt{3}$, $D(x) = 4x - \sqrt{3}$
7. $P(x) = 3x^2 + \sqrt{3}x - 20$, $D(x) = \sqrt{3}x - 4$
8. $P(x) = 2x^3 + 5x + 9\sqrt{2}$, $D(x) = \sqrt{2}x + 2$
9. $P(x) = x^3 + 2x^2 - ix^2 + 7$, $D(x) = x - 6 + i$
10. $P(x) = 16x^3 - 4ix + 41$, $D(x) = 2x - i$
11. Show that $x + 7$ is a factor of $x^3 + 6x^2 + 49$.
12. Show that $x - 7$ is not a factor of $x^3 - 6x^2 + 49$.
13. Show that $\frac{3}{2}i$ is not a root of $16x^4 + \frac{9}{4}x^2 - 1$.
14. Show that $1 - 2i$ is a root of $8x^3 + 22ix^2 - 25x + 25$.

In exercises 15–18, find $P(k)$ for the given $P(x)$ and the given k.

15. $P(x) = 9x^3 - 3x + 5$, $k = \frac{2}{3}$
16. $P(x) = 48ix^4 - 12x^2 - 2ix$, $k = -\frac{3}{2}$
17. $P(x) = ix^3 + 7x - \sqrt{2}$, $k = -6 + i$
18. $P(x) = x^2 - \frac{x}{2} + \frac{1}{\sqrt{2}}$, $k = \sqrt{2} - 3i$

In exercises 19–22, determine which of the values $-\frac{2}{3}$, 0, 1, and $2 - i$ are roots of the given polynomial function P. (Hint: review exercises 36–41 in section 3.1.)

19. $P(x) = 3x^3 - 10x^2 + 7x + 10$
20. $P(x) = 3ix^3 + 2ix^2$
21. $P(x) = x^2 + (i - 3)x + 2 - i$
22. $P(x) = x^7 - 3x^6 + 3x^5 - x^4$
23. Find all values of k so that $x - 3$ is a factor of $x^3 + k^2x^2 - 3kx - 27$.
24. Find all values of k so that $x + 3$ is a factor of $2k^2x^2 + 3kx + 9$. (Hint: k may not be a real number. Review how to use the quadratic formula.)
25. Find all values of k so that $x + k$ is a factor of $3x^2 + kx - 2$.
26. Give an example in which $P(x)$ and $D(x)$ are each complex, nonreal polynomials and $q(x) = P(x) \div D(x)$ is a real polynomial. (Hint: it's in one of the worked problems in this section.)

In exercises 27–30, use nested multiplication to rewrite the given polynomial P. Hence evaluate $P(-\frac{1}{2})$ and $P(2.7)$.

27. $P(x) = 8x^4 + 32x^3 - 2x + 5$
28. $P(x) = -2x^3 + 16x^2 - 4x - 1$
29. $P(x) = 2ix^3 + 16ix^2 + 4x - 1$
30. $P(x) = (8 + i)x^4 - 2ix + 6$

In exercises 31–34, use nested multiplication, with its associated displays, to find the quotient $q(x)$ and the remainder r when the given dividend $P(x)$ is divided by the given divisor $D(x)$.

31. $P(x) = x^4 - 2x^3 + x^2 - 7x + 6$, $D(x) = x - 3$
32. $P(x) = 2x^3 + x^2 - 18x - 9$, $D(x) = x + 3$
33. $P(x) = 4x^3 - 4x^2 - 21x - 9$, $D(x) = 2x + 3$
34. $P(x) = \frac{1}{2}x^4 + 5x^3 - 25x + 2$, $D(x) = 5x - 2$

◆ 35. (A foretaste of "continuity," discussed in the next section) Consider the polynomial function $P(x) = 7x^3 - 17x^2 + 7x - 17$. By any method, compute $P(2.3)$, $P(2.4)$, $P(2.5)$, $P(2.6)$, and $P(2.7)$. Hence estimate a root of $P(x)$ to within 0.05 unit.

36. (a) Consider the problem of dividing $2x^2 + 2x + 2$ by $2x + 1$. Show that the method of example 3.18 leads to the quotient $q(x) = x + \frac{1}{2}$ and the remainder $r = \frac{3}{2}$.
    (b) Observe that
    $$\frac{2x^2 + 2x + 2}{2x + 1} = \frac{x^2 + x + 1}{x + \frac{1}{2}}$$
    and so synthetic division leads to

    $$\begin{array}{r|rrr} -\frac{1}{2} & 1 & 1 & 1 \\ & & -\frac{1}{2} & -\frac{1}{4} \\ \hline & 1 & \frac{1}{2} & \frac{3}{4} \end{array}$$

    This result indicates a "remainder" of $\frac{3}{4}$. Explain why this does not contradict the work in (a).

3.3 GRAPHING POLYNOMIAL FUNCTIONS

We have already discussed and graphed several types of polynomial functions. At this point, you should be able to graph the types of functions summarized in table 3.1.

Chapter 3 / Polynomial and Rational Functions

Table 3.1

Degree	Type	Function	Shape of Graph
undefined	zero polynomial	$f(x) = 0$	x-axis
0	constant	$f(x) = a, a \neq 0$	horizontal line, not x-axis
1	linear, nonconstant	$f(x) = ax + b, a \neq 0$	nonhorizontal line with slope a
2	quadratic	$f(x) = ax^2 + bx + c,$ $a \neq 0$	parabola; opening up if $a > 0$, opening down if $a < 0$

In this section, we focus on graphing polynomials of degree higher than 2. Because these are more difficult to sketch, we first discuss some basic characteristics of the graphs of polynomials. Graphing uses the techniques emphasized in chapter 2, such as plotting points, finding x- and y-intercepts, and testing for symmetry with respect to the axes and the origin. This section also contains the text's second discussion of graphing calculators.

Continuity — all polynomials are continuous functions

One fundamental concept that you will study in calculus is that of continuity. Although the definition of continuity is beyond the scope of this book, the basic idea is illustrated in figures 3.4(a) and (b). In effect, a function is considered to be **continuous** if its graph can be drawn without lifting your pencil from the paper. This is the case for the function graphed in figure 3.4(a). On the other hand, the function graphed in figure 3.4(b) is **discontinuous,** or not continuous, because its graph *cannot* be drawn without lifting your pencil from the paper. Without proof, we state the theorem that **every polynomial function is continuous**. So, indeed, are most of the nonpolynomial functions that you will meet in calculus or, for that matter, in chapters 4 and 5.

One consequence of continuity is the Intermediate Value Theorem. Although its proof is deferred to a course on advanced calculus, this theorem can

Figure 3.4

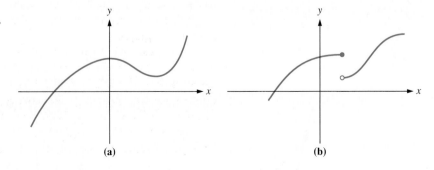

Graphing Polynomial Functions

be *used* effectively with what you know at present. Exercises 41–42 will give you practice in using the Intermediate Value Theorem to isolate and numerically approximate the real roots of a polynomial. Several computer programs have been written to carry out this important kind of numerical analysis.

Intermediate Value Theorem for Polynomials

Figure 3.5

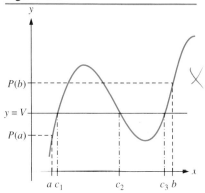

Consider a closed interval $[a, b]$ and a real polynomial function P such that $P(a) \neq P(b)$. Let V be a real number strictly between $P(a)$ and $P(b)$. Then there is at least one c strictly between a and b such that $P(c) = V$.

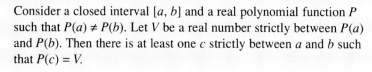

x make it equal to "0"

Smoothness

Any graph of a polynomial function is smooth. By **smooth,** we mean that the function is continuous and each curve or turn in its graph is rounded, as in figures 3.4(a) and 3.6(a). In particular, the graph of a polynomial cannot have any sharp points or cusps of the kind in figure 3.6(b) or figure 3.6(c). These two figures illustrate that not all continuous functions have smooth graphs.

Figure 3.6

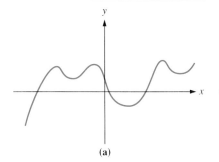
(a)

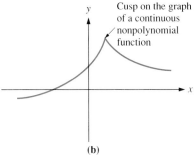
Cusp on the graph of a continuous nonpolynomial function
(b)

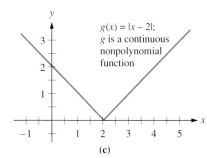
$g(x) = |x - 2|$;
g is a continuous nonpolynomial function
(c)

Turning Points

Consider the parabola $y = 4(x - 3)^2 + 1$, which is sketched in figure 3.7(a). From this graph, we see that the function $f(x) = 4(x - 3)^2 + 1$ is decreasing on the interval $(-\infty, 3)$ and increasing on $(3, \infty)$. At the point $(3, 1)$, the graph changes direction. Such a high or low point where a graph changes from increasing to decreasing or from decreasing to increasing is called a **turning point** of the graph.

In this example there was only one turning point. In general, it can be shown by the methods of calculus that the graph of an nth-degree polynomial has at most $n - 1$ turning points. Figure 3.7(b) gives an example of a fourth-degree polynomial whose graph has three turning points, and figure 3.7(c) gives an example of a fourth-degree polynomial with only one turning point.

As figures 3.7(b) and (c) show, if two polynomials have the same leading term, their graphs can have different shapes when $|x|$ is small. However, as these figures suggest, the shapes are similar when $|x|$ is large. The next result describes this as a general phenomenon.

Figure 3.7

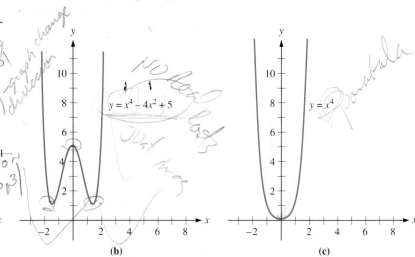

(a)   (b)   (c)

The Leading Term Test

Consider a polynomial $P(x) = a_n x^n + a_{n-1} x^{n-1} + \cdots + a_1 x + a_0$ of degree n. The **leading coefficient** of P is a_n. The sign of a_n and the parity (evenness or oddness) of n can be used to determine whether the graph of $y = P(x)$ rises or falls as values of x get increasingly large in absolute value.

Leading Term Test

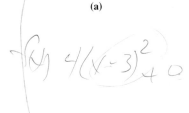

> The limiting tendencies of any real polynomial $P(x)$, as x moves without bound to either the right or the left, are the same as the corresponding tendencies of the associated power function $y = a_n x^n$ formed from the leading term of the polynomial P.

You will be asked to verify this in exercise 26.

The following two tables summarize what happens. The cases in table 3.2 are illustrated in figures 3.8(a) and (b), while the cases in table 3.3 are illustrated in figures 3.8(c) and (d). Example 3.20 shows how to use the Leading Term Test.

Table 3.2

As x increases without bound, the limiting value of $P(x)$ is	
$a_n > 0$	infinity
$a_n < 0$	negative infinity

Graphing Polynomial Functions

Table 3.3

<u>Moving w/o bound to left</u>
As x <u>decreases</u> without bound, the limiting value of $P(x)$ is

	n odd	n even
$a_n > 0$	negative infinity	infinity
$a_n < 0$	infinity	negative infinity

Figure 3.8

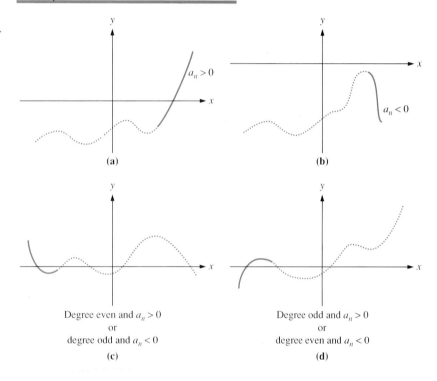

(a) Degree odd and $a_n > 0$

(b) Degree odd and $a_n < 0$

(c) Degree even and $a_n > 0$
or
degree odd and $a_n < 0$

(d) Degree odd and $a_n > 0$
or
degree even and $a_n < 0$

We have used the expression "infinity" to indicate the limiting tendency when $P(x)$ increases without bound. "Negative infinity" describes the limiting tendency when $P(x)$ decreases without bound. In calculus, you will study these concepts more rigorously.

EXAMPLE 3.20

Use the Leading Term Test to determine the nature of the graphs of the following polynomials as x moves without bound to the right and left:
(a) $f(x) = 2x^7 - 5x^5 + 3$, (b) $g(x) = \frac{1}{3}x^6 + x^2 + \frac{2}{3}x - 5$,
(c) $h(x) = -9x^8 + x^7 + 3x^6 - x + 4$, and (d) $j(x) = -\frac{2}{5}x^3 - x^2 + 1$.

Solutions

(a) The graph's limiting tendencies are the same as those of $y = 2x^7$. Here, $n = 7$ is odd and $a_n = 2 > 0$. So, by table 3.2, the limit is infinity as x moves without bound to the right. By table 3.3, the limit is negative infinity as x moves without bound to the left. The graph of f is shown in figure 3.9(a).

(b) The graph's limiting tendencies are the same as those of $y = \frac{1}{3}x^6$. By the above tables, these limits are infinity as x moves without bound to the right, and infinity as x moves without bound to the left. The graph of g is shown in figure 3.9(b).

(c) The graph's limiting tendencies are the same as those of $y = -9x^8$. These limits are negative infinity as x moves without bound to the right, and negative infinity when x moves without bound to the left. The graph of h is shown in figure 3.9(c).

(d) The limiting tendencies for the graph of j are the same as those of $y = -\frac{2}{5}x^3$. These limits are negative infinity as x moves without bound to the right, and infinity as x moves without bound to the left. The graph of j is shown in figure 3.9(d).

Figure 3.9

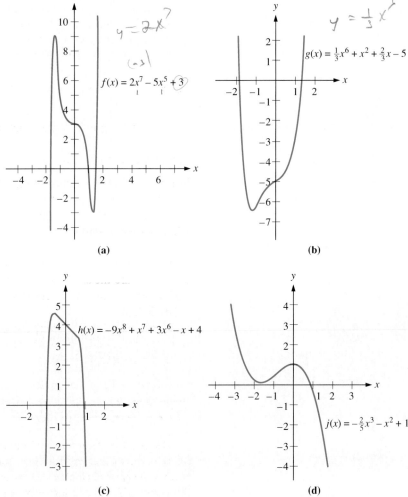

(a) $f(x) = 2x^7 - 5x^5 + 3$

(b) $g(x) = \frac{1}{3}x^6 + x^2 + \frac{2}{3}x - 5$

(c) $h(x) = -9x^8 + x^7 + 3x^6 - x + 4$

(d) $j(x) = -\frac{2}{5}x^3 - x^2 + 1$

Graphing Polynomial Functions

Limiting tendencies are just one tool for graphing polynomials. The most sophisticated tools, such as those which identify turning points, depend on calculus. We can, however, summarize some guidelines for graphing polynomials that you can use with what you know now.

Guidelines for Graphing Polynomials

1. Use the Leading Term Test to determine what happens to the graph for large positive x (as x moves to the right without bound) and for small negative x (as x moves to the left without bound).
2. Look for symmetries and transformations, as covered in chapter 2.
3. Find the y-intercept and as many x-intercepts as possible. Remember, the real zeros or roots of a real polynomial give the x-intercepts of the polynomial's graph. (The next two sections give procedures for finding some x-intercepts.)
4. Make a table of values. Synthetic division or nested multiplication can be useful here.
5. If a and b are consecutive real roots of P, compute $P(c)$ for some convenient c in (a, b), in order to determine whether the portion of the graph of P corresponding to x-values in (a, b) lies above or below the x-axis. (Notice that this step relates to the sign chart method introduced earlier.)
6. Plot the points tabulated in step 4 and connect them smoothly.

The next three examples show how to use some of these guidelines.

▶ **EXAMPLE 3.21**   Sketch the graph of $f(x) = \frac{1}{2}x^4 - 3$.

Solution   You can use the methods of section 2.2 directly, as follows. Let's begin with the graph of the function $y = x^4$, as in figure 3.7(c). If we apply a vertical shrinking by a factor of $\frac{1}{2}$ followed by a vertical translation of -3, the result is the desired graph of f.

Let's see how this graph could be found by using the above guidelines. First, for large $|x|$, the Leading Term Test tells us that the graph's limiting tendencies are the same as those of $y = \frac{1}{2}x^4$. Thus, as x moves without bound to the right or the left, the function values have limit infinity, as shown in figure 3.10(a).

Next, because the given function is even, its graph is symmetric with respect to the y-axis and so only nonnegative values of x need to be used. A table of values is given below.

x	0	0.5	1	1.5	2	2.5
$f(x)$	-3	-2.97	-2.5	-0.47	5	16.53

Notice that the graph's y-intercept is $(0, -3)$. Examination of the above table of values and application of the Intermediate Value Theorem indicate that there is a real number c between 1.5 and 2 such that $f(c) = 0$. Thus, the graph of f has an x-intercept at $(c, 0)$. We can determine this value

Chapter 3 / Polynomial and Rational Functions

of c by finding the roots of f. Now, $f(x) = \frac{1}{2}(x^4 - 6)$, and this factors as $\frac{1}{2}(x - \sqrt[4]{6})(x + \sqrt[4]{6})(x^2 + \sqrt{6})$. Thus, $f(x) = 0$ when $x = c = \sqrt[4]{6} \approx 1.5651$. (Note also that $f(x) = 0$ when $x = -\sqrt[4]{6} \approx -1.5651$.) The points tabulated above, including the x-intercept $(c, 0)$, are plotted, connected by a smooth curve, and reflected in the y-axis, to produce the graph in figure 3.10(b). ◀

Figure 3.10

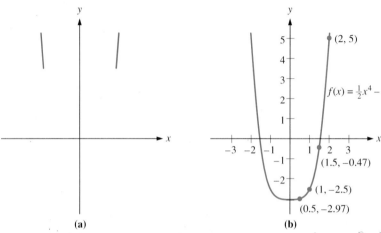

(a)   (b)

▶ **EXAMPLE 3.22**

Sketch the graph of $g(x) = -2x^3 + 4x - 1$. $= -2x(x^2 - 2) - 1$

Solution

When $|x|$ is large, this polynomial's limiting tendencies are the same as those of $-2x^3$. Thus, as x moves to the left without bound, the limit of $f(x)$ is infinity; and as x moves to the right without bound, the limit of $f(x)$ is negative infinity. These tendencies are indicated in figure 3.11(a).

The given function g is neither odd nor even. However, $h(x) = -2x^3 + 4x$ defines an odd function, whose graph therefore has the origin as a point of symmetry. As $g(x) = h(x) - 1$, the graph of g is obtained by translating the graph of h down one unit. Thus, the point $(0, -1)$ is a point of symmetry for the graph of g.

Since $h(x) = -2x^3 + 4x = -2x(x - \sqrt{2})(x + \sqrt{2})$, we can see that h has roots at $x = 0, \sqrt{2}$, and $-\sqrt{2}$. Geometrically, that means that the graph of h will meet the x-axis at the points $(0, 0)$, $(\sqrt{2}, 0)$, and $(-\sqrt{2}, 0)$. Additional points are found from the following table of values and the fact that h is an odd function.

x	0	0.5	1	$\sqrt{2}$	1.5	2
$h(x)$	0	1.75	2	0	−0.75	−8

When these tabulated points are plotted, reflected in the origin, and connected by a smooth curve, the graph of h in figure 3.11(b) results. Translating this graph vertically one unit downward results in the graph of g. ◀

Figure 3.11

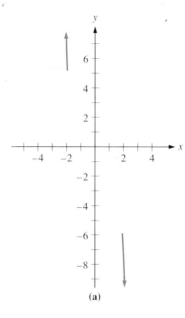
(a)

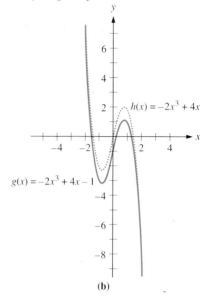
(b)

EXAMPLE 3.23 Sketch the graph of $j(x) = -x^4 + 2x^3$.

Solution The Leading Term Test directs our attention to $y = -x^4$. As shown in figure 3.12(a), when x moves without bound to either the right or the left, the limit of $j(x)$ is negative infinity.

The expression for $j(x)$ factors as $j(x) = -x^3(x - 2)$, and so we determine that $j(x)$ has roots at 0 and 2. A table of values for j is given below.

x	-2	-1	0	1	2	3
$j(x)$	-32	-3	0	1	0	-27

When these tabulated points are plotted and connected by a smooth curve, the graph in figure 3.12(b) results. However, this is *not* the graph of $y = j(x)$. The correct graph is shown in figure 3.12(c). Both graphs contain the points given in the above table. Why did our reasoning lead to the wrong graph in figure 3.12(b), and what can we do to prevent making such errors again?

In this example, if we had also checked at x-values such as $-1.5, -0.5, 0.5, 1.5,$ and 2.5, the enlarged table of values might have led us correctly to the maximum point $(\frac{3}{2}, \frac{27}{16})$. However, this "half-interval" strategy might not work on a different example.

The important point to remember about effective techniques of graphing is that they allow you to sketch a curve using a reasonably small number of points in your table of values. Calculus will allow you to analyze the graph of a function much more accurately, by identifying a "reasonably small" number of *useful* points, such as $(\frac{3}{2}, \frac{27}{16})$ in this example. We do not expect you to be able to use *pre*calculus to distinguish between subtly different graphs like those in figures 3.12(b) and (c).

Figure 3.12

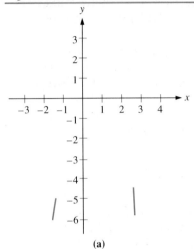
(a)

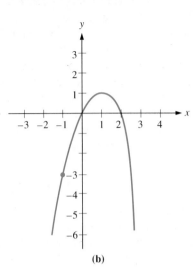
(b)

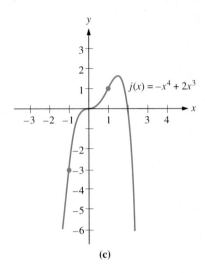
(c) $j(x) = -x^4 + 2x^3$

Technology has made it possible, at the precalculus level, to occasionally avoid the type of error mentioned in example 3.23. The next graphing calculator activity indicates how this might be accomplished.

Graphing Calculator Activity

In example 2.11, we used a graphing calculator to approximate the roots of a function. To do this, we used one of the "zoom-in" features of that calculator. Here we explore another way to zoom in, this time by restricting the domain and range of the graph on the calculator's screen either before or after the graph is displayed. The sketched graph can be used to approximate coordinates of points on the graph. We shall illustrate this capability by repeating example 3.23 using both a Texas Instruments *TI-81* and a Casio *fx-7700G* scientific calculator.

▶ **EXAMPLE 3.24** Use a graphing calculator to sketch the graph of $j(x) = -x^4 + 2x^3$. Use the sketch to approximate the graph's maximum value.

Solution  *TI-81* **Version** We begin by limiting the domain and range we want displayed on the screen. Press the RANGE key. The screen should show something like that in figure 3.13(a).

The domain to be shown on the screen is indicated by Xmin (the smallest value in the domain of interest), Xmax (the corresponding largest *x*-value), and Xscl (the *x*-axis's scale). The range is similarly restricted by Ymin, Ymax, and Yscl.

These values can be changed. Indeed, we saw in figures 3.12(b) and (c) that it is important in this example to be able to zoom in on the part of the domain corresponding to $-1 \le x \le 2$. To be safe, we focus on $[-2, 3]$ next. The cursor is initially at Xmin. To make -2 the minimum *x*-value, enter (−) 2 ENTER . Pressing the ENTER key saves the entered value and moves

Graphing Polynomial Functions

Figure 3.13

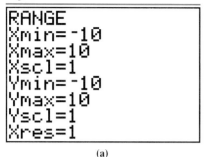

(a)

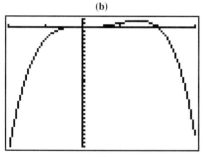

(b)

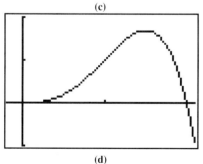

(c)

the cursor to the next location. To leave a value the same, simply press the ENTER key, and the cursor moves on. In this manner, we can make Xmax = 3, Xscl = 1, Ymin = –30, Ymax = 2, and Yscl = 1. The screen should now look like the one in figure 3.13(b).

Once the above settings have been completed, press the Y= key. If anything is written to the right of any of the four equal signs, use the ▲ and ▼ keys to move the cursor to the appropriate line, and then press the CLEAR key. When you have finished clearing, move the cursor to the top line, the line that begins ":Y1 = ". You are now able to input a formula describing the function you want graphed. To input $j(x) = -x^4 + 2x^3$, press the sequence of keys

(−) X|T ^ 4 + 2 X|T ^ 3 GRAPH

The result is the graph shown in figure 3.13(c).

If you wish to "take a closer look," the limitations on the domain and range can be changed by pressing the RANGE key and making the desired changes. Pressing GRAPH will then redraw the graph. Figure 3.13(d) shows the graph sketched with domain [–0.2, 2.1] and range limited to [–1, 2]. Thus, while figure 3.13(c) gave a hint that the graph of this function did not appear as we sketched it in figure 3.12(b), the "zoom" sketch in figure 3.13(d) clearly shows that the graph of this function is more like the one in figure 3.12(c) than the one in figure 3.12(b).

You may wish to approximate the coordinates of a particular point, such as a turning point or maximum point, on the graph. This is done by using the trace function and the cursor keys ◄ and ►. After a graph has been sketched, press TRACE and a blinking point will appear at the center of the graph. The coordinates of this point will be approximated on the bottom line of the display screen. Figure 3.13(e) shows the result of pressing TRACE after the graph in figure 3.13(d) was displayed.

To move the blinking point to the right, press the ► key; to move it to the left, press ◄. By repeatedly pressing or holding down the ► key, we can move the blinking point to, for instance, what appears to be a maximum point of the graph. There appear to be nine x-coordinates giving the same maximum y-value on this graph. Here, pressing TRACE leads to x-coordinates ranging from 1.3978947 to 1.5915789. The corresponding y-coordinates are actually different, varying from $y = 1.6447346$ for the lowest x-value to 1.6466173 for the highest x-value, with the largest y-coordinate being 1.6873759. None of these give pairs of coordinates which are exactly the same as the actual maximum point $(\frac{3}{2}, \frac{27}{16})$.

Casio fx-7700G Version

First enter Shift Cls EXE to clear the graph screen.

We begin by limiting the domain and range we want displayed on the screen. Press the Range key. Although it may show different numbers, the screen should show something like that in figure 3.13(f).

The domain to be shown on the screen is indicated by Xmin (the smallest value in the domain of interest), Xmax (the corresponding largest

(d)

160 Chapter 3 / Polynomial and Rational Functions

Figure 3.13 (continued)

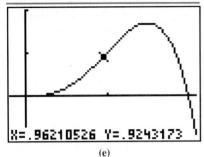
(e)

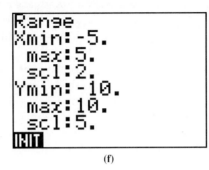
(f)

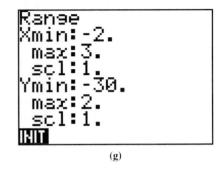
(g)

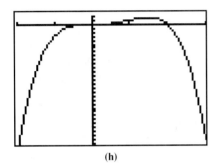
(h)

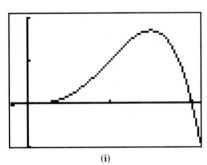
(i)

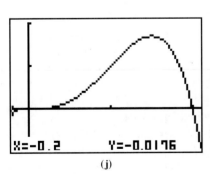
(j)

x-value), and Xscl (the x-axis's scale). The range is similarly restricted by Ymin, Ymax, and Yscl.

These values can be changed. Indeed, we saw in figures 3.12(b) and (c) that it is important in this example to be able to zoom in on the part of the domain corresponding to $-1 \leq x \leq 2$. To be safe, we focus on $[-2, 3]$ next. The cursor is initially at Xmin. To make -2 the minimum x-value, enter Shift (−) 2 EXE. Pressing the EXE key saves the entered value and moves the cursor to the next location. To leave a value the same, simply press the EXE key, and the cursor moves on. In this manner, we can make Xmax = 3, Xscl = 1, Ymin = −30, Ymax = 2, and Yscl = 1. The screen should now look like the one in figure 3.13(g).

Once the above settings have been completed, press the Range key twice. You are now able to input a formula describing the function you want graphed. To input $j(x) = -x^4 + 2x^3$, press the sequence of keys

Graph Shift (−) X,θ,T x^y 4 + 2 X,θ,T x^y 3 EXE

The result is the graph shown in figure 3.13(h).

If you wish to "take a closer look," the limitations on the domain and range can be changed by pressing the Range key and making the desired changes. Next, by pressing Range Range EXE, you can draw the graph with the new limitations on the domain and range. Figure 3.13(i) shows the graph sketched with domain $[-0.2, 2.1]$ and range limited to $[-1, 2]$. Thus, while figure 3.13(h) gave a hint that the graph of this function did not appear as we sketched it in figure 3.12(b), the "zoom" sketch in figure 3.13(i) clearly shows that the graph of this function is more like the one in figure 3.12(c) than the one in figure 3.12(b).

You may wish to approximate the coordinates of a particular point, such as a turning point or maximum point, on the graph. This is done by using the trace function and the cursor keys ◄ and ►. After a graph has been sketched, press Shift Trace and a blinking point will appear at the extreme left part of the graph. The coordinates of this point will be approximated on the bottom line of the display screen. Figure 3.13(j) shows the result of pressing Shift Trace after the graph in figure 3.13(i) was displayed. To move the blinking point to the right, press the ► key; to move it to the left, press ◄. By repeatedly pressing or holding down the ► key, we can move the blinking point to, for instance, what appears to be a maximum point of the graph.

Graphing Polynomial Functions

There appear to be two x-coordinates giving the same maximum y-value on this graph. Here, pressing Shift Trace leads to the x-coordinate 1.4882978 or 1.5127659. The corresponding y-coordinates are actually different, namely $y = 1.6868901$ and 1.6867582, respectively. Neither of these gives pairs of coordinates which are exactly the same as the actual maximum point $(\frac{3}{2}, \frac{27}{16})$. ◀

Exercises 3.3

In each of exercises 1–8, the graph of a function f is shown. State whether or not f is a continuous function. Then state whether or not the given graph is smooth.

1.

2.

3.

4.
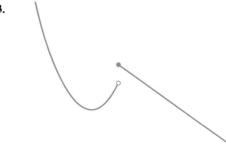

5.
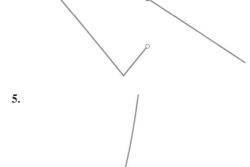

6.

7.
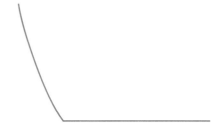

8.
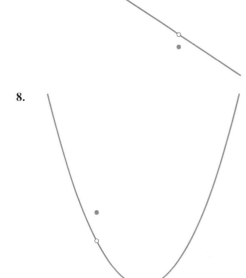

In exercises 9–12, explain why the Intermediate Value Theorem guarantees the existence of some c in the open interval (a, b) such that the given equation holds. If possible, determine how many such c there are and find them.

9. $a = 2, b = 3, 14c^3 - 7c^2 = 147$
10. $a = -1, b = 1, -2c^3 = 1$
11. $a = 0, b = 2, -3c^4 = -3$
12. $a = -2, b = -1, 6c^5 + c^4 + 170 = 0.$
13. Explain why the following information does not contradict the Intermediate Value Theorem:

$P(x) = x^2 - 4x + 4$ defines a continuous function, $P(0) = 4$, $P(6) = 16$, $P(1.5) = 0.25$, $0 < 1.5 < 6$, and 0.25 is not between 4 and 16.

14. Let $f(x) = \begin{cases} x+2 & \text{if } x \neq 3 \\ 7 & \text{if } x = 3 \end{cases}$

Verify that $f(1) < 5 < f(4)$ and that no c between 1 and 4 satisfies $f(c) = 5$. Explain why this does not contradict the Intermediate Value Theorem. Is f continuous? Why?

In exercises 15–22, use the Leading Term Test to determine the limiting tendencies of $P(x)$ as x moves without bound to the right or left.

15. $P(x) = \frac{2}{3}x^7 + \pi x^6 - \sqrt{2}x^5 + \frac{3}{7}x - 6$
16. $P(x) = -\frac{2}{3}x^9 + \pi x^8 + 4x + 1$
17. $P(x) = -\frac{1}{2}x^4 + 5x^3 - x$
18. $P(x) = \frac{3}{2}x^9 - x^5 + \frac{1}{7}x^2$
19. $P(x) = \frac{3}{2}x^8 + 3x^4 - x^2 - 3$
20. $P(x) = -6x^4 - \pi x^2 - 5$
21. $P(x) = -\frac{5}{2}x^3 - 8x^2 + 7x + \frac{1}{2}$
22. $P(x) = x^6 - x - 5$

23. Prove that if a real polynomial function P has odd degree, then P has a real root. (Hint: use the Leading Term Test to show that $P(x)$ has *different* limiting tendencies as x moves without bound to the right or the left. Then apply the Intermediate Value Theorem.)

24. Use the result in exercise 23 to show that each of the following polynomials has a real root:
  (a) $P(x) = x^5 - x + 2$ (b) $Q(x) = 7x + 5$
  (c) $R(x) = -5x + 7$
  (d) $S(x) = -5x^7 + \pi x^6 + \sqrt{2}x^5 + 3x + 1$

25. (a) (The converse of exercise 23 is false.) Find a real polynomial function P with even degree and at least one real root.
  (b) Find a real polynomial function P with even degree and no real roots.

26. Consider a real polynomial $P(x) = a_n x^n + \cdots + a_0$ of degree $n \geq 1$.
  ◆ (a) Give an intuitive argument explaining why $\dfrac{P(x)}{a_n x^n}$ has limiting tendency 1 as x increases without bound to either the right or the left. (Hint: the limiting tendency of $1/x$ in these cases is 0. What about the limit of $1/x^2$? of $1/x^3$?)
  (b) Explain how the result in (a) leads to a proof of the Leading Term Test.

27. Let $P(x) = a_n x^n + \cdots + a_0$ be an nth-degree real polynomial, with $a_0 \neq 0$, such that a_n and a_0 have opposite signs. Prove that P has at least one positive root. (Hint: adapt the hint for exercise 23. This time, you only need to consider one limit.)

28. Use the result in exercise 27 to show that each of the following polynomials has a positive root:
  (a) $P(x) = x^5 - x - 1$ (b) $R(x) = x^6 - x - 2$
  (c) $S(x) = -2x^8 + \pi x^7 + \sqrt{2}x^5 + 1$
  (d) $T(x) = -5x^7 + \pi x^6 + \sqrt{2}x^5 + 3x + 1$

29. Show that the converse of exercise 27 is false. In other words, find a real polynomial P having a positive root and such that the leading coefficient of P and the constant coefficient of P have the same sign. (Hint: think quadratic.)

30. Devise a criterion, in the spirit of exercise 27, that would guarantee the existence of a negative root. (Hint: the criterion must be given in cases, depending on whether n is odd.)

In exercises 31–40, graph each of the given polynomial functions. Assemble all available information (limiting tendencies, symmetries, transformations, intercepts) while doing so.

31. $P(x) = 2(x-1)^3 + 1$
32. $P(x) = \frac{1}{2}x^3 - x^2 + x$
33. $P(x) = -2x^3 - 2x$
34. $P(x) = \frac{1}{4}x^4 + 3$
35. $P(x) = 2x^3 - 4x - 1$
36. $P(x) = x^3 + 2x - 1$
37. $P(x) = \frac{1}{10}x^6 + \frac{1}{8}x^4 - 1$
38. $P(x) = -2x^3 + 6x^2 - 2x - 7$. (Hint: calculate $g(x-1)$, where g is as in example 3.22.)
39. $P(x) = -2x^4 + 2x^2 + 5$
40. $P(x) = \frac{1}{10}x^6 - \frac{1}{8}x^4 - 2$

In exercises 41–42, use the Intermediate Value Theorem to estimate a root of the given polynomial between 3 and 4 to within 0.005. (Hint: review exercise 35 in section 3.2 and example 3.21.)

◆ 41. $P(x) = 16x^3 - 56x^2 + x - 50$
◆ 42. $P(x) = 4x^4 - 15x^3 - x + 50$

43. One winter day, Hilkka finds that the temperature (in degrees Celsius) t hours after midnight is given by
$$T(t) = -\frac{t^3}{60} + \frac{t^2}{2} - 4t - 20.$$
  (a) What is a reasonable domain to consider for the function T?
  (b) Show that at some time between 6:00 A.M. and noon, the temperature was $-27°C$.

44. Use a graphing calculator to graph $f(x) = \frac{1}{240}(x^3 - 24x^2 - 240x)$.
  (a) Does the graph seem to be rising or falling at the right-hand side of the viewing screen?

(b) According to the Leading Term Test, what is the limiting tendency of $f(x)$ as x increases without bound?

(c) Are your answers in (a) and (b) compatible? If not, discuss how to use the graphing calculator so that a graph of the kind found in (a) can help you discover the tendency predicted in (b). (Hint: zoom out several times to get the big picture. To zoom out on a graphing calculator, change the range as described in example 3.24, or follow the following shortcuts. For a *TI-81*, press ZOOM 3 ENTER. For a Casio *fx-7700G*, press Shift Zoom F4.)

3.4 ROOTS OF A COMPLEX POLYNOMIAL

You have already learned formulas to find the roots of linear or quadratic polynomials. We shall now begin to explore methods for solving equations of the form $P(x) = 0$, where $P(x)$ is any polynomial. One reason for solving this equation is that, when P is a real polynomial, each real root r of $P(x)$ gives an x-intercept $(r, 0)$ of the graph of P. Initially, these x-intercepts will help us graph $y = P(x)$. Another reason, as you will see in section 3.5, is that x-intercepts are used as the key points for the sign charts used to solve inequalities involving P.

This text's third discussion of graphing calculators appears in this section in example 3.37.

Linear Equations and Quadratic Equations

First let's review some facts from chapter 0 and chapter 1. A linear, or first-degree, polynomial is one of the form $f(x) = ax + b$, where $a \neq 0$. If f is a real linear polynomial, its graph is a straight line with slope a and x-intercept $\left(-\frac{b}{a}, 0\right)$. The only root of such an $ax + b$ is $-\frac{b}{a}$.

A quadratic, or second-degree, polynomial is one of the form $f(x) = ax^2 + bx + c$, where $a \neq 0$. If f is a real quadratic polynomial, its graph is a parabola, and its roots r_1 and r_2 are described by the quadratic formula

$$r_1, r_2 = \frac{-b \pm \sqrt{b^2 - 4ac}}{2a}.$$

The discriminant, $b^2 - 4ac$, tells us about the nature of these roots r_1 and r_2. If the discriminant is positive, r_1 and r_2 are distinct real numbers; if the discriminant is 0, the roots r_1 and r_2 are equal real numbers; and if the discriminant is negative, r_1 and r_2 are unequal nonreal complex numbers.

The Fundamental Theorem of Algebra

Finding, or even approximating, the roots of polynomials of degree larger than 2 is often difficult. There is no single method that works well for all such polynomials. However, the next theorem states that the number of roots of a polynomial is restricted. Although this useful result was anticipated by the physicist d'Alembert, its first proof was achieved later, in 1799, by Gauss.

Chapter 3 / Polynomial and Rational Functions

The Fundamental Theorem of Algebra

> Let P be an nth-degree complex polynomial, where $n > 0$. Then P has at least one and at most n (distinct) complex roots.

Consider the quadratic polynomial $x^2 - 6x + 9$. Its discriminant is $(-6)^2 - 4(1)9 = 0$ and so it has two equal roots: $r_1 = r_2 = 3$. Notice that $x^2 - 6x + 9 = (x - 3)(x - 3)$. More generally, we shall see a benefit from counting the multiple root r each time $x - r$ appears in the factorization of $P(x)$. Namely, the Fundamental Theorem of Algebra can be restated as asserting that, *when one counts according to "multiplicities," P has exactly n roots.*

How can we find the n roots of an nth-degree polynomial? The next example illustrates a method, with $n = 3$.

▶ **EXAMPLE 3.25**

Determine the roots of $P(x) = x^3 + x^2 - 12x$.

Solution

We can factor x out of $P(x)$, getting

$$P(x) = x(x^2 + x - 12).$$

Also, $x^2 + x - 12$ factors as $(x + 4)(x - 3)$. So,

$$P(x) = x(x + 4)(x - 3).$$

By the Factor Theorem, the numbers 0, -4, and 3 are each roots of P. Since P has degree 3, the above discussion shows that these are all the roots of P. ◀

To find the n roots of any nth-degree polynomial $P(x) = a_n x^n + \cdots + a_1 x + a_0$, we may theoretically proceed as follows. Using the Factor Theorem, we know that if r_1 is a root of P, then $P(x) = (x - r_1)Q_1(x)$, where $Q_1(x)$ is the quotient obtained by dividing $P(x)$ by $x - r_1$. Remember that a_n is also the leading coefficient of $Q_1(x)$. Moreover, the degree of $Q_1(x)$ is $n - 1$. Because of this lowering of degree, it is traditional to refer to the equation $Q_1(x) = 0$ as the **depressed equation (relative to P and r_1).**

Next, suppose that r_2 is a root of $Q_1(x)$. Then $Q_1(x) = (x - r_2)Q_2(x)$, where a_n is still the leading coefficient of $Q_2(x)$, and the degree of $Q_2(x)$ is $n - 2$. Thus, we have

$$P(x) = (x - r_1)(x - r_2)Q_2(x).$$

This procedure can be repeated until we reach a quotient $Q_n(x)$ with degree 0. Since the leading coefficient of $Q_n(x)$ is a_n, we have $Q_n(x) = a_n$. We can therefore write

$$P(x) = a_n(x - r_1)(x - r_2) \cdots (x - r_n).$$

Notice that $r_1, r_2, \ldots, r_n$ are *all* the roots of $P(x)$. The point is that if r is a complex number different from the r_i's, then

$$P(r) = a_n(r - r_1)(r - r_2) \cdots (r - r_n)$$

Roots of a Complex Polynomial

is a product of $n + 1$ nonzero complex numbers and, hence, is nonzero. We have thus proved the first part, or existence assertion, in the following theorem. Exercise 53 will address the second part, or uniqueness assertion, in it.

Linear Factor Theorem

Let $P(x) = a_n x^n + a_{n-1} x^{n-1} + \cdots + a_2 x^2 + a_1 x + a_0$ be an nth-degree complex polynomial, for some $n > 0$. Then $P(x)$ can be factored into n linear factors of the form $x - r_i$ and one constant factor:

$$P(x) = a_n(x - r_1)(x - r_2)(x - r_3) \cdots (x - r_n).$$

Apart from reordering the factors, there is only one such factorization of $P(x)$. The numbers $r_1, r_2, r_3, \ldots, r_n$ appearing in it are the n roots of P.

Once we factor an nth-degree polynomial into its linear factors, it is easy to find its n roots. We shall next work two examples of this kind. You already practiced this skill in exercises 36–41 of section 3.1.

▷ **EXAMPLE 3.26**

List the roots of $P(x) = -\frac{1}{2}(x - \pi)(x + 2)(x - \frac{3}{2})^3(x + \sqrt{7})$.

Solution

Writing each linear factor in the form $x - r_i$ produces

$$P(x) = -\frac{1}{2}(x - \pi)(x - -2)(x - \frac{3}{2})^3(x - -\sqrt{7}).$$

From this, we see that the roots of P are $\pi, -2, \frac{3}{2}, \frac{3}{2}, \frac{3}{2}$, and $-\sqrt{7}$. ◁

▷ **EXAMPLE 3.27**

Given that $r_1 = -1$ is a root of $P(x) = x^3 - x^2 - 10x - 8$, find all the roots of P.

Solution

Factoring $x - r_1 = x + 1$ out of $P(x)$ produces

$$P(x) = x^3 - x^2 - 10x - 8 = (x + 1)(x^2 - 2x - 8).$$

The depressed polynomial $x^2 - 2x - 8$ factors as $(x - 4)(x + 2)$. Thus, we have $P(x) = x^3 - x^2 - 10x - 8 = (x + 1)(x - 4)(x + 2)$, from which we can see that the roots of P are -1, 4, and -2. ◁

Creating Polynomials from Known Roots

Exercises 47 and 48 of section 3.1 asked you to construct a polynomial with specified roots. Now, you can see that this amounts to using the Linear Factor Theorem in reverse. Suppose that you know that the roots of a certain nth-degree polynomial are $r_1, r_2, r_3, \ldots, r_n$. Then the polynomial must be of the form

$$a_n(x - r_1)(x - r_2) \cdots (x - r_n),$$

with only the leading coefficient a_n remaining undetermined.

▶ **EXAMPLE 3.28**   Determine a polynomial of degree 4 with roots 1, –2, 2, and $5i$.

Solution
$$P(x) = a_4(x - 1)(x + 2)(x - 2)(x - 5i).$$

The leading coefficient, a_4, can be any nonzero complex number. Each different value of a_4 produces a different satisfactory polynomial. By multiplying the linear factors, we see that

$$P(x) = a_4[x^4 + (-1 - 5i)x^3 + (-4 + 5i)x^2 + (4 + 20i)x - 20i].$$

The simplest such polynomial will be obtained by setting $a_4 = 1$. In this case,

$$P(x) = x^4 + (-1 - 5i)x^3 + (-4 + 5i)x^2 + (4 + 20i)x - 20i. \quad ◀$$

▶ **EXAMPLE 3.29**   Determine a polynomial of degree 5 with roots 2, 3, 3, $4 - i$, and $4 + i$.

Solution
$$P(x) = a_5(x - 2)(x - 3)^2(x - [4 - i])(x - [4 + i])$$
$$= a_5(x^5 - 16x^4 + 102x^3 - 322x^2 + 501x - 306), \quad \text{with } a_5 \neq 0.$$

If $a_5 = 1$, then $P(x) = x^5 - 16x^4 + 102x^3 - 322x^2 + 501x - 306$. Its graph is shown in figure 3.14. Notice that this graph has two x-intercepts, one for each of the distinct real roots of $P(x)$. ◀

Figure 3.14

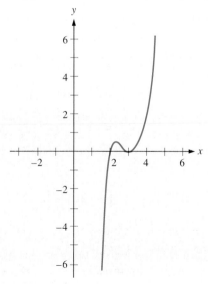

$P(x) = x^5 - 16x^4 + 102x^3 - 322x^2 + 501x - 306$

Roots of a Complex Polynomial

In example 3.29, the root 3 occurred twice. In cases such as these, we say that 3 is a root with multiplicity 2. In general, we have the following definition.

Root of Multiplicity k

If a root r of $P(x)$ "occurs" k times, in the sense that $(x - r)^k$ divides $P(x)$ but $(x - r)^{k+1}$ does not divide $P(x)$, then we say r is a **root of multiplicity k**.

The notion of multiplicity is useful in graphing. Suppose that a real number r is a root, with multiplicity k, of the real polynomial P. Then the nature of the graph of $y = P(x)$ near the x-intercept $(r, 0)$ is described in the following table:

k	Graph of $y = P(x)$
(a) k even	is "tangent" to, but does not cross, the x-axis at $(r, 0)$
(b) k odd, $k > 1$	is tangent to, and crosses, the x-axis at $(r, 0)$
(c) $k = 1$	crosses, but is not tangent to, the x-axis at $(r, 0)$

The cases indicated in the above table are illustrated in figures 3.15(a), (b), and (c).

Figure 3.15

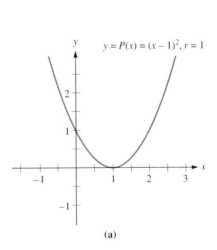
(a) $y = P(x) = (x - 1)^2$, $r = 1$

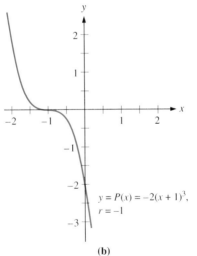
(b) $y = P(x) = -2(x + 1)^3$, $r = -1$

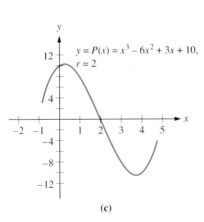
(c) $y = P(x) = x^3 - 6x^2 + 3x + 10$, $r = 2$

EXAMPLE 3.30

Determine a sixth-degree polynomial, having 1 as a root of multiplicity 4 and also having roots $2i$ and $-2i$.

Solution

$$P(x) = a_6(x - 1)^4(x - 2i)(x + 2i)$$
$$= a_6(x^6 - 4x^5 + 10x^4 - 20x^3 + 25x^2 - 16x + 4).$$

If we let $a_6 = 1$, then $P(x) = x^6 - 4x^5 + 10x^4 - 20x^3 + 25x^2 - 16x + 4$. The graph of P is shown in figure 3.16. The only x-intercept is at $(1, 0)$, corresponding to the only real root, 1.

Figure 3.16

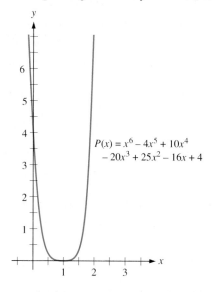

Roots of Real Polynomials

The polynomial $x^6 - 4x^5 + 10x^4 - 20x^3 + 25x^2 - 16x + 4$ in example 3.30 has real coefficients. Two of its roots, $2i$ and $-2i$, are complex conjugates of one another.

In example 3.29, the polynomial $x^5 - 16x^4 + 102x^3 - 322x^2 + 501x - 306$ has real coefficients. Two of its roots, $4 - i$ and $4 + i$, are complex conjugates of one another.

The roots of the quadratic polynomial $x^2 - 8x + 25$ are $4 + 3i$ and $4 - 3i$. This polynomial also has real coefficients, and its nonreal complex roots are conjugates of each other. We can generalize this pattern to any polynomial with real coefficients.

Conjugate Root Theorem

> Let $P(x)$ be a polynomial with real coefficients such that some nonreal complex number $z = a + bi$ is a root of P. (Here, a and b are real numbers and $b \neq 0$.) Then $\bar{z} = a - bi$, the conjugate of z, is also a root of P.

Proof: Write $P(x) = a_n x^n + a_{n-1} x^{n-1} + \cdots + a_2 x^2 + a_1 x + a_0$, where $a_0, a_1, a_2, \ldots, a_n$ are real numbers and $a_n \neq 0$. We are given that $P(z) = 0$, or

$$a_n z^n + a_{n-1} z^{n-1} + \cdots + a_2 z^2 + a_1 z + a_0 = 0.$$

If we take the conjugate of both sides using the properties of conjugation developed in section 1.5, we obtain

Roots of a Complex Polynomial

$$\overline{a_n z^n + a_{n-1} z^{n-1} + \cdots + a_2 z^2 + a_1 z + a_0} = \overline{0}$$

or

$$\overline{a_n z^n} + \overline{a_{n-1} z^{n-1}} + \cdots + \overline{a_2 z^2} + \overline{a_1 z} + \overline{a_0} = \overline{0}$$

$$\overline{a_n}\,\overline{z^n} + \overline{a_{n-1}}\,\overline{z^{n-1}} + \cdots + \overline{a_2}\,\overline{z^2} + \overline{a_1}\,\overline{z} + \overline{a_0} = \overline{0}$$

Now, the conjugate of any real number is itself. So, the preceding equation becomes

$$a_n \overline{z^n} + a_{n-1} \overline{z^{n-1}} + \cdots + a_2 \overline{z^2} + a_1 \overline{z} + a_0 = 0.$$

Since $\overline{z^n} = (\overline{z})^n$, the above equation becomes

$$a_n \overline{z}^n + a_{n-1} \overline{z}^{n-1} + \cdots + a_2 \overline{z}^2 + a_1 \overline{z} + a_0 = 0$$

or $P(\overline{z}) = 0$. In other words, $\overline{z}$ is a root of P, completing the proof.

A similar theorem concerns irrational roots of rational polynomials.

Theorem

Let $P(x)$ be a rational polynomial of positive degree such that $a + b\sqrt{c}$ is a root of P, where a, b, and c are rational numbers and $\sqrt{c}$ is irrational. Then $a - b\sqrt{c}$ is also a root of P.

▶ **EXAMPLE 3.31**

Suppose a sixth-degree polynomial with rational coefficients has $3 - 2i$, $5i$, and $6 + 2\sqrt{7}$ as three of its roots. What are its three remaining roots?

Solution

Since the coefficients are rational numbers, they are also real numbers. So, by the Conjugate Root Theorem, the conjugate of each of the nonreal roots is also a root. Hence, $3 + 2i$ and $-5i$ are roots. Also, since $\sqrt{7}$ is irrational, the previous theorem implies that $6 - 2\sqrt{7}$ is a root. We have thus found all (six) roots. ◀

▶ **EXAMPLE 3.32**

Notice that 5 is a root of $P(x) = 3x^3 - 14x^2 - 2x - 15$. What are the other roots of P?

Solution

Since 5 is a root, we know that $P(x) = (x - 5)Q(x)$, where Q is a second-degree polynomial. Using synthetic division, we obtain

```
5)   3   -14   -2   -15
          15    5    15
     ─────────────────────
     3     1    3     0
```

Thus, we see that $Q(x)$ is $3x^2 + x + 3$. Using the quadratic formula on the depressed equation, $Q(x) = 0$, we determine that $\dfrac{-1 + \sqrt{35}\,i}{6}$ and $\dfrac{-1 - \sqrt{35}\,i}{6}$ are the other two roots of P. ◀

Chapter 3 / Polynomial and Rational Functions

EXAMPLE 3.33 Notice that $4 - \sqrt{2}$ is a root of $T(x) = 3x^4 - 24x^3 + 57x^2 - 120x + 210$. What are the the other three roots of T?

Solution T is a rational polynomial and $\sqrt{2}$ is an irrational number. Since $4 - \sqrt{2}$ is a root, we know that $4 + \sqrt{2}$ is also a root. Thus,

$$T(x) = (x - 4 + \sqrt{2})(x - 4 - \sqrt{2})S(x)$$

where S is a second-degree polynomial. Multiplying $(x - 4 + \sqrt{2})$ and $(x - 4 - \sqrt{2})$, we obtain $x^2 - 8x + 14$. Thus, $S(x) = T(x) \div (x^2 - 8x + 14) = 3x^2 + 15$. The roots of S are $\pm\sqrt{5}i$. Thus,

$$T(x) = 3(x - 4 + \sqrt{2})(x - 4 - \sqrt{2})(x - \sqrt{5}i)(x + \sqrt{5}i).$$

The full list of the roots of T is $4 - \sqrt{2}, 4 + \sqrt{2}, \sqrt{5}i$, and $-\sqrt{5}i$. ◀

EXAMPLE 3.34 Two of the roots of $W(x) = 2x^6 + 9x^5 + 3x^4 + 39x^3 - 155x^2 - 378x - 180$ are $3i$ and $-3 + \sqrt{5}$. What are the other roots of W?

Solution W has rational (and, hence, real) coefficients. Thus, by the previous theorems, $-3i$ and $-3 - \sqrt{5}$ are also roots of W. This means that

$$W(x) = (x - 3i)(x + 3i)(x + 3 - \sqrt{5})(x + 3 + \sqrt{5})Y(x)$$

where Y is a second-degree polynomial. Using polynomial division, you can check that $Y(x) = 2x^2 - 3x - 5$, which factors as $(2x - 5)(x + 1)$. Thus, the roots of W are $3i, -3i, -3 - \sqrt{5}, -3 + \sqrt{5}, \frac{5}{2}$, and -1. ◀

Rational Root Test

Another useful theorem about roots reduces our search for the *rational* roots of a rational polynomial $P(x)$ to a finite list of candidates. Since multiplying by a common denominator of the coefficients of $P(x)$ does not affect the list of roots of $P(x)$, we assume that P has integral coefficients.

Rational Root Test

Let $P(x) = a_n x^n + a_{n-1} x^{n-1} + \cdots + a_2 x^2 + a_1 x + a_0$ be an nth-degree polynomial with integral coefficients such that $a_0 \neq 0$. Suppose that a rational number, written in lowest terms as $\frac{p}{q}$, is a root of P. Then

(a) p is an integral divisor of the constant term a_0, and
(b) q is an integral divisor of the leading coefficient a_n.

EXAMPLE 3.35 Find all the rational roots of $P(x) = 3x^4 - 22x^3 - 56x^2 + 102x - 27$. Graph $y = P(x)$.

Solution First, we shall find the "candidates" $\frac{p}{q}$ (in lowest terms). According to the Rational Root Test, p must be a divisor of -27 and q must be a divisor of 3.

possible values of p: ±1, ±3, ±9, ±27

possible values of q: ±1, ±3

possible rational roots: ±1, ±3, ±9, ±27, ±$\frac{1}{3}$

Let's begin testing the candidates. Using synthetic division to calculate $P(1)$, we get

$$\begin{array}{r|rrrrr} 1) & 3 & -22 & -56 & 102 & -27 \\ & & 3 & -19 & -75 & 27 \\ \hline & 3 & -19 & -75 & 27 & 0 \end{array}$$

Thus, 1 *is* a root. We shall test the next candidates on the (simpler) depressed equation, $3x^3 - 19x^2 - 75x + 27 = 0$. Notice that 1 is not a root of the depressed equation (and so 1 is not a multiple root of the given polynomial) since $3(1)^3 - 19(1)^2 - 75(1) + 27 = -64 \neq 0$. Similarly, or using synthetic division, you can check that neither -1 nor 3 is a root. However, when we test -3, we obtain

$$\begin{array}{r|rrrr} -3) & 3 & -19 & -75 & 27 \\ & & -9 & 84 & -27 \\ \hline & 3 & -28 & 9 & 0 \end{array}$$

and so -3 is a root. Now we can apply the quadratic formula to the new depressed equation, $3x^2 - 28x + 9 = 0$, to obtain the final two roots: 9 and $\frac{1}{3}$. Thus, the roots of $3x^4 - 22x^3 - 56x^2 + 102x - 27$ are 1, -3, 9, and $\frac{1}{3}$, and these all happen to be rational numbers. The Intermediate Value Theorem and the Leading Term Test permit us to sketch $y = P(x)$, as in figure 3.17.

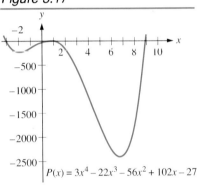

Figure 3.17

$P(x) = 3x^4 - 22x^3 - 56x^2 + 102x - 27$

EXAMPLE 3.36

Find all the rational roots of $P(x) = x^5 - \frac{11}{2}x^4 + 13x^3 - \frac{33}{2}x^2 + 11x - 3$ and graph $y = P(x)$.

Solution

P is not an integral polynomial. However, since it is a rational polynomial, we can obtain an integral polynomial having the same roots by multiplying P by 2, the lowest common denominator of its coefficients. We thus obtain the integral polynomial $T(x) = 2P(x) = 2x^5 - 11x^4 + 26x^3 - 33x^2 + 22x - 6$.

We begin by finding the "candidates" $\frac{p}{q}$ (in lowest terms). From the Rational Root Test, we know that p must be a divisor of -6 and q must be a divisor of 2. Thus we have the following:

possible values of p: ±1, ±2, ±3, ±6

possible values of q: ±1, ±2

possible rational roots: ±1, ±2, ±3, ±6, ±$\frac{1}{2}$, ±$\frac{3}{2}$

Using synthetic division to test the candidates, we obtain the following for $T(1)$:

$$\underline{1)\quad 2\quad -11\quad 26\quad -33\quad 22\quad -6}$$
$$\,\,\,\,2\quad -9\quad 17\quad -16\quad 6$$
$$2\quad -9\quad 17\quad -16\quad 6\quad 0$$

Thus, 1 *is* a root. We shall test the next candidate on the depressed equation, $2x^4 - 9x^3 + 17x^2 - 16x + 6 = 0$. We test 1 again to determine if it is a multiple root of the given polynomial, and obtain

$$\underline{1)\quad 2\quad -9\quad 17\quad -16\quad 6}$$
$$\,\,\,\,2\quad -7\quad 10\quad -6$$
$$2\quad -7\quad 10\quad -6\quad 0$$

Thus, 1 is a multiple root, with multiplicity at least 2.

Now test for roots on the new depressed equation, $2x^3 - 7x^2 + 10x - 6 = 0$. We find that 1 is not a root of *this* equation. So, the multiplicity of 1 as a root of T is exactly 2. Checking, we also determine that neither 2 nor 3 is a root. However, when we test $\frac{3}{2}$, we obtain

$$\underline{\tfrac{3}{2})\quad 2\quad -7\quad 10\quad -6}$$
$$\phantom{\tfrac{3}{2})\quad 2\quad }\,\,\,\,3\quad -6\quad 6$$
$$\phantom{\tfrac{3}{2})\,\,}2\quad -4\quad 4\quad 0$$

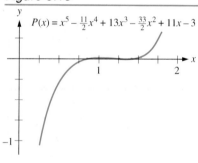

Figure 3.18

and so $\frac{3}{2}$ is a root. We now apply the quadratic formula to the newest depressed equation, $2x^2 - 4x + 4 = 0$, and determine that the final two roots are $1 + i$ and $1 - i$.

We have now established that the roots of $T(x) = 2x^5 - 11x^4 + 26x^3 - 32x^2 + 22x - 6$, and hence of $P(x)$, are $1, 1, \frac{3}{2}, 1 + i$, and $1 - i$. Since 1 is a root of multiplicity 2, the graph of $y = P(x)$ is tangent to, but does not cross, the x-axis at $(1, 0)$. The graph does cross the x-axis at $(\frac{3}{2}, 0)$. The graph of $y = P(x)$ is in figure 3.18. ◀

Section 3.5 will give a quick way to shorten the search for rational roots. This depends on a result called Descartes' Rule of Signs. For example 3.35, it would tell us that the given polynomial has one negative (real) root and either one or three positive (real) roots. For example 3.36, it would tell us that the given polynomial has either one, three, or five positive roots and no negative roots.

Graphing Calculator Activity

▶ **EXAMPLE 3.37**

Consider the functions f and g given by the cubic polynomials $f(x) = x^3 - 6x^2 + 12x - 8$ and $g(x) = x^3 - 6x^2 + 11.9975x - 7.995$. Use a graphing calculator to estimate the number and location of the real roots of f and g.

Solution

Figures 3.19(a) and (b) show the graphs of f and g obtained with a Texas Instruments *TI-81*. Figures 3.19(c) and (d) show the graphs of f and g obtained with a Casio *fx-7700G*.

Roots of a Complex Polynomial

In each case, the "flat" part of the graph near $x = 2$ makes it appear as if the function has exactly one real root. This is, in fact, the case for f, since $f(x) = (x - 2)^3$. However, g has three distinct real roots. You can check that 1.95, 2, and 2.05 are each roots of g. In effect, figures 3.19(b) and (d) fail to distinguish among these three numbers. Zoom in by making Xmin = 1.85, Xmax = 2.15, Xscl = 0.05, Ymin = –0.0001, Ymax = 0.0001, and Yscl = 0.00005. This produces the graphs of f and g, in figure 3.19(e) using a *TI-81*, and in figure 3.19(f) using a Casio *fx-7700G*. Now, you can really see how the function g oscillates as x varies from 1.9 to 2.1. ◀

Figure 3.19

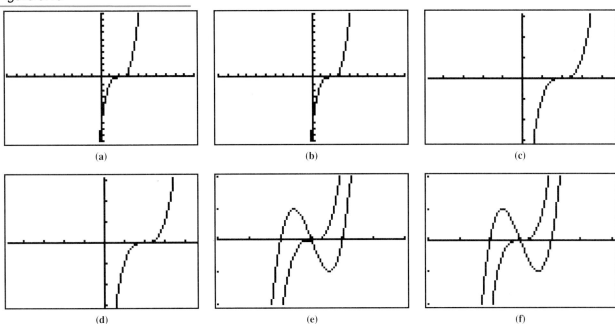

(a)   (b)   (c)

(d)   (e)   (f)

One moral to be drawn from the graphs of f and g in example 3.37 is the following. When you find a "flat" part of a graph near what appears to be a root, zooming in or tracing will help determine whether several nearby roots can be found within the flat part. According to the Fundamental Theorem of Algebra, you know at most how many real roots the given polynomial can have, but the above example of f shows that this upper bound may not be realized no matter how much you zoom in.

Exercises 3.4

In exercises 1–8, state how many complex roots the given polynomial has, when counting roots according to their multiplicities. In exercises 3–8, also determine how many distinct complex roots the given polynomial has.

1. $9x^7 - \frac{1}{2}x^6 - \pi x^5 + \sqrt{2}x^4 - \frac{3}{2}$
2. $x^4 - 4x^3 + x^2 - 1$
3. $x(x - 3)^2(x + 4)$
4. $(4x^2 - 4x + 1)^3$
5. $(4x^2 - 4x + 1)(x^2 + 4x + 4)^4$
6. $x^3 + 3x^2 + 2x$
7. $\frac{3}{2}x - \sqrt{2}$
8. $x^3(2x - 1)^2(x^2 + 1)$

Chapter 3 / Polynomial and Rational Functions

In exercises 9–12, find the depressed equation relative to the given $P(x)$ and r_1.

9. $P(x) = 2x^3 - 3x^2 + 10x - 15$, $r_1 = \frac{3}{2}$
10. $P(x) = 3x^3 + 6x^2 - 15x - 30$, $r_1 = -2$
11. $P(x) = 5x^4 - 10\sqrt{2}x^3 + 14x^2 + 2\sqrt{2}x - 3$, $r_1 = \sqrt{2} - i$
12. $P(x) = 5x^4 - 10x^3 + 26x^2 - 2x + 5$, $r_1 = 1 - 2i$

In exercises 13–18, find all the roots of the given polynomial $P(x)$ and determine each root's multiplicity. In view of these multiplicities, what conclusions can you draw about the graph of $y = P(x)$ near the corresponding x-intercepts?

13. $P(x) = (x^3 - 16x)^5$
14. $P(x) = 3(x + 4)(2x - 3)^4(x^2 + 3)$
15. $P(x) = -8(x - 2)(3x + 1)^5(x^2 + \pi)$
16. $P(x) = x^2(x - 2)(3x^2 - 7x + 6)$
17. $P(x) = (x^3 - 8)(x^2 + 1)(x^2 - 1)^2(x - 1)(x + 1)$
18. $P(x) = (x^3 + 1)(x^2 - 1)^2 x$
19. Verify that -2 is a root of $P(x) = x^3 + 3x^2 + 3x + 2$ and hence determine all the roots of $P(x)$. (Hint: think depressed.)
20. Verify that $3i$ is a multiple root of $P(x) = 2x^5 - x^4 + 36x^3 - 18x^2 + 162x - 81$ and hence determine all the roots of $P(x)$.
21. Find *the* cubic polynomial $P(x)$ with roots -2, 1, and 3 such that $P(5) = 2$. (Hint: use the Linear Factor Theorem. What is a_3?)
22. Prove that *no* fourth-degree polynomial $P(x)$ exists with all the following properties: 1 is a root, $P(0) = 4$, $P(4) = 3$, $P(-1) = 2$, $P(-2) = 6$, and $P(-3) = -8$. (Hint: if $Q(x) = 0$ is the depressed equation, try to solve a simultaneous set of five equations for the four unknown coefficients of Q.)
23. Prove that *no* fourth-degree real polynomial can have all of $1 - i$, $2 + 3i$, and -7 as roots. (Hint: too many roots!)
24. Find at least two different fourth-degree complex polynomials each having $1 - i$, $2 + 3i$, and -7 as roots. (Hint: the Conjugate Root Theorem does not hold for arbitrary complex polynomials.)

In exercises 25–32, find a real (and, if possible, rational) polynomial of degree n, having (at least) the specified numbers as roots.

25. $n = 3$; roots $-\frac{1}{2}$, $3 - \sqrt{5}$
26. $n = 4$; roots $2 + 3i$, $5 + \sqrt{3}$
27. $n = 4$; roots 2, i, $5 - \sqrt{3}$
28. $n = 3$; roots 3, -3, $\sqrt{2}$
29. $n = 4$; roots -4 (with multiplicity 2), $1 - \sqrt{5}$
30. $n = 6$; roots -2 (with multiplicity 4), $5 - i$
31. $n = 6$; roots $\frac{1}{2}$, $\sqrt{7}$, $\frac{2}{3} + i$
32. $n = 4$; roots $\frac{2}{3} + \sqrt{5}$, $1 - \frac{7}{8}i$
33. Let $P(x)$ be an nth-degree polynomial with roots $r_1, \ldots, r_n$. If k is any nonzero complex number, prove that the roots of $P\left(\frac{x}{k}\right)$ are $kr_1, \ldots, kr_n$. (Hint: write $P(x) = a_n(x - r_1)\cdots(x - r_n)$. What are the roots of $k^n P\left(\frac{x}{k}\right)$?)
34. Let $P(x) = x^3 - 4x^2 + 9$, with roots r_1, r_2, and r_3. Using exercise 33, find polynomials Q, R, S, and T:
   (a) $Q(x)$ with roots $-r_1$, $-r_2$, and $-r_3$
   (b) $R(x)$ with roots $2r_1$, $2r_2$, and $2r_3$
   (c) $S(x)$ with roots $\frac{1}{3}r_1$, $\frac{1}{3}r_2$, and $\frac{1}{3}r_3$
   (d) $T(x)$ with roots $-\frac{2}{7}r_1$, $-\frac{2}{7}r_2$, and $-\frac{2}{7}r_3$
35. Let $P(x)$ be an nth-degree polynomial with roots $r_1, \ldots, r_n$.
   (a) Show that $P(x)P(-x)$ is a $2n$th-degree polynomial whose terms of odd degree each have coefficient 0.
   (b) Using (a), write $P(x)P(-x)$ as $Q(x^2)$ for a suitable nth-degree polynomial Q. Then prove that $r_1^2, r_2^2, \ldots, r_n^2$ are the roots of Q. (Hint: if $P(x) = a_n(x - r_1)\cdots(x - r_n)$, observe that $P(x)P(-x) = (-1)^n a_n^2 (x^2 - r_1^2)\cdots(x^2 - r_n^2)$.)
36. Let r_1, r_2, and r_3 denote the roots of $x^3 - 3x^2 + 4x - 7$.
   (a) Using exercise 35, find a cubic polynomial with roots r_1^2, r_2^2, and r_3^2.
   (b) Find a cubic polynomial with roots r_1^4, r_2^4, and r_3^4.
37. Let r_1 and r_2 denote the roots of $x^2 - 6x + 13$.
   (a) Use exercise 35 to show that r_1^2 and r_2^2 are roots of $x^2 - 10x + 169$.
   (b) Use the quadratic formula to verify directly your answer to (a).
38. (a) Explain why $x^4 + 8x^2 + 16$ has no real roots. (Hint: if r is any real number, then $r^2 \geq 0$.)
   (b) Using (a) and exercise 35, show that $x^4 - 4x^3 + 8x^2 - 8x + 4$ has no real roots.
39. Suppose that $P(x)$ is a real polynomial having a nonreal root $a + bi$ (for some real numbers a and b). Prove that $x^2 - 2ax + (a^2 + b^2)$ is a factor of $P(x)$. (Hint: combine the Conjugate Root Theorem with the Linear Factor Theorem.)
40. Observe that $2 + i$ is a root of $P(x) = 2x^4 - 9x^3 + 17x^2 - 17x + 15$. Using exercise 39, find all the other roots of P.

In exercises 41–50, use the Rational Root Test to list all the "candidates" for rational roots of the given polynomial P. In exercises 47–50, actually find all the roots of P and hence graph $y = P(x)$.

41. $P(x) = 6x^3 - 8x^2 + 9x - 42$
42. $P(x) = x^4 + 72x - 2$
43. $P(x) = 3x^6 + 6x^3 + x^2 - x$. (Hint: find $Q(x)$ so that $P(x) = xQ(x)$.)
44. $P(x) = -105x^7 + 18x^5 + 27x^3 - 7x^2$

45. $P(x) = \frac{1}{15}x^6 - \frac{2}{3}x^4 + \frac{3}{5}x^2 - 5x + 2$
46. $P(x) = \frac{2}{3}x^5 + \frac{1}{2}x^4 - 14$
47. $P(x) = 3x^4 - 7x^3 + 5x^2 - 7x + 2$
48. $P(x) = 20x^5 - 52x^4 + 61x^3 - 16x^2 - 10x + 4$
49. $P(x) = 7x^6 + 18x^5 - 9x^4 - 56x^3 - 39x^2 + 6x + 9$
50. $P(x) = 3x^3 - 7x^2 - 5x + 25$
51. Give an example of a real polynomial P, having a rational root, such that at least one coefficient of P is irrational.
52. (a) Factor $P(x) = x^6 + x^4 - x^2 - 1$ "as fully as possible" using real polynomial factors (that is, find all the irreducible factors of P over the real numbers).
    (b) Use (a) to determine the "multiplicity" (in some intuitive sense) of $x^2 + 1$ as a factor of $P(x)$.
    (c) What is the "multiplicity" of $x^2 + 4$ as a factor of $x^6 + 12x^4 + 48x^2 + 64$?
    (d) What is the "multiplicity" of $x^2 + x + 1$ as a factor of $5x^3 + 3x^2 + 3x - 2$?

◆ 53. (Proof of the uniqueness assertion in the Linear Factor Theorem) Suppose that an nth-degree polynomial $P(x)$ has been factored as $P(x) = a_n(x - r_1) \cdots (x - r_n)$. Prove the following assertions.
    (a) Each r_i is a root of P.   (b) Each root of P is an r_i.
    (c) (Well-definedness of multiplicity) Suppose that $P(x) = (x - r)^k Q(x) = (x - r)^{k+1} H(x)$ for some complex number r, positive integer k, and polynomials $Q(x)$, $H(x)$. Prove that $x - r$ is a factor of $Q(x)$. (Hint: express $P(x) \div (x - r)^k$ two ways and apply the Factor Theorem.)

3.5 COUNTING AND BOUNDING REAL ROOTS; HIGH-DEGREE POLYNOMIAL INEQUALITIES

In this section, we shall explain a method for estimating the number of real roots of a real polynomial. We shall also determine when and where we can stop searching for additional positive or negative real roots. (This skill permits a sensible choice of Xmin and Xmax on a graphing calculator.) As before, we seek information about roots to help us graph a polynomial function or solve an equation or inequality.

As usual, we shall write the terms of a polynomial in descending powers of x. We shall say that a **variation of sign** occurs whenever two successive nonzero terms have opposite signs. We give two examples indicating how to count such variations and then use this counting skill to estimate numbers of positive (real) roots.

▶ **EXAMPLE 3.38**

Determine the number of variations of sign for the polynomial $P(x) = \frac{9}{5}x^7 + \pi x^6 - 12x^5 - 5x^3 + 7x - \sqrt{2}$.

Solution

$$\frac{9}{5}x^7 + \pi x^6 - 12x^5 - 5x^3 + 7x - \sqrt{2}$$

from + to −,   from − to +,   from + to −

As indicated, there are three variations of sign. Notice that since the x^7 term and the x^6 term both have positive coefficients, there is no variation of sign between these two. The x^5 and x^3 terms are consecutive nonzero terms, but they both have negative coefficients, and so there is no variation of sign between them. On the other hand, the x^3 and x terms are consecutive nonzero terms whose coefficients have opposite signs, and so there *is* a variation of sign between them. ◀

▶ **EXAMPLE 3.39**

Determine the number of variations of sign for each of the following polynomials: $W(x) = 4x^3 - 3x - 2$, $Y(x) = 5x^6 - 7x^4 - \frac{3}{7}x^2 - 2x + 1$, and $Z(x) = x^{11} - x^6 + 2x$.

Solution

$$W(x) = 4x^3 - 3x - 2 \qquad \text{1 variation of sign}$$

$$Y(x) = 5x^6 - 7x^4 - \tfrac{3}{7}x^2 - 2x + 1 \qquad \text{2 variations of sign}$$

$$Z(x) = x^{11} - x^6 + 2x \qquad \text{2 variations of sign}$$

◀

Descartes' Rule of Signs

In section 3.4, we saw how to generate polynomials having a given set of roots. That procedure is used in the next example.

EXAMPLE 3.40

Create a polynomial having the given set of roots, and determine the number of variations of sign for the created polynomial. (For convenience, let the leading coefficient of the polynomial be 1.)

Solutions

	Roots	Polynomial	Number of Variations of Sign
(a)	2, –3	$A(x) = (x-2)(x+3) = x^2 + x - 6$	1
(b)	2, –3, 1	$B(x) = (x-2)(x+3)(x-1) = x^3 - 7x + 6$	2
(c)	2, –3, 1, 3	$C(x) = x^4 - 3x^3 - 7x^2 + 27x - 18$	3
(d)	2, –3, 1, 3, –1	$D(x) = x^5 - 2x^4 - 10x^3 + 20x^2 + 9x - 18$	3
(e)	2, –3, 1, 3, –1, 1	$E(x) = x^6 - 3x^5 - 8x^4 + 30x^3 - 11x^2 - 27x + 18$	4
(f)	1, 2, i, $-i$	$F(x) = x^4 - 3x^3 + 3x^2 - 3x + 2$	4

◀

Viewing calculations like those in example 3.40, René Descartes, the man primarily responsible for developing analytic geometry, noticed a pattern between the number of variations of sign and the number of real roots of a real polynomial. This pattern is expressed in the following theorem, which bears his name.

Descartes' Rule of Signs

Let $P(x)$ be a real polynomial of degree $n > 0$. Then,

(i) Counted with multiplicity, the number of positive roots of $P(x)$ is either the same as the number of variations of sign in $P(x)$ or less than that number of variations of sign by a positive even integer.

(ii) Counted with multiplicity, the number of negative roots of $P(x)$ is either equal to the number of variations of sign in $P(-x)$ or less than that number of variations of sign by a positive even integer.

Counting and Bounding Real Roots; High-Degree Polynomial Inequalities

To illustrate Descartes' Rule of Signs, look again at example 3.40. The polynomials $A(x)$, $B(x)$, $C(x)$, $D(x)$, and $E(x)$ each have the same number of variations of sign as, counted with multiplicity, positive roots. The polynomial $F(x)$ has four variations of sign but only two positive roots.

> **EXAMPLE 3.41**

What does Descartes' Rule of Signs tell us about the number of positive roots and the number of negative roots of the equation $3x^5 + 7x^4 + 2x^3 - 5x - 4 = 0$?

Solution

The polynomial $P(x) = 3x^5 + 7x^4 + 2x^3 - 5x - 4$ exhibits one variation of sign, and so P has one positive root.

$$P(-x) = 3(-x)^5 + 7(-x)^4 + 2(-x)^3 - 5(-x) - 4$$
$$= -3x^5 + 7x^4 - 2x^3 + 5x - 4.$$

There are four variations of sign in $P(-x)$, and so $P(x)$ has either four, two, or zero negative roots.

Since $P(x)$ has 5 complex roots, there are the following possibilities:

one positive root and four negative roots;

one positive root, two negative roots, and two nonreal complex roots;

one positive root, no negative roots, and four nonreal complex roots. ◂

Advanced algebraic reasoning (Sturm's Theorem), in conjunction with the formulas of differential calculus, definitely resolves which of these possibilities holds. Our needs are generally met by the less precise, but more accessible, result of Descartes.

> **EXAMPLE 3.42**

What does Descartes' Rule of Signs tell you about the real roots of $P(x) = x^{10} + 3x^8 - 7x^6 - 9x^4 - 5x^2 = 0$?

Solution

$P(x) = x^2 Q(x)$, where $Q(x) = x^8 + 3x^6 - 7x^4 - 9x^2 - 5$ and $Q(0) = -5 \neq 0$. Hence 0 is a root of P with multiplicity 2, and the nonzero roots of P are the same as the roots of Q. As Q presents one variation of sign, Q (and, hence, P) has exactly one positive root.

$$Q(-x) = (-x)^8 + 3(-x)^6 - 7(-x)^4 - 9(-x)^2 - 5$$
$$= x^8 + 3x^6 - 7x^4 - 9x^2 - 5.$$

There is one variation of sign in $Q(-x)$, and so Q (and, hence, P) has exactly one negative root. Hence $x^{10} + 3x^8 - 7x^6 - 9x^4 - 5x^2$ has 0 as a root with multiplicity 2, one positive root, one negative root, and six nonreal complex roots. ◂

Examples 3.41 and 3.42 illustrated that if $P(x)$ has exactly one variation of sign, then P has (exactly) one positive root. Example 3.42 illustrated that if $P(-x)$ has exactly one variation of sign, then P has exactly one negative root. Although examples were not included, if $P(x)$ (or $P(-x)$) has no variations in sign, then P has no positive (or negative) roots.

Bounds for Real Roots

A real number U is an **upper bound** for the real roots of a polynomial P if no root of P is greater than U. A real number L is a **lower bound** for the real roots of P if no root of P is less than L. We shall next state, without proof, sufficient conditions for determining such U and L.

Theorem

> Let $P(x) = a_n x^n + a_{n-1} x^{n-1} + \cdots + a_2 x^2 + a_1 x + a_0$ be a real polynomial such that $a_n > 0$. Suppose you carry out the synthetic division corresponding to $P(x) \div (x - k)$. If $k \geq 0$ and each term in the bottom row of the synthetic division is nonnegative, then k is an upper bound for the real roots of $P(x)$. If $k < 0$ and the terms in the bottom row of the synthetic division alternate between positive (or zero) and negative (or zero), then k is a lower bound for the real roots of $P(x)$.

▶ **EXAMPLE 3.43** Find all the roots of the rational polynomial $P(x) = \frac{2}{3}x^3 - \frac{1}{2}x^2 + \frac{2}{3}x - \frac{1}{2}$.

Solution Before applying the Rational Root Test, we must multiply P by a common multiple of its coefficient's denominators. Multiplying by 6 produces the polynomial $Q(x) = 4x^3 - 3x^2 + 4x - 3$, with integer coefficients.

If $\frac{p}{q}$ is a rational root in lowest terms, then p is a divisor of -3 and q is a divisor of 4. Thus, we have the following possibilities:

possible values of p: $\pm 1, \pm 3$

possible values of q: $\pm 1, \pm 2, \pm 4$

possible rational roots: $\pm 1, \pm \frac{1}{2}, \pm \frac{1}{4}, \pm 3, \pm \frac{3}{2},$ and $\pm \frac{3}{4}$.

Using Descartes' Rule of Signs, we see that Q (and, hence, P) has no negative roots and either one or three positive roots. If synthetic division is used to check 1, we get

$$\begin{array}{r|rrrr} 1) & 4 & -3 & 4 & -3 \\ & & 4 & 1 & 5 \\ \hline & 4 & 1 & 5 & 2 \end{array} \leftarrow \text{all these terms are nonnegative}$$

From this, we see that 1 is not a root and, moreover, that 1 is an upper bound for the real roots. Thus, 3 and $\frac{3}{2}$ are not roots of Q. So, the only other possible positive rational roots are $\frac{1}{4}, \frac{1}{2},$ and $\frac{3}{4}$. Checking, we see that $\frac{3}{4}$ is a root. In fact,

$$4x^3 - 3x^2 + 4x - 3 = (x - \tfrac{3}{4})(4x^2 + 4)$$
$$= 4(x - \tfrac{3}{4})(x^2 + 1) = 4(x - \tfrac{3}{4})(x - i)(x + i).$$

Thus, the roots of Q (and, hence, P) are $\frac{3}{4}, i,$ and $-i$. ◀

Counting and Bounding Real Roots; High-Degree Polynomial Inequalities

Notice that, in the above example, by using the test for upper bounds of roots, we eliminated several "candidates," thus speeding up our search for roots. Although Descartes' Rule of Signs made it unnecessary to find a lower bound for the real roots of $Q(x)$, note that the synthetic division

$$
\begin{array}{r|rrrr}
-\frac{1}{2} & 4 & -3 & 4 & -3 \\
 & & -2 & \frac{5}{2} & -\frac{13}{4} \\
\hline
 & 4 & -5 & \frac{13}{2} & -\frac{25}{4}
\end{array} \leftarrow \text{terms alternate in sign}
$$

shows that $-\frac{1}{2}$ is such a lower bound. This calculation would have directly eliminated the candidates $-\frac{1}{2}, -\frac{3}{4}, -1, -\frac{3}{2}$, and -3.

It is important to notice that the criteria for the upper bounds and lower bounds are only *sufficient* conditions. This means that any number that satisfies the upper (or lower) bound criterion is an upper (or lower) bound, but there may be upper (or lower) bounds that do not satisfy the criterion. For instance, $U = 3$ does *not* satisfy the upper bound criterion for the polynomial $R(x) = x^2 - 4x + 4$ since the bottom row of the synthetic division

$$
\begin{array}{r|rrr}
3 & 1 & -4 & 4 \\
 & & 3 & -3 \\
\hline
 & 1 & -1 & 1
\end{array}
$$

has a negative entry. However, $R(x) = (x - 2)^2$ has 2 as its only real root (apart from considerations of multiplicities), and so 3 *is* an upper bound for the real roots of $R(x)$.

Polynomial Inequalities

Section 1.1 gave some techniques for solving quadratic inequalities. This subsection generalizes these to higher-degree polynomials.

▶ **EXAMPLE 3.44** Solve $(x + 4)(2 - x)(2x + 3) < 0$.

Solution In order to help us get to the new ideas, the polynomial on the left-hand side has already been factored. Its roots are -4, 2, and $-\frac{3}{2}$. We mark each of these key points on a number line. Since -4 is a root, we draw a vertical line at $x = -4$. The + and − signs in figure 3.20(a) indicate that the factor $x + 4$ is positive for $x > -4$ and negative for $x < -4$. Since -4 is not a solution of the inequality, the circle at $x = -4$ is not filled in.

Since 2 is also a root, we draw another vertical line at $x = 2$. The row of + and − signs on the line labeled $2 - x$ in figure 3.20(b) indicates that the factor $2 - x$ is negative for $x > 2$ and positive for $x < 2$.

Finally, the signs in the row labeled $2x + 3$ in figure 3.20(b) indicate that the factor $2x + 3$ is positive for $x > -\frac{3}{2}$ and negative for $x < -\frac{3}{2}$.

Since the given polynomial is the product of the three factors $x + 4$, $2 - x$, and $2x + 3$, it is negative whenever an odd number of these factors is negative. Examination of the sign chart in figure 3.20(b) shows that this occurs on the intervals $(-4, -\frac{3}{2})$ and $(2, \infty)$. Accordingly, the solution set is

Chapter 3 / Polynomial and Rational Functions

Figure 3.20

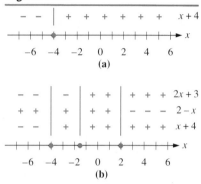

$(-4, -\frac{3}{2}) \cup (2, \infty) = \{x: -4 < x < -\frac{3}{2} \text{ or } x > 2\}.$ ◀

Example 3.44 points toward the following method for solving higher-degree polynomial inequalities. In order to determine the intervals of real numbers on which a given inequality is valid, we first solve the equation $P(x) = 0$ associated with the inequality. To do this, factor $P(x)$ into real linear and irreducible quadratic factors. (Remember that an irreducible quadratic is a real quadratic with negative discriminant—that is, a real quadratic with no real roots.) Each real root of P determines a key point of the inequality. **A sign chart is used to indicate where each factor is positive and where it is negative.** In this way, we determine where P is positive and where P is negative. The next three examples further illustrate the procedure.

▶ **EXAMPLE 3.45** Solve $(x - 3)(2x + 1)(x^2 + 3x + 5) \geq 0$.

Solution Let $P(x) = (x - 3)(2x + 1)(x^2 + 3x + 5)$. To determine the key points, we solve the associated equation $P(x) = 0$. Using the linear factors, we see that two roots of this equation are at $x = 3$ and at $x = -\frac{1}{2}$. Since the discriminant of $x^2 + 3x + 5$ is $3^2 - 4(1)5 = -11 < 0$, the quadratic factor has no real roots. In fact, this factor is always positive. (How do we know? Hint: chapter 0 and sections 2.3, 3.3.)

A sign chart is given in figure 3.21(a). The row of + signs indicates that the quadratic factor is always positive. The row labeled $x - 3$ indicates that the factor $x - 3$ is positive for $x > 3$ and negative for $x < 3$. In a similar manner, + and − signs indicate that $2x + 1$ is positive for $x > -\frac{1}{2}$ and negative for $x < -\frac{1}{2}$. Since the inequality being solved includes the case of equality, the circles at each key point are filled in. Now, $P(x)$ is positive whenever there are an even number of negative factors. By examining the sign chart in figure 3.21(a), we see that this occurs in the intervals $(-\infty, -\frac{1}{2})$ and $(3, \infty)$. Accordingly, the solution set is

$$(-\infty, -\tfrac{1}{2}] \cup [3, \infty) = \{x: x \leq -\tfrac{1}{2} \text{ or } x \geq 3\}.$$

Figure 3.21

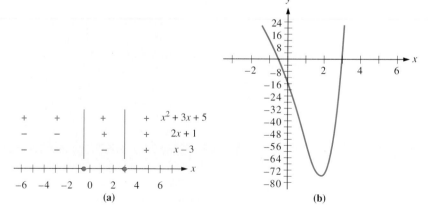

Counting and Bounding Real Roots; High-Degree Polynomial Inequalities

Using the above sign chart and key points, we obtain the graph of $y = P(x)$, as shown in figure 3.21(b).

▶ **EXAMPLE 3.46** Solve $-3x^2(x + 1)^3(5 - x)(x + 6)^4 > 0$.

Solution Let $P(x) = -3x^2(x + 1)^3(5 - x)(x + 6)^4$. To determine the key points, we solve the associated equation $P(x) = 0$. We see that the roots of this tenth-degree polynomial equation are 0, 0, −1, −1, −1, 5, −6, −6, −6, and −6.

A sign chart is given in figure 3.22(a). Notice that a row of − signs indicates that the expression $-3x^2$ is negative whenever $x \neq 0$. Next, notice that $(x + 1)^3$ has the same sign as $x + 1$. Accordingly, the + and − signs in the row labeled $(x + 1)^3$ indicate that $(x + 1)^3$ is positive if $x > -1$ and $(x + 1)^3$ is negative if $x < -1$. The next row indicates that $5 - x$ is positive for $x < 5$ and negative for $x > 5$. Finally, the row labeled $(x + 6)^4$ displays only + signs, since $(x + 6)^4$ is positive whenever $x \neq -6$. Since the inequality being solved does not include the case of equality, the circles at key points are not filled in.

$P(x)$ is positive whenever there are an even number of negative factors. By examining the sign chart in figure 3.22(a), we see that this occurs on the intervals $(-\infty, -6)$, $(-6, -1)$, and $(5, \infty)$. Notice that the signs in the row labeled $P(x)$ do *not* alternate. This is due to the fact that some roots of $P(x)$ have even multiplicity. As explained in section 3.4, we expect the graph of $y = P(x)$ to be tangent to, but not cross, the x-axis at the corresponding x-intercepts. In fact, this graph can now be sketched (see figure 3.22(b)) using the key points and sign chart in figure 3.22(a). ◀

Figure 3.22

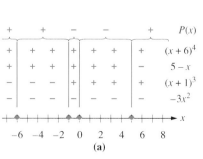

(a)

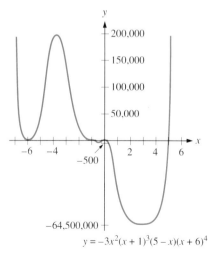

$y = -3x^2(x + 1)^3(5 - x)(x + 6)^4$

(Note: The y-axis scales for $y > 0$ and $y < 0$ are not the same.)

(b)

Chapter 3 / Polynomial and Rational Functions

EXAMPLE 3.47

Solve $2x^6 + 3x^5 - 10x^4 - 28x^3 + 23x^2 + 57x > x^6 - 18$.

Solution

We shall solve the equivalent inequality $P(x) > 0$, where $P(x) = x^6 + 3x^5 - 10x^4 - 28x^3 + 23x^2 + 57x + 18$. As usual, to find the key points, we solve $P(x) = 0$. As P has not yet been factored, we will have to use many of the techniques from the last two sections.

Using Descartes' Rule of Signs, we see that there are either two or no positive roots, and either four, two, or no negative roots. We shall next try to find rational roots. According to the Rational Root Test, the "candidate" rational roots $\frac{p}{q}$ (in lowest terms) are such that p is a divisor of 18 and q is a divisor of 1.

possible values of p: $\pm 1, \pm 2, \pm 3, \pm 6, \pm 9, \pm 18$

possible values of q: ± 1

possible rational roots: $\pm 1, \pm 2, \pm 3, \pm 6, \pm 9, \pm 18$

Some of these twelve candidates can be eliminated by the sufficient conditions for upper bounds and lower bounds discussed earlier in this section. For instance,

$$\begin{array}{r|rrrrrrr}
-6) & 1 & 3 & -10 & -28 & 23 & 57 & 18 \\
 & & -6 & 18 & -48 & 456 & -2874 & 16{,}902 \\
\hline
 & 1 & -3 & 8 & -76 & 479 & -2817 & 16{,}920
\end{array}$$

shows that -6 is a lower bound for the real roots of P. This means, in particular, that the "candidates" $-6, -9$, and -18 are not roots. Similarly,

$$\begin{array}{r|rrrrrrr}
4) & 1 & 3 & -10 & -28 & 23 & 57 & 18 \\
 & & 4 & 28 & 72 & 176 & 796 & 3412 \\
\hline
 & 1 & 7 & 18 & 44 & 199 & 853 & 3430
\end{array}$$

shows that 4 is an upper bound for the real roots of P. Thus, the "candidates" 6, 9, and 18 can be eliminated.

We now use synthetic division to test the remaining rational candidates: $\pm 1, \pm 2, \pm 3$. We determine that 1 is not a root. Next we test to see if 2 is a root, with the following result:

$$\begin{array}{r|rrrrrrr}
2) & 1 & 3 & -10 & -28 & 23 & 57 & 18 \\
 & & 2 & 10 & 0 & -56 & -66 & -18 \\
\hline
 & 1 & 5 & 0 & -28 & -33 & -9 & 0
\end{array}$$

Thus, 2 *is* a root. Notice that 2 is not a root of the depressed equation, $x^5 + 5x^4 - 28x^2 - 33x - 9 = 0$, and so 2 is not a multiple root of P. Next, we test 3 and obtain

Counting and Bounding Real Roots; High-Degree Polynomial Inequalities

$$\underline{3)} \quad \begin{array}{cccccc} 1 & 5 & 0 & -28 & -33 & -9 \\ & 3 & 24 & 72 & 132 & 297 \\ \hline 1 & 8 & 24 & 44 & 99 & 288 \end{array}$$

Thus, 3 is not a root and, in fact, 2 is the only positive rational root of P. We begin checking the possible negative rational roots with -1.

$$\underline{-1)} \quad \begin{array}{cccccc} 1 & 5 & 0 & -28 & -33 & -9 \\ & -1 & -4 & 4 & 24 & 9 \\ \hline 1 & 4 & -4 & -24 & -9 & 0 \end{array}$$

Thus, -1 *is* a root. Using the depressed equation $x^4 + 4x^3 - 4x^2 - 24x - 9 = 0$, we determine that -1 is not a multiple root of P and that -2 is not a root. We next determine that -3 is a root:

$$\underline{-3)} \quad \begin{array}{ccccc} 1 & 4 & -4 & -24 & -9 \\ & -3 & -3 & 21 & 9 \\ \hline 1 & 1 & -7 & -3 & 0 \end{array}$$

Using the depressed equation $x^3 + x^2 - 7x - 3 = 0$, we check to see if -3 is a multiple root, with the following results:

$$\underline{-3)} \quad \begin{array}{cccc} 1 & 1 & -7 & -3 \\ & -3 & 6 & 3 \\ \hline 1 & -2 & -1 & 0 \end{array}$$

Figure 3.23

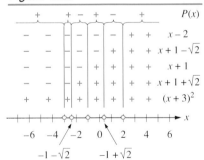

Thus, -3 *is* a multiple root. The last depressed equation, $x^2 - 2x - 1 = 0$, can be solved using the quadratic formula, giving the roots $-1 \pm \sqrt{2}$. Thus, the roots of $P(x)$ are $2, -1, -3, -3, -1 + \sqrt{2}$, and $-1 - \sqrt{2}$.

Now we are ready to solve the inequality. To do so, we use the sign chart in figure 3.23. Examining the sign chart, we see that $x^6 + 3x^5 - 10x^4 - 28x^3 + 23x^2 + 57x + 18 > 0$ occurs on the intervals $(-\infty, -3), (-3, -1 - \sqrt{2}), (-1, -1 + \sqrt{2})$, and $(2, \infty)$. ◂

Exercises 3.5

In exercises 1–10,
(a) determine the number of variations of sign in $P(x)$;
(b) use Descartes' Rule of Signs and the calculation in (a) to estimate the number of positive roots (counting multiplicities) of P;
(c) simplify the expression for, and hence determine the number of variations of sign in, $P(-x)$;
(d) use Descartes' Rule of Signs and the calculation in (c) to estimate the number of negative roots (counting multiplicities) of P;

(e) use the conclusions in (b) and (d), together with the Fundamental Theorem of Algebra, to indicate for which P you know that there exists a nonreal complex root.

1. $P(x) = 2x^5 + x^4 - \frac{3}{2}x^2 - 7$
2. $P(x) = 3x^8 - 5x^6 + 7x^4 - \frac{3}{2}x^2 + 3x$
3. $P(x) = 2x^6 + \frac{3}{2}x^5 + x^4 - 4x^3 + 5x$
4. $P(x) = 4x^6 + \frac{1}{4}x^5 + 3x^4 + 2x^3 + x^2 + 3$
5. $P(x) = \frac{2}{7}x^3 - \frac{3}{5}x^2 + 4x - \frac{2}{3}$
6. $P(x) = 2x^4 - 3x^2 + 4$
7. $P(x) = \sqrt{2}x^{10} - \sqrt{6}x^8 + 7x^4 - 7x^2 - \frac{1}{2}$
8. $P(x) = \sqrt{5}x^4 - x^3 + 4x^2 + 9$
9. $P(x) = x^6 + x^5 + 2x^4 + 6x^3$
10. $P(x) = \frac{3}{2}x^5 - 6x^4 + \frac{2}{7}x^2 + 5x + 3$
11. (An example that Descartes' Rule of Signs counts "with multiplicities")
    (a) Show that $P(x) = x^2 - 2x + 1$ exhibits two variations of sign but has only one distinct positive root. (Hint: factor $P(x)$.)
    (b) Find another polynomial which shows that Descartes' Rule of Signs must count "with multiplicity." (Hint: it's in example 3.40.)
12. Find examples of two real cubic polynomials $P(x)$ and $Q(x)$ which each exhibit three variations of sign, such that $P(x)$ has a positive root with multiplicity 3 and $Q(x)$ has one positive and two nonreal complex roots. (Hint: for $P(x)$, review the construction in exercise 11. For $Q(x)$, consider $x^3 - x^2 + x - 1$ or review example 3.40(f).)
13. (a) Explain why Descartes' Rule of Signs does not permit you to conclude directly that $P(x) = x^4 - x^2 + x - 2$ has exactly one positive root.
    (b) Does Descartes' Rule of Signs permit you to conclude directly that $Q(x) = x^5 + 2x^4 - x^3 - x^2 - 4$ has exactly one positive root?
    (c) Observe that $Q(x) = (x + 2)P(x)$. Explain why this fact, in conjunction with your answers to (b), *now* permits you to conclude that P has exactly one positive root.
14. Let $P(x) = x^5 + 2x^3 - x^2 + x - 1$ have roots $r_1, \ldots, r_5$.
    (a) Use exercise 35(c) of section 3.4 to construct a fifth-degree polynomial $Q(x)$ having roots $r_1^2, \ldots, r_5^2$.
    (b) Use Descartes' Rule of Signs to show that $Q(x)$ has only one positive root.
    (c) Use (b) and the results of section 3.4 to show that P has four nonreal complex roots. Is this sharper information than Descartes' Rule of Signs is able to provide directly?

In exercises 15–22, find an integral upper bound U and an integral lower bound L for the real roots of $P(x)$. Use these bounds in exercises 19–22 to find all the roots of $P(x)$.

15. $P(x) = 9x^6 + 6x^3 + x - 1$
16. $P(x) = x^4 + 3$
17. $P(x) = -3x^4 + 5$
18. $P(x) = \frac{1}{2}x^6 + 52x - 100$
19. $P(x) = 2x^3 + x^2 - 8x - 4$
20. $P(x) = x^3 + x^2 - 9x - 9$
21. $P(x) = x^3 - x^2 + 4x - 4$
22. $P(x) = 8x^3 - 3x^2 + 72x - 27$
23. (a) Let $P(x)$ be a real polynomial. Prove that if U is an upper bound for the real roots of P, then $-U$ is a lower bound for the real roots of $P(-x)$. (Hint: review exercise 33 of section 3.4, with $k = -1$.)
    (b) Use the text's sufficient condition for upper bounds and the result in (a) to conclude that -2 is a *lower* bound for the real roots of $R(x) = x^3 + x^2 - x - 1$. (Hint: you will eventually need to consider the polynomial $-R(-x) = x^3 - x^2 - x + 1$.)
24. Find an example of a real polynomial $P(x)$ and a real number L such that L is a lower bound for the real roots of P and the entries in the bottom line of the synthetic division corresponding to $P(x) \div (x - L)$ do *not* alternate in sign. (Hint: the text considered $R(x) = x^2 - 4x + 4$ and $U = 3$. In the spirit of exercise 23, consider $P(x) = R(-x)$ and $L = -U$.)

In exercises 25–38, solve the given inequality. In exercises 25–34, use the associated sign chart and the Leading Term Test to roughly sketch the graph of the polynomial on the left of the inequality symbol.

25. $x(x + 3)(2x - 1)(3x - 7) > 0$
26. $x^2(-x + 3)(x + \pi)(2x - 9) \geq 0$
27. $x^2(x + 1)(x - 2)(x^2 - x + 1) \leq 0$
28. $x(x - 1)(2 - x)(x^2 + x + 1)^2 < 0$
29. $(4 - 3x)(x + 3)(-x^2 + x - 1)^2 < 0$
30. $(3x + 1)(x - 4)(-x^2 + x - 1)^3 > 0$
31. $(x - 1)(x - 2)^3(x - 3)^2(x - 4) \geq 0$
32. $(x + 1)(2 - x)^4(x - 4)^5(5 - x) \leq 0$
33. $(2x - 1)(x - \pi)^4(6 - x)^5(x - 7)^2 < 0$
34. $(3x - 1)^2(x - 1)^4(x - 2)^3 > 0$
35. $8x^4 + x^3 + 20x^2 - 8 > x^4 + 6x^3 - 6x^2 + 20x$
36. $2x^4 - x^2 - 5x - 7 \geq 5x^3 + 4x^2$
37. $4x^5 + 30x^3 + 48x - 30 \leq 33x^3 - 15x^2 + 124x - 90$
38. $3x^4 - 3x^3 + 5x^2 + 2 < 2x^4 + 5x$
39. Consider a right triangle with hypotenuse of length $b + 1$, other sides of length b and c, and area 25 cm^2.
    (a) Show that b is a root of the polynomial $2x^3 + x^2 - 2500$.

(b) Show that 10 cm $< b <$ 11 cm.

40. (Proof of a result used above. It will be needed again in section 3.7.) Use the Linear Factor Theorem, the Conjugate Root Theorem, and the Fundamental Theorem of Algebra to prove that each real polynomial of positive degree is a product of real linear polynomials and irreducible real quadratics. (Hint: review exercise 39 of section 3.4.)

41. One summer day, Dwala finds that the temperature (in degrees Fahrenheit) t hours after midnight is given by

$$T(t) = -\frac{t^3}{40} + \frac{9}{10}t^2 - \frac{36}{5}t + 65.$$

(a) At which times of the day is the temperature at least 65°F?

(b) Show that at some time between 10 A.M. and noon, the temperature was 60°F.

3.6 RATIONAL FUNCTIONS AND THEIR GRAPHS

When two polynomials are multiplied, the product is also a polynomial. This is not necessarily the case when one polynomial is divided by another. The result of such a division is, however, a function. In this section, we shall examine and graph the types of functions that result when one polynomial is divided by another.

Rational Function

Suppose that $P(x)$ and $D(x)$ are two real polynomial functions, where $D(x) \neq 0$. Then the function f given by $f(x) = P(x)/D(x)$ is a **rational function**.

Notice that the (natural) domain of such a rational function f is $\{r: D(r) \neq 0\}$. For any real number r such that $D(r) = 0$, there is often a striking effect on the part of the graph of f near $x = r$.

EXAMPLE 3.48

Each of the following is a rational function. In each case, the given set is the function's natural domain.

(a) $f(x) = \dfrac{1}{x}$ \qquad $\{x: x \neq 0\}$

(b) $g(x) = \dfrac{5x^3 - 3x^2 + 7x + \frac{2}{3}}{x + 7}$ \qquad $\{x: x \neq -7\}$

(c) $h(x) = \dfrac{3x + 1}{(4x^2 + 5x + 1)}$ \qquad $\{x: x \neq -1, -\frac{1}{4}\}$

(d) $j(x) = \dfrac{2}{x^3 + 1}$ \qquad $\{x: x \neq -1\}$

To find the roots or zeros of a rational function $f = P/D$, you need only find those roots r of the numerator P such that $D(r) \neq 0$. If P and D have no common factors (except constants), the Factor Theorem implies that the roots of f are the same as the roots of P. Often, f can be written as P/D where the polynomials P and D have no common factors. For instance, $\dfrac{(x^2 - 1)(x + 1)}{(x - 1)^2(x + 2)}$ and

Chapter 3 / Polynomial and Rational Functions

$\dfrac{(x+1)^2}{(x-1)(x+2)}$ define the same rational function. The second description is preferable because its numerator and denominator have no common factor.

However, consider the rational function f given by $f(x) = \dfrac{(x^2-1)(x+1)}{(x-1)(x+2)}$. If g is the rational function given by $g(x) = \dfrac{(x+1)^2}{x+2}$, then f and g are *not* equal, since 1 is in the (natural) domain of g but not of f. As we shall show in studying vertical asymptotes, it is still useful to "simplify" $f(x)$ to $g(x)$, since $f(r)$ and $g(r)$ are equal whenever *both* are meaningfully defined real numbers.

Rational Functions of the Form $1/x^n$

Consider the graph of the rational function $h(x) = 1/x^n$, where n is a positive integer. Such graphs have two possible shapes, depending on whether n is odd or even. If n is odd, then h is an odd function, and so its graph is symmetric with respect to the origin. If n is even, then h is even, and so its graph is symmetric with respect to the y-axis. The graphs of $f(x) = 1/x$ and $g(x) = 1/x^2$ illustrate these two possibilities and will be discussed next.

▶ **EXAMPLE 3.49**　Sketch the graph of $f(x) = 1/x.$

Solution　Here, $n = 1$, making f an odd function. As its graph is therefore symmetric about the origin, we shall plot points in the first quadrant and then reflect them through the origin.

For small positive values of x, the value of $f(x)$ is large. As x increases (moves to the right) through positive values, $1/x$ decreases but is always positive. This effect is shown in table 3.4.

Table 3.4

x	0.01	0.1	0.25	0.5	1	2	4	10	100
$f(x) = \dfrac{1}{x}$	100	10	4	2	1	0.5	0.25	0.1	0.01

By plotting most of these points, connecting them to get a smooth curve, and then reflecting through the origin, we get the graph in figure 3.24(a). Notice that while the graph comes arbitrarily close to the x- and y-axes, it does not touch either of them. Thus, this graph has no x- or y-intercept. (This graph is a special case of a hyperbola. We shall return to this topic in chapter 8.)

If n is any odd positive integer, the graph of the rational function $1/x^n$ has the same general shape as the graph of $f(x) = 1/x$. Figure 3.24(b) illustrates this for $n = 3$.　◀

Rational Functions and Their Graphs

Figure 3.24

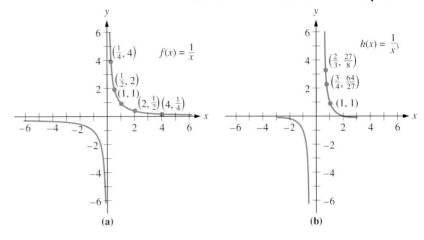

EXAMPLE 3.50

Sketch the graph of $g(x) = 1/x^2$.

Solution

Here, $n = 2$, making g an even function. As its graph is therefore symmetric about the y-axis, we shall plot points in the first quadrant and reflect them through the y-axis to obtain the final graph.

Table 3.5 contains values for x and $g(x)$. As x approaches the origin, notice that the values of $g(x)$ increase faster than did the corresponding values of $f(x)$ in example 3.49. Similarly, the values of $g(x)$ approach the x-axis more rapidly as x becomes large.

Table 3.5

x	0.01	0.1	0.25	0.5	1	2	4	5	10
$g(x) = \dfrac{1}{x^2}$	10,000	100	16	4	1	0.25	0.625	0.04	0.01

By plotting some of the points in table 3.5, connecting them with a smooth curve, and then reflecting the result in the y-axis, we produce the graph in figure 3.25(a). As was the case for the graph of f, the graph of g has no x- or y-intercepts. If n is any even positive integer, the graph of the rational function $1/x^n$ has the same general shape as the graph of $g(x) = 1/x^2$. Figure 3.25(b) illustrates this for $n = 4$.

Figure 3.25

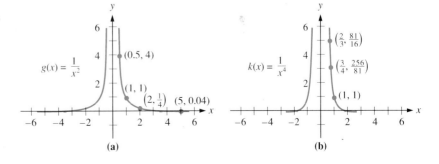

Asymptotes

In the last two examples, we saw graphs that approached but did not touch the *x*- and *y*-axes. In those examples, the axes were asymptotes. Identifying asymptotes is a helpful aid in graphing many functions—in particular, rational functions.

> **Asymptote**
>
> Let Γ be the graph of a function and let ℓ be a line. Then ℓ is an **asymptote** to Γ in case the distance between ℓ and a point $P(x, y)$ on Γ approaches 0 in at least one limiting process for which the distance from P to the origin increases without bound.

[handwritten: FUNCTION GETS CLOSER TO THE #]

This definition is admittedly rather formidable. You will come to understand it by studying the special cases in the following pages. For now, it is enough that you think of an asymptote as "a line that a graph approaches toward infinity."

Not every graph of a rational function has asymptotes, but there are three types of asymptotes that occur often enough to warrant their study. These are vertical, horizontal, and slant (also called oblique) asymptotes. An example of each type is shown in figure 3.26. Notice, in figure 3.26(b), that it is possible for a graph to cross through one of its asymptotes.

Figure 3.26

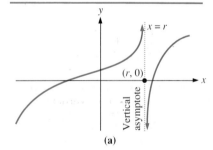
(a)

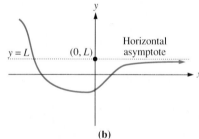
(b)

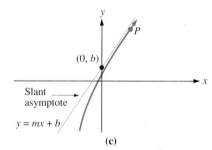
(c)

Vertical Asymptotes

The easiest asymptotes to find are the vertical asymptotes.

> **Vertical Asymptote**
>
> If the expression for a rational function f has been "simplified" to $P(x)/D(x)$ such that P and D have no common factors, then each real root r of $D(x)$ induces a **vertical asymptote** $x = r$.

Geometrically, this means that as x approaches r from one side (either the left or the right), the limiting tendency of $f(x)$ is either infinity or negative infinity.

Rational Functions and Their Graphs

EXAMPLE 3.51

Determine the vertical asymptotes of (the graph of) $f(x) = \dfrac{5x+5}{x^2+5x+4}$.

Solution

Simplify $f(x) = \dfrac{5(x+1)}{(x+4)(x+1)} = \dfrac{5}{x+4}$. The domain of f is $\{x: x \neq -4$ and $x \neq -1\}$. Here, $P(x) = 5$ and $D(x) = x + 4$ have no common factors. Notice that $r = -4$ is the only root of $D(x)$. So, $x = -4$ is the only vertical asymptote of f. Although $x = -1$ is a root of both $5x + 5$ and $x^2 + 5x + 4$, $x = -1$ is *not* an asymptote. Instead, the graph of f has a "hole" at $(-1, \frac{5}{3})$, as shown in figure 3.27. The dashed vertical line $x = -4$ is the asymptote. Looking at this figure also reveals interesting facts about limits. As x approaches -4 from the right-hand side, the limiting value of $f(x)$ is infinity; and as x approaches -4 from the left-hand side, the limiting value of $f(x)$ is negative infinity.

Figure 3.27

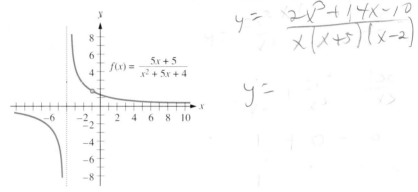

EXAMPLE 3.52

Determine the vertical asymptotes of $g(x) = \dfrac{2x^3 + 14x - 10}{x^3 + 3x^2 - 10x}$.

Solution

The denominator factors as $x(x+5)(x-2)$, with roots 0, −5, and 2. None of x, $x + 5$, and $x - 2$ is a factor of the numerator. Thus, g has three vertical asymptotes: $x = 0$, $x = -5$, and $x = 2$, as shown in figure 3.28.

Figure 3.28

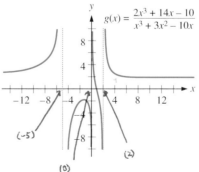

EXAMPLE 3.53

Determine the vertical asymptotes of $h(x) = \dfrac{3x^2 + 15x + 12}{x^3 - x^2 - 20x}$.

Solution

As in example 3.51, we must first simplify the functional expression by cancelling all common factors.

$$h(x) = \frac{3x^2 + 15x + 12}{x^3 - x^2 - 20x}$$

$$= \frac{3(x+4)(x+1)}{x(x+4)(x-5)}$$

$$= \frac{3(x+1)}{x(x-5)}.$$

Since $3(x+1)$ and $x(x-5)$ have no common factors, we see that $x = 0$ and $x = 5$ are the only vertical asymptotes of h. Since $x + 4$ is a common factor of both the numerator and denominator of h, the graph of h has a "hole" at $(-4, -\frac{1}{4})$, as shown in figure 3.29.

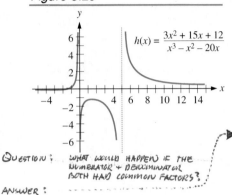

Figure 3.29

QUESTION: WHAT WOULD HAPPEN IF THE NUMERATOR + DENOMINATOR BOTH HAD COMMON FACTORS?

ANSWER:

Horizontal Asymptotes

Determining horizontal asymptotes is more difficult than finding vertical ones.

Horizontal Asymptote

— A rational function $f = P/D$ will have a horizontal asymptote if and only if the degree of P is less than or equal to the degree of D.

All this is trying to say is:
1. Find the highest polynomial degree
2. Divide — numerator and denominator by the highest degree of x
3. Solve
4. Replace all "x" denominators from equation which are positive w/ a 0 (zero)

To determine the horizontal asymptote, if any, divide the numerator and denominator each by x^n, where n is the degree of D. Now, for any positive integer k, $1/x^k$ approaches 0 when $|x|$ increases without bound. (Review exercise 26(a) of section 3.3.) Replace any term of the form $1/x^k$ by 0 and evaluate the resulting fraction. If it is a definite real number L, then $y = L$ is the only horizontal asymptote of f. Geometrically, this means that if x moves without bound either to the left or to the right, then L is the limiting tendency of $f(x)$.

▶ **EXAMPLE 3.54**

Determine the horizontal asymptote of $f(x) = \frac{2}{x+3}$.

Solution

We divide the numerator and denominator by $x^1 = x$, since the degree of $x + 3$ is 1, and obtain

$$y = \frac{\frac{2}{x}}{\frac{x}{x} + \frac{3}{x}} \qquad \text{plug in} \qquad y = \frac{\frac{2}{x}}{1 + \frac{3}{x}}.$$

As explained above, what we are really doing here is examining what happens to "y" as $|x|$ gets very large. To do this, each term with a positive power of x in its denominator is replaced with 0. The result is $\frac{0}{1+0} = 0$; thus, $y = 0$, the x-axis, is the horizontal asymptote of f. Notice also that $x = -3$ is a vertical asymptote. The graph of f is in figure 3.30.

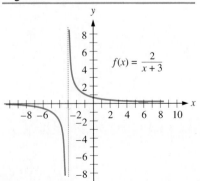

Figure 3.30

HORIZONTAL ASYMPTOTE RULES:

FORM: $\dfrac{A x^N}{B x^O}$

1. $N < O$ ⟶ $y = 0$ ASYMP.
2. $N > O$ ⟶ NO ASYMPTOTE
3. $N = O$ ⟶ $\dfrac{A}{B}$ ASYMPTOTE

WHERE'S ASYMPTOTE

Rational Functions and Their Graphs

You should also notice that this graph could be obtained directly from the graph of $y = \dfrac{1}{x}$ by means of the techniques in section 2.2: translate three units to the left and then stretch vertically by a factor of 2. ◀

EXAMPLE 3.55

Determine the horizontal asymptote of $f(x) = \dfrac{x+4}{2x-5}$. ÷ $\dfrac{x}{x}$

Solution

We divide the numerator and denominator by $x^1 = x$, since the degree of $2x - 5$ is 1, and obtain

$$y = \dfrac{1 + \dfrac{4}{x}}{2 - \dfrac{5}{x}}.$$

Figure 3.31

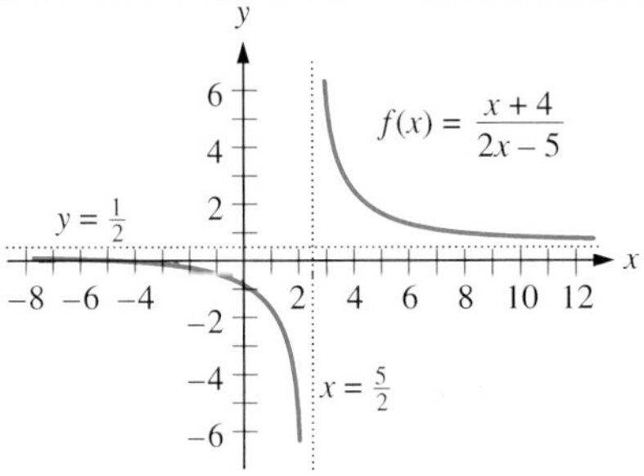

[Graph showing $f(x) = \dfrac{x+4}{2x-5}$ with horizontal asymptote $y = \dfrac{1}{2}$ and vertical asymptote $x = \dfrac{5}{2}$]

Each term with a positive power of x in its denominator is replaced with 0, resulting in

$$\dfrac{1+0}{2+0} = \dfrac{1}{2}.$$

Thus, $y = \dfrac{1}{2}$ is the horizontal asymptote of f. Notice also that $x = \dfrac{5}{2}$ is a vertical asymptote. The graph of f is in figure 3.31.

If we had divided the numerator by the denominator, $f(x)$ could have been expressed as

$$f(x) = \dfrac{x+4}{2x-5} = \dfrac{1}{2} + \dfrac{\frac{13}{2}}{2x-5} = \dfrac{1}{2} + \dfrac{\frac{13}{4}}{x-\frac{5}{2}}.$$

This indicates that the graph of f is obtained from that of $y = 1/x$ by doing the following: translate $\dfrac{5}{2}$ units horizontally to the right, stretch vertically by a factor of $\dfrac{13}{4}$, and then translate $\dfrac{1}{2}$ unit vertically upward. ◀

EXAMPLE 3.56

Determine the horizontal asymptote of $g(x) = \dfrac{x^2 + 2x - 4}{x^3 + 4x^2 - 5x}$. ÷ $\dfrac{x^3}{x^3}$

Solution

We divide the numerator and denominator by x^3, since the degree of the denominator is 3. This gives

$$y = \dfrac{\dfrac{1}{x} + \dfrac{2}{x^2} - \dfrac{4}{x^3}}{1 + \dfrac{4}{x} - \dfrac{5}{x^2}}.$$

Each term with a positive power of x in the denominator is replaced by 0, yielding

$$y = \dfrac{0 + 0 + 0}{1 + 0 + 0} = \dfrac{0}{1} = 0.$$

Figure 3.32

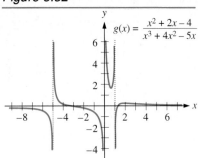

Thus, $y = 0$, the x-axis, is the horizontal asymptote of g.

Since $x^3 + 4x^2 - 5x = x(x + 5)(x - 1)$, the graph of g also has vertical asymptotes at $x = 0$, $x = -5$, and $x = 1$. Its graph is in figure 3.32.

Notice that the graph of g crosses its horizontal asymptote at two points. In general, if $y = L$ is a horizontal asymptote of a rational function f, then solving $f(x) = L$ for x will determine the points where the graph of f intersects the horizontal asymptote. In our example, the horizontal asymptote is $y = 0$, so we must solve the equation $g(x) = 0$. In other words, $x^2 + 2x - 4 = 0$ (and $x \neq 0, -5, 1$). Thus, we see that the graph of g crosses its horizontal asymptote at $x = -1 \pm \sqrt{5}$, approximately $x = -3.24$ and $x = 1.24$.

Slant Asymptotes

Suppose $f(x) = P(x)/D(x)$ is a rational function such that the degree of P is n and the degree of D is d. The above reasoning shows that f has a horizontal asymptote precisely when $n \leq d$. We shall see next that if $n = d + 1$, then f has a slant asymptote; and that if $n \geq d + 2$, then f has neither horizontal nor slant asymptotes.

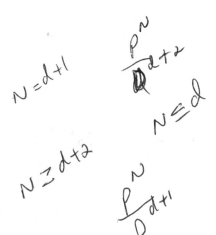

Let's consider the case $n = d + 1$, in which the degree of the numerator is exactly 1 more than the degree of the denominator. To find the slant asymptote, divide $P(x)$ by $D(x)$. The quotient is a linear polynomial $mx + b$; let $R(x)$ denote the remainder. Thus,

$$f(x) = \frac{P(x)}{D(x)} = mx + b + \frac{R(x)}{D(x)}$$

where R either is 0 or has degree less than the degree of D. Reasoning as in exercise 26(a) of section 3.3, we see that $\frac{R(x)}{D(x)}$ approaches 0 as $|x|$ increases without bound. Under these conditions, $f(x) - (mx + b)$ approaches 0; that is, $y = mx + b$ is the slant asymptote of f.

▶ **EXAMPLE 3.57** Determine the slant asymptote of $h(x) = \dfrac{6x^2 + 3x - 7}{3x + 6}$.

Solution Long division gives us

$$\begin{array}{r} 2x - 3 \\ 3x+6\overline{\smash{\big)}\,6x^2 + 3x - 7} \\ \underline{6x^2 + 12x} \\ -9x - 7 \\ \underline{-9x - 18} \\ 11 \end{array}$$

So, $h(x) = 2x - 3 + \dfrac{11}{3x + 6}$, and the slant asymptote is $y = 2x - 3$. Notice also that h has the vertical asymptote $x = -2$. The graph of h is in figure 3.33. ◀

Rational Functions and Their Graphs

Figure 3.33

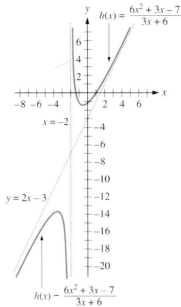

EXAMPLE 3.58   Determine the slant asymptote of $j(x) = \dfrac{2x^4 - 5x^3 + 7x^2 - 4}{2x^3 - 16}$.

Solution   Dividing, we get

$$\begin{array}{r}
x - \frac{5}{2} \\
2x^3 - 16 \overline{\smash{\big)}\, 2x^4 - 5x^3 + 7x^2 + 0x - 4} \\
\underline{2x^4 - 16x } \\
-5x^3 + 7x^2 + 16x - 4 \\
\underline{-5x^3 + 40} \\
7x^2 + 16x - 44
\end{array}$$

Figure 3.34

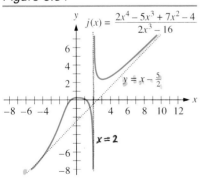

Thus, $j(x) = x - \dfrac{5}{2} + \dfrac{7x^2 + 16x - 44}{2x^3 - 16}$, and the slant asymptote is $y = x - \dfrac{5}{2}$. In addition, j has the vertical asymptote $x = 2$, as shown by the graph in figure 3.34. Notice that the graph of j crosses the asymptote $y = x - \dfrac{5}{2}$ twice. To determine these points of intersection, we solve $j(x) = x - \dfrac{5}{2}$ for x. In other words, we want the values of x that satisfy $2x^4 - 5x^3 + 7x^2 - 4 = (2x^3 - 16)(x - \dfrac{5}{2})$. These are the roots of $7x^2 + 16x - 44$, namely $\dfrac{-8 \pm 2\sqrt{93}}{7}$, approximately 1.61 and −3.90. Substituting each of these as x-values into $j(x)$, we see that the graph of j intersects its slant asymptote approximately at the points (−3.90, −6.40) and (1.61, −0.89).

Summarizing these ideas about asymptotes, we have the following.

Guidelines for Finding Asymptotes of Rational Functions

Simplify the expression for a given rational function f to the form $\frac{P(x)}{D(x)}$, where the polynomials $P(x)$ and $D(x)$ have no common factors, the degree of P is n, and the degree of D is d. Then,

1. A vertical line $x = r$ is a vertical asymptote if and only if $D(r) = 0$.
2. If $n < d$, then $y = 0$ is the horizontal asymptote.
3. If $n = d$, then $y = \frac{a_n}{b_n}$ is the horizontal asymptote, where a_n and b_n are the leading coefficients of P and D, respectively.
4. If $n = d + 1$, then $y = mx + b$ is the slant asymptote, where
$$\frac{P(x)}{D(x)} = mx + b + \frac{R(x)}{D(x)}.$$
5. If $n \geq d + 2$, then f has no horizontal or slant asymptote.

We are now able to use the ideas about asymptotes to develop the following suggestions for graphing rational functions.

Guidelines for Graphing Rational Functions

If $f(x) = \frac{P(x)}{D(x)}$, then,

1. Simplify the functional expression by canceling all common factors of P and D. Corresponding to each root of any of these common factors, the graph of f will have either a "hole" or a vertical asymptote. **For the remaining steps, we assume that P and D have no common factors.**
2. If $D(0) \neq 0$, plot the y-intercept, $(0, f(0))$.
3. Plot any x-intercepts $(r, 0)$. These arise from the real roots of P.
4. Find and graph the asymptotes, using the steps outlined above.
5. Find all points where the graph intersects an asymptote, as follows.
   - Vertical asymptote: The graph will not intersect a vertical asymptote (since the corresponding x-value is not in the domain of f).
   - Horizontal asymptote, $y = L$: Solve the equation $f(x) = L$ for x. If $x = h$ is a solution, then the graph intersects its horizontal asymptote at (h, L).
   - Slant asymptote, $y = mx + b$: Solve the equation $f(x) = mx + b$ for x. If $x = h$ is a solution, the graph intersects the slant asymptote at $(h, mh + b)$.
6. Check symmetry about the x- and y-axes and the origin, using the procedures studied in chapter 2.
7. Plot some selected points in each of the regions formed by the asymptotes as boundaries.
8. Complete the graph by connecting the plotted points in a manner consistent with the locations of intercepts, asymptotes, and additional plotted points.

Rational Functions and Their Graphs

Example 3.59

Sketch the graph of $m(x) = \dfrac{2x^2 - 2x - 4}{x^2 - 9}$.

Solution

We begin by factoring the numerator and denominator to see if they have any common factors.

$$m(x) = \frac{2x^2 - 2x - 4}{x^2 - 9} = \frac{2(x-2)(x+1)}{(x+3)(x-3)}.$$

There are no common factors.

The y-intercept is $(0, m(0)) = (0, \frac{4}{9})$. The x-intercepts arise from the real roots of $2x^2 - 2x - 4 = 2(x-2)(x+1)$; that is, $(2, 0)$ and $(-1, 0)$.

There are vertical asymptotes at $x = 3$ and $x = -3$ and a horizontal asymptote at $y = 2$. There is no slant asymptote.

To see if the graph crosses the horizontal asymptote, we solve the equation $m(x) = 2$ for x. Thus,

$$2x^2 - 2x - 4 = 2(x^2 - 9)$$
$$= 2x^2 - 18.$$
$$-2x = -14$$
$$x = 7.$$

This produces the intersection point $(7, 2)$.

Additional points between and beyond the asymptotes are

x	-6	-5	-4	-2	1	4	5	6	8	9
$m(x)$	3.20	3.50	5.14	-1.60	0.5	2.86	2.25	2.07	1.96	1.94

The graph of m is shown in figure 3.35.

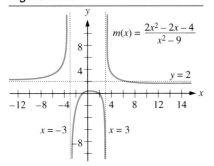

Figure 3.35

Exercises 3.6

In exercises 1–6, (a) find the natural domain of the given rational function f, (b) find the zeros of f, and (c) determine whether f equals a function that can be defined by an expression $\dfrac{P(x)}{D(x)}$ in which P and D are polynomials having no common factor.

1. $f(x) = \dfrac{x-2}{2+x}$

2. $f(x) = \dfrac{x^2 - x - 2}{x - 2}$

3. $f(x) = \dfrac{x-2}{x^2 - x - 2}$

4. $f(x) = \dfrac{5x^2 + 14x - 3}{x^2 - x - 6}$

5. $f(x) = \dfrac{x^2 - x - 2}{x^2 - 2x + 1}$

6. $f(x) = \dfrac{x^2 - 2x + 1}{x^2 - x - 2}$

In exercises 7–12, apply the techniques in section 2.2 to figures 3.24 and 3.25 in order to graph the given rational function f.

7. $f(x) = \dfrac{-1}{x+2} + 3$

8. $f(x) = \dfrac{4}{(x-2)^3}$

9. $f(x) = \dfrac{3}{x^4} - 2$

10. $f(x) = \dfrac{-2}{(x-3)^2} + 4$

11. $f(x) = \dfrac{-2}{(3x)^2} + 4$

12. $f(x) = \dfrac{1}{3x+1} - 2$

13. (a) Graph $f(x) = \dfrac{1}{x^n}$ for $n = 5$.

    (b) Graph $g(x) = \dfrac{1}{x^n}$ for $n = -5$.

14. (a) Prove that each real polynomial function is a rational function. (Hint: take $D(x)$ of low degree.)

    (b) Give an example of an increasing rational function.

In exercises 15–16, find functional expressions for the functions (a) $f + g$, (b) $f - g$, (c) $f \cdot g$, (d) f/g, and (e) $f \circ g$. Also, indicate the domains of these functions.

Chapter 3 / Polynomial and Rational Functions

15. $f(x) = \dfrac{1}{x-2}$, $g(x) = \dfrac{3x}{x^2 + x - 6}$

16. $f(x) = \dfrac{3x^2}{x+1}$, $g(x) = \dfrac{x+2}{3x-4}$

17. Prove that if f and g are rational functions, then so are $f + g, f - g, f \cdot g,$ and f/g.

18. Prove that if f and g are rational functions, then so is $f \circ g$. (Hint: use exercise 17 to reduce consideration first to the case in which f is a polynomial function and next to the case in which $f = cx^n$.)

19. Prove that figure 3.26(c) is *not* the graph of a rational function. (Hint: does the graph approach $y = mx + b$ in both cases as $|x|$ increases without bound?)

20. (a) Prove that no rational function has a graph with all three kinds of asymptotes (vertical, horizontal, and slant).
    (b) Draw the graph of a function having all three kinds of asymptotes.

21. Give an equation of a rational function whose graph has no asymptotes. (Hint: exercise 14(a).)

22. Give an example of a rational function f which is not a polynomial function and which has range $\mathbf{R}$. (Hint: check this section's figures.)

In exercises 23–30, find all the vertical asymptotes (if any exist) for the graph of the given rational function f. (Here, and for exercises 31–50, you may wish to use a graphing calculator or computer graphing program to help you or to provide a "rough check" of your work.)

23. $f(x) = \dfrac{3x + 12}{x^2 - 3x - 10}$

24. $f(x) = \dfrac{x^4 + x^3}{2x^3 - 2x}$

25. $f(x) = \dfrac{3}{x+1} - \dfrac{1}{x^2}$. (Hint: write as $f(x) = \dfrac{P(x)}{D(x)}$.)

26. $f(x) = \dfrac{-1}{x^2} + 4$

27. $f(x) = \dfrac{2x^3 - 6x^2 - 5x + 15}{x^2 - 9}$

28. $f(x) = \dfrac{2x^3 + 3x^2 - 18x - 27}{x^2 - 9}$

29. $f(x) = \dfrac{4}{x^2 + 1}$

30. $f(x) = \dfrac{-2}{(x+2)^3}$

In exercises 31–36, find the horizontal asymptote (if it exists) for the graph of the given rational function f.

31. $f(x) = \dfrac{2x^3 - \pi x + \sqrt{2}}{9x^4 + 6x^3 - \frac{3}{2}x + 4}$

32. $f(x) = \dfrac{9x^3 + 2x - 7}{14x^5 + 6x^4 - \pi x^2 + \frac{1}{2}}$

33. $f(x) = \dfrac{-x^4 + 6x^2 - 5}{-2x^3 + 4x + 3}$

34. $f(x) = \dfrac{9x^5 + 2x^3 - 7}{14x^5 - \pi x^2 + \frac{1}{2}}$

35. $f(x) = \dfrac{2x^3 - \pi x + \sqrt{2}}{9x^3 + 6x + 4}$

36. $f(x) = \dfrac{-6x^4 - 10x^2 + 1}{3x^2 + 5}$

In exercises 37–42, find the slant asymptote (if it exists) for the graph of the given rational function f.

37. $f(x) = \dfrac{9x^3 + 10x^2 - 14x - 4}{x^2 + x - 2}$

38. $f(x) = \dfrac{3x^2 + \frac{17}{2}x - 11}{2x - 1}$

39. $f(x) = \dfrac{x^2 + x - 2}{2x^2 + 2x - 4}$

40. $f(x) = \dfrac{8x^4 + x + 2}{x^3 + x^2 + 2}$

41. $f(x) = \dfrac{-6x^2 + 11x - 11}{2x - 1}$

42. $f(x) = \dfrac{x^4 + 3x - 7\pi}{7x + \pi}$

In exercises 43–50, use all the available information ("holes," intercepts, asymptotes, intersections with asymptotes, symmetry) to graph the given rational function f.

43. $f(x) = \dfrac{3x^2 + 6x - 24}{x^2 - 4}$

44. $f(x) = \dfrac{3x^2 + 10x + 11}{x + 1}$

45. $f(x) = \dfrac{2x^4 + 5x^3 + 6x^2 - 2x - 9}{x^3 - 1}$

46. $f(x) = \dfrac{8x^2 - 2x - 10}{4x^2 - 1}$

47. $f(x) = \dfrac{x - 2}{x^3 + 8}$

48. $f(x) = \dfrac{x + 2}{x^3 + 8}$

49. $f(x) = \dfrac{2x^3 + 3x^2 - x - 7}{x^2 + 3x + 4}$

50. $f(x) = \dfrac{2x^5 + 5x^4 - 3x^3 - 8x^2 + 5x - 5}{(x+2)^2 x(x-1)}$

In exercises 51–54, solve the given inequality.

51. $\dfrac{x + 3}{(x-1)(x+2)(x+4)} > 0$. (Hint: in building a sign chart, you should consider the key points $-3, 1, -2,$ and -4. You may also want to review the techniques for solving inequalities in section 1.1.)

52. $\dfrac{3x^2 - x}{(x-1)(x+2)} < \dfrac{2x}{(x-1)(x+2)}$. (Hint: obtain an equivalent inequality of the form $f(x) < 0$. Then proceed as in exercise 51.)

53. $\dfrac{3x + 2}{x - 1} \leq \dfrac{2x + 3}{x + 1}$

54. $x^2 \geq \dfrac{x^2 + 1}{x + 1}$

55. (Contrast with the Leading Term Test.) Find real polynomial functions $P(x) = a_n x^n + \cdots + a_0$ and

$D(x) = b_m x^m + \cdots + b_0$ such that $a_n \neq 0$, $b_m \neq 0$, $n > m$, P and D have no factors in common, and, as $|x|$ increases without bound, the limiting tendency of $\dfrac{P(x)}{D(x)}$ is not $\dfrac{a_n}{b_m} x^{n-m}$. (Hint: as $|x|$ increases without bound, $\dfrac{x^4+1}{x+1} - x^3$ approaches negative infinity.)

56. Prove that if n and m are distinct positive integers and c is any real number, then the expressions $1/x^n$ and c/x^m define unequal rational functions. (Hint: how many roots does the polynomial $x^m - cx^n$ have? What is its degree?)

◆ 57. (A preview of asymptotes for a hyperbola, to be studied in section 8.3) Consider the (nonrational) function f given by $f(x) = 3\sqrt{x^2 - 4}$, with its natural domain, $(-\infty, -2] \cup [2, \infty)$.
   (a) Give an intuitive argument that $y = 3x$ is a slant asymptote for the graph of f, by showing that as x increases without bound, the limiting tendency of $f(x) - 3x$ is 0. (Hint: multiply numerator and denominator each by $\sqrt{x^2 - 4} + x$.)
   (b) Give an intuitive argument, using a calculator if necessary, that as x moves to the *left* without bound, the limiting tendency of $f(x) - 3x$ is *not* 0. What is this limiting tendency?

58. Radioactive industrial waste is to be put into a lead cylindrical container as in the figure. The container's wall, top, and bottom are each 1 foot thick, its inner radius is r feet, and its height is h feet. If the volume of the entire container (including the inside, top, bottom, and wall) is 36π ft³, show that the maximum amount of cubic feet of waste that can be contained is given by the rational function

$$W(r) = \pi r^2 \left[\frac{36}{(r+1)^2} - 2 \right].$$

(Hint: $36\pi = \pi(r+1)^2 h$. You must calculate $\pi r^2 (h - 2)$.)

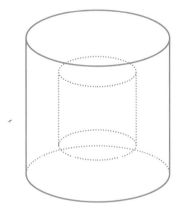

59. A chemist wishes to mix 50 ml of a solution that is 40 percent acid (by volume) with an unknown amount of pure acid, in order to produce a solution that is at least 60 percent acid. How much of the pure acid should the chemist use? (Hint: suppose x ml of the pure acid is used. Solve either a linear inequality or a rational-function inequality for x.)

*3.7 PARTIAL FRACTIONS

In your algebra courses you have had to add or subtract two rational functions. One way to do this is to find a common denominator, rewrite each function using this common denominator, add or subtract the numerators, and simplify the result. For example,

$$\frac{5}{x-1} - \frac{2x}{x^2+x+1} = \frac{5(x^2+x+1)}{(x-1)(x^2+x+1)} - \frac{2x(x-1)}{(x-1)(x^2+x+1)}$$
$$= \frac{5(x^2+x+1) - 2x(x-1)}{(x-1)(x^2+x+1)} = \frac{3x^2 + 7x + 5}{x^3 + 1}.$$

In integral calculus, it is often necessary to reverse this process—that is, to rewrite a given rational function as a sum of several "simpler" rational functions. The purpose of this section is to show you how to do so.

Advanced algebraic reasoning can be used to establish the following result.

---

*This section is optional because the material in it is not essential for later material in the book.

Chapter 3 / Polynomial and Rational Functions

Partial Fraction Decomposition

Let $f(x) = P(x)/D(x)$ be a rational function, arising from real polynomial functions P and D. Then f has a **partial fraction decomposition**

$$f(x) = \frac{P(x)}{D(x)} = Q(x) + F_1 + F_2 + F_3 + \cdots + F_n \qquad (*)$$

where the polynomial $Q(x)$ is the quotient arising when $P(x)$ is divided by $D(x)$; each F_i is a rational function taking one of the forms $\dfrac{A}{(ax+b)^m}$ or $\dfrac{Bx+C}{(ax^2+bx+c)^n}$ in which A, B, and C represent real (possibly zero) constants; m and n are positive integers; and the quadratics $ax^2 + bx + c$ that arise are irreducible over the real numbers (the criterion for which is $b^2 - 4ac < 0$).

Four steps are used to find the partial fraction decomposition of a given rational function. We shall state these next, and examine each closely by studying several cases. As you will see, it is not always necessary to use all four steps.

Method for Finding Partial Fraction Decomposition

1. Consider $f = P/D$, as above. If the degree of P is greater than or equal to the degree of D, divide $P(x)$ by $D(x)$ to get the quotient $Q(x)$ and the remainder $R(x)$. Since $P(x) = Q(x)D(x) + R(x)$, we see after dividing by $D(x)$ that $f(x) = Q(x) + R(x)/D(x)$. Hence, attention is shifted to the rational function R/D. Assume henceforth that $R = P$, that is, that the degree of P is less than the degree of D. It will also be convenient to assume that P and D have been chosen to have no common factors (besides constants).
2. Factor $D(x)$ as a product of powers of distinct real linear polynomials and real irreducible quadratic polynomials.
3. Let $(ax+b)^m$ be one of the powers appearing in the factorization obtained in step 2. The partial fraction decomposition thereby gains m terms

$$\frac{A_1}{ax+b} + \frac{A_2}{(ax+b)^2} + \frac{A_3}{(ax+b)^3} + \cdots + \frac{A_m}{(ax+b)^m}$$

for suitable (possibly zero) real constants $A_1, A_2, A_3, \ldots, A_m$.
4. Let $(ax^2 + bx + c)^n$ be one of the powers appearing in the factorization obtained in step 2. The partial fraction decomposition thereby gains n terms

$$\frac{A_1x + B_1}{ax^2 + bx + c} + \frac{A_2x + B_2}{(ax^2 + bx + c)^2} + \cdots + \frac{A_nx + B_n}{(ax^2 + bx + c)^n}$$

for suitable (possibly zero) real constants $A_1, B_1, A_2, B_2, \ldots, A_n, B_n$.

Partial Fractions

Distinct Linear Factors

EXAMPLE 3.60 Write the partial fraction decomposition for $\dfrac{x^3 - 3x^2 - 8x + 37}{x^2 - x - 12}$.

Solution The degree of the numerator is greater than the degree of the denominator, so we begin by dividing the numerator by the denominator.

$$
\begin{array}{r}
x - 2 \\
x^2 - x - 12 \overline{\smash{\big)}\, x^3 - 3x^2 - 8x + 37} \\
\underline{x^3 - x^2 - 12x} \\
-2x^2 + 4x + 37 \\
\underline{-2x^2 + 2x + 24} \\
2x + 13
\end{array}
$$

Thus, we can rewrite the given rational function as

$$\frac{x^3 - 3x^2 - 8x + 37}{x^2 - x - 12} = x - 2 + \frac{2x + 13}{x^2 - x - 12}.$$

The denominator factors as $x^2 - x - 12 = (x - 4)(x + 3)$. According to step 2, the factors $(x - 4)$ and $(x + 3)$ each contribute one term to the partial fraction decomposition:

$$\frac{2x + 13}{x^2 - x - 12} = \frac{A}{x - 4} + \frac{B}{x + 3}. \qquad (*)$$

Multiply equation (*) by the least common multiple (LCM) of the denominators, namely $(x - 4)(x + 3)$, to get

$$2x + 13 = A(x + 3) + B(x - 4). \qquad (**)$$

Notice that (**) was derived from (*), and so (**) holds for all real numbers, except possibly 4 and −3. However, a polynomial equation that holds for infinitely many numbers must hold for all real numbers, since we saw in section 3.4 that any polynomial of positive degree has only finitely many roots. Hence (**) is an "identity" and, in particular, holds for $x = 4$ and for $x = -3$.

Next we shall *carefully* select specific values of x to help solve equation (**) for A and B. The above comment permits us to let $x = 4$. If we do so, $x - 4 = 0$, and equation (**) becomes

$$2(4) + 13 = A(4 + 3) + B(0)$$
$$21 = 7A$$

or

and so $A = 3$.

To solve for B, we substitute $x = -3$ into (**). This produces

$$2(-3) + 13 = A(0) + B(-3 - 4)$$
$$7 = -7B$$

or

and so $B = -1$.

Therefore, the partial fraction decomposition is

$$\frac{x^3 - 3x^2 - 8x + 37}{x^2 - x - 12} = x - 2 + \frac{3}{x - 4} + \frac{-1}{x + 3}.$$

◀

EXAMPLE 3.61   Find the partial fraction decomposition of $\dfrac{8x + 3}{x^3 + 2x^2 - 3x}$.

Solution   The numerator's degree is less than the denominator's, and so we can proceed to step 2. Factor the denominator as $x^3 + 2x^2 - 3x = x(x + 3)(x - 1)$. So the decomposition takes the form

$$\frac{8x + 3}{x^3 + 2x^2 - 3x} = \frac{A}{x} + \frac{B}{x + 3} + \frac{C}{x - 1} \qquad (*)$$

where A, B, and C are constants that we need to determine. Multiplying by the LCM, $x(x + 3)(x - 1)$, we change equation (*) into

$$8x + 3 = A(x + 3)(x - 1) + Bx(x - 1) + Cx(x + 3). \qquad (**)$$

Just as in example 3.60, (**) can be shown to hold for *all* real x. (Henceforth, we won't pause to observe this fact explicitly, as the reasoning in example 3.60 justifies this type of step in general.)

Once again, it will be helpful to replace x by the roots of the given denominator. If we let $x = 0$ in (**), the last two terms on the right-hand side become zero. This reduces (**) to

$$8(0) + 3 = A(0 + 3)(0 - 1) + B(0) + C(0)$$

or
$$3 = -3A$$

and so $A = -1$.

Similarly, if we let $x = -3$ in (**), we can see that $B = -\frac{21}{12} = -\frac{7}{4}$; and if we let $x = 1$ in (**), we determine that $C = \frac{11}{4}$. Thus, the partial fraction decomposition is

$$\frac{8x + 3}{x^3 + 2x^2 - 3x} = \frac{-1}{x} + \frac{-\frac{7}{4}}{x + 3} + \frac{\frac{11}{4}}{x - 1}.$$

◀

Repeated Linear Factors

EXAMPLE 3.62   Determine the partial fraction decomposition of $\dfrac{2x^3 + 5x^2 - x - 8}{x^2 + 4x + 4}$.

Solution   Notice that 3, the degree of the numerator, is greater than 2, the degree of the denominator. So we begin with step 1. When $2x^3 + 5x^2 - x - 8$ is divided by $x^2 + 4x + 4$, the quotient is $2x - 3$ and the remainder is $3x + 4$. Thus, we can write the given rational function as

$$\frac{2x^3 + 5x^2 - x - 8}{x^2 + 4x + 4} = 2x - 3 + \frac{3x + 4}{x^2 + 4x + 4}.$$

Attention now shifts to finding the partial fraction decomposition of $\frac{3x + 4}{x^2 + 4x + 4}$. Its denominator factors as $(x + 2)^2$. According to step 3, this leads to the partial fraction decomposition

$$\frac{3x + 4}{(x + 2)^2} = \frac{A}{x + 2} + \frac{B}{(x + 2)^2}$$

where A and B need to be determined. Multiplying both sides of this equation by the LCM, $(x + 2)^2$, produces the identity $3x + 4 = A(x + 2) + B$. As above, we shall substitute a root of the denominator into this identity. Putting $x = -2$, we find that $B = -2$. To find A, we proceed differently. Applying exercise 43 of section 3.1 to the identity $3x + 4 = A(x + 2) + B$ allows us to conclude that the coefficients of x are equal: $3 = A$. Notice that the same information would arise from $B = -2$ and any *new* value of x. For instance, $x = 1$ leads to $3(1) + 4 = A(1 + 2) + (-2)$, or $7 = 3A - 2$, and so $A = 3$, as before. We can now conclude that

$$\frac{2x^3 + 5x^2 - x - 8}{x^2 + 4x + 4} = 2x - 3 + \frac{3}{x + 2} + \frac{-2}{(x + 2)^2}.$$

▶ **EXAMPLE 3.63**   Determine the partial fraction decomposition of $\frac{x^3 + 3x^2 - 2x + 1}{x^5 - 2x^4 + x^3}$.

Solution   The denominator factors as $x^3(x - 1)^2$. Thus, step 3 leads to

$$\frac{x^3 + 3x^2 - 2x + 1}{x^5 - 2x^4 + x^3} = \frac{A}{x} + \frac{B}{x^2} + \frac{C}{x^3} + \frac{D}{x - 1} + \frac{E}{(x - 1)^2}.$$

Multiplying both sides of the equation by the LCM, $x^3(x - 1)^2$, we obtain

$$x^3 + 3x^2 - 2x + 1 = Ax^2(x - 1)^2 + Bx(x - 1)^2 + C(x - 1)^2 + Dx^3(x - 1) + Ex^3.$$

As above, it will be helpful to substitute the roots of the denominator (namely, 0 and 1) for x. If we let $x = 0$, we see that $C = 1$. Putting $x = 1$, we obtain $E = 3$.

As in example 3.62, *one* way to find the three remaining constants A, B, and D is to equate corresponding coefficients, as you did in exercise 43 of section 3.1. Let's rewrite the above identity:

$$x^3 + 3x^2 - 2x + 1$$
$$= Ax^4 - 2Ax^3 + Ax^2 + Bx^3 - 2Bx^2 + Bx + Cx^2 - 2Cx + C + Dx^4 - Dx^3 + Ex^3$$
$$= (A + D)x^4 + (-2A + B - D + E)x^3 + (A - 2B + C)x^2 + (B - 2C)x + C.$$

Since we already know that $C = 1$ and $E = 3$, the right-hand side of the identity is

$$(A + D)x^4 + (-2A + B - D + 3)x^3 + (A - 2B + 1)x^2 + (B - 2)x + 1.$$

Equate the constant coefficients: $1 = 1$ (true, but not helpful).

Equate the coefficients of x: $-2 = B - 2$. This leads to $B = 0$.

Equate the coefficients of x^2: $3 = A - 2B + 1 = A - 2(0) + 1 = A + 1$. This gives $A = 2$.

Equate the coefficients of x^3: $1 = -2A + B - D + 3 = -2(2) + 0 - D + 3 = -1 - D$. Hence, $D = -2$.

(We could also see that $D = -2$ by equating the coefficients of x^4, for then $0 = A + D$, and so $D = -A = -2$.)

Another way to find A, B, and D is to use $C = 1$, $E = 3$ and substitute three *new* values of x. For instance,

Value of x	How the Identity Transforms
$x = -1$	$5 = 4A - 4B + 4 + 2D - 3$
$x = 2$	$17 = 4A + 2B + 1 + 8D + 24$
$x = -2$	$9 = 36A - 18B + 9 + 24D - 24$

With some persistence, you should be able to solve this system of three equations, once again finding $A = 2$, $B = 0$, and $D = -2$. (There really are systematic ways to do this. In chapter 1, you saw the substitution method. See sections 9.1 and 9.2 for other general methods.)

We can now conclude that the partial fraction decomposition is

$$\frac{x^3 + 3x^2 - 2x + 1}{x^5 - 2x^4 + x^3} = \frac{2}{x} + \frac{1}{x^3} + \frac{-2}{x - 1} + \frac{3}{(x - 1)^2}. \blacktriangleleft$$

Distinct Linear and Quadratic Factors

▶ **EXAMPLE 3.64** Find the partial fraction decomposition of $\dfrac{x^2 + 5x + 19}{(8x + 2)(x^2 + 5)}$.

Solution There is nothing to do in step 1 since 2, the numerator's degree, is less than 3, the degree of the denominator. Notice that step 2 is also trivial here. Indeed, the denominator is already fully factored: notice that the quadratic $x^2 + 5$ is irreducible since its discriminant, $0^2 - 4(1)(5) = -20$, is negative. Hence, steps 3 and 4 lead to

$$\frac{x^2 + 5x + 19}{(8x + 2)(x^2 + 5)} = \frac{A}{8x + 2} + \frac{Bx + C}{x^2 + 5}.$$

Multiplying both sides of this equation by the LCM, $(8x + 2)(x^2 + 5)$, we get

$$x^2 - 5x + 19 = A(x^2 + 5) + (Bx + C)(8x + 2). \tag{*}$$

If we substitute $x = -\frac{1}{4}$, the root of $8x + 2$, (*) becomes

$$(-\tfrac{1}{4})^2 - 5(-\tfrac{1}{4}) + 19 = A[(-\tfrac{1}{4})^2 + 5] + 0$$

or
$$\tfrac{325}{16} = \tfrac{81}{16} A$$

and so $A = \frac{325}{81}$. Next, we proceed to find B and C, by following the first method indicated in example 3.63.

We can rewrite the identity (*) as

$$x^2 - 5x + 19 = (A + 8B)x^2 + (2B + 8C)x + (5A + 2C).$$

Since we know $A = \frac{325}{81}$, this simplifies to

$$x^2 - 5x + 19 = (\tfrac{325}{81} + 8B)x^2 + (2B + 8C)x + (\tfrac{1625}{81} + 2C).$$

If we equate the constant coefficients, we obtain $19 = \frac{1625}{81} + 2C$, or $2C = 19 - \frac{1625}{81} = -\frac{86}{81}$, and so $C = -\frac{43}{81}$. Equating the coefficients of x^2, we obtain $1 = \frac{325}{81} + 8B$, and so $B = -\frac{61}{162}$.

We can now conclude that the partial fraction decomposition is

$$\frac{x^2 + 5x + 19}{(8x + 2)(x^2 + 5)} = \frac{\frac{325}{81}}{8x + 2} + \frac{-\frac{61}{162}x - \frac{43}{81}}{x^2 + 5}. \quad \blacktriangleleft$$

Repeated Quadratic Factors

▶ **EXAMPLE 3.65**   Find the partial fraction decomposition of $\dfrac{2x^4 + 7x^3 + 12x^2 + x + 3}{x(x^2 + x + 1)^2}$.

Solution   As in example 3.64, steps 1 and 2 have already been done for us, since $x^2 + x + 1$ is irreducible. Steps 3 and 4 lead to

$$\frac{2x^4 + 7x^3 + 12x^2 + x + 3}{x(x^2 + x + 1)^2} = \frac{A}{x} + \frac{Bx + C}{x^2 + x + 1} + \frac{Dx + E}{(x^2 + x + 1)^2}.$$

Multiplying both sides of this equation by the LCM, $x(x^2 + x + 1)^2$, we get the identity

$$2x^4 + 7x^3 + 12x^2 + x + 3 = A(x^2 + x + 1)^2 + (Bx + C)x(x^2 + x + 1) + (Dx + E)x.$$

Substituting the only real root of the denominator, $x = 0$, we see that $A = 3$.

As above, we may choose to find the remaining four constants by either substituting "new" values of x or by equating corresponding coefficients of powers of x. We shall do the latter. (You may wish to try the former.) Rewrite the identity as

$$2x^4 + 7x^3 + 12x^2 + x + 3 =$$
$$(A + B)x^4 + (2A + B + C)x^3 + (3A + B + C + D)x^2 + (2A + C + E)x + A.$$

Since we already know that $A = 3$, the right-hand side of this identity is

$$(3 + B)x^4 + (6 + B + C)x^3 + (9 + B + C + D)x^2 + (6 + C + E)x + 3.$$

Equate the coefficients of x^4: $2 = 3 + B$. Hence $B = -1$.

Equate the coefficients of x^3: $7 = 6 + B + C = 6 + (-1) + C = 5 + C$, and so $C = 2$.

Equate the coefficients of x^2: $12 = 9 + B + C + D = 9 + (-1) + 2 + D = 10 + D$. So, $D = 2$.

Equate the coefficients of x: $1 = 6 + C + E = 6 + 2 + E = 8 + E$, and we find that $E = -7$.

Therefore, the partial fraction decomposition is

$$\frac{2x^4 + 7x^3 + 12x^2 + x + 3}{x(x^2 + x + 1)^2} = \frac{3}{x} + \frac{-x+2}{x^2+x+1} + \frac{2x-7}{(x^2+x+1)^2}.$$

Exercises 3.7

In exercises 1–40, find the partial fraction decomposition of the given rational function.

1. $\dfrac{x+38}{(x+5)(x-6)}$

2. $\dfrac{x-41}{(x-2)(x+11)}$

3. $\dfrac{2x-11}{(x+3)^2}$

4. $\dfrac{3x^2-12x+11}{(x-2)^3}$

5. $\dfrac{-x^4-x^3+5x^2+x-2}{x^3+2x^2}$

6. $\dfrac{x^4-3}{x^4+2x^3}$

7. $\dfrac{11x^2-5\sqrt{2}x+6}{3x^2(x-\sqrt{2})}$

8. $\dfrac{3x^2-(6\sqrt{2}+1)x+6}{x(x-\sqrt{2})^2}$

9. $\dfrac{x^3+2}{x^3+x^2-6x}$

10. $\dfrac{x^3+7x^2-13x+12}{x^4+x^3-12x^2}$

11. $\dfrac{5}{6(x-6)^2}$

12. $\dfrac{2}{3(x+2)^4}$

13. $\dfrac{x^2+4x}{(x^2+4x+3)^2}$

14. $\dfrac{-7x+3}{x^2+1}$

15. $\dfrac{4x^3-6x^2+19x-31}{2x^2+8}$

16. $\dfrac{2x+6}{(x^2+4x+3)^2}$

17. $\dfrac{x^4-x^3+x^2-2x+1}{2x^3+2x^2+2x}$

18. $\dfrac{5x^3-4x^2+3x-3}{2x^4+2x^2}$

19. $\dfrac{3x^3-5x^2+27x-47}{(x^2+9)^2}$

20. $\dfrac{2x^2+7x+15}{(x^2+6)^2}$

21. $\dfrac{x^3+9x^2+x+5}{x^4-1}$

22. $\dfrac{-9x^3+18x^2+25x-5}{(x^2-x+1)(x-2)(x+3)}$

23. $\dfrac{-2x^3+7x^2+2x+9}{(x^2+1)(x+1)^2}$

24. $\dfrac{-x^5+x^4+x^3+9x^2+2x+4}{x^4-1}$

25. $\dfrac{4x^2+14\sqrt{2}x+4}{x^4+1}$ (Hint: x^4+1 has $x^2+\sqrt{2}x+1$ as a factor.)

26. $\dfrac{-2x^3+11x^2-8x+40}{x^4+7x^2+12}$

27. $\dfrac{(1-\pi)x^4+2\pi x^3+(3-12\pi)x^2+12\pi x-18(1+2\pi)}{(x^2+6)^2 x}$

28. $\dfrac{4x^4+39x^2-54x+99}{(x^2+4)^2(x-7)}$

29. $\dfrac{3x^3+19x+2}{x^4+11x^2+24}$

30. $\dfrac{4x^3-9x^2+16x-30}{(x^2+3)(x^2+4)}$

31. $\dfrac{-x^4+8x^3-3x^2-28x+68}{(x^2-4)^2}$

32. $\dfrac{-x^4+5x^3-37x^2+36x-42}{(3x^2+5)(x-1)^3}$

33. $\dfrac{-6x-29}{2x^2+6x-8}$

34. $\dfrac{-x^5-x^4-20x^3+10x^2-16x+32}{(x^3+4x)^2}$

35. $\dfrac{162x^9-243x^8+216x^7-324x^6+108x^5-162x^4+25x^3-37x^2+2x-3}{x^2(3x^2+1)^4}$

36. $\dfrac{(2+\sqrt{2})x+(2\sqrt{2}-6)}{3x^2-3x-6}$

37. $\dfrac{4x^3+3x^2-86x-27}{(3x^2-3x-6)^2}$

38. $\dfrac{7x^3+35x^2+24x-91}{(2x^2+6x-8)^2}$

39. $\dfrac{x^6-8x^5-66x^3-44x^2-130x-128}{(x^2+4)^3(x-1)^2}$

40. $\dfrac{\pi x+\sqrt{2}}{x^2+1}$

41. (Why step 1 is needed)
   (a) Consider nonzero polynomials $P_1, Q_1, P_2,$ and Q_2 such that the degree of P_i is less than the degree of Q_i (for $i = 1, 2$). Suppose that
   $$\frac{P_1}{Q_1} + \frac{P_2}{Q_2} = \frac{P}{Q}$$
   for some nonzero polynomials P and Q. Prove that the degree of P is less than the degree of Q. (Hint: you may take $P = P_1Q_2 + P_2Q_1$ and $Q = Q_1Q_2$. Apply exercise 44 of section 3.1.)
   (b) Let f be a rational function having partial fraction decomposition whose initial polynomial term is 0. Suppose $f = \frac{P}{Q}$, where P and Q are nonzero polynomials. Prove that the degree of P is less than the degree of Q. (Hint: use (a).)
   (c) Show that if P is a polynomial of positive degree, then P cannot be written as a sum of finitely many partial fractions of the form
   $$\frac{A}{(ax+b)^i} \text{ and } \frac{Bx+C}{(ax+b+c)^j}.$$

42. (Partial fractions over the complex numbers)
   (a) Find complex constants A and B so that
   $$\frac{1}{x^2+1} = \frac{A}{x-i} + \frac{B}{x+i}.$$
   (b) Find complex constants $A, B, C, D,$ and E so that
   $$\frac{x^3 - ix + 2}{(x^2+4)^2(x-3)} = \frac{A}{x-2i} + \frac{B}{(x-2i)^2} + \frac{C}{x+2i} + \frac{D}{(x+2i)^2} + \frac{E}{x-3}.$$

3.8 CHAPTER SUMMARY AND REVIEW EXERCISES

Major Concepts

- Asymptote
- Conjugate Root Theorem
- Continuous function
- Depressed equation (relative to P and r)
- Descartes' Rule of Signs
- Discontinuous function
- Division Algorithm
  - Dividend
  - Divisor
  - Quotient
  - Remainder
- Factor Theorem
- Fundamental Theorem of Algebra
- Guidelines for Graphing Polynomials
- Guidelines for Graphing Rational Functions
- Horizontal asymptote
- Intermediate Value Theorem for Polynomials
- Leading Term Test
- Linear Factor Theorem
- Lower bound
- Nested multiplication
- nth-degree polynomial
  - Coefficients
  - Complex polynomial of degree n
  - Leading coefficient
  - Real polynomial of degree n
  - Zero polynomial
- Partial fraction decomposition
- Polynomial function of degree n
- Power function of degree n
- Rational function
- Rational Root Test
- Remainder Theorem
- Root of multiplicity k
- Root of $P(x)$
- Simplified rational expression
- Slant asymptote
- Smooth function
- Synthetic division
- Upper bound
- Variation of sign
- Vertical asymptote
- Zero of $P(x)$

Review Exercises

1. Consider the functions $f, g,$ and h given by $f(x) = 2x - \pi$, $g(x) = ix^3 - ix + 1$, and $h(x) = \frac{x^3 - 8}{x - 2}$.
   (a) Which of these functions is/are a real polynomial?
   (b) Which is/are a complex polynomial?
   (c) Which is/are a rational function?

2. Determine whether each of the given functions is even, odd, or neither; graph them; and establish all valid symmetries with respect to the axes or the origin:
   (a) $f(x) = -\frac{2}{5}x^3$ (b) $g(x) = 2x^2 - 3$
   (c) $h(x) = \frac{5}{x^4} + 3$ (d) $j(x) = \frac{2}{(x-5)^3} + 4$

3. Determine the natural domain and the range for each of the functions in exercise 2. (Hint: you can use the graphs to find the ranges.)

4. Given functions f and g defined by $f(x) = \dfrac{x+1}{x-2}$ and $g(x) = \dfrac{x^3+1}{x^2-2}$, find the functional expressions for
   (a) $f + g$, (b) f/g, and (c) $g \circ f$.

5. Use the Factor Theorem to show that both $x - 2$ and $x + 3$ are factors of $P(x) = 3x^4 - x^3 - 17x^2 + 29x - 30$. Hence factor $P(x)$ as a product of real linear polynomials and an irreducible real quadratic.

6. Without substituting the value -3 into $P(x) = \frac{3}{2}x^4 + \frac{2}{3}x^3 - 3x^2 + 7$, use the Remainder Theorem to find $P(-3)$.

In exercises 7–9, find the quotient $q(x)$ and remainder $r(x)$ arising from the given dividend $P(x)$ and divisor $D(x)$.

7. $P(x) = 2x^3 - 3x + 7$, $D(x) = x + 9$
8. $P(x) = -2x^3 + ix^2 + 7$, $D(x) = 9x - 6$
9. $P(x) = ix^5 + 2x^4 + x^2 - 2ix$, $D(x) = x^4 - ix$
10. Find all values of k so that $x - 3$ is a factor of $81k^2x^2 + 108kx + 36$.
11. Use nested multiplication to rewrite $P(x) = -4x^3 - 16x^2 + 4x + 5$ and hence evaluate $P(1.2)$.

In exercises 12–13, construct a polynomial having exactly the given roots (with the indicated multiplicities).

12. $0, 0, -2, -2, -2, i, \sqrt{2}$. (Hint: the coefficients won't all be real.)
13. $2 + 3i, 2 - 3i, 5 + \sqrt{2}, 5 - \sqrt{2}, 1, 1$

In exercises 14–16, find a real (and, if possible, rational) polynomial of degree n, having (at least) the specified numbers as roots.

14. $n = 4$; roots $2 - \sqrt{3}, 3 - i$
15. $n = 4$; roots $\sqrt{3}, -i, 2$
16. $n = 5$; roots $\frac{2}{3} + \sqrt{7}, \sqrt{2}, 5$
17. In each part, the graph of a function f is shown. State whether or not f is a continuous function. Then state whether or not the given graph is smooth.

(a)

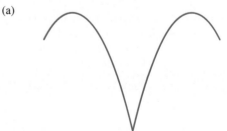

(b)

(c)

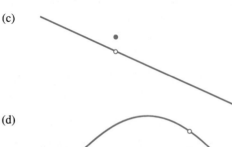

(d)

18. Use the Intermediate Value Theorem to estimate the root of $P(x) = x^4 + 6x - 42$ between 2 and 3 to within 0.005.
19. Prove that $P(x) = -7x^9 + \pi x^8 - 4$ has at least one negative root. (Hint: Descartes.)
20. Use the Leading Term Test to determine the limiting tendencies of (a) $P(x) = -\frac{2}{5}x^7 + \sqrt{2}x^6 - \pi x^3 - \frac{3}{2}x^2 + \pi$ and (b) $Q(x) = 3x^6 - 6x^3$ as x moves without bound to the right or left.

In exercises 21–25, find all the roots, with their multiplicities, of the given polynomial.

21. $(x^3 - 64)(x^2 - 16)^2(x^2 + 16)(3x - 1)$
22. $x(x - 5)^2(x^2 + 9)^3(2 + x)$
23. $4x^6 + 4x^5 - 19x^4 - 4x^3 + 15x^2$ (Hint: think rational!)
24. $3x^3 - 2x^2 + 12x - 8$
25. $5x^5 + 29x^4 + 64x^3 + 56x^2 + 11x - 5$

In exercises 26–29, what does Descartes' Rule of Signs allow you to conclude about (a) the number of positive roots (counting multiplicities) of $P(x)$, (b) the number of negative roots (counting multiplicities) of $P(x)$, and (c) whether $P(x)$ has nonreal complex roots?

26. $P(x) = \pi x^5 + x^2 + \sqrt{2}$
27. $P(x) = \frac{2}{3}x^6 - \frac{3}{2}x^5 - 4x^3 + 6x$
28. $P(x) = 2x^3 + 3x^2 + x - 6$
29. $P(x) = 9x^{10} - 8x^9 + 9x^7 - 4x^3 + 5x^2 + 6x - 3$
30. Prove that -3 is a lower bound and 3 is an upper bound for the real roots of the polynomial $4x^4 + 4x^3 - 11x^2 + 4x - 15$.

In exercises 31–34, solve the given inequality.

31. $x^2(x - 3)(x + 3)(x^2 + 2x + 3) \geq 0$

Chapter Summary and Review Exercises

32. $x(x-3)^3(x+3)^2(5-x) < 0$
33. $2x^4 + 5x^2 + 10x - 25 < 9x^3 - 3x^2 - 9x + 5$
34. $\dfrac{3x^2 + x}{(x+1)(x-2)} > \dfrac{5x}{(x+1)(2-x)}$
35. Determine the natural domain for each of the given rational functions:
    (a) $f(x) = \dfrac{(x-1)(x-2)^2(x-4)}{(x-1)^2(x-2)(x-3)^5}$
    (b) $g(x) = \dfrac{2x+6}{x^2 + 6x + 7}$ (c) $h(x) = \dfrac{-3}{(x-4)^2} + 5$
36. Determine the range of the rational function given by $f(x) = \dfrac{x^2}{x-1}$. (Hint: noticing $f(x) = x + 1 + \dfrac{1}{x-1}$, graph f. You should find a "high" point at $x = 0$ and a "low" point at $x = 2$.)

In exercises 37–40, find all the asymptotes for the graph of the given rational function.

37. $f(x) = \dfrac{(x-2)(x+4)}{(x-2)(x-3)}$
38. $f(x) = \dfrac{3x^3 - 11x^2 + 10x + 11}{x^2 - 2x - 3}$
39. $f(x) = \dfrac{\pi x^2 + 5x - 100}{x^3 - x^2 - 5x - 3}$
40. $f(x) = \dfrac{\pi x^5 + 5x^4 - 100x^3}{x^3 - x^2 - 5x - 3}$

In exercises 41–46, graph the given function. (You may wish to use a graphing calculator or computer graphing program to help you or to provide a "rough check" of your work.)

41. $P(x) = 2x^3 - 2x$
42. $P(x) = 2x^3 - x^2 + 2x - 1$
43. $P(x) = x^4 - 4x^2 - 2$
44. $f(x) = \dfrac{x^2 - 4}{x^2 - 9}$
45. $f(x) = \dfrac{x^3 + 4x - 5}{x^2 - 4}$
46. $f(x) = \dfrac{x(x-5)}{(x-1)(x-2)^2(x-3)(x-4)}$

In exercises 47–52, find the partial fraction decomposition of the given rational function.

47. $\dfrac{3x^2 - 19}{(x-3)^2(x+1)}$
48. $\dfrac{x^4 - x^3 + 2x^2 + 2x + 3}{x^4(x^2 + x + 1)}$
49. $\dfrac{2x+1}{x^2 + x + 1}$
50. $\dfrac{2x^3 + 7x^2 + 9x - 7}{x^2 + 4x + 5}$
51. $\dfrac{2x^3 - x^2 + 2x + 1}{x^4 - 1}$
52. $\dfrac{-8}{5(x-1)^3}$

53. A very large reservoir initially contains 80 gallons of pure water. Starting at time $t = 0$, a salty solution containing 0.15 pound of salt per gallon is pumped into the reservoir at a rate of 4 gallons per minute.
    (a) Show that if the reservoir never overflows, then after t minutes, the concentration of salt in the reservoir is $S(t) = \dfrac{3t}{20t + 400}$ pounds per gallon.
    (b) Show that S is an increasing function (using the domain $[0, \infty)$).
    (c) What is the limiting tendency of $S(t)$ as t increases without bound?
    (d) Give physical interpretations for your answers to (b) and (c).

54. A rectangular sheet of cardboard has dimensions 18 inches by 24 inches. Identical squares, each measuring x inches by x inches, are cut from the four corners of the sheet. The resulting cardboard is folded up to form a box with no top. If this box's volume is 460 in^3, find all possible values of x to the nearest tenth (of an inch).

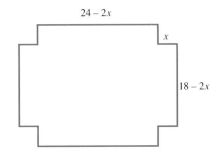

Chapter 4

Exponential and Logarithmic Functions

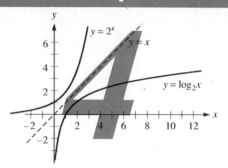

In this chapter, we shall introduce two families of functions which are very important not only for your future study of mathematics but also in view of their many business and scientific applications. We shall show some applications of these two functions in finance and in scientific areas such as growth and decay. In later chapters, you will see a few ways in which these functions are used in more advanced mathematics.

4.1 EXPONENTIAL FUNCTIONS

Many of the problems that were worked in the earlier chapters concerned a variable raised to a power. It was not unusual to consider functions like $f(x) = x^2$ or $g(t) = t^5$. What would happen if we reversed the roles of variable and constant? What if we considered the various powers of some fixed number? For example, consider the powers of 4: $4^3 = 64$, $4^4 = 256$, $4^5 = 1024$, $4^{1/2} = 2$, $4^{-1} = 0.25$, etc. For each value of x, there is a unique value for 4^x. We thus have a function that can be expressed as $f(x) = 4^x$. This is an example of an exponential function. More formally, we have the following definition.

Exponential Function with Base b

A function f is an **exponential function with base b** if c and b are constant, $c \neq 0$, $b > 0$, $b \neq 1$, and

$$f(x) = cb^x$$

Exponential Functions

for each real number x. The domain of an exponential function is the set of real numbers. Its range can be shown to be the set of positive real numbers when c is positive, and the set of negative real numbers when c is negative. The number b is called the **base**.

▶ EXAMPLE 4.1 $f(x) = 2^x$, $g(x) = 3^x$, $h(x) = \pi^x$, $j(x) = 4.2^x$, $k(x) = (\sqrt{3})^x$, and $m(x) = (\frac{3}{7})^x$ are all exponential functions. In each of these, $c = 1$.
$f(x) = (5a)^{-x}$, $g(x) = 12b^x$, $h(x) = (2.3)^{x+7}$, $j(x) = (\sqrt{5})7^x$, $k(x) = 4d^{x/5}$, and $m(x) = -2d^x$ are also exponential functions, where a, b, and d represent positive constants, $a \neq \frac{1}{5}$, and b and d are not 1.
$f(x) = (-2)^x$ and $g(x) = 0^x$ are not exponential functions because, in both cases, the "base" is not greater than 0. Also, $h(x) = 1^x$ is not an exponential function since the "base" is 1. ◀

The first examples that we shall consider have the constant $c = 1$. However, most naturally occurring examples use constants c other than one.

Graphing Exponential Functions; Asymptotes

An exponential function is graphed in the same manner as other functions. First, a set of ordered pairs in the function is determined. Next, that set of ordered pairs is plotted on a graph. Finally, these plotted points are connected in order to get a sketch of the graph. Notice that since the base b of an exponential function is positive, all powers of that base are also positive. Thus, the values of an exponential function $f(x) = cb^x$ are positive for all values of x when $c > 0$, and negative when $c < 0$. In particular, the graph of an exponential function $f(x) = cb^x$ lies entirely above the x-axis when c is positive, and entirely below the x-axis when c is negative. As noted above, it can be shown that the range of $f(x) = cb^x$ is $(0, \infty)$ for $c > 0$ and is $(-\infty, 0)$ for $c < 0$.

We shall first graph the exponential function $f(x) = 2^x$. A table of values is given below for the integer values of x from -3 to 5.

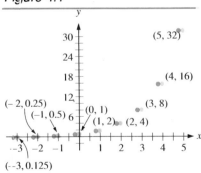
Figure 4.1

x	-3	-2	-1	0	1	2	3	4	5
$f(x) = 2^x$	0.125	0.25	0.5	1	2	4	8	16	32

These nine points are plotted in figure 4.1.

The graph of $f(x) = 2^x$ shown in figure 4.2 was formed by smoothly connecting the nine points that were plotted in figure 4.1. As x gets larger, $f(x)$ seems to increase at a faster and faster rate. As x gets smaller, $f(x)$ gets closer and closer to zero, but, since it is always positive, it never reaches the value of zero. Notice how, as x gets smaller, the curve gets closer and closer to the x-axis. In other words, as x decreases without bound, the limiting tendency of 2^x is 0. Thus, the x-axis is a horizontal asymptote to the graph of $f(x) = 2^x$.

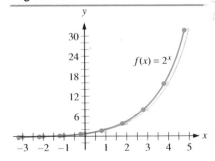
Figure 4.2

As would be expected, a calculator is helpful for finding values of an exponential function. The $\boxed{y^x}$ or $\boxed{x^y}$ key is used for most calculators; use the $\boxed{\wedge}$ key on a *TI-81*. To determine $f(4.5)$ when $f(x) = 2^x$, you press the keystroke combination below that is appropriate for your calculator. Notice that the first number is the base and the second is the exponent.

Algebraic calculator:	2 $\boxed{y^x}$ 4.5 $\boxed{=}$
RPN calculator:	2 $\boxed{\text{ENTER}}$ 4.5 $\boxed{y^x}$
TI-81 graphing calculator:	2 $\boxed{\wedge}$ 4.5 $\boxed{\text{ENTER}}$
Casio *fx-7700G* graphing calculator:	2 $\boxed{x^y}$ 4.5 $\boxed{\text{EXE}}$

In all four cases, the calculator will display 22.627417.

The most common bases are the ones larger than 1. In fact, when the base is less than 1 (and, of course, greater than 0), you get the interesting effect shown in example 4.2.

▶ **EXAMPLE 4.2**   Graph $g(x) = (\frac{1}{2})^x$.

Solution   Once again we begin with a table of values.

x	-5	-4	-3	-2	-1	0	1	2	3
$g(x) = (\frac{1}{2})^x$	32	16	8	4	2	1	0.5	0.25	0.125

Figure 4.3

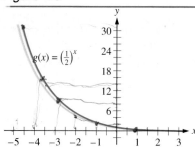

These points are plotted in figure 4.3, as is the graph that results when the points are connected. This graph would also result from reflecting the graph of $f(x) = 2^x$ in the *y*-axis.   ◀

Study the function g again. Remember that $\frac{1}{2} = 2^{-1}$, so $g(x) = (\frac{1}{2})^x = 2^{-x}$. These examples show the differences, when $b > 1$, between a function of the form $y = b^x$ and one of the form $y = b^{-x}$. If $b > 1$, then $y = b^x$ describes an increasing function, while $y = b^{-x}$ is a decreasing function. Also, for any fixed b, the graph of $y = b^{-x}$ is obtained from the graph of $y = b^x$ by reflecting about the *y*-axis.

Figure 4.4 shows the graphs of several exponential functions of the form $y = b^x$ relative to the same set of coordinate axes. These functions, indeed all exponential functions with $c = 1$, have three features in common.

Figure 4.4

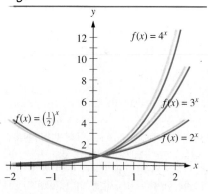

(1) The *y*-intercept is the point (0, 1).
(2) The *x*-axis is a horizontal asymptote.
(3) If $b > 1$, the function is increasing. If $0 < b < 1$, the function is decreasing.

In section 2.2, we studied horizontal stretchings and shrinkings. The next two examples show how these concepts apply to exponential functions.

Example 4.3

Consider the function $f(x) = 2^x$ from example 4.1. What can you say about the graph of $y = f(3x)$?

Solution

From section 2.2, we know that, if $|d| > 1$, the graph of $y = f(dx)$ is obtained by shrinking horizontally the graph of $y = f(x)$. In this example, $f(3x) = 2^{3x} = (2^3)^x = 8^x$. The graphs of $y = f(x) = 2^x$ and $y = f(3x) = 2^{3x}$ are shown in figure 4.5. The point $(0, 1)$ remains "fixed" on both graphs.

Figure 4.5

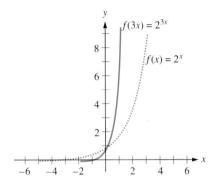

Example 4.4

If $f(x) = 2^x$, what can you say about the graph of $y = f(\tfrac{1}{4}x)$?

Solution

This is the graph of $y = f(dx)$, where $f(x) = 2^x$ and $d = \tfrac{1}{4}$. Since $|d| = \tfrac{1}{4} < 1$, the graph of $y = f(\tfrac{1}{4}x)$ is obtained by stretching horizontally the graph of $f(x) = 2^x$. The point $(0, 1)$ remains fixed. The graphs of $y = f(x) = 2^x$ and $y = f(\tfrac{1}{4}x) = (2^{1/4})^x$ are shown in figure 4.6.

Figure 4.6

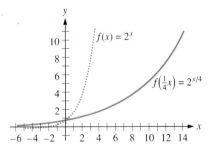

Example 4.5

If $f(x) = 2^x$, what can you say about the graph of $y = f(-x)$?

Solution

The graph of $y = f(-x)$ can be obtained from the graph of $f(x) = 2^x$ since it is of the form $y = f(dx)$ with $d = -1$. In section 2.2, you learned that $d < 0$ indicates that the graph of $y = f(-x)$ is obtained by reflecting the graph of $y = f(x)$ in the y-axis, as shown in figure 4.7. Notice that $y = f(-x)$ is the same function studied in example 4.2.

Chapter 4 / Exponential and Logarithmic Functions

Figure 4.7

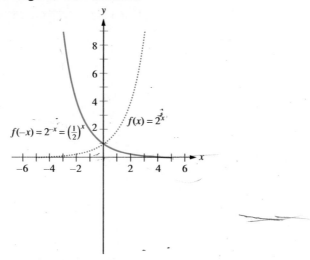

EXAMPLE 4.6

If $f(x) = 2^x$, what can you say about the graph of $y = 3 \cdot 2^x$?

Solution

From section 2.2, we know that, if $|c| > 1$, the graph of $y = cf(x)$ is obtained by stretching vertically the graph of $y = f(x)$. For example, since the y-intercept of the graph of $y = f(x)$ is 1 unit from the x-axis, the y-intercept of $y = 3f(x)$ will be 3 units from the x-axis. The graphs of $f(x) = 2^x$ and $y = 3f(x)$ are shown in figure 4.8.

Figure 4.8

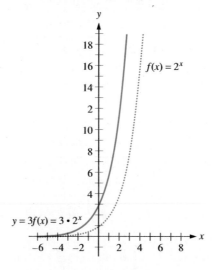

EXAMPLE 4.7

If $f(x) = 2^x$, what can you say about the graph of $y = (-\frac{3}{4})2^x$?

Solution

$y = (-\frac{3}{4})2^x$ is of the form $y = cf(x)$ with $c = -\frac{3}{4}$. Since $|c| = \frac{3}{4} < 1$, the graph of $y = (-\frac{3}{4})2^x$ is obtained from the graph of $f(x) = 2^x$ in two steps. First, shrink it vertically, to obtain the graph of $y = (\frac{3}{4})2^x$, as shown in figure 4.9(a). Next, since $c < 0$, this graph should be reflected in the x-axis to get the graph of $y = (-\frac{3}{4})2^x$, as shown in figure 4.9(b). Notice that its y-intercept is $(0, -\frac{3}{4})$.

Figure 4.9

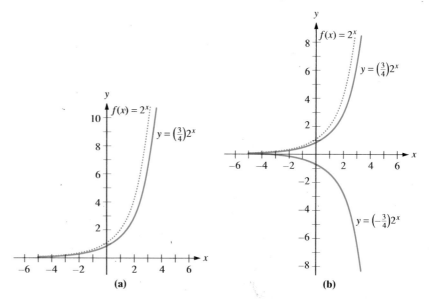

(a)

(b)

EXAMPLE 4.8

If $f(x) = 3^x$, what can you say about the graphs of (a) $g(x) = 3^x + 1$, (b) $h(x) = \frac{4}{9}(3^x) + 1$, (c) $j(x) = (-\frac{4}{9})3^x + 1$, and (d) $k(x) = (-\frac{4}{9})(3^{x+2}) + 1$?

Solutions

The graph of $f(x) = 3^x$ is shown in figure 4.10(a).

(a) $g(x) = 3^x + 1$ is of the form $g(x) = f(x) + 1$. Thus, the graph of g is obtained by translating the graph of f vertically 1 unit, as shown in figure 4.10(b).

(b) Here, $h(x) = \frac{4}{9}(3^x) + 1 = \frac{4}{9}f(x) + 1$. First, shrink the graph of $f(x) = 3^x$ vertically to obtain the graph of $y = \frac{4}{9}(3^x)$. Next, translate the resulting graph vertically 1 unit, to obtain the graph of h shown in figure 4.10(c).

(c) We see that $j(x) = (-\frac{4}{9})3^x + 1 = (-\frac{4}{9})f(x) + 1$. Three steps are needed to obtain this graph. First, shrink the graph of f vertically to obtain the graph of $y = \frac{4}{9}f(x)$. Since $-\frac{4}{9} < 0$, we next reflect the resulting graph in the x-axis to get the graph of $y = (-\frac{4}{9})3^x$. Finally, translate this graph vertically 1 unit to obtain the graph of j shown in figure 4.10(d).

Figure 4.10

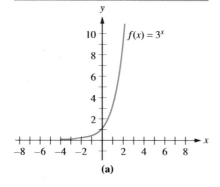

(a)

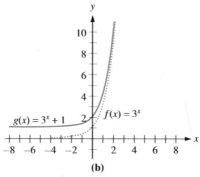

(b)

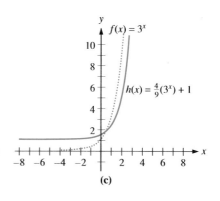

(c)

Figure 4.10 (continued)

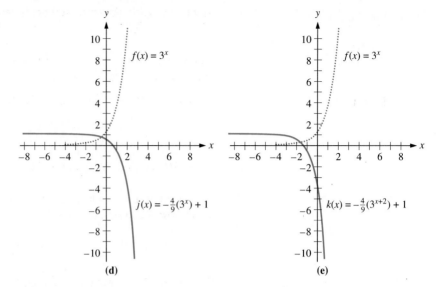

(d) We have $k(x) = (-\frac{4}{9})(3^{x+2}) + 1 = (-4)3^x + 1 = -4f(x) + 1$. Begin by stretching vertically the graph of f, reflect the result in the x-axis, and finally translate vertically 1 unit. The final graph is shown in figure 4.10(e). (Notice that you could have treated 3^{x+2} as $f(x + 2)$. Thus $k(x) = (-\frac{4}{9})f(x + 2) + 1$, and so you could have begun by translating the graph of f horizontally 2 units to the left.)

We have the following general results.

General Characteristics of the Exponential Function $f(x) = cb^x$

For any exponential function $f(x) = cb^x$ with $c \neq 0$, $b > 0$, $b \neq 1$:

1. The y-intercept is $(0, c)$.
2. The x-axis is a horizontal asymptote.
3. The domain is $\mathbf{R}$.
4. (a) For $c > 0$: The range of f is $(0, \infty)$.
   (b) For $c < 0$: The range of f is $(-\infty, 0)$.
5. (a) For $c > 0$: If $b > 1$, f is an increasing function;
              if $0 < b < 1$, f is a decreasing function.
   (b) For $c < 0$: If $b > 1$, f is a decreasing function;
              if $0 < b < 1$, f is an increasing function.

Applications

One application of exponential functions has to do with money. When an amount of money P, called the **principal,** is invested and interest is compounded at the interest rate of r per year, then the amount of money at the end of t years is given by the formula

$$S = P(1 + r)^t.$$

Exponential Functions

In this formula, **r is expressed as a decimal**. For instance, at 5 percent interest, $r = 0.05$; and at $6\frac{1}{4}$ percent interest, $r = 0.0625$.

EXAMPLE 4.9

Suppose $700 is invested at 9 percent compounded annually. How much is the total worth after 10 years?

Solution

Here $P = \$700$, $r = 0.09$, and $t = 10$. Putting these into the formula, we get

$$S = 700(1 + 0.09)^{10}$$
$$\approx \$1657.15.$$

After 10 years, the total value of this investment is $1657.15. ◀

If interest is compounded more than once a year, the formula is changed. If interest is compounded semiannually, or twice a year, the formula becomes

$$S = P\left(1 + \frac{r}{2}\right)^{2t}.$$

If interest is compounded quarterly, or four times a year, the formula is changed to

$$S = P\left(1 + \frac{r}{4}\right)^{4t}.$$

In general, if interest is compounded k times a year at an interest rate of r, the result is given by the following **compound interest formula:**

$$S = P\left(1 + \frac{r}{k}\right)^{kt}.$$

EXAMPLE 4.10

Suppose $700 is invested in a savings account paying 9 percent annual interest compounded monthly. How much is the total worth after 10 years?

Solution

As in the last example, we have $P = \$700$, $r = 0.09$, and $t = 10$. However, this time, we have $k = 12$. Putting these into the formula, we get

$$S = 700(1 + \tfrac{0.09}{12})^{12 \cdot 10}$$
$$= 700(1.0075)^{120}$$
$$\approx \$1715.95.$$

Compare this result with the result in example 4.9. Compounding monthly increases the total by $58.80. ◀

The Number e

The above formula concerned the amount of interest that has accumulated in an account after a certain period of time. An interesting variation of that formula uses $\left(1 + \frac{1}{n}\right)^n$. Let's examine the values of $\left(1 + \frac{1}{n}\right)^n$ as n gets larger and larger. You might want to use a calculator to check these figures.

Chapter 4 / Exponential and Logarithmic Functions

n	1	10	100	1,000	10,000	100,000	1,000,000
$\left(1+\dfrac{1}{n}\right)^n$	2	2.5937425	2.7048138	2.7169238	2.7181459	2.7182546	2.7182818

Mathematicians have been able to prove that as the value of n gets larger, the value of $\left(1+\dfrac{1}{n}\right)^n$ also continues to get larger, but there is an upper bound on how large it can get. We shall show in section 10.6 that $\left(1+\dfrac{1}{n}\right)^n \leq 3$ for all positive integers n. Thus, 3 is an upper bound of $\left(1+\dfrac{1}{n}\right)^n$. Of all the upper bounds for $\left(1+\dfrac{1}{n}\right)^n$, there is a smallest one, and this smallest value is such a special number that it has been given its own symbol, e. In other words, as n increases without bound, the limiting tendency of $\left(1+\dfrac{1}{n}\right)^n$ is e. The first eight digits of e are the same ones we got in the table above, and so $e \approx 2.7182818$.

The number e occurs in many scientific applications. A later section in this chapter is devoted entirely to the exponential function $f(x) = e^x$.

Approximating e^x for particular values of x is primarily done using a table, computer, or calculator. A table of exponential values is in Appendix I.

There are three possible ways to use a calculator to evaluate e^x. The first way is to use the $\boxed{y^x}$ key and let the base, y, be 2.7182818. Not only is this slow, but it is more prone to errors than are the other two methods.

The other two possible methods for calculating e^x with a calculator depend on the type of calculator that you are using. Some calculators have an $\boxed{e^x}$ key. (Notice that on some of these, you may have to press the $\boxed{\text{2nd}}$ or $\boxed{\text{Shift}}$ key before you press the $\boxed{e^x}$ key.) With one of these calculators, if you wanted to compute $e^{3.5}$, you would press either

$$3.5 \; \boxed{e^x} \quad \text{or} \quad \boxed{e^x} \; 3.5$$

depending on the type of calculator, and in a short time, something like 33.115452 would appear in the display area.

Some algebraic calculators do not have an $\boxed{e^x}$ key. Instead you will have to use a combination of two keys to calculate the value of e^x. These calculators use the $\boxed{\text{INV}}$ (or $\boxed{\text{2nd}}$) key and the $\boxed{\ln x}$ key. For example, to use an algebraic calculator to compute $e^{3.5}$, you would press

$$3.5 \; \boxed{\text{INV}} \; \boxed{\ln x} \quad \text{or} \quad 3.5 \; \boxed{\text{2nd}} \; \boxed{\ln x}$$

and once again, after a brief time, you would see 33.115452 in the display area. What is this $\boxed{\ln x}$ key? That will be explained in the next section.

Let's return to a problem involving compound interest. We know, from the compound interest formula, that if a principal amount P is invested and interest is compounded k times a year at an interest rate of r, then the amount after t years is

Exponential Functions

$$S = P\left(1 + \frac{r}{k}\right)^{kt}.$$

If we let $n = \dfrac{k}{r}$, then $k = nr$ and this formula becomes

$$S = P\left(1 + \frac{1}{n}\right)^{nrt}$$

$$= P\left[\left(1 + \frac{1}{n}\right)^{n}\right]^{rt}.$$

Now if interest is "compounded continuously," then the expression inside the brackets, $\left(1 + \dfrac{1}{n}\right)^{n}$, is essentially the number e. We thus have the following result.

Continuous Compounding of Interest

The amount accumulated after t years at the interest rate r under continuous compounding is

$$S = Pe^{rt}.$$

▶ **EXAMPLE 4.11**

Suppose the same $700 from the previous two examples is invested in a savings account paying 9 percent interest compounded continuously. How much is the total worth after 10 years?

Solution

As in the last two examples, we have $P = \$700$, $r = 0.09$, and $t = 10$. Putting these into the formula for the continuous compounding of interest, we get

$$S = 700e^{(0.09)10}$$
$$= 700e^{0.9}$$
$$\approx 700(2.4596031)$$
$$\approx 1721.72.$$

Compare this result with the result in example 4.10. Compounding continuously increased the total by another $5.77. ◀

Exercises 4.1

In exercises 1–8, use a calculator to approximate the given numbers. Round off each answer to four decimal places.

1. $6^{\sqrt{2}}$
2. 5^{π}
3. π^{7}
4. $(\sqrt{7})^{\sqrt{5}}$
5. $\sqrt{5}^{\pi}$
6. $(\sqrt{9})^{4/3}$
7. $e^{5.3}$
8. e^{π}

In exercises 9–21, (a) make a table of values and sketch the graph of the given function and (b) describe the function's domain and range.

9. $f(x) = 4^{x}$
10. $g(x) = -3^{x}$
11. $h(x) = -2 \cdot 5^{3x}$
12. $k(x) = 0.4^{2x}$
13. $j(x) = 3^{-x}$
14. $f(x) = 4^{-x}$
15. $g(x) = 3 \cdot 4^{-x}$
16. $h(x) = 5^{x/2} = \sqrt{5^{x}}$
17. $k(x) = 3^{x+1/2}$
18. $j(x) = (\frac{1}{4})^{x-1/5}$
19. $f(x) = 3^{|x|}$
20. $g(x) = 2^{[x]}$
21. $h(x) = 4^{-x} + 4^{x}$

In exercises 22–28, use the skills from the previous exercises and the transformation techniques of chapter 2 to sketch the graph of each of the following functions. (You may wish to use a graphing calculator or computer graphing program to help you or to provide a "rough check" of your work.)

22. $f(x) = 4^x - 5$
23. $g(x) = -2 \cdot 4^x$
24. $h(x) = (\frac{1}{2})4^{x-1}$
25. $j(x) = 4^{x/2} - 5$
26. $k(x) = (\frac{1}{4})^{x-1/5} - 5$
27. $m(x) = 3^{|x|} + 7$
28. $n(x) = -\frac{1}{2}(2^{[x]}) - 4$

In exercises 29–30, find the y-intercept of the graph of the given exponential function.

29. (a) $f(x) = \pi^x$ (b) $g(x) = 5 \cdot 3^{-x}$
30. (a) $h(x) = 2^{-x}$ (b) $k(x) = -3 \cdot 4^x$
31. Consider the four exponential functions studied in exercises 29 and 30.
    (a) Determine which of these is/are increasing, and which is/are decreasing.
    (b) Determine the horizontal asymptote(s) of these functions' graphs.
32. Identify the horizontal asymptotes, if any, for the graph of the given function:
    (a) $f(x) = \pi^x + 4$ (b) $g(x) = 2^{-x} + 5$
    (c) $h(x) = -3 \cdot 4^x$ (d) $k(x) = 5 \cdot 3^{-x}$
    (e) $m(x) = 5 \cdot 2^{-x} + 4$ (f) $n(x) = -3 \cdot 4^x - 2$
33. The sum of $3000 is placed in a savings account at $6\frac{1}{2}$ percent interest. What is the total after 5 years if interest is compounded (a) annually, (b) semiannually, (c) quarterly, (d) monthly, (e) daily (assume 365 days in one year), or (f) continuously?
34. The sum of $4000 is placed in a savings account at 8 percent interest. What is the total after 10 years if interest is compounded (a) annually, (b) semiannually, (c) quarterly, (d) monthly, (e) daily, or (f) continuously?
35. One bank offers 6 percent interest compounded semiannually. A second bank offers the same interest but compounded monthly. How much more income will result after 10 years from depositing $1000 in the second bank than from placing the money in the first bank?
36. A bank offers 8 percent interest compounded annually. A second bank offers 8 percent interest compounded quarterly. How much more income will result from depositing $2000 in the second bank for 10 years than from depositing it in the first bank for the same length of time?
37. A savings institution offers 7.5 percent interest compounded continuously. How much can we expect to get from an investment of $4000 after 10 years?
38. If $5000 is invested in an account that pays $6\frac{1}{8}$ percent interest compounded continuously, how much can we expect to have after 10 years? How much more is this than if the same interest had been compounded monthly instead of continuously?
39. If we have 1000 grams (= 1 kilogram) of the radioactive isotope called carbon-14, it can be shown that the amount of carbon-14 left after a decay period of t years is approximately $1000(\frac{1}{2})^{t/5700}$ grams. How many grams are left after 1 year? after 2 years? after 3 years? after 5700 years? after 11,400 years?
◆ 40. The number of bacteria in a colony increased from 200 to 1000 during the period from noon to 3 P.M. the same day. If the number of bacteria t hours after noon on that day was $200(5)^{t/3}$, what was the size of the colony at 4 P.M.? at midnight? at 8 A.M. the next day?
41. Approximate e and π by 2.781828 and 3.14159, respectively. By plotting points, sketch the graphs of e^x and π^x on the same graph paper. Which of e^π and π^e appears to be greater? Check by "finding" e^π and π^e on your calculator.
42. Using your calculator and increasingly large values of n, speculate as to the limiting tendency, if any, of
$$\left(1 + \frac{k}{n}\right)^n$$
in case (a) $k = 2$, (b) $k = 3$, or (c) $k = -1$.
43. By reviewing the text's derivation for interest accruing from continuous compounding, present a plausible argument for the conclusion suggested by exercise 42: if k is any real number, the limiting tendency of
$$\left(1 + \frac{k}{n}\right)^n,$$
as n increases without bound, is e^k. (Hint: if $k \neq 0$, consider $\left(1 + \frac{1}{m}\right)^{mk}$, with $m = \frac{n}{k}$. You may use the fact that the limiting tendency of $\left(1 + \frac{1}{m}\right)^m$ is e when m either increases without bound or decreases without bound.)

4.2 LOGARITHMIC FUNCTIONS

Turn back to the last section and look at the graphs of the exponential functions in figures 4.2 through 4.7. They each pass the horizontal line test for the existence of the inverse of a function. This is the case for the graph of any

Logarithmic Functions

exponential function. The inverse for the exponential function $y = b^x$ is called the **logarithmic function, base b.** The symbol for this logarithmic function is $\log_b$. We read $\log_b x$ as "log to the base b of x" or "logarithm of x to the base b." Note that if $f(x) = b^x$ (for some given $b > 0$, $b \neq 1$), then $f^{-1}(x) = \log_b x$. So, by the properties of inverse functions developed in chapter 2, the domain of the function $y = \log_b x$ is the range of the function $y = b^x$; and the range of $\log_b x$ is the domain of b^x. In other words, the domain of any logarithmic function is the set of positive real numbers, and its range is the set of real numbers. In particular, $\log_b x$ is not defined if x is negative or zero.

The relationship between the logarithmic and exponential functions can best be seen by the following fact.

Log$_b$ x

$$\log_b x = y \quad \text{if and only if} \quad b^y = x \quad (\text{where } b > 0, b \neq 1, x > 0).$$

Thus, each logarithm y can be expressed as an exponent and, conversely, each exponent can be expressed as a logarithm. Put differently, this says the following:

$$b^{\log_b x} = x \quad \text{for all positive real } x$$

and

$$\log_b (b^x) = x \quad \text{for all real numbers } x.$$

▶ **EXAMPLE 4.12**   Rewrite each equation using exponentials.

Solutions

	Logarithmic Form	Exponential Form
(a)	$\log_4 1024 = 5$	$4^5 = 1024$
(b)	$\log_4 \left(\frac{1}{16}\right) = -2$	$4^{-2} = \frac{1}{16}$
(c)	$\log_2 0.125 = -3$	$2^{-3} = 0.125 = \frac{1}{8}$
(d)	$\log_9 27 = \frac{3}{2}$	$9^{3/2} = 27$
(e)	$\log_b 1 = 0$	$b^0 = 1$

◀

▶ **EXAMPLE 4.13**   Compute the following logarithms (exactly): (a) $\log_3 9$, (b) $\log_3 \frac{1}{9}$, (c) $\log_{1/2} \frac{1}{16}$, and (d) $\log_{1/2} 16$.

Solutions

(a) $\log_3 9 = \log_3 3^2 = 2$

(b) $\log_3 \frac{1}{9} = \log_3 \frac{1}{3^2} = \log_3 3^{-2} = -2$

(c) $\log_{1/2} \frac{1}{16} = \log_{1/2} (\frac{1}{2})^4 = 4$

(d) $\log_{1/2} 16 = \log_{1/2} (2^4) = \log_{1/2} (\frac{1}{2})^{-4} = -4$.

▶ **EXAMPLE 4.14** Simplify: (a) $3^{\log_3 4}$, (b) $(\frac{1}{2})^{\log_{1/2} 3}$, (c) $9^{\log_3 4}$, and (d) $2^{\log_{1/2} 7}$.

Solutions

(a) $3^{\log_3 4} = 4$

(b) $(\frac{1}{2})^{\log_{1/2} 3} = 3$

(c) $9^{\log_3 4} = (3^2)^{\log_3 4} = 3^{2 \log_3 4} = (3^{\log_3 4})^2 = 4^2 = 16$

(d) $2^{\log_{1/2} 7} = ((\frac{1}{2})^{-1})^{\log_{1/2} 7} = ((\frac{1}{2})^{\log_{1/2} 7})^{-1} = 7^{-1} = \frac{1}{7}$. ◀

Graphical representations of $\log_b x$ can be made using the technique we learned for graphing inverse functions. In figure 4.11, the graph of $y = 2^x$ has been drawn and the line $y = x$ is shown as a dashed line. The reflection of $y = 2^x$ in the line $y = x$ is shown by the colored curve. This colored curve, then, is the graph of $y = \log_2 x$.

Figure 4.11

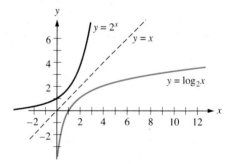

The above reflection method of graphing $y = \log_b x$ is admittedly awkward. We need a quicker way to evaluate $\log_b x$ so that we then can set up a table of values, plot some points, and connect these points in order to sketch the graph. This quicker method will be discussed after we examine logarithms for the two bases that are used most often in calculations.

Common Logs; Natural Logs

The two bases that are used most often are 10 and e. Logarithms that have a base of 10 are called **common logs** and are denoted **log**. Logarithms with a base of e are called **natural logs** and are denoted **ln**.

There are three ways in which most people find the values of a logarithm: by calculator, computer, or table. Tables of logarithms are in Appendices II and III.

Logarithmic Functions

Finding Logs Using Calculators

Your calculator probably has two keys on it that you can use to determine the logarithm of a number. These are the keys [log] and [ln x] (or [LN]). Since log x means $\log_{10} x$, the [log] key can be used to determine values of $\log_{10} x$. Similarly, the [ln x] key is used to determine values of $\log_e x$. (Some calculators require the use of the [2nd] key in combination with the [log] or [LN] keys. On a graphing calculator, you press the [log] or [LN] key before you enter numbers.)

EXAMPLE 4.15 Use a calculator to evaluate (a) log 3, (b) log 7.85, (c) ln 2, and (d) ln 13.7.

Solutions

	Press	Display
(a)	3 [log]	0.4771213
(b)	78.5 [log]	1.8948697
(c)	2 [ln x]	0.6931472
(d)	13.7 [ln x]	2.6173958

In the last section, we mentioned that some calculators use the [INV] [ln x] key combination to evaluate e^x. Since $f(x) = \ln x$ and $g(x) = e^x$ are inverse functions, you can see why they do this.

Logs Using Different Bases

You can use a calculator to determine the natural logarithm or the common logarithm of a given positive number. To find the logarithm of a number in some base other than 10 or e, use the following change of base relationship.

Change of Base Relationship

$$\log_b x = \frac{\ln x}{\ln b} \qquad (4.1)$$

This seems to be an unusual relationship. It uses some properties of logarithms that we will explain in the next section. But, it works! Since $5^4 = 625$, we know that $\log_5 625 = 4$. Let's see that formula (4.1) gives the same result. According to that formula, $\log_5 625 = \frac{\ln 625}{\ln 5}$. Using a calculator, step by step, we get the following results.

Algebraic Calculator: Press	RPN Calculator: Press	Display
625 [ln x]	625 [LN] [ENTER]	6.4377516
[÷] 5 [ln x]	5 [LN]	1.6094379
[=]	[÷]	4

Now that you know how to evaluate logarithms, you can find the values of the points on a graph of a logarithmic function.

▶ **EXAMPLE 4.16**

Plot the graphs of $f(x) = \log x$, $g(x) = \ln x$, $h(x) = \log_5 x$, and $k(x) = \log_{1/5} x$.

Solution

A table of (rounded-off functional) values is given below, and the graphs of the functions are in figure 4.12.

Figure 4.12

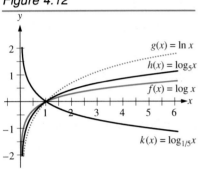

x	0.5	1	1.5	2	2.5	3	3.5	4	4.5	5	5.5	6
$f(x) = \log x$	−0.30	0	0.18	0.30	0.40	0.48	0.54	0.60	0.65	0.70	0.74	0.78
$g(x) = \ln x$	−0.69	0	0.41	0.69	0.92	1.10	1.25	1.38	1.50	1.61	1.70	1.79
$h(x) = \log_5 x$	−0.43	0	0.25	0.43	0.57	0.68	0.78	0.86	0.93	1.00	1.06	1.11
$k(x) = \log_{1/5} x$	0.43	0	−0.25	−0.43	−0.57	−0.68	−0.78	−0.86	−0.93	−1.00	−1.06	−1.11

Notice that all four curves cross the x-axis at the point $(1, 0)$. If you look back at example 4.12(e), you will see that $\log_b 1 = 0$. All graphs of logarithmic functions cross the x-axis at the point $(1, 0)$. Notice, for each of these graphs, that as x approaches 0 from the right-hand side, the limiting tendency of y is either infinity or negative infinity. This means that all such graphs have the y-axis as a vertical asymptote. A logarithmic function $f(x) = \log_b x$ is increasing if $b > 1$, and decreasing if $0 < b < 1$.

We have the following general results.

General Characteristics of Logarithmic Functions

For a logarithmic function $f(x) = \log_b x$, $b > 0$, $b \neq 1$:

1. The x-intercept is $(1, 0)$.
2. The y-axis is a vertical asymptote.
3. The domain of f is $(0, \infty)$.
4. The range of f is $\mathbf{R}$.
5. (a) If $b > 1$, f is an increasing function.
   (b) If $0 < b < 1$, f is a decreasing function.

One interesting consequence of the above change of base relationship is the fact that

$$\frac{\log_a b}{\log_a d} = \frac{\dfrac{\ln b}{\ln a}}{\dfrac{\ln d}{\ln a}} = \frac{\ln b}{\ln d}.$$

Thus, it follows that

$$\frac{\log_a b}{\log_a d} = \frac{\log_c b}{\log_c d},$$

or

Logarithmic Functions

Identity

$$\log_a b \log_c d = \log_c b \log_a d.$$

This means that a product of logarithms can be computed by exchanging bases.

EXAMPLE 4.17

Compute $\log_3 (125) \log_5 (\frac{1}{9})$.

Solution

Since 125 is 5^3 and $\frac{1}{9}$ is 3^{-2}, we can rewrite the given problem as follows:

$$\log_3 (125) \log_5 (\tfrac{1}{9}) = \log_5 (125) \log_3 (\tfrac{1}{9})$$
$$= (3)(-2) = -6.$$

Compare this solution with the following direct approach via calculator:

$$\log_3 (125) \log_5 \tfrac{1}{9} \approx (4.394920561)(-1.365212389)$$
$$\approx -5.99999999.$$

The discrepancy is due to the truncation error inherent in any calculator. This will be explored further at the end of this section. ◀

EXAMPLE 4.18

Compute $\log_4 (49) \log_7 (0.04) \log_5 (243) \log_3 (\tfrac{1}{8})$.

Solution

We rewrite the given problem as follows:

$$\log_4 (49) \log_7 (0.04) \log_5 (243) \log_3 (\tfrac{1}{8})$$
$$= \log_7 (49) \log_5 (0.04) \log_3 (243) \log_4 (\tfrac{1}{8})$$
$$= (2)(-2)(5)(-\tfrac{3}{2})$$
$$= 30.$$ ◀

Antilogarithms

We have seen how to find the logarithm of a given number. For example, we know how to determine that $\ln 16 \approx 2.7725887$. There are instances when we need to reverse this process. Suppose that you knew that $\ln n \approx 2.7725887$. How could you determine that the value of n is (approximately) 16?

The key to the solution of such problems lies in the inverse-function relationship between the natural logarithm and exponential functions. We know that if $\ln n = 2.7725887$, then $e^{2.7725887} = n$; so, using a calculator, we could find that $n \approx 16$.

We could use this process with any base. Given values of b and p and the fact that $\log_b n = p$, we would use $n = b^p$ to solve for n. Since we obtain the value of n by "eliminating" its logarithm, we call n the *antilogarithm* of p for base b. More formally, we have this definition.

Antilogarithm of p for Base b

If $\log_b n = p$, then n is called the **antilogarithm of p for base b** and has the value $n = b^p$.

Chapter 4 / Exponential and Logarithmic Functions

EXAMPLE 4.19   Find the antilogarithm corresponding to each of the following:
(a) $\log n = 2.5$, (b) $\ln n = 7.3$, and (c) $\log_7 n = 0.26$.

Solutions   Asking you to find the antilogarithm is equivalent to saying "solve for n."

(a) Since $\log n = 2.5$ is equivalent to $10^{2.5} = n$, we see that $n \approx 316.22777$.

(b) $\ln n = 7.3$ is equivalent to $e^{7.3} = n$, or $n \approx 1480.2999$.

(c) $\log_7 n = 0.26$ is equivalent to $7^{0.26} = n$, and so $n \approx 1.6585382$. ◀

EXAMPLE 4.20   Find (a) the antilogarithm of 7 for base 2, (b) the antilogarithm of $\frac{1}{3}$ for base $\frac{1}{2}$, (c) the antilogarithm of 5 for base e, and (d) the antilogarithm of 4.2 for base 10.

Solutions
(a) Let n denote the antilogarithm of 7 for base 2. In other words, $\log_2 n = 7$. This is equivalent to $n = 2^7$. Thus, $n = 128$.

(b) If we let n denote the antilogarithm of $\frac{1}{3}$ for base $\frac{1}{2}$, then $\log_{1/2} n = \frac{1}{3}$. This means that $n = \left(\frac{1}{2}\right)^{1/3} = \frac{1}{\sqrt[3]{2}}$.

(c) Here, we have $\log_e n = 5$, and so $n = e^5$.

(d) $\log_{10} n = 4.2$, and so $n = 10^{4.2}$. ◀

As indicated by our use of the $\approx$ symbol in example 4.19, a calculator will most often give only an approximate answer. It typically displays the first eight significant digits of the exact answer. This can cause "truncation" errors. For example, according to one calculator, $\ln 7.8 = 2.0541237$. This would mean that $e^{2.0541237} = 7.8$. However, using this same calculator, we determine that 2.0541237 $\boxed{e^x}$ $\boxed{=}$ yields 7.7999997, not 7.8. (A second calculator produces $\ln 7.8 = 2.054123734$, and then 2.054123734 $\boxed{e^x}$ yields 7.800000002. Both graphing calculators used for this text give the correct answer of 7.8.)

Exercises 4.2

In exercises 1–10, rewrite each equation using exponentials.

1. $\log_6 216 = 3$
2. $\log_9 6561 = 4$
3. $\log_4 16 = 2$
4. $\log_7 16{,}807 = 5$
5. $\log_{1/7}\left(\frac{1}{49}\right) = 2$
6. $\log_{1/2}\left(\frac{1}{64}\right) = 6$
7. $\log_2\left(\frac{1}{32}\right) = -5$
8. $\log_3\left(\frac{1}{243}\right) = -4$
9. $\log_9 2187 = \frac{7}{2}$
10. $\log_8 2048 = \frac{11}{3}$

In exercises 11–23, rewrite each equation using logarithms.

11. $5^3 = 125$
12. $3^7 = 2187$
13. $2^8 = 256$
14. $4.17^3 = 72.511713$
15. $7.2^4 = 2687.3856$
16. $13^2 = 169$
17. $5^{-3} = \frac{1}{125}$
18. $3^{-5} = \frac{1}{243}$
19. $4^{9/2} = 512$
20. $125^{7/3} = 78{,}125$
21. $\left(\frac{1}{2}\right)^3 = \frac{1}{8}$
22. $\left(\frac{1}{2}\right)^{-3} = 8$
23. $\left(\frac{1}{5}\right)^3 = \frac{1}{125}$

In exercises 24–26, compute the given logarithm exactly.

24. $\log_4 \frac{1}{16}$
25. $\log_{1/2} 128$
26. $\log_e e^5$

In exercises 27–29, simplify the given expression.

27. $10^{\log_{10}(0.5)}$
28. $\left(\frac{2}{3}\right)^{\log_{2/3}(3/2)}$
29. $8^{\log_2 7}$

In exercises 30–39, use a calculator or a computer to evaluate each of these logarithms.

Properties of Logarithms

30. ln 6
31. ln 3.951
32. log 28
33. log 13.57
34. $\log_5 9$
35. $\log_8 432.15$
36. ln 8
37. ln 0.1345
38. log 0.1345
39. $\log_8 0.1345$

In exercises 40–45, make a table of values and sketch the graph of each function.

40. $f(x) = \ln x$
41. $g(x) = \log x$
42. $h(x) = \log_2 x$
43. $k(x) = \log_{15} x$
44. $j(x) = \log_{1/2} x$
45. $m(x) = \log_{1/4} x$

In exercises 46–57, use the skills from the previous exercises and the transformation techniques of chapter 2 to sketch the graph of the given function. Identify the vertical asymptote(s) and x-intercept(s). State whether the given function is increasing, decreasing, or neither. (You may wish to use a graphing calculator or computer graphing program to help you or to provide a "rough check" of your work.)

46. $f(x) = \ln x + 4$
47. $g(x) = \ln (2x + 4)$
48. $h(x) = \ln |x - 3|$
49. $j(x) = \log 2x - 3$
50. $k(x) = 2 \log (x - 3)$
51. $\ell(x) = -2 \log |x - 3|$
52. $m(x) = 3 \log_{1/2} x$
53. $n(x) = -3 \ln (x + 4) - 5$
54. $p(x) = -3 \ln |x + 4| - 5$
55. $q(x) = -2 \log x$
56. $r(x) = \ln \dfrac{x}{2}$
57. $s(x) = \dfrac{1}{2} \ln (2x)$

In exercises 58–64, find the corresponding antilogarithm; that is, solve for n.

58. $\log n = 4.72$
59. $\log n = 0.05$
60. $\ln n = 8.65$
61. $\ln n = 0.75$
62. $\log_5 n = 4.65$
63. $\log_{12} n = -2.5$
64. $\log_{1/3} n = -5.75$
65. Solve for n: $\log_n 5 = 6.2$
66. Solve for n: $\log_n \left(\dfrac{1}{9}\right) = 3$
67. Find the antilogarithm of 3 for the base $\dfrac{1}{2}$.
68. Find the antilogarithm of 4.2 for the base 2.
69. Find the antilogarithm of $\dfrac{2}{3}$ for the base e.
70. Find the antilogarithm of $\dfrac{1}{2}$ for the base $\dfrac{3}{4}$.
71. The Richter scale is used to measure the magnitude of earthquakes. The formula for the Richter scale is $R = \log I$, where R is the Richter number and I is the intensity of the earthquake. Express the equation for the Richter scale in exponential form. What happens to the value of R if the value of I is multiplied by 10? by 100? by 1000?

72. The decibel (dB) scale is used for sound intensity because the response of the human ear to sound intensity is not proportional to the intensity. The intensity $I_0 = 10^{-12}$ W/m² is barely audible to the human ear. I_0 is given a value of 0 dB. A sound ten times more intense is given the value 10 dB; a sound 10^2 more intense than 0 dB is given the value 20 dB. Continuing in this manner leads to the formula

$$\beta = 10 \log \dfrac{I}{I_0},$$

where β is the intensity in decibels and I is the intensity in W/m².
(a) Express the decibel scale in exponential notation.
(b) What is the decibel value of a heavy truck passing a pedestrian at the side of a road if the truck's sound wave intensity, I, is 10^{-3} W/m²?

73. How many years will it take $2345.72 at $7\dfrac{1}{4}$ percent compounded monthly to grow to $4077.28? How long will it take if interest is compounded continuously?

74. What interest rate allows $1234.56 compounded quarterly to grow to $2345.67 after two years? What if interest is compounded continuously?

75. How long does it take 1000 grams of carbon-14 to decay down to 400 grams? to 300 grams? to 200 grams? to 100 grams? (See exercise 39, section 4.1.)

◆ 76. Consider the bacteria colony in exercise 40, section 4.1. When had the colony grown to a size of 2000? a size of 5000?

77. Simplify each of the following:
(a) $\log_2 2^x$ (b) $\log_{1/2} 2^{-x}$ (c) $\log_2 8^x$
(d) $\left(\dfrac{1}{2}\right)^{\log_2 x}$ (e) $2^{\log_2 x}$ (f) $e^{\ln x}$
(g) $e^{-\ln x}$ (h) $e^{\ln (1/x)}$

78. Without using calculator, computer, or tables, calculate each of the following exactly:
(a) $\log_2 (81) \log_9 8$
(b) $\log_3 \left(\dfrac{1}{e^2}\right) \log_{1/2} (125) \ln (243) \log_5 8$
(c) $\log_7 (64) \log_3 (9^x) \log_{1/4} (49) + 12x$

79. What is the inverse function of $2 \cdot 3^x$? of $3 \cdot 2^x$? of $c \cdot b^x$?

80. What is the inverse function of $\log (2x)$? of $\ln \left(\dfrac{x}{2}\right)$? of $\log_b (cx)$?

4.3 PROPERTIES OF LOGARITHMS

In the last section, you saw that logarithmic functions are the inverses of exponential functions. Hence, each logarithm can be written as an exponent.

Chapter 4 / Exponential and Logarithmic Functions

We shall next show how to derive properties of logarithms by using the laws of exponents that were discussed in chapter 0.

Consider any positive real numbers, x and y, and a base, b, with $b > 0$ and $b \neq 1$. We know that we can find real numbers m and n where $m = \log_b x$ and $n = \log_b y$, and so $x = b^m$ and $y = b^n$. According to a law of exponents,

$$xy = b^m b^n = b^{m+n}.$$

From the definition of $\log_b$, it follows that $\log_b xy = m + n$. Using the facts that $m = \log_b x$ and $n = \log_b y$ leads to the first property of logarithms:

$$\log_b (xy) = \log_b x + \log_b y. \qquad \textbf{Property 1}$$

▶ **EXAMPLE 4.21**

Use the first property of logarithms and other available information to rewrite (a) $\log_4 35$, (b) $\log 21$, (c) $\ln 18$, (d) $\log_2 140$, (e) $\log_3 5x$, (f) $\ln (x^2 y^3)$, (g) $\log_8 2 + \log_8 5$, (h) $\log 5 + \log 2 + \log 6$, (i) $\ln 5 + \ln y$, and (j) $3 \ln x + 2 \ln y$.

Solutions

(a) $\log_4 35 = \log_4 (5 \cdot 7) = \log_4 5 + \log_4 7$

(b) $\log 21 = \log (3 \cdot 7) = \log 3 + \log 7$

(c) $\ln 18 = \ln (2 \cdot 9) = \ln 2 + \ln (3 \cdot 3) = \ln 2 + \ln 3 + \ln 3$

(d) $\log_2 140 = \log_2 (4 \cdot 5 \cdot 7) = \log_2 4 + \log_2 5 + \log_2 7 = 2 + \log_2 5 + \log_2 7$

(e) $\log_3 5x = \log_3 5 + \log_3 x$

(f) $\ln (x^2 y^3) = \ln (x \cdot x \cdot y \cdot y \cdot y) = \ln x + \ln x + \ln y + \ln y + \ln y = 2 \ln x + 3 \ln y$

(g) $\log_8 2 + \log_8 5 = \log_8 (2 \cdot 5) = \log_8 10$

(h) $\log 5 + \log 2 + \log 6 = \log (5 \cdot 2 \cdot 6) = \log 60$

(i) $\ln 5 + \ln y = \ln 5y$

(j) $3 \ln x + 2 \ln y = \ln x + \ln x + \ln x + \ln y + \ln y = \ln (x \cdot x \cdot x \cdot y \cdot y) = \ln (x^3 y^2)$ ◀

The first property examined the log of a product. The second property looks at the log of a quotient. It is stated below without proof. You will be asked to establish it in exercise 37.

$$\log_b \frac{x}{y} = \log_b x - \log_b y. \qquad \textbf{Property 2}$$

▶ **EXAMPLE 4.22**

Use the second property of logarithms and other available information to rewrite (a) $\log \frac{3}{5}$, (b) $\ln \frac{5}{4}$, (c) $\ln \frac{7}{x}$, (d) $\log_4 \frac{x^2}{y^3}$, (e) $\log 8 - \log 2$, (f) $\ln 5 - \ln 2$, (g) $\log 700 - \log 7$, and (h) $\ln x^2 y - \ln xy^2$.

Properties of Logarithms

Solutions

(a) $\log \frac{3}{5} = \log 3 - \log 5$

(b) $\ln \frac{5}{4} = \ln 5 - \ln 4$

(c) $\ln \frac{7}{x} = \ln 7 - \ln x$

(d) $\log_4 \frac{x^2}{y^3} = \log_4 (x \cdot x) - \log_4 (y \cdot y \cdot y) = 2 \log_4 x - 3 \log_4 y$

(e) $\log 8 - \log 2 = \log \frac{8}{2} = \log 4$

(f) $\ln 5 - \ln 2 = \ln \frac{5}{2} = \ln 2.5$

(g) $\log 700 - \log 7 = \log \left(\frac{700}{7}\right) = \log 100 = 2$

(h) $\ln x^2 y - \ln xy^2 = \ln \frac{x^2 y}{xy^2} = \ln \frac{x}{y}$.

How does $\log_b x$ compare to $\log_b x^p$? Since $\log_b x = m$ means $x = b^m$, we consider $x^p = (b^m)^p = b^{mp}$. From this you can see that $\log_b x^p = mp = pm$.

$$\log_b x^p = p \log_b x. \quad \text{Property 3}$$

EXAMPLE 4.23

Use the third property of logarithms and other available information to rewrite each of the following: (a) $\log 5^3$, (b) $\ln 7^{-4}$, (c) $\log_2 16^5$, (d) $\log 100^{3.4}$, (e) $\ln \sqrt[3]{25}$, (f) $\log_4 x^{2/3}$, (g) $2 \log 5 + 3 \log 4 - 4 \log 2$, and (h) $3 \ln x + 2 \ln y - 4 \ln z$.

Solutions

(a) $\log 5^3 = 3 \log 5$

(b) $\ln 7^{-4} = -4 \ln 7$

(c) $\log_2 16^5 = 5 \log_2 16 = 5 \log_2 2^4 = 5 \cdot 4 = 20$

(d) $\log 100^{3.4} = 3.4 \log 100 = 3.4 \log 10^2 = (3.4) \cdot 2 = 6.8$

(e) $\ln \sqrt[3]{25} = \ln 25^{1/3} = \frac{1}{3} \ln 25 = \frac{1}{3} \ln 5^2 = \frac{2}{3} \ln 5$

(f) $\log_4 x^{2/3} = \frac{2}{3} \log_4 x$

(g) $2 \log 5 + 3 \log 4 - 4 \log 2 = \log 5^2 + \log 4^3 - \log 2^4$
$= \log 25 + \log 64 - \log 16$
$= \log (25 \cdot 64 \div 16)$
$= \log 100$
$= 2$

(h) $3 \ln x + 2 \ln y - 4 \ln z = \ln x^3 + \ln y^2 - \ln z^4$
$= \ln (x^3 y^2) - \ln (z^4)$
$= \ln \frac{x^3 y^2}{z^4}$.

Chapter 4 / Exponential and Logarithmic Functions

In addition to the above three properties, there are six other properties of logarithms that we have been using. All nine properties are summarized below:

Properties of Logarithms

$\log_b xy = \log_b x + \log_b y$	Property 1
$\log_b \dfrac{x}{y} = \log_b x - \log_b y$	Property 2
$\log_b x^p = p \log_b x$	Property 3
$\log_b 1 = 0$	Property 4
$\log_b b = 1$	Property 5
$\log_b b^n = n$	Property 6
$b^{\log_b n} = n$	Property 7
$\log_b m = \log_b n$ implies that $m = n$	Property 8
$b^m = b^n$ implies that $m = n$	Property 9

The properties of logarithms allow you to simplify logarithms of products, quotients, powers, and roots. This will be very useful during your later study of differential calculus. In general, there is no way to simplify logarithms of sums and differences. You cannot change $\log_b (x + y)$ to $\log_b x + \log_b y$. To see this, try $b = x = y = 10$. (Notice that log 20 ≠ 2 since 20 ≠ 100.) This is a very tempting mistake, so be alert! You should also not be tempted by other alluring varieties of properties 2 and 3.

Logarithms were developed originally to help people perform computations, and slide rules were computational tools based on logarithms. At one time every engineer, technician, scientist, and mathematician had, and knew how to use, a slide rule. The hand-held calculator and the microcomputer have replaced the slide rule and reduced the importance of logarithms as an aid in computing. However, logarithms are still important for working many theoretical and applied problems.

The remainder of this section demonstrates how to use the properties of logarithms in computations. The values that are used may be obtained from a table of logarithms or with the aid of a calculator or computer. The main purpose of these exercises is to prepare you to solve logarithmic equations such as the ones you will see later in this chapter and in differential calculus.

▶ **EXAMPLE 4.24**   If log 2 = 0.3010 and log 3 = 0.4771, determine the following: (a) log 6, (b) log 81, (c) log 5, (d) log $\sqrt{5}$, and (e) log (0.02).

Solutions

(a) log 6 = log (2 · 3) = log 2 + log 3 = 0.3010 + 0.4771 = 0.7781

(b) log 81 = log 3^4 = 4 log 3 = 4(0.4771) = 1.9084

(c) log 5 = log ($\frac{10}{2}$) = log 10 − log 2 = 1 − 0.3010 = 0.6990

(d) log $\sqrt{5}$ = log $5^{1/2}$ = $\frac{1}{2}$ log 5 = ($\frac{1}{2}$)(0.6990) = 0.3495

(e) log (0.02) = log ($\frac{2}{100}$) = log 2 − log 100 = 0.3010 − 2 = −1.6990.   ◀

Properties of Logarithms

EXAMPLE 4.25

If $\ln 2 = 0.6931$, $\ln 3 = 1.0986$, and $\ln 10 = 2.3026$, determine the following:
(a) $\ln 30$, (b) $\ln 1.5$, (c) $\ln (1.5)^3$, and (d) $\ln \sqrt{0.06}$.

Solutions

(a) $\ln (30) = \ln (3 \cdot 10) = \ln 3 + \ln 10 = 1.0986 + 2.3026 = 3.4012$

(b) $\ln 1.5 = \ln \left(\frac{3}{2}\right) = \ln 3 - \ln 2 = 1.0986 - 0.6931 = 0.4055$

(c) $\ln (1.5)^3 = 3 \ln 1.5 = 3(0.4055) = 1.2165$

(d) $\ln \sqrt{0.06} = \ln (0.06)^{1/2} = \frac{1}{2} \ln 0.06 = \frac{1}{2} \ln \frac{6}{100} = \frac{1}{2} (\ln 6 - \ln 10^2)$

$= \frac{1}{2} (\ln 2 + \ln 3 - 2 \ln 10) = \frac{1}{2} (0.6931 + 1.0986 - 4.6052)$

$= -1.4068$.

Exercises 4.3

In exercises 1–12, write the given expression as the sum or difference of two or more logarithms. Simplify further if possible. (Assume all variables represent positive numbers.)

1. $\log \frac{2}{3}$
2. $\log \frac{5}{4}$
3. $\ln 5e^2$
4. $\log 110$
5. $\log_2 \frac{92}{30y}$
6. $\log \frac{466}{5x}$
7. $\ln 15x$
8. $\ln \sqrt[4]{16t^2}$
9. $\log_2 \frac{4a^2x^3}{3b^4}$
10. $\log \frac{6r}{19uv^2}$
11. $\log \sqrt[7]{\frac{x^3y^3}{z+1}}$
12. $\ln \left(x^2 \sqrt[3]{\frac{y^2}{z^3}}\right)$

Express each quantity in exercises 13–24 as a single logarithm.

13. $\log 4 + \log 7$
14. $\ln x^2 + \ln y^5$
15. $\ln 36 - \ln 3$
16. $\log 24 - \log 8$
17. $4 \log (x + 1) - 3 \log (y - 2)$
18. $\ln 4 + \ln 7 + \ln x$
19. $\ln 16 - \ln 2 + \ln 5$
20. $4 \ln x + 2 \ln y - 3 \ln z$
21. $\frac{6}{5} \log 3 + \frac{4}{7} \log 2$
22. $3 \ln 2 - \frac{1}{5} \ln 4$
23. $\log \frac{2}{3} + \log \frac{15}{4}$
24. $\log_{16} 2 + \log_{16} (x^3)$

If $\log 2 = 0.3010$, $\log 3 = 0.4771$, and $\log 5 = 0.6990$, determine the value of each of the logarithms in exercises 25–36.

25. $\log 8$
26. $\log 9$
27. $\log 15$
28. $\log 12$
29. $\log 36$
30. $\log 40$
31. $\log \frac{2}{27}$
32. $\log \frac{45}{4}$
33. $\log 200$
34. $\log 5000$
35. $\log \sqrt{15}$
36. $\log (0.03)$

◆ 37. Prove the second property of logarithms.

38. Prove the following result that was used in section 4.2:

$\log_b x = \frac{\ln x}{\ln b}$. (Hint: convert to exponential form.)

39. Generalize exercise 38 by proving $\log_b x = \frac{\log_d x}{\log_d b}$ for any bases b and d.

Solve for x in each of exercises 40–49.

40. $\log x^2 = \log 9$
◆ 41. $\ln x + \ln (x + 4) = \ln 2$. (Hint: check your "answers." They aren't both right.)
◆ 42. $\log_2 x - 2 \log_2 (x + 1) = \log_2 4$. (Hint: check your "answer.")
43. $\log 8 - \frac{1}{2} \log 16 = \log x$
44. $3 \ln \left(\frac{\sqrt[3]{4}}{e}\right) = 1 + \ln x$
45. $2^x = 16$
46. $2^{x-3} = 32$
47. $10^{2x-4} = 10000^x$
48. $4^{2x-5} = 2^{3x+6}$
49. $e^{5x/2} = e^{2x/7} \cdot e^{3/8}$
50. Prove that $b^x = 10^{x \log b}$ for all real numbers x and all $b > 0$ such that $b \neq 1$. (Hint: the right-hand side is $(10^{\log b})^x$.)

Rewrite $f(x)^{g(x)}$ as a power of 10, using the values of $f(x)$ and $g(x)$ provided in exercises 51 and 52.

51. $f(x) = 2^{\sqrt{x}}$ and $g(x) = x^4 + 1$
52. $f(x) = x^4 + 1$ and $g(x) = 2^{\sqrt{x}}$

4.4 THE EXPONENTIAL FUNCTION e^x; APPLICATIONS INVOLVING EXPONENTIAL GROWTH AND DECAY

The number e was introduced in section 4.1 as the limiting value of $\left(1 + \frac{1}{n}\right)^n$ as n increases without bound. The value of e, approximated to the first twelve decimal places, is 2.718281828459. One application of e, shown in section 4.1, was the continuous compounding of interest. In this section, some scientific applications of the function $f(x) = e^x$ will be examined.

The function $f(x) = e^x$ is an exponential function with base e. The importance of this function can be guessed from the fact that the original scientific hand-held calculators had an $\boxed{e^x}$ key. Because such calculators also had a $\boxed{\ln x}$ key, later calculators were able to eliminate the $\boxed{e^x}$ key. Evaluations of e^x on these newer calculators require the double keystroke of the $\boxed{\text{INV}}$ $\boxed{\ln x}$ or of the $\boxed{\text{2nd}}$ $\boxed{\log}$ keys. Since the natural logarithm is the inverse function of e^x, this double keystroke has the same effect as using the old $\boxed{e^x}$ key.

Exponential Growth and Decay

The number e is used in two scientific applications known as exponential growth and exponential decay. As will be evident from their formulas, these two topics are related.

Exponential growth is often illustrated with the use of a bacterial culture. (See exercise 40, section 4.1 and exercise 76, section 4.2.) Experience has shown that, under favorable conditions, bacteria cultures tend to grow exponentially. This means that the quantity Q of bacteria at any time t is approximated by the formula

$$Q = ce^{kt},$$

where c and k are positive constants that have to be determined. Since Q depends functionally on t, this formula is better expressed by the function

$$Q(t) = ce^{kt}.$$

The original time is at $t = 0$, and at that time, $Q(0) = ce^0 = c$. If $Q(0)$ is denoted by Q_0, the exponential growth formula can be expressed as follows:

Exponential Growth Formula

$$Q(t) = Q_0 e^{kt}.$$

EXAMPLE 4.26 A bacteria culture originally numbers 600. Four hours later the culture contains 4500 bacteria. How many bacteria will there be after 5 additional hours?

Solution The original time is $t = 0$, and at that time, there are 600 bacteria. So,

The Exponential Function e^x; Applications Involving Exponential Growth and Decay

$$Q_0 = Q(0) = 600. \qquad (1)$$

Hence, the exponential growth formula for this particular culture can be expressed as

$$Q(t) = 600e^{kt}. \qquad (2)$$

We also know from the data that when $t = 4$, there are 4500 bacteria. Putting these substitutions into equation (2), we get

$$Q(4) = 600e^{k \cdot 4} = 4500$$

or

$$e^{4k} = \frac{4500}{600} = 7.5. \qquad (3)$$

After 5 additional hours, $t = 9$ (the 4 original plus the 5 additional hours), and

$$\begin{aligned} Q(9) &= 600e^{9k} \\ &= 600(e^{4k})^{9/4} \\ &= 600(7.5)^{9/4} \\ &= 600(7.5)^{2.25} \\ &\approx 600(93.086745) \\ &\approx 55852.047. \end{aligned}$$

Since you cannot have any partial bacteria, there would be approximately 55,852 bacteria in the culture at the time in question.

In solving this problem the value of k was not determined. However, there are times when one might need to find the value of the constant k. The method presented below can be used if a value for k is desired. This alternative solution picks up from the point where the value $e^{4k} = 7.5$ was found in equation (3).

Alternative strategy for finding $Q(9)$ One of the properties of logarithms (property 6 in section 4.3) states that $\ln e^x = x$. Taking the natural logarithm of the left and right sides of the equation $e^{4k} = 7.5$ produces

$$\ln (e^{4k}) = \ln (7.5)$$

$$4k = \ln 7.5.$$

Thus $k = \dfrac{\ln 7.5}{4} \approx 0.5037258$. Substituting this value of k into equation (2) yields the following formula for approximating the number of bacteria in the culture:

$$Q(t) = 600e^{0.5037258t} \qquad (4)$$

With the formula for Q known, we can now compute the approximate value for $Q(9)$:

$$Q(9) = 600e^{(0.5037258)9}$$
$$= 600e^{4.5335318}$$
$$\approx 55{,}852.047.$$

Alternative strategy for finding $Q(t)$ A general equation for $Q(t)$ can also be produced as follows.

In equation (3), you have $e^{4k} = 7.5$. Hence, $e^k = (7.5)^{1/4}$. Substituting this into equation (2) leads to

$$Q(t) = 600[(7.5)^{1/4}]^t$$
$$= 600(7.5)^{t/4},$$

which is essentially the same as (4). ◀

Exponential decay is exhibited by radioactive substances. The basic formula for radioactive decay is the exponential function

$$Q(t) = ce^{-kt},$$

where $Q(t)$ is the quantity of the substance at time t, and c and k are positive constants that need to be determined. As above, $Q(0) = ce^0 = c$, in general. If Q_0 denotes $Q(0)$, then the function can be written as

Exponential Decay Formula

$$Q(t) = Q_0 e^{-kt}.$$

A common measure of a radioactive substance's rate of decay is its half-life.

Half-Life

The **half-life** of a substance is the amount of time that is needed for half of the original atoms to transform radioactively into other kinds of matter or energy.

▶ **EXAMPLE 4.27**

The half-life of polonium-210 is 140 days. How much of 10 grams will remain after 1 year (365 days)?

Solution

Since the original quantity is 10 grams, $Q_0 = Q(0) = 10$. Thus,

$$Q(t) = 10e^{-kt}.$$

We are also given the half-life as 140 days. This means that when $t = 140$, the original quantity of 10 grams has disintegrated to 5 grams, or

$$Q(140) = 5.$$

We also know that

The Exponential Function e^x; Applications Involving Exponential Growth and Decay

$$Q(140) = 10e^{-140k}$$

and so

$$10e^{-140k} = 5$$

$$e^{-140k} = \tfrac{1}{2}.$$

Then

$$Q(365) = 10e^{-365k}$$

$$= 10(e^{-140k})^{365/140}$$

$$= 10(\tfrac{1}{2})^{365/140}$$

$$\approx 10(0.1641239)$$

$$\approx 1.64.$$

After 1 year, there are only 1.64 grams left of the original 10 grams of polonium.

As in the preceding example, one may determine k and the underlying decay function. For this equation, one finds $Q(t) = 10e^{-0.0049511t}$. ◀

In general, the exponential growth (or decay) formulas may be used when the rate of growth (or decay) of a substance is directly proportional to the existing amount of the substance. Thus, specific applications include Newton's Law of Cooling, population growth, air pressure as a function of altitude, and, of course, continuous compounding of interest. The exercises will include instances for each of these applications. One additional example will demonstrate an electrical application.

▶ **EXAMPLE 4.28**

(See figure 4.13.) Suppose that a constant electromotive force V is applied at time $t = 0$ to a direct current (DC) circuit of constant resistance R and of constant inductance L. Then, the current at any subsequent time t is given by the equation

$$I(t) = I = \frac{V}{R} - \left(\frac{V}{R} - I_0\right)e^{-\alpha t},$$

where $\alpha = \dfrac{R}{L}$. Typically, I is measured in amperes (A), V is measured in volts (V), R in ohms (Ω), and L in henries (H). Notice that the current at $t = 0$ is

$$I(0) = \frac{V}{R} - \left(\frac{V}{R} - I_0\right) = I_0.$$

If the initial current I_0 is 0, the current equation reduces to

$$I = \frac{V}{R}(1 - e^{-\alpha t}).$$

Theoretically, it takes an infinite time for the current I to reach the maximum value $\dfrac{V}{R}$. Since α is large, in practical cases it takes just a few seconds for I to be "nearly" $\dfrac{V}{R}$.

Figure 4.13

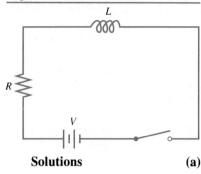

If the system is short-circuited, the short-circuit current is obtained by letting $V = 0$ in the first equation to get

$$I = I_0 e^{-\alpha t}.$$

This may be viewed as an instance of exponential decay.

Let a 0.2 H inductor whose resistance is 12 Ω be connected to a 24 V battery of negligible internal resistance. (a) What is the current in the circuit 0.05 s after the connection is made? (b) What is the current 0.1 s after the inductor has been short-circuited, assuming it had been connected to the battery for a long time?

Solutions

(a) We are given that $L = 0.2$, $V = 24$, and $R = 12$. The initial current $I_0 = 0$, and

$$\alpha = \frac{R}{L} = \frac{12 \, \Omega}{0.2 \, H} = 60 \text{ 1/s}.$$

At the time $t = 0.05$ s after the connection is made, the current is given by

$$I = \frac{V}{R}(1 - e^{-\alpha t})$$
$$= (\tfrac{24}{12})(1 - e^{-(60)(0.05)})$$
$$= 2(1 - e^{-3})$$
$$\approx 1.9004 \text{ A}.$$

So, the answer to (a) is 1.9004 A.

(b) Note next that the maximum current in this circuit before short-circuiting is $\frac{V}{R} = 2$ A. This value plays the role of I_0 in (b). Hence, the current in the inductor 0.1 s after short-circuiting is

$$I = I_0 e^{-\alpha t}$$
$$= 2e^{-(60)(0.1)}$$
$$= 2e^{-6}$$
$$\approx 0.0050 \text{ A}. \blacktriangleleft$$

Exercises 4.4

1. A bacteria colony originally consisted of 400 bacteria. Two hours later, it consisted of 3500 bacteria. How many bacteria did it contain 5 hours after the original count was made?

2. Explicitly compute the constant k figuring in the solution of the preceding exercise.

3. Consider the bacteria colony in exercise 40, section 4.1. Show that the constant k in the underlying equation for exponential growth is $k = \tfrac{1}{3} \ln 5$.

4. A herd of mammals was observed to grow exponentially. Two years after an initial count, the herd consisted of 200 mammals. One year after *that*, the herd consisted of 2000 mammals. What was the original size of the herd?

◆ 5. Suppose that a certain population of humans doubles in number every 20 years. (We then say that this population's **doubling time** is 20 years.)
   (a) Compute the constant k in the underlying equation for exponential growth.
   (b) How many years will it take this population to increase in size by a factor of 6?

6. Redo the preceding exercise, this time supposing a doubling time of 30 years.

The Exponential Function e^x; Applications Involving Exponential Growth and Decay

7. A small town's population grew exponentially from 18,000 in 1975 to 45,000 in 1985.
   (a) What would you expect its population to be in 1995?
   (b) When would you expect its population to number 300,000?
8. A radioactive substance originally consisted of 400 grams. Three hours later, it consisted of 300 grams. What was the substance's weight (more accurately, its mass) 7 hours after the original weighing?
9. Explicitly compute the constant k figuring in the solution of the preceding exercise.
10. Consider the carbon-14 decay discussed in exercise 39, section 4.1. Show that the constant k in the underlying equation for exponential decay is $k = \dfrac{\ln 2}{5700}$.
11. Let k be the constant figuring in an exponential decay equation.
    (a) Show that the associated half-life is $\dfrac{\ln 2}{k}$.
    (b) What decay phenomenon is present at time $t = \dfrac{\ln 4}{k}$? At $t = \dfrac{\ln 8}{k}$? At $t = \dfrac{\ln 12}{k}$?
12. A radioactive substance decayed exponentially. Two hours after an initial observation, 40 grams of the substance were present. Three hours after *that*, 30 grams were present. How many grams were present at the time of the initial observation?
13. Compute the constant k figuring in the exponential decay studied in the preceding exercise. At what time were there only 10 grams of the radioactive substance?
14. Let Q be the amount of a substance that is experiencing exponential growth or exponential decay. (So, $Q = ce^{\pm kt}$.)
    (a) Show that $y = \ln Q$ is a linear function of $x = t$.
    (b) Consider the graph of the linear function in (a). What is its y-intercept? When is its slope positive? When is its slope negative?
15. According to Newton's Law of Cooling, if an object with initial temperature T_0 is subject to an environment with constant temperature A, then the object's temperature T, after t units of time, is given by
    $$T = A + (T_0 - A)e^{-kt}.$$
    (Here, k is a positive parameter depending on the object and its surrounding environment.)
    (a) In the preceding equation, prove that the quantity $|T - A|$ is experiencing exponential decay.
    (b) If a bath is drawn at 112°F in a large room having constant temperature 68°F and if the water's temperature falls to 90°F after 10 minutes, when will the water's temperature be 79°F?
    (c) Compute the constant k for the situation in (b).
16. (A "heating" application of Newton's Law of Cooling; see exercise 15.) If a TV dinner having a 40°F temperature is placed in a 400°F oven and if the dinner's temperature is found to be 100°F just 7 minutes later, what is the dinner's temperature 20 minutes after being placed in the oven?
17. At an altitude h miles above sea level, the atmospheric pressure in pounds per square inch (psi) is found to be
    $$P = 14.7e^{-0.21h}.$$
    (a) What is the atmospheric pressure at sea level?
    (b) What is the atmospheric pressure at a remote village situated 1.5 miles above sea level?
    (c) At which altitude is the atmospheric pressure 7.35 psi?
    (d) What is the atmospheric pressure at the bottom of Death Valley, 282 feet *below* sea level?
18. (An application of the preceding exercise) Find the ratio of the atmospheric pressures prevailing in two towns if the first town is 0.5 mile above sea level and the second town is 0.82 mile above sea level.
19. (Review section 4.1 concerning continuous compounding, if necessary.) Let $2000 be invested in an account earning $7\frac{1}{2}$ percent compounded continuously.
    (a) How much interest has been earned 6 years later?
    (b) After approximately how many years will the investment yield $2000 in interest?
    (c) After approximately how many years will the investment yield $6000 in interest?
    (d) After approximately how many years will the investment yield $8000 in interest?
20. If an investment of $2134.75 earns interest compounded continuously and if the account (principal plus interest) totals $2817.69 after 3 years, what is the interest rate?
21. Suppose that a 0.3 H inductor whose resistance is 10 Ω is connected to a 30 V battery of negligible internal resistance.
    (a) What is the current in the circuit 0.06 s after the connection is made?
    (b) Suppose that the inductor is short-circuited after having been connected to the battery for a long time. What is the current 0.15 s later? What is the current 0.3 s later?
22. In example 4.28, show that the quantity $V/R - I$ is experiencing exponential decay.

4.5 EXPONENTIAL AND LOGARITHMIC EQUATIONS AND INEQUALITIES

Exponents and logarithms have been used in earlier sections to solve equations. This section will look at additional ways they are used as a problem-solving tool.

In this section's graphing calculator activity, we use skills developed in the text's earlier graphing calculator activities in order to approximate $e^\pi - \pi^e$. In addition to motivating the calculus topic of Newton's Method, this discussion reinforces that expressions of the form b^x and x^b determine distinctly different functions.

Exponential Equations

An **exponential equation** is an equation in which the unknown appears as part of an exponent. Several of the exercises earlier in this chapter, especially in section 4.4, involved exponential equations. For instance, we often had to solve for k in equations $Q = Q_0 e^{\pm kt}$ of exponential growth or decay. We next treat this topic of solving exponential equations systematically.

The first example is designed to illustrate the process that is used rather than actually describe a general method. In fact, this particular problem and its solution appeared as example 4.12(a).

▶ **EXAMPLE 4.29**   Solve $4^x = 1024$.

Solution   To solve this equation, take the logarithm of both sides. Any base of logarithms can be used, but, since most calculators have keys for both common and natural logarithms, it is best to use one of these bases. To illustrate that this problem can be worked using any base, it will first be worked using common logarithms and then be solved using natural logarithms. (Exercise 44 will ask you to verify that this process works in any base.) After each step a brief explanation is provided for that step. These explanations are intended only as learning aids. While they are the justifications that you should be mentally using, it is not essential that you write the process or reason when doing homework. You should, however, be able to provide the process or reason to anyone who asks.

Step	Process or Reason
$4^x = 1024$	Given
$\log 4^x = \log 1024$	Take log of both sides
$x \log 4 = \log 1024$	Property 3 of logs (sec. 4.3)
$x = \dfrac{\log 1024}{\log 4}$	Solve for x
$\approx \dfrac{3.0103}{0.60206}$	Evaluate logs
$= 5$	Divide

Exponential and Logarithmic Equations and Inequalities

Notice that $\log 1024 \div \log 4 \neq \log \frac{1024}{4} = \log 1024 - \log 4$.

This same problem, worked using natural logarithms instead of common logarithms, is shown below. As expected, the answer is 5.

$$4^x = 1024$$
$$\ln 4^x = \ln 1024$$
$$x \ln 4 = \ln 1024$$
$$x = \frac{\ln 1024}{\ln 4} \approx \frac{6.9314718}{1.3862944} \approx 5.$$

The above example shows how logarithms can be used to solve for variables appearing in exponents. In general, the process can be summarized as shown in the box.

Solving Exponential Equations

To solve an exponential equation of the form $a^{f(x)} = b^{g(x)}$, first take the logarithm of both sides and then try solving for the unknown.

EXAMPLE 4.30

Solve $5^{x+2} = 9^{2x-1}$.

Solution

Begin by taking the logarithm of both sides:

$$\ln(5^{x+2}) = \ln(9^{2x-1})$$

$(x + 2) \ln 5 = (2x - 1) \ln 9$	Property 3
$x \ln 5 + 2 \ln 5 = 2x \ln 9 - \ln 9$	Distributive law
$x \ln 5 - 2x \ln 9 = -2 \ln 5 - \ln 9$	Collect like terms
$x(\ln 5 - 2 \ln 9) = -(2 \ln 5 + \ln 9)$	Distributive law
$x = -\dfrac{2 \ln 5 + \ln 9}{\ln 5 - 2 \ln 9}$	Divide

$$\approx 1.9447.$$

EXAMPLE 4.31

Solve $\dfrac{e^x - e^{-x}}{2} = 4$.

Solution

Begin this example with some algebra:

$\dfrac{e^x - e^{-x}}{2} = 4$	Given
$e^x - e^{-x} = 8$	Multiply by 2
$e^{2x} - 1 = 8e^x$	Multiply by e^x
$(e^x)^2 - 8e^x - 1 = 0$	Quadratic equation

We've seen equations before that became manageable when reformulated as quadratic equations. Here's another instance. Substitute $u = e^x$ in the last step to obtain the quadratic equation

$$u^2 - 8u - 1 = 0.$$

This substitution might cause us to forget that u (or e^x) represents positive numbers. It is possible that the use of the quadratic formula will produce values of u that cannot be used to solve for x.

By the Quadratic Formula, the solution to this equation is

$$u = \frac{8 \pm \sqrt{64 + 4}}{2} = \frac{8 \pm \sqrt{68}}{2} = 4 \pm \sqrt{17}.$$

Since $u = e^x$, we have

$$e^x = 4 \pm \sqrt{17}.$$

Since the range of the exponential function is the set of positive real numbers and $4 - \sqrt{17} \approx -0.1231$ is negative, the only solution is $e^x = 4 + \sqrt{17}$.

Taking the natural logarithm of both sides, we obtain

$$x = \ln(4 + \sqrt{17}) \qquad \text{Remember } \ln e^x = x$$

$$\approx 2.0947.$$

◀

Logarithmic Equations

In the last three examples, logarithms were used to solve exponential equations. Similarly, exponents can be used to solve logarithmic equations. A **logarithmic equation** is an equation that contains at least one logarithm of a function of the unknown. Some examples of logarithmic equations are $\ln x = 9.4$ and $\log(6x + 5) - 3 = \log(5x - 1)$.

Solving Logarithmic Equations

> To solve a logarithmic equation, first use the properties of logarithms to combine all the logarithms into one, next rewrite this equation in exponential form, and then solve for the unknown. Remember to check whether such a solution actually is meaningful in the original equation.

In the second step, to rewrite the equation in exponential form, use the fact that $\log_b x = y$ is equivalent to $x = b^y$. Notice in this step that it is very important to *use the base of the logarithm*. Also, notice that the "checking" referred to in the final step may eliminate some purported "solutions": review exercises 41 and 42 of section 4.3.

The first example will be fairly simple in order to demonstrate the key process; later examples will be more involved.

Exponential and Logarithmic Equations and Inequalities

▶ **EXAMPLE 4.32** Solve $\ln x = 9.4$.

Solution The base in this problem is e. So

$$\log_e x = \ln x = 9.4 \qquad \text{Given}$$
$$x = e^{9.4} \qquad \text{Use definition of } \log_e$$
$$x \approx 12{,}088.38.$$ ◀

▶ **EXAMPLE 4.33** Solve $\log(6x + 5) - 3 = \log(5x - 1)$.

Solution Notice that in this problem the base of the logarithm is 10.

$$\log(6x+5) - 3 = \log(5x-1) \qquad \text{Given}$$
$$\log(6x+5) - \log(5x-1) = 3 \qquad \text{Collect log terms}$$
$$\log \frac{6x+5}{5x-1} = 3 \qquad \text{Property 2 of logs (sec. 4.3)}$$
$$\frac{6x+5}{5x-1} = 10^3 = 1000 \qquad \log_b x = y \Leftrightarrow x = b^y$$
$$6x + 5 = 1000(5x - 1) \qquad \text{Multiply by } 5x - 1$$
$$6x + 5 = 5000x - 1000 \qquad \text{Distributive law}$$
$$6x - 5000x = -1000 - 5 \qquad \text{Collect like terms}$$
$$-4994x = -1005$$
$$x = \frac{1005}{4994} \qquad \text{Solve for } x$$
$$\approx 0.2012.$$

The value of x that we found is a solution for the equation $\log \frac{6x+5}{5x-1} = 3$. To see whether it is a solution for the *given* equation we must test whether $\log(6x + 5)$ and $\log(5x - 1)$ are both meaningful when $x = \frac{1005}{4994}$. They *are*, since $6(\frac{1005}{4994}) + 5$ and $5(\frac{1005}{4994}) - 1$ are both positive. Thus, the solution set of the given equation is $\{\frac{1005}{4994}\}$. ◀

One interesting application involves the continuous compounding of interest that was studied in section 4.1. The formula stated that if a certain amount of money, P, was invested at the rate r compounded continuously, then, after t years, the investment would be worth

$$S = Pe^{rt}.$$

How long does it take for this investment to double?

If it doubles, then it is worth $2P$, and we need to solve the exponential equation

Chapter 4 / Exponential and Logarithmic Functions

$$2P = Pe^{rt}$$

or

$$2 = e^{rt}.$$

Taking the natural logarithm of both sides, we get

$$\ln 2 = rt$$

and so

$$t = \frac{\ln 2}{r}.$$

Thus, it will take $\frac{\ln 2}{r}$ years for the money to double. This same procedure can be used for anything that is growing exponentially. (Indeed, the preceding argument was anticipated in exercise 5, section 4.4.)

▶ **EXAMPLE 4.34**

How long will it take for $5000 to double at $6\frac{1}{2}$ percent interest compounded continuously?

Solution

We know $t = \dfrac{\ln 2}{r}$.

Since $r = 6\frac{1}{2}$ percent = 6.5 percent = 0.065,

$$t = \frac{\ln 2}{0.065} = 10.6638 \text{ years}$$

or about 10 years, 8 months. ◀

Exponential and Logarithmic Inequalities

Exponential and logarithmic inequalities are solved using the techniques for solving inequalities introduced in chapter 1. Using the fact that any exponential function is either increasing or decreasing, one can show that the graphs of any two exponential functions can intersect in no more than one point. For the same sort of reason, two logarithmic graphs can intersect in at most one point. This point is what we earlier called the key point and should be located by solving the equation associated with the inequality. You should then check to see on which side of the key point the inequality is valid.

▶ **EXAMPLE 4.35**

Solve $2^{5x+1} > 0.6^{2x+3}$.

Solution

To locate the key point, we shall solve the associated equation $2^{5x+1} = 0.6^{2x+3}$. Since both sides of this equation are positive, we can take the logarithm of both sides.

$$\ln 2^{5x+1} = \ln 0.6^{2x+3}$$

$$(5x + 1) \ln 2 = (2x + 3) \ln 0.6$$

$$5x \ln 2 + \ln 2 = 2x \ln 0.6 + 3 \ln 0.6$$

$$x(5 \ln 2 - 2 \ln 0.6) = 3 \ln 0.6 - \ln 2$$

Exponential and Logarithmic Equations and Inequalities

$$x = \frac{3 \ln 0.6 - \ln 2}{5 \ln 2 - 2 \ln 0.6}$$

$$\approx -0.496.$$

Thus, the key point is $x \approx -0.496$. This divides the real numbers into two intervals: $(-\infty, -0.496)$ and $(-0.496, \infty)$. We now check to see if either of these is the solution to the inequality. In order to determine if the inequality is valid on one of these intervals, we shall select one point from that interval and evaluate each side of the inequality symbol at that point.

For the interval $(-\infty, -0.496)$, we select the point $x = -1$. Substituting this value for x in the given inequality, we obtain $2^{5(-1)+1} = 2^{-4} = 0.0625$ on the left side and $0.6^{2(-1)+3} = 0.6^1 = 0.6$ on the right side. Since $0.0625 \not> 0.6$, we have determined that no point in $(-\infty, -0.496)$ can satisfy the inequality.

For the interval $(-0.496, \infty)$, we select $x = 0$ and obtain $2^{5(0)+1} = 2$ on the left side and $0.6^{2(0)+3} = 0.6^3 = 0.216$ on the right side. Since $2 > 0.216$, the numbers in the interval $(-0.496, \infty)$ all satisfy the inequality.

Graphically, you can see the validity of our conclusion by looking at figure 4.14. The figure contains the graphs of $y = 2^{5x+1}$ and $y = 0.6^{2x+3}$. These curves intersect at approximately $x = -0.496$, and you can see that the curve $y = 2^{5x+1}$ is higher than the curve $y = 0.6^{2x+3}$ for $x > -0.496$. ◀

Figure 4.14

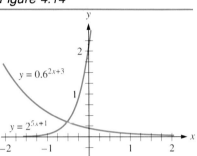

In example 4.36, we will first solve the inequality using an algebraic method. After that, we will use sign charts to solve the inequality. You may find one method easier than the other.

▷ **EXAMPLE 4.36**

Algebraic Solution

Solve $\ln (2x + 5) \geq \ln (4x - 1)$.

To locate the key point, we solve the associated equation, $\ln (2x + 5) = \ln (4x - 1)$.

$$e^{\ln(2x+5)} = e^{\ln(4x-1)}$$

$$2x + 5 = 4x - 1$$

$$-2x = -6$$

$$x = 3.$$

The key point is $x = 3$. Since the domain of $y = \ln (2x + 5)$ is $(-\frac{5}{2}, \infty)$ and the domain of $y = \ln (4x - 1)$ is $(\frac{1}{4}, \infty)$, this inequality can be considered only for x in $(\frac{1}{4}, \infty)$, which might be termed the "domain of the inequality."

The key point divides this "domain" into two intervals, $(\frac{1}{4}, 3)$ and $(3, \infty)$. We shall check one point in each interval. In the interval $(\frac{1}{4}, 3)$, we select $x = 1$, and we obtain $\ln (2 + 5) = \ln 7$ for the left side and $\ln (4 - 1) = \ln 3$ for the right side. Since $\ln 7 > \ln 3$, the inequality is valid for all values of x in the interval $(\frac{1}{4}, 3)$. In the interval $(3, \infty)$, we select $x = 4$, and we obtain $\ln (2 \cdot 4 + 5) = \ln 13$ and $\ln (4 \cdot 4 - 1) = \ln 15$. Since $\ln 13 \not> \ln 15$, no value in the interval $(3, \infty)$ can satisfy the inequality. Notice that $x = 3$ is included in the solution set. Thus, the solution set of this inequality is the interval $(\frac{1}{4}, 3]$.

Sign Chart Solution

We shall begin by using the properties of logarithms to simplify the given inequality.

$$\ln(2x+5) \geq \ln(4x-1)$$

$$\ln(2x+5) - \ln(4x-1) \geq 0$$

$$\ln\left(\frac{2x+5}{4x-1}\right) \geq 0$$

$$\ln\left(\frac{2x+5}{4x-1}\right) \geq \ln 1$$

$$\frac{2x+5}{4x-1} \geq 1$$

$$\frac{-2x+6}{4x-1} \geq 0.$$

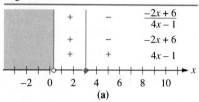

Figure 4.15

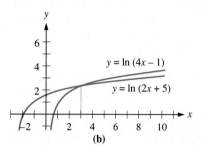

As above, this last inequality should be considered only for the "domain" $(\frac{1}{4}, \infty)$. The sign chart for this inequality is in figure 4.15(a). Examination of this sign chart indicates that the solution set of the given inequality is the interval $(\frac{1}{4}, 3]$. The shaded portion of the sign chart indicates that values of x in $(-\infty, \frac{1}{4}]$ are not in the domain of the given inequality.

Examination of the graphs of $y = \ln(2x+5)$ and $y = \ln(4x-1)$ in figure 4.15(b) confirms our solution. ◀

One tongue-in-cheek piece of wisdom often shared by mathematicians states that "If you cannot solve a problem, change it to one that you can solve. Then use the answer to the second problem in order to solve the original one." Properly applied, this is good problem-solving advice. In fact, your experience in this chapter is a case in point. Equations and inequalities in this chapter had variables in exponents or in logarithms. Earlier, in chapter 3, you had acquired a large arsenal of techniques for solving polynomials. These techniques may not work directly with exponential or logarithmic equations. However, by using logarithms or exponents, we changed each of this chapter's equations into one that could be solved by the methods used to solve polynomials. Then that solution was employed in order to solve the original problem. The underlying moral, that of identifying "type" or "paradigm" problems to which complicated situations can be reduced, should serve you well throughout your scientific studies.

Graphing Calculator Activity

You have seen that $e \approx 2.7182818$, and you know that $\pi \approx 3.1415927$. In particular, $2 < e < 3 < \pi < 4$. Consider the numbers e^π and π^e: which of these numbers is larger? The answer is not immediately obvious, since e^π and π^e are each "roughly" equal to $3^3 = 27$. However, with the help of the $\boxed{y^x}$ or $\boxed{x^y}$ key

Exponential and Logarithmic Equations and Inequalities

of a calculator (or the $\boxed{\land}$ key on a *TI-81*), you can answer the question quickly. By using the above approximations for e and π (or by using the $\boxed{e^x}$ and $\boxed{\pi}$ keys which you may have on your calculator), you find that

$$e^\pi \approx 23.14069263 \quad \text{and} \quad \pi^e \approx 22.45915772.$$

Thus, it seems safe to conclude that $e^\pi > \pi^e$ and that

$$e^\pi - \pi^e \approx 23.14069263 - 22.45915772 = 0.68153491.$$

The following discussion shows how a graphing calculator may be used to discover this same fact and, in doing so, analyze other related questions.

▶ **EXAMPLE 4.37** Consider the functions f and g defined by $f(x) = x^e$ and $g(x) = e^x$, each with domain [0, 4]. Use your graphing calculator to graph f and g on one screen. (a) At $x = \pi$, which curve lies above the other? By approximating the coordinates $(\pi, f(\pi))$ and $(\pi, g(\pi))$, approximate $e^\pi - \pi^e$. (b) The graphs of f and g appear to touch near $x = 3$. By zooming in, approximate the value of x such that $x^e = e^x$. (c) Formulate some other conjectures suggested by the graphs developed in the answers for (a) and (b).

Solutions We indicate how to proceed using a Texas Instruments *TI-81* and a Casio *fx-7700G* scientific calculator. Modify the discussion appropriately if you are using another model.

(a)
TI-81 Version First, press $\boxed{\text{Y=}}$ and input the functions f and g by pressing the sequence of keys

$$\boxed{\text{X|T}} \; \boxed{\land} \; 2.718281828 \; \boxed{\text{ENTER}} \; \boxed{\text{2nd}} \; \boxed{e^x} \; \boxed{\text{X|T}}$$

Notice that we have used $e \approx 2.7182818$ to input f and the $\boxed{\text{2nd}} \; \boxed{e^x}$ keys to input g. Next, press $\boxed{\text{RANGE}}$ and set the following values: Xmin = 0, Xmax = 4, Xscl = 1, Ymin = 0, Ymax = 45, and Yscl = 2. The screen should now look like the one in figure 4.16(a).

Next, press $\boxed{\text{GRAPH}}$. The result is the pair of graphs shown in figure 4.16(b). It's clear from figure 4.16(b) that the graphs of f and g touch to the left of $x = 3$, and that $g(x) > f(x)$ if $3 \leq x \leq 4$. In particular, $g(\pi) > f(\pi)$; that is, $e^\pi > \pi^e$.

Let's take a closer look "near" $x = \pi$. Press $\boxed{\text{RANGE}}$ to change the domain to [3.1, 3.2], with Xscl = 0.01, and range limited to [20, 26], with Yscl = 1. Press $\boxed{\text{GRAPH}}$ to redraw the graphs of f and g. The result is shown in figure 4.16(c).

Use figure 4.16(c) to approximate the coordinates $(\pi, f(\pi))$ and $(\pi, g(\pi))$; for this, you need to recall that $\pi \approx 3.141592654$. Press $\boxed{\text{TRACE}}$ and the cursor keys $\boxed{\blacktriangleleft}$ and $\boxed{\blacktriangleright}$ to move the cursor to approximately π, namely 3.1410526. By pressing the $\boxed{\blacktriangledown}$ and $\boxed{\blacktriangle}$ keys we can move the cursor from one graph to the other. Thus, we determine that $\pi^e = f(\pi) \approx 22.448665$ and $e^\pi = g(\pi) \approx 23.1282$. From this, we conclude that $e^\pi > \pi^e$, and that $e^\pi - \pi^e \approx 0.679535$. By additional zooming in, you can find better approximations.

Chapter 4 / Exponential and Logarithmic Functions

Figure 4.16

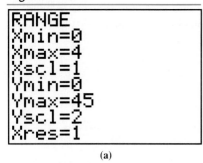
(a)

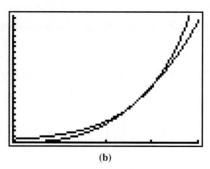
(b)

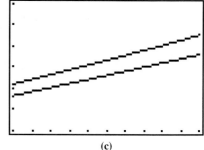
(c)

Casio fx-7700G Version

First, enter Shift Cls EXE to clear the graph screen. Next, press Range and proceed as in example 3.24, to arrange Xmin = 0, Xmax = 4, Xscl = 1, Ymin = 0, Ymax = 45, and Yscl = 2. The screen should now look like the one in figure 4.16(d).

Next, press Range twice and you are ready to input the functions f and g. Press the sequence of keys

Graph X,θ,T x^y 2.718281828 Shift ↵ Graph Shift e^x X,θ,T EXE

Notice that we have used $e \approx 2.7182818$ to input f and the Shift e^x keys to input g. The result is the pair of graphs shown in figure 4.16(e). It's clear from figure 4.16(e) that the graphs of f and g touch to the left of $x = 3$, and that $g(x) > f(x)$ if $3 \leq x \leq 4$. In particular, $g(\pi) > f(\pi)$; that is, $e^\pi > \pi^e$.

Let's take a closer look "near" $x = \pi$. Press Range to change the domain to [3.1, 3.2], with Xscl = 0.01, and range limited to [20, 26], with Yscl = 1. Press Range EXE to redraw the graphs of f and g. The result is shown in figure 4.16(f).

Use figure 4.16(f) to approximate the coordinates $(\pi, f(\pi))$ and $(\pi, g(\pi))$; for this, you need to recall that $\pi \approx 3.141592654$. Press Shift Trace and the cursor keys ◄ and ► to move the cursor to approximately π, namely 3.1414893. By pressing the ▼ and ▲ keys we can move the cursor from one graph to the other and determine that $\pi^e = f(\pi) \approx 22.457149$ and $e^\pi = g(\pi) \approx 23.138302$. From this, we conclude that $e^\pi > \pi^e$, and that $e^\pi - \pi^e \approx 0.681153$. By additional zooming in, you can find better approximations.

Figure 4.16 (continued)

(d)

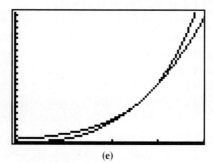
(e)

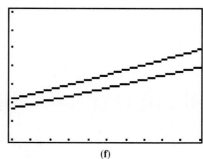
(f)

Exponential and Logarithmic Equations and Inequalities

Both TI-81 and fx-7700G
The above conclusions were based on what we saw on the calculator screen. Here, *approximations* to f and g had been graphed. (Those graphs were only approximations of the graphs of f and g because the keying-in and graphing processes for f and g rely on decimal approximations of e and π.) Therefore, our conclusions only have the status of **conjectures** (albeit very believable ones). In fact, calculus can be used to **prove** that $e^{\pi} > \pi^{e}$. Our main point in (a) has been how to use a graphing calculator to suggest what one should be trying to prove when e^{π} and π^{e} are being compared.

(b) We can do more with the data assembled in (a). Return to either figure 4.16(b) or figure 4.16(e), depending on your calculator. By using the cursor keys, move the blinking point to what appears to be the point at which the graphs of f and g first touch. Using the Trace property of your calculator leads to the x-coordinate 2.638297872. Hence, a value of x such that $x^e = e^x$ has been seen to be approximately 2.638297872. This should not surprise you, for the graphs of f and g intersect at $x = e \approx 2.7182818$; this is just the observation that $f(e) = e^e = g(e)$.

(c) The above figures suggest more than what we concluded above. For instance, figures 4.16(b) and (e) suggest that the graphs of f and g touch, but do not cross, at $x = e$. In fact, this can be proved using calculus. Sometimes, the techniques of *pre*calculus are adequate to prove what has been suggested by using a graphing calculator. See example 6.34 for one such case.

A striking feature of figures 4.16(c) and (f) is the "linear" appearance of the portions of the graphs of f and g as x varies from 3.1 to 3.2. This type of linearity is actually a rather general phenomenon. As we indicated in the introduction to section 1.3, many graphs of functions are approximately linear over small intervals. You may want to experiment with your graphing calculator to see how the appearance of linearity in figures 4.16(c) and (f) increases as the displayed interval of x-values shrinks, and how the illusion of linearity disappears when the displayed interval of x-values is expanded.

Another use of figures 4.16(b) and (e) was to suggest where to place the cursor initially if one wants to approximate a solution of the equation $x^e - e^x = 0$. In calculus, you will study Newton's Method for approximating a solution for an equation $h(x) = 0$ (in our example, $h(x) = x^e - e^x$), given a sufficiently close initial approximation for the solution (corresponding to the initial placement of the cursor). Newton's Method applies successfully to a wide class of functions h, but one needs calculus even to write the formula that describes Newton's Method. ◀

Exercises 4.5

Solve the equations or inequalities in exercises 1–42.

1. $4^x = 256$
2. $4^x = 128$
3. $4^{2x-1} = \frac{1}{128}$
4. $4^{2-x} = \frac{1}{64}$
5. $8^x = 129$
6. $5^{x^2} = 9$
7. $(\frac{2}{5})^{x-3} = 1$
8. $(\frac{2}{5})^{x-3} = \frac{2}{5}$
9. $(\frac{2}{5})^{2x+1} = \frac{3}{12345}$
10. $125^{3x-1} = 25$
11. $125^{3x-1} = 25^{6x+1}$
12. $125^{3x-1} = 26^{6x+1}$
13. $5^{2x-1} = 3^{4x+6}$
14. $125^{1-3x} = 25^{6x+1}$
15. $10^{4x+1} = 127$
16. $10^{-2x+3} = 6.17$
17. $10^{2x+3} = 0.167$
18. $e^{1-3x} = 10$
19. $e^{3x+1} = 4.472$
20. $e^{3x+1} = 1.472$

21. $e^{3x+1} = e^{5x+1}$
22. $e^{3x+1} = e^{2x+6}$
23. $e^x + e^{-x} = 3$
24. $2e^x - e^{-x} = 2\sqrt{2}$
25. $\dfrac{2e^x - 3e^{-x}}{5} = \dfrac{e^x - e^{-x}}{6}$
26. $e^x + e^{-x} = \sqrt{2}$
27. $\log_2 x = 5$
28. $\log_3 (x - 2) = -4$
29. $\ln (x^2 - 3x - 3) = 0$
30. $\ln (x^2 - 3x - 3) = 1$
31. $\log_4 (x^2 - 3x - 3) = 1$
32. $\log x = \log 72.5$
33. $\log 12x = \log (3x - 7)$ (Hint: check your "answers.")
34. $2 \log (3x - 1) - \log (2x - 7) = 1$ (Hint: check your "answers.")
35. $\log (1 - 3x) + \log (-1 - 3x) = \log 8$ (Hint: check your "answers.")
36. $\log (2x - 4) + \log (2x + 4) = 20$
37. $\ln x + 2 \ln 5 = \ln (x - 25)$
38. $\log_x 66 = 2$
39. $2^{2x+1} > 5^{5x+1}$
40. $5^{2x+1} < 5^{3x-1}$
41. $e^{x^2} \geq e^{2x-1}$
42. $2 \log x > \log 2x$
43. $\ln (2x + 1) - \ln 5 \leq \ln (x + 3)$
◆ 44. Show that $4^x = 1024$ leads to $x = 5$ by using logarithms to any base.
45. As in the preceding exercise, solve $9^{3x-5} = 343$.
46. Suppose that an amount P is invested at the rate r compounded continuously, yielding the amount A ($=$ principal plus interest) after t years.
    (a) If $P = \$2000$, $r = .5$ percent, and $t = 3$, find A.
    (b) If $P = \$2000$, $r = 6$ percent, and $A = \$3000$, find t.
    (c) If $P = \$2000$, $A = \$3167$, and $t = 4$, find r.
    (d) If $r = 7$ percent, $A = \$3167$, and $t = 4$, find P.
47. Suppose that a bacteria colony initially numbering Q_0 grows to number Q at time t, with underlying exponential growth equation having constant k.
    (a) If $Q_0 = 40$, $t = 2$, and $k = 0.56$, find Q.
    (b) If $Q_0 = 40$, $Q = 85$, and $k = 0.63$, find t.
    (c) If $Q_0 = 40$, $Q = 85$, and $t = 3$, find k.
    (d) If $Q = 85$, $t = 3$, and $k = 0.53$, find Q_0 (approximately).
48. Show that $e^x + e^{-x} = \gamma$ has a real solution if and only if $\gamma \geq 2$.
49. Let α and β be positive numbers. Show that $\alpha e^x + \beta e^{-x} = \gamma$ has a real solution if and only if $\gamma \geq 2\sqrt{\alpha\beta}$.
50. Show that the graphs of two different exponential functions $y = c_1 e^{b_1 x}$ and $y = c_2 e^{b_2 x}$ can intersect in at most one point, and that they *do* intersect precisely in the case where c_1 and c_2 have the same sign and $b_1 \neq b_2$.
51. Prove that any two different logarithmic graphs, $y = \log_{b_1} x$ and $y = \log_{b_2} x$, can intersect only at $(1, 0)$. (Hint: if $\log_{b_1} x = \log_{b_2} x$, show that either $\ln b_1 = \ln b_2$ or $\ln x = 0$. Conclude that if $x \neq 1$, then $b_1 = b_2$.)
52. (An exponential *and* logarithmic inequality can have more than one key point.) Using the idea of "continuity" introduced intuitively in chapter 3, show that the inequality

$$-6e - e\sqrt[3]{2} + 3e^x < 3e \ln x$$

has more than one key point. (Hint: build a sign diagram, emphasizing "small" positive x, $x = e^{1/e}$, and "large" x. To see how "small" and how "large" the values of x should be, you may find it helpful to use a graphing calculator or computer graphing program.)
53. Use a graphing calculator to graph $f(x) = x^\pi$ and $g(x) = \pi^x$, each with domain $[1, 5]$, on one screen.
    (a) At $x = e$, which curve lies above the other? By approximating the coordinates $(e, f(e))$ and $(e, g(e))$, approximate $e^\pi - \pi^e$. Does this approximation agree with what we found in example 4.37?
    (b) The graphs of f and g appear to cross near $x = 3$. By zooming in, approximate the value of x such that $x^\pi = \pi^x$.
    (c) By zooming, formulate a conjecture as to whether there is a real number x, other than the one you found in (b), such that $x^\pi = \pi^x$.

4.6 CHAPTER SUMMARY AND REVIEW EXERCISES

Major Concepts

- Antilogarithm of p for base b
- Change of base theorem
- Common logarithm
- Exponential decay
- Exponential equation/inequality
- Exponential function with base b
- Exponential growth
- Logarithm to the base b of x
- Logarithmic equation/inequality
- Logarithmic function with base b
- Natural logarithm
- Properties of logarithms and exponents

Review Exercises

1. Round off $e^{3.5}$ to four decimal places.
2. Round off $\sqrt{5}^{\sqrt{7}}$ to four decimal places.
3. Make a table of values for the function $f(x) = 3^{-x+2}$ and sketch its graph. Discuss this function's domain and range. Identify the y-intercept and the horizontal asymptote. Determine if this function is increasing or decreasing.
4. Repeat exercise 3 for the function given by $g(x) = -2 \cdot 4^x + 3$.
5. A bank offers 7 percent compounded semiannually. How much interest is earned after 4 years by an investment of $2963.48?
6. Write in exponential form:
   (a) $\log_3 343 = 5$ (b) $\log_{1/2} (128) = -7$
7. Write in logarithmic form:
   (a) $(\frac{1}{2})^{-4} = 16$ (b) $5^{-2} = \frac{1}{25}$
8. Compute exactly, without calculator, computer, or tables:
   (a) $\log_{1/2} (\frac{1}{8})$ (b) $9^{\log_3 2}$ (c) $\log_8 (2^x)$ (d) $e^{\ln 7}$
9. What is the inverse function of $5 \cdot 2^x$? of $\ln 2x$?
10. Use a calculator or computer to evaluate $\ln 0.182$, $\ln 1.82$, $\ln 18.2$, $\ln 182$, $\log 18.2$, and $\log_3 18.2$.
11. Make a table of values for the function $f(x) = \log_4 (2x - 1)$ and sketch its graph. Describe this function's domain and range. Identify the x-intercept and the vertical asymptote. Determine if this function is increasing or decreasing.
12. Repeat exercise 11 for the function given by $g(x) = -2 \log_{1/2} (1 + 2x) + 1$.
13. Find the antilogarithm of 2 for the base $\frac{1}{3}$.
14. Write as a sum or difference of two or more logarithms:
   (a) $\ln \frac{4x}{55}$ (b) $\log \sqrt[4]{x^2 y^3}$ (assuming $x > 0$)
15. Combine as a single logarithm:
   (a) $\log 27 - \log 4$ (b) $4 \ln x^2 - \ln y^{3/7} + 2 \ln 6$
16. If $\log 3 = 0.4771$, $\log 5 = 0.6990$, and $\log 7 = 0.8451$, determine the following:
   (a) $\log 9$ (b) $\log \sqrt{15}$ (c) $\log \frac{27}{3430}$ (d) $\log \frac{0.03}{175}$
17. If $\ln 2 = 0.6931$, $\ln 3 = 1.0986$, $\ln 5 = 1.6094$, and $\ln 7 = 1.9459$, determine the following:
   (a) $\ln 9$ (b) $\ln 2100$ (c) $\ln \sqrt{15}$
   (d) $\ln \frac{27}{3430}$ (e) $\ln \frac{0.03}{175}$
18. Write the exponential decay equation for strontium-90, given that its underlying constant is $k = 0.0246$.
19. Find the constant k underlying the radioactive decay of radium, given that the associated half-life is 1600 years.
20. What is the doubling-time of an investment at 6 percent compounded continuously?
21. What is the doubling-time of a population of mammals that took 10 years to grow from 1600 to 2500?
22. Two hours after an initial count, a bacteria colony numbered 160. One hour after *that*, the colony numbered 400. What was the original number of bacteria?
23. A chef is reheating some stale buns from room temperature (68°F) by placing them in a 450°F oven. After 1 minute, the buns' temperature has risen to 108°F. The chef intends to serve the buns as soon as their temperature reaches 148°F. How much longer must the chef wait for the buns to be hot enough?
24. (Review exercise 72, section 4.2.) What is the intensity in decibels of a thunderclap if its sound wave intensity is $I = 10^{-5}$ W/m²?

Solve the equations and inequalities in exercises 25–38.

25. $3^{2x-1} = \frac{1}{81}$
26. $3^{5x} = 9^{4x-1}$
27. $5^{2x} = 3^{4x-1}$
28. $5^{3x} = 3^{5x}$
29. $e^{-x+5} = e^{2x-3}$
30. $e^{-x} + 6 = e^x$
31. $\ln x^2 = \ln 16$
32. $2 \log_2 (x - 3) = 4$
33. $\log_2 2x = 2 + \log_2 (x - 1)$
34. $\ln (x + 1) + \ln (x - 1) = 1$
35. $e^{2x-3} \leq 4e^{x+1}$
36. $2^{1-3x} > 5^{4x+2}$
37. $\log 5x - \log (3x + 1) \leq \log (1 - 2x)$
38. $\ln (5x + 7) > \ln [(3x - 1)(x + 5)]$
39. Let $a = 4.1234567$ and $b = 3.7654321$. Use your calculator to (approximately) find $\alpha = \log a$ and $\beta = \log b$. Use the $\boxed{y^x}$ key with $y = b$ and $x = \frac{\alpha}{\beta}$. Explain why theory predicts that the calculator will display a. Did it? Discuss.
40. Calculate exactly without calculator, computer, or tables:
   (a) $\log_{1/2} (\sqrt{e}) \log_3 (8) \log (243) \ln 100$
   (b) $\dfrac{\log_2 49}{\log_3 (7) \log_2 81}$

Chapter 5

Trigonometric Functions and Their Graphs

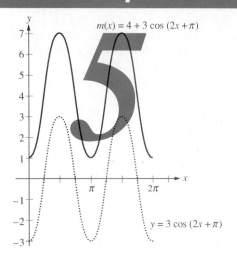

In the last two chapters, you studied polynomial, rational, exponential, and logarithmic functions. In this chapter, you will begin studying the trigonometric functions. As their name indicates, the trigonometric functions were originally developed to measure triangles. But, as so often happens in mathematics, additional uses for these functions were found. In this chapter, as well as the next two, you will study the trigonometric functions and some of their applications. Not only will you learn to use them as they were originally intended, with triangles, but you will also see them applied to areas such as surveying, navigation, astronomy, and periodic phenomena.

5.1 ANGLES AND THEIR MEASURES

For the study of trigonometry, it is of utmost importance to understand the notion of an angle. If P is a point on a line ℓ, then a **ray** is a subset of ℓ consisting of P and all the points of ℓ on one side of P; we call P the **endpoint** of such a ray. An angle is made up of two rays. More precisely, an **angle** is formed by the union of two rays that have a common endpoint, called the **vertex**. These two rays are called the **sides** of the angle; see figure 5.1.

In trigonometry, a slightly different view is sometimes taken. An angle can be formed by rotating a ray around the vertex from a starting position, called the **initial side**, to an ending position, called the **terminal side**. If the rotation is counterclockwise, the angle is considered a **positive** angle, and if the rotation is clockwise, it is a **negative** angle. (See figures 5.2(a) and (b).) If a Cartesian

Figure 5.1

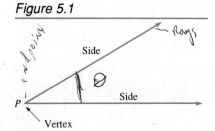

Angles and Their Measures

coordinate system is used, then an angle is in **standard position** if its vertex is at the origin and its initial side is the positive x-axis. This is shown in figure 5.2(c). If the terminal side coincides with one of the coordinate axes, as in figure 5.2(d), the angle is a **quadrantal angle.**

Figure 5.2

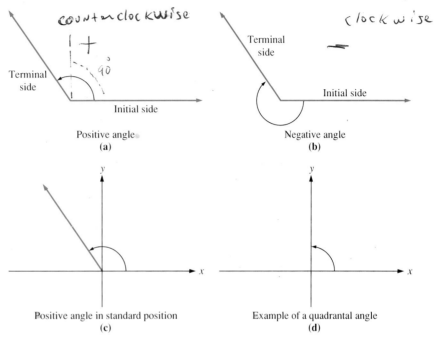

Positive angle
(a)

Negative angle
(b)

Positive angle in standard position
(c)

Example of a quadrantal angle
(d)

Degree and Radian Measure

The size of an angle can be measured using several different units of measure. The two most common are degrees and radians. A **degree** is $\frac{1}{360}$ of the angle formed when a ray is rotated through one complete revolution. The symbol ° is used to indicate degrees. Hence one degree is represented as 1°, and 15 degrees as 15°. One complete revolution in a counterclockwise direction produces an angle of 360°, and a complete revolution in a clockwise direction produces an angle of –360°. One-fourth of a complete revolution in a counterclockwise direction produces an angle of 90°. An angle of 0° results when the generating ray is not rotated at all. Two complete revolutions in a counterclockwise direction produce an angle of 720°.

Each degree is divided into 60 equal parts called **minutes,** and the symbol ′ is used to represent minutes. Thus 1° = 60′, 0.5° = 30′, and so on. Each minute is subdivided into 60 **seconds,** denoted by the symbol ″. Thus, 1′ = 60″, 2.5′ = 150″, and so on.

If two angles have the same initial and terminal sides, then they are called **coterminal angles.** One way to find the measure of a coterminal angle of a given angle is to add 360° to or subtract 360° from the size of the original angle. For example, in figure 5.3(a), the original angle has a measure of 75°. The coterminal angle shown in figure 5.3(b) would measure 75° – 360° = –285°;

and the coterminal angle in figure 5.3(c) measures 75° + 360° = 435°. In fact, you could find all possible measures of the coterminal angles of a given angle by adding all possible integral multiples of 360° to the measure of the original angle. So, 75° + 3(360°) = 75° + 1080° = 1155° is the size of another angle that is coterminal to a 75° angle. In general, all angles coterminal with a 75° angle are represented by 75° + 360°n, n any integer.

Figure 5.3

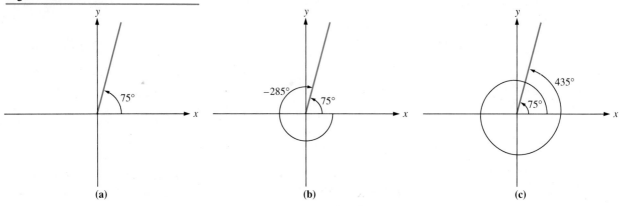

(a)   (b)   (c)

Most scientific applications require an angle measure called a radian. In order to define this term, first put the vertex of an angle at the center of a circle, as shown in figure 5.4(a). Let s be the length of the intercepted arc, and let r be the radius of the circle. Then we have the following definition.

Radian Measure of an Angle

The size of angle θ, as given by the formula

$$\theta = \frac{s}{r}, \quad \text{length of intercepted arc} \over \text{radius of circle}$$

is the **radian** measure of θ.

A useful algebraically equivalent form is $s = r\theta$. As you can see in figure 5.4(b), an angle that measures 1 radian intercepts an arc equal in length to the radius of the circle. Since the circumference of a circle with radius r is $2\pi r$, the radian measure of one complete revolution of a circle is given by

$$\theta = \frac{s}{r} = \frac{2\pi r}{r} = 2\pi.$$

It is important to notice that the above formula for θ does not depend on the chosen circle. More precisely, a theorem from geometry shows, in the notation of figure 5.4(c), that $\dfrac{s_1}{r_1} = \dfrac{s_2}{r_2}$.

Angles and Their Measures

Figure 5.4

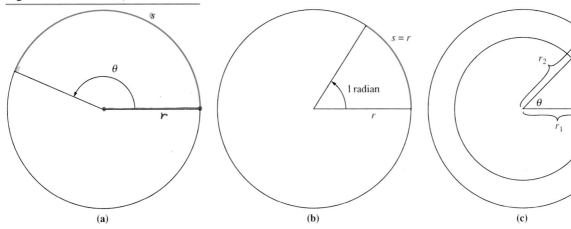

Converting Between Degrees and Radians

We just saw that 2π radians is one complete revolution of a circle. But a complete revolution is also 360°. Hence 2π radians = 360°, and we have the following equivalent relationship.

Radian-Degree Relationship

$$\pi \text{ radians} = 180°.$$

This relationship allows for an easy conversion between these two ways of measuring angles since the ratio of radians to degrees must be $\dfrac{\pi}{180}$.

▶ **EXAMPLE 5.1**   Convert 72° to radians.

Solution   Let R represent the unknown radian value. Then we have the proportion

$$\frac{R \text{ radians}}{72°} = \frac{\pi \text{ radians}}{180°}$$

$$R \text{ radians} = 72 \times \frac{\pi}{180} = \frac{2\pi}{5}.$$

Thus, $72° = \dfrac{2\pi}{5}$ radians. ◀

▶ **EXAMPLE 5.2**   Convert $\dfrac{5\pi}{6}$ radians to degrees.

Solution   If we let D represent the unknown degree value, then we have the proportion

$$\frac{\dfrac{5\pi}{6} \text{ radians}}{D°} = \frac{\pi \text{ radians}}{180°}$$

$$D° = \frac{180 \times 5\pi}{\pi \times 6} = 150°.$$

Thus, $\frac{5}{6}\pi$ radians = 150°. ◀

Using the above relationship, you can verify that 1 radian ≈ 57.3°.

With practice you will notice a quick way to do problems like example 5.2 involving a rational multiple of π radians. All you have to do is multiply the rational factor by 180°. For example, $\frac{5\pi}{6}$ radians = $\frac{5(180)°}{6}$ = 5(30°) = 150°.

It will often be helpful to refer to the entries in the following table of conversions.

Degrees	0	30	45	60	90	135	180	270	360
Radians	0	$\frac{\pi}{6}$	$\frac{\pi}{4}$	$\frac{\pi}{3}$	$\frac{\pi}{2}$	$\frac{3\pi}{4}$	π	$\frac{3\pi}{2}$	2π

Conversions Between Degrees and Radians Using Calculators

Most nongraphing calculators have a key which allows for conversion from degrees to radians or vice versa. When most of these calculators are turned on, they are in the degree mode. Many calculators indicate they are in the degree mode by showing the symbol DEG on the display screen. To get the calculator to operate in radians, you press the [DRG] key. When [DRG] is pressed, the symbol RAD will appear on the display screen. Pressing [DRG] a second time causes the calculator to operate in the GRAD mode, with the symbol GRAD appearing on the screen. (The angle measure grads will not be used in this text.) Pressing [DRG] a third time returns the calculator to the degree mode. To go through the modes in reverse order, press the [INV] [DRG] key combination.

Pressing the [2nd] [DRG▶] key combination does two things: (1) changes the mode displayed by the calculator and (2) converts the number in the display to the new unit. To convert a value from degrees to radians, you would use [2nd] [DRG▶]; and to convert from radians to degrees, you use [2nd] [INV] [DRG▶]. One difference between textbooks such as this one and calculators is that calculators express radians as decimals and textbooks often express radians as multiples of π.

On a *TI-81* calculator, you can determine whether the calculator is in radian or degree mode and, if desired, change this mode by proceeding as in the first paragraph of example 2.11.

When you turn on a Casio *fx-7700G* calculator, the third line of the display tells you whether the calculator is in radian or degree mode. If you wish to change this mode, first press [SHIFT] [DRG]. Then, press [F1] if you wish to change the calculator from radian mode to degree mode; to change from degrees to radians, press [F2]. Finally, press the [EXE] key, and you are ready to proceed.

Angles and Their Measures

EXAMPLE 5.3 Use a calculator to convert 72° to radians.

Solution On many nongraphing calculators, you may proceed along the following lines.

Press	Display
72 [2nd] [DRG▶]	1.2566371

On a Casio *fx-7700G*, first make sure that the calculator is in radian mode, and then press 72 [SHIFT] [DRG] [F4] [EXE]. The display shows 1.256637061. You would, of course, find the same result by pressing 72 [×] [SHIFT] [π] [÷] 180 [EXE].

When this same problem was worked in example 5.1, the exact answer was $\frac{2\pi}{5}$. A calculator can be used to verify that $\frac{2\pi}{5}$ is approximately 1.2566371. ◀

EXAMPLE 5.4 Use a calculator to convert $\frac{5\pi}{6}$ radians to degrees.

Solution On many nongraphing calculators, you may proceed along the following lines.

Press	Display
[DRG] 5	5
[×] π	3.1415927
[÷] 6	6
[=]	2.6179939
[2nd] [INV] [DRG▶]	150

The first four lines were used to get a decimal approximation of $\frac{5\pi}{6}$. When this problem was worked in example 5.2, the final answer was the same, 150°.

On a Casio *fx-7700G*, first make sure that the calculator is in degree mode, and then press [(] 5 [SHIFT] [π] [÷] 6 [)] [SHIFT] [DRG] [F5] [EXE]. The display shows 150. You would, of course, find the same result by pressing 5 [×] [SHIFT] [π] [÷] 6 [×] 180 [÷] [SHIFT] [π] [EXE] or, more simply, 5 [÷] 6 [×] 180 [EXE]. ◀

Most calculators represent parts of degrees as decimals rather than minutes and seconds. So, a calculator would represent 25°30′ as 25.5° and 73°50′42″ as 73.845°. Some calculators have a [DMS·DD] key that can be used to convert from degrees, minutes, and seconds to decimal degrees and vice versa. For example, to convert 32°47′6″ to decimal degrees, enter 32.4706 and press [2nd] [DMS·DD]. The number 32.875 will be displayed in the calculator screen and is understood to represent 32.875°. To change the decimal degree value of of 123.45° to degrees, minutes, and seconds, enter 123.45 and press [INV] [2nd] [DMS·DD]. Then 123.27 will be displayed; this represents 123°27′.

Applications of Radian Measure

The formula $\theta = \dfrac{s}{r}$ that was used to define radian measure has several applications. Some of these will be shown in the following examples; others will appear in the exercises.

▶ **EXAMPLE 5.5**   The vertex of a 70° angle is placed at the center of a circle with radius 9 cm. What is the length of the subtended arc?

Solution   As noted above, we may rewrite the formula $\theta = \dfrac{s}{r}$ so as to give the length of the arc as $s = r\theta$, provided that θ is in radians. Converting 70° to radians, we get $70° = \dfrac{7\pi}{18}$. Hence, $s = 9\left(\dfrac{7\pi}{18}\right) = \dfrac{7\pi}{2}$ cm ≈ 10.996 cm. ◀

Consider the shaded sector of the circle in figure 5.5. The area of this sector is proportional to the size of the central angle θ. Hence, we have $\dfrac{A_{\text{sector}}}{A_{\text{circle}}} = \dfrac{\theta}{2\pi}$. Substituting the value of πr^2 for the area of the circle produces the formula for the area of a sector.

Area of a Sector

$$A_{\text{sector}} = \dfrac{\pi r^2 \theta}{2\pi} = \dfrac{1}{2} r^2 \theta,$$

where θ is in radians.

Figure 5.5

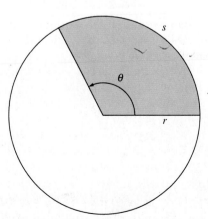

Angles and Their Measures

EXAMPLE 5.6 What is the area of the sector formed by a central angle of 150° in a circle with radius 8 inches?

Solution The above area formula is for θ in radians. Converting 150° to radians, we obtain $150° = \frac{5\pi}{6}$. (See examples 5.1 and 5.3.) Thus, $A = \frac{1}{2}8^2\left(\frac{5\pi}{6}\right) = \frac{80}{3}\pi \approx 26.67\pi \approx 83.78$ square inches. ◀

The next two applications involve linear and angular speed. The formula relating distance d, rate r, and time t specifies that $d = rt$ or $r = \frac{d}{t}$. Since rate and speed are synonymous here, this can be written as $v = \frac{d}{t}$, where v represents speed.

Consider a point P that moves at a constant speed along a circle with radius r and center O, as shown in figure 5.6. The measure of how fast the position of P is changing is called the **linear speed**.

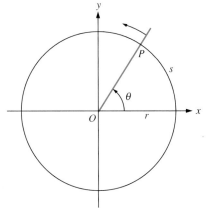

Figure 5.6

Linear Speed

If s is the length of the arc subtended by P during a time interval of duration t, then the **linear speed** is given by

$$v = \frac{s}{t}.$$ *is the same as :)*

Since $s = r\theta$, where the ray $\overrightarrow{OP}$ has moved through θ radians during the given time interval, this can also be written as

$$v = \frac{r\theta}{t} \quad \frac{s}{t} \qquad (1)$$

During the same time interval, the ray $\overrightarrow{OP}$ moves through θ angular units. The measure of how fast θ is changing is called the **angular speed** and is denoted by the Greek letter ω.

RAD/SEC

Angular Speed

Angular speed is given by

$$\omega = \frac{\theta}{t}. \qquad (2)$$

It follows from (1) and (2) that we have the following relationship between linear and angular speed when speed around a circular path is constant.

Chapter 5 / Trigonometric Functions and Their Graphs

Linear Speed–Angular Speed Relationship

$$v = r\omega.$$

The linear speed–angular speed relationship is not nearly as nice when the object is accelerating or when the path is not circular. Dealing with these less than ideal circumstances is better addressed by using the tools of calculus.

▶ **EXAMPLE 5.7**

A wheel with an 18 cm radius is rotating at 12 radians/sec. What is the speed of a point on its rim?

Solution

This problem is asking for the linear speed of this point, given $r = 18$ and $\omega = 12$. So the appropriate formula is

$$v = r\omega$$
$$= 18 \text{ cm } (12 \text{ rad/sec}) = 216 \text{ cm/sec.} \quad ◀$$

▶ **EXAMPLE 5.8**

A phonograph record has a diameter of 30 cm. If a point on the rim moves at 52.36 cm/sec, what is the angular speed of the record in (a) radians per second and (b) rpm (revolutions per minute)?

Solutions

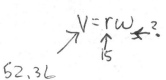

(a) Since the diameter is 30 cm, the radius is 15 cm. We are told that the linear speed, v, is 52.36 cm/sec. Solving the formula $v = r\omega$ for ω determines that

$$\omega = \frac{v}{r} = \frac{52.36}{15} \approx 3.49 \text{ radians/sec.}$$

(b) To get the angular speed in rpms, we must convert the value of 3.49 radians/sec to revolutions per minute. One revolution is 2π radians; so 1 rpm is 2π radians/min. As there are 60 seconds in one minute, 1 rpm is $\frac{2\pi}{60}$ radians/sec. Hence 3.49 rad/sec = $3.49 \times \frac{60}{2\pi} \approx \frac{104.7}{\pi} \approx 33\frac{1}{3}$ rpm. ◀

Exercises 5.1

In exercises 1–4, draw each of the given angles in standard position. Use a protractor if necessary. (Unspecified angular units are *always* radians.)

1. (a) 30° (b) −60° (c) −750° (d) 180°
2. (a) −135° (b) 120° (c) 405° (d) −270°
3. (a) $-\frac{2\pi}{3}$ (b) $\frac{3\pi}{4}$ (c) $\frac{37\pi}{6}$ (d) $-\pi$
4. (a) $\frac{\pi}{6}$ (b) $-\frac{\pi}{3}$ (c) $-\frac{7\pi}{3}$ (d) $\frac{3\pi}{2}$

In exercises 5–6, convert to radian measure. (Round off to nearest hundredths.)

5. (a) 135° (b) −30°20′ (c) 20″ (d) 904°
6. (a) −18° (b) 45°30′ (c) 1°2′3″ (d) 180°

In exercises 7–8, convert the given radian measure to degree measure. (Round off to nearest hundredths of a degree.)

7. (a) π (b) 1 (c) −7 (d) $-\frac{\pi}{6}$
8. (a) 2.5 (b) $\frac{7\pi}{4}$ (c) -7π (d) 7

In exercises 9–10, convert to degrees, minutes, and seconds. (Round off seconds to nearest hundredths.)

9. (a) 2.5° (b) −30.125°
   (c) $\frac{4\pi}{11}$ radians (d) 8 radians
10. (a) 46.48219° (b) 2.583°
    (c) −5.2 radians (d) 4 radians

Angles and Their Measures

◆ 11. Give the degree measure of the smallest positive angle that is coterminal with
   (a) −90°   (b) 135°   (c) 370°
   (d) −180°   (e) 904°   (f) −760°

◆ 12. Determine which of the following, when placed in standard position, is/are a pair of coterminal angles.
   (a) 731°, 11°   (b) −30°, 330°   (c) 120°, 1020°
   (d) π radians, $-\pi$ radians   (e) $\frac{5\pi}{2}$ radians, 2π radians
   (f) $-\pi$ radians, 7π radians   (g) $-30°$, $\frac{35\pi}{6}$ radians

13. The minute hand of a clock is 5 cm long. How far does its tip travel in 20 minutes?

14. How many revolutions per minute (rpms) are made by a truck's tire if the truck is traveling 60 miles per hour (= 88 feet per second) and the tire's diameter is 3 feet? (You may assume that the truck rolls without slipping. This means, in particular, that, in one second, a point on the outer rim of the tire travels a path measuring 88 feet.)

15. A pulley is driven at 500 rpm by a belt which (does not slip and) travels at the rate of 75 feet per second. Find the radius of the pulley.

16. Find the radius of a circle in which a central angle of 235° subtends an arc of 10 feet.

17. Find the radian measure of a central angle θ in a circle of radius 4 inches if θ subtends an arc of 3π inches.

18. Find the arc length subtended by a central angle of $\frac{\pi}{6}$ radians in a circle of radius 2. How does the answer change if the central angle's measure is doubled? How does the first answer change if the circle's radius is doubled?

19. A clock's pendulum is 40 cm long and sweeps out an angle of 20°. Find the length of the arc traveled by the pendulum's tip.

20. A circular portion of a roller coaster has length 175 feet and radius 50 feet. What is the radian measure of the angle through which a person traveling on this portion of the roller coaster is turned?

For simplicity, in exercises 21 and 22, you may assume that the Earth is a sphere with radius 3960 miles.

21. Let C denote the Earth's center. A nautical mile is the shortest distance along the Earth's surface between points P and Q such that the measure of angle PCQ is 1′. Explain why a nautical mile is approximately 1.15 miles.

22. Consider two points Q and K on the Earth's equator. Q is at 78° west longitude (near Quito), and K is at 32° east longitude (near Kampala). Find the approximate distance in miles between Q and K.

23. A sector of a circle with radius 5 feet has area 8 square feet. What is the radian measure of the sector's angle?

24. A cherry pie has a circular top with diameter 20 cm. If Jacques cuts a slice of pie whose top forms a sector with angle $\frac{\pi}{6}$ radians, what is the area of the top of Jacques's piece of pie?

25. Jeanne plans to order a slice of Capriciossa pizza. She has three options corresponding to the slice's radii: 6 inches, 8 inches, or 11 inches. All slices have tops in the shape of sectors having angle $\frac{\pi}{4}$ radians. If Jeanne wants a slice whose top has area approximately 24 square inches, which slice should Jeanne choose? What is the actual area of the top of the chosen slice? (Hint: the second answer is what Jeanne then did.)

26. Two concentric circles have radii 5 feet and 18 feet. The region between these circles is to be planted with flowers. What is the area of the (top of the) proposed garden?

27. Consider the hour hand and the minute hand of the face of a (nondigital!) clock. Suppose that the time is x minutes after 1:00, where $0 < x < 60$.
   (a) Explain why the hour hand has moved $\frac{60 + x}{720}$ of one (clockwise!) revolution since the most recent 12:00.
   (b) What fraction of a revolution has the minute hand moved since 1:00?
   (c) Show that the hour hand and the minute hand point in the same direction precisely when $x = \frac{60}{11}$.

28. At which time shortly after 7:00 do the hour hand and the minute hand of a nondigital clock point in the same direction?

29. A wheel is rotating with an angular speed of 50 radians per second.
   (a) How many revolutions per second is the wheel making?
   (b) If the wheel's diameter is 6 feet, find the linear speed of a point on the wheel's rim.

30. Find the angular speed in radians per second of a phonograph record playing at 45 rpm. Does your answer depend on the record's radius?

31. A centrifuge in a chemistry laboratory spins around at 100 rpm. A solution in a test tube at the edge of the centrifuge has linear speed 140 cm per second. Find the centrifuge's radius.

32. A child is securely attached to the outer rim of a merry-go-round which is 50 feet in diameter. If the ride

makes 12 revolutions in 5 minutes, determine the child's angular speed and linear speed.

33. A fly lands on the outer rim of a roulette wheel just before the wheel begins to rotate. Because of a new nearly frictionless mechanism, this wheel is able to maintain a nearly constant angular speed of 5 radians per second for several seconds. During that period, the fly's linear speed is 150 cm per second. Find the wheel's radius.

34. A wheel with diameter 28 feet is to revolve in a vertical plane at 6 radians per second. The Human Fly, a daredevil, plans to attach herself to the wheel at a point 12 feet from the wheel's center. She can hold on only if her speed does not exceed 75 feet per second.

(a) Will she be able to hold on once the revolution comes?
(b) If so, what is the area of the sector formed by her arc during one second of motion?

35. (a) Ignoring the units of measurement, you found in exercise 34 that your answer in (b) was the product of the angular speed with your answer in (a). Does this also happen if that exercise is changed by positioning the Human Fly just 10 feet from the wheel's center?
(b) What if the angular speed is reduced to 4 radians per second and the Human Fly is positioned 8 feet from the wheel's center?

5.2 DEFINING THE TRIGONOMETRIC FUNCTIONS

Let θ be any angle in standard position with vertex O, and let $P(x, y)$ be another fixed point on the terminal side of θ, as shown in figure 5.7. If r is the distance from O to P, then $r = OP = \sqrt{x^2 + y^2}$, and if $P'(x_1, y_1)$ is any other point on the terminal side of θ, we shall let $r_1 = OP' = \sqrt{x_1^2 + y_1^2}$. If Q and Q' are points on the initial side of θ such that $\overline{PQ} \perp \overrightarrow{OQ}$ and $\overline{P'Q'} \perp \overrightarrow{OQ'}$, then $\triangle OPQ$ and $\triangle OP'Q'$ are similar right triangles. Hence, these triangles have corresponding sides proportional. In particular, $\dfrac{y}{r} = \dfrac{y_1}{r_1}, \dfrac{x}{r} = \dfrac{x_1}{r_1}$, and $\dfrac{x}{y} = \dfrac{x_1}{y_1}$. Thus, these ratios depend only on the angle θ, and not on the choice of P or P'.

There are six of these ratios possible: the above three and their reciprocals. Because these ratios are determined by the size of the angle (that is, they are functions of the angle θ), they are called the trigonometric functions of the angle θ. The names of these six trigonometric functions and their abbreviations are sine (sin), cosine (cos), tangent (tan), cotangent (cot), secant (sec), and cosecant (csc). A formal definition of the six trigonometric functions is given below.

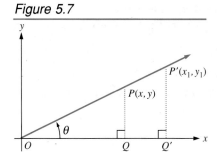

Figure 5.7

Trigonometric Functions of θ

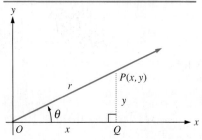

Figure 5.8

Let θ be the radian measure of an angle placed in standard position. Let $P(x, y)$ be any point, other than the origin, on the terminal side of the angle. Let r be the distance between the origin and P. (See figure 5.8.) Then the six **trigonometric functions of θ** are defined as

$$\sin \theta = \frac{y}{r} \qquad \cos \theta = \frac{x}{r} \qquad \tan \theta = \frac{y}{x}$$

$$\cot \theta = \frac{x}{y} \qquad \sec \theta = \frac{r}{x} \qquad \csc \theta = \frac{r}{y}$$

Remember that an expression involving division by 0 is undefined.

Defining the Trigonometric Functions

The trigonometric functions, as defined above, are functions of a standard-position angle θ.

It is often convenient to adopt an equivalent point of view in which the trigonometric functions are functions of a real number (representing the radian measure of θ).

This idea, developed further in section 5.3, leads to viewing the trigonometric functions as **circular functions**. One benefit of this second point of view will be our ability to graph the trigonometric functions.

In section 5.6, we shall graph the trigonometric functions and, in particular, tabulate their domains and ranges. These depend on the function being considered. For example, both sine and cosine have domain equal to the set of real numbers. However, the domains of the other four trigonometric functions are more restricted. Both tangent and secant have domain consisting of all real numbers θ such that $x \neq 0$. Thus, $\tan \theta$ and $\sec \theta$ are not defined if θ is $\frac{\pi}{2}$ radians (90°), $\frac{3\pi}{2}$ radians (270°), and so on. Similarly, the cotangent and cosecant are not defined at $\theta = 0$, π (180°), 2π (360°), and so on.

Since $|x| \leq r$ and $|y| \leq r$, sin and cos each have range $\{y: -1 \leq y \leq 1\}$; and so their reciprocals, csc and sec, each have range $\{y: y \leq -1 \text{ or } y \geq 1\}$. Similarly, tan and cot each have range consisting of all real numbers.

The above definitions lead to some valuable equations connecting some of the trigonometric functions. For instance, $\sin \theta = \frac{1}{\csc \theta}$ since $\frac{y}{r} = 1 / \left(\frac{r}{y}\right)$. Similar relationships also hold for the other trigonometric functions. These relationships, known as the **reciprocal identities**, are shown in the following box.

Reciprocal Identities

$$\sin \theta = \frac{1}{\csc \theta} \qquad \cos \theta = \frac{1}{\sec \theta} \qquad \tan \theta = \frac{1}{\cot \theta}$$

$$\csc \theta = \frac{1}{\sin \theta} \qquad \sec \theta = \frac{1}{\cos \theta} \qquad \cot \theta = \frac{1}{\tan \theta}$$

These are interpreted so that the left-hand side is undefined when the corresponding right-hand side's denominator is zero.

Next, consider dividing $\sin \theta$ by $\cos \theta$: $\frac{y/r}{x/r} = \frac{y}{x}$. Since this is the definition of $\tan \theta$, one gets the following identity.

Quotient or Ratio Identity

$$\frac{\sin \theta}{\cos \theta} = \tan \theta.$$

Then, using the reciprocal identity $\cot \theta = \frac{1}{\tan \theta}$, we find the identity shown in the box below.

Chapter 5 / Trigonometric Functions and Their Graphs

Quotient or Ratio Identity

$$\frac{\cos \theta}{\sin \theta} = \cot \theta.$$

The above two identities are called the **quotient** or **ratio identities**. In them, denominators of zero are interpreted as for the reciprocal identities. In particular, $\tan \theta$ is undefined if $\theta = \frac{\pi}{2}$.

You will find an excellent use for the above identities in section 5.4. Meanwhile, they should be memorized.

▶ **EXAMPLE 5.9**

Given the point $(-15, 8)$ on the terminal side of an angle θ in standard position, determine the six trigonometric functions of θ.

Solution

Here $x = -15$ and $y = 8$, so $r = \sqrt{(-15)^2 + 8^2} = \sqrt{289} = 17$, as shown in figure 5.9. The values of the trigonometric functions at this θ are

$$\sin \theta = \frac{8}{17} \qquad \cos \theta = \frac{-15}{17} \qquad \tan \theta = \frac{-8}{15}$$

$$\cot \theta = \frac{-15}{8} \qquad \sec \theta = \frac{-17}{15} \qquad \csc \theta = \frac{17}{8} \quad ◀$$

Figure 5.9

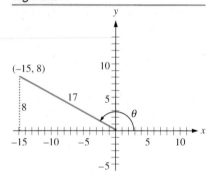

The trigonometric functions can be grouped into three sets of cofunctions. For instance, sine and cosine are **cofunctions** (of one another). Similarly, tan and cot are cofunctions; and sec and csc are cofunctions. Later in this section, we shall introduce the cofunction identities. They will be very useful in chapter 6.

You may remember that an identity is an equation that is true for all reasonable replacements of the variable. Once an identity has been established, it can be used to either develop or verify other identities. These identities can then be used to help solve equations or inequalities. For this reason, skill at establishing identities is helpful later in this chapter and in subsequent chapters.

▶ **EXAMPLE 5.10**

Use the reciprocal and ratio identities to establish the following identity:

$$\frac{\cot \theta + \cos \theta}{\cos \theta} = 1 + \csc \theta.$$

Solution

We begin by using the reciprocal and ratio identities to rewrite the left-hand side in terms of the sine and cosine functions.

$$\frac{\cot \theta + \cos \theta}{\cos \theta} = \frac{\dfrac{\cos \theta}{\sin \theta} + \cos \theta}{\cos \theta}$$

$$= \frac{\dfrac{\cos \theta + \cos \theta \sin \theta}{\sin \theta}}{\cos \theta}$$

Defining the Trigonometric Functions

$$= \frac{\cos\theta + \cos\theta\sin\theta}{\sin\theta\cos\theta}$$

$$= \frac{\cos\theta}{\sin\theta\cos\theta} + \frac{\cos\theta\sin\theta}{\sin\theta\cos\theta}$$

$$= \frac{1}{\sin\theta} + 1$$

$$= \csc\theta + 1 = 1 + \csc\theta.$$

We have converted the left-hand side into the right-hand side and thus have established that the two sides are equivalent. ◀

Right Triangles and the Trigonometric Functions

Calculating the trigonometric functions for an angle θ in standard position is not always the most convenient procedure to follow. Consider an angle θ in standard position and construct a triangle as in figure 5.8. Focusing on $\triangle POQ$ produces a figure like figure 5.10. Let's say that the length of the hypotenuse is r, y is the length of the side opposite angle θ, and x is the length of the side adjacent to angle θ. This notation permits the following alternative, and very useful, definitions for the trigonometric functions of acute angles.

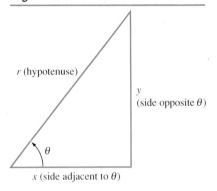

Figure 5.10

Trigonometric Functions Defined Using a Right Triangle

If θ is an acute angle of a right triangle, r is the length of the hypotenuse, y is the length of the side opposite angle θ, and x is the length of the side adjacent to angle θ, then the six **trigonometric functions of** θ are defined as

$$\sin\theta = \frac{y}{r} = \frac{\text{side opposite }\theta}{\text{hypotenuse}} \qquad \cos\theta = \frac{x}{r} = \frac{\text{side adjacent to }\theta}{\text{hypotenuse}}$$

$$\tan\theta = \frac{y}{x} = \frac{\text{side opposite }\theta}{\text{side adjacent to }\theta} \qquad \cot\theta = \frac{x}{y} = \frac{\text{side adjacent to }\theta}{\text{side opposite }\theta}$$

$$\sec\theta = \frac{r}{x} = \frac{\text{hypotenuse}}{\text{side adjacent to }\theta} \qquad \csc\theta = \frac{r}{y} = \frac{\text{hypotenuse}}{\text{side opposite }\theta}$$

These new definitions allow us to determine the trigonometric functions of any acute angle, once it is given as part of a right triangle, without making reference to standard position or to any coordinate system.

▶ **EXAMPLE 5.11**

Determine the values of the six trigonometric functions for an angle θ of a right triangle with hypotenuse of 53 and an opposite side of length 28, as shown in figure 5.11.

Solution

The length of the adjacent side, x, can be determined by using Pythagoras' Theorem:

$$x = \sqrt{53^2 - 28^2} = \sqrt{2025} = 45.$$

Figure 5.11

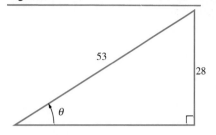

The trigonometric functions for θ are then

$$\sin \theta = \frac{y}{r} = \frac{\text{side opposite } \theta}{\text{hypotenuse}} = \frac{28}{53} \qquad \cos \theta = \frac{x}{r} = \frac{\text{side adjacent to } \theta}{\text{hypotenuse}} = \frac{45}{53}$$

$$\tan \theta = \frac{y}{x} = \frac{\text{side opposite } \theta}{\text{side adjacent to } \theta} = \frac{28}{45} \qquad \cot \theta = \frac{x}{y} = \frac{\text{side adjacent to } \theta}{\text{side opposite } \theta} = \frac{45}{28}$$

$$\sec \theta = \frac{r}{x} = \frac{\text{hypotenuse}}{\text{side adjacent to } \theta} = \frac{53}{45} \qquad \csc \theta = \frac{r}{y} = \frac{\text{hypotenuse}}{\text{side opposite } \theta} = \frac{53}{28}$$

▶ **EXAMPLE 5.12**

Suppose θ is an acute angle of a right triangle and $\tan \theta = \frac{7}{9}$. What are the values of the other trigonometric functions of θ?

Solution

Since $\tan \theta = \frac{7}{9}$, we can find a similar triangle in which the length of the side opposite θ is 7 and the length of the side adjacent to θ is 9, as shown in figure 5.12. Using Pythagoras' Theorem, it can be determined that the hypotenuse is $\sqrt{130}$. The values for the other five trigonometric functions of θ are

$$\sin \theta = \frac{7}{\sqrt{130}} \qquad \cos \theta = \frac{9}{\sqrt{130}}$$

$$\cot \theta = \frac{9}{7} \qquad \sec \theta = \frac{\sqrt{130}}{9} \qquad \csc \theta = \frac{\sqrt{130}}{7}$$

Figure 5.12

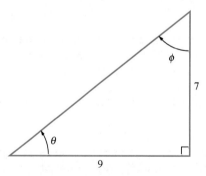

▶ **EXAMPLE 5.13**

In figure 5.12, ϕ is the complement of the angle θ. What are the values of the six trigonometric functions of ϕ?

Solution

The length of the side opposite angle ϕ is 9, and the length of the side adjacent to angle ϕ is 7. The length of the hypotenuse is still $\sqrt{130}$. The values for the six trigonometric functions of ϕ are

$$\sin \phi = \frac{9}{\sqrt{130}} \qquad \cos \phi = \frac{7}{\sqrt{130}} \qquad \tan \phi = \frac{9}{7}$$

$$\cot \phi = \frac{7}{9} \qquad \sec \phi = \frac{\sqrt{130}}{7} \qquad \csc \phi = \frac{\sqrt{130}}{9}$$

Defining the Trigonometric Functions

Examine figure 5.12 again. Angles θ and ϕ are complementary. Look at the results of examples 5.12 and 5.13, and note that

$$\sin \theta \underset{\substack{\uparrow \\ \text{complementary} \\ \text{angles}}}{= \overset{\overset{\text{cofunctions}}{\overbrace{}}}{\frac{7}{\sqrt{130}} = \cos \phi}}.$$

Similar results will be noticed by examining the other pairs of cofunctions; in particular, $\tan \theta = \cot \phi$ and $\sec \theta = \csc \phi$. These results indicate that the cofunctions of complementary angles are equal. Specifically,

Cofunction Identities

If θ and ϕ are complementary angles with radian measure, the **cofunction identities** state that

$$\sin \theta = \cos\left(\frac{\pi}{2} - \theta\right) = \cos \phi \qquad \sin \phi = \cos\left(\frac{\pi}{2} - \phi\right) = \cos \theta$$

$$\tan \theta = \cot\left(\frac{\pi}{2} - \theta\right) = \cot \phi \qquad \tan \phi = \cot\left(\frac{\pi}{2} - \phi\right) = \cot \theta$$

$$\sec \theta = \csc\left(\frac{\pi}{2} - \theta\right) = \csc \phi \qquad \sec \phi = \csc\left(\frac{\pi}{2} - \phi\right) = \csc \theta$$

We have seen how to prove the cofunction identities when θ is acute, by noting that the side opposite θ is the side adjacent to ϕ. In exercise 27, you will be asked to establish these identities in general.

Trigonometric functions were originally used to solve problems that involved triangles. The next example shows how one such problem may be solved.

▶ **EXAMPLE 5.14**

Four cables are needed to support a sign. If each cable forms an angle of 55° with the ground, and is anchored 8 feet from the base of the sign along a line perpendicular to the sign's base, how far up the sign does each cable reach?

Solution

A three-dimensional, or "real-life," drawing of the problem is in figure 5.13(a). The essential facts needed to solve the problem are in the two-dimensional drawing in figure 5.13(b), where y denotes the required distance. Notice that y is opposite the 55° angle, while the given side is adjacent to that angle.

Since the opposite and adjacent sides are involved, the tangent function will be used. (The cotangent function could also be used.) Thus, we have

$$\tan 55° = \frac{y}{8}$$

which means that $y = 8 \tan 55°$.

We leave the answer in this form since you do not yet know how to evaluate tan 55°. In the next section, we shall develop the techniques that will allow you to determine that $y \approx 11.43$ feet.

Figure 5.13

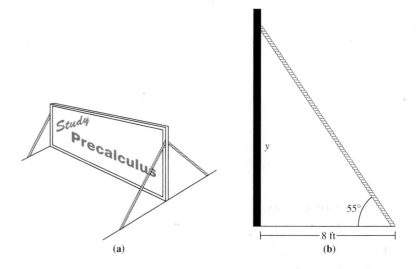

One relationship that can be deduced from the definitions of the trigonometric functions is a trigonometric version of Pythagoras' Theorem. If a right triangle such as the one in figure 5.8 has legs of lengths $|x|$ and $|y|$ and a hypotenuse of length r, then Pythagoras' Theorem states $x^2 + y^2 = r^2$. Dividing both sides by r^2, we get $\left(\dfrac{x}{r}\right)^2 + \left(\dfrac{y}{r}\right)^2 = 1$. Since $\dfrac{x}{r} = \cos \theta$ and $\dfrac{y}{r} = \sin \theta$, this can be written as $(\cos \theta)^2 + (\sin \theta)^2 = 1$, or, equivalently, $(\sin \theta)^2 + (\cos \theta)^2 = 1$. Mathematicians prefer to write $\sin^2 \theta$ instead of $(\sin \theta)^2$, and $\cos^2 \theta$ instead of $(\cos \theta)^2$. Using this notation, we have another identity.

Pythagorean Identity

$$\sin^2 \theta + \cos^2 \theta = 1.$$

This identity will be used in chapter 7 to develop the Law of Cosines, a useful method for studying triangles when they have no right angles. In chapter 6, we will find two other Pythagorean identities involving the remaining four trigonometric functions.

Exercises 5.2

In exercises 1–2, assume that the given point is on the terminal side of an angle θ in standard position. Evaluate the six trigonometric functions at θ, provided they are defined.

1. (a) $(3, 4)$ (b) $(-\sqrt{2}, 4)$ (c) $(-\tfrac{3}{4}, 0)$
   (d) $(0, \sqrt{2})$ (e) $(2, 3)$
2. (a) $(-\sqrt{2}, \pi)$ (b) $(-12, -5)$
   (c) $(0, -\tfrac{4}{3})$ (d) $(2, 0)$

In exercises 3–4, determine the algebraic sign of the value of the given trigonometric function of θ, if θ is an angle in standard position whose terminal side lies in the given quadrant.

3. (a) sin, I (b) csc, II (c) tan, III (d) cot, IV
4. (a) cos, II (b) sec, IV (c) sin, II (d) cos, IV

5. Evaluate the six trigonometric functions of an angle θ in standard position if the terminal side of θ lies along the line $y = -3x$ in the second quadrant.

6. Find a point on the terminal side of an acute angle θ in standard position such that $\sin \theta = \frac{1}{\sqrt{3}}$.

7. Establish that $-\sqrt{2}$ is in the range of csc.

8. Determine all values of θ between 0 and $\frac{\pi}{2}$ such that $\cot \theta > 1$.

◆ 9. Simplify each of the following expressions:
   (a) $\sin(\theta + 2\pi)$ (b) $\cos(\theta - 2\pi)$ (c) $\tan(\theta + 6\pi)$

◆ 10. Let θ and ϕ be coterminal angles and let T be one of the six trigonometric functions. Explain why $T(\theta) = T(\phi)$.

◆ 11. (a) Give examples of angles θ and ϕ, both in standard position, such that $\sin \theta = \sin \phi$ although θ and ϕ are not coterminal.
   (b) If θ and ϕ are angles in standard position such that $\sin \theta = \sin \phi$ and $\cos \theta = \cos \phi$, show that θ and ϕ are coterminal.

12. Determine which of the following real numbers is/are in the domain of tan:
   (a) 0   (b) $-\frac{\pi}{4}$ (c) $\frac{\pi}{2}$ (d) π
   (e) $-\frac{3\pi}{2}$ (f) $\frac{5\pi}{4}$ (g) $\frac{5\pi}{2}$ (h) 8π

13. Determine which of the real numbers listed in exercise 12 is/are in the domain of csc.

◆ 14. (a) Use similar triangles to prove that $\sin(-\theta) = -\sin \theta$. (Hint: place θ in standard position. There are several cases, depending on which quadrant contains the terminal side of θ.)
   (b) What conclusion may be drawn from (a) in regard to symmetry of the graph of the sine function?
   (c) Is your conclusion in (b) supported if you "check" using a graphing calculator or a computer graphing program?

◆ 15. Show that $\cos(-\theta) = \cos \theta$ and, hence, draw a conclusion regarding symmetry of the graph of the cosine function. (Hint: review the hint given for exercise 14.)

◆ 16. (a) Using the reciprocal and ratio identities (and exercises 14 and 15), conclude that sec is an even function and that tan, cot, and csc are each odd functions.
   (b) Draw conclusions regarding symmetry from (a).

17. Use the reciprocal and ratio identities to establish that
$$1 + \sec \theta = \frac{\tan \theta + \sin \theta}{\sin \theta}$$
is an identity (in the sense that the equation is valid whenever both its sides are meaningful).

18. Establish the identity $\dfrac{\sin \theta}{1 + \cos \theta} = \dfrac{1 - \cos \theta}{\sin \theta}$.

◆ 19. Establish that $\dfrac{\sin \theta}{1 - \cos \theta} = \dfrac{\tan \theta}{2 \sin \theta - 1}$ is *not* an identity. (You must find at least one θ for which the sides of this equation are meaningful and unequal. Try $\theta = -\frac{\pi}{4}$.)

20. Prove that the equation in exercise 19 is valid in case $\theta = \frac{\pi}{4}$. (Hint: $\sin \frac{\pi}{4} = \frac{1}{\sqrt{2}} = \cos \frac{\pi}{4}$.)

21. In the given right triangle, find all the trigonometric functions of both θ and ϕ.

In exercises 22–24, θ is an acute angle with the given trigonometric behavior. Find the other five trigonometric functional values of θ. (Hint: it may be helpful to place θ into a right triangle as in example 5.12.)

22. (a) $\sin \theta = \frac{5}{13}$ (b) $\cos \theta = \frac{4}{5}$

23. (a) $\sec \theta = \frac{2}{\sqrt{3}}$ (b) $\csc \theta = \frac{2}{\sqrt{3}}$

24. (a) $\tan \theta = \frac{1}{2}$ (b) $\cot \theta = \frac{1}{3}$

◆ 25. Suppose one knows that $\sin(\theta - \phi) = \sin \theta \cos \phi - \cos \theta \sin \phi$ for all θ and ϕ. Use the cofunction identities to conclude that $\cos(\theta + \phi) = \cos \theta \cos \phi - \sin \theta \sin \phi$.

(Hint: $\frac{\pi}{2} - (\theta + \phi) = \left(\frac{\pi}{2} - \theta\right) - \phi$.)

26. Combine the facts in exercises 14, 15, and 25 to show that $\cos(\theta - \phi) = \cos \theta \cos \phi + \sin \theta \sin \phi$ and $\sin(\theta + \phi) = \sin \theta \cos \phi + \cos \theta \sin \phi$.

◆ 27. Recall that we proved the cofunction identities for acute θ. Using expansion formulas of the type in exercises 25 and 26, one may derive the cofunction identities in general. For instance, show $\cos\left(\frac{\pi}{2} - \theta\right) = \sin \theta$ for any angle θ. (Hint: determine $\cos \frac{\pi}{2}$ and $\sin \frac{\pi}{2}$.)

28. Let ℓ be a nonvertical line, with slope m. Let θ be the angle between ℓ and the positive direction of the x-axis. (In radian measure, $0 \le \theta < \pi$; θ is called the **angle of inclination** of ℓ.) Prove that $m = \tan\theta$.

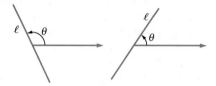

◆ 29. Let ℓ be a nonvertical, nonhorizontal line that does not pass through the origin. Let the line through the origin and perpendicular to ℓ intersect ℓ at P. Let $d = OP$, the distance from the origin to ℓ. Let ϕ be the angle of inclination of OP (in the sense defined in exercise 28). Show that $x\cos\phi + y\sin\phi - d = 0$ is an equation of ℓ. (Hint: apply the point-slope form, noting P is $(d\cos\phi, d\sin\phi)$ and ℓ has slope $-\cot\phi$.)

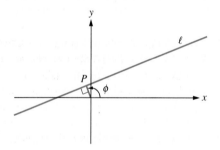

30. By suitably reinterpreting the parameters d and ϕ in exercise 29, discuss the possible validity of $x\cos\phi + y\sin\phi - d = 0$ as an equation of ℓ, if the line ℓ is vertical or horizontal or passes through the origin.

In exercise 31–32, use exercise 29 to find the distance from the origin to the given line.

31. (a) $\frac{3}{5}x + \frac{4}{5}y - 2 = 0$
    (b) $4x + 3y - 7 = 0$ (Hint: divide through by $\sqrt{4^2 + 3^2}$.)
32. (a) $12x - 5y + 4 = 0$ (Hint: divide through by -13. Why?)
    (b) $12x - 5y - 4 = 0$
33. Using exercise 29, find the distance from $(-1, 2)$ to the line $3x - 4y + 5 = 0$. (Hint: first translate axes to $(-1, 2)$, as in section 3.2.)
◆ 34. A stone block of weight w rests on a horizontal board. As one end of the board is lifted, we have the accompanying figure, with $AB = w$.

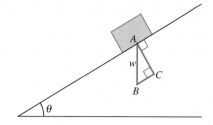

(a) Prove that $\angle BAC$ is congruent to θ, $AC = w\cos\theta$, and $BC = w\sin\theta$.
(b) As the board is raised, θ reaches a critical value θ_c. There, if the block is nudged downward, it will descend at constant speed. This occurs when $BC = \mu(AC)$, where μ is a physical constant (called the coefficient of kinetic friction) depending on the board and the stone. Prove that $\mu = \tan\theta_c$.

In exercises 35–39, you will need to leave your answers in terms of trigonometric functional values. In section 5.4, you will learn how to compute such values. Such problems recur in section 7.1.

◆ 35. Determine the height of a totem pole if it casts a shadow 75 feet long when the angle of elevation of the top of the totem pole, viewed from the far end of the shadow, is 30°. (By definition, the **angle of elevation** is the angle, measured from the horizontal, through which an observer must raise his or her line of sight in order to see an object. We shall return to this topic in section 7.1.)

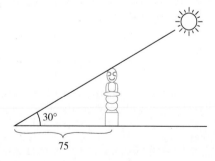

◆ 36. A short child named Petra is playing with a kite. Just as she has let out 120 feet of string and the kite's angle of elevation is 60°, a safety pin falls off the kite. Assuming the pin falls straight down and there is no sag in the string, determine how far it lands from Petra's feet.
◆ 37. A secret new aircraft is able to take off at an angle of elevation of 70° and travel 600 miles per hour. How

high above the ground is this plane 4 minutes after takeoff?

◆ 38. An observer positions herself 500 meters from a rocket launch pad. Shortly after a vertical launch, she finds that the tail of the rocket has an angle of elevation of 17° and the nose of the rocket has an angle of elevation of 20°. What is the length of the rocket?

◆ 39. A UFO passes 1000 feet directly over an observer and flies away without changing its speed or altitude. After 1.16 seconds have elapsed, the observer measures the UFO's angle of elevation as 60°. How fast is the UFO moving?

40. Romeo wishes to elope with Rowena by using a 25 foot ladder. Her bedroom window is 24 feet above the ground. Romeo cannot place the base of his ladder closer than 8 feet from Rowena's house. Is Romeo's ladder long enough to reach Rowena's window?

41. After nearly colliding, two ships steam off at right angles, with speeds of 18 mph and 24 mph. How far apart are the ships 12 minutes later?

42. On each step of a staircase, the horizontal board is 14 inches deep and the vertical board is 8 inches high. If the staircase rises 10 feet, how long is its banister railing?

5.3 CIRCULAR FUNCTIONS

In section 5.2 we developed one standard method for introducing the six trigonometric functions. In that method, these functions involved the independent variable θ. We considered θ to be a standard-position angle or the measure of such an angle. In particular, by taking θ to be the radian measure of a standard-position angle, we were able to view the trigonometric functions as functions of real numbers.

In this section, we shall introduce another useful way to regard the trigonometric functions as functions whose domains (and ranges) are certain sets of real numbers. In presenting this point of view, we shall take the opportunity to preview certain topics which will be discussed more fully later in this chapter. As you will see in exercises 19 and 20, this new point of view also yields new insights about the approach in section 5.2.

Let's begin with another geometric interpretation for the definition of the trigonometric functions. Consider the graph of $x^2 + y^2 = r^2$, where r is a positive constant. As you saw in section 1.4, this graph is the circle with center at the origin O and radius r. Let $P(x, y)$ be any point on this circle, and let θ denote the standard-position angle whose terminal side contains P, as shown in figure 5.14. From the definitions in section 5.2, we have $\cos \theta = \dfrac{x}{r}$ and $\sin \theta = \dfrac{y}{r}$. Thus, we may conclude that $x = r \cos \theta$ and $y = r \sin \theta$.

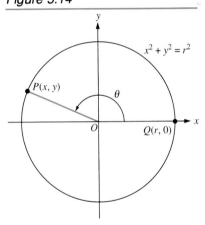

Figure 5.14

We can extend this idea. Once again, consider distinct points $P(x, y)$ and $Q(r, 0)$ on the circle $x^2 + y^2 = r^2$. Let α be the real number such that the arclength from Q to P counterclockwise along the circle is $r\alpha$. (Thus, $0 < \alpha < 2\pi$.) Let θ be the (radian) measure of the standard-position angle $\angle QOP$. The data are summarized in figure 5.15. Using the formula $s = r\theta$ from section 5.1, we see that the arc QP has length $r\theta$. Thus $r\alpha = r\theta$, and so $\theta = \alpha$. From the above, we therefore have $x = r \cos \alpha$ and $y = r \sin \alpha$.

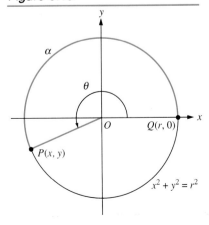

Figure 5.15

Now, let's specialize the constructions in the preceding paragraph to the unit circle, $x^2 + y^2 = 1$. This amounts to taking $r = 1$ for those constructions. As above, θ denotes the radian measure of the standard-position angle $\angle QOP$, and

268 Chapter 5 / Trigonometric Functions and Their Graphs

Figure 5.16

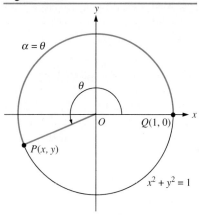

α is the arclength from $Q(1, 0)$ to $P(x, y)$. Since $r = 1$, the formula $s = r\theta$ tells us that $\alpha = \theta$. Thus, **if $0 < \alpha < 2\pi$, then the real number α can be regarded either as the radian measure of the angle θ or as the length of arc QP on the unit circle.** The data are summarized in figure 5.16.

In the above discussion, we will denote P by $P(\alpha)$. This indicates that P is the unique point on the unit circle such that the arclength, measured counterclockwise, from $(1, 0)$ to P is α. Thus far, the discussion assumed that $0 < \alpha < 2\pi$. However, as we shall show next, we may extend the discussion to *any* real number α.

If α is any real number, let $\angle QOP$ be the standard-position angle whose measure is α radians. (As usual, we take $Q(1, 0)$ and $P(x, y)$ on the unit circle, $x^2 + y^2 = 1$.) Note the following physical interpretation. As the line segment $\overline{OQ}$ is revolved α radians about the nonnegative x-axis, the point Q travels along the unit circle, finally coming to rest at P. During this process, Q travels a total distance of $|\alpha|$. It is customary to let $P(\alpha)$ denote P. Several examples illustrating this notation appear in figure 5.17.

Figure 5.17

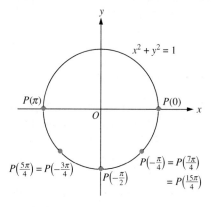

We have seen how each real number α determines a point $P(\alpha)$ on the unit circle. Just as in section 5.2, the rectangular coordinates (x, y) of $P(\alpha)$ may be used to find the six trigonometric functions of α. For instance,

$$\cos \alpha = \frac{x}{1} = x.$$

Using the same procedure for the remaining five trigonometric functions produces the following results.

Circular Functions

If α is a real number and $P(x, y)$ is the point on the unit circle that is determined by α, then

$\sin \alpha = y$ $\qquad \cos \alpha = x$ $\qquad \tan \alpha = \frac{y}{x}$ (if $x \ne 0$)

$\csc \alpha = \frac{1}{y}$ $\qquad \sec \alpha = \frac{1}{x}$ $\qquad \cot \alpha = \frac{x}{y}$

Circular Functions

Since the above formulas express the trigonometric functions in terms of the coordinates of a point on the unit circle, the trigonometric functions are sometimes referred to as the **circular functions**. This name is well chosen for another reason as well, which we address in the next paragraph.

The above discussion revealed that the unit circle may be described as the set $\{(\cos \theta, \sin \theta): 0 \leq \theta < 2\pi\}$. Moreover, if a and b are any real numbers such that $b - a \geq 2\pi$, then $\{(\cos \alpha, \sin \alpha): a \leq \alpha < b\}$ is also the unit circle. Such "parametric" descriptions of geometric objects are frequently very useful, as will be noted in section 10.4. A preview of this topic from the point of view of the unit circle will be given in exercises 25–30.

▷ **EXAMPLE 5.15**

Use the unit circle to find the following trigonometric values:

(a) $\sin \dfrac{\pi}{2}$, (b) $\cos \pi$, (c) $\sec \dfrac{\pi}{4}$, (d) $\tan \pi$, and (e) $\sin \left(-\dfrac{\pi}{2}\right)$.

Solutions

(a) If $\alpha = \dfrac{\pi}{2}$, then $P(\alpha)$ has the coordinates $(0, 1)$, as shown in figure 5.18(a). From this, we see that $x = 0$ and $y = 1$ (and, of course, $r = 1$). Hence, $\sin \dfrac{\pi}{2} = \sin \alpha = y = 1$. Thus, we see that $\sin \dfrac{\pi}{2} = 1$.

(b) $P(\pi)$ has the coordinates $(-1, 0)$, as shown in figure 5.17 and again in figure 5.18(b). Accordingly, $x = -1$ and $y = 0$. Thus, $\cos \pi = \cos \alpha = x = -1$, and we conclude that $\cos \pi = -1$.

(c) The terminal ray of the standard-position angle with radian measure $\dfrac{\pi}{4}$ intersects the unit circle at a point $P(x, x)$, as shown in figure 5.18(c). Since P is on the unit circle, we see that

$$x^2 + x^2 = 1, \quad \text{or} \quad 2x^2 = 1. \rightarrow x^2 = \dfrac{1}{2} \rightarrow \boxed{x = \dfrac{1}{\sqrt{2}}}$$

Figure 5.18

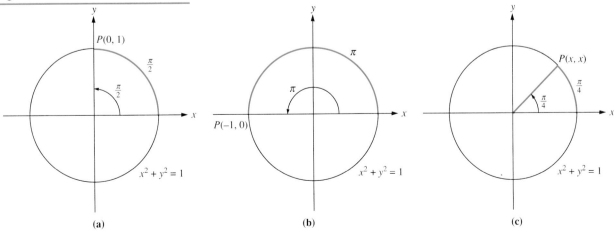

(a) (b) (c)

Solving for x produces $x = \frac{1}{\sqrt{2}}$. Therefore, $P\left(\frac{\pi}{4}\right)$ has the coordinates $\left(\frac{1}{\sqrt{2}}, \frac{1}{\sqrt{2}}\right)$. Since $x \neq 0$, $\sec \frac{\pi}{4} = \sec \alpha = \frac{1}{x} = \frac{1}{1/\sqrt{2}} = \sqrt{2}$, and we conclude that $\sec \frac{\pi}{4} = \sqrt{2}$.

(d) From (b), we see that $x = -1$ and $y = 0$. Hence $\tan \pi = \tan \alpha = \frac{y}{x} = \frac{0}{1} = 0$.

(e) As you can see from figure 5.17, $P\left(-\frac{\pi}{2}\right)$ has coordinates $(0, -1)$. Hence $\sin\left(-\frac{\pi}{2}\right) = \sin \alpha = y = -1$.

▶ **EXAMPLE 5.16**

Use the circular function approach to the trigonometric functions to simplify the expression $\tan(\alpha + \pi)$.

Solution

Let $P(x, y)$ be $P(\alpha)$, the point on the unit circle that is determined by the real number α. As in section 1.2, we can show by comparing slopes that $R(-x, -y)$ is collinear with P and O. Thus, R is the point on the unit circle that is determined by $\alpha + \pi$; in other words, $P(\alpha + \pi) = R$. From the above formulas, we see that $\tan(\alpha + \pi) = \frac{-y}{-x} = \frac{y}{x} = \tan \alpha$. Thus, $\tan(\alpha + \pi) = \tan \alpha$.

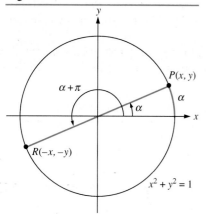

Figure 5.19

The circular function approach can also be used to make some observations about the variation of the trigonometric functions. Once again, consider the unit circle in figure 5.16. Recall that we may interpret the real number α as either the radian measure of θ or the arclength traveled from Q to P. If we let α increase from 0 to 2π, then the point $P(x, y)$ that is determined by α travels around the unit circle once in a counterclockwise direction. Since $\cos \alpha = x$, we can see that the range of the cosine function is $[-1, 1]$. Similarly, since $\sin \alpha = y$, one sees that the range of the sine function is also $[-1, 1]$. Although we shall not stress how to determine a variety of values of sin and cos until section 5.4, we can now make the following observations.

As α increases from 0 to $\frac{\pi}{2}$, $\cos \alpha$ decreases from 1 to 0; during this process, $\cos \alpha$ actually takes on every value between 1 and 0. As α increases from $\frac{\pi}{2}$ to π, $\cos \alpha$ decreases from 0 to -1, taking on every value between 0 and -1. As α increases from π to 2π, $\cos \alpha$ increases from -1 to 1.

Next, as α increases from 2π to 4π, an interesting phenomenon occurs. $P(x, y)$ simply retraces the unit circle, pointing out the "periodicity" results that $\cos(\alpha + 2\pi) = \cos \alpha$ and $\sin(\alpha + 2\pi) = \sin \alpha$. We shall return to this subject of periodicity in section 5.4. Meanwhile, exercise 19 will carry matters further.

By reasoning as in examples 5.15 and 5.16, we can produce the following table.

Circular Functions

α	0	$\dfrac{\pi}{4}$	$\dfrac{\pi}{2}$	$\dfrac{3\pi}{4}$	π	$\dfrac{5\pi}{4}$	$\dfrac{3\pi}{2}$	$\dfrac{7\pi}{4}$	2π
$\cos \alpha$	1	$\dfrac{1}{\sqrt{2}}$	0	$-\dfrac{1}{\sqrt{2}}$	-1	$-\dfrac{1}{\sqrt{2}}$	0	$\dfrac{1}{\sqrt{2}}$	1

The above observations about increasing/decreasing behavior, periodicity, and particular values of cos establish the foundation for the graph of the cosine function. In section 5.4, you will see that this graph has the shape shown in figure 5.20(a). In exercise 21, you will be asked to similarly sketch the graph of the sine function.

The remaining material in this section is for readers with a graphing calculator. If you do not have a graphing calculator, you may want to skip to the exercise set.

Your graphing calculator could also be used to obtain the result shown in figure 5.20(a). To do this, first make sure the calculator is in radian mode. Next, set Xmin = $\dfrac{-7\pi}{4}$ (≈ -5.5), Xmax = $\dfrac{7\pi}{2}$ (≈ 11), Xscl = $\dfrac{\pi}{4}$ (≈ 0.785), Ymin = -1, Ymax = 1, and Yscl = 1. Then, on a *TI-81*, you can obtain the desired graph by pressing Y= COS X|T GRAPH ; the result is shown in figure 5.20(b). On a Casio *fx-7700G*, you can press Graph cos X,θ,T EXE to get the graph; the result is shown in figure 5.20(c).

Notice that the vertical dimension on these two graphs seems stretched in relation to the horizontal. This happens because the unit length on the horizon-

Figure 5.20

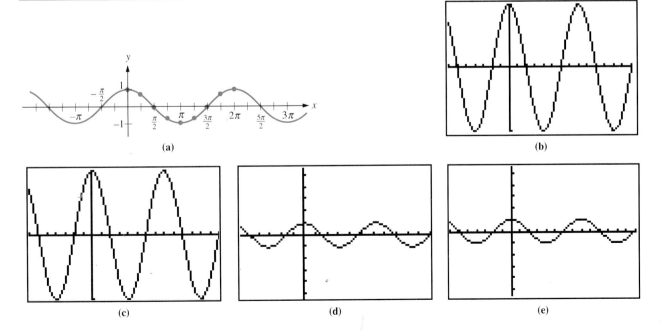

tal axis is not the same as the unit length on the vertical axis. (For reasons of engineering and economics, these unit lengths are in the ratio of 3:2.) If you want graphs that appear more realistic, then some adjustments are necessary in order to arrange the same unit length on both axes.

What are the appropriate adjustments for the above calculator-generated graphs of the cosine function? For a more realistic graph, we need to arrange $\frac{\text{Xmax} - \text{Xmin}}{\text{Ymax} - \text{Ymin}} = \frac{3}{2}$. We had $\frac{\text{Xmax} - \text{Xmin}}{\text{Ymax} - \text{Ymin}} = \frac{7\pi/2 - (-7\pi/4)}{1 - (-1)} = \frac{21\pi/4}{2} = \frac{21\pi}{8} \neq \frac{3}{2}$. Since we want the display to be based on the interval $\left[-\frac{7\pi}{4}, \frac{7\pi}{2}\right]$ along the x-axis, we need to adjust Ymax $-$ Ymin in order to get the desired ratio of 3:2. Solving $\frac{\text{Xmax} - \text{Xmin}}{\text{Ymax} - \text{Ymin}} = \frac{21\pi/4}{\text{Ymax} - \text{Ymin}} = \frac{3}{2}$, we determine that Ymax $-$ Ymin $= \frac{7\pi}{2}$. So, if you want the x-axis to pass through the center of the screen, then you should set Ymin $= -\frac{7\pi}{4}$ and Ymax $= \frac{7\pi}{4}$. Using these values on a Casio fx-7700G produces the graph in figure 5.20(d). The TI-81 will do all this work automatically if you press ZOOM 5 after obtaining figure 5.20(b); this gives the graph in figure 5.20(e).

Exercises 5.3

In exercises 1–12, (a) find the coordinates of the point of the unit circle that is determined by the given real number α, and (b) hence find the values of the trigonometric functions at α.

1. 0
2. π
3. $\frac{\pi}{2}$
4. $\frac{3\pi}{2}$
5. $\frac{5\pi}{4}$
6. $\frac{7\pi}{4}$
7. 3π
8. 7π
9. $-\pi$
10. -3π
11. $-\frac{5\pi}{4}$
12. $-\frac{11\pi}{4}$

In exercises 13–16, simplify the given expressions.

13. (a) $\cos(\alpha + \pi)$. (Hint: review example 5.16.)
    (b) $\cot(\alpha + \pi)$ (c) $\sin(\alpha + \pi)$
14. (a) $\sec(\alpha + \pi)$ (b) $\csc(\alpha + \pi)$ (c) $\cos(\alpha - \pi)$
15. (a) $\cos\left(\alpha + \frac{\pi}{2}\right)$. (Hint: consider figure 1.20.)
    (b) $\cos\left(\alpha - \frac{\pi}{2}\right)$. (Hint: use (a) and exercise 13(a).)
16. (a) $\sin\left(\alpha + \frac{\pi}{2}\right)$. (Hint: adapt the hint for exercise 15(a).)
    (b) $\csc\left(\alpha + \frac{\pi}{2}\right)$

17. If $\sin \alpha = \frac{1}{2}$, determine (a) $\sin(\alpha + \pi)$ and
    (b) $\sin\left(\alpha + \frac{\pi}{2}\right)$.
18. If $\cos \alpha = \frac{\sqrt{3}}{2}$, determine (a) $\cos(\alpha + \pi)$ and
    (b) $\cos\left(\alpha + \frac{\pi}{2}\right)$.

♦ 19. Use this section's approach to the trigonometric functions to redo exercise 9 of section 5.2.
♦ 20. Use this section's approach to the trigonometric functions to redo exercise 7 of section 5.2. (Hint: where does the line $y = x$ intersect the unit circle in quadrant III?)
♦ 21. Adapt the discussion in the text concerning the graph of the cosine function in order to graph the sine function. (Hint: you may find that exercise 15(a) or 16(a) is relevant.)
22. In the figure, θ is an acute central angle of the unit circle and $\overrightarrow{QT} \perp \overrightarrow{OQ}$. Explain why arc $QP = \theta$, $OR = \cos \theta$, $RP = \sin \theta$, and $QT = \tan \theta$.

Values of the Trigonometric Functions

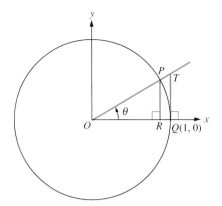

23. On one graph, sketch the following points on the unit circle and find their coordinates:

(a) $P\left(\dfrac{3\pi}{2}\right)$ (b) $P\left(-\dfrac{5\pi}{4}\right)$ (c) $P\left(-\dfrac{5\pi}{4}\right)$

(d) $P\left(-\dfrac{5\pi}{2}\right)$ (e) $P(2\pi)$ (f) $P(-3\pi)$

24. Use the "distance traveled" interpretation for $|\alpha|$ to give an intuitive argument for the conclusions in exercises 14 and 15 of section 5.2—namely that sin is an odd function and cos is even.

◆ In exercises 25–30, geometrically identify the given set. (Hint: $\{(\cos\theta, \sin\theta): 0 \le \theta < 2\pi\}$ is the unit circle—that is, the graph of $x^2 + y^2 = 1$.)

25. $\{(\cos\theta, \sin\theta): 0 \le \theta \le \pi\}$
26. $\{(\cos\theta, \sin\theta): \pi \le \theta \le 2\pi\}$
27. $\left\{(\cos\theta, \sin\theta): \dfrac{\pi}{2} \le \theta \le \dfrac{3\pi}{2}\right\}$
28. $\left\{(\cos\theta, \sin\theta): -\dfrac{\pi}{2} \le \theta \le \dfrac{\pi}{2}\right\}$
29. $\{(\cos\theta, \sin\theta): \pi \le \theta \le 4\pi\}$
30. $\left\{(\cos\theta, \sin\theta): \dfrac{\pi}{4} \le \theta < \dfrac{9\pi}{4}\right\}$

5.4 VALUES OF THE TRIGONOMETRIC FUNCTIONS

In this section, the values for the trigonometric functions at some of the more common angles will be calculated, and techniques for finding the trigonometric functions for any angle will be examined.

The $\dfrac{\pi}{4}$ Radians (45°) Angle

Consider a right triangle with both legs of length 1 unit. The hypotenuse then has length $\sqrt{2}$, as shown in figure 5.21. As this right triangle is isosceles, its acute angles are congruent; that is, $\theta = \phi = \dfrac{\pi}{4}$ radians (or 45°). Using the trigonometric definitions based on a right triangle, we find the following values for the trigonometric functions of a $\dfrac{\pi}{4}$ radians, or 45°, angle.

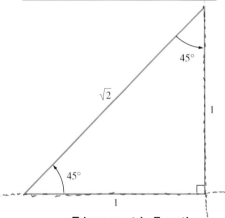

Figure 5.21

Trigonometric Functions Evaluated at $\dfrac{\pi}{4}$ Radians or 45°

$\sin\dfrac{\pi}{4} = \sin 45° = \dfrac{1}{\sqrt{2}} = \dfrac{\sqrt{2}}{2}$ $\cos\dfrac{\pi}{4} = \cos 45° = \dfrac{1}{\sqrt{2}} = \dfrac{\sqrt{2}}{2}$

$\tan\dfrac{\pi}{4} = \tan 45° = 1$ $\cot\dfrac{\pi}{4} = \cot 45° = 1$

$\sec\dfrac{\pi}{4} = \sec 45° = \sqrt{2}$ $\csc\dfrac{\pi}{4} = \csc 45° = \sqrt{2}$

The $\frac{\pi}{6}$ Radians (30°) and $\frac{\pi}{3}$ Radians (60°) Angles

Next, construct an equilateral triangle whose sides are each 2 units long. This triangle is also equiangular, and hence each of its angles measures $\frac{\pi}{3}$ radians (or $\frac{180°}{3} = 60°$). From one of the vertices, drop a perpendicular to the opposite side, as shown in figure 5.22. From geometry, it is known that this perpendicular bisects both the opposite side and the vertex angle. Thus, the equilateral triangle has been bisected into two congruent triangles. One of these smaller triangles is shown at the right of figure 5.22. Its hypotenuse is 2 units long, the side opposite the 30° angle is 1 unit, and so the remaining side is $\sqrt{3}$ units. Again using right triangle trigonometry, we now find the following values for the trigonometric functions of $\frac{\pi}{6}$ radians (30°) and of $\frac{\pi}{3}$ radians (60°).

Figure 5.22

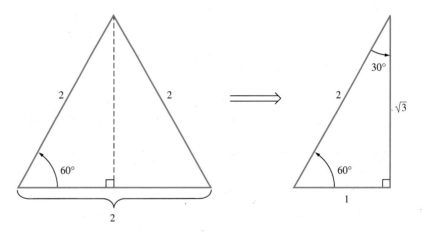

Trigonometric Functions Evaluated at $\frac{p}{6}$ Radians or 30°

$\sin \frac{\pi}{6} = \sin 30° = \frac{1}{2}$ $\cos \frac{\pi}{6} = \cos 30° = \frac{\sqrt{3}}{2}$

$\tan \frac{\pi}{6} = \tan 30° = \frac{1}{\sqrt{3}} = \frac{\sqrt{3}}{3}$ $\cot \frac{\pi}{6} = \cot 30° = \sqrt{3}$

$\sec \frac{\pi}{6} = \sec 30° = \frac{2}{\sqrt{3}} = \frac{2\sqrt{3}}{3}$ $\csc \frac{\pi}{6} = \csc 30° = 2$

Trigonometric Functions Evaluated at $\frac{p}{3}$ Radians or 60°

$\sin \frac{\pi}{3} = \sin 60° = \frac{\sqrt{3}}{2}$ $\cos \frac{\pi}{3} = \cos 60° = \frac{1}{2}$

$\tan \frac{\pi}{3} = \tan 60° = \sqrt{3}$ $\cot \frac{\pi}{3} = \cot 60° = \frac{1}{\sqrt{3}} = \frac{\sqrt{3}}{3}$

$\sec \frac{\pi}{3} = \sec 60° = 2$ $\csc \frac{\pi}{3} = \csc 60° = \frac{2}{\sqrt{3}} = \frac{2\sqrt{3}}{3}$

Values of the Trigonometric Functions

The Quadrantal Angles

As stated earlier, a **quadrantal angle** is an angle in standard position whose terminal side falls along an axis. Each quadrantal angle is coterminal with one of the following: 0, $\frac{\pi}{2}$, π, or $\frac{3\pi}{2}$. Consider the $\frac{\pi}{2}$ radians (90°) angle in figure 5.23. With $r = 1$, the corresponding point on the terminal side is $(0, 1)$. Since a triangle cannot be formed, we cannot use right triangle trigonometry. Instead, we will use the original definitions of the trigonometric functions in terms of x, y, and r. The relevant values are $x = 0$, $y = 1$, and $r = 1$. Thus the values of the trigonometric functions for a $\frac{\pi}{2}$ radians (90°) angle are

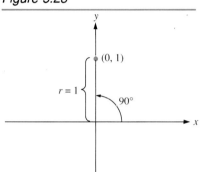

Figure 5.23

$$\sin\frac{\pi}{2} = \frac{y}{r} = \frac{1}{1} = 1 \qquad \cos\frac{\pi}{2} = \frac{x}{r} = \frac{0}{1} = 0$$

$$\tan\frac{\pi}{2} = \frac{y}{x} = \frac{1}{0}: \text{undefined} \qquad \cot\frac{\pi}{2} = \frac{x}{y} = \frac{0}{1} = 0$$

$$\sec\frac{\pi}{2} = \frac{r}{x} = \frac{1}{0}: \text{undefined} \qquad \csc\frac{\pi}{2} = \frac{r}{y} = \frac{1}{1} = 1.$$

A similar procedure can be followed to find the values of the trigonometric functions of any quadrantal angle. Table 5.1 summarizes these values, as well as the ones discussed earlier in this section.

Table 5.1 Exact Values at Special Angles

				Angle θ				
Degrees	0°	30°	45°	60°	90°	180°	270°	360°
Radians	0	$\frac{\pi}{6}$	$\frac{\pi}{4}$	$\frac{\pi}{3}$	$\frac{\pi}{2}$	π	$\frac{3\pi}{2}$	2π
$\sin\theta$	0	$\frac{1}{2}$	$\frac{1}{\sqrt{2}}$	$\frac{\sqrt{3}}{2}$	1	0	-1	0
$\cos\theta$	1	$\frac{\sqrt{3}}{2}$	$\frac{1}{\sqrt{2}}$	$\frac{1}{2}$	0	-1	0	1
$\tan\theta$	0	$\frac{1}{\sqrt{3}}$	1	$\sqrt{3}$	Undefined	0	Undefined	0
$\csc\theta$	Undefined	2	$\sqrt{2}$	$\frac{2}{\sqrt{3}}$	1	Undefined	-1	Undefined
$\sec\theta$	1	$\frac{2}{\sqrt{3}}$	$\sqrt{2}$	2	Undefined	-1	Undefined	1
$\cot\theta$	Undefined	$\sqrt{3}$	1	$\frac{1}{\sqrt{3}}$	0	Undefined	0	Undefined

Values for "Common" Angles That Are Not Acute

Values of the trigonometric functions of acute angles can be used to find the values of the trigonometric functions of *any* angle. To do this, we use the concept of a reference angle.

Reference Angle

If an angle θ is in standard position and is not a quadrantal angle, its **reference angle** θ_{ref} is defined to be the (positive) acute angle formed by the terminal ray of θ and the horizontal axis.

For example, the reference angle for a 120° angle is 60°, as shown in figure 5.24(a). If θ is 295°, then θ_{ref} is 65° (figure 5.24(b)). If $\theta = -\frac{9\pi}{5}$, $\theta_{ref} = \frac{\pi}{5}$, as shown in figure 5.24(c).

Figure 5.24

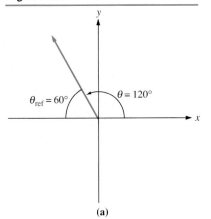
(a)

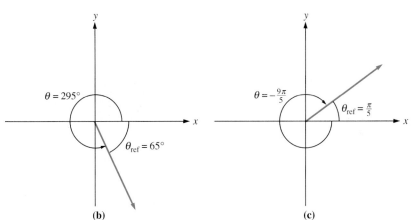
(b)   (c)

Consider the angles $\frac{\pi}{3}$ (60°), $\frac{2\pi}{3}$ (120°), $\frac{4\pi}{3}$ (240°), and $\frac{5\pi}{3}$ (300°), each in standard position. These all have the same reference angle, namely $\frac{\pi}{3}$ (60°). In each case, choose the point P on the terminal side of the angle where $r = 2$. Since similar triangles have proportional corresponding sides, the 30°–60° right triangle in figure 5.22 can be used to determine the coordinates of P. These are shown in figures 5.25(a), (b), (c), and (d), respectively. Notice that the only differences in the coordinates of the various points P stem from the signs of the x and y coordinates. Hence, at any two of these angles, the values of any given trigonometric function, which is after all just a ratio involving x, y, and r, will differ at most by sign. For instance, it can be determined from the four parts of figure 5.25 that $\sin \frac{\pi}{3} = \frac{\sqrt{3}}{2}$, $\sin \frac{2\pi}{3} = \frac{\sqrt{3}}{2}$, $\sin \frac{4\pi}{3} = -\frac{\sqrt{3}}{2}$, and $\sin \frac{5\pi}{3} = -\frac{\sqrt{3}}{2}$.

Values of the Trigonometric Functions

Figure 5.25

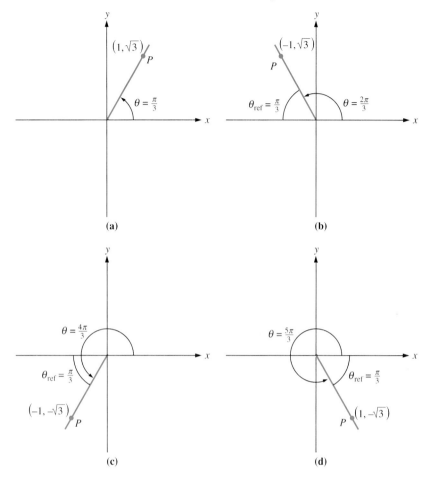

Figure 5.26

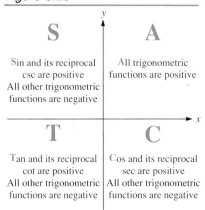

The reference angle of a given nonquadrantal angle θ is an acute angle that may be used to find the absolute values of trigonometric functions of θ, regardless of the quadrant of θ. The algebraic signs of these trigonometric values depend, however, on the quadrant of θ. For instance, if θ is in standard position and $P(x, y)$ is a second-quadrant point on the terminal side of θ, then $\sin \theta = \sin \theta_{\text{ref}}$, which is positive, and so $\csc \theta = \csc \theta_{\text{ref}} > 0$. But $\cos \theta = -\cos \theta_{\text{ref}} < 0$. In the same way, $\tan \theta = -\tan \theta_{\text{ref}} < 0$, $\cot \theta = -\cot \theta_{\text{ref}} < 0$, and $\sec \theta = -\sec \theta_{\text{ref}} < 0$.

A memory aid for the signs of the trigonometric functions is shown in figure 5.26. Starting with quadrant I and moving through quadrants II, III, and IV in order are the letters A, S, T, and C, which correspond to the mnemonic "All Students Take Calculus." Pictorially, this looks like $\dfrac{S \mid A}{T \mid C}$. The meanings of these letters are indicated in figure 5.26.

▶ **EXAMPLE 5.17**   Determine the values of the six trigonometric functions for a 210° angle.

Solution

Consider an angle of 210° in standard position, as shown in figure 5.27(a). Its reference angle is 30°. Since the 210° angle is in the third quadrant, we know from the above mnemonic that $\tan \theta$ and $\cot \theta$ are the only positive trigonometric values. Using this information and the trigonometric functions of 30°, we determine the required values as

$$\sin 210° = -\sin 30° = -\frac{1}{2} \qquad \cos 210° = -\cos 30° = -\frac{\sqrt{3}}{2}$$

$$\tan 210° = \tan 30° = \frac{1}{\sqrt{3}} \qquad \cot 210° = \cot 30° = \sqrt{3}$$

$$\sec 210° = -\sec 30° = -\frac{2}{\sqrt{3}} \qquad \csc 210° = -\csc 30° = -2$$

▶ **EXAMPLE 5.18**   Determine the values of the six trigonometric functions for a $\frac{5\pi}{3}$ angle.

Solution

An angle of $\frac{5\pi}{3}$ radians in standard position is shown in figure 5.27(b). Its reference angle is $2\pi - \frac{5\pi}{3} = \frac{\pi}{3}$. Since the $\frac{5\pi}{3}$ angle is in quadrant IV, we know from the above mnemonic that $\cos \theta$ and $\sec \theta$ are positive and the other four trigonometric values are negative. Using this information and the trigonometric functions of $\frac{\pi}{3}$, we find

$$\sin \frac{5\pi}{3} = -\frac{\sqrt{3}}{2} \qquad \cos \frac{5\pi}{3} = \frac{1}{2} \qquad \tan \frac{5\pi}{3} = -\sqrt{3}$$

$$\cot \frac{5\pi}{3} = -\frac{1}{\sqrt{3}} \qquad \sec \frac{5\pi}{3} = 2 \qquad \csc \frac{5\pi}{3} = -\frac{2}{\sqrt{3}}$$

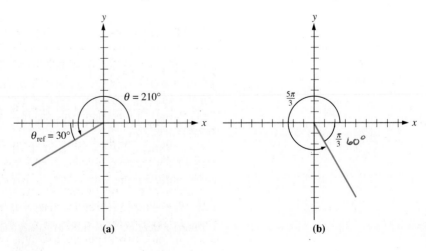

Figure 5.27

(a)   (b)

▶ **EXAMPLE 5.19**   Find an acute angle θ for which $\sin \theta = \cos 3\theta$.

Solution

We only want to find one such angle θ and so we may use anything that has set sine and cosine values equal in the past.

Since the sine and cosine are cofunctions, we know that if angles θ and 3θ are complementary, then $\sin \theta = \cos 3\theta$. Thus it is enough to solve $\theta + 3\theta = \dfrac{\pi}{2}$:

$$4\theta = \dfrac{\pi}{2}$$

$$\theta = \dfrac{\pi}{8} \text{ radians } (22.5°).$$

◀

Trigonometric Values on a Calculator

Calculators economize by having only three trigonometric function keys: $\boxed{\sin}$, $\boxed{\cos}$, and $\boxed{\tan}$. To find the values of the other three trigonometric functions, you need to use the $\boxed{1/x}$ key in conjunction with one of the trigonometric function keys. To understand why this works, consider the reciprocal identities discussed in section 5.2.

▶ **EXAMPLE 5.20**

Determine the values of the trigonometric functions for an angle of 73.8°.

Solution

Function	Press	Display
sin 73.8°	73.8 $\boxed{\sin}$	0.9602937
cos 73.8°	73.8 $\boxed{\cos}$	0.2789911
tan 73.8°	73.8 $\boxed{\tan}$	3.4420226
cot 73.8°	73.8 $\boxed{\tan}$ $\boxed{1/x}$	0.2905269
sec 73.8°	73.8 $\boxed{\cos}$ $\boxed{1/x}$	3.5843437
csc 73.8°	73.8 $\boxed{\sin}$ $\boxed{1/x}$	1.0413481

On graphing calculators, you press the function key before the measure of the angle. For example, you would enter cos 73.8° and cot 73.8°, respectively, by pressing the key sequences $\boxed{\cos}$ 73.8 and $\boxed{\tan}$ $\boxed{\text{2nd}}$ (or $\boxed{\text{SHIFT}}$) $\boxed{x^{-1}}$ 73.8. ◀

Naturally, if the angle is measured in radians, you must first change your calculator from degree mode to radian mode by pressing the appropriate keys.

One nice aspect of using a calculator is that there is no need to use reference angles. For example, to determine tan 287°, you need to press only 287 $\boxed{\tan}$, and the result −3.2708526 is shown on the display screen.

Some calculators restrict the size of the angle that can be entered into the calculator when you are evaluating a trigonometric function. For example, some calculators require that $|x| < 1440°$ or, in radians, $|x| \leq 8\pi$ when sin x, cos x, or tan x is being evaluated.

You should recognize the limitations inherent in using calculators as above. For instance, sin 73.8° is a perfectly respectable real number. It is *not*

0.9602937. However, the latter decimal expression is the best seven-place approximation of sin 73.8°. Such approximations are what we can get from technology, and often they are good enough for our purposes.

Trigonometric Values on a Computer

A computer operates in much the same way as a calculator with two main differences. (1) A computer normally operates in radians. If you want to work in degrees, you have to write a procedure for converting degrees to radians into the program. (2) Parentheses must be placed around the size of the angle. Like calculators, computers do not have the secant, cosecant, and cotangent functions. In example 5.21 below, we have used the same angle that was used in example 5.20 except that 73.8° ≈ 1.288053 radians has been rounded off to 1.288 radians.

EXAMPLE 5.21

Use a computer to determine the sin, tan, and csc of 1.288 radians.

Solution

Function	Press	Display
sin 1.288	PRINT SIN(1.288)	.960278901
tan 1.288	PRINT TAN(1.288)	3.44134194
csc 1.288	PRINT 1/SIN(1.288)	1.04136413

Trigonometric Values Using Tables

Until the recent advent of hand-held calculators and personal computers, the main way in which people determined the value of trigonometric functions was by using trigonometric tables. Many mathematics books, including this one, have a set of these tables in the appendices. The table in Appendix IV is incremented for every 0.1 degree, and the table in Appendix V is incremented in 0.01 radian. Tables for smaller increments, or in minutes and seconds, can be found in several books devoted to mathematical tables. Often tables, such as the ones in this book, are given only for angles between 0° and 90°. If you intend to use a table to find trigonometric functions of a nonacute angle, you must determine its reference angle and then proceed as above. As we've noted, calculators and computers eliminate the need to use reference angles.

Frankly, we expect that you will use a calculator or a computer (or both) to determine the values of trigonometric functions when an approximate value is wanted. (An exception occurs in exercises instructing you to "calculate exactly.")

Exercises 5.4

In exercises 1–2, find θ_{ref} for each of the given values of θ.

1. (a) 122° (b) –864° (c) 14° (d) –41°
2. (a) –147° (b) 205°6′ (c) 349° (d) 369°5′2″

In exercises 3–4, find θ_{ref} (in radians) for each of the given values of θ (in radians).

3. (a) $\dfrac{7\pi}{15}$ (b) $-\dfrac{4\pi}{3}$ (c) $-\dfrac{11\pi}{5}$

Values of the Trigonometric Functions

(d) $\dfrac{5\pi}{3}$ (e) $\dfrac{5\pi}{8}$ (f) -2.9

4. (a) $-\dfrac{5\pi}{3}$ (b) $\dfrac{19\pi}{3}$ (c) $\dfrac{6\pi}{5}$

   (d) $-\dfrac{5\pi}{9}$ (e) $-\dfrac{\pi}{6}$ (f) 4

In exercises 5–6, use reference angles to find exact values (without calculator, computer, or tables).

5. (a) $\sin\dfrac{7\pi}{6}$ (b) $\cos\left(\dfrac{2\pi}{3}\right)$

   (c) $\sec\left(-\dfrac{2\pi}{3}\right)$ (d) $\cot\left(\dfrac{\pi}{4}\right)$

6. (a) $\cos\dfrac{5\pi}{4}$ (b) $\tan\dfrac{5\pi}{3}$

   (c) $\sin\left(-\dfrac{\pi}{6}\right)$ (d) $\csc\left(-\dfrac{4\pi}{3}\right)$

In exercises 7–8, θ is an angle in standard position. Compute exactly.

7. (a) Compute $\sin\theta$ if $\cot\theta = -\dfrac{3}{2}$ and the terminal side of θ lies in quadrant IV.
   (b) Compute $\cot\theta$ if $\sin\theta = -\dfrac{5}{12}$ and $\cos\theta < 0$.

8. (a) Compute $\tan\theta$ if $\sec\theta = -\sqrt{2}$ and $\sin\theta < 0$.
   (b) Compute $\csc\theta$ if $\tan\theta = \sqrt{3}$ and $\cos\theta < 0$.

9. In which quadrant does the terminal side of a standard-position angle θ lie in the following cases?
   (a) $\cos\theta > 0$, $\sin\theta < 0$ (b) $\sec\theta < 0$, $\csc\theta > 0$
   (c) $\tan\theta > 0$, $\csc\theta < 0$ (d) $\sec\theta < 0$, $\csc\theta < 0$
   (e) $\cot\theta < 0$, $\sin\theta > 0$ (f) $\cos\theta < 0$, $\cot\theta < 0$

10. Algebraically simplify each of the following to an exact value:
    (a) $\sin^2\dfrac{\pi}{3} + \cos^2\dfrac{\pi}{3}$ (b) $\sin\left(-\dfrac{7\pi}{6}\right)\csc\left(-\dfrac{7\pi}{6}\right)$
    (c) $\sin\dfrac{\pi}{4}\cos\dfrac{\pi}{3} + \cos\dfrac{\pi}{4}\sin\dfrac{\pi}{3}$ (d) $\dfrac{2\tan\dfrac{\pi}{6}}{1 - \tan^2\dfrac{\pi}{6}}$
    (e) $2\sin 45° - 3\cos 60°$ (f) $5\tan 30° + \sqrt{2}\sin 210°$

11. Verify the following equations without using approximations by calculator, computer, or table:
    (a) $\sin 60° = 2\sin 30°\cos 30°$
    (b) $\cos 240° = 2\cos^2(120°) - 1$
    (c) $\tan\dfrac{2\pi}{3} = \dfrac{2\tan(\pi/3)}{1 - \tan^2(\pi/3)}$
    (d) $\sin\dfrac{\pi}{6} = \sqrt{\dfrac{1 - \cos(\pi/3)}{2}}$

In exercises 12–13, find the values of the six trigonometric functions for each of the given angles. (Use calculator/computer/tables as needed to obtain approximate answers.)

12. (a) $37°$ (b) $-137.21°$
    (c) $-214°30'$ (d) $37°19'18''$

13. (a) $-37.21°$ (b) $305°20''$
    (c) $-181°$ (d) $716°30'$

In exercises 14–15, find the values of the six trigonometric functions at each of the given real numbers.

14. (a) -1.23 (b) $\dfrac{9\pi}{8}$ (c) -0.87 (d) $\dfrac{7\pi}{8}$

15. (a) $-\dfrac{7\pi}{9}$ (b) 2.13 (c) -7 (d) 30

◆ 16. (a) Find $\tan 89°$, $\tan 89.9°$, $\tan 89.99°$, and $\tan 89.999°$.
   (b) Find $\tan 91°$, $\tan 90.1°$, $\tan 90.01°$, and $\tan 90.001°$.
   (c) What are you led to suspect about the "limiting tendency," if any, of $\tan\theta$ as θ "approaches" $90°$?

17. Draw a conclusion about the limiting tendencies of $\sin\theta$ and $\cos\theta$ as θ approaches 0. (Use the following data values for θ: 1, 0.1, 0.01, 0.001, 0.0001, -1, -0.1, -0.01, -0.001, and -0.0001.)

◆ 18. Develop enough data to permit you to draw a conclusion about the limiting tendency of $\dfrac{\sin\theta}{\theta}$ as θ approaches 0 radians.

◆ 19. One way to prove the assertion in exercise 18 depends upon showing that $\cos\theta < \dfrac{\sin\theta}{\theta} < \dfrac{1}{\cos\theta}$ if $-\dfrac{\pi}{2} < \theta < \dfrac{\pi}{2}$, $\theta \neq 0$.
   (a) By comparing the areas of the sector and triangles in the figure, prove this inequality for $0 < \theta < \dfrac{\pi}{2}$.

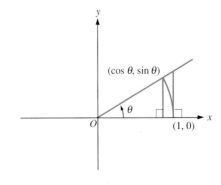

(b) Hence use exercises 14 and 15 of section 5.2 to derive the inequality for all nonzero θ strictly between $-\frac{\pi}{2}$ and $\frac{\pi}{2}$.

20. (a) Find an acute angle θ (equivalently, a real number θ so that $0 < \theta < \frac{\pi}{2}$) with the property that $\tan \theta = \cot 4\theta$.

    (b) Is your conclusion in (a) supported if you "check" using a graphing calculator or computer graphing program?

21. Find an acute angle θ such that $\tan \theta = 2 \sin^2 \theta$.

22. Find two values of θ such that $\pi < \theta < \frac{3\pi}{2}$ and $\sec \theta = \csc 5\theta$. (Hint: 5θ and $\frac{\pi}{2} - \theta$ must differ by $2\pi n$, an integral multiple of 2π. Try $n = 3, 4$.)

23. A daredevil named Sinus plans to swim directly across a crocodile-infested river. After doing so, he would signal his success by waving a flag. Sinus knows of a bridge that is 700 yards long and crosses the river at a 40° angle. Sinus expects that his crocodile repellent will protect him in the water for 450 yards. Will Sinus wave?

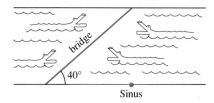

Sinus

In exercises 35–39 of section 5.2, you were asked to leave your answers in terms of uncalculated trigonometric functional values. Exercises 24–28 below are the same five exercises; now you should obtain numerical answers (exact, if possible) for each of them.

24. Determine the height of a totem pole if it casts a shadow 75 feet long when the angle of elevation of the top of the pole, viewed from the far end of the shadow, is 30°.

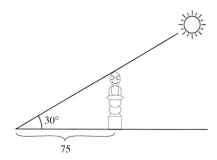

25. A short child named Petra is playing with a kite. Just as she has let out 120 feet of string and the kite's angle of elevation is 60°, a safety pin falls off the kite. Assuming the pin falls straight down and there is no sag in the string, determine how far the pin lands from Petra's feet.

26. A secret new aircraft is able to take off at an angle of elevation of 70° and travel 600 miles per hour. How high above the ground is this plane 4 minutes after takeoff?

27. An observer positions herself 500 meters from a rocket launch pad. Shortly after a vertical launch, she finds that the tail of the rocket has an angle of elevation of 17° and the nose of the rocket has an angle of elevation of 20°. What is the length of the rocket?

28. A UFO passes 1000 feet directly overhead an observer and flies away without changing its speed or altitude. After 1.16 seconds have elapsed, the observer measures the UFO's angle of elevation as 60°. How fast is the UFO moving?

5.5 GRAPHS OF THE SINE AND COSINE FUNCTIONS

With each new function, we have studied its graph. The graph of a function helps us to summarize the information about that function, get a feel for its behavior, and communicate this to each other. In this section and the next, we shall examine the graphs of the six trigonometric functions. It will be convenient, and traditional, to denote the independent variable by x rather than θ.

Graphing $y = \sin x$

As usual, we begin the graphing procedure for a new function by building a table of its functional values. In section 5.2, we indicated that the domain of

Graphs of the Sine and Cosine Functions

sin is **R** and its range is [−1, 1]. Table 5.2 contains the value of $f(x) = \sin x$ at each real number or radian measure between 0 and 2π corresponding to the common angles. The exact values of sin x are given in the third line of the table, and decimal approximations for them are given in the bottom row.

Table 5.2

x (radians)	0	$\frac{\pi}{6}$	$\frac{\pi}{4}$	$\frac{\pi}{3}$	$\frac{\pi}{2}$	$\frac{2\pi}{3}$	$\frac{3\pi}{4}$	$\frac{5\pi}{6}$	π	$\frac{7\pi}{6}$	$\frac{5\pi}{4}$	$\frac{4\pi}{3}$	$\frac{3\pi}{2}$	$\frac{5\pi}{3}$	$\frac{7\pi}{4}$	$\frac{11\pi}{6}$	2π
x (degrees)	0	30	45	60	90	120	135	150	180	210	225	240	270	300	315	330	360
y = sin x (exact)	0	$\frac{1}{2}$	$\frac{1}{\sqrt{2}}$	$\frac{\sqrt{3}}{2}$	1	$\frac{\sqrt{3}}{2}$	$\frac{1}{\sqrt{2}}$	$\frac{1}{2}$	0	$\frac{-1}{2}$	$\frac{-1}{\sqrt{2}}$	$\frac{-\sqrt{3}}{2}$	−1	$\frac{-\sqrt{3}}{2}$	$\frac{-1}{\sqrt{2}}$	$\frac{-1}{2}$	0
y = sin x (approx.)	0	0.5	0.71	0.87	1	0.87	0.71	0.5	0	−0.5	−0.71	−0.87	−1	−0.87	−0.71	−0.5	0

As we have seen in exercises 9(a) and 10 of section 5.2, the trigonometric functions repeat their values on a regular basis. For example, $\sin \frac{\pi}{6} = \sin\left(2\pi + \frac{\pi}{6}\right) = \sin\left(4\pi + \frac{\pi}{6}\right) = \cdots$. This repetition means that once we have completed table 5.2 for x varying from 0 to 2π (or 0° to 360°), we can determine the entries elsewhere in the table. Any function that exhibits this repeating behavior is called a periodic function.

Periodic Function

A function f is **periodic** if there exists a positive number c such that

$$f(x + c) = f(x)$$

for all x in the domain of f. If a least such number c exists, this least value of c is called the **period** of f.

The graph formed by smoothly connecting the points in table 5.2 is indicated by the solid curve in figure 5.28. If this graph were continued for values of x larger than 2π (radians) or for values of x less than 0, the points indicated by the dotted curve in figure 5.28 would result.

The part of this graph corresponding to the domain interval [0, 2π) is called the **sine wave**. A graph with the general shape of $y = \sin x$ is said to be **sinusoidal**.

Since we have seen that $\sin \theta = \sin(2\pi + \theta)$, we know that the sine function is periodic. (Indeed, so is every trigonometric function.) Examination of the graph of $f(x) = \sin x$ shows that 2π is the length of the smallest interval over which the curve must be graphed before its shape repeats. Thus, **the period of the function given by $f(x) = \sin x$ is 2π.** Later, we will see functions that have periods other than 2π.

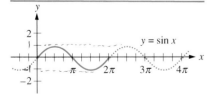

Figure 5.28

Further examination of the graph of sin x corroborates that its range is [−1, 1]. In particular, the maximum value of $f(x) = \sin x$ is 1, and the minimum value is −1. We can also see that this graph appears to be symmetric with respect to the origin. As we saw in section 2.2, this means that the sine function is an odd function; that is, $\sin(-x) = -\sin x$. A direct proof of this fact was sketched in exercise 14(a) of section 5.2.

Graphing y = cos x

A procedure similar to the one above may be used to graph the cosine function. As $\cos \theta = \cos(\theta + 2\pi)$, the cosine function is also periodic. Table 5.3 contains values for the cosine function at the common angles from 0 to 2π. The graph formed by smoothly connecting the corresponding points is in figure 5.29. Examination of the graph shows that **the period of the cosine function is 2π** and that its range is [−1, 1].

Table 5.3

x (radians)	0	$\frac{\pi}{6}$	$\frac{\pi}{4}$	$\frac{\pi}{3}$	$\frac{\pi}{2}$	$\frac{2\pi}{3}$	$\frac{3\pi}{4}$	$\frac{5\pi}{6}$	π	$\frac{7\pi}{6}$	$\frac{5\pi}{4}$	$\frac{4\pi}{3}$	$\frac{3\pi}{2}$	$\frac{5\pi}{3}$	$\frac{7\pi}{4}$	$\frac{11\pi}{6}$	2π
x (degrees)	0	30	45	60	90	120	135	150	180	210	225	240	270	300	315	330	360
y = cos x (exact)	1	$\frac{\sqrt{3}}{2}$	$\frac{1}{\sqrt{2}}$	$\frac{1}{2}$	0	$\frac{-1}{2}$	$\frac{-1}{\sqrt{2}}$	$\frac{-\sqrt{3}}{2}$	−1	$\frac{-\sqrt{3}}{2}$	$\frac{-1}{\sqrt{2}}$	$\frac{-1}{2}$	0	$\frac{1}{2}$	$\frac{1}{\sqrt{2}}$	$\frac{\sqrt{3}}{2}$	1
y = cos x (approx.)	1	0.87	0.71	0.5	0	−0.5	−0.71	−0.87	−1	−0.87	−0.71	−0.5	0	0.5	0.71	0.87	1

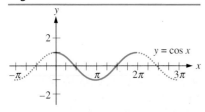

Figure 5.29

Notice that the graph of cos appears to be symmetric with respect to the y-axis. Hence, by section 2.2, cos is an even function; that is, $\cos(-x) = \cos x$. A direct proof of this was sketched in exercise 15 of section 5.2.

If you compare the graph of y = cos x in figure 5.29 with the graph of y = sin x in figure 5.28, you will see that it seems as if the graph of the cosine function can be obtained by translating the graph of y = sin x to the left $\frac{\pi}{2}$ units. In other words, $\cos x = \sin\left(x + \frac{\pi}{2}\right)$. Using a cofunction identity and the fact that cos is an even function, we may algebraically verify the connection between the graphs of cos and sin. Indeed, $\cos x = \cos(-x) = \sin\left(\frac{\pi}{2} - (-x)\right) = \sin\left(x + \frac{\pi}{2}\right)$, as the graphs suggested.

Amplitude

In section 2.2, we developed the ideas of vertical stretching and vertical shrinking. In fact, we stated that

Graphs of the Sine and Cosine Functions

The graph of the function given by $y = cf(x)$, where c is a constant, $c \neq 0$ and $c \neq 1$, is obtained by doing the following to the graph of $y = f(x)$:

(i) if $|c| > 1$, stretch vertically;
(ii) if $|c| < 1$, shrink vertically;
(iii) if $c < 0$, also reflect across the x-axis.

Now let's apply the idea of vertical stretching and shrinking to the graphs of the sine and cosine functions. We know that both these functions have a maximum functional value of 1 and a minimum of -1. Thus, it should be clear that the graphs of $y = A \sin x$ and $y = A \cos x$ are obtained by vertically stretching/shrinking the graphs of $y = \sin x$ and $y = \cos x$, respectively, by a factor of $|A|$ units. Both $y = A \sin x$ and $y = A \cos x$ have a maximum of $|A|$ and a minimum of $-|A|$.

Amplitude

The number $|A|$ is called the **amplitude** of the functions given by $y = A \sin x$ and $y = A \cos x$.

Until now, we have studied trigonometric functions whose graphs are "centered" about the x-axis. In these cases, the amplitude is simply the maximum height of the graph above the horizontal axis. A more general definition of amplitude is needed for some periodic functions, and so we have the following general definition.

Amplitude

If f is a periodic function and m and M are the minimum and maximum values of f, respectively, then the **amplitude** of f is defined as

$$\frac{M - m}{2}.$$

EXAMPLE 5.22

On the same coordinate axis, sketch the graphs of $f(x) = \sin x$ and $g(x) = 3 \sin x$ for $0 \leq x \leq 4\pi$.

Solution

The graph of $f(x) = \sin x$ is shown by the dotted curve in figure 5.30. To sketch the graph of $g(x) = 3 \sin x$, simply stretch the graph of f vertically by a factor of 3. Since $\sin 0 = \sin \pi = \sin 2\pi = \sin 3\pi = \sin 4\pi = 0$, the graph of $g(x) = 3 \sin x$ also crosses the x-axis at $x = 0, \pi, 2\pi, 3\pi,$ and 4π. The maximum value of 3 will occur at $x = \dfrac{\pi}{2}$ and $x = \dfrac{\pi}{2} + 2\pi = \dfrac{5\pi}{2}$. The minimum value of -3 will be at $x = \dfrac{3\pi}{2}$ and $x = \dfrac{3\pi}{2} + 2\pi = \dfrac{7\pi}{2}$. Here, $M = 3$ and $m = -3$, and so the amplitude is $\dfrac{3 - (-3)}{2} = 3$.

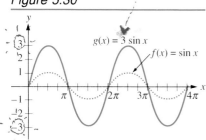

Figure 5.30

286 Chapter 5 / Trigonometric Functions and Their Graphs

▶ **EXAMPLE 5.23** Sketch the graph of $h(x) = \frac{1}{2}\sin x$.

Solution As a guide, the graph of $f(x) = \sin x$ is shown as a dotted curve in figure 5.31. Using the same reasoning as in the previous example, we see that $h(x) = 0$ when $x = 0, \pi, 2\pi$, and so on. The maximum value of $\frac{1}{2}$ for $h(x)$ will occur at $x = \frac{\pi}{2}, \frac{\pi}{2} + 2\pi = \frac{5\pi}{2}$, and so on; the minimum value of $-\frac{1}{2}$ occurs at $x = \frac{3\pi}{2}, \frac{7\pi}{2}$, and so on. The graph of h is the solid curve in figure 5.31. ◀

Figure 5.31

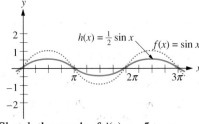

▶ **EXAMPLE 5.24** Sketch the graph of $j(x) = -5\cos x$.

Solution The graph of j is obtained by using the transformation methods of section 2.2. First, we shall use the technique of stretching/shrinking. The amplitude of j is 5, and so we can obtain the graph of $y = 5\cos x$ by stretching the graph of $y = \cos x$ vertically by a factor of 5. The result is shown by the dotted curve in figure 5.32. But the negative sign in the formula for j means that this dotted curve must be reflected in the x-axis in order to get the desired graph. The graph of $j(x) = -5\cos x$ is shown by the solid curve in figure 5.32. ◀

Figure 5.32

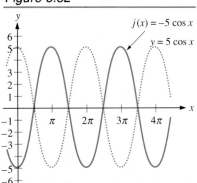

Period

Once again, we shall apply the general theory of functions. This time, we refer to the topic of horizontal stretching and shrinking, as discussed in section 2.2. There we found that:

> The graph of the function given by $y = f(dx)$, where d is a constant, $d \neq 0$ and $d \neq 1$, is obtained by doing the following to the graph of $y = f(x)$:
>
> (i)   if $|d| < 1$, stretch horizontally;
> (ii) if $|d| > 1$, shrink horizontally;
> (iii) if $d < 0$, also reflect across the y-axis.

In chapter 2, we found that the graph of $y = f(dx)$ was a stretched or shrunk version of the graph of $y = f(x)$, but we did not determine the amount of that

Graphs of the Sine and Cosine Functions

stretching or shrinking. We shall now attempt to do this for the functions given by $y = \sin Bx$ and $y = \cos Bx$. We know that $y = \sin x$ and $y = \cos x$ each have a period of 2π, and so each of these functions completes one cycle or wave as x varies from 0 to 2π. Thus, $\sin Bx$ and $\cos Bx$ will each complete a cycle from $Bx = 0$ to $Bx = 2\pi$. Solving each of these equations for x, we see that the functions $\sin Bx$ and $\cos Bx$ complete a cycle or wave as x varies from 0 to $\frac{2\pi}{|B|}$. Hence we conclude that **the functions $\sin Bx$ and $\cos Bx$ each have period $\frac{2\pi}{|B|}$**. The multiplier B acts as a horizontal stretcher if $0 < |B| < 1$ and as a horizontal shrinker if $|B| > 1$.

▶ **EXAMPLE 5.25** Sketch the graphs of $f(x) = \sin x$ and $g(x) = \sin 2x$.

Solution The graph of $f(x) = \sin x$ is shown by the dotted curve in figure 5.33. For the graph of $g(x) = \sin 2x$, the period is $\frac{2\pi}{|B|} = \frac{2\pi}{|2|} = \pi$ units. Thus, the graph of g will complete one cycle every π units, as shown by the solid curve in figure 5.33. Notice that the maximum and minimum functional values of g are 1 and -1, respectively. Thus, the graph of g has amplitude $\frac{1-(-1)}{2} = 1$. ◀

Figure 5.33

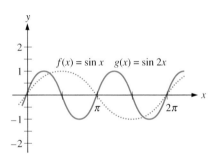

▶ **EXAMPLE 5.26** Sketch the graph of $h(x) = 3 \cos \frac{1}{2}x$.

Solution The period is $\frac{2\pi}{1/2} = 4\pi$, and so the graph of h will complete one wave every 4π units. This graph is shown in figure 5.34. The amplitude of $3 \cos \frac{1}{2}x$ is 3 since h has maximum value 3 and minimum value -3. ◀

Figure 5.34

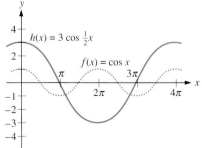

Phase Shift or Horizontal Translation

Next, let's consider a function given by an expression of the form $f(x) = A \sin(Bx - C)$, with $B \neq 0$. (A similar analysis would apply to $A \cos(Bx - C)$.) We know that the amplitude of this function is $|A|$. From our study of horizontal translations in section 2.2, we know that the graph of $g(x) = \sin(Bx - C)$ is a horizontal translation of the graph of $h(x) = \sin Bx$. As $Bx - C = B\left(x - \dfrac{C}{B}\right)$, it is seen that the graph of $y = \sin Bx$ must first be translated horizontally $\left|\dfrac{C}{B}\right|$ units. This translation is to the right if $\dfrac{C}{B} > 0$ and to the left if $\dfrac{C}{B} < 0$.

Phase Shift

The number $\dfrac{C}{B}$ is called the **phase shift.**

As we have seen, phase shift can be thought of as a horizontal translation. Finally, to get the graph of f, vertically expand the graph of g by a factor of $|A|$, reflecting through the x-axis if $A < 0$.

▶ **EXAMPLE 5.27**

Sketch the graph of $k(x) = 3 \cos (2x + \pi)$.

Solution

We see that $A = 3$, $B = 2$, and $C = -\pi$. Hence, the amplitude is 3, the period is $\dfrac{2\pi}{|B|} = \pi$, and the phase shift is $\dfrac{C}{B} = -\dfrac{\pi}{2}$. The graphs of $y = 3 \cos 2x$ and of $k(x) = 3 \cos (2x + \pi)$ are both shown in figure 5.35. ◀

It should be recognized that phase shift depends on the expression being used to generate a given function. For instance, we have just seen that the expression $k(x) = 3 \cos (2x + \pi)$ has a phase shift of $-\dfrac{\pi}{2}$. However, since $k(x) = 3 \sin \left(2x + \pi + \dfrac{\pi}{2}\right) = 3 \sin \left(2x + \dfrac{3\pi}{2}\right)$, the same function arises from an expression giving a phase shift of $-\dfrac{3\pi}{4}$. Moreover, the same function also arises from the expression $p(x) = -3 \cos 2x$, which gives a phase shift of 0. The moral is that the graph of a given function is obtained from the graph of any of its generating expressions $A \sin (Bx - C)$ (or $A \cos (Bx - C)$) by translating the graph of $\sin Bx$ (or $\cos Bx$) by $\dfrac{C}{B}$ units, vertically expanding by a factor of $|A|$, and finally reflecting in the x-axis if $A < 0$.

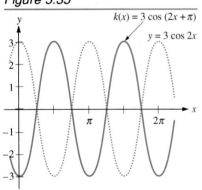

Figure 5.35

Vertical Translations

From our study of vertical translations in section 2.2, we know that replacing y in an equation by $y - D$, where D is a constant, produces a vertical translation of the original equation's graph by $|D|$ units. The translation is upward if $D > 0$ and downward if $D < 0$.

Graphs of the Sine and Cosine Functions

Summary

For Sin + Cos only
TAN on page 295

A function given by either $D + A \sin(Bx - C)$ or $D + A \cos(Bx - C)$ has

- an amplitude of $|A|$,
- a period of $\dfrac{2\pi}{|B|}$, and
- a vertical translation of D.

- The given expression has a phase shift of $\dfrac{C}{B}$.

The graph of such a function is achieved as follows:

1. Graph $\sin Bx$ (or $\cos Bx$) using the graph of $\sin x$ (or $\cos x$) and the above information about period.
2. Horizontally translate using the phase shift.
3. Expand vertically using the amplitude; reflect through the x-axis if $A < 0$.
4. Translate vertically $|D|$ units.

EXAMPLE 5.28

Sketch two cycles of the graph of $m(x) = 4 + 3\cos(2x + \pi)$.

Solution

We have $A = 3$, $B = 2$, $C = -\pi$, and $D = 4$. Thus, the amplitude is $|A| = |3| = 3$, the period is $\dfrac{2\pi}{|B|} = \dfrac{2\pi}{|2|} = \pi$, and the phase shift or horizontal translation is $\dfrac{C}{B} = -\dfrac{\pi}{2}$. Thus, the horizontal translation is to the left $\dfrac{\pi}{2}$ units, while $D = 4$ produces a vertical translation of 4 units upward. Figures 5.36(a)–(d) show how to graph m by starting with the graph of $y = \cos x$ and progressively changing the function until the graph of m results, as in figure 5.36(d).

Figure 5.36

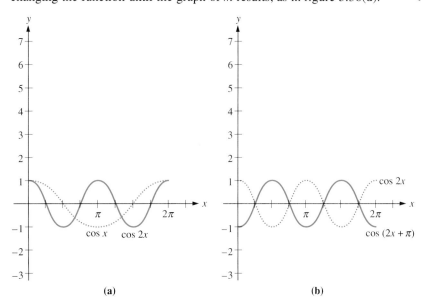

Chapter 5 / Trigonometric Functions and Their Graphs

Figure 5.36 (continued)

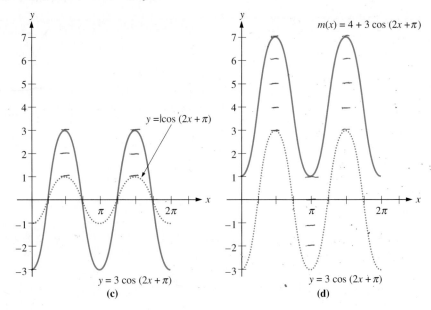

$y = |\cos(2x + \pi)|$

$y = 3\cos(2x + \pi)$
(c)

$m(x) = 4 + 3\cos(2x + \pi)$

$y = 3\cos(2x + \pi)$
(d)

▶ **EXAMPLE 5.29**

Sketch the graph of $n(x) = -\dfrac{\pi}{4} - 2\sin\left(\dfrac{1}{2}x - 1\right)$.

Solution

We have $A = -2$, $B = \dfrac{1}{2}$, $C = 1$, and $D = -\dfrac{\pi}{4}$. Thus, the amplitude is $|A| = |-2| = 2$, and since $A < 0$, sketching the curve involves a reflection in the x-axis. The period is $\dfrac{2\pi}{|B|} = \dfrac{2\pi}{1/2} = 4\pi$, and the phase shift or horizontal translation is $\dfrac{C}{B} = \dfrac{1}{1/2} = 2$. Hence, the horizontal translation is to the right 2 units, and since $D = -\dfrac{\pi}{4}$, the vertical translation is downward $\dfrac{\pi}{4}$ units.

The above recipe yields in turn the graphs of the functions $\sin x$, $\sin \tfrac{1}{2}x$, $\sin(\tfrac{1}{2}x - 1)$, $2\sin(\tfrac{1}{2}x - 1)$, $-2\sin(\tfrac{1}{2}x - 1)$, and n. One cycle of the graph of n is shown in figure 5.37(e). Figures 5.37(a)–(e) show how to graph n by starting with the graph of $y = \sin x$ and progressively changing the function until the result shown in figure 5.37(e) is reached. ◀

Figure 5.37

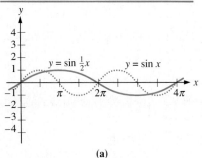
(a)

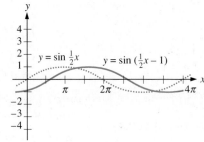
(b)

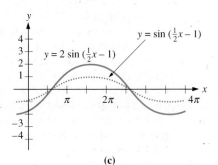
(c)

Graphs of the Sine and Cosine Functions

Figure 5.37 (continued)

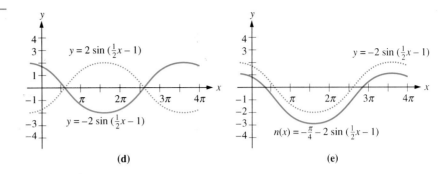

(d)   (e)

Exercises 5.5

1. Which of the following functions is/are periodic?
   (a) $f(x) = \sin(-2x)$ (b) $g(x) = -3\cos x$
   (c) $h(x) = 4\sin 2x$ (d) $j(x) = 5 - \frac{1}{2}\cos 3x$
   (e) $k(x) = 8 - \frac{1}{2}\sin(3x - \frac{1}{7})$ (f) $m(x) = x^2$
   (g) $n(x) = e^x$ (h) $p(x) = \pi$

2. For each of the periodic functions identified in exercise 1, determine its period or indicate why it does not have a period.

3. (a) Is the function f given by $f(x) = 1$ for all x a periodic function? If so, does it have a period?
   (b) Prove that the function $\cos 0x$ is periodic. Does it have a period?
   (c) Prove that each constant function is periodic but has no period.
   (d) Prove that $\sin 0x$ is a periodic function. Does it have a period?
   (e) Explain why no increasing function is periodic (unless its domain consists of at most one point). (Hint: think of its graph.)

4. Consider the functions f and g given by $f(x) = \sin x$ and $g(x) = 2x$. Let $F = f \circ g$ and $G = g \circ f$.
   (a) Show that $F(x) = \sin 2x$ and $G(x) = 2\sin x$ for all real numbers x.
   (b) Show that F and G have the same domain.
   (c) Show that F and G do not have the same range.
   ◆ (d) Show that $F \neq G$ by finding a specific x in $[0, 2\pi)$ such that $\sin 2x \neq 2\sin x$.
   (e) Find two (that's all there are) distinct values of x in $[0, 2\pi)$ such that $\sin 2x = 2\sin x$.

5. If f and g are functions such that g is periodic and $f \circ g$ is defined, show that $f \circ g$ is periodic. (Hint: if $g(x + d) = g(x)$ for all x, consider $(f \circ g)(x + d)$.)

6. Give an example of functions f and g each having domain $\mathbf{R}$ such that f is periodic and $f \circ g$ is not periodic. Justify your assertions intuitively with a graph. (Hint: consider $\sin(x^2)$.)

7. (a) Determine over which of the following intervals sin is an increasing function: $\left(0, \frac{\pi}{2}\right), \left(\frac{\pi}{2}, \pi\right),$ $\left(\pi, \frac{3\pi}{2}\right), \left(\frac{3\pi}{2}, 2\pi\right), (0, \pi), \left(\frac{\pi}{2}, \frac{3\pi}{2}\right), (\pi, 2\pi),$ and $\left(\frac{\pi}{4}, \frac{3\pi}{4}\right)$.
   (b) Determine over which of the following intervals sin is a decreasing function: $\left(0, \frac{\pi}{2}\right), \left(\frac{\pi}{2}, \pi\right),$ $\left(\pi, \frac{3\pi}{2}\right), \left(\frac{3\pi}{2}, 2\pi\right), (0, \pi), \left(\frac{\pi}{2}, \frac{3\pi}{2}\right), (\pi, 2\pi),$ and $\left(\frac{\pi}{4}, \frac{3\pi}{4}\right)$.

8. (a) Determine over which of the following intervals cos is an increasing function: $\left(0, \frac{\pi}{2}\right), \left(\frac{\pi}{2}, \pi\right),$ $\left(\pi, \frac{3\pi}{2}\right), \left(\frac{3\pi}{2}, 2\pi\right), (0, \pi), \left(\frac{\pi}{2}, \frac{3\pi}{2}\right), (\pi, 2\pi),$ and $\left(\frac{\pi}{4}, \frac{3\pi}{4}\right)$.
   (b) Determine over which of the following intervals cos is a decreasing function: $\left(0, \frac{\pi}{2}\right), \left(\frac{\pi}{2}, \pi\right),$ $\left(\pi, \frac{3\pi}{2}\right), \left(\frac{3\pi}{2}, 2\pi\right), (0, \pi), \left(\frac{\pi}{2}, \frac{3\pi}{2}\right), (\pi, 2\pi),$ and $\left(\frac{\pi}{4}, \frac{3\pi}{4}\right)$.

In exercises 9–10, prove that each of the given equations has its graph symmetric to the origin.

9. (a) $y = \sin 2x$ (b) $4x = \sin 3y$

10. (a) $x = \sin y$ (b) $-2y = \sin 3x$
11. Prove that the graph of the sine function is not symmetric to the y-axis. (You must find a specific ordered pair (a, b) in sin such that $(-a, b)$ is not in sin.)
12. Prove that the graph of the cosine function is not symmetric to the origin.

In exercises 13–14, prove that each of the given equations has its graph symmetric to the x-axis.

13. (a) $x = \cos y$ (b) $2y^2 = \cos 3x$
14. (a) $y^4 = \cos \frac{x}{2}$ (b) $-4x = \cos 5y$
15. Graph the relations $x = \sin y$ and $x = \cos y$.
16. Let f and g be functions each having domain $\mathbf{R}$. Suppose that f is even and there exists a real number d such that $g(d - x) = f(x)$ for all real numbers x. Show that $f(x) = g(d + x)$ for all real numbers x. (Hint: you've already seen this in this text in case $f = \cos$, $g = \sin$, and $d = \frac{\pi}{2}$.)
17. Prove that $|\sin x|$ is a periodic function, with period π.
    (Hint: $\sin (x + \pi) = \sin \left(\left(x + \frac{\pi}{2}\right) + \frac{\pi}{2}\right)$. Review exercises 15(a) and 16(a) of section 5.2.)
18. For each of the following functions, identify the domain, range, maximum functional value, and minimum value:
    (a) $f(x) = 3 \sin x$ (b) $g(x) = \cos 2x$
    (c) $h(x) = \sin (-3x + 4)$ (d) $j(x) = 2 \sin (3x - 4)$
    (e) $k(x) = -2 \cos (-3x - 4)$
    (f) $m(x) = 7 - 2 \sin (3x + 4)$
19. For each of the six (periodic) functions in exercise 18, determine the period and the amplitude.
20. (a) Explain geometrically why a horizontal translation of the graph of a periodic function does not change its period or its amplitude.
    (b) Explain geometrically why a vertical translation of the graph of a periodic function does not change its period or its amplitude.

In exercises 21–22, calculate the phase shift for each of the given expressions.

21. (a) $-\frac{1}{2} \sin (x + 3)$ (b) $4 \cos (-2x)$
    (c) $-3 + \sin (2x - 6)$
22. (a) $2 \cos (x - 5)$ (b) $5 + \frac{2}{3} \sin x$
    (c) $1 - \sin (2x + 1)$

◆ 23. (a) Let f be a function given by an expression of the form $f(x) = A \cos (Bx - C)$, for suitable nonzero constants A, B, and C. Show that f can also be given by an expression having phase shift $\left(C - \frac{\pi}{2}\right) \Big/ B$.

  (b) Find an expression with phase shift $\frac{\pi}{12}$ which describes the function given by $f(x) = \frac{2}{3} \cos \left(-9x + \frac{\pi}{4}\right)$.

◆ 24. Find an expression $A \sin (Bx - C)$ with phase shift $\frac{7\pi}{8}$ which describes the function given by $f(x) = -\frac{2}{3} \sin \left(2x + \frac{\pi}{4}\right)$.

In exercises 25–34, graph each of the given functions. (You may wish to use a graphing calculator or computer graphing program to help you or to provide a "rough check" of your work.)

25. $f(x) = -3 \cos x$
26. $g(x) = -\sin 2x$
27. $h(x) = 3 \sin \frac{x}{2}$
28. $j(x) = 2 \cos (4x - \pi)$
29. $k(x) = 5 - 2 \sin (2x + \pi)$
30. $m(x) = -\pi - 3 \cos \left(x + \frac{\pi}{2}\right)$
31. $p(x) = -2 + 3 \cos (x - \pi)$
32. $r(x) = 1 + \frac{1}{2} \sin (4x - \pi)$
33. $s(x) = |3 \cos x| - 2$
34. $t(x) = 3|\cos x - 2|$

In exercises 35–38, find constants A, B, C, and D such that the given curve is the graph of $y = D + A \sin (Bx - C)$.

35.

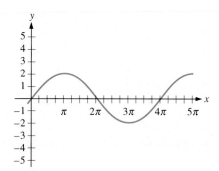

Graphs of the Sine and Cosine Functions

36.
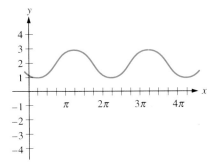

37.
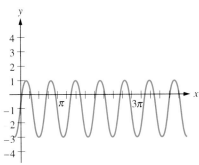

38.
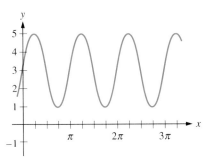

In exercises 39–42, find constants A, B, C, and D such that the given curve is the graph of $y = D + A \cos(Bx - C)$.

39.
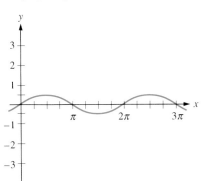

40.
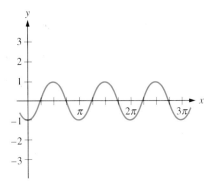

41.
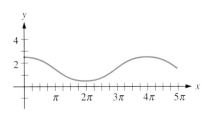

42.
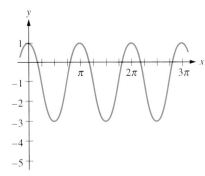

Numerous instances of sinusoidal functions arise in practice. These are used to model phenomena as diverse as body rhythms, ocean waves, and various mechanical or musical vibrations. We shall illustrate these in section 7.6. The next two exercises give a foretaste of such applications. You can work them with what you already know.

43. The systolic phase of the heart's activity pumps blood into the aorta from the left ventricle. One arch of Ed's systolic phase is pictured below:

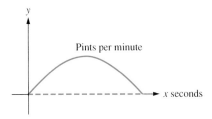

For x in the shaded interval, this arch is the graph of $y = A \sin Bx$, where A and B are suitable constants.

If Ed's systolic phase lasts 0.26 second and has a maximum flow rate of 16.73 pints per minute, find A and B.

44. It is found experimentally that on the xth day of a nonleap year, the number of hours of daylight in Washington, D.C. is roughly

$$f(x) = 12 + \pi \cos\left(-\frac{2\pi}{365}x + \frac{137\pi}{146}\right).$$

(a) Explain why, as x ranges from 1 to 365,

$$\theta = -\frac{2\pi}{365}x + \frac{137\pi}{146}$$ decreases from approximately 0.93π to approximately -1.06π.

(b) Conclude that the number of hours of daylight in Washington is greatest when $\theta = 0$, namely on about June 20.
(c) Explain why the number of hours of daylight in Washington is least on about December 20.
(d) Show that the above formula for $f(x)$ describes a periodic function, with natural domain R and period 365.
(e) Show that the above expression for $f(x)$ has phase shift $\frac{50005}{292}$. In view of your answer to (b), give a physical interpretation for this value of "phase shift."

5.6 GRAPHS OF THE OTHER TRIGONOMETRIC FUNCTIONS

In this section we graph the remaining trigonometric functions, starting with the graph of the tangent function. Once again, we begin with a table of values. Table 5.4 contains the values of $y = \tan x$ for the common angles from 0 to 2π. In this table, we see that $\tan \frac{\pi}{2}$ and $\tan \frac{3\pi}{2}$ are each undefined.

Table 5.4

x (radians)	0	$\frac{\pi}{6}$	$\frac{\pi}{4}$	$\frac{\pi}{3}$	$\frac{\pi}{2}$	$\frac{2\pi}{3}$	$\frac{3\pi}{4}$	$\frac{5\pi}{6}$	π	$\frac{7\pi}{6}$	$\frac{5\pi}{4}$	$\frac{4\pi}{3}$	$\frac{3\pi}{2}$	$\frac{5\pi}{3}$	$\frac{7\pi}{4}$	$\frac{11\pi}{6}$	2π
x (degrees)	0	30	45	60	90	120	135	150	180	210	225	240	270	300	315	330	360
$y = \tan x$ (exact)	0	$\frac{1}{\sqrt{3}}$	1	$\sqrt{3}$	Undef	$-\sqrt{3}$	-1	$\frac{-1}{\sqrt{3}}$	0	$\frac{1}{\sqrt{3}}$	1	$\sqrt{3}$	Undef	$-\sqrt{3}$	-1	$\frac{-1}{\sqrt{3}}$	0
$y = \tan x$ (approx.)	0	0.58	1	1.73	Undef	-1.73	-1	-0.58	0	0.58	1	1.73	Undef	-1.73	-1	-0.58	0

We turn next to a calculator experiment of the type suggested in exercise 16 of section 5.4. Table 5.5 takes a closer look at the values of tan x when x is close to $\frac{\pi}{2} \approx 1.5708$. Plotting the points in tables 5.4 and 5.5 and smoothly connecting them produces the graph of the tangent function in figure 5.38. The dashed lines in figure 5.38 represent vertical asymptotes.

Graphs of the Other Trigonometric Functions

Table 5.5 RADIAN MODE

x (radians)	1.50	1.52	1.54	1.56	1.57	$\frac{\pi}{2}$	1.58	1.59	1.60	1.62
y = tan x (approx.)	14.10	19.67	32.46	92.62	1255.8		−108.65	−52.07	−34.23	−20.31

Figure 5.38

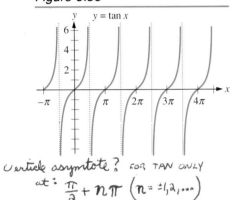

vertical asymptote? FOR TAN ONLY
at: $\frac{\pi}{2} + n\pi$ ($n = \pm 1, 2, ...$)

Studying the above graph and tables suggests some conclusions about the tangent function. The period of the tangent function is π. There are vertical asymptotes at $\frac{\pi}{2} + n\pi$, where $n = 0, \pm 1, \pm 2, \pm 3, \ldots$. (Given the identity tan x = $\frac{\sin x}{\cos x}$, we can show that the vertical asymptotes of tan occur when cos x = 0 and sin x ≠ 0, namely when $x = \frac{\pi}{2} + n\pi$.) There is an x-intercept midway between each pair of consecutive asymptotes. The function is increasing on the interval between consecutive asymptotes. For example, tan is increasing over each of the intervals $\left(-\frac{\pi}{2}, \frac{\pi}{2}\right)$, $\left(\frac{\pi}{2}, \frac{3\pi}{2}\right)$, and $\left(\frac{3\pi}{2}, \frac{5\pi}{2}\right)$. As you showed in exercise 16 of section 5.2, the graph of y = tan x is symmetric with respect to the origin. In other words, tan is an odd function: tan (−x) = −tan x. Notice that **the tangent function does not have an amplitude** since its range is unbounded.

Graphing Hint

Determines how wide "^" gets: phase shift, period, vert. are all same thing

To graph a function of the form $f(x) = D + A \tan(Bx - C)$, it is helpful to know that its [period is $\frac{\pi}{|B|}$], the [phase shift is $\frac{C}{B}$, RIGHT/LEFT $\frac{C}{B} > 0 \to$ right, $\frac{C}{B} < 0 \to$ left], and its [vertical translation is D. up/down "D"].

As usual, the number A has the effect of stretching or shrinking the graph vertically. As in section 5.5 (and chapter 2), the graph is achieved by considering successively the following transformations of the function tan x: $\tan\left(x - \frac{C}{B}\right)$, tan (Bx − C), A tan (Bx − C), and D + A tan (Bx − C).

▶ **EXAMPLE 5.30**

Sketch the graph of $g(x) = \tan\frac{x}{2}$.

Solution

$\tan\frac{x}{2}$ is of the form tan Bx, with $B = \frac{1}{2}$. Hence its period is $\frac{\pi}{|B|} = \frac{\pi}{1/2} = 2\pi$. One complete cycle occurs from $-\pi$ to π. There are vertical asymptotes at $x = -\pi$ and $x = \pi$. The graph of g(x) is shown in figure 5.39; for reference purposes, tan x appears as the dotted graph in the figure. ◀

Figure 5.39

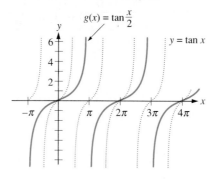

EXAMPLE 5.31

Sketch the graph of $h(x) = -2 \tan 3x$.

Solution

The period is $\dfrac{\pi}{|B|} = \dfrac{\pi}{3}$. The effect of $A = -2$ is a vertical stretching, followed by reflection in the x-axis. The graph of h is shown in figure 5.40.

Figure 5.40

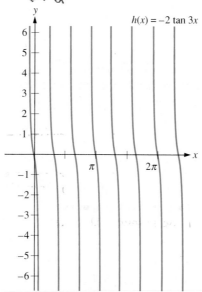

EXAMPLE 5.32

Sketch the graph of $j(x) = \pi + \tfrac{3}{5} \tan (2x - \tfrac{1}{2})$.

Solution

We have $A = \tfrac{3}{5}$, $B = 2$, $C = \tfrac{1}{2}$, and $D = \pi$. Thus we see that the period is $\dfrac{\pi}{2}$, the phase shift is $\dfrac{C}{B} = \dfrac{1/2}{2} = \dfrac{1}{4}$ unit to the right, the vertical translation is $D = \pi$ units upward, and the graph is shrunk vertically since $|A| < 1$. The graph of $y = j(x)$ is in figure 5.41.

Graphs of the Other Trigonometric Functions

Figure 5.41

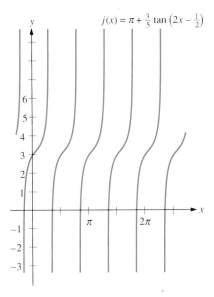
$j(x) = \pi + \frac{3}{5}\tan\left(2x - \frac{1}{2}\right)$

Graphs of Reciprocal Functions

The graphs of the three remaining trigonometric functions can be obtained via the reciprocal identities from the three that we have already graphed. For example, since $\csc x = \frac{1}{\sin x}$, we can find values for $y = \csc x$ by computing the reciprocals of the values given in table 5.2 for $y = \sin x$. Similarly, we can use the facts that $\sec x = \frac{1}{\cos x}$ and $\cot x = \frac{1}{\tan x}$ to obtain the graphs of the secant and cotangent functions. The graph of $y = \csc x$ is shown in figure 5.42 as a solid curve, with the reference graph of $y = \sin x$ as a dotted curve. Similarly, figure 5.43 shows the graph of the secant function as a solid curve, with $y = \cos x$ as a dotted curve; and figure 5.44 shows the graph of the cotangent function as solid, with its reciprocal, the tangent function, as dotted. In each graph the vertical asymptotes are indicated by dotted lines. These graphs

Figure 5.42

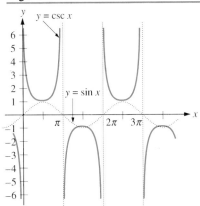

Figure 5.43

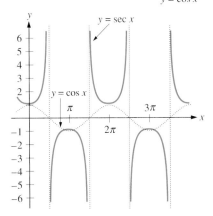

indicate the conclusions about symmetry and evenness/oddness that we already addressed in exercise 16 of section 5.2. They give new information about vertical asymptotes, summarized at the end of the section.

Figure 5.44

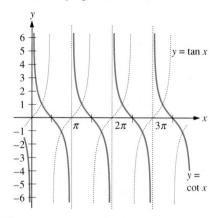

▶ EXAMPLE 5.33

Sketch one cycle of the graph of $k(x) = 3 \sec\left(x + \frac{\pi}{4}\right)$.

Solution

In this formula $A = 3$, $B = 1$, $C = -\frac{\pi}{4}$, and $D = 0$. Because of the value of A, graphing involves a vertical stretching. So the range of k is $\{y: y \leq -3 \text{ or } y \geq 3\}$. The period of k is $\frac{2\pi}{|B|} = 2\pi$. There is a phase shift of $\frac{C}{B} = -\frac{\pi}{4}$, indicated by a horizontal translation of $\frac{\pi}{4}$ units to the left, as shown in figure 5.45. ◀

Figure 5.45

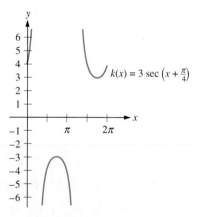

▶ EXAMPLE 5.34

Sketch two cycles of the graph of $m(x) = -1.5 + \frac{1}{3}\cot(2x - 1)$.

Solution

Here $A = \frac{1}{3}$, $B = 2$, $C = 1$, and $D = -1.5$. This function has a period of $\frac{\pi}{|B|} = \frac{\pi}{2}$. Graphing involves a phase shift of $\frac{C}{B} = \frac{1}{2}$ units (to the right), a vertical

Graphs of the Other Trigonometric Functions

shrinking, and a vertical translation 1.5 units downward. The sketch of the graph of this function is in figure 5.46.

Figure 5.46

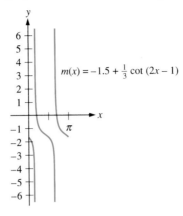

$m(x) = -1.5 + \frac{1}{3}\cot(2x - 1)$

A summary of the significant graphical properties of the six basic trigonometric functions is in table 5.6.

Table 5.6

Function	Figure	Domain	Range	Period	x-Intercepts	Asymptotes
sin x	5.28	R	$\{y: |y| \leq 1\}$	2π	$(n\pi, 0)$	None
cos x	5.29	R	$\{y: |y| \leq 1\}$	2π	$\left(n\pi + \frac{\pi}{2}, 0\right)$	None
tan x	5.38	$\{x: x \neq \frac{\pi}{2} + n\pi\}$	R	π	$(n\pi, 0)$	$x = n\pi + \frac{\pi}{2}$
cot x	5.44	$\{x: x \neq n\pi\}$	R	π	$\left(n\pi + \frac{\pi}{2}, 0\right)$	$x = n\pi$
sec x	5.43	$\{x: x \neq \frac{\pi}{2} + n\pi\}$	$\{y: |y| \geq 1\}$	2π	None	$x = n\pi + \frac{\pi}{2}$
csc x	5.42	$\{x: x \neq n\pi\}$	$\{y: |y| \geq 1\}$	2π	None	$x = n\pi$

Exercises 5.6

In exercises 1–2, determine the period for each of the given functions.

1. (a) $\tan 2x$ (b) $\tan(x - \frac{1}{2})$ (c) $5 + \cot x$
   (d) $\pi - 2\cot(-2x)$ (e) $-3 + 2\sec(4x)$
   (f) $4\sec(x - \pi)$ (g) $\csc 2x$
   (h) $\csc(3 - 2x)$ (i) $-4 + 3\cot(4x - \pi)$

2. (a) $\tan(2x - 3)$ (b) $4\tan\left(x + \frac{\pi}{2}\right)$
   (c) $-2 + \cot\left(x - \frac{\pi}{2}\right)$ (d) $2\cot(-x + \pi)$
   (e) $-5 + \sec x$ (f) $2 + \sec(2x + 3\pi)$
   (g) $\csc(2x - 1)$ (h) $3\csc\left(2x - \frac{\pi}{4}\right)$
   (i) $6 - 2\sec(3x - 4\pi)$

3. Show that each of the following functions is even and hence has graph symmetric to the y-axis:
   (a) $\sec x$ (b) $5 - 3\sec(2x)$ (c) $-1 + 4\sec(-2x)$
   (d) $2\sec(x^2 + 4)$ (e) $\csc(x^2 + 4)$ (f) $\csc(|x|)$

4. Show that each of the following functions is odd and hence has graph symmetric to the origin:

(a) $\tan(-3x)$ (b) $2\cot 4x$ (c) $-3\csc\dfrac{x}{2}$
(d) $-3\tan x$ (e) $-4\cot(-2x)$ (f) $-2\csc(-3x)$

5. Show that none of the following functions is even:
(a) $\tan x$ (b) $\sec\left(x-\dfrac{\pi}{4}\right)$
(c) $\sec\left(\left(x-\dfrac{\pi}{4}\right)^2\right)$ (d) $2x - \sec x$

6. Show that none of the following functions is odd:
(a) $1 + \tan x$ (b) $-1 + 2\cot x$
(c) $-2 - \csc 3x$ (d) $\pi - 2\csc x$

In exercises 7–8, find the natural domain for each of the given functions.

7. (a) $\tan 2x$ (b) $3\sec(-x)$ (c) $4 - 2\csc\left(x+\dfrac{\pi}{4}\right)$

(Hint: in (a), you are finding the real numbers x such that $2x \neq \dfrac{\pi}{2} + n\pi$, for $n = 0, \pm 1, \pm 2, \pm 3, \ldots$)

8. (a) $\cot(2x - 1)$ (b) $-3\tan 2x$
(c) $-4 + \csc(3x - \pi)$

In exercises 9–10, find the range for each of the given functions.

9. (a) $\cot\dfrac{x}{2}$ (b) $-1 + \tan(2x - 1)$
(c) $-1 + 2\csc(-2x)$

10. (a) $-2\tan 3x$ (b) $-\cot(|x|)$ (c) $2\sec\left(x-\dfrac{\pi}{4}\right)$

11. (a) Use the graph of $y = \csc x$ to determine over which of the following intervals csc is an increasing

function: $\left(0, \dfrac{\pi}{2}\right), \left(\dfrac{\pi}{2}, \pi\right), \left(\pi, \dfrac{3\pi}{2}\right), (0, \pi)$.

(b) Use the graph of $y = \csc x$ to determine over which of the following intervals csc is a decreasing

function: $\left(0, \dfrac{\pi}{2}\right), \left(\dfrac{\pi}{2}, \pi\right), \left(\pi, \dfrac{3\pi}{2}\right), (0, \pi)$.

12. (a) Use the graph of $y = \sec x$ to determine over which of the following intervals sec is an increasing

function: $\left(0, \dfrac{\pi}{2}\right), \left(\dfrac{\pi}{2}, \pi\right), \left(\pi, \dfrac{3\pi}{2}\right), (0, \pi)$.

(b) Use the graph of $y = \sec x$ to determine over which of the following intervals sec is a decreasing

function: $\left(0, \dfrac{\pi}{2}\right), \left(\dfrac{\pi}{2}, \pi\right), \left(\pi, \dfrac{3\pi}{2}\right), (0, \pi)$.

13. (a) Find a specific value of x in $[0, \pi)$ such that $\tan 2x$ and $2\tan x$ are defined and unequal.
(b) Find one (that's all there is) value of x in $[0, \pi)$ such that $\tan 2x = 2\tan x$.

14. Explain why none of the graphs of tan, cot, sec, csc has an amplitude.

15. Use your calculator to investigate values of the function $\dfrac{\tan x}{x}$ for x near 0. What are you led to suspect about the "limiting tendency," if any, of $\dfrac{\tan x}{x}$ as x approaches 0?

16. (a) Use your calculator to investigate the "limiting tendency," if any, of $\dfrac{\cos x}{x}$ as x approaches 0.
(b) Using your conclusions in (a), explain why the y-axis is a vertical asymptote of the graph of $y = \dfrac{\cos x}{x}$.
(c) Explain why the y-axis is the only vertical asymptote of $y = \dfrac{\cos x}{x}$.
(d) Are your answers in (b) and (c) supported if you "check" using a graphing calculator or computer graphing program?

17. Explain why the y-axis is not a vertical asymptote for the graph of $y = \dfrac{\sin x}{x}$. (Hint: review exercise 18 of section 5.4.)

18. Find equations of the (vertical) asymptotes for the graphs of each of the following functions:
(a) $\tan 2x$ (b) $\cot(2x - 1)$ (c) $3\sec(-x)$
(d) $4 - 2\csc\left(x+\dfrac{\pi}{4}\right)$ (e) $-4 + \csc(3x - \pi)$
(f) $-3\tan 2x$

In exercises 19–20, calculate the phase shift for each of the given expressions:

19. (a) $-2\tan(-x + 5)$ (b) $3 + \sec(6x - 2)$
(c) $2\csc(-4x)$

20. (a) $\dfrac{1}{2}\cot(3x + 1)$ (b) $4 - \tan(x - \pi)$
(c) $-5 + \dfrac{3}{2}\csc x$

21. Find an expression with phase shift $\dfrac{\pi}{4} + \dfrac{3}{2}$ which describes the same function as $f(x) = 5 - 4\tan(2x - 3)$. (Hint: use a cofunction identity.)

◆ **22.** Find an expression with phase shift $\dfrac{\pi}{4}$ which describes the function given by $f(x) = \dfrac{3}{2}\sec\left(-9x + \dfrac{\pi}{4}\right)$.

In exercises 23–28, graph each of the given functions.

23. $f(x) = -2\tan\dfrac{x}{2}$ **24.** $g(x) = 1 + \cot\left(x+\dfrac{\pi}{4}\right)$

25. $h(x) = -1 + 2\sec x$ **26.** $j(x) = -\csc(4x + \pi)$

27. $k(x) = 1 + \dfrac{1}{2}\tan(4x - \pi)$ **28.** $m(x) = 1 + |\csc x|$

Graphs of the Other Trigonometric Functions

In exercises 29–32, find constants A, B, C, and D such that the given curve is the graph of a function of the given form.

29. $D + A \tan (Bx - C)$

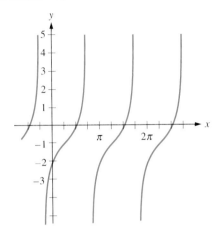

30. $D + A \cot (Bx - C)$

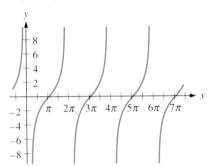

31. $D + A \sec (Bx - C)$

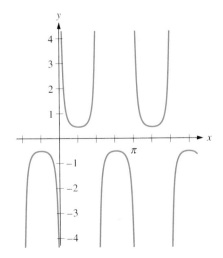

32. $D + A \csc (Bx - C)$

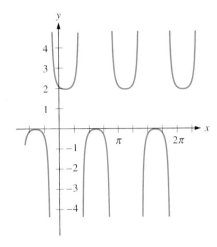

33. (A geometric way to find sec x) We return to the figure in exercise 22 of section 5.3, with $0 < \theta < \dfrac{\pi}{2}$. In this unit circle, the arc QP has length θ. Prove that the line passing through Q and parallel to $\overleftrightarrow{RS}$ meets the y-axis at $(0, \sec \theta)$.

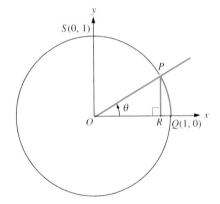

5.7 THE INVERSE TRIGONOMETRIC FUNCTIONS

In the previous chapters, our study of a specific function often led to considering the inverse relation of that function. For instance, in chapter 4, after studying exponential functions, we considered their inverse relations and found that these too were functions, the logarithmic functions. Based on the first six sections of this chapter, we now have on hand a six-member family of functions called the trigonometric functions. This section investigates the inverse relations of these trigonometric functions.

Section 2.4 discussed ways in which new functions could be created from old ones. One of these procedures involved the inverse of a function. The easiest way to tell if a function had an inverse function was to apply the Horizontal Line Test to its graph. We also stated that a function f has an inverse function if and only if f is one-to-one. From our experience, we know that trigonometric functions are not one-to-one and that their graphs fail to pass the Horizontal Line Test. Hence the inverses of the trigonometric functions are relations but not functions.

It will nevertheless be useful to identify inverse functions for trigonometric functions with suitably restricted domains. Consider the sine function. It is not one-to-one since, for instance, $\sin \frac{\pi}{6} = \sin \frac{5\pi}{6} = \sin\left(-\frac{7\pi}{6}\right) = \sin\left(-\frac{11\pi}{6}\right) = \frac{1}{2}$. What are we to understand by the "inverse sine" of $\frac{1}{2}$? All of $\frac{\pi}{6}, \frac{5\pi}{6}, -\frac{7\pi}{6}, -\frac{11\pi}{6}, \ldots$ seem equally entitled to this designation. A systematic way to deal with such repetition of functional values is shown next. It is graphically intuitive for periodic functions.

Inverse Sine Function

Consider the graph of the sine function in figure 5.47. The portion of the graph over the interval $\left[-\frac{\pi}{2}, \frac{\pi}{2}\right]$ is shown by a solid curve and passes the Horizontal Line Test. Hence this part of the sine function has a inverse function. We shall therefore use this interval as the restricted domain for the sine function when we define the inverse function, $\sin^{-1}$. Strictly speaking, $\sin^{-1}$ is not the inverse function of $\sin$, for we have seen that $\sin$ does not have an inverse function. Indeed, $\sin^{-1}$ is the inverse function of the function obtained from $\sin$ by restricting its domain to $\left[-\frac{\pi}{2}, \frac{\pi}{2}\right]$.

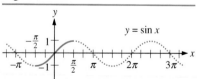

Figure 5.47

Inverse Sine Function

The **inverse sine function**, denoted by **sin⁻¹**, is defined by

$$y = \sin^{-1} x \quad \text{if and only if} \quad \sin y = x,$$

where $-1 \leq x \leq 1$ and $-\frac{\pi}{2} \leq y \leq \frac{\pi}{2}$ (or $-90° \leq y \leq 90°$).

The Inverse Trigonometric Functions

In other words, the inverse sine of x is the unique real number in $\left[-\frac{\pi}{2}, \frac{\pi}{2}\right]$ whose sine is x. The notation **arcsin** x is also used to denote $\sin^{-1} x$. (In exercise 34, you will find an explanation why the "arc" notation is used for this inverse function.) To denote $\frac{1}{\sin x}$, you may write $(\sin x)^{-1}$. You should note that, in general, $\sin^{-1} x \neq (\sin x)^{-1}$.

The fact that $f \circ f^{-1}$ and $f^{-1} \circ f$ are identity functions for any one-to-one function f specializes to the following useful rules:

$$\sin(\sin^{-1} x) = x \quad \text{whenever} \quad -1 \leq x \leq 1.$$

$$\sin^{-1}(\sin x) = x \quad \text{whenever} \quad -\frac{\pi}{2} \leq x \leq \frac{\pi}{2}.$$

EXAMPLE 5.35

(a) $\sin^{-1} \frac{\sqrt{3}}{2} = \frac{\pi}{3}$ because $\frac{\pi}{3}$ is the unique value of y in $\left[-\frac{\pi}{2}, \frac{\pi}{2}\right]$ such that $\sin y = \frac{\sqrt{3}}{2}$.

(b) $\arcsin\left(\frac{-1}{\sqrt{2}}\right) = -\frac{\pi}{4}$ because $-\frac{\pi}{4}$ is in the interval $\left[-\frac{\pi}{2}, \frac{\pi}{2}\right]$ and $\sin\left(-\frac{\pi}{4}\right) = \frac{-1}{\sqrt{2}}$.

(c) $\sin^{-1}(-1) = -\frac{\pi}{2}$.

(d) $\sin^{-1}(0) = 0$.

(e) $\sin(\sin^{-1}(0.8)) = 0.8$.

(f) $\sin(\sin^{-1}(2))$ is not defined.

(g) $\sin^{-1}\left(\sin \frac{\pi}{3}\right) = \frac{\pi}{3}$.

(h) $\sin^{-1}\left(\sin \frac{5\pi}{6}\right) = \sin^{-1}\left(\frac{1}{2}\right) = \frac{\pi}{6} \neq \frac{5\pi}{6}$.

Graphing the Inverse Sine Function

We shall apply the technique for graphing inverse functions that was used in chapters 2 and 4. In figure 5.48, the graph of $y = \sin x$ has been drawn over the restricted domain of $\left[-\frac{\pi}{2}, \frac{\pi}{2}\right]$ and the line $y = x$ is shown as a dashed line. The reflection of $y = \sin x$ in the line $y = x$ is shown by the colored curve. This colored curve is then the graph of $y = \sin^{-1} x$.

Figure 5.48

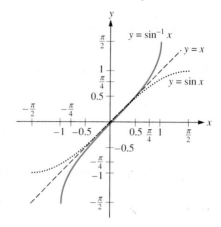

The above procedure for graphing $\sin^{-1}$ took advantage of our knowledge about the graph of sin and the relationship between the graph of a function and the graph of its inverse. What we did was different from the direct procedure of graphing a function by making a table of its functional values, plotting those values in the table, and then smoothly connecting the corresponding points. As a result, we have a graph of $y = \sin^{-1} x$ although we do not have a table of values for the $\sin^{-1}$ function. We can get these values yielding "common" angles by using the definition of $y = \sin^{-1} x$ and reading table 5.2 backwards. Doing this results in the values for $y = \sin^{-1} x$ that are in table 5.7.

Table 5.7

x	-1	$-\frac{\sqrt{3}}{2}$	$-\frac{1}{\sqrt{2}}$	-0.5	0	0.5	$\frac{1}{\sqrt{2}}$	$\frac{\sqrt{3}}{2}$	1
$y = \sin^{-1} x$	$-\frac{\pi}{2}$	$-\frac{\pi}{3}$	$-\frac{\pi}{4}$	$-\frac{\pi}{6}$	0	$\frac{\pi}{6}$	$\frac{\pi}{4}$	$\frac{\pi}{3}$	$\frac{\pi}{2}$

Another way to graph $\sin^{-1}$ is to use a graphing calculator. On a *TI-81*, access the "equation screen," and press [2nd] [sin⁻¹] [X|T] [GRAPH]. On a Casio *fx-7700G*, input [Graph] [SHIFT] [sin⁻¹] [X,θ,T] [EXE].

▶ **EXAMPLE 5.36**

Evaluate $\sin^{-1}\left(\cos\frac{\pi}{6}\right)$.

Solution

Since $\cos\frac{\pi}{6} = \frac{\sqrt{3}}{2}$,

$$\sin^{-1}\left(\cos\frac{\pi}{6}\right) = \sin^{-1}\left(\frac{\sqrt{3}}{2}\right).$$

Consulting table 5.7, we see that $\sin^{-1}\left(\frac{\sqrt{3}}{2}\right) = \frac{\pi}{3}$.

◀

The Inverse Trigonometric Functions

▶ **EXAMPLE 5.37** Evaluate $\sin^{-1}\left(\cos\frac{\pi}{5}\right)$.

Solution Since we do not yet know the precise value for $\cos\frac{\pi}{5}$, we cannot proceed as in example 5.36. But we can use the cofunction identity $\cos\theta = \sin\left(\frac{\pi}{2} - \theta\right)$. In fact, $\sin^{-1}\left(\cos\frac{\pi}{5}\right) = \frac{\pi}{2} - \frac{\pi}{5}$ since $-\frac{\pi}{2} < \frac{\pi}{2} - \frac{\pi}{5} < \frac{\pi}{2}$ and $\sin\left(\frac{\pi}{2} - \frac{\pi}{5}\right) = \cos\frac{\pi}{5}$. ◀

▶ **EXAMPLE 5.38** Evaluate $\tan\left(\sin^{-1}\left(-\frac{1}{3}\right)\right)$.

Solution We shall work this example two ways. A triangle with a hypotenuse of 3 and leg of 1 has been drawn in figure 5.49 so that $\theta = \sin^{-1}\left(-\frac{1}{3}\right)$. As you can see, θ is in the fourth quadrant. Using Pythagoras' Theorem, we determine that the length of the other leg is $\sqrt{8}$. Hence, we obtain $\tan\theta = -\frac{1}{\sqrt{8}}$, and so
$$\tan\left(\sin^{-1}\left(-\frac{1}{3}\right)\right) = -\frac{1}{\sqrt{8}}.$$

Figure 5.49

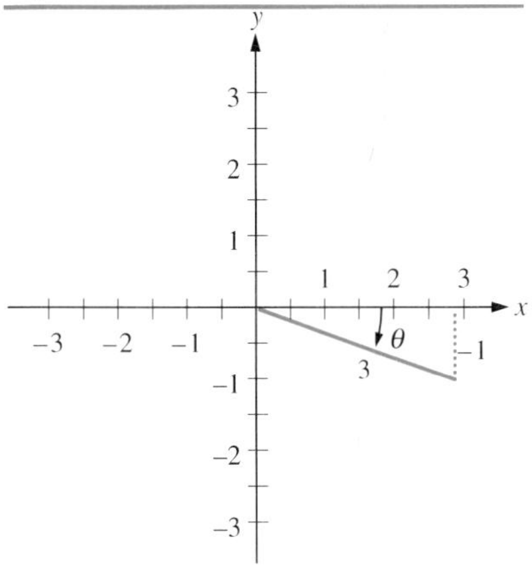

The second method uses the Pythagorean identity, $\sin^2\theta + \cos^2\theta = 1$. Since $\sin\theta = -\frac{1}{3}$, we can determine that $\cos^2\theta = \frac{8}{9}$; and with θ in quadrant IV, that means $\cos\theta = \frac{\sqrt{8}}{3}$. Thus, $\tan\theta = \frac{\sin\theta}{\cos\theta} = -\frac{1}{\sqrt{8}}$. ◀

Inverse Cosine Function

We shall restrict the domain of the cosine function in order to obtain a one-to-one function f whose inverse function can then be called the inverse cosine function. Define f by

$$f(x) = \cos x, \quad \text{for all } x \text{ in the domain } [0, \pi].$$

Then f has the same range as cos, namely $[-1, 1]$, but in an "optimal" way. Specifically, f assumes each of these function values exactly once. Consequently, f is a one-to-one function, and so has an inverse function.

Inverse Cosine Function

The **inverse cosine function**, denoted by **$\cos^{-1}$** or **arccos,** is defined by

$$y = \cos^{-1} x \quad \text{if and only if} \quad \cos y = x,$$

where $-1 \leq x \leq 1$ and $0 \leq y \leq \pi$ (or $0° \leq y \leq 180°$).

The graph of the inverse cosine function is shown by the solid curve in figure 5.50.

Chapter 5 / Trigonometric Functions and Their Graphs

Figure 5.50

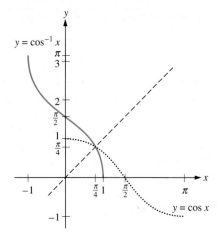

> **EXAMPLE 5.39** Evaluate (a) $\cos^{-1}\left(-\frac{1}{2}\right)$, (b) $\arccos\left(\sin\frac{5\pi}{4}\right)$, and (c) $\cot\left(\cos^{-1}\left(-\frac{3}{5}\right)\right)$.

Solutions

(a) If $y = \cos^{-1}\left(-\frac{1}{2}\right)$, then y must be in the interval $[0, \pi]$ and $\cos y = -\frac{1}{2}$. Hence, $y = \frac{2\pi}{3}$ is the uniquely determined answer.

(b) Since $\sin\frac{5\pi}{4} = -\frac{\sqrt{2}}{2}$, we are seeking $y = \arccos\left(-\frac{\sqrt{2}}{2}\right)$. Then $y \in [0, \pi]$ and $\cos y = -\frac{\sqrt{2}}{2}$. The only correct answer is $\frac{3\pi}{4}$.

I think the got the answer by using cofunction identities P263

(c) If $\theta = \cos^{-1}\left(-\frac{3}{5}\right)$, then θ is in quadrant II and $\sin\theta = \frac{4}{5}$. Since $\cot\theta = \frac{\cos\theta}{\sin\theta}$, *126.87°*

we have $\cot\theta = -\frac{3/5}{4/5} = -\frac{3}{4}$, which is the desired answer. ◀

Inverse Tangent Function

We next define an inverse function for a suitable restriction of the tangent function. We cannot use closed intervals as above since the tangent function is not defined at odd-integral multiples of $\frac{\pi}{2}$. However, over the open interval $\left(-\frac{\pi}{2}, \frac{\pi}{2}\right)$, tan assumes each value in its range exactly once.

Inverse Tangent Function

> The **inverse tangent function**, denoted by **arctan** or **tan^{-1}**, is defined by
> $$y = \tan^{-1} x \quad \text{if and only if} \quad \tan y = x,$$
> where x is a real number and $-\frac{\pi}{2} < y < \frac{\pi}{2}$ (or $-90° < y < 90°$).

The Inverse Trigonometric Functions

Figure 5.51

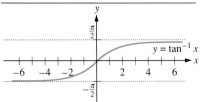

The graph of the inverse tangent function is in figure 5.51.

The Other Inverse Trigonometric Functions

Inverse functions also exist in the above sense for each of the three other trigonometric functions. You would use similar procedures to establish their domains and choose their ranges. Table 5.8 contains the domains and ranges of each of the six inverse trigonometric functions. The graphs of the $\sec^{-1}$, $\csc^{-1}$, and $\cot^{-1}$ functions are in figure 5.52(a), (b), and (c). Other texts may choose different ranges for their $\sec^{-1}$ and $\csc^{-1}$ functions; in such cases, their graphs are naturally different from ours. One benefit that you will derive from this text's choice of a range for $\sec^{-1}$ is a slightly easier formula that you will study later in integral calculus.

Figure 5.52

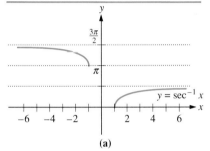

(a)

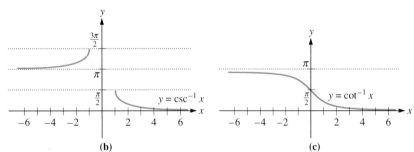

(b)   (c)

Table 5.8

Function	Alternative Notation	Domain	Range
$\sin^{-1} x$	arcsin x	$[-1, 1]$	$\left[-\dfrac{\pi}{2}, \dfrac{\pi}{2}\right]$
$\cos^{-1} x$	arccos x	$[-1, 1]$	$[0, \pi]$
$\tan^{-1} x$	arctan x	$\mathbf{R}$	$\left(-\dfrac{\pi}{2}, \dfrac{\pi}{2}\right)$
$\sec^{-1} x$	arcsec x	$(-\infty, -1] \cup [1, \infty)$	$\left[0, \dfrac{\pi}{2}\right) \cup \left[\pi, \dfrac{3\pi}{2}\right)$
$\csc^{-1} x$	arccsc x	$(-\infty, -1] \cup [1, \infty)$	$\left(0, \dfrac{\pi}{2}\right] \cup \left(\pi, \dfrac{3\pi}{2}\right]$
$\cot^{-1} x$	arccot x	$\mathbf{R}$	$(0, \pi)$

Until now, we have evaluated the inverse trigonometric functions at values associated with trigonometric functions of the common angles. But, what should you do if you need to find the inverse trigonometric function of some other value? Here, as before, you have three options for determining such inverse values: a calculator, a computer, or a trigonometric table. The use of the first two of these is explained below.

Finding Angles with a Calculator

In the last chapter, we learned that we could evaluate e^x by using the [INV] [ln x] keystroke combination. Similarly, we shall use the [INV] key in combination with a trigonometric function key to determine values of the inverse trigonometric functions. (While many calculators use the [INV] key, some use the [sin⁻¹] or [arc] key. Be sure to consult the manual for your calculator.) Suppose you wanted to find a value of θ for which tan θ = 2.5. On an algebraic calculator, you use the keystroke combination 2.5 [INV] [tan]; use [2nd] [tan] 2.5 on a graphing calculator. You get the value 68.19859° or 1.1902899 radians, depending on the mode in which your calculator is working. If you are given the value of one of the reciprocal functions, then you must first use the [1/x] key. For instance, to find cot⁻¹ 0.4 on an algebraic calculator, use the combination 0.4 [1/x] [INV] [tan]; use [2nd] [tan] 0.4 [2nd] [x⁻¹] on a graphing calculator. Here we use the reciprocal identity cot $\theta = \dfrac{1}{\tan \theta}$; since $0.4 = \dfrac{1}{2.5}$, we determine that cot⁻¹ 0.4 = tan⁻¹ 2.5.

▶ **EXAMPLE 5.40**

Use a calculator to determine a possible measure, in degrees, of each of the indicated angles with the given properties: sin α = 0.85, cos β = 0.325, and sec γ = 5.625.

Solution

Given	Press	Display
sin α = 0.85	.85 [INV] [sin]	58.211669
cos β = 0.325	.325 [INV] [cos]	71.034425
sec γ = 5.625	5.625 [1/x] [INV] [cos]	79.759652

Thus, we get that $\alpha \approx 58.21°$, $\beta \approx 71.03°$, and $\gamma \approx 79.76°$. ◀

▶ **EXAMPLE 5.41**

Use a calculator to determine a possible radian measure of each of the indicated angles with the given properties: tan α = −0.625 and csc β = 1.372.

Solution

First, we make sure the calculator is in the radian mode by pressing the [DRG] key.

Given	Press	Display
tan α = −0.625	.625 [+/−] [INV] [tan]	−0.5585993
csc β = 1.372	1.372 [1/x] [INV] [sin]	0.8166598

Thus, we get that $\alpha \approx -0.559$ and $\beta \approx 0.817$ (radian). ◀

Finding Angles Using a Computer

While most computer programming languages, such as BASIC, have the sine, cosine, and tangent functions, they have only one inverse trigonometric function, namely the inverse tangent function. Most languages use the abbreviation ATN as the name of the inverse trigonometric function. (ATN is an abbreviation for arctan, which is a synonym of inverse tangent.) You also need to remember that computers normally work in radians, and hence your result will be in radians unless you do something to convert it to degrees. For example, if you wanted the value of an angle θ where $\tan \theta = 3.5$, you would enter PRINT ATN (3.5), press the <RETURN> key, and the computer would display 1.29249667.

Since a computer does not have inverse functions built in for five of the six trigonometric functions, a user-defined function can be used as part of the program. We next give defined BASIC functions for finding the inverses of the trigonometric functions not included in the BASIC language. (In exercise 27, you will be asked to justify the BASIC function for ARCCOS(X).) If you let PI = 3.141592653, then PI/2 = 1.57079633. (For greater precision, you may prefer to let PI = 4 * ATN(1), and then PI/2 = 2 * ATN(1).)

$$\text{ARCSIN}(X) = \text{ATN}(X/\text{SQR}(-X * X + 1))$$

$$\text{ARCCOS}(X) = -\text{ATN}(X/\text{SQR}(-X * X + 1)) + \text{PI}/2$$

$$\text{ARCCSC}(X) = \text{ATN}(1/\text{SQR}(X * X - 1)) + (1 - \text{SGN}(X)) * \text{PI}/2$$

$$\text{ARCSEC}(X) = \text{ATN}(\text{SQR}(X * X - 1)) + (1 - \text{SGN}(X)) * \text{PI}/2$$

$$\text{ARCCOT}(X) = -\text{ATN}(X) + \text{PI}/2$$

Here the BASIC function SGN(X) returns –1 if X is negative, 0 if X is zero, and 1 if X is positive. In the above formulas for ARCSIN(X), ARCCOS(X), and ARCCSC(X), you should be careful not to divide by zero.

Exercises 5.7

In exercises 1–2, compute exactly.

1. (a) $\sin^{-1}\left(-\frac{\sqrt{3}}{2}\right)$ (b) $\arccos 0$ (c) $\text{arccot } \sqrt{3}$
   (d) $\tan^{-1} 0$ (e) $\arccos\left(-\frac{\sqrt{3}}{2}\right)$

2. (a) $\arcsin\left(\frac{\sqrt{2}}{2}\right)$ (b) $\cos^{-1}\left(\frac{\sqrt{2}}{2}\right)$ (c) $\cot^{-1}\left(-\frac{1}{\sqrt{3}}\right)$
   (d) $\arctan(-1)$ (e) $\cos^{-1} \frac{1}{2}$

In exercises 3–4, find the given values (approximately, if necessary).

3. (a) $\arcsin(-0.82)$ (b) $\tan^{-1} 6$ (c) $\text{arccsc } \frac{2}{\sqrt{3}}$

4. (a) $\cos^{-1} 0.41$ (b) $\text{arccot}(-1.28)$
   (c) $\sec^{-1}\left(-\frac{2}{\sqrt{3}}\right)$

In exercises 5–10, evaluate each of the given expressions.

5. (a) $\sin(\sin^{-1} \frac{2}{3})$ (b) $\cos(\arccos(-0.8))$
   (c) $\tan(\arctan \pi)$ (d) $\cot\left(\cot^{-1} \frac{2}{\pi}\right)$
   (e) $\sec\left(\sec^{-1} \frac{\pi}{2}\right)$ (f) $\csc(\text{arccsc}(-2))$

6. (a) $\sin^{-1}\left(\sin\left(-\frac{\pi}{5}\right)\right)$ (b) $\arccos\left(\cos \frac{9\pi}{4}\right)$
   (c) $\cot^{-1}\left(\cot \frac{4\pi}{3}\right)$

7. (a) $\arcsin\left(\sin \frac{2\pi}{7}\right)$ (b) $\tan^{-1}\left(\tan\left(-\frac{3\pi}{4}\right)\right)$
   (c) $\sec^{-1}\left(\sec \frac{\pi}{6}\right)$

310 Chapter 5 / Trigonometric Functions and Their Graphs

8. (a) $\sin^{-1}\left(\cos\dfrac{5\pi}{3}\right)$ (b) $\arcsin\left(\cos\dfrac{\pi}{6}\right)$
   (c) $\arctan\left(\cot\dfrac{\pi}{3}\right)$ (d) $\cot^{-1}\left(\tan\left(-\dfrac{\pi}{6}\right)\right)$

9. (a) $\tan(\sin^{-1}(-\tfrac{1}{3}))$ (b) $\cos(\cot^{-1}(-0.01))$

10. (a) $\sin(\arctan(-\tfrac{1}{3}))$ (b) $\cot(\sin^{-1}\tfrac{3}{5})$
    (c) $\cot(\text{arcsec}\,\tfrac{3}{2})$

11. Prove or disprove: $\sec^{-1}\dfrac{1}{x} = \cos^{-1} x$ whenever $-1 \le x \le 1,\ x \ne 0$.

12. Simplify each of the following:
    (a) $\sin^{-1} 1 - \sin^{-1}(-1)$ (b) $\arccos 1 - \arccos(-1)$

In exercises 13–14, use a calculator or computer to approximate a standard-position angle θ with the given trigonometric functional value and with terminal side in the given quadrant.

13. (a) $\sin\theta = 0.29$, I   (b) $\cos\theta = -0.92$, II
    (c) $\csc\theta = -7.01$, III   (d) $\tan\theta = -4$, IV

14. (a) $\cot\theta = 0.25$, I   (b) $\cos\theta = -0.92$, III
    (c) $\sin\theta = -0.92$, III   (d) $\sec\theta = 1.07$, I

In exercises 15–16, find the natural domain for each of the given functions.

15. (a) $f(x) = \sin^{-1}\dfrac{x}{2}$ (Hint: solve the inequality $-1 \le \dfrac{x}{2} \le 1$.)
    (b) $g(x) = -2\cot^{-1} 2x$ (c) $h(x) = -2 - \tfrac{1}{2}\csc^{-1}(x-1)$

16. (a) $k(x) = 2\arccos x$
    (b) $m(x) = \pi + \tfrac{1}{2}\arctan\left(x + \dfrac{\pi}{4}\right)$
    (c) $n(x) = 1 + \sec^{-1} x$

17. Find the range for each function listed in exercises 15 and 16.

18. Find the interval(s) over which each of the following functions is increasing:
    (a) $f(x) = \sin^{-1} 3x$ (Hint: look at the graph.)
    (b) $g(x) = -3\arccos x$ (c) $h(x) = -3 + \arctan(-3x)$
    (d) $j(x) = 2\tan^{-1}(x-3)$ (e) $k(x) = 2 - \text{arcsec}\,x$
    (f) $m(x) = 2\csc^{-1} x$

19. Find the interval(s) over which each of the following functions is decreasing:
    (a) $f(x) = \sin^{-1} 3x$ (b) $g(x) = -3\arccos x$
    (c) $h(x) = -3 + \arctan(-3x)$ (d) $j(x) = 2\tan^{-1}(x-3)$
    (e) $k(x) = 2 - \text{arcsec}\,x$ (f) $m(x) = 2\csc^{-1} x$

20. Find all x in the given interval such that the given equation holds:

(a) $\sin^2 x = \dfrac{1}{2},\ 0 \le x \le \dfrac{\pi}{2}$ (b) $\sin^2 x = \dfrac{1}{2},\ 0 \le x \le \dfrac{5\pi}{2}$
(c) $2\sin^2 x - 2\sqrt{2}\sin x + 1 = 0,\ \pi \le x \le 2\pi$
(d) $2\cos^2 x - 3\cos x + 1 = 0,\ 0 \le x \le \dfrac{\pi}{2}$

21. Find all x in the given interval such that the given equation holds:
    (a) $\sin^3 x = \dfrac{3}{8}\sqrt{3},\ \dfrac{\pi}{2} \le x \le \pi$
    (b) $\sin x \tan x = \cos x,\ 0 \le x \le \pi$
    (c) $\sin^2 x = 2,\ 0 \le x \le 2\pi$

22. (a) Without appealing to a graph, prove that $\sin^{-1}$ is an odd function. (Hint: if $-\dfrac{\pi}{2} \le y \le \dfrac{\pi}{2}$ and $\sin y = x$, show that $-\dfrac{\pi}{2} \le -y \le \dfrac{\pi}{2}$ and $\sin(-y) = -x$.)
    (b) What conclusion about symmetry of the graph of $y = \sin^{-1} x$ can you draw from (a)?

23. Prove that arccos is not an odd function.

24. Prove that arctan is an odd function. Hence infer a fact about symmetry of the graph of $y = \arctan x$.

25. Prove that $\cos^{-1} x = \dfrac{\pi}{2} - \sin^{-1} x$ for all x in $[-1, 1]$.
    (Hint: use a cofunction identity and solve some inequalities.)

26. Prove that $\text{arccot}\,x = \dfrac{\pi}{2} - \arctan x$ for all real numbers x. (Hint: adapt the hint for exercise 25.)

♦ 27. Use exercise 25 and known trigonometric identities to justify the text's BASIC function for ARCCOS(X).

In exercises 28–32, graph the given function.

28. $f(x) = 2\sin^{-1} x$ 29. $g(x) = \sin^{-1} 2x$
30. $h(x) = 1 + \arccos(x - \tfrac{1}{2})$
31. $j(x) = -2\arctan\dfrac{x}{2}$ 32. $k(x) = \pi - \cot^{-1}(x+1)$

33. Using a calculator to find various inverse sine values as needed, calculate the area of the shaded sector.

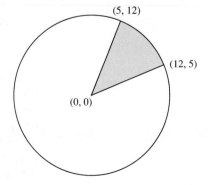

◆ 34. (An explanation of the "arc" notation for inverse trigonometric functions) Let $P(a, b)$ be a point on the unit circle. Let s be the length of the arc from $Q(1, 0)$ counterclockwise to P. Prove that $s = \arccos a$ if $b \geq 0$; that $s = \pi - \arcsin b$ if $b < 0$ and $a \leq 0$; and that $s = 2\pi + \arcsin b$ if $b < 0$ and $a > 0$.

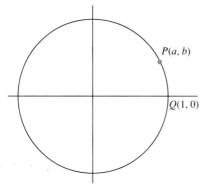

35. Tabby is stuck in an old tree 22 meters above the level of a rescue truck. The rescue squad hopes to save Tabby with a ladder that can extend from the truck to a maximum length of 25 meters, at an angle of elevation θ that can take on any value such that $0 \leq \theta \leq \frac{\pi}{3}$. If the ladder is not long enough, the rescue squad will have to call the Human Fly. Will the call be necessary?

5.8 TRIGONOMETRIC EQUATIONS AND INEQUALITIES

A **trigonometric equation** is an equation involving at least one trigonometric expression. Trigonometric identities, which we have encountered repeatedly in this chapter, are types of trigonometric equations that are true for all values of the variables in the joint domains of the expressions involved. In this section, we focus more generally on **conditional equations**, or equations that may be true for only certain values of their variables. The equations in exercises 11 and 19 of section 5.2 were conditional equations. In those exercises, you were asked to find counterexamples showing that given statements were not identities. As with any equation, any value of the unknown variable for which a given trigonometric equation is valid is called a **solution** of the equation and is said to **satisfy** the equation. To **solve** a conditional equation means to find all its solutions.

The periodic nature of trigonometric functions means that in most cases if there is at least one solution to a trigonometric equation, then there are infinitely many solutions. This does not prove to be much of a complication. What we can do is illustrated by the equation $\sin \theta = \frac{1}{2}$. One of its solutions in $[0, 2\pi)$ is found by using the inverse sine function to obtain $\theta = \sin^{-1} \frac{1}{2} = \frac{\pi}{6}$. The other solution in this interval is $\theta = \frac{5\pi}{6}$. All other solutions of $\sin \theta = \frac{1}{2}$ take one of the forms $\frac{\pi}{6} + 2k\pi$ and $\frac{5\pi}{6} + 2k\pi$, where k is an integer.

The simplest type of trigonometric equation is one in which a single trigonometric function of an unknown variable is set equal to a constant, and hence is called a **simple** trigonometric equation. Examples of simple trigonometric equations include $\sin \theta = \frac{1}{2}$, $\cos x = -1$, $\tan \theta = 5$, and $\sec \phi = \sqrt{2}$. In solving any conditional trigonometric equation, one aims to reduce the problem to solving one or more simple equations. So, we shall begin by illustrating this central idea.

Chapter 5 / Trigonometric Functions and Their Graphs

▶ **EXAMPLE 5.42**   Solve the simple trigonometric equation $2 \cos \theta = -\sqrt{2}$.

Solution   We first solve for $\cos \theta$, obtaining the equivalent equation

$$\cos \theta = -\frac{\sqrt{2}}{2}.$$

The only values of θ in $[0, 2\pi)$ for which $\cos \theta = -\frac{\sqrt{2}}{2}$ are $\frac{3\pi}{4}$ and $\frac{5\pi}{4}$.

Then every solution for θ differs from either $\frac{3\pi}{4}$ or $\frac{5\pi}{4}$ by an integral multiple of 2π. These solutions can be described by

$$\theta = \frac{3\pi}{4} + 2k\pi \quad \text{or} \quad \theta = \frac{5\pi}{4} + 2k\pi, \quad k \text{ an integer} \quad ◀$$

No one general method can be used to solve all trigonometric equations. One frequently useful method is to express all trigonometric functions occurring in the equation in terms of one function and then solve algebraically for that function. For instance, the conditional equation $\sin x \cot x = -\frac{\sqrt{2}}{2}$ is equiv-

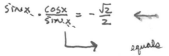

equals alent, thanks to a reciprocal identity, to $\cos x = -\frac{\sqrt{2}}{2}$, and so has the solutions found in example 5.42.

In reworking trigonometric equations algebraically you will use many of the same problem-solving techniques used throughout this book. These techniques included factoring, squaring both sides, and multiplying both sides by an expression containing the variable. Remember that if you use either of these last two techniques (squaring both sides or multiplying both sides by an expression containing the variable), then it is very important to check all alleged solutions in the original equation.

▶ **EXAMPLE 5.43**   Solve $2 \sin^2 \alpha - 3 \sin \alpha + 1 = 0$.   *BULLCRAP! THE ENTIRE EQUATION CAN BE FACTORED.*

Solution   The left-hand side of the given equation can be factored:

$$(2 \sin \alpha - 1)(\sin \alpha - 1) = 0.$$

Setting each factor equal to 0, we have

$$2 \sin \alpha - 1 = 0 \quad \text{or} \quad \sin \alpha - 1 = 0$$

which is equivalent to

$$\sin \alpha = \tfrac{1}{2} \quad \text{or} \quad \sin \alpha = 1.$$

We have seen that the solutions of $\sin \alpha = \tfrac{1}{2}$ in the interval $[0, 2\pi)$ are $\alpha = \frac{\pi}{6}$ and $\alpha = \frac{5\pi}{6}$. The only solution of $\sin \alpha = 1$ in the interval $[0, 2\pi)$ is $\frac{\pi}{2}$. Thus, the solutions of the given equation are

Trigonometric Equations and Inequalities

$$\alpha = \frac{\pi}{6} + 2k\pi, \frac{5\pi}{6} + 2k\pi, \text{ and } \frac{\pi}{2} + 2k\pi, \quad k \text{ any integer.}$$

Trigonometric identities will be studied in the next chapter. We shall see that identities may often be used to simplify a given trigonometric equation or inequality before attempting to solve it. The next three examples offer a brief exposure to this strategy.

EXAMPLE 5.44 Solve $3 \sin^2 \beta = \cos^2 \beta$ in the interval $[0, 2\pi)$.

Solution Substituting the Pythagorean identity $\cos^2 \beta = 1 - \sin^2 \beta$, we get

$$3 \sin^2 \beta = 1 - \sin^2 \beta$$
$$4 \sin^2 \beta = 1$$
$$\sin^2 \beta = \tfrac{1}{4}$$
$$\sin \beta = \tfrac{1}{2} \quad \text{or} \quad \sin \beta = -\tfrac{1}{2}.$$

The values of β in $[0, 2\pi)$ which result are

$$\beta = \frac{\pi}{6}, \frac{5\pi}{6}, \frac{7\pi}{6}, \text{ and } \frac{11\pi}{6}.$$

EXAMPLE 5.45 Solve $\sec \gamma - 2 \cos \gamma - \tan \gamma = 0$ in the interval $[0, 2\pi)$.

Solution We shall employ some reciprocal identities, replacing $\sec \gamma$ with $\dfrac{1}{\cos \gamma}$ and $\tan \gamma$ with $\dfrac{\sin \gamma}{\cos \gamma}$. Next we shall eliminate denominators by multiplying the equation through by $\cos \gamma$. This, of course, introduces a risk, so it will be very important to check all alleged solutions.

$$\sec \gamma - 2 \cos \gamma - \tan \gamma = 0$$
$$\frac{1}{\cos \gamma} - 2 \cos \gamma - \frac{\sin \gamma}{\cos \gamma} = 0$$
$$1 - 2 \cos^2 \gamma - \sin \gamma = 0$$
$$1 - 2(1 - \sin^2 \gamma) - \sin \gamma = 0 \quad \text{(by the Pythagorean identity)}$$
$$2 \sin^2 \gamma - \sin \gamma - 1 = 0.$$

We may now proceed as in example 5.43. By factoring, we have

$$(2 \sin \gamma + 1)(\sin \gamma - 1) = 0.$$

Thus $2 \sin \gamma + 1 = 0$ or $\sin \gamma - 1 = 0$; that is, $\sin \gamma = -\tfrac{1}{2}$ or $\sin \gamma = 1$. From $\sin \gamma = -\tfrac{1}{2}$ we get $\gamma = \dfrac{7\pi}{6}$ and $\gamma = \dfrac{11\pi}{6}$; from $\sin \gamma = 1$ we obtain $\gamma = \dfrac{\pi}{2}.$

By substitution, we find that $\frac{7\pi}{6}$ and $\frac{11\pi}{6}$ each satisfy the original equation. However, $\frac{\pi}{2}$ does not, since neither $\sec \frac{\pi}{2}$ nor $\tan \frac{\pi}{2}$ is defined. Therefore, the required solutions in the interval $[0, 2\pi)$ are $\frac{7\pi}{6}$ and $\frac{11\pi}{6}$. ◀

▶ **EXAMPLE 5.46**   Solve $\tan \delta + 1 - \sqrt{3} = \sqrt{3} \cot \delta$ in the interval $[0, 2\pi)$.

Solution   We shall replace $\cot \delta$ by $\frac{1}{\tan \delta}$ and then multiply through by $\tan \delta$ to clear denominators.

$$\tan \delta + 1 - \sqrt{3} = \sqrt{3} \cot \delta$$

$$\tan \delta + 1 - \sqrt{3} = \frac{\sqrt{3}}{\tan \delta}$$

$$\tan^2 \delta + (1 - \sqrt{3}) \tan \delta = \sqrt{3}$$

$$\tan^2 \delta + (1 - \sqrt{3}) \tan \delta - \sqrt{3} = 0.$$

This does not appear to factor nicely using integer coefficients, but we can regard it as a quadratic equation with $\tan \delta$ as the variable. Solving by the quadratic formula, we obtain

$$\tan \delta = \frac{-(1 - \sqrt{3}) \pm \sqrt{(1 - \sqrt{3})^2 - 4(1)(-\sqrt{3})}}{2}$$

$$= \frac{-1 + \sqrt{3} \pm \sqrt{(1 + \sqrt{3})^2}}{2}$$

$$= \frac{-1 + \sqrt{3} \pm (1 + \sqrt{3})}{2}.$$

From this, we see that $\tan \delta = -1$ or $\tan \delta = \sqrt{3}$. If $\tan \delta = -1$, we obtain $\delta = \frac{3\pi}{4}$ and $\delta = \frac{7\pi}{4}$. From $\tan \delta = \sqrt{3}$, we get $\delta = \frac{\pi}{3}$ and $\delta = \frac{4\pi}{3}$. These four values of δ satisfy the original equation and, hence, are the required solutions. ◀

Solving Trigonometric Equations with "Multiple Angles"

To solve trigonometric equations involving a compound angle $n\theta$, it is usually best to first substitute x for the compound angle (that is, $x = n\theta$), next solve for x by the above methods, and finally rewrite the solutions in terms of θ. (Notice that $\theta = \frac{x}{n}$ since $x = n\theta$.)

Trigonometric Equations and Inequalities

EXAMPLE 5.47 Solve $\cos 3\theta = \frac{1}{2}$ for θ in the interval $[0, 2\pi)$.

Solution Let $x = 3\theta$. Then the given equation may be rewritten as the simple trigonometric equation $\cos x = \frac{1}{2}$. The condition $0 \leq \theta < 2\pi$ is equivalent to $0 \leq \frac{x}{3} < 2\pi$ (that is, $0 \leq x < 6\pi$). The solutions of $\cos x = \frac{1}{2}$ for x in $[0, 2\pi)$ are $x = \frac{\pi}{3}$ and $x = \frac{5\pi}{3}$. Adding 2π to each of these, we obtain two more solutions:

$$x = \frac{\pi}{3} + 2\pi = \frac{7\pi}{3} \quad \text{and} \quad x = \frac{5\pi}{3} + 2\pi = \frac{11\pi}{3}$$

which are in the interval $[2\pi, 4\pi)$. Adding 2π a second time, we obtain two additional solutions:

$$x = \frac{7\pi}{3} + 2\pi = \frac{13\pi}{3} \quad \text{and} \quad x = \frac{11\pi}{3} + 2\pi = \frac{17\pi}{3}$$

which are in the interval $[4\pi, 6\pi)$. Thus, the solutions for $0 \leq x < 6\pi$ are

$$x = \frac{\pi}{3}, \frac{5\pi}{3}, \frac{7\pi}{3}, \frac{11\pi}{3}, \frac{13\pi}{3}, \text{ and } \frac{17\pi}{3},$$

and so the desired values of $\theta = \frac{x}{3}$ in $[0, 2\pi)$ are

$$\theta = \frac{\pi}{9}, \frac{5\pi}{9}, \frac{7\pi}{9}, \frac{11\pi}{9}, \frac{13\pi}{9}, \text{ and } \frac{17\pi}{9}.$$

EXAMPLE 5.48 Solve $4 + \sqrt{3} \sec\left(2x + \frac{\pi}{4}\right) = 2$ in the interval $[0, 2\pi)$.

Solution By substituting $\theta = 2x + \frac{\pi}{4}$, we can rewrite the equation as the trigonometric equation $4 + \sqrt{3} \sec \theta = 2$. An equivalent equation is

$$\sec \theta = -\frac{2}{\sqrt{3}}$$

or

$$\cos \theta = -\frac{\sqrt{3}}{2}.$$

The condition $0 \leq x < 2\pi$ is equivalent to $\frac{\pi}{4} \leq \theta < \frac{17\pi}{4}$.

The solutions for θ in this interval are $\theta = \frac{5\pi}{6}, \frac{7\pi}{6}, \frac{17\pi}{6}, \text{ and } \frac{19\pi}{6}$.

Thus, the desired solutions for $x = \left(\theta - \frac{\pi}{4}\right)/2$ are

$$x = \frac{7\pi}{24}, \frac{11\pi}{24}, \frac{31\pi}{24}, \text{ and } \frac{35\pi}{24}.$$

Chapter 6 will include additional exercises involving trigonometric equations with multiple angles. There, new identities will permit us to solve much more complicated multiple-angle equations.

Solving Trigonometric Inequalities

A trigonometric inequality may be solved using sign charts in much the same manner as any other inequality. First, locate the key points. These are the points where either the equality holds or at least one member of the inequality is undefined. Next, determine whether the inequality is valid between successive key points. The periodicity of the trigonometric functions must be considered when completing a sign chart.

▶ **EXAMPLE 5.49** Solve $2 \sin \theta < \sqrt{3}$ in the interval $[0, 2\pi)$.

Solution We shall find the key points by solving the associated equation $2 \sin \theta = \sqrt{3}$, or, equivalently, $\sin \theta = \frac{\sqrt{3}}{2}$. In the interval $[0, 2\pi)$, we find $\theta = \frac{\pi}{3}, \frac{2\pi}{3}$.

These solutions divide $[0, 2\pi)$ into the subintervals $\left[0, \frac{\pi}{3}\right)$, $\left(\frac{\pi}{3}, \frac{2\pi}{3}\right)$, and $\left(\frac{2\pi}{3}, 2\pi\right)$. We shall select one point in each interval to determine if the inequality is valid for that interval.

For $\left[0, \frac{\pi}{3}\right)$, if we select $\theta = \frac{\pi}{6}$, we see that $2 \sin \theta = 2 \sin \frac{\pi}{6} = 2\left(\frac{1}{2}\right) = 1$. Since $1 < \sqrt{3}$, all the values in $\left[0, \frac{\pi}{3}\right)$ satisfy the given inequality.

Next, select a point in $\left(\frac{\pi}{3}, \frac{2\pi}{3}\right)$, say $\theta = \frac{\pi}{2}$. Then $2 \sin \theta = 2 \sin \frac{\pi}{2} = 2(1) = 2 \not< \sqrt{3}$. Hence the given inequality holds for no value in the interval $\left(\frac{\pi}{3}, \frac{2\pi}{3}\right)$.

Finally, select a point in $\left(\frac{2\pi}{3}, 2\pi\right)$, say $\theta = \pi$. Then $2 \sin \pi = 2(0) = 0 < \sqrt{3}$. Thus, the given inequality is valid on the interval $\left(\frac{2\pi}{3}, 2\pi\right)$.

Therefore, the solutions for the inequality $2 \sin \theta < \sqrt{3}$ in $[0, 2\pi)$ consist of the union of the intervals $\left[0, \frac{\pi}{3}\right)$ and $\left(\frac{2\pi}{3}, 2\pi\right)$. As with many inequalities, a convenient way to organize our reasoning is to use a sign chart, as shown in figure 5.53(a).

Trigonometric Equations and Inequalities

The solution set $\left[0, \frac{\pi}{3}\right) \cup \left(\frac{2\pi}{3}, 2\pi\right)$ can also be found geometrically by studying the graph of $y = 2 \sin \theta$ in figure 5.53(b). A dashed horizontal line is drawn at $y = \sqrt{3}$. Solving the inequality $2 \sin \theta < \sqrt{3}$ on $[0, 2\pi)$ is equivalent to asking for the values of θ for which $y < \sqrt{3}$. These are the values of θ for which the graph of $y = 2 \sin \theta$ is below the dashed horizontal line. By studying figure 5.53(b), you can verify that these are the intervals $\left[0, \frac{\pi}{3}\right)$ and $\left(\frac{2\pi}{3}, 2\pi\right)$.

Figure 5.53

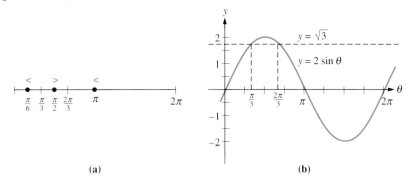

(a)   (b)

You may find it more convenient to organize your work as in the following table.

Interval	Test Point θ	Evaluation at θ	$2 \sin \theta < \sqrt{3}$
$\left[0, \frac{\pi}{3}\right)$	$\frac{\pi}{6}$	$2\left(\frac{1}{2}\right) = 1 < \sqrt{3}$	True
$\left(\frac{\pi}{3}, \frac{2\pi}{3}\right)$	$\frac{\pi}{2}$	$2(1) = 2 \not< \sqrt{3}$	False
$\left(\frac{2\pi}{3}, 2\pi\right)$	π	$2(0) = 0 < \sqrt{3}$	True

▶ **EXAMPLE 5.50**   Solve $\sin^2 x + 5 \cos^2 x \geq 3$ for the interval $[0, 2\pi)$.

Solution   First we find the key points by solving the associated equation $\sin^2 x + 5 \cos^2 x = 3$. Using the Pythagorean identity $\sin^2 x + \cos^2 x = 1$ in its equivalent form $1 - \cos^2 x = \sin^2 x$, we may rewrite the associated equation as follows.

$$(1 - \cos^2 x) + 5 \cos^2 x = 3$$

$$4 \cos^2 x = 2$$

$$\cos^2 x = \tfrac{1}{2}$$

$$\cos x = \pm \frac{\sqrt{2}}{2}.$$

Thus, the key points are $\frac{\pi}{4}, \frac{3\pi}{4}, \frac{5\pi}{4}$, and $\frac{7\pi}{4}$. Therefore we need to test the inequality over each of the intervals $\left[0, \frac{\pi}{4}\right), \left(\frac{\pi}{4}, \frac{3\pi}{4}\right), \left(\frac{3\pi}{4}, \frac{5\pi}{4}\right), \left(\frac{5\pi}{4}, \frac{7\pi}{4}\right)$, and $\left(\frac{7\pi}{4}, 2\pi\right]$. This work may be summarized in a table, just as in the preceding example. We have left the completion of the table to you.

Interval	Test Point θ	Evaluation at θ	$\sin^2 x + 5 \cos^2 x \geq 3$
$\left[0, \frac{\pi}{4}\right)$	$\frac{\pi}{6}$	$\left(\frac{1}{2}\right)^2 + 5\left(\frac{\sqrt{3}}{2}\right)^2 = \frac{1}{4} + \frac{15}{4} = 4 \geq 3$	True
$\left(\frac{\pi}{4}, \frac{3\pi}{4}\right)$	$\frac{\pi}{2}$	—	False
$\left(\frac{3\pi}{4}, \frac{5\pi}{4}\right)$	π	—	—
$\left(\frac{5\pi}{4}, \frac{7\pi}{4}\right)$	—	—	—
$\left(\frac{7\pi}{4}, 2\pi\right)$	—	—	—

Once you have completed the table and taken account of the key points, you should see that the solution set for the inequality is $\left[0, \frac{\pi}{4}\right] \cup \left[\frac{3\pi}{4}, \frac{5\pi}{4}\right] \cup \left[\frac{7\pi}{4}, 2\pi\right)$. ◂

Exercises 5.8

In exercises 1–38, solve the given conditional equation in the interval $[0, 2\pi)$. Whenever possible, find the exact solutions; elsewhere, use a calculator to approximate the solutions.

1. $\sin \theta = \frac{\sqrt{3}}{2}$
2. $3 \cos x = -1$
3. $2 - 3 \tan x = 0$
4. $3 \cot \theta = -2$
5. $2 \sec \theta = -1$
6. $\csc x = 2$
7. $(\sec x - 2)(\sqrt{3} \sec x - 2) = 0$
8. $\sin \theta \tan \theta = 0$
9. $\sin^2 \theta = \sin \theta$
10. $\tan^2 x = \tan x$
11. $\tan^2 x + 9 \tan x = 0$
12. $\tan^2 \theta + 6 \tan \theta + 9 = 0$
13. $\tan^2 \theta + \tan \theta - 2 = 0$
14. $\tan^2 x + \tan x + 2 = 0$
15. $|\tan x| = \sqrt{3}$
16. $\sqrt{\tan^2 \theta} = \sqrt{3}$
17. $\tan^2 \theta = 3$
18. $2 \sin x + \tan x = 0$
19. $\sin\left(\alpha - \frac{\pi}{4}\right) = \frac{1}{2}$
20. $\sin\left|\beta - \frac{\pi}{4}\right| = \frac{1}{2}$
21. $\left|\sin\left(\gamma - \frac{\pi}{4}\right)\right| = \frac{1}{2}$
22. $3 \cos^2 \beta = \sin^2 \beta$
23. $\sin^2 x - \cos x = 1$
24. $4 \sin^2 \theta + 2(1 + \sqrt{3}) \cos \theta = 4 + \sqrt{3}$
25. $2 \cos \theta = \sec \theta + \sqrt{17}$
26. $3 \tan x - \cot x + 1 = 0$
27. $(\sec x + \tan x) \sec x = 2$
28. $\sec^4 x = 16$
29. $\sin \frac{\theta}{2} = \frac{1}{2}$
30. $2 \sin 4x = 1$
31. $2 \cos 2\left(x - \frac{\pi}{4}\right) = 1$
32. $1 - 2 \cos(2x - \pi) = 2$
33. $2 - \tan 3x = 0$
34. $\cot 3x = -2$
35. $1 + \sec 2\theta = 1$

36. $(\csc 2\theta - 1)(\sqrt{2} \csc \theta + 2) = 0$

37. $\sqrt{|\sin \theta|} = \dfrac{1}{\sqrt[4]{2}}$
38. $\sqrt{\sin \theta} = \dfrac{1}{\sqrt[4]{2}}$

39. (a) Use the Pythagorean identity and a formula in exercise 25 of section 5.2 to show that $\cos 2\theta = 1 - 2\sin^2 \theta$ is an identity.
    (b) Using (a) and example 5.43, solve the conditional equation $\cos 2x + 3 \sin x - 2 = 0$ in the interval $[0, 2\pi)$.

40. (a) Use the Pythagorean identity and reciprocal and ratio identities to prove that $\sec^2 x = 1 + \tan^2 x$ is an identity.
    (b) Using (a), solve the conditional equation $\sec^2 x + \tan x - 1 = 0$ in the interval $[0, 2\pi)$.
    (c) Using (a), prove that the substitution $\theta = \tan^{-1} \dfrac{3x}{5}$ changes $\sqrt{25 + 9x^2}$ into $5 \sec \theta$.

In exercises 41–48, solve the given inequality in the interval $[0, 2\pi)$.

41. $2 \sin \theta < 1$
42. $2 \cos 2x > \sqrt{3}$
43. $\tan \dfrac{x}{2\pi} \geq 1$
44. $2 \sec(\theta + \pi) \leq 1$
45. $2 \sin^2 \theta + 3 \cos^2 \theta < \dfrac{5}{2}$
46. $2 \cos^2 \theta + 3 \sin^2 \theta < \dfrac{5}{2}$
47. $\cos^2 x - \sin x \geq 1$
48. $\cos^2 x + 2 \cos x + 1 < 0$
49. $\tan^2 x + \sec^2 x \leq 3$ (Hint: use exercise 40(a).)
50. Rowena has been ordered by her parents "to find any boy except Romeo." She misunderstood, and is swimming towards a buoy, bobbing in a local river. The buoy's height above the surface of the water at time t seconds is given by $y = 3 \sin \dfrac{\pi t}{28}$ feet. Rowena reaches a position directly above the buoy at $t = 91$.
    (a) How far beneath Rowena is this buoy?
    (b) How long must Rowena wait for it to surface?

5.9 CHAPTER SUMMARY AND REVIEW EXERCISES

Major Concepts

- Amplitude
- Angle
  - Coterminal angles
  - Initial side
  - Negative angle
  - Positive angle
  - Quadrantal angle
  - Sides
  - Standard position
  - Terminal side
  - Vertex
- Angular speed
- Circular functions
- Cofunctions
- Conditional equations
- Degree
- Horizontal translation
- Linear speed
- Minutes
- Period
- Periodic function
- Phase shift
- Pythagorean Identity
- Quotient or ratio identities
- Radian
- Reciprocal identities
- Reference angle θ_{ref}
- Satisfy an equation
- Seconds
- Simple trigonometric equation
- Sine wave
- Sinusoidal graph
- Solution of a trigonometric equation
- Solve a conditional equation
- Trigonometric equation
- Trigonometric functions of θ
- Vertical translation

Review Exercises

1. Draw each of the following angles θ in standard position. Convert each θ to radian measure and identify θ_{ref}.
   (a) $60°$ (b) $-30°$ (c) $-210°$ (d) $930°$

2. Convert each of the following measures to degrees, minutes, and seconds:
   (a) $\dfrac{7\pi}{11}$ radians   (b) $19°$
   (c) $-60.321°$ (d) -5 radians

3. To the nearest hundredth, calculate the arc length along the unit circle from $(1, 0)$ counterclockwise to $\left(\dfrac{3}{5}, \dfrac{4}{5}\right)$.

4. Austin, Texas and Winnipeg, Canada lie approximately on a line of longitude ($97°$ W). If their respective latitudes are $30.15°$ N and $49.55°$ N and if the Earth's radius is approximately 3960 miles, estimate the distance from Austin to Winnipeg.

5. A point P is located on the outer rim of a rotating disk having radius 18 cm. In 10 seconds, the line connecting the disk's center to P sweeps out an angle of $60°$.
   (a) What is the area of the sector swept out by P during 10 seconds? During 15 seconds?
   (b) What is the disk's angular speed?
   (c) What is the linear speed of P?

Chapter 5 / Trigonometric Functions and Their Graphs

6. Evaluate the six trigonometric functions at a standard-position angle θ in each of the following situations:
   (a) $(-5, 12)$ is on the terminal side of θ.
   (b) $\cos \theta = \frac{2}{3}$, $\tan \theta < 0$.
   (c) The terminal side of θ lies along the line $y = \frac{x}{2}$ in the third quadrant.

7. Prove that each of the following functions is even, and draw a conclusion about its graph's symmetry:
   (a) $\cos x$ (b) $\sin x^2$
   (c) $3 - 2 \sec 5x$ (d) $3 \tan |x|$

8. Prove that the graph of each of the following functions is symmetric about the origin:
   (a) $\sin 5x$ (b) $2 \tan \frac{x}{2}$
   (c) $-3 \csc x$ (d) $2 \sin^{-1} 3x$

9. In the given right triangle, find all the trigonometric functions of α.

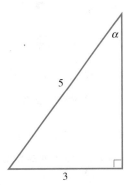

10. Find a value of k such that $(0, -1)$ is on the graph of $y = \cos(x + k)$.

11. Prove that $\frac{\pi}{2}$ is the smallest positive real number a such that $\sin(x + a) = \cos x$ for all real numbers x.

12. Prove that $\sec(\pi - x) = -\sec x$ for all real numbers x.

    (Hint: consider $\dfrac{1}{\cos\left(\frac{\pi}{2} - \left(x - \frac{\pi}{2}\right)\right)}$.)

13. Simplify $\cos\left(x + \frac{\pi}{6}\right) \cos x + \sin\left(x + \frac{\pi}{6}\right) \sin x$. (Hint: apply exercise 26 of section 5.2.)

14. Geometrically identify $\left\{(\cos \theta, \sin \theta) : \frac{\pi}{2} \le \theta \le \frac{3\pi}{2}\right\}$.

15. Find the coordinates of the point on the unit circle determined by the real number $\alpha = -\frac{3\pi}{4}$, and hence determine the trigonometric functions of α.

16. If $\sin \alpha = \frac{\sqrt{3}}{2}$, determine (a) $\sin(\alpha + \pi)$ and
    (b) $\sin\left(\alpha + \frac{\pi}{2}\right)$.

17. Frankie wishes to swim from a point A on a lake shore directly across to Johnnie, who is waiting 100 yards away at point B. Calvin is observing the action from a point C on land nearby. If $\angle ACB$ is a right angle and $\angle ABC$ measures $60°$, how far is Calvin from Frankie? from Johnnie?

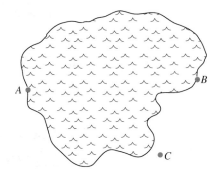

18. A guy wire that is 21 meters long is holding a radio-transmitter antenna and is attached to the antenna 15 meters from the ground. Find θ, the angle of elevation of the guy wire.

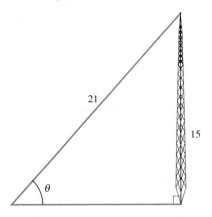

19. Find numerical values (exact, if possible) for each of the following:

(a) $\sin \dfrac{11\pi}{6}$ (b) $\arccos \dfrac{1}{2}$ (c) $\tan^{-1}(-2)$

(d) $\sec \dfrac{3\pi}{4}$ (e) $\operatorname{arcsec}\left(-\dfrac{2\sqrt{3}}{3}\right)$

(f) $\csc(27°12'13'')$

20. Evaluate each of the following (exactly, if possible):
 (a) $\arctan(\tan(-\pi))$ (b) $\sin(\sin^{-1}(-\tfrac{2}{3}))$
 (c) $\sin(\arctan(-2))$ (d) $\cos^{-1}\left(\tan \dfrac{\pi}{6}\right)$

21. (a) Determine which of the six trigonometric functions is increasing over the interval (2, 3).
 (b) Determine which of the functions $\sec^{-1}$, $\csc^{-1}$, $\tan^{-1}$, and $\cot^{-1}$ is increasing over the interval (2, 3).
 (c) Determine which of the functions $\sin^{-1}$ and $\cos^{-1}$ is increasing over the interval $(-\tfrac{1}{2}, -\tfrac{1}{3})$.

22. Determine the period and, if it exists, the amplitude for each of the following periodic functions:
 (a) $\cos \dfrac{x}{2}$ (b) $3 - \sin 2x$
 (c) $\cos^2 x$ (d) $-2 + 7\tan(3x - \pi)$
 (e) $\sec \dfrac{x}{2}$ (f) $-5 + 2\csc(x - \pi)$

23. Determine the domain, range, and, if they exist, the maximum and minimum functional values for each of the following functions:
 (a) $7 - 2\cos(-4x + 2)$ (b) $3\tan^{-1}\dfrac{x}{2}$
 (c) $-2\arcsin(x+2)$ (d) $\dfrac{2}{3} + \dfrac{1}{2}\tan(2x - 4)$
 (e) $\tan |x|$ (f) $\operatorname{arccot}\dfrac{x}{2}$

24. Calculate the phase shift for each of the following expressions:
 (a) $4\sin\left(-\dfrac{x}{2}\right)$ (b) $-5 + 2\tan(x + \pi)$
 (c) $3 - \sec(1 - 2x)$ (d) $2 - \cos(2x + 3)$

◆ 25. Find an expression with phase shift $\dfrac{\pi}{4} - \dfrac{3}{2}$ which describes the same function as $f(x) = 4 + 2\sin(2x + 3)$.

26. (a) Explain why an acceptable inverse function for (a suitably restricted) sin could be developed with range $\left[\dfrac{3\pi}{2}, \dfrac{5\pi}{2}\right]$.
 (b) Speculate as to any possible advantages of the text's arcsin function over the function developed in (a).

27. Show that the substitution $\theta = \arcsin \dfrac{5x}{6}$ changes $\sqrt{36 - 25x^2}$ into $6\cos\theta$.

28. Prove that $\tan(\sin^{-1} x) = \dfrac{x}{\sqrt{1-x^2}}$ whenever $|x| < 1$.

In exercises 29–34, graph each of the given functions. (You may wish to use a graphing calculator or computer graphing program to help you or to provide a "rough check" of your work.)

29. $f(x) = \cos(2x - \pi)$
30. $g(x) = 1 - \dfrac{1}{2}\tan(2x + 3\pi)$
31. $h(x) = 2\left|\sec \dfrac{x}{2}\right| + 1$ ◆32. $k(x) = x\sin x$
33. $m(x) = \sin x - \cos x$ ◆34. $n(x) = \dfrac{\sin x}{e^x}$

35. Find constants A, B, C, and D such that the given curve is the graph of $y = A\cos(Bx - C) + D$.

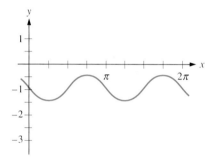

In exercises 36–46, solve the given equations or inequalities in the interval $[0, 2\pi)$.

36. $\sin 2\theta = \dfrac{\sqrt{3}}{2}$ 37. $\cos x \sec x = 1$
38. $3\tan x - 4 = 0$ 39. $\cos^2\theta + \cos\theta = 0$
40. $\cos^2\theta = 1 - \sin^2\theta$ 41. $\sin x + \cos^2 x = \dfrac{5}{4}$
42. $2\csc \dfrac{x}{2} - \cot \dfrac{x}{2} = 0$
43. $\sin x + \cos x = 1$ (Hint: solve for $\sin x$, square both sides, and check by substituting possible solutions in the given equation.)
44. $2\sin \dfrac{\theta}{2} \le \sqrt{3}$ 45. $7 - 3\cos(2\theta - \pi) > 5$
46. $\sin x < \cos x$

47. Verify that $\dfrac{\pi}{4}$ and $\dfrac{5\pi}{4}$ are solutions in $[0, 2\pi)$ for the conditional equation $\sin x + \cot x = \cos x + \tan x$. Then prove that any *other* solution satisfies $\sin x + \cos x = \sin x \cdot \cos x$. (Remark: such values of x do exist!)

48. In the accompanying figure, show that
 (a) D has coordinates $(\tfrac{3}{2}, \tfrac{1}{2})$.

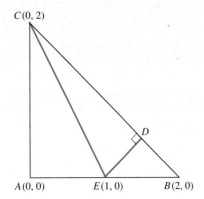

(b) $\tan(\angle ACE) = \frac{1}{2}$ and $\tan(\angle ECB) = \frac{1}{3}$

(c) $\arctan \frac{1}{2} + \arctan \frac{1}{3} = \frac{\pi}{4}$

49. Hoping to win the approval of Rowena's parents, Romeo sends them a letter containing an equation,

$$(\sin x - \cos x)(\sin x \sec x + 1) = \sec x - 2 \cos x,$$

that only he has seen. Has Romeo managed to establish his own identity?

Chapter 6

Analytic Trigonometry

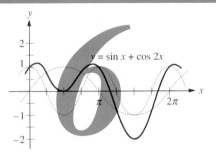

In chapter 5, we discussed the reciprocal identities, the ratio or quotient identities, and a Pythagorean identity. Additional material on identities appeared in the exercises of sections 2, 7, 8, and 9 of chapter 5. In this chapter we shall establish even more trigonometric identities. These will include identities for the sums, differences, and multiples of angles. This work will expand our ability to graph trigonometrically defined functions and to solve trigonometric equations and inequalities. It will also be used in sections 6.6 and 6.7 to expand our skills in performing the arithmetic operations on complex numbers.

6.1 BASIC IDENTITIES

An **identity** is an equation that holds for all values of the variables for which all the expressions in the equation are meaningful. Identities that we have studied before include the following three trigonometric identities from chapter 5:

$$\sin(-\theta) = -\sin\theta$$
$$\cos(-\theta) = \cos\theta$$

and
$$\sin\theta = \frac{1}{\csc\theta}.$$

An example of an identity that is not a trigonometric identity is $\ln xy = \ln x + \ln y$, developed in section 4.3.

Let's look again at the Pythagorean identity developed in section 5.2:

$$\sin^2 \theta + \cos^2 \theta = 1.$$

Divide both sides of this equation by $\cos^2 \theta$ to get

$$\frac{\sin^2 \theta}{\cos^2 \theta} + \frac{\cos^2 \theta}{\cos^2 \theta} = \frac{1}{\cos^2 \theta}$$

$$\tan^2 \theta + 1 = \sec^2 \theta.$$

If we had divided both sides by $\sin^2 \theta$, we would have found

$$\frac{\sin^2 \theta}{\sin^2 \theta} + \frac{\cos^2 \theta}{\sin^2 \theta} = \frac{1}{\sin^2 \theta}$$

$$1 + \cot^2 \theta = \csc^2 \theta.$$

Adding these last two identities to the ones developed in the last chapter gives us a list of eight basic identities. These are listed below as a handy reference.

this is nothing new!! :)

Reciprocal Identities

$$\csc \theta = \frac{1}{\sin \theta} \quad \text{or} \quad \sin \theta = \frac{1}{\csc \theta}$$

$$\sec \theta = \frac{1}{\cos \theta} \quad \text{or} \quad \cos \theta = \frac{1}{\sec \theta}$$

$$\cot \theta = \frac{1}{\tan \theta} \quad \text{or} \quad \tan \theta = \frac{1}{\cot \theta}.$$

Quotient or Ratio Identities

$$\frac{\sin \theta}{\cos \theta} = \tan \theta$$

$$\frac{\cos \theta}{\sin \theta} = \cot \theta.$$

REMEMBER: **Pythagorean Identities**

$$\sin^2 \theta + \cos^2 \theta = 1$$

$$\tan^2 \theta + 1 = \sec^2 \theta$$

$$\cot^2 \theta + 1 = \csc^2 \theta.$$

Proving Identities

The above eight basic identities can be used to establish other identities. Your ability to prove identities will depend to a great extent on your familiarity with the eight basic ones. The majority of this chapter will be spent developing other identities or verifying that some identities are true. While this may seem to be

Basic Identities

just a mental exercise, many of these additional identities will be used throughout the chapter to help solve trigonometric equations and inequalities. As you continue through this book, and then through calculus, these later identities will also enable you to solve problems that would otherwise have been extremely difficult, if not impossible, to solve.

A traditional method for proving an identity is to transform one side of the equation into the other. This method will be used whenever practical. However, you can always simplify by passing to an algebraically equivalent problem in any of the following ways. You may add or subtract the same quantity or term to both sides; and you may clear denominators by multiplying through by some or all of the denominators. One reason that some texts or teachers frown on these methods of simplifying the problem is that, while our advice will often lead to quicker *proofs* of suggested identities, only the traditional method of systematically altering one side at a time will lead you to *discover* new identities.

EXAMPLE 6.1   Prove the identity $\cos\theta \csc\theta = \cot\theta$.

Solution   **Hint: it is often helpful to write all the trigonometrically defined expressions in terms of the sine and cosine functions.**

We write the left-hand side of this identity in terms of the sine and cosine functions:

$$\cos\theta \csc\theta = \cos\theta \, \frac{1}{\sin\theta}$$

$$= \frac{\cos\theta}{\sin\theta}$$

$$= \cot\theta$$

and we have simplified the left-hand side to the right-hand side.   ◀

EXAMPLE 6.2   Prove the identity $|\tan\theta| = \sqrt{\dfrac{1 + \tan^2\theta}{1 + \cot^2\theta}}$.

Solution   **Hint: begin with the more complicated side and simplify it.**

The right-hand side of this identity is more complicated than the left-hand side. We shall use the Pythagorean identities, replacing $1 + \tan^2\theta$ with $\sec^2\theta$ and $1 + \cot^2\theta$ with $\csc^2\theta$.

$$\sqrt{\frac{1 + \tan^2\theta}{1 + \cot^2\theta}} = \sqrt{\frac{\sec^2\theta}{\csc^2\theta}}$$

$$= \sqrt{\left(\frac{\sec\theta}{\csc\theta}\right)^2}$$

$$= \left|\frac{\sec\theta}{\csc\theta}\right|$$

NEXT PAGE...

$$= \left|\frac{1/\cos\theta}{1/\sin\theta}\right|$$

$$= \left|\frac{\sin\theta}{\cos\theta}\right|$$

$$= |\tan\theta|. \blacktriangleleft$$

EXAMPLE 6.3   Prove the identity $\dfrac{1}{1-\sin x} - \dfrac{1}{1+\sin x} = 2\sin x \sec^2 x$.

Solution   **Hint: begin with the more complicated side and complete the indicated operations.**

The left-hand side of this identity is the more complicated side. We begin by rewriting the fractions with a common denominator and then combining the two terms.

$$\frac{1}{1-\sin x} - \frac{1}{1+\sin x} = \frac{1}{(1-\sin x)}\frac{(1+\sin x)}{(1+\sin x)} - \frac{1}{(1+\sin x)}\frac{(1-\sin x)}{(1-\sin x)}$$

$$= \frac{1+\sin x}{1-\sin^2 x} - \frac{1-\sin x}{1-\sin^2 x}$$

$$= \frac{2\sin x}{\cos^2 x} \quad \longrightarrow \text{remember: } \cos^2\theta + \sin^2\theta = 1$$
$$\text{well, } \cos^2 x \text{ is same as } 1-\sin^2 x$$

$$= 2\sin x\left(\frac{1}{\cos x}\right)^2$$

$$= 2\sin x \sec^2 x. \blacktriangleleft$$

EXAMPLE 6.4   Prove the identity $\sec^2 \alpha = \dfrac{\sin^4 \alpha - \cos^4 \alpha}{\sin^2 \alpha - \cos^2 \alpha} + \tan^2 \alpha$.

Solution   **Hint: factoring will often simplify an expression.**

The numerator of the rational expression can be factored:

$$\sin^4 \alpha - \cos^4 \alpha = (\sin^2 \alpha - \cos^2 \alpha)(\sin^2 \alpha + \cos^2 \alpha).$$

Since one of these factors is the same as the denominator, the rational expression can next be simplified by cancellation.

$$\frac{\sin^4 \alpha - \cos^4 \alpha}{\sin^2 \alpha - \cos^2 \alpha} + \tan^2 \alpha = \frac{(\sin^2 \alpha - \cos^2 \alpha)(\sin^2 \alpha + \cos^2 \alpha)}{\sin^2 \alpha - \cos^2 \alpha} + \tan^2 \alpha$$

$$= (\sin^2 \alpha + \cos^2 \alpha) + \tan^2 \alpha$$

$$= 1 + \tan^2 \alpha$$

$$= \sec^2 \alpha. \blacktriangleleft$$

Basic Identities

> **EXAMPLE 6.5**

Prove the identity $\dfrac{\cos \beta}{1 + \sin \beta} = \sec \beta - \tan \beta$.

Solution

Hint: multiply the numerator and denominator of a rational expression by the same nonzero quantity.

Notice that $\cos \beta \neq 0$ for any relevant value of β, since $\sec \beta$ is defined. In particular, $\sin \beta \neq 1$, and so $\dfrac{1}{1 - \sin \beta}$ makes sense.

$$\dfrac{\cos \beta}{1 + \sin \beta} = \dfrac{\cos \beta}{1 + \sin \beta} \cdot \dfrac{1 - \sin \beta}{1 - \sin \beta}$$

$$= \dfrac{\cos \beta (1 - \sin \beta)}{1 - \sin^2 \beta}$$

$$= \dfrac{\cos \beta (1 - \sin \beta)}{\cos^2 \beta}$$

$$= \dfrac{1 - \sin \beta}{\cos \beta}$$

$$= \dfrac{1}{\cos \beta} - \dfrac{\sin \beta}{\cos \beta}$$

$$= \sec \beta - \tan \beta.$$

One may attack a trigonometric identity with any reversible operation, such as multiplying by a nonzero quantity. Using this idea, we next give another way to do example 6.5. In this example, $\sin \beta$ is never -1, so the problem is equivalent to proving that, if $1 + \sin \beta \neq 0$, then

$$\cos \beta = (\sec \beta - \tan \beta)(1 + \sin \beta).$$

The right-hand side reduces to

$$\sec \beta + \sec \beta \sin \beta - \tan \beta - \tan \beta \sin \beta = \dfrac{1}{\cos \beta} + \dfrac{\sin \beta}{\cos \beta} - \dfrac{\sin \beta}{\cos \beta} - \dfrac{\sin^2 \beta}{\cos \beta}$$

$$= \dfrac{1 - \sin^2 \beta}{\cos \beta}$$

$$= \dfrac{\cos^2 \beta}{\cos \beta}$$

$$= \cos \beta.$$ ◀

> **EXAMPLE 6.6**

Prove the identity $\dfrac{\csc \gamma}{\csc \gamma + 1} = \dfrac{1}{1 + \sin \gamma}$.

Solution

This identity will be proved using two methods. The first method is the more "traditional" one, since it works with only the expression on the left-hand side.

First Method: We shall rewrite each expression on the left-hand side in terms of sin γ and use properties of rational expressions to simplify the resulting quantity.

$$\frac{\csc \gamma}{\csc \gamma + 1} = \frac{\dfrac{1}{\sin \gamma}}{\dfrac{1}{\sin \gamma} + 1}$$

$$= \frac{\dfrac{1}{\sin \gamma}}{\dfrac{1 + \sin \gamma}{\sin \gamma}}$$

$$= \frac{1}{\sin \gamma} \cdot \frac{\sin \gamma}{1 + \sin \gamma}$$

$$= \frac{1}{1 + \sin \gamma}.$$

Second Method: In the second method, we shall use any reversible operation. We obtain an equivalent problem by multiplying through by $(\csc \gamma + 1)(1 + \sin \gamma)$. Our new task is to prove that if $\csc \gamma + 1$ and $1 + \sin \gamma$ are nonzero, then

$$\csc \gamma \, (1 + \sin \gamma) = \csc \gamma + 1.$$

The new left-hand side is just $\csc \gamma \, (1 + \sin \gamma) = \csc \gamma + \csc \gamma \sin \gamma = \csc \gamma + 1$, which is the right-hand side. ◀

The traditional method for verifying an identity is to transform one side of the equation into the other. There are times when both sides involve complicated expressions and this approach may not seem practical. We may then transform each side of the equation into the same expression, being careful that each step is reversible.

As we saw in the last chapter, especially in section 5.8, many trigonometric equations are not identities. While they may be valid for some values of the variable, they are not true for all values. To show that a trigonometric equation is not an identity, it is necessary to find values of the variables which do not satisfy the equation. Such concrete values serve as a **counterexample.** When searching for a value to construct a counterexample, it is best to avoid the quadrantal angles because they often result in undefined trigonometric functions. You had practice in finding counterexamples in exercise 19 of section 5.2 and exercise 4(d) of section 5.5. Here is another example of that type.

▶ **EXAMPLE 6.7**

Show that $\dfrac{\sin \delta - \cos \delta}{\cos \delta} = 1 - \tan \delta$ is not an identity.

Solution

This equation is true for some values. For example, if $\delta = \frac{\pi}{4}$, then $\sin \delta = \cos \delta = \frac{\sqrt{2}}{2}$ and $\tan \delta = 1$, so both sides of the equation are equal to 0 in this case. However, let's see what happens if we put $\delta = \frac{\pi}{3}$. Then $\sin \delta = \frac{\sqrt{3}}{2}$, $\cos \delta = \frac{1}{2}$, and $\tan \delta = \sqrt{3}$. In this case, the equation in question would give

$$\frac{\sin \delta - \cos \delta}{\cos \delta} \stackrel{?}{=} 1 - \tan \delta$$

$$\frac{\sqrt{3}/2 - 1/2}{1/2} \stackrel{?}{=} 1 - \sqrt{3}$$

$$\sqrt{3} - 1 \neq 1 - \sqrt{3}.$$

Notice that in the first two lines, a question mark was put over the "is equal to" symbol to indicate that the equality was being questioned. In the last line, we had two unequal values. Thus, we can conclude that $\frac{\sin \delta - \cos \delta}{\cos \delta} = 1 - \tan \delta$ is not an identity. (You can check that choosing $\delta = 0$ leads to another counterexample.) ◀

Exercises 6.1

In exercises 1–4, rewrite the given expression as a fraction involving $\sin x$ and $\cos x$.

1. $\sin x + \tan x \cos x$
2. $\csc x + \tan x$
3. $\sec^2 x + \tan^2 x$
4. $\tan^2 x + \cot^2 x$

In exercises 5–39, prove that the given equation is an identity.

5. $\cos x \tan x = \sin x$
6. $\tan \theta = \sin \theta \sec \theta$
7. $\cos \theta \sec \theta = 1$
8. $1 = \sin x \cot x \sec x$
9. $\cos x + \sin x \tan x = \sec x$
10. $\cot \alpha - \tan \beta = (\cos \alpha \cos \beta - \sin \alpha \sin \beta) \csc \alpha \sec \beta$
11. $\frac{1 + \sin \alpha}{\cos \alpha} = \frac{\cos \alpha}{1 - \sin \alpha}$
12. $2 \cos^2 \theta - 1 = 1 - 2 \sin^2 \theta$
13. $\frac{1}{\sec \theta - \tan \theta} = \sec \theta + \tan \theta$
14. $\frac{1 + \cos x}{\sin x} = \frac{1}{\csc x - \cot x}$
15. $\frac{1 + \tan x}{1 - \tan x} = \frac{1 + \cot x}{-1 + \cot x}$
16. $\frac{\sin \alpha - \cos \alpha}{\sin \alpha \cos \alpha} = \frac{\tan \alpha - \cot \alpha}{\cos \alpha + \sin \alpha}$
17. $\csc \alpha - \sin \alpha = \cos \alpha \cot \alpha$
18. $\frac{\tan^2 \theta}{1 + \sec \theta} = \frac{1 - \cos \theta}{\cos \theta}$

19. $(\sec \theta - \tan \theta)^2 = \frac{1 - \sin \theta}{1 + \sin \theta}$
20. $1 = \left(\frac{\sin^3 x}{\tan^6 x}\right)^2 \left(\frac{\csc x}{\cot^2 x}\right)^6$
21. $(\csc x - \sin x)^2 = \cot^2 x - \cos^2 x$
22. $\frac{2 \sin \beta \cos \beta}{\sin \beta + \cos \beta + 1} = \sin \beta + \cos \beta - 1$
23. $\cos^2 \beta = \frac{1 + \cos^2 \beta}{1 + \sec^2 \beta}$
24. $\frac{1}{1 - \cos \alpha} + \frac{1}{1 + \cos \alpha} = 2 \csc^2 \alpha$
25. $2 \csc \alpha = \frac{\sin \alpha}{1 + \cos \alpha} + \frac{1 + \cos \alpha}{\sin \alpha}$
26. $\frac{1}{\sec \theta} - \frac{1}{\csc \theta} = \frac{\cot^2 \theta - \tan^2 \theta}{(\csc \theta + \sec \theta) \csc \theta \sec \theta}$
27. $\frac{1}{\tan \gamma + \sec \gamma} - \frac{1}{\tan \gamma - \sec \gamma} = 2 \sec \gamma$
28. $\tan^2 \theta - \cot^2 \theta = \frac{\sec \theta}{\cos \theta} - \frac{\csc \theta}{\sin \theta}$
29. $\frac{\csc^4 x - 1}{\cot^2 x + 2} = \cot^2 x$
30. $\tan^4 x = \sec^4 x - 2 \sec^2 x + 1$
31. $\sqrt{\cos \beta} (\sec \beta - \cos \beta) = |\sin \beta|$

32. $|\sin \beta| = \sqrt{\dfrac{\sec^2 \beta - 1}{\sec^2 \beta}}$

33. $1 + \tan^2(-\gamma) = \sec^2(-\gamma)$

34. $\tan \gamma^2 = \dfrac{\sec \gamma^2}{\csc \gamma^2}$

35. $\ln \sin \theta = -\ln \csc \theta$

36. $\ln(\sec \theta + \tan \theta) + \ln(\sec \theta - \tan \theta) = 0$

37. $\ln |\csc x - \cot x| = \ln |1 - \cos x| + \ln |\csc x|$

38. $\log_{10}(10^{\sin x}) = \sin x$

39. $2^{\log_2 (\cos^2 \alpha)} = \cos^2 \alpha$

40. (a) Show that the substitution $\theta = \arcsin \dfrac{x}{2}$ changes $\sqrt{4 - x^2}$ into $2 \cos \theta$. (Hint: first show that $x = 2 \sin \theta$. Is $\cos \theta$ positive?)

   (b) If A and B are positive real numbers, prove that the substitution $\theta = \sin^{-1}\left(\dfrac{Bx}{A}\right)$ changes $\sqrt{A^2 - B^2 x^2}$ into $A \cos \theta$.

41. (a) Show that the substitution $\theta = \tan^{-1}\left(\dfrac{u}{\sqrt{7}}\right)$ changes $\sqrt{7 + u^2}$ into $\sqrt{7} \sec \theta$.

   (b) If A and B are positive real numbers, prove that the substitution $\theta = \tan^{-1}\left(\dfrac{Bu}{A}\right)$ changes $\sqrt{A^2 + B^2 u^2}$ into $A \sec \theta$.

42. Find a substitution $\theta = \ldots$ that changes $9u^2 - 4$ into $4 \tan^2 \theta$.

43. (a) Prove that there are no positive constants a, b, and c for which $a^2 \tan^2 \theta + c^2 \cos^2 \theta = b^2$ is an identity.

   (b) Using (a), prove that there are no positive constants a, b, and c for which the substitution $\theta = \tan^{-1}\dfrac{bx}{a}$ changes $\sqrt{1 - x^2}$ into $c \cos \theta$.

In exercises 44–48, prove that each given equation is *not* an identity by producing a counterexample to it.

44. $\cos 2\theta = 2 \cos \theta$

45. $\sqrt{1 - \sin^2 x} = \cos x$

46. $\cos(\alpha - \beta) = \cos \alpha - \cos \beta$

47. $\tan \theta = \dfrac{\cos \theta}{\sin \theta}$

48. $\csc^2 x + 1 = \cot^2 x$

In exercises 49–53, solve the given conditional equations or inequalities in $[0, 2\pi)$.

49. $\tan^2 x + \sec x - 1 = 0$

50. $\sec^2 \theta + \tan \theta - 1 = 0$

51. $\sin\left(\dfrac{\theta}{2}\right) \sec\left(\dfrac{\theta}{2}\right) > \dfrac{1}{2}$

52. $3 \csc^2 x - 2 \cot^2 x > \dfrac{5}{2}$

53. $\tan^2 x - \sec x \le 1$

54. Criticize the following *reasoning*, which attempts to show that the equation $\dfrac{\sqrt{1 + \tan^2 \theta}}{\tan \theta} = \csc \theta$ is an identity.

$$\dfrac{\sqrt{1 + \tan^2 \theta}}{\tan \theta} = \dfrac{\sqrt{\sec^2 \theta}}{\tan \theta} = \dfrac{\sec \theta}{\tan \theta} = \dfrac{1}{\cos \theta} \cdot \dfrac{\cos \theta}{\sin \theta}$$

$$= \dfrac{1}{\sin \theta} = \csc \theta.$$

55. Show by counterexample that the equation considered in exercise 54 is not an identity.

56. Prove that the equation

$$3 \sin^4 \theta \tan \theta - 6 \sin^2 \theta + 3 \sin^2 \theta \tan \theta \cos^2 \theta$$
$$= 6 \cos^2 \theta + 3 \sin^2 \theta \tan \theta - 6$$

is an identity.

6.2 THE ADDITION AND SUBTRACTION IDENTITIES

In exercises 44 and 46 at the end of the last section, you saw that $\cos 2\theta \ne 2 \cos \theta$ and $\cos(\alpha - \beta) \ne \cos \alpha - \cos \beta$. In this section, we shall develop some identities that answer the underlying question posed by those two exercises, namely how to simplify (or rewrite) a trigonometric expression of a sum or difference.

cos $(\alpha + \beta)$ and cos $(\alpha - \beta)$

We begin by developing an identity for $\cos(\alpha - \beta)$, and then use it to develop an identity for $\cos(\alpha + \beta)$. Consider the unit circle—that is, the circle with radius 1 and center at the origin, as shown in figure 6.1(a). In figure 6.1(a), point A is a point on the circle, B is any other point on the circle, and the measure of $\angle AOB$ is $\alpha - \beta$. If we rotate $\triangle AOB$ about O through the angle $-\beta$, we obtain $\triangle COD$, as shown in figure 6.1(b).

Figure 6.1

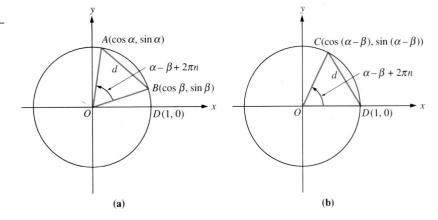

We first examine figure 6.1(a). We see that A has the coordinates $(\cos \alpha, \sin \alpha)$ and $B = (\cos \beta, \sin \beta)$. Using the distance formula, we find

$$d(A, B) = \sqrt{(\cos \beta - \cos \alpha)^2 + (\sin \beta - \sin \alpha)^2}.$$

Squaring both sides, we obtain

$$\begin{aligned}[d(A, B)]^2 &= (\cos \beta - \cos \alpha)^2 + (\sin \beta - \sin \alpha)^2 \\ &= \cos^2 \beta - 2 \cos \alpha \cos \beta + \cos^2 \alpha + \sin^2 \beta - 2 \sin \alpha \sin \beta + \sin^2 \alpha \\ &= (\cos^2 \beta + \sin^2 \beta) + (\cos^2 \alpha + \sin^2 \alpha) - 2 \cos \alpha \cos \beta - 2 \sin \alpha \sin \beta \\ &= 1 + 1 - 2 \cos \alpha \cos \beta - 2 \sin \alpha \sin \beta \\ &= 2 - 2(\cos \alpha \cos \beta + \sin \alpha \sin \beta).\end{aligned}$$

Next, we examine figure 6.1(b). We see that C has the coordinates $(\cos (\alpha - \beta), \sin (\alpha - \beta))$ and D the coordinates $(1, 0)$. Once again, using the distance formula produces

$$\begin{aligned}[d(C, D)]^2 &= [\cos (\alpha - \beta) - 1]^2 + [\sin (\alpha - \beta)]^2 \\ &= \cos^2 (\alpha - \beta) - 2 \cos (\alpha - \beta) + 1 + \sin^2 (\alpha - \beta) \\ &= [\sin^2 (\alpha - \beta) + \cos^2 (\alpha - \beta)] - 2 \cos (\alpha - \beta) + 1 \\ &= 1 - 2 \cos (\alpha - \beta) + 1 \\ &= 2 - 2 \cos (\alpha - \beta).\end{aligned}$$

Since $\triangle AOB \cong \triangle COD$ (by the SAS congruence criterion), we know that $AB = CD$ and so

$$[d(C, D)]^2 = [d(A, B)]^2.$$

Thus,
$$2 - 2\cos(\alpha - \beta) = 2 - 2(\cos\alpha\cos\beta + \sin\alpha\sin\beta)$$
$$-2\cos(\alpha - \beta) = -2(\cos\alpha\cos\beta + \sin\alpha\sin\beta).$$

Dividing both sides by -2 results in the following identity:

Subtraction Identity for Cosine

$$\cos(\alpha - \beta) = \cos\alpha\cos\beta + \sin\alpha\sin\beta.$$

Many proofs of the above formula for $\cos(\alpha - \beta)$ have hidden assumptions. Our proof is no different, for we assumed implicitly that AOB gives a genuine triangle. What if it doesn't? In that case, $\angle AOD$ and $\angle BOD$ are coterminal, so that $\alpha - \beta = 2\pi n$ for some integer n. Then we find $\sin\alpha = \sin\beta$, $\cos\alpha = \cos\beta$, and so

$$\cos\alpha\cos\beta + \sin\alpha\sin\beta = \cos^2\alpha + \sin^2\alpha = 1 = \cos(\alpha - \beta).$$

Thus, the above formula for $\cos(\alpha - \beta)$ is valid for *all* real numbers α and β.

In order to determine a formula for $\cos(\alpha + \beta)$, we replace β with $-\beta$ in the above formula, thus obtaining

$$\cos(\alpha + \beta) = \cos[\alpha - (-\beta)] = \cos\alpha\cos(-\beta) + \sin\alpha\sin(-\beta).$$

Since the sine function is odd, $\sin(-\beta) = -\sin\beta$, and since the cosine function is even, $\cos(-\alpha) = \cos\alpha$. With these substitutions, the previous equation becomes

$$\cos(\alpha + \beta) = \cos\alpha\cos\beta + \sin\alpha(-\sin\beta)$$
$$= \cos\alpha\cos\beta - \sin\alpha\sin\beta.$$

This yields the second important identity of this section.

Addition Identity for Cosine

$$\cos(\alpha + \beta) = \cos\alpha\cos\beta - \sin\alpha\sin\beta.$$

▶ **EXAMPLE 6.8** Verify that $\cos\dfrac{\pi}{3}\cos\dfrac{\pi}{6} + \sin\dfrac{\pi}{3}\sin\dfrac{\pi}{6} = \cos\dfrac{\pi}{6}$.

Solution We shall verify this using two methods: a slow way and a fast way.

The slow way uses the trigonometric values from the last chapter. Proceeding in this manner, we obtain

$$\cos\dfrac{\pi}{3}\cos\dfrac{\pi}{6} + \sin\dfrac{\pi}{3}\sin\dfrac{\pi}{6} = \left(\dfrac{1}{2}\right)\left(\dfrac{\sqrt{3}}{2}\right) + \left(\dfrac{\sqrt{3}}{2}\right)\left(\dfrac{1}{2}\right)$$
$$= \dfrac{\sqrt{3}}{4} + \dfrac{\sqrt{3}}{4} = \dfrac{\sqrt{3}}{2} = \cos\dfrac{\pi}{6}.$$

The fast method recognizes that

$$\cos\dfrac{\pi}{3}\cos\dfrac{\pi}{6} + \sin\dfrac{\pi}{3}\sin\dfrac{\pi}{6} = \cos\left(\dfrac{\pi}{3} - \dfrac{\pi}{6}\right) = \cos\dfrac{\pi}{6}.$$ ◀

The Addition and Subtraction Identities

EXAMPLE 6.9   Find the exact value of $\cos\left(\dfrac{7\pi}{12}\right)$.

Solution   After a little calculation we notice that $\dfrac{\pi}{3} + \dfrac{\pi}{4} = \dfrac{7\pi}{12}$. Hence,

$$\cos\frac{7\pi}{12} = \cos\left(\frac{\pi}{3}+\frac{\pi}{4}\right) = \cos\frac{\pi}{3}\cos\frac{\pi}{4} - \sin\frac{\pi}{3}\sin\frac{\pi}{4}$$

$$= \frac{1}{2}\cdot\frac{\sqrt{2}}{2} - \frac{\sqrt{3}}{2}\cdot\frac{\sqrt{2}}{2} = \frac{(1-\sqrt{3})\sqrt{2}}{4}. \blacktriangleleft$$

EXAMPLE 6.10   Find the exact value of $\cos\left(\dfrac{\pi}{12}\right)$.

Solution   This time, we notice that $\dfrac{\pi}{3} - \dfrac{\pi}{4} = \dfrac{\pi}{12}$. Hence,

$$\cos\frac{\pi}{12} = \cos\left(\frac{\pi}{3}-\frac{\pi}{4}\right) = \cos\frac{\pi}{3}\cos\frac{\pi}{4} + \sin\frac{\pi}{3}\sin\frac{\pi}{4}$$

$$= \frac{1}{2}\cdot\frac{\sqrt{2}}{2} + \frac{\sqrt{3}}{2}\cdot\frac{\sqrt{2}}{2} = \frac{(1+\sqrt{3})\sqrt{2}}{4}. \blacktriangleleft$$

EXAMPLE 6.11   Verify the identity $\cos\left(\dfrac{\pi}{2} - \theta\right) = \sin\theta$.

Solution   Using the formula for the cosine of a difference, we see that

$$\cos\left(\frac{\pi}{2} - \theta\right) = \cos\frac{\pi}{2}\cos\theta + \sin\frac{\pi}{2}\sin\theta$$

$$= 0\cdot\cos\theta + 1\cdot\sin\theta$$

$$= \sin\theta. \blacktriangleleft$$

Example 6.11 establishes one of the cofunction identities that were stated in section 5.2. All the cofunction identities may now be proved in the above style (and you will be asked to prove them in the exercises). A clumsier, but complete, proof of them was sketched in exercise 27 of section 5.2.

You should verify, reasoning as in example 6.11, that $\cos(\theta + \pi) = -\cos\theta = \cos(\theta - \pi)$.

sin $(\alpha + \beta)$ and sin $(\alpha - \beta)$

We proceed in the spirit of exercise 25 in section 5.2. First, note that if we had let $\theta = \dfrac{\pi}{2} - \phi$ in example 6.11, we would have found the cofunction identity

$$\sin\left(\frac{\pi}{2} - \phi\right) = \cos\phi.$$

Now, suppose we let $\theta = \alpha + \beta$. Then, according to the cofunction identities and the formula for the cosine of a difference, we have

$$\sin(\alpha + \beta) = \cos\left(\frac{\pi}{2} - [\alpha + \beta]\right)$$

$$= \cos\left(\left[\frac{\pi}{2} - \alpha\right] - \beta\right)$$

$$= \cos\left(\frac{\pi}{2} - \alpha\right)\cos\beta + \sin\left(\frac{\pi}{2} - \alpha\right)\sin\beta$$

$$= \sin\alpha\cos\beta + \cos\alpha\sin\beta.$$

We have thus derived the third important identity of this section.

Addition Identity for Sine

$$\sin(\alpha + \beta) = \sin\alpha\cos\beta + \cos\alpha\sin\beta.$$

In order to determine the sine of the difference of two angles, we rewrite $\alpha + \beta$ as $\alpha - (-\beta)$ and use the fact that sine is an odd function and cosine is an even function, to obtain

$$\sin(\alpha - [-\beta]) = \sin\alpha\cos(-\beta) + \cos\alpha\sin(-\beta)$$

$$= \sin\alpha\cos\beta - \cos\alpha\sin\beta.$$

Thus, we have a fourth identity.

Subtraction Identity for Sine

$$\sin(\alpha - \beta) = \sin\alpha\cos\beta - \cos\alpha\sin\beta.$$

Special cases of these formulas include $\sin(\theta - \pi) = -\sin\theta = \sin(\theta + \pi)$. Some additional applications are indicated next.

▶ **EXAMPLE 6.12** Find the exact value of $\sin\left(\frac{7\pi}{12}\right)$.

Solution We write $\frac{7\pi}{12}$ as $\frac{\pi}{3} + \frac{\pi}{4}$. Hence,

$$\sin\frac{7\pi}{12} = \sin\left(\frac{\pi}{3} + \frac{\pi}{4}\right)$$

$$= \sin\frac{\pi}{3}\cos\frac{\pi}{4} + \cos\frac{\pi}{3}\sin\frac{\pi}{4}$$

$$= \frac{\sqrt{3}}{2} \cdot \frac{\sqrt{2}}{2} + \frac{1}{2} \cdot \frac{\sqrt{2}}{2} = \frac{(\sqrt{3} + 1)\sqrt{2}}{4}.$$

◀

The Addition and Subtraction Identities

EXAMPLE 6.13 Show that $\sin(x+y)\cos y - \cos(x+y)\sin y = \sin x$.

Solution One way to proceed is to work with the left-hand side of the equation and use the expansion identities for $\sin(\alpha+\beta)$ and $\cos(\alpha+\beta)$.

$$\begin{aligned}\sin(x+y)&\cos y - \cos(x+y)\sin y \\ &= (\sin x \cos y + \cos x \sin y)\cos y - (\cos x \cos y - \sin x \sin y)\sin y \\ &= \sin x \cos^2 y + \cos x \sin y \cos y - (\cos x \cos y \sin y - \sin x \sin^2 y) \\ &= \sin x \cos^2 y + \sin x \sin^2 y \\ &= \sin x(\cos^2 y + \sin^2 y) \\ &= \sin x.\end{aligned}$$

The above proof was rather long and arduous and certainly provided practice with the expansion identities for $\sin(\alpha+\beta)$ and $\cos(\alpha+\beta)$. It was not, however, a very sophisticated or elegant proof. A much quicker proof results if you recognize that the left-hand side of the equation is just the expansion of $\sin[(x+y)-y] = \sin x$.

EXAMPLE 6.14 Prove that $\sin\left(\cot^{-1} 2 + \sin^{-1}\dfrac{5}{13}\right) = \dfrac{22}{13\sqrt{5}}$.

Solution Let $\alpha = \cot^{-1} 2$ and $\beta = \sin^{-1}\dfrac{5}{13}$. Then the left-hand side of the given identity is $\sin(\alpha+\beta)$.

From the definition of α, we have $\cot \alpha = 2$. Using this, we can sketch a right triangle containing α, as in figure 6.2. Then, using Pythagoras' Theorem and the definitions of sin and cos, we find that $\sin \alpha = \dfrac{1}{\sqrt{5}}$ and $\cos \alpha = \dfrac{2}{\sqrt{5}}$.

Similarly, the definition of β leads to $\sin \beta = \dfrac{5}{13}$, and using the Pythagorean identity, we find that $\cos \beta = \dfrac{12}{13}$. Thus,

$$\begin{aligned}\sin(\alpha+\beta) &= \sin\alpha\cos\beta + \cos\alpha\sin\beta \\ &= \dfrac{1}{\sqrt{5}}\cdot\dfrac{12}{13} + \dfrac{2}{\sqrt{5}}\cdot\dfrac{5}{13} \\ &= \dfrac{22}{13\sqrt{5}}.\end{aligned}$$

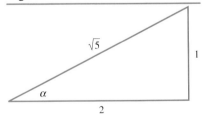

Figure 6.2

Thus, we have verified that $\sin\left(\cot^{-1} 2 + \sin^{-1}\dfrac{5}{13}\right) = \dfrac{22}{13\sqrt{5}}$.

$\tan(\alpha+\beta)$ and $\tan(\alpha-\beta)$

For the final two identities in this section, we shall develop identities for $\tan(\alpha+\beta)$ and $\tan(\alpha-\beta)$.

In order to develop $\tan(\alpha + \beta)$, we use the quotient identity $\tan\theta = \dfrac{\sin\theta}{\cos\theta}$ and the above addition identities for sin and cos.

$$\tan(\alpha + \beta) = \frac{\sin(\alpha + \beta)}{\cos(\alpha + \beta)}$$

$$= \frac{\sin\alpha\cos\beta + \cos\alpha\sin\beta}{\cos\alpha\cos\beta - \sin\alpha\sin\beta}.$$

Dividing each term of the numerator and denominator of this fraction by $\cos\alpha\cos\beta$, we get

$$\frac{\dfrac{\sin\alpha\cos\beta}{\cos\alpha\cos\beta} + \dfrac{\cos\alpha\sin\beta}{\cos\alpha\cos\beta}}{\dfrac{\cos\alpha\cos\beta}{\cos\alpha\cos\beta} - \dfrac{\sin\alpha\sin\beta}{\cos\alpha\cos\beta}} = \frac{\tan\alpha + \tan\beta}{1 - \tan\alpha\tan\beta}$$

provided, of course, that the denominator is not zero.

Thus, we have the formula for $\tan(\alpha + \beta)$.

Addition Identity for Tangent

$$\tan(\alpha + \beta) = \frac{\tan\alpha + \tan\beta}{1 - \tan\alpha\tan\beta}.$$

In exercise 19, you will be asked to show that the following is a valid identity.

Subtraction Identity for Tangent

$$\tan(\alpha - \beta) = \frac{\tan\alpha - \tan\beta}{1 + \tan\alpha\tan\beta}.$$

In order to save you some page flipping, we repeat the six addition and subtraction formulas from this section.

The Addition and Subtraction Identities

$$\sin(\alpha + \beta) = \sin\alpha\cos\beta + \cos\alpha\sin\beta$$
$$\sin(\alpha - \beta) = \sin\alpha\cos\beta - \cos\alpha\sin\beta$$
$$\cos(\alpha + \beta) = \cos\alpha\cos\beta - \sin\alpha\sin\beta$$
$$\cos(\alpha - \beta) = \cos\alpha\cos\beta + \sin\alpha\sin\beta$$
$$\tan(\alpha + \beta) = \frac{\tan\alpha + \tan\beta}{1 - \tan\alpha\tan\beta}$$
$$\tan(\alpha - \beta) = \frac{\tan\alpha - \tan\beta}{1 + \tan\alpha\tan\beta}.$$

Exercises 6.2

In exercises 1–6, find the exact value of each given expression.

1. $\sin \dfrac{7\pi}{12}$
2. $\sin \dfrac{\pi}{12}$
3. $\cos \dfrac{13\pi}{12}$
4. $\cos \dfrac{5\pi}{12}$
5. $\tan \dfrac{5\pi}{12}$
6. $\tan \dfrac{23\pi}{12}$

In exercises 7–14, write each given expression as a trigonometrically defined function evaluated at one angle.

7. $\dfrac{\sin x}{2} - \dfrac{\sqrt{3}}{2} \cos x$ (Hint: $\dfrac{1}{2} = \cos \dfrac{\pi}{3}$ and $\dfrac{\sqrt{3}}{2} = \sin \dfrac{\pi}{3}$. The angle you are looking for is $x - \dfrac{\pi}{3}$.)
8. $\cos x - \sqrt{3} \sin x$
9. $\sqrt{3} \cos x + \sin x$
10. $\dfrac{2 \tan x}{1 - \tan^2 x}$
11. $\cos x \cos \alpha + \sin x \sin \alpha$
12. $\cos^2 \theta - \sin^2 \theta$
13. $\sin x \cos \dfrac{\pi}{5} + \cos x \sin \dfrac{\pi}{5}$
14. $\cos x \cos \dfrac{\pi}{5} + \sin x \sin \dfrac{\pi}{5}$
15. If θ is a standard-position angle whose terminal side is in the second quadrant and $\sin \theta = \frac{5}{13}$, calculate $\cos \left(\theta - \dfrac{\pi}{3} \right)$. (Hint: first calculate $\cos \theta$.)
16. If θ is a standard-position angle whose terminal side is in the second quadrant and $\cos \left(\theta - \dfrac{\pi}{3} \right) = 0.6$, calculate $\cos \theta$ and $\sin \theta$. (Hint: $\left(\theta - \dfrac{\pi}{3} \right) + \dfrac{\pi}{3} = \theta$ and $\sin \left(\theta - \dfrac{\pi}{3} \right) > 0$.)
17. If $\tan x = 2$ and $\tan y = -\frac{1}{2}$, calculate $\tan (x + y)$.
18. If x and y are acute angles such that $\sin x = 0.51$ and $\sin y = 0.69$, determine whether $x + y$ is an acute angle.

In exercises 19–42, prove that each given equation is an identity.

◆ 19. $\tan (\alpha - \beta) = \dfrac{\tan \alpha - \tan \beta}{1 + \tan \alpha \tan \beta}$
20. $\tan \left(\dfrac{\pi}{2} - \theta \right) = \cot \theta$. (Hint: write the left-hand side in terms of sin and cos, and then use cofunction identities.)
21. $\tan (\pi + \beta) = \tan \beta$
22. $\cot \left(\dfrac{\pi}{2} - x \right) = \tan x$
23. $\sec (x + y) = \dfrac{\csc x \csc y}{\cot x \cot y - 1}$
24. $\sec (x + y) = \dfrac{\sec x \sec y}{1 - \tan x \tan y}$
25. $\sec \left(\dfrac{\pi}{2} - \theta \right) = \csc \theta$
26. $\csc \left(\dfrac{\pi}{2} - \theta \right) = \sec \theta$
27. $\cos^2 \alpha + \cos^2 \left(\dfrac{\pi}{2} - \alpha \right) = 1$
28. $\sec^2 \alpha = 1 + \cot^2 \left(\alpha - \dfrac{\pi}{2} \right)$
29. $\tan \alpha \sin \left(\alpha - \dfrac{\pi}{2} \right) + \sin \alpha = 0$
30. $\cos (\alpha + \beta) \cos \beta + \sin (\alpha + \beta) \sin \beta = \cos \alpha$
31. $\tan x = \dfrac{\tan (x + y) - \tan y}{1 + \tan(x + y) \tan y}$
32. $\cot (x + y) = \dfrac{\cot x \cot y - 1}{\cot x + \cot y}$
33. $\cot (\theta - \pi) = \cot \theta$
34. $\sin (\theta + \phi) + \sin (\phi - \theta) = 2 \sin \phi \cos \theta$
35. $\sin (\alpha + \beta) - \sin (\alpha - \beta) = 2 \cos \alpha \sin \beta$
36. $\sin (\alpha + \beta) \sin (\alpha - \beta) = \sin^2 \alpha \cos^2 \beta - \cos^2 \alpha \sin^2 \beta$
37. $\dfrac{\sin (x + y)}{\sin (x - y)} = \dfrac{\tan x + \tan y}{\tan x - \tan y}$
38. $2 \sin \left(x + \dfrac{\pi}{6} \right) = \sqrt{3} \sin x + \cos x$
39. $\tan \left(\theta - \dfrac{\pi}{4} \right) = \dfrac{\tan^2 \theta - 1}{(1 + \tan \theta)^2}$
40. $\sin (\theta + \phi + \varphi) = \sin \theta \cos \phi \cos \varphi + \cos \theta \sin \phi \cos \varphi + \cos \theta \cos \phi \sin \varphi - \sin \theta \sin \phi \sin \varphi$
41. $\sin 3\alpha = 3 \sin \alpha - 4 \sin^3 \alpha$ (Hint: let $\theta = \phi = \varphi$ in exercise 40.)
◆ 42. $\cos 3\alpha = 4 \cos^3 \alpha - 3 \cos \alpha$ (Hint: mimic exercises 40 and 41.)
43. Prove by counterexample that $\sin (x - y) = \sin x - \sin y$ is not an identity.
44. Show that if x and y are real numbers such that $\sin (x + y) = \sin x + \sin y$, then $\cot x + \cot y = \csc x + \csc y$, provided that both these expressions are defined.
45. (a) If $h \neq 0$, prove that

$$\dfrac{\sin (x + h) - \sin x}{h} = -\sin x \left(\dfrac{1 - \cos h}{h} \right) + \cos x \left(\dfrac{\sin h}{h} \right).$$

(b) In view of (a) and exercise 18 in section 5.4, what "limiting tendency" do you expect $\dfrac{\sin (x + h) - \sin x}{h}$ to exhibit as h approaches 0 and

Chapter 6 / Analytic Trigonometry

x is held fixed? (Hint: first consider the tendency of $\frac{1 - \cos h}{h}$.)

46. Use exercise 28 of section 5.2 and exercise 19 above to obtain a new proof that nonvertical lines ℓ_1 and ℓ_2, with respective slopes m_1 and m_2, are perpendicular if and only if $m_1 m_2 = -1$.

47. Show that $\sin x - \cos x = \sqrt{2} \sin \left(x - \frac{\pi}{4} \right)$ for all real numbers x.

48. (a) Find a real number C such that $3 \sin x - 4 \cos x = 5 \sin (x - C)$. (Hint: arrange that $\cos C = \frac{3}{5}$ and $\sin C = \frac{4}{5}$.)
    (b) Does there exist a real number C such that $3 \sin x - 4 \cos x = 5 \cos (x - C)$?
    (c) Graph $y = 3 \sin x - 4 \cos x$.

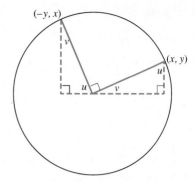

In exercises 49–52, solve the given conditional equations or inequalities in $[0, 2\pi)$.

49. $\sin 3x \cos x = \cos x \sin 3x$

50. $\cos^2 \frac{\theta}{2} - \sin^2 \frac{\theta}{2} \leq \frac{1}{2}$

51. $\frac{\tan \theta}{1 - \tan^2 \theta} > 1$

52. $3 \sin \alpha - 4 \sin^3 \alpha = 1$ (Hint: use exercise 41.)

53. Prove that $\sin \left(\tan^{-1} 3 - \cos^{-1} \frac{4}{5} \right) = \frac{9\sqrt{10}}{50}$.

54. Prove that $\sin (\sin^{-1} x + \cos^{-1} x) - 1 = 0$ if $-1 \leq x \leq 1$.

55. Use the unit circle approach (see section 5.3) to prove

$$\cos \left(x + \frac{\pi}{2} \right) = -\sin x \text{ and } \sin \left(x + \frac{\pi}{2} \right) = \cos x.$$

56. Romeo intends to climb a defective 60 foot pole/ladder running from ground level to Rowena's window, which is 24 feet above the ground. Rowena's brother will support the defective ladder by bracing its base with another pole extending from a window 12 feet below Rowena's. To the nearest hundredth, estimate the radian measure of the angle keeping the poles apart.

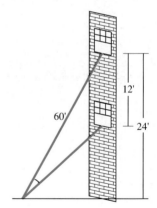

6.3 THE DOUBLE- AND HALF-ANGLE IDENTITIES

There are two main objectives for this section. The first one is to develop the double-angle formulas which rewrite expressions such as $\sin 2\alpha$ and $\cos 2\alpha$. The second is to develop the half-angle formulas for expressions such as $\sin \frac{1}{2}\alpha$ and $\tan \frac{1}{2}\alpha$.

Double-Angle Formulas

Consider the formula $\sin (\alpha + \beta) = \sin \alpha \cos \beta + \cos \alpha \sin \beta$ which was derived in section 6.2. If we replace β with α, we get

$$\sin 2\alpha = \sin (\alpha + \alpha) = \sin \alpha \cos \alpha + \cos \alpha \sin \alpha$$
$$= 2 \sin \alpha \cos \alpha.$$

The Double- and Half-Angle Identities

Dealing similarly with the other addition formulas in section 6.2, we can derive formulas for $\cos 2\alpha$ and $\tan 2\alpha$, with the following results.

Double-Angle Identities

$$\sin 2\alpha = 2 \sin \alpha \cos \alpha$$

$$\cos 2\alpha = \cos^2 \alpha - \sin^2 \alpha$$

$$\tan 2\alpha = \frac{2 \tan \alpha}{1 - \tan^2 \alpha}, \quad \text{if } \tan \alpha \neq \pm 1.$$

There are two other useful ways to rewrite $\cos 2\alpha$. First, recall that we may restate the Pythagorean identity $\sin^2 \alpha + \cos^2 \alpha = 1$ as $\sin^2 \alpha = 1 - \cos^2 \alpha$ and as $\cos^2 \alpha = 1 - \sin^2 \alpha$. If the first of these is substituted in the above identity for $\cos 2\alpha$, we get

$$\cos 2\alpha = \cos^2 \alpha - (1 - \cos^2 \alpha) = 2 \cos^2 \alpha - 1.$$

Similarly, if the second identity is used, we get

$$\cos 2\alpha = (1 - \sin^2 \alpha) - \sin^2 \alpha = 1 - 2 \sin^2 \alpha.$$

Summarizing these results, we have the following alternative identities for $\cos 2\alpha$.

Double-Angle Identities for Cosine

$$\cos 2\alpha = \cos^2 \alpha - \sin^2 \alpha$$
$$= 2 \cos^2 \alpha - 1$$
$$= 1 - 2 \sin^2 \alpha.$$

A phrase such as "θ is an angle in standard position whose terminal side is in the third quadrant" is too long and cumbersome to keep repeating. From now on, we shall shorten this type of phrase to a phrase such as "θ is in the third quadrant."

▷ **EXAMPLE 6.15**

If $\cos \alpha = -\frac{4}{5}$ and $\frac{\pi}{2} < \alpha < \pi$, find $\sin 2\alpha$, $\cos 2\alpha$, and $\tan 2\alpha$. Determine which quadrant 2α is in.

Solution

From $\cos \alpha = -\frac{4}{5}$ and α in the second quadrant, we find as in section 5.4 that $\sin \alpha = \frac{3}{5}$ and $\tan \alpha = -\frac{3}{4}$. Then

$$\sin 2\alpha = 2 \sin \alpha \cos \alpha = 2(\tfrac{3}{5})(-\tfrac{4}{5}) = -\tfrac{24}{25}$$

$$\cos 2\alpha = \cos^2 \alpha - \sin^2 \alpha = (-\tfrac{4}{5})^2 - (\tfrac{3}{5})^2 = \tfrac{7}{25}.$$

By dividing the first result by the second, we get

$$\tan 2\alpha = \frac{\sin 2\alpha}{\cos 2\alpha} = \frac{-24/25}{7/25} = -\frac{24}{7}.$$

As $\pi \leq 2\alpha \leq 2\pi$ and $\cos 2\alpha > 0$, 2α is in the fourth quadrant. ◀

We know that $\cos 2\theta = 2\cos^2 \theta - 1 = 1 - 2\sin^2 \theta$. Hence, we have the following variants of the double-angle identities for cosine.

$$\cos^2 \theta = \tfrac{1}{2}(1 + \cos 2\theta)$$
$$\sin^2 \theta = \tfrac{1}{2}(1 - \cos 2\theta).$$

These formulas will be very helpful when you study integral calculus.

▶ **EXAMPLE 6.16**

Express $\cos^4 \theta$ in terms of $\cos 2\theta$ and $\cos 4\theta$.

Solution

As we saw above, $\cos^2 \theta = \tfrac{1}{2}(1 + \cos 2\theta)$.
Squaring and then substituting, we see that

$$\cos^4 \theta = (\cos^2 \theta)^2 = [\tfrac{1}{2}(1 + \cos 2\theta)]^2$$
$$= \tfrac{1}{4}(1 + \cos 2\theta)^2$$
$$= \tfrac{1}{4}(1 + 2\cos 2\theta + \cos^2 2\theta).$$

This may be rewritten further as

$$\frac{1}{4} + \frac{\cos 2\theta}{2} + \frac{\tfrac{1}{2}(1 + \cos 4\theta)}{4} = \frac{3 + 4\cos 2\theta + \cos 4\theta}{8}. \quad ◀$$

We are next able to show a most efficient way to obtain the conclusion of exercise 42 in section 6.2.

▶ **EXAMPLE 6.17**

Express $\cos 3\theta$ in terms of $\cos \theta$.

Solution

Write 3θ as $2\theta + \theta$. Then

$$\cos 3\theta = \cos(2\theta + \theta)$$
$$= \cos 2\theta \cos \theta - \sin 2\theta \sin \theta$$
$$= (\cos^2 \theta - \sin^2 \theta)\cos \theta - (2\sin \theta \cos \theta)\sin \theta$$
$$= \cos^3 \theta - \sin^2 \theta \cos \theta - 2\sin^2 \theta \cos \theta$$
$$= \cos^3 \theta - 3\sin^2 \theta \cos \theta$$
$$= \cos^3 \theta - 3(1 - \cos^2 \theta)\cos \theta$$
$$= \cos^3 \theta - 3\cos \theta + 3\cos^3 \theta$$
$$= 4\cos^3 \theta - 3\cos \theta. \quad ◀$$

Trigonometric equations and inequalities can often be more easily solved by first simplifying some expressions in them by using trigonometric identities. We shall illustrate this in the next two examples by using the formulas for $\cos 2\theta$ and $\sin 2\theta$.

The Double- and Half-Angle Identities

EXAMPLE 6.18 Solve $\cos 2\theta = \sin \theta$ in the interval $[0, 2\pi)$.

Solution By using the double-angle formula $\cos 2\theta = 1 - 2\sin^2 \theta$, we may rewrite the given conditional equation as

$$1 - 2\sin^2 \theta = \sin \theta$$

or

$$2\sin^2 \theta + \sin \theta - 1 = 0.$$

Factoring, we get

$$(2\sin \theta - 1)(\sin \theta + 1) = 0,$$

and so

$$\sin \theta = \tfrac{1}{2} \quad \text{or} \quad \sin \theta = -1.$$

The values in $[0, 2\pi)$ which satisfy $\sin \theta = \dfrac{1}{2}$ are $\theta = \dfrac{\pi}{6}$ and $\theta = \dfrac{5\pi}{6}$. The only value in $[0, 2\pi)$ which satisfies $\sin \theta = -1$ is $\theta = \dfrac{3\pi}{2}$. So the solution set consists of $\dfrac{\pi}{6}, \dfrac{5\pi}{6},$ and $\dfrac{3\pi}{2}$. ◀

EXAMPLE 6.19 Solve $\cos 3x + \cos x + 1 = -\cos 2x$ in the interval $[0, 2\pi)$.

Solution From example 6.17, we know that $\cos 3x = 4\cos^3 x - 3\cos x$. Since we also know that $\cos 2x = 2\cos^2 x - 1$, we can rewrite the original equation as

$$4\cos^3 x - 3\cos x + \cos x + 1 = -(2\cos^2 x - 1)$$

$$4\cos^3 x + 2\cos^2 x - 2\cos x = 0.$$

Factoring produces first

$$2\cos x(2\cos^2 x + \cos x - 1) = 0$$

and then

$$2\cos x(2\cos x - 1)(\cos x + 1) = 0.$$

By setting the factors separately equal to 0, we have either $\cos x = 0$ or $\cos x = \tfrac{1}{2}$ or $\cos x = -1$.

Thus, the solutions are $\dfrac{\pi}{2}, \dfrac{3\pi}{2}; \dfrac{\pi}{3}, \dfrac{5\pi}{3};$ and π. ◀

EXAMPLE 6.20 Solve $2\cot 2x \geq \sec x$ in the interval $[0, 2\pi)$.

Solution As in section 5.8, we shall find the key points by solving the associated equation $2\cot 2x = \sec x$ in the interval $[0, 2\pi)$.

Rewriting this equation in terms of sines and cosines, we obtain

$$\frac{2\cos 2x}{\sin 2x} = \frac{1}{\cos x} \quad \text{or} \quad \frac{2\cos 2x}{\sin 2x} - \frac{1}{\cos x} = 0.$$

The double-angle formulas are used to rewrite the left-hand side as follows:

$$\frac{2(1 - 2\sin^2 x)}{2\sin x \cos x} - \frac{1}{\cos x} = 0.$$

This simplifies to

$$\frac{1 - \sin x - 2\sin^2 x}{\sin x \cos x} = 0$$

or

$$\frac{(1 - 2\sin x)(1 + \sin x)}{\sin x \cos x} = 0.$$

From this equation, we obtain the key points $x = 0, \frac{\pi}{6}, \frac{\pi}{2}, \frac{5\pi}{6}, \pi$, and $\frac{3\pi}{2}$. These points are marked on the number line in figure 6.3(a). (Why are the points at $x = \frac{\pi}{6}$ and $\frac{5\pi}{6}$ the only ones filled in?) Using the sign chart in figure 6.3(b), we determine that the inequality's solution is

$$\left(0, \frac{\pi}{6}\right] \cup \left(\frac{\pi}{2}, \frac{5\pi}{6}\right] \cup \left(\pi, \frac{3\pi}{2}\right).$$

Figure 6.3

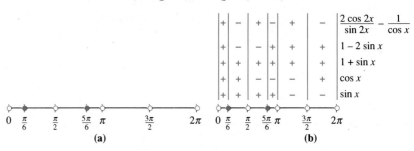

Half-Angle Formulas

Let's take another look at the two alternative identities for $\cos 2\alpha$:

$$\cos 2\alpha = 1 - 2\sin^2 \alpha = 2\cos^2 \alpha - 1.$$

If we use $\alpha = \frac{1}{2}\theta$ or, equivalently, $\theta = 2\alpha$ in these formulas, we get

$$\cos \theta = 1 - 2\sin^2 \tfrac{1}{2}\theta$$

and

$$\cos \theta = 2\cos^2 \tfrac{1}{2}\theta - 1.$$

The Double- and Half-Angle Identities

Solving each of these for the squared term, we get

$$\sin^2 \tfrac{1}{2}\theta = \tfrac{1}{2}(1 - \cos \theta)$$

and

$$\cos^2 \tfrac{1}{2}\theta = \tfrac{1}{2}(1 + \cos \theta).$$

We have already seen an argument of this kind in example 6.16, but next we go further. Taking a square root of both sides of each equation, we obtain the following results.

Half-Angle Identities

$$\sin \tfrac{1}{2}\theta = \pm\sqrt{\frac{1 - \cos \theta}{2}}$$

$$\cos \tfrac{1}{2}\theta = \pm\sqrt{\frac{1 + \cos \theta}{2}}$$

$$\tan \tfrac{1}{2}\theta = \pm\sqrt{\frac{1 - \cos \theta}{1 + \cos \theta}}, \quad \text{unless } \cos \theta = -1.$$

The appropriate sign (+ or −) to use in each of these equations depends upon the quadrant that contains the angle $\frac{\theta}{2}$ in standard position.

▶ **EXAMPLE 6.21**

Determine the exact value of $\cos \frac{\pi}{8}$.

Solution

Applying the second half-angle formula to $\theta = \frac{\pi}{4}$ yields

$$\cos \frac{\pi}{8} = \cos \tfrac{1}{2}\left(\frac{\pi}{4}\right)$$

$$= \sqrt{\frac{1 + \cos \frac{\pi}{4}}{2}}$$

$$= \sqrt{\frac{1 + \sqrt{2}/2}{2}}$$

$$= \frac{\sqrt{2 + \sqrt{2}}}{2}.$$

Notice that the positive square root was chosen because $\frac{\pi}{8}$ is in the first quadrant and the cosine function is positive in the first quadrant. ◀

Chapter 6 / Analytic Trigonometry

EXAMPLE 6.22 Suppose that $\pi < x < \frac{3\pi}{2}$ and $\cos x = -\frac{5}{13}$. Find $\sin \frac{x}{2}$.

Solution Because $\pi < x < \frac{3\pi}{2}$, we have $\frac{1}{2}(\pi) < \frac{1}{2}x < \frac{1}{2}\left(\frac{3\pi}{2}\right)$ or $\frac{\pi}{2} < \frac{x}{2} < \frac{3\pi}{4}$. Therefore $\frac{x}{2}$ is a quadrant II angle, and so $\sin \frac{x}{2} > 0$.

$$\sin \frac{x}{2} = \sqrt{\frac{1 - \cos x}{2}} = \sqrt{\frac{1 - (-5/13)}{2}} = \sqrt{\frac{18}{26}} = \frac{3}{13}\sqrt{13}.$$

EXAMPLE 6.23 Prove that $\cot \frac{1}{2}\theta = \frac{\sin \theta}{1 - \cos \theta}$.

Solution We know that

$$\cot \frac{1}{2}\theta = \frac{\cos(\theta/2)}{\sin(\theta/2)} = \frac{\cos^2(\theta/2)}{\sin(\theta/2)\cos(\theta/2)}$$

$$= \frac{2\cos^2(\theta/2)}{2\sin(\theta/2)\cos(\theta/2)}$$

$$= \frac{1 + \cos \theta}{\sin \theta}$$

$$= \frac{(1 + \cos \theta)(1 - \cos \theta)}{\sin \theta (1 - \cos \theta)}$$

$$= \frac{\sin^2 \theta}{\sin \theta (1 - \cos \theta)}$$

$$= \frac{\sin \theta}{1 - \cos \theta}.$$

The preceding argument involved division by $\cos \frac{\theta}{2}$. What if $\cos \frac{\theta}{2} = 0$? In this case $\cot \frac{\theta}{2}$ also equals zero, since $\cot \frac{\theta}{2} = \frac{\cos(\theta/2)}{\sin(\theta/2)}$. We still need to show that if $\cos \frac{\theta}{2} = 0$, then $\frac{\sin \theta}{1 - \cos \theta} = 0$. This *does* follow from a double-angle formula because $\sin \theta = 2 \sin \frac{\theta}{2} \cos \frac{\theta}{2}$. Therefore, we have proved that $\cot \frac{1}{2}\theta = \frac{\sin \theta}{1 - \cos \theta}$ in all cases.

EXAMPLE 6.24 Solve the equation $\cos \frac{1}{2}\theta = \sin \theta$ in the interval $[0, 2\pi)$.

Solution

We need to analyze $\cos \frac{1}{2}\theta = \pm\sqrt{\frac{1+\cos\theta}{2}} = \sin\theta$. Squaring leads to the equation $\frac{1+\cos\theta}{2} = \sin^2\theta$. Substituting $1 - \cos^2\theta$ for $\sin^2\theta$ and multiplying by 2, we obtain

$$1 + \cos\theta = 2(1 - \cos^2\theta)$$
$$2\cos^2\theta + \cos\theta - 1 = 0$$
$$(2\cos\theta - 1)(\cos\theta + 1) = 0.$$

Hence either $\cos\theta = \frac{1}{2}$ or $\cos\theta = -1$. This means that θ is one of $\frac{\pi}{3}, \frac{5\pi}{3}$, and π. We need to check each of these in the original equation because we have used the "squaring both sides" technique. In fact, all these values of θ do satisfy the original equation, and so these values are the desired solutions.

Alternative solution

Recall the identity $\sin\theta = 2\sin\frac{\theta}{2}\cos\frac{\theta}{2}$ which was used in example 6.23. This identity permits us to rewrite the given equation as

$$\cos\frac{\theta}{2} = 2\sin\frac{\theta}{2}\cos\frac{\theta}{2} \quad \text{or} \quad \left(\cos\frac{\theta}{2}\right)\left(1 - 2\sin\frac{\theta}{2}\right) = 0.$$

Thus, for θ in $[0, 2\pi)$, we have

$$\cos\frac{\theta}{2} = 0 \quad \text{or} \quad 1 - 2\sin\frac{\theta}{2} = 0$$

$$\frac{\theta}{2} = \frac{\pi}{2} \qquad\qquad \sin\frac{\theta}{2} = \frac{1}{2}$$

$$\theta = \pi \qquad\qquad \frac{\theta}{2} = \frac{\pi}{6}, \frac{5\pi}{6}$$

$$\theta = \frac{\pi}{3}, \frac{5\pi}{3}.$$

Thus the required solution set is $\left\{\frac{\pi}{3}, \pi, \frac{5\pi}{3}\right\}$.

Exercises 6.3

In exercises 1–4, use the given information to calculate the exact values of $\sin 2\theta$, $\cos 2\theta$, and $\tan 2\theta$, and hence determine the quadrant of 2θ.

1. $\sin\theta = 0.7$, θ is in quadrant I
2. $\cos\theta = -\frac{1}{2}$, θ is in quadrant III
3. $\tan\theta = 3$, $\sin\theta < 0$
4. $\cos\theta = \frac{12}{13}$, $\sin\theta = -\frac{5}{13}$

In exercises 5–8, use the given information and the fact that 2θ is in quadrant I to calculate $\sin\theta$, $\cos\theta$, and $\tan\theta$. Notice that there will be an ambiguity in the algebraic signs of $\sin\theta$ and $\cos\theta$, since θ is in either quadrant I or quadrant III.

5. $\cos 2\theta = 0.6$
6. $\sin 2\theta = 0.4$
7. $\tan 2\theta = \frac{1}{3}$
8. $\sec 2\theta = \sqrt{2}$

9. Find $\sin\frac{x}{2}$, given that $\cos x = \frac{1}{3}$ and $\frac{3\pi}{2} < x < 2\pi$.

10. Find $\sin\frac{\alpha}{2}$, given that $\tan\alpha = 2$ and $2\pi < \alpha \leq \frac{5\pi}{2}$.

11. Find $\cos\frac{\alpha}{2}$, given that $\tan\alpha = -2$ and $\frac{\pi}{2} < \alpha < \pi$.

12. Find $\cos\frac{\theta}{2}$, given that $\cos\theta = \frac{1}{3}$ and $\frac{3\pi}{2} < \theta < 2\pi$.

13. Find $\tan\frac{\theta}{2}$, given that $\cos\theta = \frac{1}{3}$ and $\frac{3\pi}{2} < \theta < 2\pi$.

14. Find $\tan\frac{x}{2}$, given that $\sin x = -\frac{1}{3}$ and $\pi < x < \frac{3\pi}{2}$.

15. Find $\tan\frac{x}{2}$, given that $\tan x = 3$ and $\pi < x < \frac{3\pi}{2}$.

In exercises 16–22, use double- or half-angle formulas to find the exact value of each of the given expressions.

16. $\tan 15°$
17. $\sin\frac{2\pi}{3}$
18. $\cos\frac{10\pi}{3}$
19. $\cos 67°30'$
20. $\sin 202°30'$
21. $\tan\frac{\pi}{12}$
22. $\csc 292°30'$
23. Express $\sin^4\theta$ in terms of $\cos 2\theta$ and $\cos 4\theta$.
24. Express $\cos 4\theta$ in terms of $\cos\theta$. (Hint: $4\theta = 2\phi$, where $\phi = 2\theta$.)

In exercises 25–39, prove that each given equation is an identity.

25. $\sin 4x = 4\sin x \cos x - 8\sin^3 x \cos x$
26. $\cos^4\alpha - \sin^4\alpha = \cos 2\alpha$
27. $1 + \frac{\sin 2\alpha}{2} = \frac{\sin^3\alpha - \cos^3\alpha}{\sin\alpha - \cos\alpha}$
28. $2\csc 2\theta = \sec\theta\csc\theta$
29. $\sec 2\theta = \frac{\csc^2\theta}{\cot^2\theta - 1}$ (Hint: see exercise 23 in section 6.2.)
30. $\sec 2x = \frac{\sec^2 x}{2 - \sec^2 x}$
31. $2\sin 4x \cos 4x = \sin 8x$
32. $(\sin\alpha + \cos\alpha)^2 = 1 + \sin 2\alpha$
33. $\tan\frac{\alpha}{2} = \frac{1 - \cos\alpha}{\sin\alpha}$
34. $\sec\frac{\theta}{2} = \pm\sqrt{\frac{2}{1 + \cos\theta}}$
35. $\csc\frac{\theta}{2} = \pm\sqrt{\frac{2\sec\theta}{\sec\theta - 1}}$
36. $\sqrt{\frac{1 - \cos 8x}{2}} = |\sin 4x|$
37. $\sin x = \frac{2\tan(x/2)}{1 + \tan^2(x/2)}$
38. $\frac{1 - \tan^2(\alpha/2)}{1 + \tan^2(\alpha/2)} = \cos\alpha$
39. $\tan\alpha + \sec\alpha = \frac{1 + \tan(\alpha/2)}{1 - \tan(\alpha/2)}$

In exercises 40–54, solve the given conditional equations or inequalities in $[0, 2\pi)$.

40. $2\cos 2x - 2(1 + \sqrt{3})\cos x + 2 + \sqrt{3} = 0$ (Hint: see exercise 24 of section 5.8.)
41. $\sin\frac{\theta}{2} = \cos\theta$
42. $\cos 2\theta = \cos\theta$
43. $\cos 2\alpha + \cos\alpha + 1 = 0$
44. $\sin\alpha = 1 - \cos 2\alpha$
45. $\tan 2x + \tan x = 0$
46. $|\tan 2x - 2\cos x| = 0$
47. $\sin\theta\cos 2\theta = \cos\theta\sin 2\theta$
48. $2\sin 2\theta - 2\sin\theta - 2\cos\theta + 1 = 0$
49. $\sin\frac{\alpha}{2} = \sin\alpha$
50. $\cos 2\alpha + 4\cos\alpha + 3 \geq 0$ (Hint: see exercise 48 of section 5.8.)
51. $\cos 2x \leq \cos x$
52. $\cos 2x + \cos x < 0$
53. $2\cot 2\theta > \csc\theta$
54. $2\tan x \leq \cos\frac{x}{2}$

55. Graph $y = \cos^2 x$. (Hint: rewrite y in terms of $\cos 2x$ and apply section 5.4.)

56. Graph $y = \sin^2\left(x - \frac{\pi}{4}\right)$.

57. The Human Fly is asked to climb a 70 foot long ladder whose top is 60 feet above the ground. (See the figure.) She agrees to do so, stipulating that she will charge a fee if and only if $2\theta > \arccos 0.47$. The deal is struck. Will the Human Fly be climbing for a fee or has she been too obtuse?

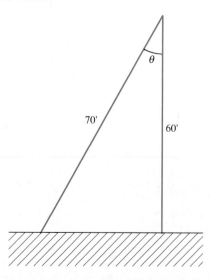

58. (a) Use a graphing calculator to sketch the graphs of the functions $f_1(x) = \sin kx$ and $f_2(x) = k\sin x$ for each of the following values of k: $-2, -1.5, -1, -0.5, 0, 0.5, 1, 1.5, 2,$ and 2.5.

(b) Use your results from (a) to compare $f_1(x)$ and $f_2(x)$. (Hint: graph $y = f_1(x) - f_2(x)$ for the above values of k.) Using this work, make a conjecture about the values of k for which $\sin kx = k \sin x$ is an identity.

(c) Prove the conjecture you made in (b). (Hint: if $f_1 = f_2$, the set of zeros of f_1 is equal to the set of zeros of f_2.)

6.4 GRAPHS OF COMBINATIONS OF SINE AND COSINE CURVES

In chapter 2, we looked at ways to make new functions from old ones by addition, subtraction, multiplication, division, or composition. At that time, the data functions were all "algebraic." For example, given $f(x) = 2x$ and $g(x) = x^2 + 1$, we could understand $f + g$ by graphing $y = (f + g)(x) = x^2 + 2x + 1 = (x + 1)^2$. As we shall see, algebraic simplification is not as straightforward when we try to graph combinations of trigonometric functions or of algebraic and trigonometric functions, but some useful techniques can still be identified. These techniques will remain useful even after you have learned calculus.

In this section's graphing calculator activity, we hunt for roots in a "flat" part of a graph of a trigonometrically defined function. This builds on the graphing calculator activity in section 3.4. The new discussion also highlights the role of graphing calculators (or, for that matter, computer graphing programs) as tools for discovery. This section's discussion includes the text's most substantial graphing calculator activity. Because of this, we have divided that activity into two parts.

Addition of Ordinates

In this procedure, two functions are being combined by addition or subtraction. The easiest method is first to make a sketch, on the same set of axes, of each of the data functions. Next, add or subtract the ordinates (or y-values) for "representative" x-values in the domain. It is a good idea to select these representative x-values to be the **special points**. These are the x-values where at least one of the graphs of the data functions crosses the x-axis or has a maximum or minimum point. Finally, smoothly connect this new set of points. A ruler or compass can be used to help with the addition of ordinates procedure, as is shown in the following example.

EXAMPLE 6.25

Use addition of ordinates to sketch the graph of $y = \sin x + \cos 2x$.

Solution

We can think of y as $f(x) + g(x)$, where $f(x) = \sin x$ and $g(x) = \cos 2x$. Begin by sketching the graphs of f and g on the same set of axes. (You should be able to do this part quickly. Review sections 5.4 and 5.5 if necessary.) The graphs of f and g are shown as dashed curves in figure 6.4(a). The period of f is 2π and of g is π. The period of $f + g$ will be no larger than the "least common multiple" of 2π and π, which is 2π. On $[0, 2\pi]$, the special points of the function f are $0, \dfrac{\pi}{2}, \pi, \dfrac{3\pi}{2}$, and 2π. For the function g, the special points on $[0, 2\pi]$ are these same five points, $0, \frac{\pi}{2}, \pi, \frac{3\pi}{2}$, and 2π, and also $\frac{\pi}{4}, \frac{3\pi}{4}, \frac{5\pi}{4}$, and $\frac{7\pi}{4}$.

Adding the ordinates at these special points, we get the indicated points in figure 6.4(a). As you can see, a compass is a useful tool for geometrically adding (or subtracting) ordinates. Smoothly connecting these points, we obtain the curve in figure 6.4(b).

In calculus, you will learn how to determine that the maximum value, M, of this curve is 1.125 and that the minimum, m, is -2. Hence, the amplitude is $\frac{1}{2}(M - m) = 1.5625$. ◀

Figure 6.4

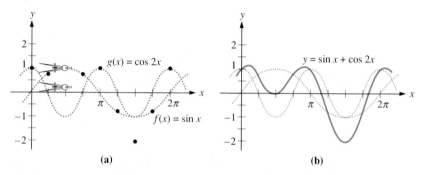

(a) (b)

Computer programs and some calculators will graph trigonometric functions for you. Thus, you might ask why it is necessary to perform this graphing by hand. What we wish to instill here is, first, the ability to quickly sketch a graph as an aid to solving a problem and, second, the ability to look at a computer- or calculator-produced graph and see if it appears to be correct. It is very easy to accept and rely on computers and calculators without questioning their results. One of the abilities you must maintain and/or acquire is that of estimating an answer and of checking your work. Estimation and checking are essential so that you do not blindly accept the results produced by a computer or a calculator.

▶ **EXAMPLE 6.26**

Sketch the graph of $y = 2 \cos x + \frac{1}{3} \sin 4x$.

Solution

First we make a sketch, on the same set of axes, of the graphs of the functions given by $h(x) = 2 \cos x$ and $j(x) = \frac{1}{3} \sin 4x$. These are represented by the dashed curves in figure 6.5(a). The period of h is 2π and of j is $\frac{2\pi}{4} = \frac{\pi}{2}$. Hence the period of $h + j$ will be no larger than the "least common multiple" of 2π and $\frac{\pi}{2}$, which is 2π. For h, the special values on $[0, 2\pi]$ are $x = 0, \frac{\pi}{2}, \pi, \frac{3\pi}{2}$, and 2π. The graph of the function j crosses the x-axis at 0, $\frac{\pi}{4}, \frac{\pi}{2}, \frac{3\pi}{4}, \pi, \frac{5\pi}{4}, \frac{3\pi}{2}, \frac{7\pi}{4}$, and 2π, and has either a maximum or a

Graphs of Combinations of Sine and Cosine Curves

minimum midway between each consecutive pair of these points. By adding the ordinates of h and j at each of these special points and smoothly connecting the resulting points, we obtain the curve in figure 6.5(b). Its period is 2π.

Figure 6.5

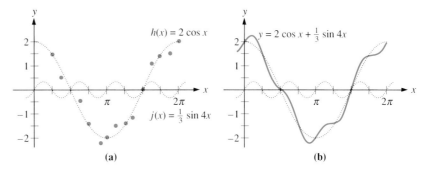

EXAMPLE 6.27

Sketch the graph of $y = 2 \sin 3x - 5 \cos 3x$.

Solution

We shall think of this as $y = f(x) - g(x)$, where $f(x) = 2 \sin 3x$ and $g(x) = 5 \cos 3x$. As usual, we begin by sketching the graphs of f and g on the same set of axes, as shown by the dashed curves in figure 6.6(a). Both f and g have period $\dfrac{2\pi}{3}$, and the period of $f + g$ is also $\dfrac{2\pi}{3}$. The special points in the interval $\left[0, \dfrac{2\pi}{3}\right]$ are $0, \dfrac{\pi}{6}, \dfrac{\pi}{3}, \dfrac{\pi}{2}$, and $\dfrac{2\pi}{3}$. By subtracting the ordinates of f and g at these five points and smoothly connecting the resulting points, we obtain the curve in figure 6.6(b).

Figure 6.6

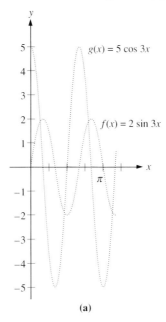
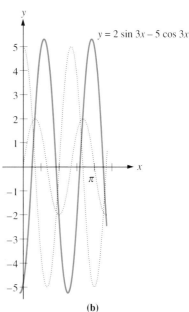

Chapter 6 / Analytic Trigonometry

EXAMPLE 6.28 Sketch the graph of $f(x) = \sin \frac{3}{2}x + 2 \sin 4x$.

Solution We shall think of this as $f(x) = g(x) + h(x)$, where $g(x) = \sin \frac{3}{2}x$ and $h(x) = 2 \sin 4x$. The graphs of g and h are the dashed curves in figure 6.7(a). The period of g is $\frac{2\pi}{3/2} = \frac{4\pi}{3}$ and the period of h is $\frac{\pi}{2}$. Since 4π is an integral multiple of $\frac{4\pi}{3}$ and $\frac{\pi}{2}$, $f(x + 4\pi) = f(x)$. (In fact, the period of f is 4π.) The special points have been graphed in figure 6.7(a). Smoothly connecting these results in the curve in figure 6.7(b).

Figure 6.7

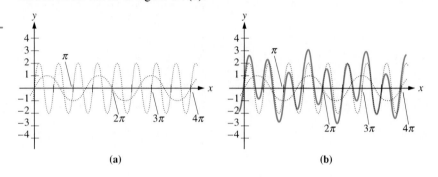

In the next example, we prove an interesting result about two periodic functions.

EXAMPLE 6.29 Prove that if f and g are periodic functions having the same period, then $f + g$ is a periodic function.

Proof Let c denote the period of f. This means that c is the smallest positive number such that $f(x + c) = f(x)$ for all x in D_f, the domain of f. The fact that g is also periodic and has period c means that $g(x + c) = g(x)$ for all x in D_g, the domain of g.

Consider $f + g$. By definition, $(f + g)(x) = f(x) + g(x)$ for all $x \in D_f \cap D_g$, the domain of $f + g$. So, if $x \in D_f \cap D_g$, then

$$(f + g)(x + c) = f(x + c) + g(x + c) = f(x) + g(x) = (f + g)(x).$$

Thus, $f + g$ is a periodic function. (As you will see in the exercises, $f + g$ need not have a period.)

In examples 6.25–6.28, we added or subtracted two trigonometric functions. In the next example, we shall subtract a trigonometric function from an algebraic function.

EXAMPLE 6.30 Use the addition of ordinates method to graph $f(x) = x - \cos x$.

Solution We view this as $f(x) = g(x) - h(x)$, where $g(x) = x$ and $h(x) = \cos x$. The graphs of g and h are sketched as dashed curves in figure 6.8(a). The special

Graphs of Combinations of Sine and Cosine Curves

points of h on $[0, 2\pi]$ are $0, \dfrac{\pi}{2}, \pi, \dfrac{3\pi}{2}$, and 2π. By (geometrically) subtracting the ordinates of h from those of g at each of these special points and then smoothly connecting the resulting points, we obtain the curve in figure 6.8(b). Notice that f is not periodic; of course, neither was g.

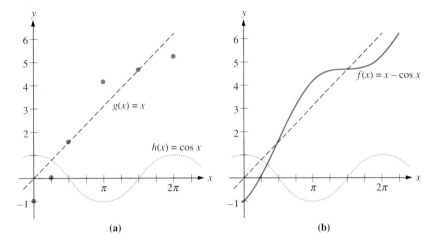

Figure 6.8

Products of Functions

In exercises 32 and 34 of section 5.9, you were asked to graph the functions $x \sin x$ and $\dfrac{\sin x}{e^x}$. More generally, one could be asked to graph a product $f(x) = g(x) \cdot h(x)$ or a quotient $\dfrac{g(x)}{h(x)}$. As we shall show next, this is done by using properties of the individual functions g and h. As was the case in addition of ordinates, you will want to consider what happens at the special points of g or h. We illustrate this now by reconsidering $x \sin x$.

EXAMPLE 6.31

Sketch the graph of $f(x) = x \sin x$.

Solution

We shall consider $f(x)$ as $g(x) \cdot h(x)$, where $g(x) = x$ and $h(x) = \sin x$. The graphs of the functions g and h are sketched as dashed curves in figure 6.9(a). As noted in example 6.26, the special points of h are $0, \dfrac{\pi}{2}, \pi, \dfrac{3\pi}{2}, 2\pi$, and so on.

There is an additional fact that will help us to sketch the required graph. Recall from section 5.2 that the range of sin is $[-1, 1]$. In particular, $|\sin x| \leq 1$ for all x. So, using the properties of absolute value developed in chapter 0, we have

$$|f(x)| = |g(x)| \cdot |h(x)| = |x| \cdot |\sin x| \leq |x|.$$

Consequently,

$-x \leq f(x) \leq x$.

This means that the graph of $f(x) = x \sin x$ lies between the graphs of $y = x$ and $y = -x$. The graph of $y = -x$ has also been included as a dashed curve in figure 6.9(a).

As for the behavior of $f(x)$ at the special points of g or h, we may notice the following. For any integer n, $f(n\pi) = 0$ and $f\left(\dfrac{\pi}{2} + n\pi\right) = \pm\left(\dfrac{\pi}{2} + n\pi\right)$. The points on the graph of f which correspond to these special points have been plotted in figure 6.9(a). When they are smoothly connected, we obtain the required curve in figure 6.9(b).

Figure 6.9

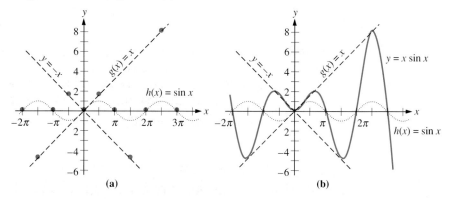

In engineering and physics, an oscillation whose maximum displacement from equilibrium decreases exponentially with the passage of time is said to be damped. An example of a damped function (viewing x as time) is given in example 6.32. Damped functions will be studied further in section 7.6.

▶ **EXAMPLE 6.32**

Sketch the graph of the "damped cosine" $g(x) = \dfrac{\cos x}{e^x}$.

Solution

Consider g as the product of two functions in the following way. Write $g(x) = j(x) \cdot k(x)$, where $j(x) = e^{-x}$ and $k(x) = \cos x$. As in example 6.31, information about ranges is useful. Recall that $e^{-x} > 0$ and $|\cos x| \leq 1$ for any real number x. Hence, again using a fact about absolute value, we have

$$|g(x)| = |j(x)| \cdot |k(x)| \leq e^{-x}.$$

Consequently,

$$-e^{-x} \leq g(x) \leq e^{-x}.$$

This means that the graph of g is between the graphs of $y = -e^{-x}$ and $y = e^{-x}$. The graphs of these two curves are shown dashed in figure 6.10.

Notice that the function j has no special points. Examining the special points of k, we see, for any integer n, that

$$g(x) = e^{-x} \cos x = 0 \qquad \text{at } x = \dfrac{\pi}{2} + n\pi,$$

Figure 6.10

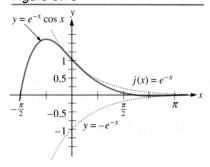

Graphs of Combinations of Sine and Cosine Curves

and $\quad\quad\quad\quad\quad\quad g(x) = e^{-x} \cos x = \pm e^{-x} \quad$ at $x = n\pi$.

The graph of g crosses the x-axis at $x = \dfrac{\pi}{2} + n\pi$, and touches one of the curves $y = e^{-x}$ and $y = -e^{-x}$ whenever x is an integral multiple of π. A sketch of the graph of g is shown in figure 6.10. ◀

The next example that we shall develop in this section concerns the function given by $\dfrac{1}{x}\sin x$. You should become very familiar with its graph, for such knowledge is fundamental to studying the differential calculus of trigonometric functions.

EXAMPLE 6.33

Sketch the graph of $h(x) = \dfrac{1}{x}\sin x$.

Solution

As before, range information will be useful. Since $|\sin x| \leq 1$,

$$|h(x)| = \left|\frac{1}{x}\sin x\right| = \left|\frac{1}{x}\right| \cdot |\sin x| \leq \left|\frac{1}{x}\right|.$$

Equivalently,

$$-\frac{1}{x} \leq \frac{1}{x}\sin x \leq \frac{1}{x} \quad \text{if } x \neq 0.$$

As a result, the graph of $h(x) = \dfrac{1}{x}\sin x$ lies between the graph of $y = \dfrac{1}{x}$ and the graph of $y = -\dfrac{1}{x}$. These last two graphs are shown as dashed curves in figure 6.11.

Notice that $\dfrac{1}{x}$ has no special points. Examining the special points of the sine function, we see, for any integer n, that

$$h(x) = \frac{1}{x}\sin x = 0 \quad \text{at } x = n\pi \quad \text{if } n \neq 0,$$

and $\quad\quad\quad h(x) = \dfrac{1}{x}\sin x = \pm\dfrac{1}{x} \quad$ at $x = \dfrac{\pi}{2} + n\pi$.

Thus, the graph of h crosses the x-axis whenever x is a nonzero integral multiple of π, and touches one of the curves $y = -\dfrac{1}{x}$ and $y = \dfrac{1}{x}$ at $x = \dfrac{\pi}{2} + n\pi$.

Although 0 is not in the domain of $\dfrac{1}{x}$, 0 is a special point of the sine function. What happens when x is close to 0? (You have already been asked, in exercise 18 of section 5.4, to answer this question!) Using a calculator, you should be able to verify the following values of $\dfrac{1}{x}\sin x$.

Chapter 6 / Analytic Trigonometry

x	−0.5	−0.4	−0.3	−0.2	−0.1	0.1	0.2	0.3	0.4	0.5
$\frac{1}{x}\sin x$	0.959	0.974	0.985	0.993	0.998	0.998	0.993	0.985	0.974	0.959

The values in the table indicate that $\frac{1}{x}\sin x$ approaches 1 as x approaches 0. (In fact, we indicated in exercise 19 of section 5.4 how to carry out a formal proof of this fact.) This conclusion is reinforced by the additional data in the following table.

Figure 6.11

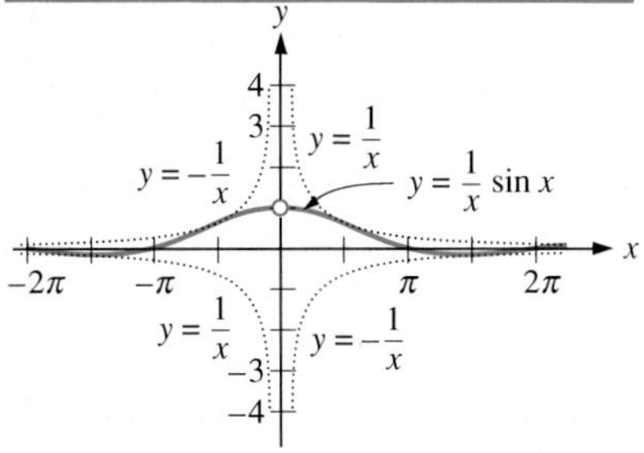

x	−0.05	−0.01	−0.001	0.001	0.01	0.05
$\frac{1}{x}\sin x$	0.9995833	0.9999833	0.9999998	0.9999998	0.9999833	0.9995833

Now, with the behavior of h understood at or near the special points, we can obtain a sketch of its graph. This is shown in figure 6.11.

Graphing Calculator Activity

In exercise 58 of section 6.3, you showed that the functions $\sin kx$ and $k \sin x$ are equal for only certain values of k (namely 0, 1, and −1). The development of that exercise revealed that graphing calculators can play a role in "discovering" facts. After we use the graphing calculators, the discovered "facts" actually only have the status of conjectures, which we must then try to prove by traditional mathematical methods. The following discussion expands upon this theme of "discovery by calculators, then traditional proof."

▶ **EXAMPLE 6.34**

Consider the function h defined by $h(x) = \tan x - x$, with domain $\left(-\frac{\pi}{2}, \frac{\pi}{2}\right)$.

Use your graphing calculator to graph h. (Make sure your calculator is in radian mode.) What conjectures can you make about the function h? Try to prove these conjectures.

Solution

Part One: Your calculator should produce a graph of h like that in figure 6.12(a), which we found with a Casio *fx-7700G* graphing calculator, or a graph like the one in figure 6.12(b), produced by a *TI-81*. (We set Xmin = −1.57079632, Xmax = 1.57079632, Xscl = 1.57079632, Ymin = −1, Ymax = 1, and Yscl = 0.5. Note that 1.57079632 ≈ $\frac{\pi}{2}$.)

Notice that the graph of h seems to have a "flat" part near the origin. What's happening in this flat part? Certainly, 0 is a root of h since $h(0) = \tan(0) - 0 = 0 - 0 = 0$. Are there other zeros of h which are close enough to 0 so that they are "hidden inside" the flat part? In the graphing calculator activity of section 3.4, you saw an example (called g there) where a flat part contained "hidden" zeros, which could be found by repeatedly zooming in

Graphs of Combinations of Sine and Cosine Curves

Figure 6.12

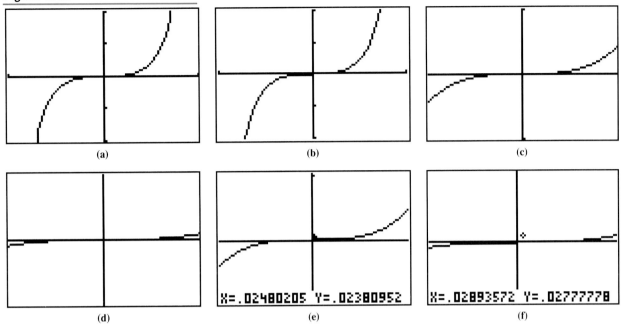

with a graphing calculator. In that same example, you also saw a function f whose graph's flat part had only one zero (and no other to be found regardless of how often you zoomed in). Is h like f or like g?

Let's zoom in twice on figures 6.12(a) or (b), first with domain $\left[-\frac{\pi}{4}, \frac{\pi}{4}\right]$, and next with domain $\left[-\frac{\pi}{8}, \frac{\pi}{8}\right]$. The results for a Casio *fx-7700G* are in figures 6.12(c) and (d); the results for a *TI-81* are in figures 6.12(e) and (f). From these figures, it seems safe to **conjecture** that h has only one root, namely 0.

The above "conjecture" is a fact, and you have already seen it. In exercise 19 of section 5.4, you showed that if $-\frac{\pi}{2} < \theta < \frac{\pi}{2}$ and $\theta \neq 0$, then $\cos \theta < \frac{\sin \theta}{\theta} < \frac{1}{\cos \theta}$. Replacing θ with x and using the ratio identity $\frac{\sin x}{\cos x} = \tan x$, we see that $\tan x < x$ if $-\frac{\pi}{2} < x < 0$; and $\tan x > x$ if $0 < x < \frac{\pi}{2}$. (To verify this using the multiplication properties of inequalities from section 1.1, you also need to recall that $\cos x > 0$ whenever $-\frac{\pi}{2} < x < \frac{\pi}{2}$: review figure 5.29.) In particular, $\tan x \neq x$ if $-\frac{\pi}{2} < x < \frac{\pi}{2}$ and $x \neq 0$. Thus $h(x) = \tan x - x \neq 0$ if $-\frac{\pi}{2} < x < \frac{\pi}{2}$ and $x \neq 0$. This completes the proof of the conjecture.

Part Two: Are any other conjectures suggested by the above graphs of h? Don't all those graphs suggest that h is an increasing function? On the basis of these graphs, we can surely **conjecture** that h is increasing. Can we prove *this* new conjecture?

In a later course, you will learn a fast way to use differential calculus to show that the new conjecture is valid: *h is* increasing. A proof using *pre*calculus is much more taxing, but it is possible, and we give it next.

We must show that if $-\frac{\pi}{2} < x_1 < x_2 < \frac{\pi}{2}$, then $h(x_1) < h(x_2)$. It can be assumed that x_1 and x_2 have the same sign, since we saw above, via exercise 19 of section 5.4, that $h(x) < 0$ if $-\frac{\pi}{2} < x < 0$; and $h(x) > 0$ if $0 < x < \frac{\pi}{2}$.

Moreover, we can actually assume $0 < x_1 < x_2 < \frac{\pi}{2}$. The reason involves the fact that h is an odd function. (Did you notice this fact, namely that $h(-x) = -h(x)$? Check this, recalling from exercise 16(a) of section 5.2 that tan is odd.) If fact, if $-\frac{\pi}{2} < x_1 < x_2 < 0$, then $0 < -x_2 < -x_1 < \frac{\pi}{2}$; and if the case of positive variables is known, we have $h(-x_2) < h(-x_1)$, and so $-h(x_2) < -h(x_1)$, which leads to $h(x_1) < h(x_2)$, the desired conclusion. Thus, we may assume $0 < x_1 < x_2 < \frac{\pi}{2}$ for the rest of the proof.

Since $\tan(x_2)$ and $\tan(x_1)$ are each positive, $1 < 1 + \tan(x_2)\tan(x_1)$, and so $\frac{1}{1 + \tan(x_2)\tan(x_1)} < 1$. Moreover, $\tan(x_1) < \tan(x_2)$ since tan is increasing on $\left(0, \frac{\pi}{2}\right)$: recall figure 5.38 in section 5.6. Hence $\tan(x_2) - \tan(x_1) > 0$ and so, by the positive multiplication property of inequalities,

$$\frac{\tan(x_2) - \tan(x_1)}{1 + \tan(x_2)\tan(x_1)} < \tan(x_2) - \tan(x_1).$$

The subtraction formula for the tangent function from section 6.2 can now be applied. The result is

$$\tan(x_2 - x_1) < \tan(x_2) - \tan(x_1).$$

However $\tan(x_2 - x_1) > x_2 - x_1$ by the above discussion of exercise 19 of section 5.4, since $0 < x_2 - x_1 < \frac{\pi}{2}$. Hence

$$x_2 - x_1 < \tan(x_2 - x_1) < \tan(x_2) - \tan(x_1)$$
$$\tan(x_1) - x_1 < \tan(x_2) - x_2$$
$$h(x_1) < h(x_2). \quad \blacktriangleleft$$

At least four morals can be drawn from part two of example 6.34. First, you should appreciate how much you already know and are able to use: tan is

Sum and Product Identities

EXAMPLE 6.35   Express $\sin 5\theta \cos 3\theta$ as a sum or difference.

Solution   Let $\alpha = 5\theta$ and $\beta = 3\theta$ in identity (5). The result is

$$\sin 5\theta \cos 3\theta = \tfrac{1}{2}[\sin(5\theta + 3\theta) + \sin(5\theta - 3\theta)]$$
$$= \tfrac{1}{2}[\sin 8\theta + \sin 2\theta]$$
$$= \tfrac{1}{2}\sin 8\theta + \tfrac{1}{2}\sin 2\theta.$$

◀

EXAMPLE 6.36   Evaluate $\cos 75° \cos 45°$ exactly.

Solution   We let $\alpha = 75°$ and $\beta = 45°$ in identity (7). The result is

$$\cos 75° \cos 45° = \tfrac{1}{2}[\cos(75° + 45°) + \cos(75° - 45°)]$$
$$= \tfrac{1}{2}(\cos 120° + \cos 30°)$$
$$= \tfrac{1}{2}\left[\left(-\tfrac{1}{2}\right) + \tfrac{\sqrt{3}}{2}\right]$$
$$= \tfrac{1}{4}(\sqrt{3} - 1).$$

◀

Changing Sums and Differences of Sines and Cosines into Products

If we let $x = \alpha + \beta$ and $y = \alpha - \beta$, then we see that $\alpha = \tfrac{1}{2}(x + y)$ and $\beta = \tfrac{1}{2}(x - y)$. Substituting these values into the sum and difference identities (identities (5)–(8)), we get the following four identities:

$$\sin \tfrac{1}{2}(x + y) \cos \tfrac{1}{2}(x - y) = \tfrac{1}{2}(\sin x + \sin y)$$
$$\cos \tfrac{1}{2}(x + y) \sin \tfrac{1}{2}(x - y) = \tfrac{1}{2}(\sin x - \sin y)$$
$$\cos \tfrac{1}{2}(x + y) \cos \tfrac{1}{2}(x - y) = \tfrac{1}{2}(\cos x + \cos y)$$
$$\sin \tfrac{1}{2}(x + y) \sin \tfrac{1}{2}(x - y) = \tfrac{1}{2}(\cos y - \cos x).$$

Multiplying both sides by 2 and reversing the sides produces four "sum" identities.

Sum-to-Product Identities

$$\sin x + \sin y = 2 \sin \tfrac{1}{2}(x + y) \cos \tfrac{1}{2}(x - y) \quad (9)$$
$$\sin x - \sin y = 2 \cos \tfrac{1}{2}(x + y) \sin \tfrac{1}{2}(x - y) \quad (10)$$
$$\cos x + \cos y = 2 \cos \tfrac{1}{2}(x + y) \cos \tfrac{1}{2}(x - y) \quad (11)$$
$$\cos x - \cos y = -2 \sin \tfrac{1}{2}(x + y) \sin \tfrac{1}{2}(x - y). \quad (12)$$

You are strongly cautioned not to confuse (9)–(12) with the identities in section 6.2. For instance, $\sin x + \sin y$ is not the same as $\sin(x + y)$, and so the identities rewriting them are not interchangeable.

EXAMPLE 6.37   Write $\cos 2\phi - \cos 6\phi$ as a product of two functions.

Solution

We shall use identity (12),

$$\cos x - \cos y = -2 \sin \tfrac{1}{2}(x + y) \sin \tfrac{1}{2}(x - y),$$

with $x = 2\phi$ and $y = 6\phi$. We find

$$\cos 2\phi - \cos 6\phi = -2 \sin \tfrac{1}{2}(2\phi + 6\phi) \sin \tfrac{1}{2}(2\phi - 6\phi)$$
$$= -2 \sin 4\phi \sin (-2\phi)$$
$$= 2 \sin 4\phi \sin 2\phi.$$

▶ **EXAMPLE 6.38** Prove that $\dfrac{\sin 5\theta + \sin 3\theta}{\sin 4\theta} = 2 \cos \theta$ is an identity.

Solution

We shall use identity (9) to rewrite the numerator of the left-hand side. Thus, with $x = 5\theta$ and $y = 3\theta$, we reduce the left side to the right side as follows:

$$\frac{\sin 5\theta + \sin 3\theta}{\sin 4\theta} = \frac{2 \sin \left(\dfrac{5\theta + 3\theta}{2}\right) \cos \left(\dfrac{5\theta - 3\theta}{2}\right)}{\sin 4\theta}$$

$$= \frac{2 \sin 4\theta \cos \theta}{\sin 4\theta}$$

$$= 2 \cos \theta.$$

▶ **EXAMPLE 6.39** Prove the identity $\dfrac{\sin 5x + \sin 3x}{\cos 5x + \cos 3x} = \tan 4x.$

Solution

$$\frac{\sin 5x + \sin 3x}{\cos 5x + \cos 3x} = \frac{2 \sin 4x \cos x}{2 \cos 4x \cos x} = \frac{\sin 4x}{\cos 4x} = \tan 4x.$$

▶ **EXAMPLE 6.40** Solve $\cos 3x + \cos x = -\cos 2x$ in the interval $[0, 2\pi)$.

Solution

We shall use identity (11), in the form

$$\cos \alpha + \cos \beta = 2 \cos \tfrac{1}{2}(\alpha + \beta) \cos \tfrac{1}{2}(\alpha - \beta),$$

to change the left-hand side of the given equation to a product. This leads to

$$2 \cos 2x \cos x = -\cos 2x$$

or

$$\cos 2x(2 \cos x + 1) = 0.$$

Equivalently, either $\cos 2x = 0$ or $\cos x = -\tfrac{1}{2}$.

The values of x in $[0, 2\pi)$ which satisfy $\cos 2x = 0$ are given by $2x = \dfrac{\pi}{2}$, $\dfrac{3\pi}{2}$, $\dfrac{5\pi}{2}$, and $\dfrac{7\pi}{2}$, or $x = \dfrac{\pi}{4}$, $\dfrac{3\pi}{4}$, $\dfrac{5\pi}{4}$, and $\dfrac{7\pi}{4}$. The values of x in $[0, 2\pi)$ which satisfy $\cos x = -\dfrac{1}{2}$ are $x = \dfrac{2\pi}{3}$ and $\dfrac{4\pi}{3}$. Hence, the required solutions are

$$x = \frac{\pi}{4}, \frac{2\pi}{3}, \frac{3\pi}{4}, \frac{5\pi}{4}, \frac{4\pi}{3}, \text{ and } \frac{7\pi}{4}.$$

Expressions of the Form $a \cos \theta + b \sin \theta$

In a number of scientific applications, it is very useful to rewrite a sum of the form $a \cos \theta + b \sin \theta$ as either a single sine or a single cosine term. We shall show next how this can be done. Of course, there will be no harm in assuming that the constants a and b are nonzero.

If we choose α so that $\tan \alpha = \frac{b}{a}$, then

$$\cos \alpha = \frac{a}{\pm\sqrt{a^2 + b^2}} \quad \text{and} \quad \sin \alpha = \frac{b}{\pm\sqrt{a^2 + b^2}}$$

or

$$a = \pm\sqrt{a^2 + b^2} \cos \alpha \quad \text{and} \quad b = \pm\sqrt{a^2 + b^2} \sin \alpha.$$

Substituting these values for a and b into the expression $a \cos \theta + b \sin \theta$, we get

$$a \cos \theta + b \sin \theta = \pm\sqrt{a^2 + b^2} \cos \alpha \cos \theta \pm \sqrt{a^2 + b^2} \sin \alpha \sin \theta$$
$$= \pm\sqrt{a^2 + b^2} (\cos \alpha \cos \theta + \sin \alpha \sin \theta)$$
$$= \pm\sqrt{a^2 + b^2} \cos (\theta - \alpha).$$

If we had chosen β so that $\tan \beta = \frac{a}{b}$ and proceeded in a similar manner, we would have found that $a \cos \theta + b \sin \theta = \pm\sqrt{a^2 + b^2} \sin (\theta + \beta)$.

In summary, we have the following results.

Expressing $a \cos u + b \sin u$ as a Sine or Cosine

If θ is any angle, then

$$a \cos \theta + b \sin \theta = c \cos (\theta - \alpha)$$
$$= c \sin (\theta + \beta),$$

where $c = \pm\sqrt{a^2 + b^2}$, $\tan \alpha = \frac{b}{a}$, and $\tan \beta = \frac{a}{b}$. (The $\pm$ sign depends on the quadrant of α or β.)

Before illustrating these formulas, we advise you to review portions of chapter 5. In particular, review how to graph functions given by expression of the forms $A \sin (Bx - C)$ and $A \cos (Bx - C)$. We also saw that such a function had several corresponding such expressions, exhibiting different phase shifts. This latter fact was borne out through exercises 23 and 24 of section 5.5, exercise 22 of section 5.6, and exercise 25 of section 5.9. Let's apply this information from chapter 5 in the following example.

▶ **EXAMPLE 6.41**   Transform $3 \cos \theta + 4 \sin \theta$ to the form $c \sin (\theta + \beta)$.

Solution

Here $a = 3$ and $b = 4$, so $\tan \beta = \dfrac{a}{b} = \dfrac{3}{4}$ and $c = \pm\sqrt{a^2 + b^2} = \pm 5$. Since $\tan \beta$ is positive, β is in quadrant I or quadrant III. Restricting β to $[0, 2\pi)$, we have that either $\beta = \arctan \tfrac{3}{4} \approx 0.6435$ radian or $\beta \approx 3.7851$ radians.

If we take $\beta \approx 0.6435$, then $\sin \beta = \tfrac{3}{5}$ and $\cos \beta = \tfrac{4}{5}$ and we must take the plus sign with $\sqrt{a^2 + b^2}$. If we take $\beta \approx 3.7851$ radians, then $\sin \beta = -\tfrac{3}{5}$, $\cos \beta = -\tfrac{4}{5}$, and we take the minus sign with $\sqrt{a^2 + b^2}$.

As a result, we get

$$3 \cos \theta + 4 \sin \theta = 5 \sin (\theta + \arctan \tfrac{3}{4})$$
$$= -5 \sin (\theta + \pi + \arctan \tfrac{3}{4}).$$

The graph of the function $3 \cos \theta + 4 \sin \theta$ is shown in figure 6.13. It is obtained either as in section 5.5 or by addition of ordinates, as in section 6.3.

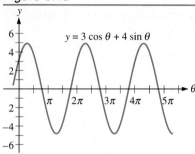

Figure 6.13

$y = 3 \cos \theta + 4 \sin \theta$

Example 6.42

Transform $y = 2 \sin 3x - 5 \cos 3x$ to the form $c \cos (\theta - \alpha)$.

Solution

Here $\theta = 3x$, $a = -5$, and $b = 2$, so $\tan \alpha = \dfrac{b}{a} = -0.4$ and $c = \pm\sqrt{a^2 + b^2} = \pm\sqrt{29}$. Restricting α to $[0, 2\pi)$, we have that α is in either quadrant II or quadrant IV; that is, either $\alpha = \pi + \arctan(-0.4) \approx 2.7611$ radians or $\alpha = 2\pi + \arctan(-0.4) \approx 5.9027$ radians.

If we take $\alpha \approx 2.7611$ radians (in quadrant II), then $\sin \alpha = \dfrac{2}{\sqrt{29}}$, $\cos \alpha = -\dfrac{5}{\sqrt{29}}$, and we must take the positive sign with $\sqrt{a^2 + b^2}$. If $\alpha \approx 5.9027$ radians (in quadrant IV), then $\sin \alpha = -\dfrac{2}{\sqrt{29}}$, $\cos \alpha = \dfrac{5}{\sqrt{29}}$, and we take the negative sign with $\sqrt{a^2 + b^2}$.

As a result,

$$2 \sin 3x - 5 \cos 3x = \sqrt{29} \cos (3x - \pi - \arctan(-0.4))$$
$$= -\sqrt{29} \cos (3x - \arctan(-0.4)).$$

This function *was* originally graphed as part of example 6.27, with the result shown again in figure 6.14.

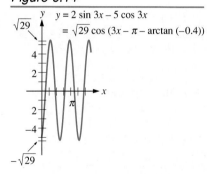

Figure 6.14

$y = 2 \sin 3x - 5 \cos 3x$
$= \sqrt{29} \cos (3x - \pi - \arctan(-0.4))$

Exercises 6.5

In exercises 1–6, rewrite the given expression as a sum or difference.

1. $\cos 5\theta \sin 3\theta$
2. $\sin 4x \cos x$
3. $\tfrac{1}{4} \cos 3x \cos 2x$
4. $3 \sin 6\alpha \sin 4\alpha$
5. $5 \sin 7\alpha \sin 3\alpha$
6. $2 \cos 3\theta \cos 8\theta$

In exercises 7–14, rewrite the given expression as a product.

7. $\cos 2\theta + \cos 6\theta$
8. $\sin 6x + \sin 2x$
9. $\sin 4x - \sin 5x$
10. $6 \sin 3\alpha - 6 \sin \alpha$
11. $\cos \tfrac{6}{5}\alpha - \cos \tfrac{4}{5}\alpha$
12. $2 \cos \tfrac{5}{6}\theta - 2 \cos \theta$
13. $4 \sin \tfrac{\theta}{2} + 4 \sin 4\theta$
14. $\cos x + \cos \pi x$

In exercises 15–18, evaluate each given expression exactly.

15. $\sin 45° \cos 75°$
16. $\cos \dfrac{7\pi}{12} \cos \dfrac{13\pi}{12}$

17. $2 \sin \dfrac{11\pi}{12} \sin \dfrac{\pi}{4}$ 18. $\cos \dfrac{23\pi}{24} \sin \dfrac{17\pi}{24}$

19. Derive formula (8) from (7), cofunction identities, an "addition" formula from section 6.2, and the fact that cos is an even function.

In exercises 20–26, prove that each given equation is an identity.

20. $\dfrac{\cos 5\theta + \cos 7\theta}{\sin 5\theta + \sin 7\theta} = \cot 6\theta$

21. $-\tan 4\theta = \dfrac{\cos 3\theta - \cos 5\theta}{\sin 3\theta - \sin 5\theta}$

22. $-\cot \dfrac{11x}{2} = \dfrac{\sin 8x - \sin 3x}{\cos 8x - \cos 3x}$

23. $\sin \left(x + \dfrac{\pi}{6}\right) + \sin \left(x - \dfrac{\pi}{6}\right) = \sqrt{3} \sin x$

24. $\cos \left(\alpha - \dfrac{\pi}{3}\right) - \cos \left(\alpha + \dfrac{\pi}{3}\right) = \sqrt{3} \sin \alpha$

25. $\tan \dfrac{\alpha + \beta}{2} = \dfrac{\sin \alpha + \sin \beta}{\cos \alpha + \cos \beta}$

26. $4 \cos x \cos 2x \sin 3x = \sin 2x + \sin 4x + \sin 6x$ (Hint: express $\sin 2x + \sin 4x$ using this section and express $\sin 6x$ using section 6.3.)

In exercises 27–34, solve the given conditional equations or inequalities in $[0, 2\pi)$.

27. $\sin 3x + \sin x = 0$ 28. $\sin 3x - \sin x = 0$
29. $\cos 3\theta + \cos \theta = 0$
30. $\sin \theta + \sin 2\theta + \sin 3\theta = 0$ (Hint: use exercise 26.)
31. $\sin \alpha + \sin 3\alpha + \sin 5\alpha = 0$
32. $\cos \alpha \cos 2\alpha \geq 0$
33. $\sin 3x - \sin x < 0$ (Hint: express the left-hand side as a product.)
34. $\cos x + \cos 2x + \cos 3x \leq 0$ (Hint: review example 6.40.)

In exercises 35–40, transform each given expression to the form $c \sin (\theta + \beta)$.

35. $3 \cos \theta + 5 \sin \theta$ 36. $3 \cos \theta - 4 \sin \theta$
37. $4 \cos \theta - 5 \sin \theta$ 38. $5 \cos \theta + 4 \sin \theta$
39. $5 \cos \theta - 4 \sin \theta$
40. $-5 \cos \theta + 4 \sin \theta$ (Hint: do this directly or use exercise 39.)

In exercises 41–46, transform each given expression to the form $c \cos (\theta - \alpha)$.

41. $\cos \theta + 2 \sin \theta$ 42. $\cos \theta - 2 \sin \theta$
43. $2 \cos \theta + \sin \theta$ 44. $-\cos \theta + \dfrac{1}{2} \sin \theta$
45. $2 \cos \theta - \sin \theta$ 46. $\cos \theta + \sin \theta$

Notice that α and β in the above exercises could always be chosen in $\left(-\dfrac{\pi}{2}, \dfrac{\pi}{2}\right)$, but that c might be negative. The next exercises use the "circular functions" approach to replace c with a nonnegative number, at the expense of permitting α and β to be replaced by a number chosen from $[0, 2\pi)$.

47. Given constants a and b, find a constant $d \geq 0$ and a constant γ in $[0, 2\pi)$ so that $a \cos \theta + b \sin \theta = d \cos (\theta - \gamma)$. (Hint: take $d = \sqrt{a^2 + b^2}$ and, reasoning as in section 5.3, find γ satisfying $a = d \cos \gamma$ and $b = d \sin \gamma$.)

48. Apply exercise 47 to rewrite each of the following as $d \cos (\theta - \gamma)$, with $d \geq 0$, $\gamma \in [0, 2\pi)$:
    (a) $\cos \theta - \sin \theta$ (b) $-4 \cos \theta + 3 \sin \theta$
    (c) $\cos \theta + 2 \sin \theta$ (d) 0

49. Given constants a and b, find a constant $d \geq 0$ and a constant δ in $[0, 2\pi)$ so that $a \cos \theta + b \sin \theta = d \sin (\theta + \delta)$. (Hint: use the addition formula for sin and reason as in exercise 47.)

50. Apply exercise 49 to rewrite each of the expressions in exercise 48 as $d \sin (\theta + \delta)$, with $d \geq 0$, $\delta \in [0, 2\pi)$.

6.6 POLAR FORM OF A COMPLEX NUMBER

In this section, you will see how trigonometry can be used to develop a new description of complex numbers. Next, you will see how this trigonometric description is a useful tool in performing the various arithmetic operations on complex numbers. Additional applications of this kind will appear in section 6.7.

Trigonometric or Polar Form of Complex Numbers

In section 1.5, you found that a complex number z can be written in the form

$$z = a + bi,$$

364
Chapter 6 / Analytic Trigonometry

where a and b are real numbers. This complex number can be also represented by the ordered pair (a, b), where a is the real part of z and b is the imaginary part. As in chapter 0 and section 1.2, this ordered pair forms the coordinates of a point in the plane. When we use rectangular coordinates in a plane to represent the complex numbers in this way, the underlying rectangular coordinate system is called the **complex plane**; the horizontal axis is called the **real axis,** and the vertical axis is the **imaginary axis.** In figure 6.15, several points have been plotted in the complex plane.

Figure 6.15

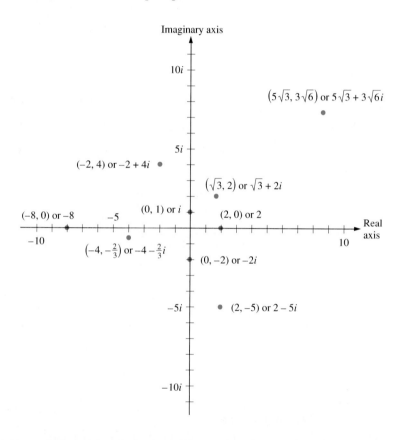

The absolute value function can be extended to the complex numbers in a geometrically natural way. Recall from chapter 0 that if x is any real number, then $|x|$ is the distance from the origin to the point x on the real number line. More generally, $|a + bi|$ is defined to be the distance in the complex plane from the origin to the point representing $a + bi$. By applying the distance formula (see figure 6.16), we see the following.

Absolute Value of a Complex Number $a + bi$

$$|a + bi| = \sqrt{a^2 + b^2}.$$

Polar Form of a Complex Number

In particular, $|a + bi|$ is a nonnegative real number; it is 0 if and only if $a + bi = 0$.

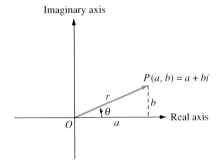

Figure 6.16

> **EXAMPLE 6.43**
>
> Find the absolute value of (a) $\sqrt{3} + 2i$, (b) $-4 - \frac{2}{3}i$, (c) $-2i$, (d) -8, and (e) 0.
>
> **Solutions**
>
> Using $|a + bi| = \sqrt{a^2 + b^2}$, we get the following results.
>
> (a) $|\sqrt{3} + 2i| = \sqrt{(\sqrt{3})^2 + 2^2} = \sqrt{3 + 4} = \sqrt{7}$
>
> (b) $|-4 - \frac{2}{3}i| = \sqrt{16 + (-\frac{2}{3})^2} = \sqrt{16 + \frac{4}{9}} = \frac{2}{3}\sqrt{37}$
>
> (c) $|-2i| = \sqrt{0 + 4} = 2$
>
> (d) $|-8| = \sqrt{64 + 0} = 8$
>
> (e) $|0| = 0$ ◀

Look again at figure 6.16. If $a + bi$ is any nonzero complex number, the ray $\overrightarrow{OP}$ can be considered the terminal side of angle θ in standard position. From the very definitions of the trigonometric functions in section 5.2, we know that $a = r \cos \theta$ and $b = r \sin \theta$. Thus, we can write $a + bi = (r \cos \theta) + (r \sin \theta)i$. Simplifying this, we obtain the desired trigonometric description of a complex number.

Trigonometric or Polar Form of a Complex Number

> For any complex number $z = a + bi$, the **trigonometric form** or **polar form of the complex number** z is
>
> $$z = r(\cos \theta + i \sin \theta),$$
>
> where $a = r \cos \theta$, $b = r \sin \theta$, and $0 \le \theta < 2\pi$ (or $0° \le \theta < 360°$). The expression $\cos \theta + i \sin \theta$ is sometimes abbreviated **cis** θ, and so the trigonometric or polar form of z can be written as
>
> $$z = r \text{ cis } \theta.$$
>
> The real number r is called the **modulus** of z, and is uniquely determined as $r = |z| = \sqrt{a^2 + b^2}$. The measure (or angle) θ, when restricted as above, is called the **principal argument** of z; θ is uniquely determined if $z \ne 0$.

In attempting to find the polar form of a given complex number z, one is often first led to $z = r \text{ cis } \alpha$, where $r = |z|$ but α is not the principal argument θ of z. By adding (or subtracting) a suitable integral multiple of 2π to α, one can, however, recover θ from α, and thus obtain the polar form of z. You developed the underlying skills in exercises 11 and 12 of section 5.1. The next example will illustrate how they can be implemented.

▶ **EXAMPLE 6.44**

Write the complex number $2 - 4i$ in trigonometric (or polar) form.

Solution

The geometric representation of $2 - 4i$ is shown in figure 6.17. As you can see, it is in the fourth quadrant, so the principal argument θ is between $\frac{3\pi}{2}$ and 2π. As for the modulus, we see from the given information that $r = \sqrt{a^2 + b^2} = \sqrt{4 + 16} = 2\sqrt{5}$. Thus, $\cos\theta = \frac{a}{r} = \frac{\sqrt{5}}{5}$ and $\sin\theta = \frac{b}{r} = -\frac{2\sqrt{5}}{5}$.

In view of the ranges of the arcsin and arccos functions (see section 5.7) and the $\begin{array}{c|c} S & A \\ \hline T & C \end{array}$ mnemonic (see section 5.4), it follows that $2 - 4i = 2\sqrt{5} \text{ cis } \alpha$, where $\alpha = \arcsin\left(-\frac{2\sqrt{5}}{5}\right) \approx -63.4349° \approx -1.1071$ radians. Thus, $\theta = \alpha + 2\pi \approx 5.1760$ radians, giving the desired polar form

$$2 - 4i = 2\sqrt{5} \text{ cis }\left(\arcsin\left(-\frac{2\sqrt{5}}{5}\right) + 2\pi\right)$$

$$\approx 2\sqrt{5} \text{ cis } 5.1760. \quad ◀$$

Figure 6.17

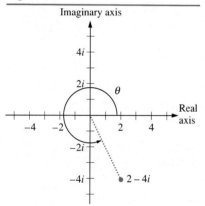

▶ **EXAMPLE 6.45**

Express $-2 + 2\sqrt{3}i$ in trigonometric form.

Solution

In this case, $r = |-2 + 2\sqrt{3}i| = \sqrt{4 + 12} = \sqrt{16} = 4$. Moreover, $\cos\theta = \frac{-2}{4} = -\frac{1}{2}$ and $\sin\theta = \frac{2\sqrt{3}}{4} = \frac{\sqrt{3}}{2}$. As shown in figure 6.18, the point

Figure 6.18

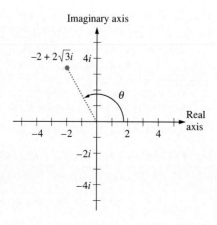

Polar Form of a Complex Number

representing $-2 + 2\sqrt{3}\,i$ is in quadrant II, and so we find θ as $\arccos\left(-\frac{1}{2}\right) = \frac{2\pi}{3}$. Thus, we have

$$-2 + 2\sqrt{3}\,i = r \text{ cis } \theta = 4 \text{ cis } \frac{2\pi}{3}.$$

◀

EXAMPLE 6.46 Express $-2 - 2\sqrt{3}\,i$ in trigonometric form.

Solution In this case, $r = |-2 - 2\sqrt{3}\,i| = \sqrt{4 + 12} = \sqrt{16} = 4$. This time, $\cos\theta = \frac{-2}{4} = -\frac{1}{2}$ and $\sin\theta = \frac{-2\sqrt{3}}{4} = -\frac{\sqrt{3}}{2}$. As the point representing $-2 - 2\sqrt{3}\,i$ is in quadrant III, we find θ as $\pi - \arcsin\left(-\frac{\sqrt{3}}{2}\right) = \pi - \left(-\frac{\pi}{3}\right) = \frac{4\pi}{3}$. Thus, we have

$$-2 - 2\sqrt{3}\,i = r \text{ cis } \theta = 4 \text{ cis } \frac{4\pi}{3}.$$

◀

EXAMPLE 6.47 Express the complex number $8 \text{ cis } \frac{7\pi}{6}$ in the standard or rectangular form $a + bi$.

Solution Since $\cos\frac{7\pi}{6} = -\frac{1}{2}\sqrt{3}$ and $\sin\frac{7\pi}{6} = -\frac{1}{2}$, we have

$$8 \text{ cis } \frac{7\pi}{6} = 8\left(\cos\frac{7\pi}{6} + i\sin\frac{7\pi}{6}\right)$$
$$= 8\left(-\tfrac{1}{2}\sqrt{3} - \tfrac{1}{2}i\right)$$
$$= -4\sqrt{3} - 4i.$$

◀

Multiplication and Division of Complex Numbers

Multiplication or division of complex numbers is often easier to carry out when the numbers are written in trigonometric form.

Let $z_1 = r_1(\cos\theta_1 + i\sin\theta_1)$ and $z_2 = r_2(\cos\theta_2 + i\sin\theta_2)$ be two complex numbers expressed in trigonometric form. By definition of the product of complex numbers, we get

$$z_1 \cdot z_2 = r_1 r_2(\cos\theta_1 \cos\theta_2 + i\cos\theta_1 \sin\theta_2 + i\sin\theta_1 \cos\theta_2 + i^2 \sin\theta_1 \sin\theta_2)$$
$$= r_1 r_2[(\cos\theta_1 \cos\theta_2 - \sin\theta_1 \sin\theta_2) + i(\sin\theta_1 \cos\theta_2 + \cos\theta_1 \sin\theta_2)].$$

However, we can now apply the addition formulas

$$\cos\theta_1 \cos\theta_2 - \sin\theta_1 \sin\theta_2 = \cos(\theta_1 + \theta_2)$$
$$\sin\theta_1 \cos\theta_2 + \cos\theta_1 \sin\theta_2 = \sin(\theta_1 + \theta_2)$$

from section 6.2. Hence we obtain the following.

Chapter 6 / Analytic Trigonometry

Multiplication of Complex Numbers

> The product of two complex numbers $z_1 = r_1(\cos\theta_1 + i\sin\theta_1) = r_1 \text{ cis } \theta_1$ and $z_2 = r_2(\cos\theta_2 + i\sin\theta_2) = r_2 \text{ cis } \theta_2$ is given by
>
> $$z_1 \cdot z_2 = r_1 r_2 \left[\cos(\theta_1 + \theta_2) + i\sin(\theta_1 + \theta_2)\right] = r_1 r_2 \text{ cis } (\theta_1 + \theta_2).$$

By uniqueness of the modulus in polar form, this gives an easy proof that $|z_1 z_2| = |z_1||z_2|$, thus extending a fact obtained in chapter 0 about the absolute value function of real numbers. In addition, if θ_1 is the principal argument of z_1 and θ_2 is the principal argument of z_2, then the principal argument of $z_1 z_2$ is either $\theta_1 + \theta_2$ or $\theta_1 + \theta_2 - 2\pi$.

This result can be extended to three or more factors.

Similar reasoning leads to the following formula for the quotient of complex numbers.

Division of Complex Numbers

> The quotient of two complex numbers $z_1 = r_1(\cos\theta_1 + i\sin\theta_1) = r_1 \text{ cis } \theta_1$ and $z_2 = r_2(\cos\theta_2 + i\sin\theta_2) = r_2 \text{ cis } \theta_2 \neq 0$ is given by
>
> $$\frac{z_1}{z_2} = \frac{r_1}{r_2}\left[\cos(\theta_1 - \theta_2) + i\sin(\theta_1 - \theta_2)\right] = \frac{r_1}{r_2} \text{ cis } (\theta_1 - \theta_2).$$

▶ **EXAMPLE 6.48** Multiply $-1 + \sqrt{3}i$ and $2\sqrt{3} + 2i$.

Solution As in example 6.45, we find the polar form $z_1 = -1 + \sqrt{3}i = 2 \text{ cis } \frac{2\pi}{3} = r_1 \text{ cis } \theta_1$. Moreover for $z_2 = 2\sqrt{3} + 2i = r_2 \text{ cis } \theta_2$, we have $r_2 = \sqrt{12 + 4} = 4$, $\cos\theta_2 = \frac{\sqrt{3}}{2}$, $\sin\theta_2 = \frac{1}{2}$, and $0 < \theta_2 < \frac{\pi}{2}$. Hence $\theta_2 = \arcsin\frac{1}{2} = \frac{\pi}{6}$, and so $z_2 = 4 \text{ cis } \frac{\pi}{6}$.

Consequently, the product is

$$(-1 + \sqrt{3}i)(2\sqrt{3} + 2i) = z_1 z_2 = (r_1 \text{ cis } \theta_1)(r_2 \text{ cis } \theta_2)$$
$$= r_1 r_2 \text{ cis } (\theta_1 + \theta_2)$$
$$= 2 \cdot 4 \text{ cis }\left(\frac{2\pi}{3} + \frac{\pi}{6}\right)$$
$$= 8 \text{ cis } \frac{5\pi}{6}.$$

If this is converted to standard form, we get

$$z_1 z_2 = 8\left(\cos\frac{5\pi}{6} + i\sin\frac{5\pi}{6}\right) = 8\left(-\frac{\sqrt{3}}{2} + \frac{1}{2}i\right) = -4\sqrt{3} + 4i.$$

You should check that the methods of section 1.5 would lead to the same answer. ◀

Polar Form of a Complex Number

EXAMPLE 6.49 Multiply $1 + \sqrt{3}i$ and $3\sqrt{2} - 3\sqrt{2}i$.

Solution Let $z_1 = 1 + \sqrt{3}i$. Then its absolute value is $r_1 = \sqrt{1 + 3} = 2$ and you can check that its principal argument $\theta_1 = \arcsin\left(\frac{\sqrt{3}}{2}\right) = \frac{\pi}{3}$. As a result, $z_1 = 2\left(\cos\frac{\pi}{3} + i\sin\frac{\pi}{3}\right) = 2 \operatorname{cis}\frac{\pi}{3}$.

For $z_2 = 3\sqrt{2} - 3\sqrt{2}i$, we proceed as in example 6.44, finding $r_2 = 6$, $\alpha = \arcsin\left(-3\frac{\sqrt{2}}{6}\right) = -\frac{\pi}{4}$, and $\theta_2 = \alpha + 2\pi = \frac{7\pi}{4}$. So, $z_2 = r_2 \operatorname{cis}\theta_2 = 6 \operatorname{cis}\frac{7\pi}{4}$.

Consequently, the product is

$$(1 + \sqrt{3}i)(3\sqrt{2} - 3\sqrt{2}i) = z_1 z_2 = r_1 r_2 \operatorname{cis}(\theta_1 + \theta_2)$$

$$= 2 \cdot 6 \operatorname{cis}\left(\frac{\pi}{3} + \frac{7\pi}{4}\right)$$

$$= 12 \operatorname{cis}\frac{25\pi}{12} = 12 \operatorname{cis}\frac{\pi}{12}$$

$$= 12\left(\cos\frac{\pi}{12} + i\sin\frac{\pi}{12}\right).$$

By the half-angle identities in section 6.3,

$$\cos\frac{\pi}{12} = \sqrt{\frac{1 + \cos(\pi/6)}{2}} = \sqrt{\frac{1 + \sqrt{3}/2}{2}} = \frac{\sqrt{2 + \sqrt{3}}}{2}$$

and, similarly, $\sin\frac{\pi}{12} = \frac{\sqrt{2 - \sqrt{3}}}{2}$. Thus, we may convert to standard form, getting $z_1 z_2 = 6\sqrt{2 + \sqrt{3}} + 6\sqrt{2 - \sqrt{3}}\,i$. Notice that if you also work the example by the methods of section 1.5, then comparing answers leads to the additional facts $6\sqrt{2 + \sqrt{3}} = 3\sqrt{2} + 3\sqrt{6}$ and $6\sqrt{2 - \sqrt{3}} = 3\sqrt{6} - 3\sqrt{2}$. ◀

EXAMPLE 6.50 Divide $2 + 2\sqrt{3}i$ by $\sqrt{3} + i$.

Solution Since $z_1 = 2 + 2\sqrt{3}i = 4 \operatorname{cis}\frac{\pi}{3} = r_1 \operatorname{cis}\theta_1$ and $z_2 = \sqrt{3} + i = 2 \operatorname{cis}\frac{\pi}{6} = r_2 \operatorname{cis}\theta_2$, we have

$$\frac{2 + 2\sqrt{3}i}{\sqrt{3} + i} = \frac{z_1}{z_2} = \frac{r_1}{r_2}[\operatorname{cis}(\theta_1 - \theta_2)]$$

$$= \frac{4}{2}\left[\operatorname{cis}\left(\frac{\pi}{3} - \frac{\pi}{6}\right)\right]$$

$$= 2 \operatorname{cis}\frac{\pi}{6} = 2\left(\cos\frac{\pi}{6} + i\sin\frac{\pi}{6}\right)$$

$$= 2\left(\tfrac{1}{2}\sqrt{3} + \tfrac{1}{2}i\right) = \sqrt{3} + i.$$ ◀

Chapter 6 / Analytic Trigonometry

Exercises 6.6

In exercises 1 and 2, plot each of the given points in the complex plane.

1. (a) $-4i$ (b) 7 (c) $\frac{3}{2} + 2i$
   (d) $-2 + \frac{i}{\sqrt{2}}$ (e) $\frac{3}{i}$

2. (a) $3 - \frac{2}{5}i$ (b) $-1 - 3i$ (c) π
   (d) $3i$ (e) $(3 - i)(4i)$

In exercises 3 and 4, calculate the absolute value of each of the given complex numbers.

3. (a) $-1 + \sqrt{2}i$ (b) $\frac{5}{13} - \frac{12}{13}i$
   (c) -2 (d) 0 (e) $-3 - 4i$ (f) $2i^{407}$

4. (a) $(\sqrt{3} + i)(4i)$ (b) $\frac{5}{i}$
   (c) $4 + 3i$ (d) $4 - 3i$ (e) $-6i$

5. Let r be any real number, viewed as a complex number. Prove that the modulus $|r|$ is the same as the absolute value of r (as defined in chapter 0). (Hint: just *how* is r viewed as a complex number?)

6. Prove, for any complex number z, that $|-z| = |z|$. (Hint: first consider the case $z = 1$.)

7. Prove directly from the definition of modulus that $|z_1 z_2| = |z_1||z_2|$ for all complex numbers z_1 and z_2.

8. Use exercise 7 to explain why $z_1 z_2 = 0$, for complex numbers z_1 and z_2, entails that either $z_1 = 0$ or $z_2 = 0$ (or both).

In exercises 9–20, find the trigonometric (polar) form of each of the given complex numbers. (Approximate the principal argument only if necessary.)

9. $2 + 4i$
10. $-2 + 4i$
11. $-2 - 4i$
12. $1 - i$
13. r, where r is a positive real number
14. $-r$, where r is a positive real number
15. $-i$
16. $2i$
17. $3 - 4i$
18. $-3 - 3i$
19. $-3 + 4i$
20. $4 + 4\sqrt{3}i$

21. Find the radius of the circle in the complex plane that is centered at the origin and passes through $-12 + 5i$.

22. Prove that the graph of $|z| = 5$ in the complex plane (that is, the set of complex numbers z such that $|z| = 5$) is the circle that is centered at the origin and has radius 5.

23. (a) If z_1 is a complex number and r is a positive real number, prove that the graph of $|z - z_1| = r$ in the complex plane is the circle that is centered at z_1 and has radius r. (Hint: convert to rectangular coordinates or interpret modulus geometrically.)

(b) Using (a), identify the graph of $|z - 2 + 3i| = 4$ in the complex plane.
(c) Using (a), identify the graph of $|2z - i| = 3$ in the complex plane. (Hint: first find an equivalent equation of the form $|z - z_1| = r$.)

24. Explain why cis $(\alpha + n2\pi)$ = cis α for each real number α and each integer n.

25. If α and β are real numbers such that cis α = cis β, prove that $\alpha = \beta + n2\pi$ for some integer n. (Hint: see exercise 11(b) in section 5.2.)

26. Show that cis $(\theta + \pi) = -$cis θ for each real number θ. (Hint: first consider cos $(\theta + \pi)$ and sin $(\theta + \pi)$.)

27. (a) Prove that $|\text{cis } \theta| = 1$ for each real number θ.
    (b) Prove that if z is a complex number such that $|z| = 1$, then there exists a real number θ such that $z = $ cis θ.

In exercises 28 and 29, write each of the given complex numbers in rectangular form. (Use your calculator to approximate only if necessary.)

28. (a) $4 \text{ cis } \frac{\pi}{6}$ (b) $3 \text{ cis } \frac{\pi}{2}$
    (c) $\frac{3}{2} \text{ cis } \pi$ (d) $5 \text{ cis } \frac{13\pi}{12}$

29. (a) $\frac{1}{2} \text{ cis } 0$ (b) $4 \text{ cis } \frac{4\pi}{3}$
    (c) $\sqrt{2} \text{ cis } \frac{7\pi}{4}$ (d) $-4 \text{ cis } \frac{13\pi}{2}$

In exercises 30–37, find both the polar form and the rectangular form for each of the given products and quotients. (Approximate these forms by calculator only when necessary.)

30. $(1 - \sqrt{3}i)(-5 + 5i)$
31. $\frac{\sqrt{3} - i}{1 + i}$
32. $\frac{2i}{-1 + \sqrt{3}i}$
33. $(3 + 4i)(-12 - 5i)$
34. $\left(2 \text{ cis } \frac{\pi}{6}\right)\left(3 \text{ cis } \frac{\pi}{3}\right)$
35. $\frac{\sqrt{6} \text{ cis } \frac{\pi}{4}}{\sqrt{2} \text{ cis } \frac{\pi}{6}}$
36. $\frac{\sqrt{6} \text{ cis } \frac{\pi}{6}}{\sqrt{2} \text{ cis } \frac{\pi}{4}}$
37. $\left(3 \text{ cis } \frac{11\pi}{6}\right)\left(4 \text{ cis } \frac{\pi}{4}\right)$

38. If z_1 and z_2 are nonzero complex numbers, prove that $\left|\frac{z_1}{z_2}\right| = \frac{|z_1|}{|z_2|}$. (Hint: review the text's discussion of $|z_1 z_2|$.)

◆ 39. If z is a nonreal complex number with polar form $r \text{ cis } \theta$, prove that z^{-1} has the polar form $r^{-1} \text{ cis } (2\pi - \theta)$.

40. Given nonzero complex numbers with polar forms $z_1 = r_1 \text{ cis } \theta_1$ and $z_2 = r_2 \text{ cis } \theta_2$, prove that the principal argument of $\dfrac{z_1}{z_2}$ is either $\theta_1 - \theta_2$ or $\theta_1 - \theta_2 + 2\pi$. Determine under precisely what conditions this principal argument is $\theta_1 - \theta_2$.

41. (a) Show that
$$(r_1 \text{ cis } \theta_1)(r_2 \text{ cis } \theta_2)(r_3 \text{ cis } \theta_3) = r_1 r_2 r_3 \text{ cis } (\theta_1 + \theta_2 + \theta_3)$$
for all real numbers $r_1, r_2, r_3, \theta_1, \theta_2$, and θ_3.
(b) Using (a), show that if z has polar form $r \text{ cis } \theta$, then $z^3 = r^3 \text{ cis } 3\theta$.
(c) Using (b) with $r = 1$, recover the identities describing $\sin 3\theta$ and $\cos 3\theta$ in exercises 41 and 42 of section 6.2.

42. Give a direct proof of the equation $6\sqrt{2 + \sqrt{3}} = 3\sqrt{2} + 3\sqrt{6}$ which was established in example 6.49.

43. Give a direct proof of the equation $6\sqrt{2 - \sqrt{3}} = 3\sqrt{6} - 3\sqrt{2}$ which was established in example 6.49. (Hint: you'll need to notice that $\sqrt{3} > 1$.)

44. (Together with exercises 22 and 23, this helps to motivate the study of polar coordinates in chapter 10.)
(a) Prove that the graph in the complex plane of $\{z: z$ is a complex number and the principal argument of z is $\dfrac{\pi}{4}\}$ consists of all the points $x + xi$ such that $x > 0$.
(b) Prove that the graph in the complex plane of $\{z: z$ is a complex number and $z = r \text{ cis } \dfrac{\pi}{4}$ for some real number $r\}$ consists of all the points $x + xi$ (such that x is a real number). (Hint: combine (a) with exercise 26.)

45. Let z be a nonzero complex number viewed in the complex plane. If z is rotated 90° counterclockwise about the origin, show that the resulting point is iz. (Hint: use the polar form of z, and recall that
$$i = \text{cis } \frac{\pi}{2}.)$$

46. (This generalizes exercise 45.) Let z_1 and z_2 be nonzero complex numbers, with $r_1 \text{ cis } \theta_1$ as the polar form of z_1. Explain why $z_1 z_2$ may be obtained in the complex plane from z_2 as follows: rotate z_2 counterclockwise θ_1 radians about the origin and then alter the resulting point's modulus (leaving its principal argument alone) by a factor of r_1.

47. In exercise 46, does $z_1 z_2$ necessarily result if you first alter the modulus of z_2 by a factor of r_1 and then rotate the resulting point counterclockwise θ_1 radians about the origin?

48. Do the analogues of exercises 46 and 47 for $\dfrac{z_1}{z_2}$ instead of $z_1 z_2$.

49. Let z_1, with polar form $r_1 \text{ cis } \theta_1$, be a fixed complex number. Let $\mathcal{R}$ be a rectangle in the complex plane having vertices 0, a, $a + bi$, and bi, for suitable positive (real) numbers a and b. Using exercise 46 or exercise 47, explain intuitively why multiplication by z_1 converts $\mathcal{R}$ into another rectangle, say $\mathcal{S}$. Then show that the area of $\mathcal{S}$ is the product of r_1^2 and the area of $\mathcal{R}$.

50. (a) Let z_1 be as in exercise 49, but replace $\mathcal{R}$ with a triangle $\mathcal{T}$ having vertices 0, b, and $c + hi$ for suitable positive numbers b, c, and h. Explain why multiplication by z_1 converts $\mathcal{T}$ into a triangle whose area is r_1^2 times the area of $\mathcal{T}$.
(b) Determine whether the "r_1^2 phenomenon" noticed in (a) and exercise 49 persists if $\mathcal{R}$ is replaced with a circle centered at the origin. (You may wish to use a graphing calculator or computer graphing program to see if the phenomenon persists for other geometric figures, such as parallelograms or trapezoids.)

6.7 POWERS AND ROOTS OF COMPLEX NUMBERS

In the last section, we learned the trigonometric form for complex numbers and how this form could expedite multiplication and division of complex numbers. In this section, we continue investigating uses of the trigonometric form of complex numbers and we show how it can speed the process of finding powers and roots of complex numbers. All this follows from a theoretical result, De Moivre's Theorem, to which we now turn.

Powers of Complex Numbers and De Moivre's Theorem

If $z = r(\cos \theta + i \sin \theta)$ in trigonometric form, then using last section's rule for multiplying complex numbers, we see that $z^2 = r^2(\cos 2\theta + i \sin 2\theta)$. In the same way,

$$z^3 = z^2 \cdot z = (r^2 \text{ cis } 2\theta)(r \text{ cis } \theta)$$
$$= r^3(\text{cis } 3\theta) = r^3(\cos 3\theta + i \sin 3\theta).$$

Similar computations would show that $z^4 = r^4(\cos 4\theta + i \sin 4\theta)$ and $z^5 = r^5(\cos 5\theta + i \sin 5\theta)$. A fairly obvious pattern has emerged. This pattern, known as **De Moivre's Theorem,*** is expressed formally below.

De Moivre's Theorem

If $z = r \text{ cis } \theta = r(\cos \theta + i \sin \theta)$ is the polar form of a complex number z and if n is a positive integer, then

$$z^n = r^n \text{ cis } n\theta = r^n (\cos n\theta + i \sin n\theta).$$

In other words, the modulus of the nth power of z is the nth power of the modulus of z; and if $z \neq 0$, the principal argument of the nth power of z differs by an integral multiple of 2π from the product of n and the principal argument of z.

▶ **EXAMPLE 6.51**

Find $(-1 + \sqrt{3}i)^8$ in trigonometric form and express the result in standard or rectangular form.

Solution

We begin by converting $-1 + \sqrt{3}i$ to trigonometric form. Its modulus is $r = \sqrt{1+3} = \sqrt{4} = 2$. Its principal argument θ satisfies $\cos \theta = -\frac{1}{2}$ and $\sin \theta = \frac{\sqrt{3}}{2}$. Hence $\theta = \frac{2\pi}{3}$; it follows that $-1 + \sqrt{3}i = 2 \text{ cis } \frac{2\pi}{3}$.

By De Moivre's Theorem, we have

$$(-1 + \sqrt{3}i)^8 = \left(2 \text{ cis } \frac{2\pi}{3}\right)^8$$

$$= 2^8 \text{ cis } \left(8 \cdot \frac{2\pi}{3}\right)$$

$$= 256 \text{ cis } \frac{16\pi}{3}$$

$$= 256 \text{ cis } \left[2(2\pi) + \frac{4\pi}{3}\right]$$

$$= 256 \text{ cis } \frac{4\pi}{3}$$

---

*A proof of De Moivre's Theorem by means of mathematical induction is contained in section 10.5.

Powers and Roots of Complex Numbers

$$= 256\left(\cos\frac{4\pi}{3} + i\sin\frac{4\pi}{3}\right)$$
$$= 256\left(-\frac{1}{2} - \frac{1}{2}\sqrt{3}\,i\right)$$
$$= -128 - 128\sqrt{3}\,i. \quad \blacktriangleleft$$

▶ **EXAMPLE 6.52** Find $[\sqrt{5}(\cos 15° + i\sin 15°)]^{12}$ and express the result in standard form.

Solution By De Moivre's Theorem,

$$(\sqrt{5}\operatorname{cis} 15°)^{12} = (\sqrt{5})^{12}[\operatorname{cis}(12)15°]$$
$$= 15{,}625 \operatorname{cis} 180°$$
$$= 15{,}625(\cos 180° + i\sin 180°)$$
$$= -15{,}625. \quad \blacktriangleleft$$

Notice that the result in example 6.52 was a real number. That should not come as a surprise. After all, $i^2 = -1$ is also real. In general, if θ is the principal argument of z, then z^n is real if and only if $n\theta$ is an integral multiple of π.

De Moivre's Theorem also holds when n is a negative integer. This can be derived from the case stated above. To see this, we let $n = -m$, where m is a positive integer. Then $(r \operatorname{cis} \theta)^n = (r \operatorname{cis} \theta)^{-m} = [(r \operatorname{cis} \theta)^m]^{-1} = (r^m \operatorname{cis} m\theta)^{-1}$. By exercise 39 of section 6.6, this is just $(r^m)^{-1} \operatorname{cis}(2\pi - m\theta) = r^{-m}\operatorname{cis}(-m\theta) = r^n \operatorname{cis} n\theta = r^n(\cos n\theta + i\sin n\theta)$.

▶ **EXAMPLE 6.53** Find $(-1 + \sqrt{3}\,i)^{-8}$ in trigonometric form and express the result in standard form.

Solution From example 6.51, we know that $-1 + \sqrt{3}\,i = 2 \operatorname{cis} \frac{2\pi}{3}$. Using the above-noted extension of De Moivre's Theorem to negative exponents, we obtain

$$(-1 + \sqrt{3}\,i)^{-8} = \left(2 \operatorname{cis} \frac{2\pi}{3}\right)^{-8}$$
$$= 2^{-8} \operatorname{cis}\left[(-8)\frac{2\pi}{3}\right]$$
$$= \frac{1}{256} \operatorname{cis}\left(-\frac{16\pi}{3}\right)$$
$$= \frac{1}{256} \operatorname{cis}\left[-3(2\pi) + \frac{2\pi}{3}\right]$$
$$= \frac{1}{256} \operatorname{cis} \frac{2\pi}{3}$$

$$= \frac{1}{256}\left(\cos\frac{2\pi}{3} + i\sin\frac{2\pi}{3}\right)$$

$$= \tfrac{1}{256}\left(-\tfrac{1}{2} + \tfrac{1}{2}\sqrt{3}\,i\right)$$

$$= -\tfrac{1}{512} + \tfrac{1}{512}\sqrt{3}\,i. \quad \blacktriangleleft$$

Roots of Complex Numbers

If $z = r(\cos\theta + i\sin\theta)$ is the polar form of a nonzero complex number and if $w = R(\cos\phi + i\sin\phi)$ is the polar form of a complex number w such that $w^n = z$, then w is called a **complex nth root of z**. (Here, as usual, n is a positive integer.) Now, by De Moivre's Theorem,

$$w^n = (R\operatorname{cis}\phi)^n = R^n \operatorname{cis} n\phi = r\operatorname{cis}\theta \qquad (*)$$

Now, two nonzero complex numbers are equal if and only if their absolute values are equal and their principal arguments are equal. From (*), we see that the condition $w^n = z$ is equivalent to

$$R^n = r \quad\text{and}\quad n\phi = \theta + 2k\pi,$$

where k is an integer. Equivalently,

$$R = \sqrt[n]{r} \quad\text{and}\quad \phi = \frac{\theta}{n} + \frac{2k\pi}{n},$$

where k is an integer and $\sqrt[n]{r}$ denotes the principal nth root of the positive number r.

By substituting these values of R and ϕ into the polar form of w, we obtain the nth roots of z. Stated formally, this says the following.

---

nth Roots of a Complex Number

Let z be any nonzero complex number and n any positive integer. Then z has exactly n distinct nth roots. If $z = r\operatorname{cis}\theta$ in polar form, then the nth roots of z are given by

$$w_k = \sqrt[n]{r}\,\operatorname{cis}\left(\frac{\theta}{n} + \frac{2k\pi}{n}\right),$$

where k takes the values $0, 1, 2, 3, \ldots, n-1$. (If θ is measured in degrees rather than radians, then replace $\dfrac{\theta + 2k\pi}{n}$ by $\dfrac{\theta + 360°k}{n}$.)

▶ **EXAMPLE 6.54**

Find all fourth roots of $z = 8 - 8\sqrt{3}\,i$.

Solution

We begin by writing $8 - 8\sqrt{3}\,i$ in trigonometric form as

$$z = 8 - 8\sqrt{3}\,i = 16\operatorname{cis}\frac{5\pi}{3}.$$

Here $r = 16$ and $\theta = \dfrac{5\pi}{3}$. Each of the fourth roots of z has modulus $\sqrt[4]{16} = 2$. These roots have principal arguments as follows:

$$\text{for } k = 0: \quad \frac{5\pi/3}{4} + \frac{2\pi(0)}{4} = \frac{5\pi}{12}$$

$$\text{for } k = 1: \quad \frac{5\pi/3}{4} + \frac{2\pi(1)}{4} = \frac{11\pi}{12}$$

$$\text{for } k = 2: \quad \frac{5\pi/3}{4} + \frac{2\pi(2)}{4} = \frac{17\pi}{12}$$

$$\text{for } k = 3: \quad \frac{5\pi/3}{4} + \frac{2\pi(3)}{4} = \frac{23\pi}{12}.$$

Using these angles, we find that the fourth roots of $8 - 8\sqrt{3}\,i$ are

$$w_0 = 2\left(\cos\frac{5\pi}{12} + i\sin\frac{5\pi}{12}\right)$$

$$w_1 = 2\left(\cos\frac{11\pi}{12} + i\sin\frac{11\pi}{12}\right)$$

$$w_2 = 2\left(\cos\frac{17\pi}{12} + i\sin\frac{17\pi}{12}\right)$$

$$w_3 = 2\left(\cos\frac{23\pi}{12} + i\sin\frac{23\pi}{12}\right).$$

As shown in figure 6.19, these four roots all lie on the circle that is centered at the origin and has radius 2. Notice that the roots are equally spaced about the circle, and are $\dfrac{2\pi}{4} = \dfrac{\pi}{2}$ radians apart. You will be asked to show in exercise 28(a) that this is a general phenomenon. ◀

Figure 6.19

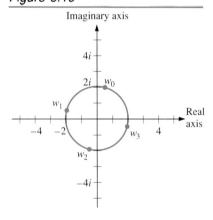

▶ **EXAMPLE 6.55**   Find the (five) fifth roots of 1.

Solution   First, let's write 1 in trigonometric form:

$$1 = 1(\cos 0° + i\sin 0°).$$

Then, by the nth root formula, with $n = 5$, $r = 1$, and $\theta = 0°$, the requested roots have the form

$$\sqrt[5]{1}\left[\cos\frac{(0° + 360°k)}{5} + i\sin\frac{(0° + 360°k)}{5}\right]$$

or simply cis $72°k$, for $k = 0, 1, 2, 3, 4$.

Thus, the fifth roots of 1 are

$$w_0 = 0$$

$$w_1 = \text{cis } 72°$$

Chapter 6 / Analytic Trigonometry

Figure 6.20

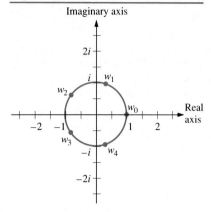

$w_2 = \text{cis } 144°$

$w_3 = \text{cis } 216°$

$w_4 = \text{cis } 288°$.

These five roots are plotted in figure 6.20. Notice again the phenomenon that we pointed out in example 6.54: these roots are equally spaced (this time, 72° apart) around a circle centered at the origin (this time, the unit circle). You can also see that the nonreal complex roots occur in conjugate pairs. This example illustrates an important special case: the (n distinct) nth roots of 1 are called the **nth roots of unity.**

Let $w_1, w_2, \ldots, w_n$ be the (n distinct) nth roots of a nonzero complex number z. In other words, the w_k's are the roots of the polynomial $x^n - z$. As we saw in section 3.4, this means that

$$x^n - z = (x - w_1)(x - w_2) \cdots (x - w_n).$$

You will be requested to find applications of this factorization in exercises 32 and 33.

Now, recall that the fundamental theorem of algebra (as in section 3.4) guarantees that any nth degree complex polynomial has n complex roots, counting multiplicities. We have seen that in the case of the polynomial $x^n - z$, each of these roots has multiplicity 1, and we have even found formulas for these roots w_k. Having a formula for the roots of a given polynomial is, of course, quite rare. One formula that applies to appropriate (that is, second degree) complex polynomials is the quadratic formula (from section 1.5). You will be asked to find complex applications of it in exercises 46–50, using the material of this section.

EXAMPLE 6.56

Solve $x^5 - i = 0$.

Solution

Let us consider the equivalent equation $x^5 = i$. Its roots (and so the roots of the given equation) are the fifth roots of i. To find them, we first write i in trigonometric form:

$$i = 0 + i = 1 \text{ cis } \frac{\pi}{2}.$$

The modulus of any fifth root of i is $\sqrt[5]{1} = 1$. The principal arguments of these fifth roots are given by

$$\frac{\pi/2 + 2k\pi}{5}, \quad k = 0, 1, 2, 3, 4.$$

By using these data, we see that the fifth roots of i are cis $\frac{\pi}{10}$, cis $\frac{\pi}{2}$, cis $\frac{9\pi}{10}$, cis $\frac{13\pi}{10}$, and cis $\frac{17\pi}{10}$. The second of these roots is i. The others

Powers and Roots of Complex Numbers

cannot be expressed easily in standard form. However, with the aid of a calculator, we can approximate them as $0.951 + 0.309i$, $-0.951 + 0.309i$, $-0.588 - 0.809i$, and $0.588 - 0.809i$. ◀

Exercise 24 of section 6.3 expressed $\cos 4\theta$ in terms of $\cos \theta$. Earlier, example 6.17 had expressed $\cos 3\theta$ as $4\cos^3 \theta - 3 \cos \theta$. In fact, one can develop a general method for expressing $\sin n\theta$ and $\cos n\theta$ in terms of $\sin \theta$ and $\cos \theta$, respectively. This method depends on De Moivre's Theorem and the ability to expand $(\cos \theta + i \sin \theta)^n$. A quick way to do the latter via the Binomial Theorem is discussed in section 10.6. For the moment, however, we illustrate the general method in the next example.

▶ **EXAMPLE 6.57**   Prove the identity $\sin 3\theta = 3 \sin \theta - 4 \cos^3 \theta$.

Solution   From De Moivre's Theorem, $\cos 3\theta + i \sin 3\theta = (\cos \theta + i \sin \theta)^3$. In chapter 0, we saw that $(c + d)^3 = c^3 + 3c^2d + 3cd^2 + d^3$. If we replace c with $\cos \theta$ and d with $i \sin \theta$, then

$$(\cos \theta + i \sin \theta)^3 = \cos^3 \theta + 3i \cos^2 \theta \sin \theta + 3i^2 \cos \theta \sin^2 \theta + i^3 \sin^3 \theta.$$

Using the identities $\cos^2 \theta = 1 - \sin^2 \theta$ and $\sin^2 \theta = 1 - \cos^2 \theta$ and the facts that $i^2 = -1$ and $i^3 = -i$, we can rewrite this as

$$(\cos \theta + i \sin \theta)^3$$
$$= \cos^3 \theta + 3i(1 - \sin^2 \theta) \sin \theta - 3 \cos \theta (1 - \cos^2 \theta) - i \sin^3 \theta$$
$$= \cos^3 \theta + 3i \sin \theta - 3i \sin^3 \theta - 3 \cos \theta + 3 \cos^3 \theta - i \sin^3 \theta$$
$$= (4 \cos^3 \theta - 3 \cos \theta) + i(3 \sin \theta - 4 \sin^3 \theta).$$

Thus, $\cos 3\theta + i \sin 3\theta = (4 \cos^3 \theta - 3 \cos \theta) + i(3 \sin \theta - 4 \sin^3 \theta)$.
Equating the corresponding real and imaginary parts, we have established not only the desired identity

$$\sin 3\theta = 3 \sin \theta - 4 \sin^3 \theta$$

but also the earlier identity from example 6.17,

$$\cos 3\theta = 4 \cos^3 \theta - 3 \cos \theta.$$

◀

Exercises 6.7

For each of the numbers in exercises 1–10, find the polar form and hence (approximating via calculator only if necessary) find the standard form.

1. $\left(3 \text{ cis } \dfrac{\pi}{4}\right)^5$
2. $(\sqrt{3} - i)^6$
3. $(-1 + \sqrt{3}i)^5$
4. $(2 \cos 120° - i2 \sin 120°)^3$
5. $(1 - i)^{18}$
6. $(-3 - 4i)^{-7}$
7. $\dfrac{81}{(4 + 3i)^5}$
8. $\dfrac{(1 + i)^3}{(8 - 8\sqrt{3}i)^4}$
9. $(\text{cis } 2\theta)^{-2}$, where θ is a real number
10. $\left(2 \text{ cis } \dfrac{\pi}{6}\right)^{-3}\left(3 \text{ cis } \dfrac{\pi}{2}\right)^2$

11. Prove the following assertion that was made in the text. If n is a positive integer and if θ is the principal argument of a nonzero complex number z, then z^n is real if and only if $n\theta$ differs from either 0 or π by an integral multiple of 2π. (Hint: review exercise 25 in section 6.6.)

12. Find a nonreal complex number z and a positive integer n such that z^n is a positive (real) number. (Hint: if necessary, apply exercise 11.)

In exercises 13–16, find the polar form of z^{-1} for each of the given numbers z.

13. $\sqrt{3} - i$

14. i^7

15. $3 \operatorname{cis} \dfrac{7\pi}{6}$

16. $4 + 3i$

In exercises 17–26, find the polar forms of all the nth roots of z, for the given values of n and z. Hence find the standard forms of these roots, approximating via calculator if necessary.

17. $n = 3, z = 2$
18. $n = 2, z = -1$
19. $n = 2, z = i$
20. $n = 3, z = 1 + i$
21. $n = 4, z = 1 - i$
22. $n = 6, z = 32\sqrt{3} + 32i$
23. $n = 8, z = -128 - 128\sqrt{3}\, i$
24. $n = 4, z = -i$
25. $n = 4, z = 2 \operatorname{cis} 5\pi$
26. $n = 5, z = (8 \cos 210° + 8i \sin 210°)\left(4 \operatorname{cis} \dfrac{5\pi}{6}\right)$

27. Let w be an nth root of a complex number z. Show that $\overline{w}$, the conjugate of w, is an nth root of $\overline{z}$.

28.◆(a) Let n be a positive integer. Prove that all the nth roots of any given nonzero complex number z lie on a circle, with center at the origin and radius $\sqrt[n]{|z|}$, and that these roots are spaced every $\dfrac{2\pi}{n}$ radians.

 (b) Let n be a positive integer. Using (a), draw an inference about the nth roots of unity.

29. Let n be a positive integer. Prove that there exists a complex number w such that $\{w^k : k = 0, 1, 2, \ldots, n-1\}$ is the set of (n distinct) nth roots of unity.

30. (This addresses the converse of exercise 29.) If n is a positive integer and z and w are nonzero complex numbers such that both w and w^2 are nth roots of z, prove that $z = 1$. (Hint: what are the roots of the polynomial $x^2 - x$?)

31. Show that $-\dfrac{1}{2} + \dfrac{i\sqrt{3}}{2}$ is a cube (nth, with $n = 3$) root of unity. Then find the other two cube roots of unity.

◆32. If w_1, w_2, and w_3 are the cube roots of unity, show that $w_1 + w_2 + w_3 = 0$ and $w_1 w_2 w_3 = 1$.

◆33. (A generalization of exercise 32) If $w_1, w_2, \ldots, w_n$ are the nth roots of unity for some positive integer n, prove that $w_1 + w_2 + \cdots + w_n = 0$ and $w_1 w_2 \cdots w_n = (-1)^{n-1}$.

34. If w is both a seventh root of unity and an eleventh root of unity, show that $w = 1$.

35. Find the polar forms of the sixth roots of unity.

36. Find the square (nth, with $n = 2$) roots of unity.

37. Find the fourth roots of unity.

38. Using a calculator, approximate the standard forms of the seventh roots of unity.

In exercises 39–50, solve the given equations by finding all complex numbers x satisfying them. Exhibit these roots in both trigonometric form and standard form, approximating via calculator if necessary.

39. $x^5 = 32$
40. $x^5 + 32 = 0$
41. $x^6 + 64i = 0$
42. $x^6 = 64i$
43. $x^3 - 8i = 0$
44. $x^5 + 32 - 32i = 0$
45. $ix^3 + 8\sqrt{3} = 8i$
◆46. $2x^2 - 2\sqrt{2}x + 3 = 0$
◆47. $4x^2 + 4x - i = 0$ (Hint: it would help to find the square roots of $16 + 16i$.)
◆48. $ix^2 - 2x = i$
◆49. $2x^4 - 2x^2 + 1 = 0$ (Hint: first find the numbers z such that $2z^2 - 2z + 1 = 0$.)
◆50. $\left(x - \dfrac{1}{x}\right)^2 + i\left(x - \dfrac{1}{x}\right) + 1 = 0$

51. Prove the identity $\cos 5\theta = 16 \cos^5 \theta - 20 \cos^3 \theta + 5 \cos \theta$.

52. A car on a carnival ride travels counterclockwise along the unit circle, beginning at the point 1. Shortly after the ride begins, the car becomes stuck, after having rotated through only $\dfrac{10e}{\pi}$ degrees (which is less than 9°). Inexplicably, the ride begins to run again, stopping after another rotation of $\dfrac{10e}{\pi}$ degrees. This pattern continues, until the ride mechanism breaks utterly just as the eighteenth rotation ends. The Human Fly sets out to rescue the stranded riders, declaring "This is no job for sissies." At what complex point on the unit circle will she find the riders?

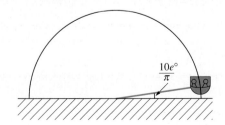

6.8 CHAPTER SUMMARY AND REVIEW EXERCISES

Major Concepts

- Absolute value
- Addition and Subtraction Identities
- Complex nth root of z
- Complex plane
- Counterexample
- De Moivre's Theorem
- Double-Angle Identities
- Half-Angle Identities
- Identity
- Imaginary axis
- Modulus
- n distinct nth roots
- nth roots of unity
- Polar form of a complex number
- Principal argument
- Product-to-Sum Identities
- Product of two complex numbers
- Pythagorean Identities
- Quotient of two complex numbers
- Quotient or Ratio Identities
- Real axis
- Reciprocal Identities
- Sum-to-Product Identities
- Trigonometric form of a complex number

Review Exercises

1. Rewrite the expression $\sec x + \tan^2 x \sin x$ as an expression involving $\sin x$ and $\cos x$.
2. Show that the substitution $\theta = \sin^{-1} 2x$ changes $\sqrt{9 - 36x^2}$ into $3 \cos \theta$.
3. Find the exact values of (a) $\cos \dfrac{\pi}{8}$, (b) $\sin 195°$, and (c) $\cot 337°30'$.
4. Express $\sqrt{3} \sin x + \cos x$ as a trigonometrically defined function evaluated at one angle.
5. If θ is a third-quadrant angle such that $\cos \theta = -\dfrac{4}{5}$, calculate $\tan\left(\theta + \dfrac{\pi}{6}\right)$.
6. If θ is a second-quadrant angle such that $\cos \theta = -0.55$, calculate $\sin 2\theta$, $\cos 2\theta$, and $\tan 2\theta$, and hence determine the quadrant of 2θ.
7. If 2θ is in quadrant I (so that θ is either first- or third-quadrant) and $\sin 2\theta = 0.55$, calculate $\sin \theta$, $\cos \theta$, and $\tan \theta$.
8. Find $\sin \dfrac{x}{2}$, given that $\cot x = 4$ and $\pi < x < \dfrac{3\pi}{2}$.
9. Express (a) $\cos 5x \sin 2x$ and (b) $\sin 5x \cos 2x$ as sums or differences.
10. Rewrite (a) $\sin 2\theta - \sin 5\theta$ and (b) $7 \cos 2\theta + 7 \cos 5\theta$ as products.
11. Transform both (a) $5 \cos \theta - 4 \sin \theta$ and (b) $-\dfrac{1}{4} \cos \theta - \sin \theta$ into the forms $c \sin (\theta + \beta)$ and $c \cos (\theta - \alpha)$.
12. Show by counterexample that $\tan^{-1} x = \dfrac{\sin^{-1} x}{\cos^{-1} x}$ is not an identity.

In exercises 13–28, prove that the given equation is an identity.

13. $\sin x - \cos x \cot x = -\cos 2x \csc x$
14. $-\tan \theta \sin \theta = \cos \theta - \sec \theta$
15. $\dfrac{-1}{\tan \theta - \sec \theta} = \tan \theta + \sec \theta$
16. $\ln |\tan \alpha - \sec \alpha| - \ln |\sec \alpha| = \ln |1 - \sin \alpha|$
17. $\tan\left(\alpha + \dfrac{7\pi}{4}\right) = \dfrac{\tan \alpha - 1}{\tan \alpha + 1}$
18. $\dfrac{1 - \tan x \tan y}{1 + \tan x \tan y} = \dfrac{\cos (x+y)}{\cos (x-y)}$
19. $\sin\left(\dfrac{\pi}{2} - x\right) = \sin\left(\dfrac{\pi}{2} + x\right)$
20. $\sin 2\theta = \dfrac{2 \cos \theta}{\csc \theta}$
21. $\cot 2\theta = 2 \cos^2 \theta \csc 2\theta - \csc 2\theta$
22. $(\sin \alpha + \cos \alpha - 1)(\sin \alpha + \cos \alpha + 1) \csc 2\alpha = 1$
23. $\tan \dfrac{\alpha}{2} = \dfrac{\sin \alpha}{1 + \cos \alpha}$
24. $\sec \dfrac{x}{2} = \pm \sqrt{\dfrac{2 \sec x}{1 + \sec x}}$
25. $\sqrt{\dfrac{2}{1 - \cos x}} = \left|\csc \dfrac{x}{2}\right|$
26. $\dfrac{\sin 2\theta + \sin 6\theta}{\cos 2\theta + \cos 6\theta} = \tan 4\theta$
27. $\dfrac{\sin 3\theta}{\sin \theta} = 2 + \dfrac{\cos 3\theta}{\cos \theta}$
28. $\tan\left(\dfrac{\alpha}{2} + \dfrac{\pi}{4}\right) = \tan \alpha + \sec \alpha$ (Hint: use exercise 39 of section 6.3.)
29. (A generalization of exercise 48(c) in section 5.9) Prove that

$$\tan^{-1} x + \tan^{-1} y = \tan^{-1}\left(\dfrac{x+y}{1-xy}\right)$$

if x and y are any positive numbers such that $xy < 1$.

In exercises 30–40, solve the given conditional equations or inequalities in $[0, 2\pi)$.

30. $\sec^2 x - \tan x + 1 = 0$
31. $\sin 2\theta \cos 4\theta = \cos 4\theta \sin 2\theta$
32. $2 \sin 2\theta + 2\sqrt{3} \sin \theta = 2 \cos \theta + \sqrt{3}$

33. $|\sin 2\alpha + \sin \alpha| = 0$ **34.** $\cos \dfrac{\alpha}{2} = \cos \alpha$

35. $\cos 3x + \cos x = 0$

36. $\cos\left(2x - \dfrac{\pi}{4}\right) = \sin x$ (Hint: sketching the graph indicates four solutions. Try to find them using cofunction or addition identities.)

37. $\dfrac{\cos(\theta/4)}{\csc(\theta/4)} \leq \dfrac{1}{2}$ **38.** $\cos^2 \dfrac{\theta}{2} - \sin^2 \dfrac{\theta}{2} > \dfrac{\sqrt{3}}{2}$

39. $2 \sin \alpha > \cos \dfrac{\alpha}{2}$ **40.** $\sin \alpha \sin 2\alpha \leq 0$

In exercises 41–46, graph the given function over the interval $[-\pi, 3\pi]$.

41. $2 \cos x - 3 \sin 2x$ **42.** $2 \sin(2x - \pi) - \cos x$
43. $|\cos x| + \sin |x|$ **44.** $x - \sin x$
45. $\cos^2\left(x + \dfrac{\pi}{4}\right)$ **46.** $\cos \dfrac{x}{2} \sin x$

47. What is the period of the function $3 \cos 6x - 2 \sin 4x$?
48. Identify the graph of $|z + 3 - 2i| = 1$ in the complex plane.
49. Write each of the following complex numbers in rectangular form:
(a) $2 \operatorname{cis} \dfrac{5\pi}{3}$ (b) $3 \operatorname{cis} \pi$ (c) $(1 - i)^{10}$

For each of the numbers in exercises 50–59, plot it in the complex plane, find its absolute value and principal argument, and also find its trigonometric (polar) form.

50. $-12 - 5i$
52. $-6i$
54. $\dfrac{\sqrt{3} + i}{-2i}$
56. $\left(3 \operatorname{cis} \dfrac{\pi}{6}\right)^2 \left(2 \operatorname{cis} \dfrac{\pi}{4}\right)^3$
58. $(1 - i)^{17}$

51. $-8 + 8\sqrt{3}\,i$
53. $-2i^{411}$
55. $\dfrac{[\sqrt{2} \operatorname{cis}(\pi/6)]^2}{[\sqrt{6} \operatorname{cis}(\pi/24)]^4}$
57. $\left(\dfrac{1}{2} \operatorname{cis} \dfrac{7\pi}{4}\right)^{-1}$
59. $\dfrac{(3 \cos 60° - i4 \sin 60°)^5}{(3^{10} + 4^{10})^{1/2}}$

60. Find polar forms and (approximate, if necessary) rectangular forms of the fifth roots of unity, and plot these five points in the complex plane.
61. Find the polar forms of the fourth roots of $2 - 2\sqrt{3}\,i$.
62. If w is both a sixth and a seventeenth root of unity, show that $w = 1$.

In exercises 63–66, find the polar forms of all solutions of the given equations.

63. $x^3 - 1 + \sqrt{3}\,i = 0$ **64.** $2x^2 + \sqrt{7}x + 2 = 0$
65. $x^2 - 2ix = 1$ **66.** $\left(x + \dfrac{1}{x}\right)^2 + 1 = 0$

67. Prove the identity $\cos 4\theta = 8 \cos^4 \theta - 8 \cos^2 \theta + 1$.
68. (a) Use exercise 51 of section 6.7 to show that $\cos^2(18°)$ is a root of the equation $16x^2 - 20x + 5 = 0$. Conclude that $\cos^2 18° = \dfrac{5 + \sqrt{5}}{8}$.

(b) Use (a) and exercise 67 to show that $\cos 72° = \dfrac{-1 + \sqrt{5}}{4}$.

Chapter 7

Applications of Trigonometry

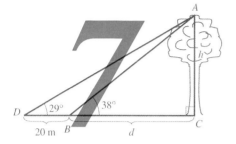

In the last two chapters, you learned about trigonometrically defined functions and their graphs. In this chapter, we first use trigonometry to solve right triangles and later use it to solve arbitrary triangles. You should not be surprised that trigonometry enjoys such uses. In fact, historically, the earliest applications of trigonometry involved solving right triangles. Some examples of this kind have already been given in the exercise sets of sections 5.2, 5.4, and 5.9. Other applications of trigonometry in this chapter involve harmonic motion and vectors. The former concerns motion of the kind found in simple pendulums or vibrating strings. Vector quantities, such as force, have both magnitude and direction and may therefore be studied both algebraically and geometrically.

7.1 RIGHT TRIANGLES

In sections 5.2 and 5.4, you learned how to determine the values of the trigonometric functions of an acute angle θ, given the lengths of two sides of a right triangle one of whose angles is θ. In section 5.7, you learned how to find any such θ from the value of one of its trigonometric functions by using the corresponding inverse trigonometric function. These skills will be used next to show how knowing the values of three parts of a right triangle (sides, angles other than the right angle) lets us determine the values of the remaining parts of the triangle.

Consider the right triangle in figure 7.1. In keeping with standard notation, the vertex of the right angle has been labeled C and the other two vertices are

Figure 7.1

381

382 Chapter 7 / Applications of Trigonometry

labeled A and B. The length of the side opposite each of the vertices has been labeled with a lower-case version of the letter for the vertex. The angles at A, B, and C are denoted by α, β, and γ, respectively. Since α and β are complementary angles, we can apply some of our knowledge to find the values of unknown sides and angles in terms of known ones. The next two examples show how.

▶ **EXAMPLE 7.1**

Given $\alpha = 65°$ and $b = 9.65$ in figure 7.2, determine the values of the remaining parts of the right triangle $\triangle ABC$.

Solution

We can display the given information in tabular form as follows:

$$\alpha = 65° \qquad a = \underline{}$$
$$\beta = \underline{} \qquad b = 9.65$$
$$\gamma = 90° \qquad c = \underline{}.$$

Figure 7.2

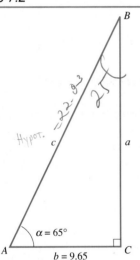

Since $\alpha + \beta + \gamma = 180°$, we have $\beta = 180° - (\alpha + \gamma) = 180° - 155° = 25°$. By examining the drawing of this triangle in figure 7.2 and applying the definition of the tangent function, we also have $\tan \alpha = \dfrac{a}{b}$. Thus, we can determine b as follows.

$$\tan \alpha = \frac{a}{b}$$
$$\tan 65° = \frac{a}{9.65}$$
$$a = 9.65 \tan 65°$$
$$\approx (9.65)(2.1445069)$$
$$\approx 20.694492$$
$$\approx 20.69.$$

One way to find the length c is to use the fact that $\sec \alpha = \dfrac{c}{b}$. So,

$$c = b \sec \alpha$$
$$\approx (9.65)(2.3662016)$$
$$\approx 22.83.$$

Summarizing our findings, we have determined that the parts of $\triangle ABC$ have the following measures:

$$\alpha = 65° \qquad a \approx 20.69$$
$$\beta = 25° \qquad b = 9.65$$
$$\gamma = 90° \qquad c \approx 22.83.$$

In example 7.1, we could also have found the length c in several other ways. One of these applies Pythagoras' Theorem: $c = \sqrt{a^2 + b^2}$. Other ways use the derived value for a and another trigonometric function: for instance, $c = a \sec \beta$. However, **in order to reduce round-off errors, it is usually preferable to adopt methods using the given values rather than derived values**.

Example 7.2

If $\triangle ABC$ is a right triangle with $a = 25$ and $b = 32$ (and right angle at C), determine the values of the other parts of the triangle.

Solution

The given information can be displayed as follows:

$\alpha =$ _____ $a = 25$

$\beta =$ _____ $b = 32$

$\gamma = 90°$ $c =$ _____.

From Pythagoras' Theorem, $c^2 = a^2 + b^2 = 25^2 + 32^2 = 625 + 1024 = 1649$. Hence, $c = \sqrt{1649} \approx 40.61$.

We know that $\tan \alpha = \dfrac{a}{b} = \dfrac{25}{32} = 0.78125$, and so $\alpha = \tan^{-1}\left(\dfrac{25}{32}\right) \approx 38°$ and $\beta \approx 90° - 38° = 52°$.

As a result, the parts of $\triangle ABC$ have the following measures:

$\alpha \approx 38°$ $a = 25$

$\beta \approx 52°$ $b = 32$

$\gamma = 90°$ $c \approx 40.61$.

Angles of Elevation or Depression

Frequently, in solving problems via trigonometry, we refer to the **angle of elevation**. This is the angle, measured from the horizontal, through which an observer must raise his or her line of sight in order to see an object. (See figure 7.3(a).) Similarly, the **angle of depression** is the angle, measured from the horizontal, through which an observer must lower his or her line of sight in order to see an object, as shown in figure 7.3(b).

Figure 7.3

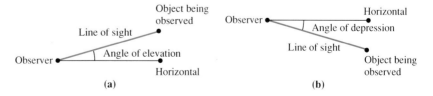

In section 5.2, exercises 35–39 introduced you to solving right triangles in situations involving angles of elevation. Here is a similar example, to refresh your memory.

Chapter 7 / Applications of Trigonometry

▶ **EXAMPLE 7.3**

A person is standing 65 meters from the base of a vertical cliff. The angle of elevation to the top of the cliff is 70°. How high is the cliff?

Solution

A right triangle has been sketched in figure 7.4 with the height of the cliff labeled x. We know that $\tan 70° = \dfrac{x}{65}$, so that

$$x = 65 \tan 70°$$
$$\approx 65(2.7474774)$$
$$\approx 178.6.$$

Hence, the cliff rises about 178.6 meters above the observer's eye-level. ◀

Figure 7.4

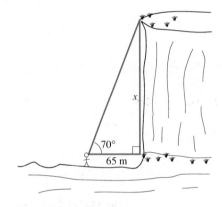

▶ **EXAMPLE 7.4**

A surveyor gathered the data shown on the right triangle in figure 7.5. Find the side lengths a and b and the angle β.

Solution

The data gathered by the surveyor can be organized as follows:

$\alpha = 27°15'$ $a = $ _____

$\beta = $ _____ $b = $ _____

$\gamma = 90°$ $c = 36.8$ ft.

Figure 7.5

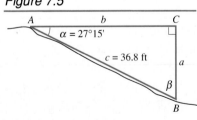

We shall begin with β. As α and β are complementary,

$$\beta = 90° - \alpha = 90° - 27°15'$$
$$= 89°60' - 27°15' \qquad 27.25$$
$$= 62°45'.$$

To find the length a, we use the facts that $\sin \alpha = \dfrac{a}{c}$ and $27°15' = 27.25°$.

$$\sin 27.25° = \dfrac{a}{36.8}$$
$$a = 36.8(\sin 27.25°)$$

$$\approx 16.85 \text{ ft.}$$

To find the length b, we shall use $\cos \alpha = \dfrac{b}{c}$.

$$\cos 27.25° = \frac{b}{36.8}$$

$$b = 36.8(\cos 27.25°)$$

$$\approx 36.8(0.8890)$$

$$\approx 32.72 \text{ ft.}$$

The earliest applications of trigonometry were to astronomy, surveying, and navigation. The next example gives an application to navigation.

EXAMPLE 7.5

In aerial navigation, directions are given in degrees, clockwise from the north. Thus, east is 90°, south is 180°, west is 270°, and north is 0°. An airplane leaves an airport and travels 150 miles in the direction 237°. How far south of the airport is the airplane at this time? How far west?

Solution

The direction of the flight is shown in figure 7.6(a), with the triangle in figure 7.6(b) indicating the necessary information. In this triangle, s is the southerly distance we are to find, and w is the requested westerly distance. By definition of sine and cosine, we have

$$\sin 33° = \frac{s}{150} \quad \text{and} \quad \cos 33° = \frac{w}{150}.$$

Hence

$$s = 150 \sin 33° \approx 150(0.544639) \approx 81.70$$

$$w = 150 \cos 33° \approx 150(0.8386706) \approx 125.80.$$

The airplane is approximately 81.7 miles south and 125.8 miles west of the airport.

Figure 7.6

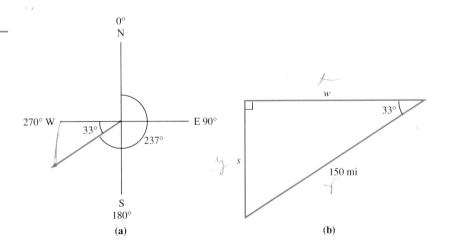

Chapter 7 / Applications of Trigonometry

▶ **EXAMPLE 7.6** A forest ranger wants to determine the height of a tree. Because the tree is surrounded by other trees and undergrowth, she is unable to simply measure the tree. Instead, she determines that the angle of elevation to the top of the tree is 38°. She then backs away an additional 20 meters from the base of the tree and finds that the treetop's angle of elevation is now 29°. What is the height of the tree?

Solution This problem involves two right triangles instead of the usual one. The diagram in figure 7.7 summarizes the information that the forest ranger was able to gather. She wants to determine h.

From $\triangle ACD$, we have

$$\tan 29° = \frac{h}{d + 20}$$

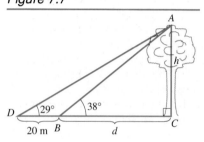

Figure 7.7

or

$$h = (d + 20) \tan 29°;$$

a. all we did was solve for H

b. use H solution for second equat.

and from $\triangle ABC$, we have

1. $\quad \tan 38° = \frac{h}{d}$

or

2. $\quad h = d \tan 38°.$

used from this

Equating the above two expressions for h, we obtain

$$(d + 20) \tan 29° = d \tan 38°$$

$$20 \tan 29° = d(\tan 38° - \tan 29°)$$

$$d = \frac{20 \tan 29°}{\tan 38° - \tan 29°}$$

$$\approx 48.842842 \text{ m.}$$

From this, we see that

$$h = d \tan 38°$$

$$\approx (48.842842)(0.7812856)$$

$$\approx 38.16021.$$

Thus, the height of the tree is about 38.16 meters. ◀

Exercises 7.1

Note: Many of the answers to exercises in chapter 7 are given as rounded-off decimal approximations.

In exercises 1–14, assume that $\triangle ABC$ is a right triangle with right angle at C, and use the given data to solve for the other parts of $\triangle ABC$.

1. $\alpha = 37°, b = 12$
2. $\alpha = 62°30', c = 1.2$
3. $\alpha = 67°30', c = 1.2$
4. $\alpha = 15°, a = 1.53$
5. $\beta = \frac{\pi}{12}$ radians, $b = 15.3$
6. $\beta = 0.925$ radians, $a = 12$
7. $a = 3, b = 4$
8. $a = 5.1, c = 13.1$
9. $b = 6, c = 13$
10. $a = 4, b = 5$
11. $\alpha = 48.24°, b = 23.7$
12. $\beta = 55.52°, c = 9.24$
13. $a = 20.4, c = 22.1$
14. $a = 9.18, b = 40.8$
15. Prove or disprove: given the parts α, β, and γ of a right triangle $\triangle ABC$, the remaining parts (sides a, b, and c) can be uniquely determined.

16. Prove or disprove: given the parts a, b, and c of a right triangle, the remaining parts (angles α, β, and γ) can be uniquely determined.
17. Prove or disprove: given any three parts other than the right angle of a right triangle, one can then solve uniquely for the remaining parts of the triangle.
18. Determine the height of a tree if it casts a shadow 36 meters long when the sun is 29° above the horizon.
19. A guy wire that is 19 meters long is holding a transmitter antenna and is attached to the antenna 13 meters from the ground. Find the angle of elevation of the guy wire. (Hint: review exercise 16 in section 5.9.)
20. While a group of shoppers are driving toward a new mall, they look up and, at a 5° angle of elevation, they see a promotional balloon flying above the mall.
    (a) If the balloon is 370 feet above ground level, how far are the shoppers from the mall?
    (b) At 50 miles per hour, will the shoppers reach the mall in less than one minute?
21. As Romeo approaches Rowena's house, he observes that the angle of elevation to Rowena's window is 30°. After he moves 40 feet closer, Romeo finds that the window's angle of elevation is 58°12′. How high above Romeo's eye-level is Rowena's window?
22. From a tower 40 meters above the ground, a fire ranger notices, at an angle of depression of 30°, a small fire at ground level. She is permitted to handle a fire by herself if the fire is at most 70 meters from the tower. Is this a job for a lone ranger?
23. A surveyor is planning a bridge to connect the tops of two vertical cliffs having the same height. Standing on the edge of one cliff, he looks 2250 feet directly across to the other cliff, then straight down to the base of that other cliff. He determines that the angle of depression to that base is 2°. How high are the cliffs?
24. A lighthouse is 250 feet tall and is located on a straight shoreline. Two boats are approaching the lighthouse on a path perpendicular to the shoreline. From the top of the lighthouse, an observer measures the boats' angles of depression as 2°15′ and 3°. How far apart are the boats?
25. A mathematician visiting New York City looks out a window 175 feet above the ground. He finds that the angle of elevation to a point P on the top of the Empire State Building is 61°18′ and the angle of depression to a ground-level point directly beneath P is 16°30′. How high does he calculate the Empire State Building to be?
♦ 26. An airplane leaves an airport and travels 200 miles in the direction 32°. How far north of the airport is the airplane at this time? How far east?
♦ 27. An airplane flies east at a rate which, in the absence of wind, would be 325 miles per hour. Wind, however, causes a southward drift of 32 miles per hour. Determine the airplane's effective speed and direction.
♦ 28. Two aerial scouts leave the same point and each travels 200 miles. If the scouts travel in the directions 150° and 60°, how far apart are they after completing their travels?
29. One bird watcher is 100 yards directly north of another bird watcher. Both see a UFO (unidentified fowl overhead) directly to their south. The bird watchers report angles of elevation of 56°18′ and 57°4′ for the UFO. Determine the altitude of the UFO.
30. The mathematician in exercise 25 later visits the Empire State Building and, from the 86th floor, looks out a window and gazes toward the horizon. At that moment, he is about 1032 feet above ground level. Assuming the radius of the earth is 3960 miles, how far can he see? (Hint: a tangent to a circle is perpendicular to the "radius vector" at the point of tangency.)

7.2 THE LAW OF SINES AND THE AMBIGUOUS CASE

In section 7.1, we used trigonometry to solve for the parts of a right triangle. In this section and the next, we shall develop two trigonometric laws that can then be applied to solve for the parts of any triangle, including an "oblique" triangle having no right angle.

The first of these is the **Law of Sines**.

Law of Sines

In any triangle $\triangle ABC$,

$$\frac{a}{\sin \alpha} = \frac{b}{\sin \beta} = \frac{c}{\sin \gamma}.$$

We shall now prove the Law of Sines. Since each triangle has at least two acute angles, there is no harm in assuming that α is acute. By interchanging the roles of B and C (and hence of β and γ) in the argument, we see that it will be enough to prove that

$$\frac{a}{\sin \alpha} = \frac{b}{\sin \beta}.$$

There are three cases to consider, depending on where the perpendicular from C meets the line $\overleftrightarrow{AB}$. Let h denote the length of this perpendicular and let D denote the point at which this perpendicular meets $\overleftrightarrow{AB}$. In the first case, β is acute and D is between A and B. In the second case, β is a right angle and $B = D$. In the third case, β is obtuse and D lies outside the line segment $\overline{AB}$. These cases are depicted in figures 7.8(a), (b), and (c).

Figure 7.8

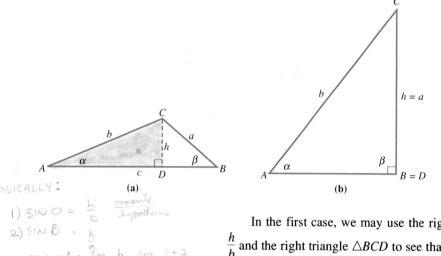
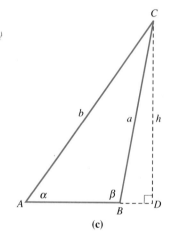
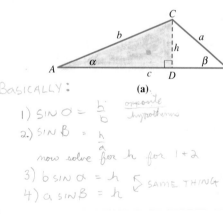

In the first case, we may use the right triangle $\triangle ACD$ to see that $\sin \alpha = \frac{h}{b}$ and the right triangle $\triangle BCD$ to see that $\sin \beta = \frac{h}{a}$. Thus, $h = b \sin \alpha$ and $h = a \sin \beta$. In particular, $a \sin \beta = b \sin \alpha$, and we conclude that

$$\frac{a}{\sin \alpha} = \frac{b}{\sin \beta}.$$

In the second case, the argument is, if anything, simpler. Indeed, we have $\sin \alpha = \frac{h}{b} = \frac{a}{b}$ and $\sin \beta = 1$. Thus

$$\frac{a}{\sin \alpha} = \frac{b \sin \alpha}{\sin \alpha} = b = \frac{b}{1} = \frac{b}{\sin \beta}$$

in this case as well.

The third case is, perhaps, the most interesting. Working with the right triangles $\triangle ACD$ and $\triangle BCD$, we see that $\sin \alpha = \frac{h}{b}$ and $\sin (\angle CBD) =$

The Law of Sines and the Ambiguous Case

$\sin(\pi - \beta) = \dfrac{h}{a}$. (Notice that, for the moment, β is considered in radian measure.) However, a subtraction identity in section 6.2 leads to

$$\sin(\pi - \beta) = \sin \pi \cos \beta - \cos \pi \sin \beta = 0(\cos \beta) - (-1)\sin \beta = \sin \beta.$$

Thus, $h = b \sin \alpha$ and $h = a \sin \beta$. Hence, $\dfrac{a}{\sin \alpha} = \dfrac{b}{\sin \beta}$. This completes the proof of the Law of Sines.

When to Use the Law of Sines

> The Law of Sines can be used to solve a triangle when the known parts of the triangle fit either of the following two cases:
>
> 1. Two angles and one side: AAS or ASA;
> 2. Two sides and the angle opposite one of these sides: SSA.
>
> Case 2, the "ambiguous case," **may** lead to two incongruent triangles.

The ambiguous case will be discussed in detail later in this section.

The first two examples will consider case 1: solving a triangle when two angles and one side are known.

▶ **EXAMPLE 7.7** Solve $\triangle ABC$ if $\alpha = 76.4°$, $\beta = 57.5°$, and $b = 8.92$ cm.

Solution It is often helpful to begin by drawing a triangle approximately to scale and labeling the given parts, as in figure 7.9. As in section 7.1, we organize the known and unknown information in a table:

$\alpha = 76.4°$ $a =$ ____

$\beta = 57.5°$ $b = 8.92$ cm

$\gamma =$ ____ $c =$ ____ .

Figure 7.9

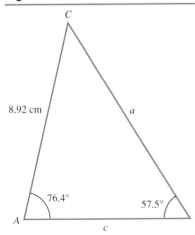

Since the values of α, β, and b are known, begin with the part of the Law of Sines that involves these three variables:

$$1. \quad \dfrac{a}{\sin \alpha} = \dfrac{b}{\sin \beta}.$$

Substituting the known values gives

$$2. \quad \dfrac{a}{\sin 76.4°} = \dfrac{8.92}{\sin 57.5°}$$

which, when solved for a, produces

$$3. \quad a = \dfrac{8.92 \sin 76.4°}{\sin 57.5°}.$$

remember: you are using this equation because you have no right triangle

If you use a calculator to find the answer, do not round off your answers until the final result is found. At that time, round the answer to the desired number of significant digits.

On many algebraic calculators, you would press the key sequence

$$8.92 \; \boxed{\times} \; 76.4 \; \boxed{\sin} \; \boxed{\div} \; 57.5 \; \boxed{\sin} \; \boxed{=}$$

to get 10.279796. Rounding this off to 3 significant digits, we see that

$$a \approx 10.3 \text{ cm}.$$

On an RPN calculator, press the sequence

$$8.92 \; \boxed{\text{ENTER}} \; 76.4 \; \boxed{\sin} \; \boxed{\times} \; 57.5 \; \boxed{\sin} \; \boxed{\div}$$

to get the same value for a as above.

On a Casio *fx-7700G* or a *TI-81*, press the sequence

$$8.92 \; \boxed{\text{SIN}} \; 76.4 \; \boxed{\div} \; \boxed{\text{SIN}} \; 57.5$$

followed by either $\boxed{\text{EXE}}$ or $\boxed{\text{ENTER}}$, to get the value 10.27979614 for a. We round this off to 10.3 centimeters.

Since the sum of the degree measures of the angles of a triangle is 180°, we see that

$$\gamma = 180° - \alpha - \beta$$
$$= 180° - 76.4° - 57.5°$$
$$= 46.1°.$$

Now use the Law of Sines again in order to find the length c.

$$\frac{b}{\sin \beta} = \frac{c}{\sin \gamma}$$

$$\frac{8.92}{\sin 57.5°} = \frac{c}{\sin 46.1°}$$

$$c = \frac{8.92 \sin 46.1°}{\sin 57.5°} \approx 7.6207981.$$

The length c is about 7.62 centimeters.

The measures of the angles and the sides of this triangle are

$$\alpha = 76.4° \qquad a \approx 10.3 \text{ cm}$$
$$\beta = 57.5° \qquad b = 8.92 \text{ cm}$$
$$\gamma = 46.1° \qquad c \approx 7.62 \text{ cm}.$$

◀

The Law of Sines and the Ambiguous Case

EXAMPLE 7.8

A high tension wire is to be strung across a river from tower A to tower C. A point B is located on the same side of the river as tower A, 450 meters from it. If $\angle ABC$ is $62.8°$ and $\angle BAC$ is $36.7°$, what is the distance between towers A and C?

Solution

Start by drawing a triangle approximately to scale, labeling the given parts as in figure 7.10. The known parts of this triangle are angles α and β and side c. In order to use the Law of Sines, it is first necessary to find the measure of angle γ.

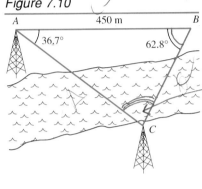

Figure 7.10

■ASA

$$\gamma = 180° - \alpha - \beta$$
$$= 180° - 36.7° - 62.8°$$
$$= 80.5°.$$

Now we can find b, the length of $\overline{AC}$, by applying the Law of Sines.

$$\frac{b}{\sin \beta} = \frac{c}{\sin \gamma}$$

$$\frac{b}{\sin 62.8°} = \frac{450}{\sin 80.5°}$$

$$b = \frac{450 \sin 62.8°}{\sin 80.5°} \approx 405.80271.$$

The distance between the towers is about 406 meters. ◀

The next three examples show how the Law of Sines can be used to solve case 2, the SSA case, when two sides and the angle opposite one of these sides are known.

EXAMPLE 7.9

Solve $\triangle ABC$ if $a = 22$, $b = 25$, and $\alpha = 55°$.

Solution

First, sketch the triangle as in figure 7.11. We shall find β by applying the Law of Sines.

■SSA

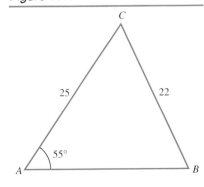

Figure 7.11

$$\frac{a}{\sin \alpha} = \frac{b}{\sin \beta}$$

$$\frac{22}{\sin 55°} = \frac{25}{\sin \beta}$$

$$\sin \beta = \frac{25 \sin 55°}{22} \approx 0.9308546.$$

We now have two possibilities for the angle β since the sine function takes on the value 0.9308546 once in quadrant I and once in quadrant II. Using the inverse sine function, we can determine the reference angle for these two possible solutions:

Chapter 7 / Applications of Trigonometry

AAS
ASA } Law of Sines
SSA

SAS } Law of Cos
SSS

$$\beta_{ref} \approx \sin^{-1}(0.9308546) \approx 68.6°.$$

We shall denote these two possible solutions as β_1 and β_2. The corresponding values of γ will be denoted γ_1 and γ_2, respectively.

Possible Solution 1: β_1 is in quadrant I, and so $\beta_1 = \beta_{ref} \approx 68.6°$. Since $\gamma_1 = 180° - \beta_1 - \alpha$, we see that $\gamma_1 \approx 180° - 68.6° - 55° = 56.4°$. We shall determine the side length c_1 shortly.

Possible Solution 2: β_2 is in quadrant II, and so $\beta_2 = 180° - \beta_{ref} \approx 180° - 68.6° = 111.4°$. Thus, $\gamma_2 = 180° - \beta_2 - \alpha \approx 180° - 111.4° - 55° = 13.6°$. (Notice that in this example, "possible solution 2" has not led to a contradiction.) We shall determine the side length c_2 shortly.

Now, we shall use the above information to find the length of side c. For solution 1, $\gamma_1 \approx 56.4°$. Then, applying the Law of Sines, we have

$$\frac{a}{\sin \alpha} = \frac{c_1}{\sin \gamma_1}$$

$$\frac{22}{\sin 55°} \approx \frac{c_1}{\sin 56.4°}$$

$$c_1 \approx \frac{22 \sin 56.4°}{\sin 55°} \approx 22.37.$$

The length c_1 is about 22.37.

For solution 2, $\gamma_2 \approx 13.6°$. Then

$$\frac{a}{\sin \alpha} = \frac{c_2}{\sin \gamma_2}$$

$$\frac{22}{\sin 55°} \approx \frac{c_2}{\sin 13.6°}$$

$$c_2 \approx \frac{22 \sin 13.6°}{\sin 55°} \approx 6.32.$$

The length of side c_2 is about 6.32.

The two possible solutions from the given data are shown as $\triangle AB_1C$ and $\triangle AB_2C$ in figure 7.12. In tabular form, these two solutions are

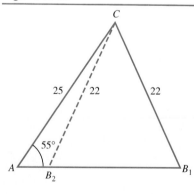

Figure 7.12

$\triangle AB_1C$		$\triangle AB_2C$	
$\alpha = 55°$	$a = 22$	$\alpha = 55°$	$a = 22$
$\beta \approx 68.6°$	$b = 25$	$\beta \approx 111.4°$	$b = 25$
$\gamma \approx 56.4°$	$c \approx 22.37$	$\gamma \approx 13.6°$	$c \approx 6.32$

▶ **EXAMPLE 7.10**   Solve $\triangle ABC$ if $c = 5.74$ cm, $\beta = 51.2°$, and $b = 9.85$ cm.

The Law of Sines and the Ambiguous Case

Solution

First, sketch the triangle as in figure 7.13. We proceed to find γ by applying the Law of Sines.

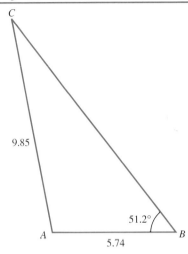

Figure 7.13

$$\frac{b}{\sin \beta} = \frac{c}{\sin \gamma}$$

$$\frac{9.85}{\sin 51.2°} = \frac{5.74}{\sin \gamma}$$

$$\sin \gamma = \frac{5.74 \sin 51.2°}{9.85} \approx 0.4541523.$$

There are two possibilities for γ since the sine function takes on the value 0.4541523 once in the first quadrant and once in the second quadrant. As in example 7.9, we shall use the inverse sine function to determine the reference angle for these two possible solutions:

$$\gamma_{\text{ref}} \approx \sin^{-1}(0.4541523) \approx 27°.$$

We shall denote the two possible solutions as γ_1 and γ_2. The corresponding values of α will be denoted α_1 and α_2, respectively.

Possible Solution 1: γ_1 is in quadrant I, and so $\gamma_1 = \gamma_{\text{ref}} \approx 27°$. Since $\alpha_1 = 180° - \beta - \gamma_1$, we see that $\alpha_1 \approx 180° - 51.2° - 27° = 101.8°$.

Possible Solution 2: γ_2 is in quadrant II, and so $\gamma_2 = 180° - \gamma_{\text{ref}} \approx 180° - 27° = 153°$. Thus, $\beta + \gamma_2 \approx 51.2° + 153° = 204.2°$. Since each angle in a triangle has positive measure, $\alpha + \beta + \gamma_2 > \beta + \gamma_2$. In particular, $\alpha + \beta + \gamma_2 > 180°$, contradicting the fact that the sum of the angle measures in any triangle is 180°. Thus, "possible solution 2" via γ_2 is *not* possible in *this* example.

We return to the only solution in this example, namely "possible solution 1," and complete it by finding the side length a. Apply the Law of Sines.

$$\frac{a}{\sin \alpha} = \frac{b}{\sin \beta}$$

$$\frac{a}{\sin 101.8°} \approx \frac{9.85}{\sin 51.2°}$$

$$a \approx \frac{9.85 \sin 101.8°}{\sin 51.2°} \approx 12.37.$$

The length a is about 12.37 centimeters. ◀

There is one other problem that can develop when data fit case 2, SSA. It is possible that there will be no solution that satisfies the given conditions. Consider example 7.11.

▶ **EXAMPLE 7.11**

Solve $\triangle ABC$ if $a = 20$, $b = 30$, and $\alpha = 70°$.

Chapter 7 / Applications of Trigonometry

Solution Applying the Law of Sines, we see that

$$\frac{a}{\sin \alpha} = \frac{b}{\sin \beta}$$

$$\frac{20}{\sin 70°} = \frac{30}{\sin \beta}$$

$$\sin \beta = \frac{30 \sin 70°}{20} \approx 1.4095389.$$

But we know that the range of the sine function is $[-1, 1]$. So the sine of an angle is never larger than 1. Thus, it is not possible for *any* triangle to satisfy the given conditions. Figure 7.14 illustrates why no such triangle exists. ◂

Figure 7.14

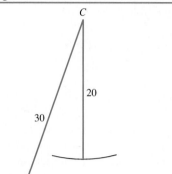

The SSA case is often referred to as the **ambiguous case** because three possible situations can arise from such data: no such triangle exists (as in example 7.11), one such triangle exists (as in example 7.10), or two distinct (incongruent) triangles satisfy the given conditions (see example 7.9).

In the ambiguous case, we can determine which of the possible outcomes to expect. Suppose we are given sides a and b and an acute angle α. Construct α and locate the vertex C on one ray of α so that $AC = b$. Let the other ray of α be $\overrightarrow{AE}$. From C drop a perpendicular, meeting $\overrightarrow{AE}$ at D. The length of $\overline{CD}$ is then $b \sin \alpha$. Of course, $\overline{CD}$ is the shortest distance from C to $\overrightarrow{AE}$. Thus if $a < b \sin \alpha$, no triangle meeting the given conditions is possible. (See figure 7.15(a).)

In figure 7.15(b), $a = CD$. This illustrates that if $a = b \sin \alpha$ and α is acute, then there is (apart from congruent copies) exactly one triangle that meets the given data on a, b, and α, and this triangle is a right triangle.

In figure 7.15(c), $b > a > b \sin \alpha$ and two noncongruent oblique triangles are formed that fit the given information. These two triangles are $\triangle AB_1C$ and $\triangle AB_2C$.

Figure 7.15

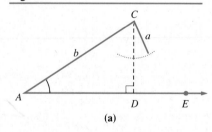
(a)

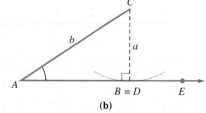
(b)

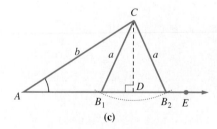
(c)

We shall leave it for you to show what happens when either α is acute and $a \geq b$ or $\alpha \geq 90°$. What we have shown and what we have left for you to prove are summarized in the following table of the ambiguous case.

The Law of Sines and the Ambiguous Case

The Ambiguous Case of the Sine Law

IN OTHER WORDS:
a) $b \sin \alpha$ (and)
$a < b$

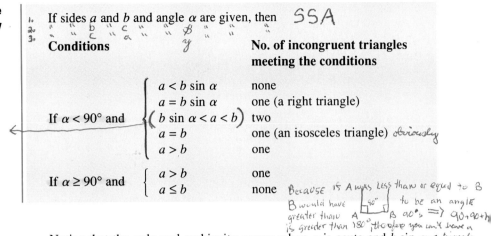

If sides a and b and angle α are given, then SSA

Conditions		No. of incongruent triangles meeting the conditions
If $\alpha < 90°$ and	$a < b \sin \alpha$	none
	$a = b \sin \alpha$	one (a right triangle)
	$b \sin \alpha < a < b$	two
	$a = b$	one (an isosceles triangle) obviously
	$a > b$	one
If $\alpha \geq 90°$ and	$a > b$	one
	$a \leq b$	none — Because is A was less than or equal to B, B would have to be an angle greater than A, 90°'s => 90+90+ is greater than 180°, therefore you can't have a triangle = NONE

Notice that the only real ambiguity occurs when α is acute and $b \sin \alpha < a < b$. You can deal with any such ambiguous conditions as in example 7.9.

Exercises 7.2

In exercises 1–6, use the given parts of $\triangle ABC$ to solve for the unknown parts.

1. $\alpha = 62.3°$, $\beta = 44°$, $b = 9.28$ cm
2. $\alpha = 36°$, $\beta = 40.5°$, $a = 5.03$ in.
3. $\beta = \frac{\pi}{6}$ (radians), $\gamma = \frac{2\pi}{3}$, $c = 3\sqrt{2}$ ft
4. $\alpha = \frac{3\pi}{4}$, $\gamma = \frac{\pi}{6}$, $c = 2\sqrt{3}$ m
5. $\alpha = 38°$, $\gamma = 51.7°$, $b = 4.7$ m
6. $\beta = 2°$, $\gamma = 93.5°$, $a = 7.41$ yd

In exercises 7–20, determine how many pairwise incongruent triangles $\triangle ABC$ fit the given conditions. For each such triangle, solve for the unknown parts.

7. $a = 4$ cm, $b = 6$ cm, $\alpha = 30°$
8. $a = 6$ in., $b = 4$ in., $\alpha = \frac{\pi}{3}$
9. $a = 6$ ft, $c = 8$ ft, $\gamma = \frac{\pi}{4}$
10. $b = 6$ m, $c = 8$ m, $\beta = 46.5°$
11. $a = 2$ m, $b = 6$ m, $\alpha = 30°$
12. $a = 4$ yd, $c = 2$ yd, $\gamma = \frac{5\pi}{8}$
13. $a = 6$ in., $b = 4$ in., $\alpha = \frac{2\pi}{3}$
14. $b = 3.58$ cm, $c = 4.17$ cm, $\beta = 1.05°$
15. $b = 3$ m, $c = 3$ m, $\beta = 37°$
16. $a = 1.79$ yd, $b = 1.79$ yd, $\alpha = 125°$
17. $a = 2.19$ ft, $c = 9.12$ ft, $\alpha = \frac{7\pi}{12}$
18. $b = 2.19$ m, $c = 9.12$ m, $\gamma = \frac{7\pi}{12}$
19. $a = 2$, $c = 2$, $\gamma = 94.5°$
20. $a = 2$, $c = 2$, $\gamma = 73°$
21. Tom and Dick are forest rangers, whose observation posts are at points A and B, respectively, 10 miles apart. As their friend, Harriet, is walking from A to B, she sees a forest fire at point C, and reports this via walkie-talkie to Tom and Dick. If the rangers measure $\angle CAB$ as $42°$ and $\angle CBA$ as $58°$, how far is the fire from Tom? From Dick?
22. The first annual Mathlete Trot will be a road race along a triangular path from A to B and then to C and back to A. Part of John's strategy is to lay out the course to thwart David, whose usual race plan wins only for distances not exceeding 8 miles. John arranges $AB = 2$ mi, $\angle ABC = 65°$, and $\angle ACB = 46°$. Knowing this, David still believes that following his usual race plan will give him a reasonable chance to outtrot the other mathletes. Is David right?
23. A tree grows tilted away from the sun, forming a $5°$ angle with the vertical. This tree casts a 40 foot long shadow and, according to an observer at the end of the shadow, the treetop's angle of elevation is $6°$. How tall is the tree? (Ignore the observer's height.)

24. Tom, Dick, and Harriet are equally fast swimmers. The boys are swimming and Harriet is on the shoreline, with Tom to her left and Dick to her right. She is 500 yards from Tom, and the boys are 750 yards apart. At a point between Tom and Dick, a boating accident places a nonswimmer, Ken, in jeopardy. Harriet measures $\angle THK$ as 50° and estimates $\angle TKH$ as 85°. She concludes that she can reach Ken more quickly than either Tom or Dick can. Is Harriet correct?

25. The Human Fly awakens in her tree house to find that an earthquake has caused the tree where she resides to form an angle of inclination of 80° with a straight sidewalk. At an angle of depression of 60°, she sees Frank, who has run along the sidewalk 40 feet away from the base of the falling tree. How far is Frank from the Human Fly?

26. A sloping hillside makes a 12° angle with the horizontal. A vertical flagpole is erected at a point P on the hillside 40 feet from a point Q at the bottom of the hill. If R is the point at the top of the flagpole and $\angle PQR$ measures 30°, determine the height of the flagpole. (Hint: first determine the measure of $\angle QRP$.)

27. An engineer surveys the area prior to building a bridge connecting two points A and B on opposite sides of a river. From a point C on the same side of the river as A, he determines that $\angle ACB$ measures 51°. Moving to A, he finds that $\angle CAB$ measures 58° and $AC = 412$ ft. How long should the bridge be?

28. On a playground, a children's slide is 35 feet long and is inclined 30° from the horizontal. Access to the slide is gained via an 18 foot long ladder. Determine the angle of inclination of the ladder. (Hint: there are two possible answers.)

29. (We return to the Human Fly who was left hanging in exercise 25.) The tree has continued to tilt and now is inclined 30° to the horizontal. The Human Fly has crawled along the tree to a point 40 feet from the tree's base. Frank has not moved since we left him. Standing on the sidewalk, Lee offers to rescue the Human Fly with his 35 foot long ladder. How far is Lee from Frank? (Hint: frankly, there are two possible answers.)

30. (A continuation of exercise 29) Before the Human Fly can respond to Lee, Scarlett arrives, stands next to Frank, and offers to save the Human Fly with a 20 foot long ladder.
    (a) Must Scarlett move along the sidewalk in order to be able to act on her offer?
    (b) If the Human Fly decides to honor Scarlett's offer, where must Scarlett position herself in order to carry out the rescue?

31. Show that if $\triangle ABC$ is an equilateral triangle, then it satisfies the "cosine version of the Law of Sines," namely $\dfrac{a}{\cos \alpha} = \dfrac{b}{\cos \beta} = \dfrac{c}{\cos \gamma}$. (Hint: $\triangle ABC$ does satisfy the Law of Sines and you know α, β, and γ.)

32. (The converse of exercise 31) If $\triangle ABC$ is a triangle such that $\dfrac{a}{\cos \alpha} = \dfrac{b}{\cos \beta} = \dfrac{c}{\cos \gamma}$, prove that $\triangle ABC$ is equilateral. (Hint: if k denotes the above ratio, use the Pythagorean identity to calculate $\dfrac{1}{k^2} + \dfrac{\sin^2 \alpha}{a^2}$, and apply the Law of Sines.)

33. Prove the text's assertions for the ambiguous case when either α is acute and $a \geq b$ or $\alpha \geq 90°$.

34. Having established his own identity (in exercise 47 of section 5.9), Romeo now claims to have discovered a new trigonometric law: for any triangle $\triangle ABC$,
$$b \cos \alpha + a \cos \beta = c.$$
Luckless Romeo has lost his notes. Help him prove this law. (Hint: besides the Law of Sines, you'll need to recall the formulas for $\sin(x \pm y)$ in section 6.2 and notice that $\gamma = \pi - \alpha - \beta$.)

35. (An interpretation for the ratios in the statement of the Law of Sines) Prove that the diameter of the circumscribed circle of a triangle $\triangle ABC$ is $\dfrac{a}{\sin \alpha}$. (Hint: the center, D, of the circumscribed circle is the intersection of the perpendicular bisectors of the sides of $\triangle ABC$. Extend $\overline{BD}$ so as to meet the circle at E. If α is as shown in the figure, explain from your earlier study of geometry why $\angle BEC$ is congruent to $\angle BAC$ and why $\angle BCE$ is a right angle. Use $\triangle BEC$ to compute $\sin \alpha$.)

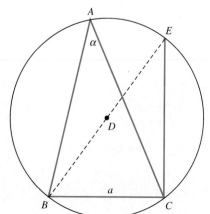

7.3 THE LAW OF COSINES

In the last section, you learned that the Law of Sines could be used to solve for the unknown parts of a triangle in two cases: (1) when the measures of two angles and one side are known, the AAS or ASA case, and (2) when the measures of two sides and the angle opposite one of them are known, the SSA case.

There are two other cases in which one can solve a triangle. In both of these cases, whether the triangle is a right triangle or an oblique triangle, the Law of Cosines is used.

When to Use the Law of Cosines

The Law of Cosines is used to help solve a triangle in the following two cases:

3. When the measures of two sides and the included angle are known: SAS;
4. When the lengths of all three sides are known: SSS.

In order to develop the Law of Cosines, let's consider an arbitrary triangle $\triangle ABC$. We shall place it on a coordinate system with vertex B at the origin and side $\overline{BC}$ along the positive x-axis, as shown in figure 7.16. Because the length of $\overline{BC}$ is a, C has the coordinates $(a, 0)$. Let (x, y) denote the coordinates of vertex A. Then $\sin \beta = \dfrac{y}{c}$ and $\cos \beta = \dfrac{x}{c}$, or $y = c \sin \beta$ and $x = c \cos \beta$. Hence, the coordinates of vertex A can be written as $(c \cos \beta, c \sin \beta)$.

The length of side $\overline{AC}$ is b. By the distance formula, as in section 1.2,

$$b = \sqrt{(c \cos \beta - a)^2 + (c \sin \beta)^2}.$$

Squaring both sides and then simplifying, we are led to the following sequence of steps.

$$\begin{aligned} b^2 &= (c \cos \beta - a)^2 + (c \sin \beta)^2 \\ &= c^2 \cos^2 \beta - 2ac \cos \beta + a^2 + c^2 \sin^2 \beta \\ &= c^2(\cos^2 \beta + \sin^2 \beta) + a^2 - 2ac \cos \beta \\ &= c^2 + a^2 - 2ac \cos \beta. \end{aligned}$$

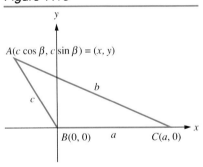

Figure 7.16

This result is one form of the Law of Cosines. It would have been just as easy to place A or C at the origin in figure 7.16. Each of these options provides another variation of the Law of Cosines. In summary, we have

Law of Cosines

SAS   SSS

If ABC is any triangle, then   I   II

$$a^2 = b^2 + c^2 - 2bc \cos \alpha, \quad \cos\alpha =$$
$$b^2 = a^2 + c^2 - 2ac \cos \beta \quad \cos\beta =$$
$$c^2 = a^2 + b^2 - 2ab \cos \gamma. \quad \to \cos\gamma = \dfrac{c^2-a^2-b^2}{2ab}$$

Notice that Pythagoras' Theorem is the special case of the Law of Cosines in which the known angle is a right angle.

EXAMPLE 7.12

Solve $\triangle ABC$ if $a = 12.8$, $b = 15.9$, and $\gamma = 107.3°$.

Solution

We can display the given information in tabular form as follows:

$$\alpha = \underline{\hspace{1cm}} \qquad a = 12.8$$
$$\beta = \underline{\hspace{1cm}} \qquad b = 15.9$$
$$\gamma = 107.3° \qquad c = \underline{\hspace{1cm}}.$$

Figure 7.17 shows a sketch of a triangle fitting the given information. Since we are given sides a and b and angle γ, we shall use the following version of the Law of Cosines:

$$c^2 = a^2 + b^2 - 2ab \cos \gamma.$$

Substituting the given information, we see that

$$c^2 = 12.8^2 + 15.9^2 - 2(12.8)(15.9) \cos 107.3°$$
$$\approx 163.84 + 252.81 - 2(12.8)(15.9)(-0.2973749)$$
$$\approx 537.69347$$
$$c \approx 23.2.$$

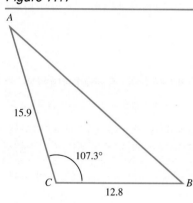

Figure 7.17

At this point, you must choose which method to use to solve the remainder of the problem. You now know the lengths of the three sides, so you may use the Law of Cosines. However, you also know the measure of one angle, so you may instead use the Law of Sines. Since this section concerns the Law of Cosines, that is the method used below.

To use the Law of Cosines to find the measure of angle α, we first need to rewrite the Law of Cosines as

$$\cos \alpha = \frac{b^2 + c^2 - a^2}{2bc}.$$

Substituting the given information, we get

$$\cos \alpha \approx \frac{15.9^2 + 23.2^2 - 12.8^2}{2(15.9)(23.2)}$$
$$= \frac{252.81 + 538.24 - 163.84}{2(15.9)(23.2)}$$
$$\approx 0.8501545.$$
$$\alpha = \cos^{-1} (\cos \alpha) \approx \cos^{-1} (0.8501545)$$
$$\approx 31.8°.$$

Since $\alpha \approx 31.8°$ and $\gamma = 107.3°$, $\beta \approx 180° - 31.8° - 107.3° = 40.9°$. So, in tabular form, the solution may be summarized as follows:

$$\alpha \approx 31.8° \qquad a = 12.8$$
$$\beta = 40.9° \qquad b = 15.9$$
$$\gamma = 107.3° \qquad c \approx 23.2.$$
◀

EXAMPLE 7.13

An electric transmission line is planned to go over a heavily wooded area from a tower to be placed at A to a tower located at B, as indicated in figure 7.18. A surveyor cannot see the location of either of the proposed towers from the location of the other. She does find a place, marked C, from which she can see the locations of the two towers. If the distance from A to C is 372 meters and the distance from B to C is 432 meters and $\angle ACB = 79.6°$, what is the distance between the proposed locations of the two towers?

Solution

Figure 7.18 shows a sketch of the situation. This is an SAS type of problem in which we want to find $\overline{AB} = c$. Using the Law of Cosines, we have

$$c^2 = a^2 + b^2 - 2ab \cos \gamma$$
$$= 432^2 + 372^2 - 2(432)(372) \cos 79.6°$$
$$\approx 266\,987.7$$
$$c \approx 516.7.$$

Figure 7.18

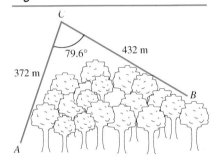

The distance from the proposed site of tower A to the proposed site of tower B is approximately 517 meters. ◀

The SSS criterion for congruence of triangles assures us that we can solve for the unknown parts of a triangle if we are given its side lengths a, b, and c. This is done by the Law of Cosines, at least initially, as in the solution for α in example 7.12. We next turn to another useful application of knowing all three side lengths.

The Law of Cosines can be used to establish several interesting formulas that involve triangles. One of the most useful is **Heron's** (or Hero's) **formula** for the area of a triangle. A proof of this result was indicated in exercise 60 of section 1.4. More elementary trigonometric formulas for area will be developed and used in exercises 24–30.

Heron's Formula

> The area, K, of $\triangle ABC$ is
> $$K = \sqrt{s(s-a)(s-b)(s-c)},$$
> where $s = \frac{1}{2}(a + b + c)$.

The quantity s, which is one-half the perimeter of $\triangle ABC$, is called the **semiperimeter** of $\triangle ABC$. Another mathematically useful role for semiperimeter will be indicated in exercise 35.

EXAMPLE 7.14

Find the area of the triangle with side lengths $a = 96$ yd, $b = 73$ yd, and $c = 51$ yd.

Solution First, let's find the semiperimeter:

$$s = \tfrac{1}{2}(a + b + c) = \tfrac{1}{2}(96 + 73 + 51) = 110.$$

Then, apply Heron's formula:

$$\begin{aligned} K &= \sqrt{s(s-a)(s-b)(s-c)} \\ &= \sqrt{110(110-96)(110-73)(110-51)} \\ &= \sqrt{110(14)(37)(59)} \\ &= \sqrt{3361820} \\ &\approx 1{,}833.5 \text{ yd}^2. \end{aligned}$$

◀

Since antiquity, many trigonometric laws have been developed. In addition to those introduced in sections 7.2 and 7.3, you should know that there is also a Law of Tangents. It was often used in conjunction with logarithm tables to solve (and check) triangles in various situations. Given the advent of calculators, such a role is much less important today. You may also be interested in another law, usually attributed to Mollweide (1808) but already known to Newton. This result facilitates checking, as it involves all six parts of a triangle in a single equation.

Mollweide's Theorem

If ABC is any triangle, then

$$\frac{a+b}{c} = \frac{\cos\left[\tfrac{1}{2}(\alpha - \beta)\right]}{\sin\tfrac{\gamma}{2}}.$$

You know enough now to prove this result, but you are warned that its proof is rather difficult.

Exercises 7.3

In exercises 1–10, use the given parts of $\triangle ABC$ to solve for the unknown parts.

1. $a = 11.07$ cm, $b = 3.15$ cm, $\gamma = 24°$
2. $b = 5$ in., $c = 7.2$ in., $\alpha = \dfrac{\pi}{3}$
3. $b = 1.08$ ft, $c = 3$ ft, $\alpha = \dfrac{4\pi}{7}$
4. $a = 2.6$ m, $c = 1.75$ m, $\beta = 107.5°$
5. $a = 38$, $c = 12$, $\beta = 30°$
6. $a = 3$, $b = 4$, $\gamma = \dfrac{2\pi}{5}$
7. $a = 2.6$ m, $b = 3.7$ m, $c = 4$ m
8. $a = 3$ in., $b = 2$ in., $c = 4$ in.
9. $a = 1.1$, $b = 2.3$, $c = 2.55$
10. $a = 2.2$, $b = 4.1$, $c = 3$

◆ 11. After refueling in mid-flight, a fighter plane heads due east for three hours at 500 miles per hour, and the plane that had carried the extra fuel to the fighter flies 1050 miles in a 120° direction. How far apart are the two planes three hours after refueling?

◆ 12. Upon leaving an airport, an airplane flies due east for 200 miles and then, changing to a 120° direction, flies another 150 miles. How far is the airplane then from this airport?

◆ 13. Upon leaving an airport, an airplane flies in a 60° direction for 20 minutes and then, changing to a 40° direction, flies another 30 minutes. As a result of the 50 minute flight, the plane is 210 miles from the airport. If the plane's speed has been constant, determine this constant speed.

The Law of Cosines

14. Tom asks Dick and Harriet to help him survey a proposed wildlife preserve. If Tom is 3 kilometers from Dick and 4 kilometers from Harriet, and if ∠DTH measures 72°, how far apart are Dick and Harriet?

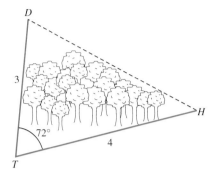

15. A vertical antenna 89 feet high is erected on a hillside that is inclined 12° to the horizontal. The antenna is held in place by four guy wires. Two of the wires are anchored on the hillside 40 feet from the antenna's base, as shown in the figure. Determine the lengths of each of these two guy wires.

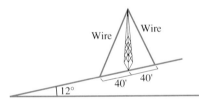

16. Repeat exercise 15 with the following change in data: assume now that the antenna is not vertical, but instead has an angle of inclination (with the horizontal) of 82°.

For exercises 17–18, you will need the following sketch of a baseball "diamond."

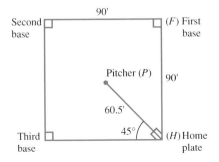

17. (a) How far is the pitcher from first base?
    (b) Determine the measure of ∠FPH.

18. A runner has traveled three-fifths of the way from second base to third base. At this moment, how far is the runner from the pitcher?

19. A straight racetrack runs from point P to point Q. Directly over the racetrack, an observer in a weather balloon views P and Q, with angles of depression 45° and 30°, respectively. The observer is 75 meters from P and $75\sqrt{2}$ meters from Q.
    (a) How long is the racetrack?
    (b) How high above the racetrack is the weather balloon?

20. A parallelogram has sides of length 6 inches and 8 inches. (Since "opposite" parts of a parallelogram are congruent, you know that the other sides of the given parallelogram have lengths 6 inches and 8 inches.) One of its diagonals is 9 inches long.
    (a) Determine the measures of the parallelogram's angles. (Hint: the sum of these four measures is 360°.)
    (b) Determine the length of the parallelogram's other diagonal.

21. Prove that, in any triangle $\triangle ABC$,
$$a^2 + b^2 + c^2 = 2(bc \cos \alpha + ca \cos \beta + ab \cos \gamma).$$

22. Prove that the "sine version of the Law of Cosines,"
$$\begin{cases} a^2 = b^2 + c^2 - 2bc \sin \alpha \\ b^2 = a^2 + c^2 - 2ac \sin \beta \\ c^2 = a^2 + b^2 - 2ab \sin \gamma \end{cases}$$
is false for each triangle $\triangle ABC$. (Hint: the Law of Cosines *does* hold. Deduce that $\sin \alpha = \cos \alpha$. So?)

23. Prove that a triangle $\triangle ABC$ satisfies two of the three equations $a^2 = b^2 + c^2 - 2bc \sin \alpha$, $b^2 = a^2 + c^2 - 2ac \sin \beta$, and $c^2 = a^2 + b^2 - 2ab \sin \gamma$ if and only if $\triangle ABC$ is an isosceles right triangle. (Hint: review exercise 22.)

◆ 24. Prove that the area, K, of any triangle $\triangle ABC$ is given by each of the following equal expressions: $\frac{1}{2}ac \sin \beta$, $\frac{1}{2}bc \sin \alpha$, and $\frac{1}{2}ab \sin \gamma$. (Hint: K is one-half the product of a base length and altitude h. Review the formulas for h in the derivation of the Law of Sines in section 7.2.)

◆ In exercises 25–30, use exercise 24 or Heron's formula (if it may be applied) to find the area of a triangle $\triangle ABC$ meeting the given conditions.

25. $a = 4\sqrt{3}$ cm, $c = 2.01$ cm, $\beta = 35°$

26. $a = 2\sqrt{3}$ in., $b = 3\sqrt{2}$ in., $\gamma = \dfrac{2\pi}{3}$

27. $A(1, 2)$, $B(3, 4)$, $C(5, 7)$

28. $A(-1, -1)$, $B(2, 2)$, $C(1, 3)$
29. $a = 4$, $b = 8$, $c = 6$
30. $a = 2.1$ cm, $b = 3.9$ cm, $c = 5.6$ cm
31. Find the area of the parallelogram $PQRS$ having vertices $P(-1, 2)$, $Q(4, 8)$, $R(14, 13)$, $S(9, 7)$. (Hint: the required area is twice that of $\triangle PQR$.)

Exercises 32–34 refer to a regular n-gon F ($n \geq 4$) inscribed in a circle of radius r. The figure depicts the case $n = 6$.

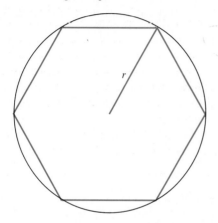

32. Show that the perimeter p of the regular n-gon F is given by $p = n\sqrt{2}r \sqrt{1 - \cos\left(\frac{2\pi}{n}\right)}$.

33. Show that the area A of the regular n-gon F is given by $A = \frac{nr^2}{2} \sin\left(\frac{2\pi}{n}\right)$. (Hint: use exercise 24.)

34. Let p and A be the perimeter and area, respectively, of the regular n-gon F, as in exercises 32–33.
    (a) Show that $A = \frac{p^2}{4}\left(\frac{\sin\left(\frac{2\pi}{n}\right)}{n\left(1 - \cos\left(\frac{2\pi}{n}\right)\right)}\right) = \frac{p^2}{4}\left(\frac{1 + \cos\left(\frac{2\pi}{n}\right)}{n \sin\left(\frac{2\pi}{n}\right)}\right)$.
    (b) Consider the limiting tendency of A as n "approaches infinity" (that is, as n increases without bound). Why is it intuitively obvious that p "approaches" $2\pi r$? Using exercises 18 and 19 of section 5.4, conclude that the limiting tendency of A is πr^2. What does this tell you about the circle of radius r?

◆ 35. Let r be the radius of the inscribed circle of a triangle $\triangle ABC$. (The center, D, of this circle is the intersection of the angle-bisectors of $\triangle ABC$.) Let s be as in the statement of Heron's formula. Prove that $\tan\frac{\alpha}{2} = \frac{r}{s - a}$, $\tan\frac{\beta}{2} = \frac{r}{s - b}$, and $\tan\frac{\gamma}{2} = \frac{r}{s - c}$. (Hint: $\angle AED$ is a right angle. Show that $AE = s - a$.)

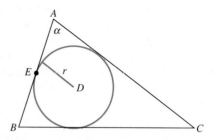

36. Use the Law of Cosines to obtain a new proof of the subtraction identity,
$$\cos(x - y) = \cos x \cos y + \sin x \sin y,$$
that was first derived in section 6.2. (Hint: consider the triangle with vertices at the origin, $P(\cos y, \sin y)$, and $Q(\cos x, \sin x)$. Compute the distance PQ two different ways. Remember to discuss "degenerate" cases, such as $x - y = 2\pi n$, n an integer.)

37. (a) Explain why no triangle $\triangle ABC$ can satisfy the data $a = 2$, $b = 3$, $c = 7$.
    (b) Intuitively, derive the "geometric triangle inequality": for any triangle, the sum of lengths of any two of its sides is greater that the length of the other side of the triangle. (A proof of this result was given in section 1.1.)
    (c) Using (b), explain why $s > a$, $s > b$, and $s > c$ in the context of Heron's formula.

38. (A converse of exercise 37(b)) Given positive (real) numbers) a, b, and c such that $a + b > c$, $a + c > b$, and $b + c > a$, give an intuitive argument that there exists a triangle with sides having lengths a, b, and c.

39. (a) Show that the following data satisfy Mollweide's equation: $a = 0.89$, $b = 0.11$, $c = \sqrt{2 - \sqrt{2}}$, $\alpha = 127.5°$, $\beta = 7.5°$, and $\gamma = 45°$.
    (b) Show that, despite (a), there is no triangle meeting the data listed in (a). (Hint: see exercise 37(b).) What can you conclude?

40. Use Mollweide's equation to prove the "geometric triangle inequality" stated in exercise 37(b). (Hint: show that $\sin\left(\frac{\gamma}{2}\right) = \cos\left(\frac{\alpha + \beta}{2}\right)$ and recall that $\cos$ is a decreasing function on $\left[0, \frac{\pi}{2}\right]$.)

7.4 VECTORS IN THE PLANE

Many applications of mathematics to the physical, biological, and social sciences are concerned with quantities that have both magnitude and direction. Any quantity that involves only magnitude, such as 12 ft, 25 kg, or 7 s, is called a **scalar quantity** or **scalar.** Any quantity that has both magnitude and direction, such as a northwest wind of 35 mph, is called a **vector quantity** or **vector.**

A vector quantity is normally represented by a **directed line segment**. This is a line segment to which a direction has been assigned by considering one of the segment's endpoints as the vector's **initial point**. The other endpoint of the segment is the vector's **terminal point**. We write $\overrightarrow{PQ}$ (not to be confused with the notation for a ray) to denote the vector with initial point P and terminal point Q. To draw $\overrightarrow{PQ}$, we draw the segment $\overline{PQ}$ and then place an arrowhead at Q, to indicate that the associated vector quantity is applied at P and acts in the direction from P to Q. The length of a directed line segment is called the **magnitude** of the associated vector quantity.

We suggested above that many physical concepts may be represented by vectors. For example, the vector $\overrightarrow{PQ}$ in figure 7.19(a) represents a force of 10 pounds which is applied at P and acts at an angle of $\dfrac{\pi}{4}$ to the positive x-axis. Other examples, as shown in figure 7.19(b), include the **displacement vector**

Figure 7.19

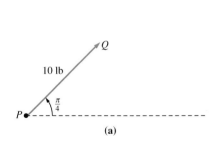

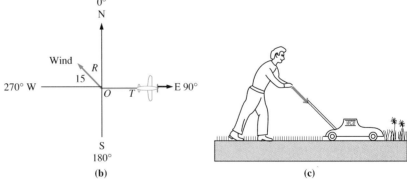

$\overrightarrow{OT}$ used to represent the change in position of an airplane that has traveled east from O to T; and the **velocity vector** $\overrightarrow{OR}$ which indicates a wind blowing from the southeast at 15 km/h. Yet another physical example of a vector quantity is given by the **force vector** in figure 7.19(c) representing the force being exerted on the lawnmower handle.

The symbol for a vector is often printed in boldface type. Since one cannot write in boldface, hand-written vectors are often represented with an arrow over the letter or letters. Thus, **v** and $\vec{v}$ represent the same vector, as do **PQ** and $\overrightarrow{PQ}$. Of course, **QP** means the same as $\overrightarrow{QP}$, as shown in figure 7.20. The magnitude of a vector may be denoted by absolute value notation. For instance, the magnitude of **v** is |**v**|, while the magnitude of **PQ** is |**PQ**| = $|\overrightarrow{PQ}|$.

Figure 7.20

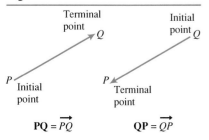

A **zero vector**, denoted by **0** or $\vec{0}$, is a vector having length 0. This useful concept corresponds to the situation in which the associated line segment degenerates to a point. In other words, $\vec{PQ} = \mathbf{0}$ if and only if and only if $P = Q$.

A **unit vector** is any vector having magnitude 1.

The vectors that we have discussed so far have been **bound,** in the sense that the quantity represented by $\vec{PQ}$ acts at the specific point P. A more useful concept is that of a **free vector**. We often drop the qualifier "free."

Free Vector

> A free vector consists of all bound vectors which have the same magnitude and direction as a given bound vector.

In order to avoid the "bound" and "free" terminology, we agree to the following definition.

Equal Vectors

> Call two vectors **equal** if and only if they have the same magnitude and the same direction.

Figure 7.21

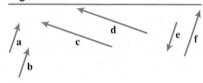

The "direction" of a zero vector is not defined, and so we also agree that **v** is equal to the zero vector if and only if $\mathbf{v} = \mathbf{0}$.

In particular, **the directed line segments representing two equal vectors do not need to have the same initial point**. In figure 7.21, vectors **a** and **b** are equal and so we write $\mathbf{a} = \mathbf{b}$; similarly, $\mathbf{c} = \mathbf{d}$. However, vectors **a** and **e** are unequal, since they have different directions. Similarly, $\mathbf{a} \neq \mathbf{f}$, since **a** and **f** have different magnitudes.

Figure 7.22

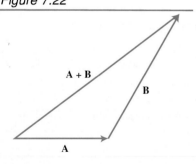

Addition of Vectors

Let **A** and **B** be two vectors. Their sum, denoted by $\mathbf{A} + \mathbf{B}$, is found by placing the initial point of **B** at the terminal point of **A**, as shown in figure 7.22. Then the vector with the same initial point as **A** and the same terminal point as **B** is the sum $\mathbf{A} + \mathbf{B}$. This sum is also called the **resultant** of **A** and **B** because $\mathbf{A} + \mathbf{B}$ physically represents a vector that exerts the same effect as **A** and **B** acting together. Notice that the zero vector is well named, for $\mathbf{A} + \mathbf{0} = \mathbf{A}$ for any vector **A**.

If **A** and **B** do not have the same line of action, there is another method to find the vectorial sum $\mathbf{A} + \mathbf{B}$. First, place **A** and **B** so that their initial points coincide. Next, complete the parallelogram that has **A** and **B** as adjacent sides, as shown in figure 7.23. Then the diagonal of this parallelogram that has the same initial point as **A** and **B** is $\mathbf{A} + \mathbf{B}$. The "parallelogram law of vector addition" was discovered empirically by considering resultants of displacement vectors, or velocity vectors, or force vectors, and so on.

Figure 7.23

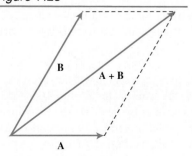

It can be shown by careful geometric reasoning that vector addition is commutative and associative. We omit this verification, as it would take us too far afield.

Vectors in the Plane

Opposite of a Vector

If **V** is a nonzero vector, then its opposite, −**V**, is defined as a vector with the same magnitude as **V** but with opposite direction. Moreover, the opposite of a zero vector is taken to be a zero vector.

In all cases, **V** + (−**V**) = **0**.

By analogy with the elementary algebra of real numbers, we define the **difference A − B** of two given vectors **A**, **B** in the following box.

Difference of Vectors

$$\mathbf{A} - \mathbf{B} = \mathbf{A} + (-\mathbf{B}).$$

Figure 7.24

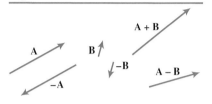

These concepts are illustrated in figure 7.24.

Scalar Multiplication

There is a useful way to multiply a vector **V** by a scalar (real number) r.

Scalar Multiple

The **scalar multiple** $r\mathbf{V}$ is a vector whose magnitude is $|r|$ times $|\mathbf{V}|$. If $\mathbf{V} \neq \mathbf{0}$ and r is positive, then $r\mathbf{V}$ has the same direction as **V**; if $\mathbf{V} \neq \mathbf{0}$ and $r < 0$, the direction of $r\mathbf{V}$ is opposite to the direction of **V**. As degenerate cases, we define $r\mathbf{0}$ and $0\mathbf{V}$ both to be **0**.

For example, if $\mathbf{A} \neq \mathbf{0}$, $2\mathbf{A}$ has the same direction as **A** and is twice as long as **A**.

▶ **EXAMPLE 7.15**

For vectors **A** and **B** as in figure 7.25(a), draw the vector $3\mathbf{A} + 2\mathbf{B}$.

Solution

Using **A**, we construct the vector $3\mathbf{A}$ having the same direction as **A** and having length 3 times that of **A**. Using the same initial point, we next draw $2\mathbf{B}$, the vector in the same direction as **B** but having twice the length of **B**. Construct the parallelogram with adjacent sides $3\mathbf{A}$ and $2\mathbf{B}$. As shown in figure 7.25(b), the resultant vector $3\mathbf{A} + 2\mathbf{B}$ is the diagonal having the same initial point as $3\mathbf{A}$ and $2\mathbf{B}$. ◀

Figure 7.25

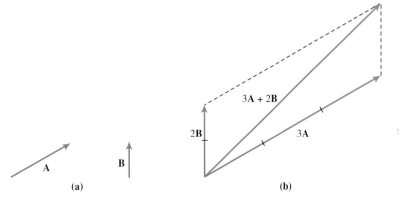

Vector Components

Performing vector addition by using diagrams is rather arduous and inexact. Consequently, a more arithmetic method is appropriate for precise scientific computations. This is accomplished by resolving each vector into its "components" and then performing suitable arithmetic with those components. To explain these concepts and the procedure, we shall next relate (free) vectors to Cartesian coordinates.

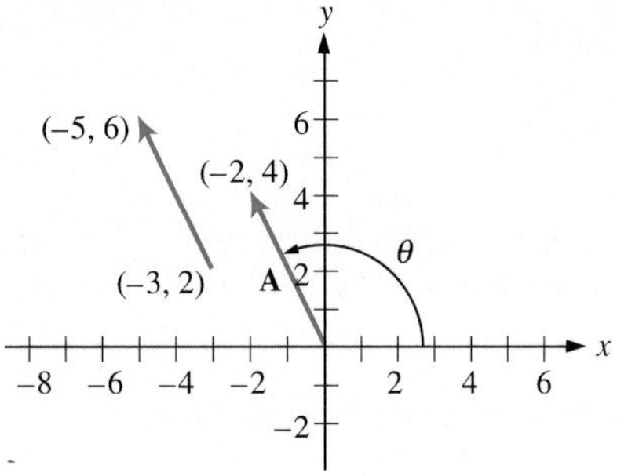

Each (bound) vector has an initial and a terminal point. Consider the vector with initial point $(-3, 2)$ and terminal point $(-5, 6)$, as in figure 7.26. As a free vector, this equals the vector with initial point $(0, 0)$ and terminal point $(-2, 4)$. In the same way, each (free) vector corresponds to exactly one bound vector with initial point at the origin. Then the coordinates of the terminal point of that bound vector uniquely determine the free vector. Thus, the vector in figure 7.26 can be specified as $\langle -2, 4 \rangle$. The magnitude of $\langle -2, 4 \rangle$, as determined by Pythagoras' Theorem, is $\sqrt{(-2)^2 + 4^2} = \sqrt{20} = 2\sqrt{5}$. The direction, as indicated by the angle θ in standard position, is approximately 2.0344 radians.

We turn to the general situation. First we define **i** and **j** to be the unit vectors along the positive coordinate axes, as shown in figure 7.27(a). In terms of coordinates, these unit vectors are defined as follows.

$$\mathbf{i} = \langle 1, 0 \rangle \quad \text{and} \quad \mathbf{j} = \langle 0, 1 \rangle.$$

Using **i** and **j**, we can describe the vector $\langle -2, 4 \rangle$ as $-2\mathbf{i} + 4\mathbf{j}$. (See figure 7.27(b).) More generally, one may show the following.

Figure 7.26

Figure 7.27

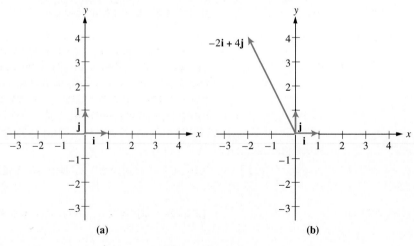

Describing a Vector in Terms of i and j

Given points $P_1(x_1, y_1)$ and $P_2(x_2, y_2)$,
$$\overrightarrow{P_1 P_2} = \langle x_2 - x_1, y_2 - y_1 \rangle = (x_2 - x_1)\mathbf{i} + (y_2 - y_1)\mathbf{j}.$$

Vectors in the Plane

In particular, if **V** is a vector with initial point at the origin and terminal point (V_x, V_y), then we can give the following description of **V**.

$$\mathbf{V} = \langle V_x, V_y \rangle = V_x \mathbf{i} + V_y \mathbf{j}.$$

The numbers V_x and V_y are the **components** of **V**. The magnitude of **V** is, by Pythagoras' Theorem, given as follows.

Magnitude of Vector V

$$|\mathbf{V}| = \sqrt{V_x^2 + V_y^2}.$$

If $\mathbf{V} \neq \mathbf{0}$, then the **direction** of **V** is the standard-position angle θ, $0 \leq \theta < 2\pi$, determined by **i** and **V**. If $V_x \neq 0$, it is often convenient to find θ by using the equation

$$\tan \theta = \frac{V_y}{V_x}.$$

The definitions of the sine and cosine functions in section 5.2 help us relate V_x, V_y, $|\mathbf{V}|$, and θ.

$$V_x = |\mathbf{V}| \cos \theta \quad \text{and} \quad V_y = |\mathbf{V}| \sin \theta.$$

Do not be alarmed if you feel that you have seen similar reasoning and formulas earlier. You *are* experiencing *déjà vu!* Finding a vector's magnitude and direction is essentially the same process as finding a complex number's modulus and argument, a skill you acquired in section 6.6. This analogy will be made explicit and pursued in exercise 52. A lot of scientific and mathematical progress depends on detecting and exploiting such formal similarities.

▶ **EXAMPLE 7.16**   Write **a** in terms of its components if $|\mathbf{a}| = 6$ and the direction of **a** is $\theta = \frac{5\pi}{3}$.

Solution   Since $a_x = |\mathbf{a}| \cos \theta$, we have $a_x = 6 \cos \frac{5\pi}{3} = 6 \left(\frac{1}{2} \right) = 3$. Similarly,

$a_y = |\mathbf{a}| \sin \theta = 6 \sin \frac{5\pi}{3} = 6 \left(-\frac{\sqrt{3}}{2} \right) = -3\sqrt{3}$. Thus,

$$\mathbf{a} = \langle a_x, a_y \rangle = \langle 3, -3\sqrt{3} \rangle = 3\mathbf{i} - 3\sqrt{3}\mathbf{j}.$$ ◀

▶ **EXAMPLE 7.17**   Find the magnitude and the direction of $\mathbf{b} = 4\mathbf{i} - 4\mathbf{j}$.

Solution

The magnitude is $|\mathbf{b}| = \sqrt{b_x^2 + b_y^2} = \sqrt{4^2 + (-4)^2} = 4\sqrt{2}$.

The direction θ satisfies $\tan \theta = \dfrac{b_y}{b_x} = \dfrac{-4}{4} = -1$. Since $(4, -4)$ is in the fourth quadrant, so is θ. Hence, $\theta = \dfrac{7\pi}{4}$.

▶ **EXAMPLE 7.18**

Is $\mathbf{c} = \dfrac{\sqrt{3}}{2}\mathbf{i} - \dfrac{1}{2}\mathbf{j}$ a unit vector?

Solution

The question is whether $|\mathbf{c}|$ equals 1. We have

$$|\mathbf{c}| = \sqrt{\left(\dfrac{\sqrt{3}}{2}\right)^2 + \left(\dfrac{1}{2}\right)^2} = \sqrt{\dfrac{3}{4} + \dfrac{1}{4}} = \sqrt{1} = 1.$$

Hence, $\mathbf{c}$ is a unit vector.

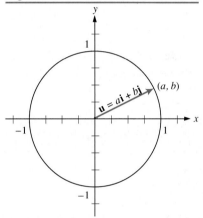

Figure 7.28

If $\mathbf{u} = a\mathbf{i} + b\mathbf{j}$ is a unit vector, then $|\mathbf{u}| = \sqrt{a^2 + b^2} = 1$. This means that the point (a, b) lies on the unit circle, as shown in figure 7.28. If θ is the direction of $\mathbf{u}$, then $a = u_x = |\mathbf{u}| \cos \theta = 1 \cos \theta = \cos \theta$. Similarly, $b = \sin \theta$. Thus (compare exercise 25 of section 5.3), we have the following characterization of unit vectors.

> Any unit vector can be written in the form
> $$\mathbf{u} = (\cos \theta)\mathbf{i} + (\sin \theta)\mathbf{j},$$
> where θ is the direction of $\mathbf{u}$.

▶ **EXAMPLE 7.19**

Find the vector $\mathbf{a}$ with direction $\dfrac{3\pi}{4}$ and magnitude 9.

Solution

You may proceed as in example 7.16. We shall use another method, in order to show a new role for unit vectors. The unit vector with direction $\dfrac{3\pi}{4}$ is

$$\mathbf{u} = \left(\cos \dfrac{3\pi}{4}\right)\mathbf{i} + \left(\sin \dfrac{3\pi}{4}\right)\mathbf{j} = -\dfrac{\sqrt{2}}{2}\mathbf{i} + \dfrac{\sqrt{2}}{2}\mathbf{j}.$$

Thus,

$$\mathbf{a} = 9\mathbf{u} = 9\left(-\dfrac{\sqrt{2}}{2}\mathbf{i} + \dfrac{\sqrt{2}}{2}\mathbf{j}\right) = -\dfrac{9\sqrt{2}}{2}\mathbf{i} + \dfrac{9\sqrt{2}}{2}\mathbf{j}.$$

The reasoning in example 7.19 establishes the following general fact.

> If $\mathbf{v}$ is a nonzero vector, then $\mathbf{u} = \dfrac{\mathbf{v}}{|\mathbf{v}|}$ is the unit vector having the same direction as $\mathbf{v}$.

Vectors in the Plane

Let θ be the angle that the resultant vector makes with the positive x-axis. Now, $\tan\theta \approx \dfrac{182.1346}{150.6418} \approx 1.2091$, and so $\theta \approx 50.4062°$. This means that the actual heading relative to the ground is about $90° - 50.4062° \approx 39.59°$. We saw earlier that the speed relative to the ground is approximately 236.36 mph. ◀

Figure 7.29

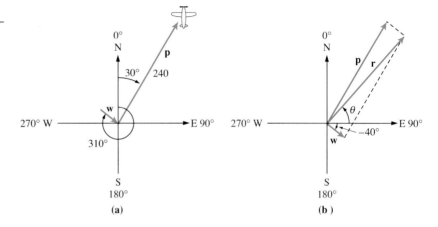

If more than one force is applied to an object, then the **resultant** of the forces applied to the object is the vector sum of these forces. It often helps to think of the resultant force as the net applied force. An object is said to be in **equilibrium** if the net force acting on it is the zero vector. This means that the sum of the x-components of the forces and the sum of the y-components of the forces are both zero.

▶ EXAMPLE 7.24

Two cables support a 500 kilogram load, as shown in figure 7.30(a). Find the tension in each cable.

Figure 7.30

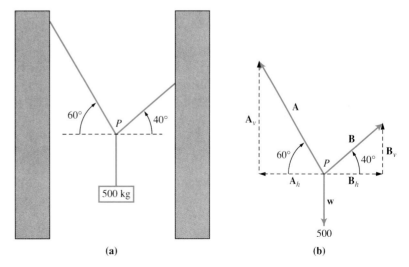

Solution

At point P, there are three forces acting: the load is pulling down, while the two cables are pulling upward and outward. A diagram of the forces is in figure 7.30(b). The forces exerted by the cables are indicated by vectors $\mathbf{A}$ and $\mathbf{B}$, while the 500 kilogram load is indicated by vector $\mathbf{W}$. The horizontal components of $\mathbf{A}$ and $\mathbf{B}$ are $A_h = -|\mathbf{A}| \cos 60°$ and $B_h = |\mathbf{B}| \cos 40°$. Of course, the horizontal component of $\mathbf{W}$ is 0. Since there is equilibrium, their sum must be 0. So,

$$|\mathbf{A}| \cos 60° = |\mathbf{B}| \cos 40°.$$

The vertical components of the forces in the cables are $A_v = |\mathbf{A}| \sin 60°$ and $B_v = |\mathbf{B}| \sin 40°$. By the equilibrium condition,

$$|\mathbf{A}| \sin 60° + |\mathbf{B}| \sin 40° + (-500) = 0.$$

Thus, we have two equations in two variables:

$$|\mathbf{A}| \cos 60° - |\mathbf{B}| \cos 40° = 0 \qquad (1)$$

$$|\mathbf{A}| \sin 60° + |\mathbf{B}| \sin 40° = 500. \qquad (2)$$

If we solve equation (1) for $|\mathbf{A}|$, we obtain

$$|\mathbf{A}| = \frac{|\mathbf{B}| \cos 40°}{\cos 60°}. \qquad (3)$$

Next, use (3) to substitute for $|\mathbf{A}|$ in equation (2), with the result

$$\frac{|\mathbf{B}| \cos 40°}{\cos 60°} \cdot \sin 60° + |\mathbf{B}| \sin 40° = 500$$

or

$$\frac{|\mathbf{B}| \cos 40° \cdot \sin 60°}{\cos 60°} + \frac{|\mathbf{B}| \sin 40° \cdot \cos 60°}{\cos 60°} = 500$$

and so $|\mathbf{B}| \cos 40° \cdot \sin 60° + |\mathbf{B}| \sin 40° \cdot \cos 60° = 500 \cos 60°$.

In general, you would proceed to solve this linear equation for $|\mathbf{B}|$, perhaps using your calculator to approximate the trigonometric values. However, in this example, it is faster to use an addition formula from section 6.2.

$$|\mathbf{B}| \sin (40° + 60°) = 500 \cos 60° = 500(\tfrac{1}{2}) = 250,$$

and so

$$|\mathbf{B}| = \frac{250}{\sin 100°} \approx 253.85665 \text{ kg}.$$

Substituting this value in equation (3), we obtain

$$|\mathbf{A}| \approx 388.93096 \text{ kg}. \qquad \blacktriangleleft$$

▶ **EXAMPLE 7.25** A force of 5 N is applied to the left side of an object, a force of 12 N is applied from the bottom, and a force of 17 N is applied at a direction of $\frac{4\pi}{3}$ to the horizontal. What is the resultant of the forces applied to the object? Find its magnitude and direction.

Vectors in the Plane

Solution

The situation is depicted in figure 7.31(a), with the corresponding force vector diagram in figure 7.31(b). Thus, we have $\mathbf{F}_1 = 5\mathbf{i}$, $\mathbf{F}_2 = 12\mathbf{j}$, and $\mathbf{F}_3 = -\frac{17}{2}\mathbf{i} - \frac{17}{2}\sqrt{3}\,\mathbf{j}$. The resultant is given by

$$\mathbf{F} = \mathbf{F}_1 + \mathbf{F}_2 + \mathbf{F}_3 = (5 - \tfrac{17}{2})\mathbf{i} + (12 - \tfrac{17}{2}\sqrt{3}\,)\mathbf{j} = -\tfrac{7}{2}\mathbf{i} + (12 - \tfrac{17}{2}\sqrt{3}\,)\mathbf{j}.$$

The magnitude of $\mathbf{F}$ is $|\mathbf{F}| = \sqrt{373 - 204\sqrt{3}} \approx \sqrt{19.661635} \approx 4.4341443$.

The direction θ of $\mathbf{F}$ can be calculated by first finding the unit vector in the direction of $\mathbf{F}$. Proceed as in example 7.20:

$$(\cos\theta)\mathbf{i} + (\sin\theta)\mathbf{j} = \frac{\mathbf{F}}{|\mathbf{F}|} \approx \frac{-\tfrac{7}{2}}{\sqrt{19.661635}}\mathbf{i} + \frac{12 - \tfrac{17}{2}\sqrt{3}}{\sqrt{19.661635}}\mathbf{j}$$

$$\approx -0.7893293\mathbf{i} - 0.6139701\mathbf{j}.$$

Thus, $\cos\theta \approx -0.7893293$ and $\sin\theta \approx -0.6139701$. This means that θ is in the third quadrant, and so $\theta \approx 3.8026732$. In other words, $\theta \approx 217.87712°$, as shown in figure 7.31(b).

Figure 7.31

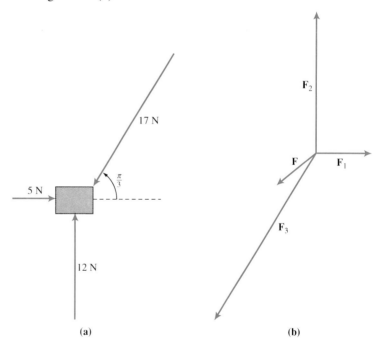

EXAMPLE 7.26

What additional force must be applied in order for the object in example 7.25 to remain at rest?

Solution

From example 7.25, we know that the resultant $\mathbf{F}$ of $\mathbf{F}_1$, $\mathbf{F}_2$, and $\mathbf{F}_3$ is $-\frac{7}{2}\mathbf{i} + (12 - \frac{17}{2}\sqrt{3}\,)\mathbf{j}$. A force $\mathbf{F}_4$ must be applied so that the resultant of all four forces is $\mathbf{0}$. We write $\mathbf{F}_4 = a\mathbf{i} + b\mathbf{j}$. The desired condition is

$$\mathbf{0} = \mathbf{F} + \mathbf{F}_4 = (-\tfrac{7}{2} + a)\mathbf{i} + (12 - \tfrac{17}{2}\sqrt{3} + b)\mathbf{j}.$$

This means that

$$F_4 = -F = \tfrac{7}{2}i + (\tfrac{17}{2}\sqrt{3} - 12)j,$$

as shown in figure 7.32.

Figure 7.32

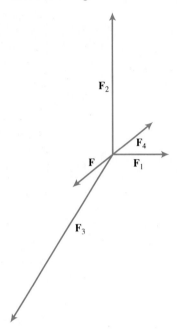

Exercises 7.4

In exercises 1–2, determine which of the given quantities are scalars and which are vectors.

1. (a) 5 centimeters
   (b) a northeast wind of 20 kilometers per hour
   (c) a vertically upward force of 160 pounds
   (d) 14 hours
2. (a) an arrow pointing from (–2, 3) to (4, –1)
   (b) 8π square feet   (c) 51 grams
   (d) a vertically downward acceleration of 32 feet per second per second

In exercises 3–4, draw each of the given quantities as a directed line segment.

3. (a) a displacement vector describing linear travel from (4, –1) to (–2, 3)
   (b) a force vector of 5 pounds, applied at (1, 2) and acting at an angle of $\dfrac{\pi}{6}$ to the positive x-axis
4. (a) a force vector of 6 N, applied at (1, 0) and acting at an angle of $\dfrac{5\pi}{6}$ to the positive x-axis

   (b) a velocity vector of 14 mph due west at (2, 1)
5. Explain geometrically why $\vec{v} + \vec{0} = \vec{v}$ for all vectors **v**.
6. What may you conclude if $\overrightarrow{PQ} = \overrightarrow{QP}$? Why?

In exercises 7–8, determine which of the vectors **B**, **C**, **D**, **E** is/are equal to the given vector **A**.

7.

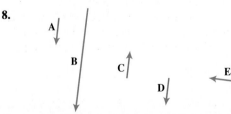

8.

9. **A** and **B** both have initial point P. Draw and label –**A** and –**B** as directed line segments with initial point P.

Vectors in the Plane

27. $|\mathbf{v}| = \pi;\ \theta = 4.24$ 28. $|\mathbf{v}| = \sqrt{2};\ \theta = 0$

In exercises 29–36, find the magnitude and the direction of the vector **V** having the given components or description. Approximate by calculator if necessary.

29. $V_x = 10;\ V_y = 10\sqrt{3}$ 30. $V_x = -\frac{2}{3};\ V_y = -\frac{2}{3}$
31. $\mathbf{V} = \langle 5, -10 \rangle$ 32. $\mathbf{V} = \langle -2\pi, 0 \rangle$
33. $\mathbf{V} = -5\mathbf{j}$ 34. $\mathbf{V} = -3\sqrt{2}\mathbf{i} + \sqrt{2}\mathbf{j}$
35. $\mathbf{V} = \overrightarrow{P_1P_2},\ P_1(-3, -2),\ P_2(1, -2)$
36. $\mathbf{V} = \overrightarrow{P_1P_2},\ P_1(\sqrt{3}, 0),\ P_2(0, 1)$
37. If a particle moves from $P_1(-1, 7)$ to P_2 and $\overrightarrow{P_1P_2} = 4\mathbf{i} - 3\mathbf{j}$, find the coordinates of P_2. (Hint: express the components of $\overrightarrow{P_1P_2}$ in terms of the coordinates of P_1 and P_2.)
38. If a particle moves from P_1 to $P_2(-1, 7)$ and $\overrightarrow{P_1P_2} = 4\mathbf{i} - 3\mathbf{j}$, find the coordinates of P_1.
39. Determine which of the following is/are a unit vector:
    (a) $-\frac{3}{5}\mathbf{i} + \frac{4}{5}\mathbf{j}$ (b) $-\mathbf{j}$ (c) $-2\mathbf{i}$ (d) $0.67\mathbf{i} - 0.75\mathbf{j}$
40. If $a\mathbf{i} + b\mathbf{j} = c\mathbf{i} + d\mathbf{j}$, explain why $a = c$ and $b = d$. (Hint: you may argue algebraically or geometrically.)

10. Explain geometrically why $\mathbf{v} + (-\mathbf{v}) = \mathbf{0}$ for all vectors **v**.

In exercises 11–12, **A** and **B** both have initial point P. Draw and label each of the following as directed line segments with initial point P: $\mathbf{A} + \mathbf{B},\ \mathbf{A} - \mathbf{B},\ 2\mathbf{A} - 3\mathbf{B}$.

11.

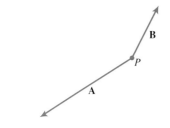

12.

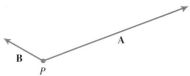

13. If **V** represents a northwest wind blowing at 15 mph, what sort of wind does $-\mathbf{V}$ represent? (Hint: the wind is blowing *from*, not toward, the northwest.)
14. The solution to example 7.26 implicitly used the following fact. If $\mathbf{u} + \mathbf{v} = \mathbf{0}$, then $\mathbf{v} = -\mathbf{u}$. Prove this fact. (Hint: one method starts by adding $-\mathbf{u}$ to both sides.)

In exercises 15–16, express $\overrightarrow{P_1P_2}$ as $\langle a, b \rangle$.

15. $P_1(6, 9),\ P_2(2, 7)$ 16. $P_1(0, \sqrt{2}),\ P_2(3, 1)$

In exercises 17–18, express $\overrightarrow{P_1P_2}$ as $a\mathbf{i} + b\mathbf{j}$.

17. $P_1(4, -6),\ P_2(e, -2)$ 18. $P_1(-4, 2),\ P_2(0, 5)$

In exercises 19–28, **v** has the given magnitude $|\mathbf{v}|$ and direction θ. Sketch **v** as a directed line segment with initial point at the origin. Also, write **v** in terms of its components. Approximate the components by calculator if necessary.

19. $|\mathbf{v}| = 2;\ \theta = \frac{\pi}{3}$ 20. $|\mathbf{v}| = \frac{1}{2};\ \theta = \frac{\pi}{6}$
21. $|\mathbf{v}| = \frac{3}{4};\ \theta = 45°$ 22. $|\mathbf{v}| = 3;\ \theta = 90°$
23. $|\mathbf{v}| = \frac{3}{2};\ \theta = \frac{5\pi}{6}$ 24. $|\mathbf{v}| = 2;\ \theta = \frac{3\pi}{4}$
25. $|\mathbf{v}| = 1;\ \theta = 180°$ 26. $|\mathbf{v}| = \frac{3}{4};\ \theta = 300°$

In exercises 41–44, determine (a) $\mathbf{u} + \mathbf{v}$, (b) $\mathbf{u} - \mathbf{v}$, (c) $5\mathbf{u}$, (d) $-\frac{2}{3}\mathbf{v}$, and (e) $\frac{3}{4}\mathbf{u} + 2\mathbf{v}$.

41. $\mathbf{u} = \langle -2, -8 \rangle,\ \mathbf{v} = \langle 7, 2 \rangle$ 42. $\mathbf{u} = \langle -12, -2 \rangle,\ \mathbf{v} = \langle 6, 8 \rangle$
43. $\mathbf{u} = \frac{3}{2}\mathbf{i} + 6\mathbf{j},\ \mathbf{v} = 2\mathbf{i} - 3\mathbf{j}$ 44. $\mathbf{u} = 9\mathbf{i} - \mathbf{j},\ \mathbf{v} = -3\mathbf{i} + \frac{1}{2}\mathbf{j}$

In exercises 45–48, find a unit vector with the same direction as the given vector.

45. $3\mathbf{i} - 5\mathbf{j}$ 46. $\langle -\frac{1}{2}, \frac{1}{2} \rangle$
47. $\frac{2}{3}(6\mathbf{i} + \mathbf{j}) - (5\mathbf{i} - \frac{1}{3}\mathbf{j})$ 48. $3(-\mathbf{i} + \mathbf{j}) + 2(\mathbf{i} - \frac{3}{2}\mathbf{j})$
49. Prove the following facts for all scalars r, r_1, r_2 and all vectors $\mathbf{v}_1, \mathbf{v}_2, \mathbf{v}$.
    (a) $r(\mathbf{v}_1 + \mathbf{v}_2) = r\mathbf{v}_1 + r\mathbf{v}_2$ (b) $(r_1 + r_2)\mathbf{v} = r_1\mathbf{v} + r_2\mathbf{v}$
    (c) $r_1(r_2\mathbf{v}) = (r_1r_2)\mathbf{v}$ (d) $1\mathbf{v} = \mathbf{v}$
    (Hint: view $\mathbf{v}_1$ as $\langle a_1, b_1 \rangle$ and $\mathbf{v}_2$ as $\langle a_2, b_2 \rangle$.)
50. The solution to example 7.24 implicitly used the following fact. If r is a scalar and **v** a vector such that $r\mathbf{v} = \mathbf{0}$, then either $r = 0$ or $\mathbf{v} = \mathbf{0}$. Prove this fact. (Hint: if $r \ne 0$, compute $r^{-1}(r\mathbf{v})$ two different ways. You'll need some parts of exercise 49.)
51. (a) In figure 7.19(c), the force vector has a magnitude of 45 pounds and a direction of $\theta = 5.323$ radians. Determine the components of this force vector.
    (b) Just after a centerfielder throws a baseball, the ball's velocity vector has a magnitude of 60 mph and makes an angle of 40° with the horizon. Determine the components of this velocity vector.

Chapter 7 / Applications of Trigonometry

◆ **52.** Consider the following "dictionary" connecting the set of complex numbers and the set of planar vectors: $z = a + bi$ (with a and b real) corresponds to $\mathbf{v} = a\mathbf{i} + b\mathbf{j}$.
  (a) If z corresponds to $\mathbf{v}$, prove that the modulus of z equals the magnitude of $\mathbf{v}$ and that the argument of z equals the direction of $\mathbf{v}$.
  (b) Show that a complex number z corresponds to a unit vector if and only if there exists a real number θ such that $z = \text{cis } \theta$. (Hint: review exercise 27(b) in section 6.6.)
  (c) If z_1 corresponds to $\mathbf{v}_1$ and z_2 corresponds to $\mathbf{v}_2$, show that $z_1 + z_2$ corresponds to $\mathbf{v}_1 + \mathbf{v}_2$.
  (d) (Viewing scalar multiplication via complex numbers) Let r be a real number and let $z = a + bi$ be a complex number. Let $\mathbf{w}$ be the vector $r(a\mathbf{i} + b\mathbf{j})$ that results when the vector $a\mathbf{i} + b\mathbf{j}$ is multiplied by the scalar r. Show that $\mathbf{w}$ is the vector corresponding to the complex product of $(r + 0i)$ and z.

53. A plane is flying with an airspeed of 280 mph and a compass bearing of 290°. The wind is blowing from 35° at 30 mph. Find the direction and speed of the plane relative to the ground.

54. Find the airspeed and compass bearing of a plane if its encounter with a northwest wind of 35 mph produces a path in the direction of 274° at a speed of 240 mph relative to the ground. (Hint: you can determine $\mathbf{p}$ by subtracting $\mathbf{w}$ from $\mathbf{p} + \mathbf{w}$.)

55. An airplane leaves an airport and flies at 260 mph in the direction 60° for 90 minutes. The plane then changes course and flies at 250 mph in the direction 110° for 30 minutes. At this moment,
  (a) How far is the plane from the airport?
  (b) How far north of the airport is the plane?
  (c) How far east of the airport is the plane?
  (Hint for all parts: find the resultant of the two displacement vectors. Then find this resultant's components and magnitude.)

56. As a ship steams due north at 30 mph, little Petra runs directly across the ship's deck from port to starboard (left to right) at 2 mph.
  (a) Determine Petra's velocity vector relative to the Earth's surface.
  (b) What is her speed relative to the Earth?

57. Two cables support a 102 pound load, as shown in the figure. Find the tension in each cable.

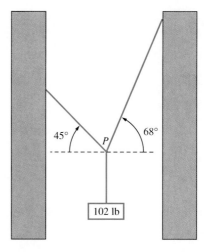

58. A force of 15 N is applied to the right side of an object, a force of 2 N is applied from the top, and a force of 20 N is applied at a direction of 330°. What is the resultant of the forces applied to the object?

59. As shown in the figure, two thieves are dragging a box containing John and David's valuable library of mathematics books. If the thieves are pulling with the indicated forces, determine (a) the resultant force on the box of books and (b) the force that the Human Fly would need to apply to the box in order for it to come to rest. (Carry your calculations to at least four decimal places.)

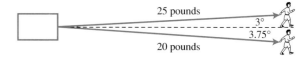

60. As in the figure, a car is on a 20° ramp. The car must exert 1060 pounds of force to keep from sliding down the ramp. Determine the weight of the car to the nearest pound. (Hint: review exercise 34 in section 5.2. Find the "component" of the car's weight in the direction of the ramp.)

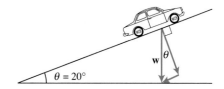

7.5 DOT PRODUCT

In the last section, we introduced the multiplication of a vector by a scalar. This had a geometric interpretation in terms of directed line segments (and, in

Dot Product

exercise 52 of section 7.4, another interpretation in terms of multiplication of complex numbers). There are other types of products involving vectors, and this section will introduce the one that is most useful in the plane. Like the material of section 7.4, it has a natural extension to three-dimensional space, which you will study in a calculus course.

We shall consider two types of applications for this new product. The first involves what engineers and physicists call "work." To study this, we shall consider "projections," which may be viewed as a generalization of the last section's method of resolving a vector into its horizontal and vertical components. The second type of application is to deriving formulas and results in geometry. Optional enrichment material is placed at the end of the section to help indicate the power and flexibility of vectorial methods.

Dot Product

Let $\mathbf{u} = a_1\mathbf{i} + b_1\mathbf{j}$ and $\mathbf{v} = a_2\mathbf{i} + b_2\mathbf{j}$ be two vectors. The **dot product** of $\mathbf{u}$ and $\mathbf{v}$, denoted $\mathbf{u} \cdot \mathbf{v}$, is defined by

$$\mathbf{u} \cdot \mathbf{v} = a_1 a_2 + b_1 b_2.$$

Notice that the dot product of two vectors is a scalar. For this reason, some people refer to the dot product as the **scalar product**. Others refer to the dot product as the **inner product.**

▶ **EXAMPLE 7.27**   If $\mathbf{a} = 2\mathbf{i} + 4\mathbf{j}$ and $\mathbf{b} = 3\mathbf{i} - 5\mathbf{j}$, compute $\mathbf{a} \cdot \mathbf{b}$.

Solution   Here, $a_1 = 2$, $b_1 = 4$, $a_2 = 3$, and $b_2 = -5$. Thus,

$$\mathbf{a} \cdot \mathbf{b} = 2(3) + 4(-5) = 6 - 20 = -14. \quad ◀$$

Properties of the Dot Product

If $\mathbf{u}$, $\mathbf{v}$, and $\mathbf{w}$ are vectors and r, s are scalars, then

1. $\mathbf{u} \cdot \mathbf{v} = \mathbf{v} \cdot \mathbf{u}$ commutative law for dot product
2. $\mathbf{u} \cdot (\mathbf{v} + \mathbf{w}) = \mathbf{u} \cdot \mathbf{v} + \mathbf{u} \cdot \mathbf{w}$ distributive law for dot product
3. $r(\mathbf{u} \cdot \mathbf{v}) = (r\mathbf{u}) \cdot \mathbf{v} = \mathbf{u} \cdot (r\mathbf{v})$
4. (a generalization of Property 3) $r\mathbf{u} \cdot s\mathbf{v} = rs(\mathbf{u} \cdot \mathbf{v})$
5. $\mathbf{0} \cdot \mathbf{v} = 0$
6. $\mathbf{v} \cdot \mathbf{v} = |\mathbf{v}|^2$; equivalently, $|\mathbf{v}| = \sqrt{\mathbf{v} \cdot \mathbf{v}}$
7. $\mathbf{i} \cdot \mathbf{i} = \mathbf{j} \cdot \mathbf{j} = 1; \mathbf{i} \cdot \mathbf{j} = 0$

We shall prove (1) and assign the proofs of the other properties in exercises 5–6.

Proof of 1: Write $\mathbf{u} = a_1\mathbf{i} + b_1\mathbf{j}$ and $\mathbf{v} = a_2\mathbf{i} + b_2\mathbf{j}$. Then $\mathbf{u} \cdot \mathbf{v} = a_1 a_2 + b_1 b_2$. Now, $a_1 a_2 = a_2 a_1$ and $b_1 b_2 = b_2 b_1$ by the commutative law for multiplication of real numbers. So, $\mathbf{u} \cdot \mathbf{v} = a_2 a_1 + b_2 b_1$. But, $a_2 a_1 + b_2 b_1 = \mathbf{v} \cdot \mathbf{u}$. Hence, $\mathbf{u} \cdot \mathbf{v} = \mathbf{v} \cdot \mathbf{u}$.

We shall use the above properties frequently, along with minor extensions such as "generalized distributivity." This law implies, in particular, that

$$(\mathbf{u} + \mathbf{v}) \cdot (\mathbf{w} + \mathbf{p}) = \mathbf{u} \cdot \mathbf{w} + \mathbf{u} \cdot \mathbf{p} + \mathbf{v} \cdot \mathbf{w} + \mathbf{v} \cdot \mathbf{p}$$

for all vectors **u**, **v**, **w**, and **p**.

Angle Between Two Vectors

Suppose **u** and **v** are nonzero vectors, arranged to have the same initial point, as in figure 7.33. The dot product can be used to calculate the angle between **u** and **v**, as shown in the following box.

Angle Between Two Vectors

If θ is the angle between the nonzero vectors **u** and **v**, taken so that $0 \leq \theta \leq \pi$, then

$$\cos \theta = \frac{\mathbf{u} \cdot \mathbf{v}}{|\mathbf{u}||\mathbf{v}|}.$$

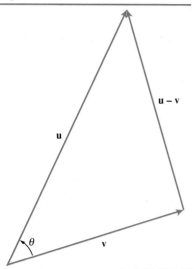

Figure 7.33

Proof: According to the Law of Cosines,

$$|\mathbf{u} - \mathbf{v}|^2 = |\mathbf{u}|^2 + |\mathbf{v}|^2 - 2|\mathbf{u}||\mathbf{v}| \cos \theta.$$

By Properties (1) and (6) and generalized distributivity,

$$|\mathbf{u} - \mathbf{v}|^2 = (\mathbf{u} - \mathbf{v}) \cdot (\mathbf{u} - \mathbf{v})$$
$$= \mathbf{u} \cdot \mathbf{u} - 2\mathbf{u} \cdot \mathbf{v} + \mathbf{v} \cdot \mathbf{v}$$
$$= |\mathbf{u}|^2 - 2\mathbf{u} \cdot \mathbf{v} + |\mathbf{v}|^2.$$

Thus,

$$|\mathbf{u}|^2 - 2\mathbf{u} \cdot \mathbf{v} + |\mathbf{v}|^2 = |\mathbf{u}|^2 + |\mathbf{v}|^2 - 2|\mathbf{u}||\mathbf{v}| \cos \theta$$

or

$$\mathbf{u} \cdot \mathbf{v} = |\mathbf{u}||\mathbf{v}| \cos \theta.$$

Since neither $|\mathbf{u}|$ nor $|\mathbf{v}|$ is 0, by hypothesis, we conclude

$$\cos \theta = \frac{\mathbf{u} \cdot \mathbf{v}}{|\mathbf{u}||\mathbf{v}|}.$$

If θ is the angle between two nonzero vectors **u** and **v**, consider the angle $2\pi - \theta$. Although not necessarily "between" **u** and **v**, $2\pi - \theta$ is certainly determined by **u** and **v**. Thus, it is natural to seek a formula for $\cos (2\pi - \theta)$. Fortunately, it follows from a subtraction formula in section 6.2 that $\cos (2\pi - \theta) = \cos \theta$. (In order to verify this, you'll need to review the values of the cosine and sine functions at 2π.) Thus

$$\cos(2\pi - \theta) = \frac{\mathbf{u} \cdot \mathbf{v}}{|\mathbf{u}||\mathbf{v}|} = \cos \theta.$$

EXAMPLE 7.28   Find the angle θ between $\mathbf{a} = 4\mathbf{i} + 3\mathbf{j}$ and $\mathbf{b} = 2\mathbf{i} - 5\mathbf{j}$.

Solution   We need to know $|\mathbf{a}| = \sqrt{4^2 + 3^2} = 5$, $|\mathbf{b}| = \sqrt{2^2 + (-5)^2} = \sqrt{29}$, and $\mathbf{a} \cdot \mathbf{b} = 4(2) + 3(-5) = -7$.

Thus, $\cos \theta = \frac{\mathbf{a} \cdot \mathbf{b}}{|\mathbf{a}||\mathbf{b}|} = \frac{-7}{5\sqrt{29}}$. Since $0 \leq \theta \leq \pi$, the angle between $\mathbf{a}$ and $\mathbf{b}$ is $\pi - \arccos\left(\frac{7}{5\sqrt{29}}\right) \approx 1.8338$ radians. ◀

EXAMPLE 7.29   In example 7.23, a plane was flying at an airspeed of 240 mph with a compass bearing of 30° and the wind was blowing from 310° at 40 mph. Find the angle by which the plane's actual course deviates from its compass heading.

Solution   We found in example 7.23 that the actual velocity with respect to the ground is approximately

$$\mathbf{r} = 150.6418\mathbf{i} + 182.1346\mathbf{j}.$$

If θ is the angle between $\mathbf{r}$ and the vector for the plane $\mathbf{p} = 120\mathbf{i} + 120\sqrt{3}\,\mathbf{j}$, then

$$\cos \theta = \frac{\mathbf{r} \cdot \mathbf{p}}{|\mathbf{r}||\mathbf{p}|} \approx \frac{(150.6418)(120) + (182.1346)(120\sqrt{3})}{(236.3598)(240)}$$

$$\approx 0.9860141$$

and so $\theta \approx \arccos(0.9860141) \approx 9.5938°$.

Another way to work this example is to recall from example 7.23 that the compass bearing of the plane is 30° and the plane's actual bearing is about 39.59°. The difference between these bearings is about 9.59°, just as the above method showed. ◀

Parallel and Perpendicular Vectors

It is natural to say that directed line segments are **parallel** if and only if their lines of action are parallel lines. Accordingly, we have the following definition.

Parallel Vectors

Vectors $\mathbf{u}$ and $\mathbf{v}$ are parallel if and only if one of the following two situations holds:

1. $\mathbf{u} \neq \mathbf{0}$ and $\mathbf{v} = r\mathbf{u}$ for some scalar r;
2. $\mathbf{u} = \mathbf{0}$.

Nonzero vectors **u** and **v** are **perpendicular,** or **orthogonal,** if and only if the angle between them is $\frac{\pi}{2}$, namely if and only if $\mathbf{u} \cdot \mathbf{v} = 0$. We also agree to say that **0** is perpendicular to any vector. Thus, it is correct in all cases to say:

Perpendicular Vectors

> **a** and **b** are perpendicular if and only if $\mathbf{a} \cdot \mathbf{b} = 0$.

It is important to notice that any scalar multiple of **i** is perpendicular to any scalar multiple of **j**.

▶ **EXAMPLE 7.30**

Show that the vectors $\mathbf{a} = 3\mathbf{i} + 5\mathbf{j}$ and $\mathbf{b} = 10\mathbf{i} - 6\mathbf{j}$ are orthogonal.

Solution

$$\mathbf{a} \cdot \mathbf{b} = 3(10) + 5(-6) = 30 + (-30) = 0.$$

Since $\mathbf{a} \cdot \mathbf{b} = 0$, **a** and **b** are orthogonal. ◀

▶ **EXAMPLE 7.31**

Find a real number s so that $\mathbf{c} = s\mathbf{i} + \mathbf{j}$ and $\mathbf{d} = 3\mathbf{i} - 7\mathbf{j}$ are (a) perpendicular or (b) parallel.

Solutions

(a) **c** and **d** are perpendicular if and only if $\mathbf{c} \cdot \mathbf{d} = 0$. Now, $\mathbf{c} \cdot \mathbf{d} = 3s - 7$, and so we must solve the equation $3s - 7 = 0$. The solution is $s = \frac{7}{3}$.

(b) **c** and **d** are parallel if and only if $\mathbf{d} = r\mathbf{c}$ for some real number r. Thus, we must solve the equation $3\mathbf{i} - 7\mathbf{j} = r(s\mathbf{i} + \mathbf{j})$. By equating corresponding components (after using the distributive law), we see that this is equivalent to solving the pair of equations $3 = rs$ and $-7 = r$. These yield

$$s = \frac{3}{r} = \frac{3}{-7} = -\frac{3}{7}.$$ ◀

Additional geometric applications of these ideas will be given in examples 7.36–7.39.

Projections of Vectors

Some applications require you to find the effective force of a vector **v** in the direction of another vector **u**. Suppose that **u** and **v** are vectors with the same initial point, as shown in figure 7.34. Also suppose $\mathbf{u} \neq \mathbf{0}$; so **u** has a direction. From the terminal point of **v**, drop a perpendicular meeting the line of action of **u** at Q. The vector with the same initial point as **u** and with terminal point Q is the **projection of v onto u,** denoted $\text{proj}_\mathbf{u} \mathbf{v}$.

The magnitude of $\text{proj}_\mathbf{u} \mathbf{v}$ is $|\mathbf{v}| \cos \theta$, and its direction is either the same as or opposite to the direction of **u**. Thus,

$$\text{proj}_\mathbf{u} \mathbf{v} = |\mathbf{v}| \cos \theta \frac{\mathbf{u}}{|\mathbf{u}|}$$

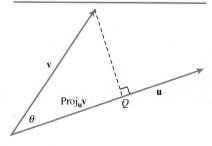

Figure 7.34

since $\frac{\mathbf{u}}{|\mathbf{u}|}$ is the unit vector determined by $\mathbf{u}$. We know that $\cos\theta = \frac{\mathbf{u}\cdot\mathbf{v}}{|\mathbf{u}||\mathbf{v}|}$, and so we have the following description.

$$\text{proj}_{\mathbf{u}}\,\mathbf{v} = \left(\frac{\mathbf{u}\cdot\mathbf{v}}{|\mathbf{u}|}\right)\frac{\mathbf{u}}{|\mathbf{u}|}$$

$$= \left(\frac{\mathbf{u}\cdot\mathbf{v}}{|\mathbf{u}|^2}\right)\mathbf{u}.$$

Notice that the quantity in parentheses is a scalar.

EXAMPLE 7.32

If $\mathbf{a} = -4\mathbf{i} + 5\mathbf{j}$ and $\mathbf{b} = 2\mathbf{i} + 4\mathbf{j}$, calculate $\text{proj}_{\mathbf{b}}\,\mathbf{a}$.

Solution

$$\text{proj}_{\mathbf{b}}\,\mathbf{a} = \left(\frac{\mathbf{a}\cdot\mathbf{b}}{|\mathbf{b}|^2}\right)\mathbf{b}$$

$$= \left(\frac{(-4)\cdot 2 + 5\cdot 4}{2^2 + 4^2}\right)(2\mathbf{i} + 4\mathbf{j})$$

$$= \tfrac{12}{20}(2\mathbf{i} + 4\mathbf{j})$$

$$= \tfrac{6}{5}\mathbf{i} + \tfrac{12}{5}\mathbf{j}.$$

Figure 7.35

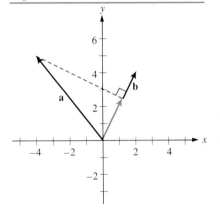

This is illustrated in figure 7.35.

It is important to notice that $\mathbf{v}$ may be "resolved" as $\text{proj}_{\mathbf{u}}\,\mathbf{v} + \mathbf{w}$, where (as we have seen) $\text{proj}_{\mathbf{u}}\,\mathbf{v}$ is parallel to $\mathbf{u}$ and (as we shall next verify) $\mathbf{w}$ is orthogonal to $\mathbf{u}$. The point is that

$$\mathbf{w}\cdot\mathbf{u} = (\mathbf{v} - \text{proj}_{\mathbf{u}}\,\mathbf{v})\cdot\mathbf{u} = \mathbf{v}\cdot\mathbf{u} - \text{proj}_{\mathbf{u}}\,\mathbf{v}\cdot\mathbf{u}$$

$$= \mathbf{v}\cdot\mathbf{u} - \left(\frac{\mathbf{u}\cdot\mathbf{v}}{|\mathbf{u}|^2}\mathbf{u}\right)\cdot\mathbf{u} = \mathbf{v}\cdot\mathbf{u} - \frac{\mathbf{u}\cdot\mathbf{v}}{|\mathbf{u}|^2}\mathbf{u}\cdot\mathbf{u} = \mathbf{v}\cdot\mathbf{u} - \frac{\mathbf{u}\cdot\mathbf{v}}{|\mathbf{u}|^2}|\mathbf{u}|^2$$

$$= \mathbf{v}\cdot\mathbf{u} - \mathbf{u}\cdot\mathbf{v} = 0.$$

EXAMPLE 7.33

Resolve $\mathbf{a} = 5\mathbf{i} + 2\mathbf{j}$ as a sum of component vectors, one parallel and the other perpendicular to $\mathbf{b} = -2\mathbf{i} - 6\mathbf{j}$.

Solution

The vectors $\mathbf{a}$ and $\mathbf{b}$ are shown in figure 7.36(a). The vector $\mathbf{a}$ will be resolved as $\mathbf{a} = \mathbf{a}_1 + \mathbf{a}_2$, where $\mathbf{a}_1$ is parallel to $\mathbf{b}$ and $\mathbf{a}_2$ is perpendicular to $\mathbf{b}$. This is done by taking

$$\mathbf{a}_1 = \text{proj}_{\mathbf{b}}\,\mathbf{a} = \frac{\mathbf{a}\cdot\mathbf{b}}{|\mathbf{b}|^2}\mathbf{b} \quad\text{and}\quad \mathbf{a}_2 = \mathbf{a} - \text{proj}_{\mathbf{b}}\,\mathbf{a}.$$

Substituting, we obtain $\mathbf{a}_1 = \dfrac{-10 - 12}{4 + 36}(-2\mathbf{i} - 6\mathbf{j}) = \dfrac{-22}{40}(-2\mathbf{i} - 6\mathbf{j}) = 1.1\mathbf{i} + 3.3\mathbf{j}$. Then $\mathbf{a}_2 = (5\mathbf{i} + 2\mathbf{j}) - (1.1\mathbf{i} + 3.3\mathbf{j}) = 3.9\mathbf{i} - 1.3\mathbf{j}$. Figure 7.36(b) shows vectors $\mathbf{a}$, $\mathbf{b}$, $\mathbf{a}_1$, and $\mathbf{a}_2$. ◀

Figure 7.36

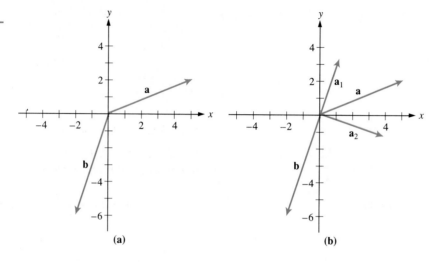

When a force $\mathbf{F}$ moves an object a distance D and $\mathbf{F}$ acts in the same direction in which the object moves, then the **work** done is defined to be $W = |\mathbf{F}|D$. For instance, if a force of $20\mathbf{i}$ moves an object from $(-7, 2)$ to $(14, 2)$, we find that the work done is $|\mathbf{F}|D = 20(14 - (-7)) = 420$ units. Typical units of work are joules (which uses the symbol J) and foot-pounds (ft-lb)*. Notice that if a "holding action" effort causes no motion, then it produces zero "work," according to the above definition.

What should the definition of work be when the force $\mathbf{F}$ does *not* act in the exact direction $\mathbf{d}$ of motion? The idea is to replace $\mathbf{F}$ in the above definition with the effective force it generates in the direction of $\mathbf{d}$. In other words, we replace $\mathbf{F}$ in the above definition with $\text{proj}_{\mathbf{d}}\ \mathbf{F}$. So, if a force $\mathbf{F}$ moves an object from the point P to the point Q, we consider the **displacement vector** $\mathbf{d} = \overrightarrow{PQ}$ and find that the work done is $|\text{proj}_{\mathbf{d}}\ \mathbf{F}|\ |\mathbf{d}| = \dfrac{|\mathbf{d} \cdot \mathbf{F}|}{|\mathbf{d}|}|\mathbf{d}| = |\mathbf{d} \cdot \mathbf{F}| = |\mathbf{F} \cdot \mathbf{d}|$. In short, we have the following result.

Work

The work done is the absolute value of the dot product of the force and the displacement vector:

$$W = |\mathbf{F} \cdot \mathbf{d}|.$$

---

*A joule is defined as 1 newton-meter—that is, the amount of work done by the force of one newton moving a particle a distance of 1 meter in the direction of the force. The joule is named after James Prescott Joule.

Dot Product

EXAMPLE 7.34 Find the work done in moving an object along the vector $\mathbf{a} = 5\mathbf{i} + 4\mathbf{j}$ if the force applied is $\mathbf{b} = 2\mathbf{i} + 3\mathbf{j}$.

Solution

$$\text{Work} = |\text{Force} \cdot \text{displacement vector}|$$
$$= |\mathbf{b} \cdot \mathbf{a}|$$
$$= |(2\mathbf{i} + 3\mathbf{j}) \cdot (5\mathbf{i} + 4\mathbf{j})|$$
$$= |2(5) + 3(4)| = 22 \text{ units.}$$

EXAMPLE 7.35 A force of 8 N acts in the direction $\dfrac{\pi}{6}$. What is the work done if this force moves an object from the point $(2, -1)$ to the point $(6, 5)$? Assume all distances are measured in meters.

Solution A unit vector with direction $\dfrac{\pi}{6}$ is given by $\mathbf{u} = \left(\cos\dfrac{\pi}{6}\right)\mathbf{i} + \left(\sin\dfrac{\pi}{6}\right)\mathbf{j} = \left(\dfrac{\sqrt{3}}{2}\right)\mathbf{i} + \dfrac{1}{2}\mathbf{j}$. Thus, $\mathbf{F} = 8\mathbf{u} = 4\sqrt{3}\,\mathbf{i} + 4\mathbf{j}$. The displacement vector $\mathbf{d}$ is given by $(6 - 2)\mathbf{i} + (5 - -1)\mathbf{j} = 4\mathbf{i} + 6\mathbf{j}$. As a result,

$$W = |\mathbf{F} \cdot \mathbf{d}| = |(4\sqrt{3}\,\mathbf{i} + 4\mathbf{j}) \cdot (4\mathbf{i} + 6\mathbf{j})| = 16\sqrt{3} + 24 \approx 51.7128 \text{ J.}$$

Geometric Applications

Our first geometric application will be a vectorial proof of the formula for the distance from a point to a line. Other ways to derive this formula were given in section 1.3 and exercise 29 in section 5.2. In a calculus course, you will see how the vectorial method easily extends to give a formula for the distance from a point to a plane.

The result states that the distance D from a point $P(x_1, y_1)$ to a line $\ell : ax + by + c = 0$ is given by

$$D = \dfrac{|ax_1 + by_1 + c|}{\sqrt{a^2 + b^2}}.$$

Proof: Consider the data in figure 7.37.

Since P_2 and P_3 are points on ℓ, their coordinates satisfy the given equation for ℓ. In particular, $ax_2 + by_2 + c = 0$, and so $ax_2 + by_2 = -c$. Similarly, $ax_3 + by_3 = -c$. It follows that the vector $\mathbf{v} = a\mathbf{i} + b\mathbf{j}$ is perpendicular to $\overrightarrow{P_2P_3}$, for

$$\mathbf{v} \cdot \overrightarrow{P_2P_3} = (a\mathbf{i} + b\mathbf{j}) \cdot [(x_3 - x_2)\mathbf{i} + (y_3 - y_2)\mathbf{j}]$$
$$= a(x_3 - x_2) + b(y_3 - y_2)$$
$$= ax_3 - ax_2 + by_3 - by_2$$
$$= (ax_3 + by_3) - (ax_2 + by_2) = -c - (-c) = 0.$$

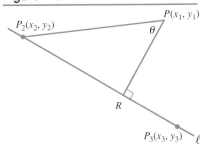

Figure 7.37

If **u** is a unit vector in the direction of $\overrightarrow{PR}$, then

$$D = PR = |\overrightarrow{PP_2}| \cos \theta = |\overrightarrow{PP_2}| |\mathbf{u}| \cos \theta = \overrightarrow{PP_2} \cdot \mathbf{u}.$$

However, **u** is parallel to **v**, since both are perpendicular to $\overrightarrow{P_2P_3}$. It follows that if $\mathbf{w} = \dfrac{1}{|\mathbf{v}|}\mathbf{v}$, the unit vector in the direction of **v**, then $\overrightarrow{PP_2} \cdot \mathbf{w}$ is either $|\overrightarrow{PP_2}| \cos \theta$ or $|\overrightarrow{PP_2}| \cos(\pi - \theta)$. Since $\cos(\pi - \theta) = -\cos \theta$, $\overrightarrow{PP_2} \cdot \mathbf{w} = \pm D$. Hence,

$$D = |\overrightarrow{PP_2} \cdot \mathbf{w}| = \left|((x_1 - x_2)\mathbf{i} + (y_1 - y_2)\mathbf{j}) \cdot \dfrac{a\mathbf{i} + b\mathbf{j}}{\sqrt{a^2 + b^2}}\right|$$

$$= \dfrac{|(x_1 - x_2)a + (y_1 - y_2)b|}{\sqrt{a^2 + b^2}}$$

$$= \dfrac{|ax_1 + by_1 - (ax_2 + by_2)|}{\sqrt{a^2 + b^2}}$$

$$= \dfrac{|ax_1 + by_1 - (-c)|}{\sqrt{a^2 + b^2}} = \dfrac{|ax_1 + by_1 + c|}{\sqrt{a^2 + b^2}},$$

completing the proof.

▶ **EXAMPLE 7.36**

Find the distance from the point $P(-3, 4)$ to the line given by (a) $12x - 5y + 4 = 0$, (b) $y = \frac{2}{3}x + 9$, and (c) $x = \pi$.

Solutions

(a) Here, $a = 12$, $b = -5$, and $c = 4$. Thus,

$$D = \dfrac{|12(-3) + (-5)4 + 4|}{\sqrt{12^2 + (-5)^2}} = \dfrac{|-52|}{\sqrt{169}} = \dfrac{52}{13} = 4.$$

(b) The equation of ℓ can be taken, equivalently, as $\frac{2}{3}x - y + 9 = 0$, with $a = \frac{2}{3}$, $b = -1$, and $c = 9$. Thus,

$$D = \dfrac{\left|\frac{2}{3}(-3) + (-1)4 + 9\right|}{\sqrt{\left(\frac{2}{3}\right)^2 + (-1)^2}} = \dfrac{|3|}{\sqrt{\frac{13}{9}}} = \dfrac{9}{\sqrt{13}}.$$

(c) Rewrite the equation of ℓ as $x - \pi = 0$, with $a = 1$, $b = 0$, and $c = -\pi$. Thus,

$$D = \dfrac{|1(-3) + 0(4) + (-\pi)|}{\sqrt{1^2 + 0^2}} = \dfrac{|-3 - \pi|}{1} = 3 + \pi. \quad \blacktriangleleft$$

You have probably seen a proof in a plane geometry course that any angle inscribed in a semicircle is a right angle. (This fact is also needed in the hint for exercise 35 in section 7.2.) We now give a vectorial proof of this.

▶ **EXAMPLE 7.37**

If $\overline{AB}$ is the diameter of a circle and C is any point on the circle but not on $\overline{AB}$, then $\angle ACB$ is a right angle.

Proof

Figure 7.38

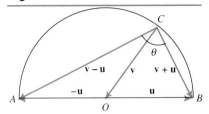

Let the circle have center O and radius r, as in figure 7.38. It will be helpful to let $\mathbf{u} = \overrightarrow{OB}$ and $\mathbf{v} = \overrightarrow{CO}$. Then $\overrightarrow{OA} = -\mathbf{u}$ since $\overrightarrow{OA}$ is the opposite of $\overrightarrow{OB}$. To show that $\angle ACB$ is a right angle or, equivalently, that $\overrightarrow{CA}$ is orthogonal to $\overrightarrow{CB}$, we shall show that $\overrightarrow{CA} \cdot \overrightarrow{CB} = 0$. To do this, first notice that $\overrightarrow{CA} = \overrightarrow{CO} + \overrightarrow{OA} = \mathbf{v} - \mathbf{u}$ and, similarly, $\overrightarrow{CB} = \mathbf{v} + \mathbf{u}$. Thus,

$$\overrightarrow{CA} \cdot \overrightarrow{CB} = (\mathbf{v} - \mathbf{u}) \cdot (\mathbf{v} + \mathbf{u})$$
$$= \mathbf{v} \cdot \mathbf{v} + \mathbf{v} \cdot \mathbf{u} - \mathbf{u} \cdot \mathbf{v} - \mathbf{u} \cdot \mathbf{u}$$
$$= \mathbf{v} \cdot \mathbf{v} - \mathbf{u} \cdot \mathbf{u}$$
$$= |\mathbf{v}|^2 - |\mathbf{u}|^2 = r^2 - r^2 = 0. \quad \blacktriangleleft$$

The next application has several proofs. (One of these was sketched in exercise 60 of section 1.2.) We will give a vectorial proof of it using dot products, much in the spirit of example 7.37.

EXAMPLE 7.38

The altitudes of any triangle are concurrent.

Proof

In figure 7.39, the altitudes from B and C meet at the point D. It will be convenient to let $\mathbf{u} = \overrightarrow{BA}$ and $\mathbf{v} = \overrightarrow{AC}$. Then $\overrightarrow{BC} = \mathbf{u} + \mathbf{v}$. To show that the altitude from A passes through D, we shall show that $\overrightarrow{AD}$ is perpendicular to $\overrightarrow{BC}$ or, equivalently, $\overrightarrow{AD} \cdot \overrightarrow{BC} = 0$.

Figure 7.39

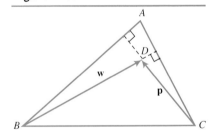

Now, using the two given altitudes, we have that $\overrightarrow{BD} = \mathbf{w}$ is perpendicular to $\overrightarrow{AC}$ and $\mathbf{p} = \overrightarrow{CD}$ is perpendicular to $\overrightarrow{BA}$. In other words,

$$\mathbf{w} \cdot \mathbf{v} = 0 \quad \text{and} \quad \mathbf{p} \cdot \mathbf{u} = 0.$$

Since $\overrightarrow{AD} = \overrightarrow{AB} + \overrightarrow{BD} = -\mathbf{u} + \mathbf{w}$ and, similarly, $\mathbf{p} = -\mathbf{u} - \mathbf{v} + \mathbf{w}$, we may use the properties of dot product for the following computation.

$$\overrightarrow{AD} \cdot \overrightarrow{BC} = (-\mathbf{u} + \mathbf{w}) \cdot (\mathbf{u} + \mathbf{v}) = -\mathbf{u} \cdot \mathbf{u} - \mathbf{u} \cdot \mathbf{v} + \mathbf{w} \cdot \mathbf{u} + \mathbf{w} \cdot \mathbf{v}$$
$$= -\mathbf{u} \cdot \mathbf{u} - \mathbf{v} \cdot \mathbf{u} + \mathbf{w} \cdot \mathbf{u} + \mathbf{w} \cdot \mathbf{v}$$
$$= (-\mathbf{u} - \mathbf{v} + \mathbf{w}) \cdot \mathbf{u} + \mathbf{w} \cdot \mathbf{v}$$
$$= \mathbf{p} \cdot \mathbf{u} + \mathbf{w} \cdot \mathbf{v} = 0 + 0 = 0. \quad \blacktriangleleft$$

Our final application of vectors does *not* concern dot product, but is a fundamental geometric fact that you might not have seen before.

EXAMPLE 7.39

Thales' Theorem: The line segment connecting the midpoints of two sides of a triangle is parallel to the third side and is half the length of that third side.

Proof

In figure 7.40, D is the midpoint of $\overline{AB}$, and E is the midpoint of $\overline{AC}$. We are to show that $\overline{DE} \parallel \overline{BC}$ and $DE = \frac{1}{2}BC$. This is **equivalent** to establishing the vector equation $\overrightarrow{DE} = \frac{1}{2}\overrightarrow{BC}$.

It will be convenient to let $\mathbf{u} = \overrightarrow{DA}$ and $\mathbf{v} = \overrightarrow{AE}$. Since D is the midpoint of $\overline{AB}$, we have $\overrightarrow{BD} = \overrightarrow{DA}$. In other words, $\overrightarrow{BD} = \mathbf{u}$; similarly, $\overrightarrow{EC} = \overrightarrow{AE} = \mathbf{v}$.

Figure 7.40

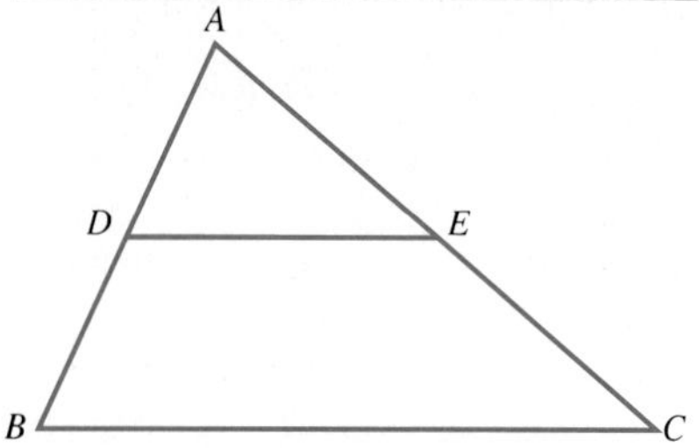

Now,
$$\vec{DE} = \vec{DA} + \vec{AE} = \mathbf{u} + \mathbf{v}$$

and
$$\vec{BC} = \vec{BD} + \vec{DA} + \vec{AE} + \vec{EC}$$
$$= \mathbf{u} + \mathbf{u} + \mathbf{v} + \mathbf{v}$$
$$= 2\mathbf{u} + 2\mathbf{v} = 2(\mathbf{u} + \mathbf{v}).$$

Hence, $\vec{DE} = \mathbf{u} + \mathbf{v} = \frac{1}{2}(2(\mathbf{u} + \mathbf{v})) = \frac{1}{2}\vec{BC}$, as asserted. ◂

Exercises 7.5

In exercises 1–4, use the definition of dot product to compute $\mathbf{u} \cdot \mathbf{v}$ for the given $\mathbf{u}$ and $\mathbf{v}$.

1. $\mathbf{u} = 2\mathbf{i} - \frac{3}{2}\mathbf{j}$, $\mathbf{v} = -\frac{3}{5}\mathbf{i} + 6\mathbf{j}$ (Hint: $b_1 = -\frac{3}{2}$.)
2. $\mathbf{u} = -4\mathbf{i}$, $\mathbf{v} = -6\mathbf{i} + 20\mathbf{j}$ (Hint: $b_1 = 0$.)
3. $\mathbf{u} = 3\mathbf{i} + 12\mathbf{j}$, $\mathbf{v} = \frac{1}{4}\mathbf{j}$ 4. $\mathbf{u} = -\frac{2}{3}\mathbf{i} + \frac{7}{2}\mathbf{j}$, $\mathbf{v} = \frac{5}{2}\mathbf{i} + \frac{8}{3}\mathbf{j}$
♦ 5. Prove Properties 2 and 3 of the dot product. (Hint: take $\mathbf{u} = a_1\mathbf{i} + b_1\mathbf{j}$, $\mathbf{v} = a_2\mathbf{i} + b_2\mathbf{j}$, and $\mathbf{w} = a_3\mathbf{i} + b_3\mathbf{j}$.)
♦ 6. Prove Properties 4–7 of the dot product.
7. Find the vector $\mathbf{u}$ such that $(2\mathbf{i} - 3\mathbf{j}) \cdot \mathbf{u} = 7$ and $(-4\mathbf{i} + 5\mathbf{j}) \cdot \mathbf{u} = 6$. (Hint: write $\mathbf{u} = a\mathbf{i} + b\mathbf{j}$. Solve a system of two equations for a and b.)
8. Prove that if a vector $\mathbf{u}$ satisfies $\mathbf{u} \cdot \mathbf{u} = 0$, then $\mathbf{u} = \mathbf{0}$. (Hint: one way involves writing $\mathbf{u} = a\mathbf{i} + b\mathbf{j}$. Another way involves finding $|\mathbf{u}|$.)

In exercises 9–12, find the radian measure of the angle between the given vectors $\mathbf{u}$ and $\mathbf{v}$. Approximate by calculator if necessary.

9. $\mathbf{u} = 3\mathbf{i} - 4\mathbf{j}$, $\mathbf{v} = 5\mathbf{i} + 6\mathbf{j}$ 10. $\mathbf{u} = \sqrt{2}\mathbf{i} + \sqrt{2}\mathbf{j}$, $\mathbf{v} = -2\mathbf{i}$
11. $\mathbf{u} = \frac{3}{5}\mathbf{j}$, $\mathbf{v} = -\frac{5}{3}\mathbf{j}$ 12. $\mathbf{u} = \frac{4}{5}\mathbf{i} - \frac{3}{5}\mathbf{j}$, $\mathbf{v} = \frac{12}{13}\mathbf{i} + \frac{5}{13}\mathbf{j}$

In exercises 13–14, use the given information to find $|\mathbf{u}|$.

13. $\mathbf{u} \cdot \mathbf{v} = 2$, $|\mathbf{v}| = 5$, and the angle between $\mathbf{u}$ and $\mathbf{v}$ is $\frac{\pi}{3}$.
14. $\mathbf{u} \cdot \mathbf{v} = -3$, $\mathbf{v} \cdot \mathbf{v} = 81$, and the angle between $\mathbf{u}$ and $\mathbf{v}$ is $\frac{3\pi}{4}$.

15. A plane is flying with a compass bearing of 135° and an airspeed of 280 mph. If the wind is blowing from 270° at 20 mph, find the angle by which the plane's actual course deviates from its compass bearing.
16. Rework exercise 15, with the following new data. The plane has an airspeed of 290 mph and the wind is blowing at 30 mph. (The compass bearing and the direction of the wind remain as in exercise 15.)

In exercises 17–18, show that the given vectors are orthogonal.

17. $\frac{3}{2}\mathbf{i} - \sqrt{7}\mathbf{j}$ and $8\sqrt{7}\mathbf{i} + 12\mathbf{j}$
18. $-6\mathbf{i} + 21\mathbf{j}$ and $14\mathbf{i} + 4\mathbf{j}$

In exercises 19–20, show that the given vectors are parallel.

19. $-36\mathbf{i} + 48\mathbf{j}$ and $3\mathbf{i} - 4\mathbf{j}$ 20. $2\sqrt{7}\mathbf{i} - \frac{3}{2}\mathbf{j}$ and $12\mathbf{i} - \frac{9}{\sqrt{7}}\mathbf{j}$

21. In each case, determine whether the given vectors are orthogonal:
    (a) $-8\mathbf{i} - \sqrt{5}\mathbf{j}$ and $\sqrt{3}\mathbf{i} - 6\mathbf{j}$ (b) $-3\mathbf{i} + 4\mathbf{j}$ and $4\mathbf{i} + 3\mathbf{j}$
22. In each case, determine whether the given vectors are parallel:
    (a) $-12\mathbf{i} - 30\mathbf{j}$ and $-10\mathbf{i} - 25\mathbf{j}$
    (b) $\frac{25}{6}\mathbf{i} - \frac{7}{3}\mathbf{j}$ and $-\frac{3}{5}\mathbf{i} + \frac{1}{3}\mathbf{j}$

In exercises 23–24, find a real number s so that the given vectors are (a) orthogonal or (b) parallel.

23. $s\mathbf{i} - 7\mathbf{j}$ and $2\mathbf{i} + \mathbf{j}$ 24. $-2\mathbf{i} + s\mathbf{j}$ and $\frac{3}{2}\mathbf{i} - 4\mathbf{j}$

In exercises 25–28, calculate $\text{proj}_\mathbf{u} \mathbf{v}$ for the given vectors $\mathbf{u}$ and $\mathbf{v}$.

25. $\mathbf{u} = -\frac{3}{5}\mathbf{i} + 2\mathbf{j}$, $\mathbf{v} = 5\mathbf{i} - 12\mathbf{j}$
26. $\mathbf{u} = 4\mathbf{i} + 2\mathbf{j}$, $\mathbf{v} = 4\mathbf{i} - 5\mathbf{j}$
27. $\mathbf{u} = -30\mathbf{i} + 20\mathbf{j}$, $\mathbf{v} = 36\mathbf{i} - 24\mathbf{j}$
28. $\mathbf{u} = \frac{2}{3}\mathbf{i} + \frac{3}{5}\mathbf{j}$, $\mathbf{v} = 9\mathbf{i} - 10\mathbf{j}$

In exercises 29–32, write the given vector $\mathbf{a}$ as $\mathbf{a}_1 + \mathbf{a}_2$, where $\mathbf{a}_1$ is parallel to $\mathbf{b}$ and $\mathbf{a}_2$ is perpendicular to $\mathbf{b}$.

29. $\mathbf{a} = 5\mathbf{i} + 3\mathbf{j}$, $\mathbf{b} = 2\mathbf{i} - 4\mathbf{j}$ 30. $\mathbf{a} = \frac{5}{3}\mathbf{i} - 2\mathbf{j}$, $\mathbf{b} = 12\mathbf{i} + 5\mathbf{j}$
31. $\mathbf{a} = 10\mathbf{i} - 9\mathbf{j}$, $\mathbf{b} = \frac{2}{7}\mathbf{i} + \frac{20}{63}\mathbf{j}$
32. $\mathbf{a} = -3\mathbf{i} + 2\mathbf{j}$, $\mathbf{b} = 9\mathbf{i} - 6\mathbf{j}$

In exercises 33–38, find the work done under the given conditions.

33. A force of 12 N acts in the direction $\frac{\pi}{6}$ and moves an object along the displacement vector $4\mathbf{i} + 5\mathbf{j}$. Assume distances are measured in meters.
34. A force of 64 pounds acts in the direction $\frac{7\pi}{4}$ and moves an object along the displacement vector $\frac{3}{4}\mathbf{i} - \frac{5}{8}\mathbf{j}$. Assume distances are measured in feet.
35. The force $\frac{3}{5}\mathbf{i} - \frac{4}{5}\mathbf{j}$ moves an object from the point $(-\frac{1}{5}, \frac{2}{5})$ to the point $(1, -\frac{6}{5})$.
36. The force $-3\mathbf{i} - 8\mathbf{j}$ moves an object from the point $(5, 6)$ to the point $(2, -1)$.
37. The boy pulling the wagon in the figure exerts 40 pounds of force and the wagon moves 7 feet horizontally. (Hint: first find **F** and **d**.)

38. The girl pushing the lawnmower in the figure exerts 35 pounds of force and the mower moves 4 feet horizontally.

In exercises 39–42, find the distance from the given point P to the line ℓ having the given equation.

39. $P(3, -4)$; ℓ: $14x + 13y + 8 = 0$
40. $P(-4, 5)$; ℓ: $y = -x + \frac{3}{2}$
41. $P(-\frac{2}{3}, 4)$; ℓ: $y = 3x + 6$ 42. $P(-3, \pi)$; ℓ: $x = -4$
43. (a) Suppose that a nonzero vector $\mathbf{n} = a\mathbf{i} + b\mathbf{j}$ is perpendicular to (a vector in either direction of) a line ℓ. Prove that there exists a real number c such that ℓ has an equation of the form $ax + by + c = 0$. (Hint: let **n** have its initial point $P_1(x_1, y_1)$ on ℓ. Then a point $P(x, y)$ is on ℓ if and only if $\mathbf{n} \cdot \overrightarrow{PP_1} = 0$.)

(b) Use (a) to prove that $5x + 7y - 1 = 0$ is an equation for the line which passes through $(3, -2)$ and is perpendicular to the vector $5\mathbf{i} + 7\mathbf{j}$. (Hint: plug in $x = 3$ and $y = -2$ in order to find c.)

44. (a) If the vectors **u** and **v** are perpendicular to **w** and r, s are any real numbers, prove that $r\mathbf{u} + s\mathbf{v}$ is also perpendicular to **w**. (Hint: use the properties of the dot product to rewrite $(r\mathbf{u} + s\mathbf{v}) \cdot \mathbf{w}$.)
    (b) Redo (a) with "perpendicular" replaced throughout by "parallel."
    (c) If a vector **u** is perpendicular to itself, show that $\mathbf{u} = \mathbf{0}$. (Hint: compare exercise 8.)
    (d) Suppose that $\mathbf{b} \neq \mathbf{0}$. If $\mathbf{a} = \mathbf{a}_1 + \mathbf{a}_2$, where $\mathbf{a}_1$ is parallel to **b** and $\mathbf{a}_2$ is perpendicular to **b**, show that $\mathbf{a}_1 = \text{proj}_\mathbf{b}\, \mathbf{a}$. (Hint: use (a)–(c) to argue that $\mathbf{a}_1 - \text{proj}_\mathbf{b}\, \mathbf{a}$ is both parallel and perpendicular to **b**, and hence must be **0**.)
45. Suppose that P_1 and P_2 are distinct points on a line having slope m. Show that $\overrightarrow{P_1P_2}$ is parallel to $\mathbf{i} + m\mathbf{j}$. (Hint: consider $P_1(x_1, y_1)$ and $P_2(x_2, y_2)$. What is $(x_2 - x_1)(\mathbf{i} + m\mathbf{j})$?)
46. Use exercise 45 to show that
    (a) given $P_1(-2, 3)$ and $P_2(-4, 21)$, then $\overrightarrow{P_1P_2}$ is parallel to $\mathbf{i} - 9\mathbf{j}$.
    (b) $\mathbf{i} + \frac{2}{3}\mathbf{j}$ has one of the directions of the line $2x - 3y + 4 = 0$.
47. Let ℓ_1 and ℓ_2 be nonvertical lines with slopes m_1 and m_2, respectively. If θ is an angle between ℓ_1 and ℓ_2, show that

$$\frac{1 + m_1 m_2}{\sqrt{(1 + m_1^2)(1 + m_2^2)}} = \pm\cos\theta.$$

(Hint: using exercise 45, interpret $(\mathbf{i} + m_1\mathbf{j}) \cdot (\mathbf{i} + m_2\mathbf{j})$. Recall that $\cos(\pi - \theta) = -\cos\theta$.)

48. Use exercise 47 to obtain this book's third proof of the following fact. If ℓ_1 and ℓ_2 are nonvertical lines with respective slopes m_1 and m_2, then ℓ_1 is perpendicular to ℓ_2 if and only if $m_1 m_2 = -1$.
49. Prove that the diagonals of any square meet at right angles. (Hint: if adjacent sides of the square are represented by **u** and **v**, then $\mathbf{u} + \mathbf{v}$ and $\mathbf{u} - \mathbf{v}$ are diagonals. Model your argument after example 7.37.)
50. Let E, F, G, and H be the midpoints of the sides of a quadrilateral, as in the figure. Show that $EFGH$ is a parallelogram. (Hint: one way is to let $\mathbf{u} = \overrightarrow{AB}$, $\mathbf{v} = \overrightarrow{AD}$, and $\mathbf{w} = \overrightarrow{DC}$, and then use some of the ideas in example 7.38. A quicker way is via several applications of Thales' Theorem (example 7.39) to appropriate triangles.) See the figure on the next page.

428
Chapter 7 / Applications of Trigonometry

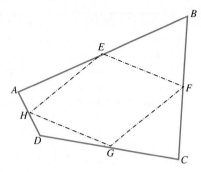

51. In the figure, D is the midpoint of $\overline{AB}$, F is between A and C, and $\overline{DF} \parallel \overline{BC}$. Prove that F is the midpoint of $\overline{AC}$. (Hint: if E is the midpoint of $\overline{AC}$, Thales' Theorem allows you to conclude $\overline{DE} \parallel \overline{BC}$. Why is $F = E$?)

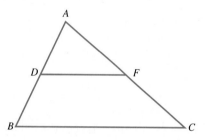

52. Consider the dictionary established in exercise 52 of section 7.4.
  (a) Suppose that complex numbers z_1, z_2 correspond respectively to vectors $\mathbf{v}_1$, $\mathbf{v}_2$. Prove that the dot product $\mathbf{v}_1 \cdot \mathbf{v}_2$ equals the real part of the complex number $\overline{z_1}z_2$. (Hint: write $z_1 = a_1 + b_1 i$ and $z_2 = a_2 + b_2 i$.)
  (b) Suppose that the complex number z corresponds to the vector $\mathbf{v}$. Use (a) to prove again that the modulus of z equals the magnitude of $\mathbf{v}$. (Hint: put $z_1 = z_2 = z$ in (a).)

53. (Cauchy-Schwarz inequality) Prove that $|\mathbf{u} \cdot \mathbf{v}| \leq |\mathbf{u}||\mathbf{v}|$ for all vectors $\mathbf{u}$ and $\mathbf{v}$. (Hint: let θ be the angle between $\mathbf{u}$ and $\mathbf{v}$. Recall that $-1 \leq \cos \theta \leq 1$. You'll need a fact about absolute value from chapter 0.)

54. (A continuation of exercise 53) Let $\mathbf{u}$ and $\mathbf{v}$ be vectors. Prove that $|\mathbf{u} \cdot \mathbf{v}| = |\mathbf{u}||\mathbf{v}|$ if and only if at least one of the following three conditions holds: $\mathbf{u} = \mathbf{0}$; $\mathbf{v} = \mathbf{0}$; $\mathbf{u}$ and $\mathbf{v}$ are nonzero and parallel. (Hint: when is $\cos \theta = \pm 1$?)

55. (a) Prove the following form of the "geometric triangle inequality": $|\mathbf{u} + \mathbf{v}| \leq |\mathbf{u}| + |\mathbf{v}|$ for all vectors $\mathbf{u}$ and $\mathbf{v}$. (Hint: $f(x) = \sqrt{x}$ defines an increasing function. So, it is enough to show that $|\mathbf{u} + \mathbf{v}|^2 \leq (|\mathbf{u}| + |\mathbf{v}|)^2$. Expand the dot product of $\mathbf{u} + \mathbf{v}$ with itself. Exercise 54 is relevant.)
  (b) Prove that $|\mathbf{u} + \mathbf{v}| = |\mathbf{u}| + |\mathbf{v}|$ if and only if at least one of the following three conditions holds: $\mathbf{u} = \mathbf{0}$; $\mathbf{v} = \mathbf{0}$; $\mathbf{u}$ and $\mathbf{v}$ are nonzero and have the same direction. (Hint: when does $\mathbf{u} \cdot \mathbf{v} = |\mathbf{u}||\mathbf{v}|$ hold?)

56. Let $\mathbf{u}$, $\mathbf{v}$ be two nonzero nonparallel vectors. If $\mathbf{w}$ is any vector, prove that there exists a uniquely determined pair of real numbers (r, s) such that $\mathbf{w} = r\mathbf{u} + s\mathbf{v}$. (Hint: if you argue algebraically to find (r, s), take $\mathbf{u} = \langle a_1, b_1 \rangle$, $\mathbf{v} = \langle a_2, b_2 \rangle$, and $\mathbf{w} = \langle c, d \rangle$. In this case, you will solve a system of two equations in two variables, and you will need to explain why the hypotheses guarantee that $a_1 b_2 - a_2 b_1 \neq 0$. On the other hand, you may argue geometrically by analyzing the parallelogram in the accompanying figure.)

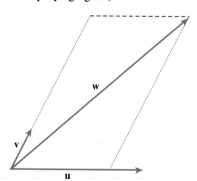

*7.6 HARMONIC MOTION

One type of periodic motion is **simple harmonic motion**. Examples of simple harmonic motion include a light weight bouncing on a spring, a simple pendulum, a boat bobbing on the water, the vibration of a string or air column as in a musical instrument, and the vibration of a tuning fork. Equations similar to those for simple harmonic motion arise in electricity in the study of alternating current.

---

*This section is optional because the material in it is not essential for later material in the book.

Harmonic Motion

Consider an object suspended on a spring. At rest, the object is at the equilibrium position shown in figure 7.41(a). When the object is pulled down to position A, the spring is stretched as shown in figure 7.41(b). When the object is released, the spring pulls the object past the equilibrium position to position B. This causes the spring to be compressed as in figure 7.41(c), and so the spring then pushes the object downward toward the equilibrium position.

Figure 7.41

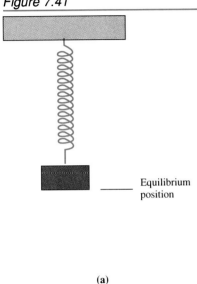
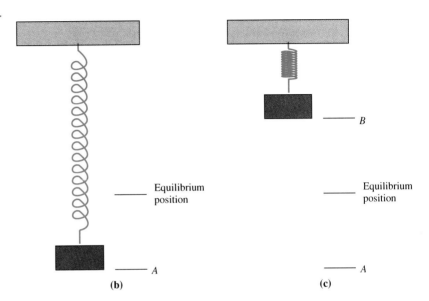

Assume that the object is very light, that the spring is not stretched "too far" initially, and that neither the air nor the internal mechanism of the system exerts any friction forces. Then, under these hypotheses, the object will oscillate up and down in simple harmonic motion. The position $s(t)$ of the object at time t is given by the following formulas:

Simple Harmonic Motion

$$s(t) = a \sin(2\pi f t + \phi)$$

or

$$s(t) = a \sin(\omega t + \phi),$$

where $\omega = 2\pi f$.

Here, a is a positive number whose physical interpretation is the **amplitude of the motion**; this amounts to the maximum displacement from the equilibrium position. Moreover, f is the **frequency** of the motion, the number of complete oscillations the object makes in one time unit (for example, seconds, minutes, hours, or days); ω is the **angular speed**; and ϕ is the **phase constant**. (Note that the phase constant is not the same as the phase shift that was introduced in section 5.5. The phase shift of $s(t)$ is $-\dfrac{\phi}{2\pi f}$, the value of t when $2\pi f t + \phi = 0$.)

Chapter 7 / Applications of Trigonometry

The above equation for simple harmonic motion may also be expressed in terms of the cosine function. Using a cofunction identity, we see that the equation would be $s(t) = a \cos(2\pi f t + \theta)$, where $\theta = \phi - \frac{\pi}{2}$.

The frequency f depends on the mass of the suspended object and the stiffness of the spring. The phase constant ϕ depends on the position and speed of the object at time $t = 0$. The **period** T of the motion, in time units, is defined as $T = \frac{1}{f}$. You will be asked to show in exercise 30 that T is the time needed for one complete oscillation.

▶ **EXAMPLE 7.40**

An object is suspended from a spring. The object is pulled down 15 centimeters from the equilibrium position, released, and then oscillates in simple harmonic motion. If it takes 0.4 second for it to complete one oscillation, write an equation to describe the motion.

Solution

We are given that the amplitude is $a = 15$ cm and that the period $T = 0.4$ s. Since the frequency $f = \frac{1}{T}$, we can determine $f = \frac{1}{0.4} = 2.5$ oscillations per second. Thus, we can write the equation for the position of the object relative to its equilibrium as

$$s(t) = a \sin(2\pi f t + \phi) = 15 \sin(5\pi t + \phi).$$

Let us agree that the object is released at time $t = 0$. As the object was then 15 centimeters below the equilibrium position, $s(0) = -15$. According to the above equation, $s(0) = 15 \sin \phi$, and so $\sin \phi = -\frac{15}{15} = -1$. Thus we can take ϕ to be $\phi = \sin^{-1}(-1)$, or $-\frac{\pi}{2}$. Hence, an equation that describes the motion of this object is

$$s(t) = 15 \sin\left(5\pi t - \frac{\pi}{2}\right).$$ ◀

▶ **EXAMPLE 7.41**

An object with simple harmonic motion has its position at time t seconds given by $s(t) = 9 \cos\left(\frac{3\pi}{4}t + \frac{\pi}{6}\right)$ cm. Determine (a) the maximum displacement, (b) the frequency, (c) the position at $t = 0$, (d) the position at $t = 3.5$, and (e) the amount of time it takes to first return to the equilibrium position.

Solutions

The given expression has the form $s(t) = a \cos(2\pi f t + \theta)$, with $a = 9$, $2\pi f = \frac{3\pi}{4}$, and $\theta = \frac{\pi}{6}$. Therefore, we have the following results:

(a) Maximum displacement = amplitude = $a = 9$ cm.

(b) Frequency = $f = \frac{3\pi/4}{2\pi} = \frac{3}{8}$ oscillations per second.

Harmonic Motion

(c) Position at $t = 0$ is $s(0) = 9 \cos \dfrac{\pi}{6} = \dfrac{9}{2}\sqrt{3} \approx 7.7942$ cm above the equilibrium position.

(d) Position at $t = 3.5$ is $s(3.5) = 9 \cos \left(\dfrac{3\pi}{4}\left(\dfrac{7}{2}\right) + \dfrac{\pi}{6} \right) = 9 \cos \dfrac{67\pi}{24} \approx -7.1402$, or about 7.1 centimeters below the equilibrium position.

(e) The object is at the equilibrium position when $s(t) = 0$ or when $9 \cos \left(\dfrac{3\pi}{4}t + \dfrac{\pi}{6} \right) = 0$. This occurs when $\dfrac{3\pi}{4}t + \dfrac{\pi}{6} = \dfrac{\pi}{2}, \dfrac{3\pi}{2}, \dfrac{5\pi}{2}, \ldots$. If $\dfrac{3\pi}{4}t + \dfrac{\pi}{6} = \dfrac{\pi}{2}$, then $t = \dfrac{4}{9}$. This is the smallest positive value of t for which $s(t) = 0$. So, the first time that this object returns to the equilibrium position is $t = \dfrac{4}{9}$ second. ◂

Another example of simple harmonic motion arises in the analysis of musical sounds. The "pure" musical sound (or sound wave) produced by a tuning fork can be described using the formulas for simple harmonic motion. If two (or more) tuning forks are vibrating simultaneously, the resultant sound is described by the algebraic sum of the forks' sound waves. The resulting motion is called a **superposition** of the original motions. The resulting graph may be periodic, but not necessarily a simple harmonic motion.

It can be shown that the superposition of two simple harmonic motions with periods T_1 and T_2 is a periodic motion only if $\dfrac{T_1}{T_2}$ is a rational number. If $\dfrac{T_1}{T_2} = \dfrac{n}{m}$, where n and m are positive integers and $\dfrac{n}{m}$ is in lowest terms, then the period of the superposition is given by

$$T = mT_1 = nT_2.$$

▸ **EXAMPLE 7.42**

Two tuning forks are vibrating simultaneously. The sound waves they produce are described by $y_1 = 2 \sin 600\pi t$ cm and $y_2 = 2 \sin \left(800\pi t + \dfrac{\pi}{2} \right)$ cm, respectively. (a) Write an equation for the superposition of the sound waves produced by the two tuning forks. (b) What is the frequency of the superposition sound? (c) Sketch the graph of one cycle of the superposition sound wave.

Solutions

(a) The combined sound wave y is the algebraic sum of the individual sound waves. Thus

$$y = y_1 + y_2 = 2 \sin 600\pi t + 2 \sin \left(800\pi t + \dfrac{\pi}{2} \right)$$

$$= 2 \left[\sin 600\pi t + \sin \left(800\pi t + \dfrac{\pi}{2} \right) \right].$$

We shall rewrite this using formula (9) in section 6.5, $\sin \alpha + \sin \beta = 2 \sin \frac{1}{2}(\alpha + \beta) \cos \frac{1}{2}(\alpha - \beta)$:

$$y = 2\left[2 \sin \frac{1}{2}\left(600\pi t + 800\pi t + \frac{\pi}{2}\right) \cos \frac{1}{2}\left(600\pi t - 800\pi t - \frac{\pi}{2}\right)\right]$$

$$= 4 \sin\left(700\pi t + \frac{\pi}{4}\right) \cos\left(-100\pi t - \frac{\pi}{4}\right) \text{ cm.}$$

(b) The periods of y_1 and y_2 are $T_1 = \frac{2\pi}{\omega_1} = \frac{2\pi}{600\pi} = \frac{1}{300}$ and $T_2 = \frac{2\pi}{\omega_2} = \frac{2\pi}{800\pi} = \frac{1}{400}$, respectively. The ratio of these two periods is $\frac{T_1}{T_2} = \frac{1/300}{1/400} = \frac{4}{3} = \frac{n}{m}$. This is in lowest terms, and so the period of the superposition is $T = mT_1 = 3T_1 = \frac{1}{100}$. From this, we can determine the superposition's frequency, $f = \frac{1}{T} = 100$ Hz.

(c) The graph of one cycle of the superposition sound wave is given in figure 7.42. In calculus, you will learn methods for verifying that the maximum value of y is 4 (when $t = \frac{3}{400}$ s) and the minimum value of y is approximately -3.6039 (at $t = \frac{17}{2800}$ and $t = \frac{25}{2800}$ s). Thus, the amplitude of the superposition sound wave is approximately $\frac{1}{2}(4 + 3.6039) = 3.8020$ cm. ◀

Figure 7.42

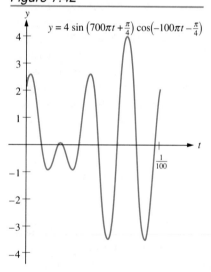

A generator converts mechanical energy to electrical energy by rotating a wire loop about a diameter in a magnetic field. Each half of the loop moves up and then down through the field. As a result, the current induced in the loop is always changing in magnitude and reverses (or alternates) its direction twice during each complete rotation of the loop. This explains the name alternating current (ac). In a simple ac circuit, both the voltage V and the current I vary with time according to formulas of "simple harmonic motion" type:

$$V = V_{\max} \sin(2\pi ft + \phi) = V_{\max} \sin(\omega t + \phi)$$

and

$$I = I_{\max} \sin(2\pi ft + \phi) = I_{\max} \sin(\omega t + \phi).$$

The parameters in the above equations can be interpreted as follows. Time t is typically measured in seconds. Then the unit of frequency is the hertz (Hz), where 1 Hz = 1 cycle per second. The quantity $\omega = 2\pi f$ is the angular speed of the wire loop. Finally, $V_{\max}$ is the maximum value of the voltage and $I_{\max}$ is the maximum value of the current.

▶ **EXAMPLE 7.43**

Consider an ac circuit for which $I_{\max} = 6.5$ A, $\omega = 120\pi$, and $\phi = \frac{\pi}{6}$.

(a) Write an equation describing the current at any time t, $t \geq 0$. Determine (b) the current at $t = 0$, (c) the frequency, and (d) the current at $t = 2.25$ s. Finally, (e) sketch one cycle of the graph of the current.

Harmonic Motion

Solutions

(a) Using the given data and the equation $I = I_{max} \sin(\omega t + \phi)$, we obtain $I = 6.5 \sin\left(120\pi t + \dfrac{\pi}{6}\right)$ A.

(b) When $t = 0$, the current is $I = 6.5 \sin\left(0 + \dfrac{\pi}{6}\right) = 6.5\left(\dfrac{1}{2}\right) = 3.25$ A.

(c) The frequency f is obtained from the relationship $\omega = 2\pi f$, from which we conclude that $f = \dfrac{\omega}{2\pi}$. As $\omega = 120\pi$, $f = \dfrac{120\pi}{2\pi} = 60$ Hz.

(d) When $t = 2.25$ s, the current is $I = 6.5 \sin\left[120\pi(2.5) + \dfrac{\pi}{6}\right] = 6.5 \sin\left(300\pi + \dfrac{\pi}{6}\right) = 6.5\left(\dfrac{1}{2}\right) = 3.25$ A.

(e) The sketch of one cycle of the current's graph is given in figure 7.43.

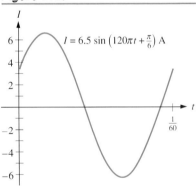

Figure 7.43

The above equations for simple harmonic motion are particular cases of functional expressions already studied in sections 5.5 and 6.5. You should review the appropriate exercises in chapters 5 and 6 to reacquaint yourself with the associated graphs and ways to rewrite these expressions.

The above vibrating springs were not very realistic since their oscillations did not decrease with the passage of time, as experience would predict. In theory, ideal springs of the above kind continue to vibrate forever. In practice, however, internal friction and other forces such as air resistance cause a motion's amplitude to lessen with time, so that eventually a spring stops vibrating. This more realistic model is known as (exponentially) **damped simple harmonic motion**. Here, as time passes, the vibrations continue with constant frequency f but with decreasing amplitude, as indicated in the following formulas.

Damped Simple Harmonic Motion

A mathematical model for damped simple harmonic motion is

$$y = ae^{-mt} \sin(2\pi ft + \phi)$$

or

$$y = ae^{-mt} \cos(2\pi ft + \theta),$$

where the positive constant m is called the **damping factor**.

We call $\dfrac{1}{f}$ the **quasiperiod** or, simply, the **period**.

▶ **EXAMPLE 7.44**

A weight is attached to a spring, pulled below the "equilibrium" position $y = 0$, and then released at $t = 0$. The position $s(t)$ of this object at time t ($t \geq 0$) is given by the equation $s(t) = -3e^{-0.5t} \cos 2\pi t$ cm. Determine (a) the maximum displacement from $y = 0$, (b) the frequency, and (c) the displacement from "equilibrium" after one cycle.

Chapter 7 / Applications of Trigonometry

Solutions

The graph of s lies between the graphs of $y = -3e^{-0.5t}$ and $y = 3e^{-0.5t}$. These two bounding curves are shown dashed in figure 7.44, while the graph of s is shown as a solid curve. Notice that s is physically "damped" in the sense that $s(t)$ has the limiting tendency of 0 as t "approaches infinity." In fact, $|s(t)| < 0.001$ if $t > 16.02$.

Figure 7.44

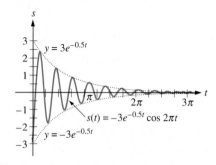

(a) The maximum displacement occurs at $t = 0$, the time the object is released. As $s(0) = -3$, the maximum displacement is 3 centimeters below the "equilibrium" position.

(b) We can see that $2\pi f = 2\pi$, and so the frequency, f, is 1 oscillation per second. As time passes, successive maxima (or minima) of $s(t)$ are not the same, even though the (quasi) period of time between them is constant. In this example, the period is $\dfrac{1}{f} = 1$ s.

(c) Since the frequency is 1 oscillation per second, the displacement after one cycle is found at $t = 1$. It is $s(1) = -3e^{-0.5} \cos 2\pi = -3e^{-0.5} \approx -1.8196$ cm, or about 1.82 centimeters below "equilibrium." ◀

EXAMPLE 7.45

An object with damped simple harmonic motion has its position at time t given by $s(t) = 5e^{-\sqrt{2}t} \sin\left(\dfrac{2\pi}{3}t + \dfrac{\pi}{6}\right)$ in. (a) Graph the curve that describes this motion. Also, determine (b) the quasiperiod, (c) the position at $t = 0$, and (d) the position at $t = 2$.

Solutions

(a) The graph of s lies between the graphs of $y = 5e^{-\sqrt{2}t}$ and $y = -5e^{-\sqrt{2}t}$. These two graphs are shown as dashed curves in figure 7.45; the graph of s is the solid curve between them. Once again, notice the physical damping. In this example, $|s(t)| < 0.001$ if $t > 6.03$.

Figure 7.45

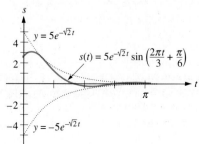

Harmonic Motion

(b) Since $2\pi f = \frac{2\pi}{3}$, we see that the frequency $f = \frac{1}{3}$. Thus the quasiperiod is $\frac{1}{f} = 3$ s.

(c) At $t = 0$, the position is $s(0) = 5 \sin \frac{\pi}{6} = \frac{5}{2}$ in. above "equilibrium."

(d) At $t = 2$, the position is $s(2) = 5e^{-2\sqrt{2}} \sin\left(\frac{4\pi}{3} + \frac{\pi}{6}\right) = 5e^{-2\sqrt{2}} \sin \frac{3\pi}{2} = -5e^{-2\sqrt{2}} \approx -0.2955$, or about 0.3 inch below "equilibrium." ◀

Exercises 7.6

In exercises 1–4, an object is suspended from a spring. The object is pulled down a units from the equilibrium position, released, and then oscillates in simple harmonic motion. If it takes t seconds to complete one oscillation, write an equation to describe the motion.

1. $a = 10$ in., $t = 0.65$
2. $a = 8$ cm, $t = 0.4$
3. $a = 25$ cm, $t = 0.8$
4. $a = 3.5$ in., $t = 0.27$

In exercises 5–8, proceed as in exercises 1–4, except that now you should assume that the suspended object is initially pushed up a units.

5. $a = 12$ cm, $t = 1.3$
6. $a = 4$ in., $t = 0.33$
7. $a = 6.2$ in., $t = 0.72$
8. $a = 7.5$ cm, $t = 0.25$

In exercises 9–10, an object moving in simple harmonic motion has its position at time t described by the given expression for $s(t)$. Determine (a) the maximum displacement, (b) the frequency, (c) the period, (d) the position at $t = 0$, (e) the position at $t = 3.5$, and (f) the amount of time it takes to first return to the equilibrium position.

9. $5 \cos\left(\frac{2\pi t}{7} + \frac{\pi}{4}\right)$ cm
10. $\frac{1}{2} \sin(3t + 1)$ in.

11. Rowena, having discovered that she likes buoys, is watching one bob up and down in simple harmonic motion. Its maximum displacement from equilibrium is 15 inches and it takes 0.4 minute to complete one oscillation. Rowena calculates that the buoy's motion is described by $s(t) = 15 \cos(5\pi t - 3\pi)$ in. Pointing to example 7.40, Rowena's father asserts that $s(t)$ should instead be $15 \sin 5\pi t$ inches. Disappointed, he orders Rowena to give up buoys. How should she respond? (Hint: she begins with "Dad, we're out of sync.")

12. Explain why the maximum displacement from equilibrium of an object moving in simple harmonic motion according to either $s(t) = a \sin(2\pi f t + \phi)$ or $s(t) = a \cos(2\pi f t + \theta)$ is $|a|$. (Hint: you'll need to remember a fact about absolute value from chapter 0 and the ranges of sine and cosine from section 5.2.)

13. Two sound waves are described by the equations $y_1 = 0.005 \sin(400\pi t)$ and $y_2 = 0.005 \sin(1200\pi t)$.
    (a) Write an equation for the superposition sound wave produced by combining these two waves.
    (b) What are the frequency and period of the superposition wave?
    (c) Sketch the graphs of one cycle of y_1, y_2, and $y_1 + y_2$.

14. Two tuning forks produce the sound waves given by $y_1 = 5 \sin 600\pi t$ and $y_2 = 5 \sin\left(600\pi t + \frac{\pi}{2}\right)$.
    (a) Write an equation for the superposition sound wave produced by combining these two waves.
    (b) What is the frequency of the superposition wave?
    (c) Sketch the graphs of one cycle of y_1 and y_2.
    (d) Sketch the graph of $y_1 + y_2$.

15. The sound wave from a string instrument is made from three component sound waves: $y_1 = 6 \sin 300\pi x$, $y_2 = 3 \sin 600\pi x$, and $y_3 = \sin 900\pi x$.
    (a) Sketch the graph of the composite sound wave.
    (b) What is the period of the composite sound wave?

16. At a frequency of 60 Hz, the maximum voltage across a power line is 170 V. What are the amplitude, period, and angular speed of this voltage?

17. Write the equation for the current in an ac circuit for which $I_{\max} = 6.8$ A, $f = 80$ Hz, and the phase constant is $\frac{\pi}{3}$ radians.

18. Explain why, as $t \geq 0$, the maximum absolute value of an expression of the form $ae^{-mt} \cos 2\pi f t$ is attained at $t = 0$. (Hint: this fact was needed in example 7.44. Review the hint to exercise 12, recall the graph of the exponential function from section 4.1, and remember that $m > 0$ by convention.)

In exercises 19–22, an object moving in damped simple harmonic motion has its position described by the given expression for $s(t)$, $t \geq 0$. Determine (a) the frequency, (b) the quasiperiod, (c) the position at $t = 4$, and (d) displacement from "equilibrium" after one cycle.

19. $2e^{-t} \cos\left(3\pi t + \dfrac{\pi}{6}\right)$ ft
20. $2e^{-0.25t} \sin\left(\dfrac{\pi}{3}t - \dfrac{\pi}{4}\right)$ m
21. $-2e^{-1.5t} \sin(4t + 1)$ m
22. $-0.5e^{-3t} \cos(-4t + 3)$ ft

In exercises 23–26, an object is moving in damped simple harmonic motion according to the position function $s(t) = ae^{-mt} \sin\left(2\pi ft + \dfrac{\pi}{6}\right)$ cm, $t \geq 0$. Use the given values of a, the quasiperiod T, and displacement D from "equilibrium" after one "cycle" to determine m.

23. $T = 4.25$ s, $a = 3$ cm, $D = 1.25$ cm
24. $T = 2$ s, $a = -6$ cm, $D = -2$ cm
25. $T = 1$ s, $a = -0.5$ cm, $D = -0.2$ cm
26. $T = 0.75$ s, $a = \sqrt{5}$ cm, $D = 1$ cm
27. Prove that no damped simple harmonic motion modeled by $s(t) = 3e^{-mt} \cos(2\pi ft + \phi)$, $t \geq 0$, can satisfy all the following conditions: the quasiperiod $T = 2$ s; displacement from "equilibrium" after one "cycle" is $D = 1.5$; $s(5) = -1$; and $s(6) = 0.8$.
28. (a) Justify the assertion in example 7.44 that $|s(t)| < 0.001$ if $t > 16.02$ by showing that $|s(t)| < 0.001$ if $t > 2 \ln 3000$. (Hint: review exercise 18 and the material on inequalities involving exponentials in section 4.5.)
    (b) Justify the analogous assertion in example 7.45.
29. (a) (A generalization of exercise 28) If a damped simple harmonic motion is described by $s(t) = ae^{-mt} \cos(2\pi ft + \theta)$ as usual and if k is a positive integer, prove that $|s(t)| < 10^{-k}$ whenever $t > \dfrac{\ln(10^k |a|)}{m}$.
    (b) Prove the analogue of (a) in case "cos" is replaced with "sin" in the formula for $s(t)$.
30. Prove the assertion made in the text that $T = \dfrac{1}{f}$ for simple harmonic motion. (Hint: consider time values t_0 and t_1 such that $\omega t_0 + \phi = 0$ and $\omega t_1 + \phi = 2\pi$. Observe that $\omega = 2\pi f$ and $T = t_1 - t_0$.)
31. Having won the argument in exercise 11, Rowena gains her father's permission to study real buoys. She concludes that a real buoy *could* undergo a damped simple harmonic motion modeled by $s(t) = 3e^{-mt} \cos(2\pi ft + \phi)$, $t \geq 0$, satisfying all the conditions in exercise 27 except $s(6) = 0.8$. She bets her father a new ladder that she can describe such an $s(t)$ in which ϕ is within one degree of 50°. Who wins the bet?

7.7 CHAPTER SUMMARY AND REVIEW EXERCISES

Major Concepts

- Amplitude of the motion
- Angle of depression
- Angle of elevation
- Angular speed
- Bound vector
- Components of a vector
- Damped simple harmonic motion
- Damping factor
- Difference of two vectors
- Directed line segment
- Dot product
  Properties
- Equal vectors
- Equilibrium
- Free vector
- Frequency
- Heron's formula
- Hero's formula
- i
- Initial point
- Inner product
- j
- Law of Cosines
- Law of Sines
  Ambiguous case
- Magnitude of a vector
- Opposite of a vector
- Orthogonal vectors
- Parallel vectors
- Period
- Perpendicular vectors
- Phase constant
- Projection of **v** onto **u**
- proj$_{\mathbf{u}}$ **v**
- Quasiperiod
- Resultant vector
- Scalar
- Scalar product
- Scalar quantity
- Semiperimeter
- Simple harmonic motion
- Sum of two vectors
- Terminal point
- Unit vector
- Vector
  Quantity
- Vector addition
  Associative law
  Commutative law
- Zero vector

Chapter Summary and Review Exercises

Review Exercises

In exercises 1–18, determine how many pairwise incongruent triangles $\triangle ABC$ fit the given conditions. For each such triangle, solve for the unknown parts.

1. $\beta = 73°$, $\gamma = 90°$, $c = 5$ cm
2. $a = 5$ in., $b = 5$ in., $\alpha = 84°$
3. $\alpha = 36°$, $\gamma = 27°$, $a = 8$ ft
4. $\alpha = 25.5°$, $\beta = 64.5°$, $a = 4$ m
5. $a = 3.7$, $b = 9$, $\alpha = 35°$
6. $a = 3.7$, $b = 9$, $\alpha = 20°$
7. $a = \sqrt{13}$, $b = 5$, $c = 2\sqrt{3}$
8. $a = 5$, $b = 6$, $\beta = 111.5°$
9. $a = \sqrt{13}$ in., $b = 5$ in., $c = \sqrt{3}$ in.
10. $b = 3.6$ cm, $c = 2.7$ cm, $\gamma = \dfrac{\pi}{4}$
11. $\beta = \dfrac{5\pi}{6}$, $\gamma = \dfrac{\pi}{12}$, $b = 3$
12. $b = 12.04$ ft, $c = 3.17$ ft, $\alpha = \dfrac{\pi}{3}$
13. $a = 4.1$ yd, $c = 1.8$ yd, $\beta = \dfrac{3\pi}{4}$
14. $a = 3.1$ mi, $b = 2.7$ mi, $\alpha = \dfrac{\pi}{6}$
15. $a = 5$ in., $b = 5$ in., $\beta = \dfrac{4\pi}{7}$
16. $a = 14.2$ m, $b = 9.6$ m, $c = 7.3$ m
17. $a = 5.2$ cm, $c = 2.5$ cm, $\gamma = 25.5°$
18. $a = 4.01$ cm, $b = 6.12$ cm, $c = 11.69$ cm
19. Building A is 120 feet away from building B. From the roof of building A, the angle of elevation to the roof of building B is 34° and the angle of depression to the bottom of B is 48°. Determine the height of building B.

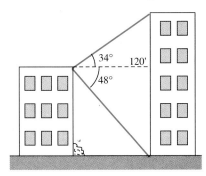

20. A hillside slopes at 13° to the horizontal. On top of the hill stands a 500 foot tall vertical tower. For an observer on the hill, the angle of elevation of the tower's top is 51°. How far is the observer from the tower's bottom? (Ignore the observer's height.)

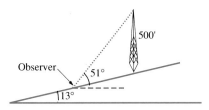

21. On top of a hillside that is inclined 10° to the horizontal, a vertical tower has been erected. For an observer on the hill 60 feet from the tower's bottom, the angle of elevation of the tower's top is 41°. Determine the height of the tower.
22. A pilot intends to fly 200 miles due east from point A to point B. However, she flies 100 miles from point A in a 110° direction before realizing that she is (20°) off course.
    (a) How should she then change the plane's direction in order to arrive at B?
    (b) If she changes direction as indicated in (a), determine the total distance she has flown while traveling from A to B.
23. A river flows from west to east at 6 miles per hour. Aiming to go straight across to the north bank, a motorboat leaves the south bank at 25 miles per hour with respect to still water. Determine the actual speed and direction of the motorboat.
24. If the motorboat in exercise 23 expects to go directly across the river, in which direction should the motorboat's skipper aim?
25. If a plane flies in the direction 15° at 360 miles per hour for 20 minutes and then flies in the direction 8° at 320 miles per hour for 30 minutes, how far is the plane from its point of origination?
26. Two lighthouses are located 20 miles apart on a straight shoreline. During bad weather, observers at the lighthouses see a ship in distress. As in the figure, the angles involved in these sightings are 37° and 45°.

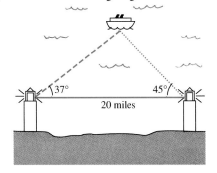

Determine the ship's distance to each lighthouse and its distance to the shore.

27. A surveyor intends to string high-tension wire from point A to point B. Using a point C as in the figure, he finds $AC = 105$ m, $\angle BAC = 87°$, and $\angle ACB = 40°$. How long should his wire be?

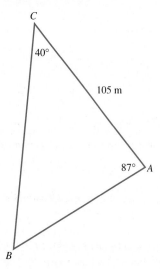

28. As he approaches a target, a bomber passes 5000 feet directly over an observer on the ground. At that moment, the target presents a 10° angle of depression to the bomber pilot and a 30° angle of elevation to the observer. How far is the bomber from the target?
29. A children's slide is 30 feet long and is inclined 35° from the horizontal. Determine the possible angles of inclination of an 18 foot long access ladder.
30. Repeat exercise 29, this time supposing that the access ladder is 17 feet long.
31. Repeat exercise 29, this time supposing that the access ladder is 39 feet long.
32. A railroad track crosses a highway at a 58° angle. A train is 275 feet from the intersection and is moving away from it at 60 miles per hour. At this time, 4 miles from the intersection, a car on the highway is approaching the intersection at 30 miles per hour. How far apart are the car and the train 3 minutes later? (Hint: there are two possible answers, but *not* because of the "ambiguous case.")
33. Suppose that the triangle $\triangle ABC$ has vertices $A(3, 4)$, $B(1, 1)$, and $C(4, 3)$. Find the radius of the circumscribed circle of $\triangle ABC$. (Hint: one way is to use exercise 35 of section 7.2. Another way follows from material in chapter 1.)

34. Find the diameter of the circumscribed circle of a triangle $\triangle ABC$ such that $a = 40$, $b = 50$, and $\beta = \frac{\pi}{6}$. (Hint: see exercise 33.)
35. Find the radius of the inscribed circle of a triangle $\triangle ABC$ such that $a = 3$, $b = 4$, and $c = 6$. (Hint: use exercise 35 of section 7.3.)
36. Suppose that the triangle $\triangle ABC$ has vertices $A(0, 0)$, $B\left(\frac{\sqrt{81095}}{8}, \frac{43}{8}\right)$, and $C(0, 4)$.
    (a) Find equations for the angle-bisectors of $\angle BAC$ and $\angle BCA$.
    (b) Using (a), find the center of the inscribed circle of $\triangle ABC$.
    (c) Using (b), find the radius of the inscribed circle of $\triangle ABC$.
37. Prove that the area of any triangle $\triangle ABC$ is given by each of the following equal expressions:

$$\frac{a^2 \sin \beta \sin \gamma}{2 \sin \alpha}, \quad \frac{b^2 \sin \alpha \sin \gamma}{2 \sin \beta}, \quad \frac{c^2 \sin \alpha \sin \beta}{2 \sin \gamma}.$$

(Hint: apply exercise 24 of section 7.3 and the Law of Sines.)

In exercises 38–41, find the area of a triangle $\triangle ABC$ meeting the given conditions.

38. $a = 4$ cm, $\beta = \frac{\pi}{3}$, $\gamma = \frac{\pi}{4}$
39. $\alpha = 37°$, $\beta = 72.5°$, $c = 5$ in.
40. $A(2, 1)$, $B(3, 4)$, $C(6, 5)$
41. $a = 4$ m, $b = 5.2$ m, $c = 7.3$ m
42. For the triangle $\triangle ABC$ in the figure, $a = 11$, $b = 8$, and $c = 6$. Find $\cos \beta$, β, and the length h of the altitude from A to $\overline{BC}$.

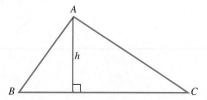

43. An isosceles triangle has a vertical angle of 62° and a base of 36 feet. Find the perimeter of this triangle. (Hint: first calculate the measures of the unknown congruent sides and the unknown congruent angles.)
44. A right triangle has sides of length 5 inches, $2\sqrt{6}$ inches, and 7 inches. Find the length of the median to the hypotenuse. (Hint: a median of a triangle is a line

segment from a vertex to the midpoint of the side opposite that vertex.)

45. Consider the box in the figure. Determine the angle formed by the diagonals $\overline{AB}$ and $\overline{AC}$. (Remark: this exercise encourages you to find planar phenomena such as right triangles lurking in our three-dimensional spatial world. Such skills will be of use when you study calculus.)

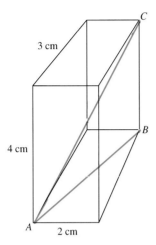

46. From afar, a high-strung surveyor nervously estimates the distance AB as 33.5 meters, after noting that $\angle ACB = 45°$, $\angle ACD = 66°$, $\angle ADB = 55°$, $\angle BDC = 30°$, and $CD = 21$ m. Explain how he uses his data, in conjunction with the Law of Sines and the Law of Cosines, in order to actually calculate AB.

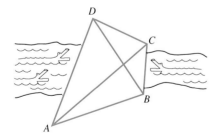

47. (a) Draw a directed line segment representing a force vector of 4 N, applied at $(-2, 1)$ and acting at an angle of $\dfrac{\pi}{3}$ to the positive x-axis.

(b) Vectors **A** and **B** both have initial point P. Draw and label $\mathbf{A} + \mathbf{B}$ and $-3\mathbf{A} + 2\mathbf{B}$ as directed line segments with initial point P.

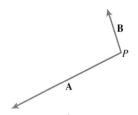

48. If $\mathbf{u} + \mathbf{v} = \mathbf{u} + \mathbf{w}$, show that $\mathbf{v} = \mathbf{w}$. (Hint: this generalizes exercise 14 of section 7.4. The hint for that exercise works here, too.)

49. Given $P_1(-4, -2)$ and $P_2(-9, 7)$, express $\overrightarrow{P_1P_2}$ in the forms $\langle a, b \rangle$ and $a\mathbf{i} + b\mathbf{j}$.

50. A vector **V** has magnitude $|\mathbf{V}| = 3$ and direction $\theta = \dfrac{11\pi}{6}$. Sketch **V** as a directed line segment with initial point at the origin. Also, write **V** in terms of its components.

51. Find the magnitude and direction of the vector **v** having components $v_x = -2\sqrt{3}$ and $v_y = 2$.

52. If $\mathbf{u} = -2\mathbf{i} + 3\mathbf{j}$ and $\mathbf{v} = 6\mathbf{i} - \dfrac{3}{2}\mathbf{j}$, determine (a) $\mathbf{u} + \mathbf{v}$, (b) $\mathbf{u} - \mathbf{v}$, (c) $-2\mathbf{u} + \dfrac{4}{3}\mathbf{v}$, and (d) the magnitude and the direction of $3\mathbf{u} + 4\mathbf{v}$.

53. Find a unit vector with the same direction as $\dfrac{-1}{5}\mathbf{i} + \dfrac{1}{5}\sqrt{3}\,\mathbf{j}$.

54. A car drives at 30 mph up a ramp that is inclined 15° to the positive x-axis. Determine the car's velocity vector and that vector's components.

55. As in the figure, two cables are supporting a metal box containing the Human Fly, who is attempting her famous handcuff escape. The box and its contents together weigh 300 pounds. Find the tension in each cable.

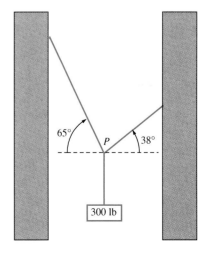

Chapter 7 / Applications of Trigonometry

56. Forces of 25 pounds and 35 pounds are applied at the same point. If the angle between these two force vectors is 30°, determine the magnitude of the resultant force vector. (Hint: one way uses the Law of Cosines. For another approach, view the first force as 25**i**.)

57. Compute the following dot products:
    (a) $(3\mathbf{i} + 2\mathbf{j}) \cdot (5\mathbf{i} + 4\mathbf{j})$ (b) $(3\mathbf{i} + 2\mathbf{j}) \cdot (3\mathbf{i} - 2\mathbf{j})$
    (c) $(\frac{3}{2}\mathbf{i} - \frac{2}{3}\mathbf{j}) \cdot (4\mathbf{i} + 12\mathbf{j})$ (d) $(\frac{3}{2}\mathbf{i} - \frac{2}{3}\mathbf{j}) \cdot (\frac{2}{3}\mathbf{i} + \frac{3}{2}\mathbf{j})$

58. Find the vector **u** such that $(2\mathbf{i} - 3\mathbf{j}) \cdot \mathbf{u} = 6$ and $(-4\mathbf{i} + 5\mathbf{j}) \cdot \mathbf{u} = 7$.

59. In the following cases, use a calculator to find the radian measure of the angle between the given pair of vectors:
    (a) $-4\mathbf{i} + 5\mathbf{j}$ and $6\mathbf{i} + 7\mathbf{j}$ (b) $2\mathbf{i} + 5\mathbf{j}$ and $-3\mathbf{j}$

60. Given that $\mathbf{u} \cdot \mathbf{v} = -3$, $|\mathbf{v}| = 4$, and the angle between **u** and **v** is $\frac{5\pi}{6}$, find $|\mathbf{u}|$.

61. Find a real number s so that the vectors $7\mathbf{i} - s\mathbf{j}$ and $\frac{3}{2}\mathbf{i} + 4\mathbf{j}$ are (a) orthogonal or (b) parallel.

62. Calculate $\text{proj}_\mathbf{u} \mathbf{v}$ in each of the following cases:
    (a) $\mathbf{u} = 4\mathbf{i} - 5\mathbf{j}$, $\mathbf{v} = 4\mathbf{i} + 2\mathbf{j}$
    (b) $\mathbf{u} = 4\mathbf{i} - 5\mathbf{j}$, $\mathbf{v} = -10\mathbf{i} - 8\mathbf{j}$
    (c) $\mathbf{u} = 4\mathbf{i} - 5\mathbf{j}$, $\mathbf{v} = 12\mathbf{i} - 15\mathbf{j}$

63. Consider the vectors $\mathbf{a} = 3\mathbf{i} - 2\mathbf{j}$ and $\mathbf{b} = 5\mathbf{i} + 4\mathbf{j}$. Write **a** as $\mathbf{a}_1 + \mathbf{a}_2$, where $\mathbf{a}_1$ is parallel to **b** and $\mathbf{a}_2$ is perpendicular to **b**.

64. Find the work done if a force of 8 pounds acting in the direction $\frac{\pi}{3}$ moves an object from $(-2, 11)$ to $(11, 12)$. Assume distances are measured in feet.

65. Prove that the diagonals of any parallelogram bisect each other. (Hint: in the figure, $\mathbf{u} + \mathbf{v}$ and $\mathbf{u} - \mathbf{v}$ are diagonals. By writing $\overrightarrow{PQ}$ in two different ways, we have real numbers r, s such that $r(\mathbf{u} + \mathbf{v}) = \mathbf{v} + s(\mathbf{u} - \mathbf{v})$. Explain why $r - s = 0 = r - 1 + s$. Solve these two equations for r, s and interpret geometrically.)

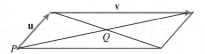

66. A triathlon swimmer intends to swim due north across an inlet, but will encounter a 0.4 mph current from the northwest. If he can swim 4 mph in the absence of a current, in what direction should he point himself? (Hint: if **u** is a unit vector having the desired direction, then $\frac{0.4}{\sqrt{2}}\mathbf{i} - \frac{0.4}{\sqrt{2}}\mathbf{j} + 4\mathbf{u} = r\mathbf{j}$ for some positive real number r.)

67. Why do we not try to solve for the unknown parts of a triangle $\triangle ABC$ in an "AAA" context, given α, β, and γ? (Hint: exercise 15 in section 7.1 is similar.)

68. Explain why no triangle $\triangle ABC$ satisfies $a = 2$, $b = 3$, and $c = 7$.

69. An object is suspended from a spring. The object is pulled down 5 centimeters from the equilibrium position, released, and then oscillates in simple harmonic motion. If it takes 0.37 second to complete one oscillation, write an equation to describe the motion.

70. Voltage V in a particular alternating current is given by $V = 130 \sin(ht + 4)$. If the circuit is carrying a 50 cycles-per-second current, determine the constant h. (Hint: first determine the frequency, f.)

71. An object moving in simple harmonic motion has its position at time t given by $s(t) = 4 \cos\left(\frac{2\pi}{5}t + \frac{\pi}{6}\right)$ in. Determine (a) the maximum displacement, (b) the frequency, (c) the period, (d) the position at $t = 0$, (e) the position at $t = 5.3$, and (f) the amount of time it takes to first return to the equilibrium position.

72. Two tuning forks are struck at approximately the same time and produce the sound waves given by $y_1 = 2 \sin 400\pi t$ and $y_2 = 2 \sin(300\pi t + \pi)$.
    (a) Write an equation for the superposition sound wave produced by combining these two waves.
    (b) What are the frequency and period of the superposition wave?
    (c) Sketch the graphs of one cycle of y_1, y_2, and $y_1 + y_2$.

73. An object moves in damped simple harmonic motion, with its position described by $s(t) = 3e^{-2t} \cos(4\pi t)$ m, $t \geq 0$. Determine (a) the maximum displacement from the "equilibrium" position $y = 0$, (b) the frequency, (c) the quasiperiod, and (d) the displacement from "equilibrium" after one "cycle."

74. For the motion in exercise 73, show that $|s(t)| < 0.0016$ m whenever $t > 3.8$.

75. (With this exercise, we bid farewell to Rowena and Romeo.) Rowena intends to elope with Romeo by using the ladder that she won from her father in exercise 31 of section 7.6. When she holds the ladder at a 40° angle of inclination, it falls 4 meters short of Romeo's window. By moving closer to his house, Rowena is able to hold the ladder at an angle of inclination of 60°. Then the ladder just reaches Romeo's window, and so they elope. How long is Rowena's ladder?

Chapter 8

Conic Sections and the General Second-Degree Equation in Two Variables

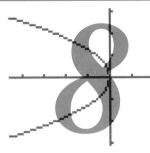

This chapter is devoted to the analytic geometry and physical applications of the conic sections. The history and study of the conic sections date back over 2200 years to the early Greeks. The Greek mathematician Apollonius, a contemporary of Archimedes, made a thorough study of the conic sections and is the first person known to have used the terms parabola, ellipse, and hyperbola.

A **conic section** is formed by the intersection of a plane and a double-napped right circular cone. If we assume that this intersection is not empty, then seven possibilities result: point, line, two intersecting lines, circle, ellipse, parabola, and hyperbola. The last four are shown in figures 8.1(a)–(d). You have already studied the analytic geometry asso-

Figure 8.1

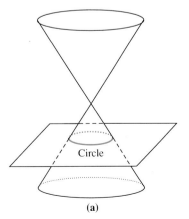
(a)

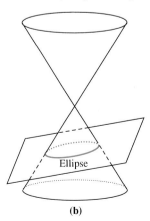
(b)

442 Chapter 8 / Conic Sections and the General Second-Degree Equation in Two Variables

Figure 8.1 (continued)

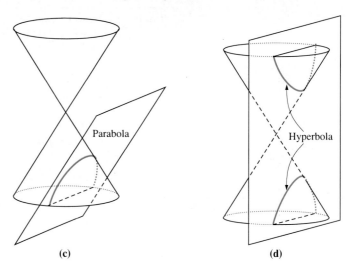

(c)   (d)

ciated with the "degenerate" conic sections (point, line, two intersecting lines, circle) in chapter 1; and in section 2.3, you studied quadratic polynomials having parabolic graphs.

In this chapter, we focus on the parabola, ellipse, and hyperbola. This will culminate in section 8.4, with the result that **conic sections are the only possible graphs of equations of the form $f(x, y) = 0$, where f is a second-degree or quadratic polynomial expression in x and y.** Although most of this chapter can be read directly after chapter 1, section 8.4 depends on the trigonometric addition formulas developed in chapter 6.

8.1 PARABOLAS

Definitions

> A **parabola** is the set of points equidistant from a given line, the **directrix,** and a given point, the **focus,** not on the directrix. The line through the focus perpendicular to the directrix is called the **axis.** The point midway between the focus and the directrix is called the **vertex.**

To sketch the parabola having a given focus and directrix, one may follow the procedure outlined in figure 8.2. Begin by drawing the directrix, the focus F, and the axis. Select a positive real number r. Draw the circle with center F and radius r. Next, draw the line parallel to the directrix, situated r units from the directrix and on the same side of the directrix as the focus. If this line and the circle intersect, as in figure 8.2(a), the point(s) of intersection is(are) on the parabola. Plot additional points on the parabola by repeating the procedure for different positive values of r. Figures 8.2(b)–(d) indicate how the procedure is repeated with a second, third, and fourth circle. The final result, shown in figure 8.2(e), is the set of points forming the parabola (which is the colored figure "opening upward" in the diagram). Notice in this figure that the parabola is

Parabolas

symmetric about its axis and that the vertex is on the parabola. These observations hold true for any parabola.

Figure 8.2

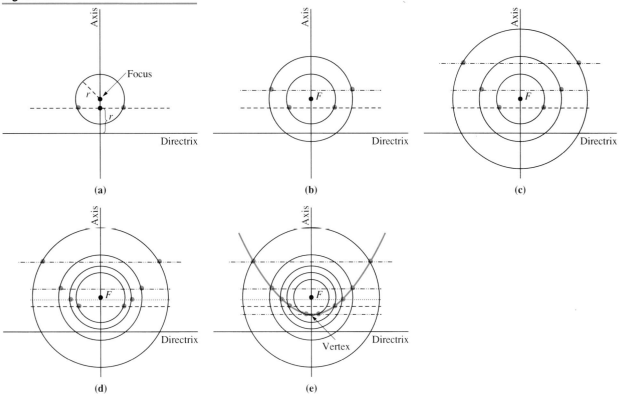

Standard Form with Vertex at the Origin

Let's see how we can express this algebraically. On a Cartesian coordinate system, let the vertex of the parabola be at the origin and suppose that the parabola's axis is the y-axis. Then the focus is at $(0, p)$ for some real number p, in which case the directrix must be $y = -p$. Now, consider any point $P(x, y)$ on the parabola. By the distance formula, $d(P, F) = \sqrt{x^2 + (y - p)^2}$; and the distance from P to the directrix is $|y + p|$. Since P is on the parabola, these distances are equal, and so

$$\sqrt{x^2 + (y - p)^2} = |y + p|,$$
$$x^2 + (y - p)^2 = (y + p)^2$$
$$x^2 + y^2 - 2py + p^2 = y^2 + 2py + p^2$$
$$x^2 = 4py.$$

Thus, we have the following standard form equation of a parabola.

Standard Form Equation of a Parabola with Horizontal Directrix

$x^2 = 4py$ is the "standard form" of an equation for the parabola with focus at $(0, p)$ and directrix $y = -p$. This parabola opens upward if $p > 0$ and downward if $p < 0$. Its vertex is the origin, $(0, 0)$. The y-axis is its only line of symmetry.

The equation $x^2 = 4py$ essentially defines y as a function (of x). The inverse relation of this function is described by $y^2 = 4px$. Although this relation is not a function, its graph is still a parabola. Indeed, in this case, we get the following standard form equation.

Standard Form Equation of a Parabola with Vertical Directrix

$y^2 = 4px$ is the "standard form" of an equation for the parabola with focus at $(p, 0)$ and directrix $x = -p$. This parabola opens to the right if $p > 0$ and to the left if $p < 0$. Its vertex is the origin, $(0, 0)$. The x-axis is its only line of symmetry.

▶ **EXAMPLE 8.1**

Sketch the graph of $y^2 = -8x$. Identify its focus and directrix.

Solution

This equation has the form $y^2 = 4px$, with $4p = -8$, or $p = -2$. Thus, its graph is a parabola, opening to the left, with focus at $(p, 0) = (-2, 0)$, directrix $x = -p = 2$, and vertex $(0, 0)$.

Figure 8.3

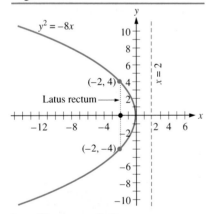

Two additional points on a parabola that are easy to plot are the endpoints of the chord which passes through the focus F and is parallel to the directrix. This chord is called the parabola's **latus rectum.** Any chord of the parabola that passes through the focus is a **focal chord.** So, the latus rectum is the particular focal chord that is parallel to the directrix (that is, perpendicular to the axis). You will be asked to prove in exercise 9 that the length of the latus rectum is $4|p|$.

In this example, the latus rectum is $4|p| = 8$ units long. Notice that F is the midpoint of the latus rectum. Thus, for this example, the endpoints of the latus rectum are $(-2, 4)$ and $(-2, -4)$. Using the vertex and these endpoints as a guide, we can sketch the graph shown in figure 8.3. ◀

▶ **EXAMPLE 8.2**

Sketch the graph of $x^2 = 2y$. Identify its focus and directrix.

Solution

This equation has the form $x^2 = 4py$, with $4p = 2$ or $p = \frac{1}{2}$. So, its graph is a parabola opening upward, with vertex $(0, 0)$, focus $(0, p) = (0, \frac{1}{2})$, and directrix $y = -p = -\frac{1}{2}$. The endpoints of the latus rectum are at $(1, \frac{1}{2})$ and $(-1, \frac{1}{2})$. The focus and these endpoints lead to the sketch of the parabola in figure 8.4(a).

You could have used a graphing calculator to sketch the graph of the above parabola. To do this, you must first rewrite the equation of the parabola so that only "y" appears on the left-hand side of "=". For this example, you obtain $y = \frac{1}{2}x^2$ or $y = 0.5x^2$. To graph this using a Casio *fx-7700G*, first press [Range] [SHIFT] [F1] to set the display to the calculator's default scale, and then press

Parabolas

Figure 8.4

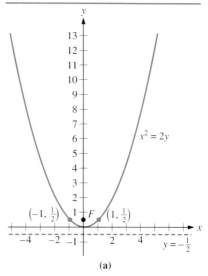
(a)

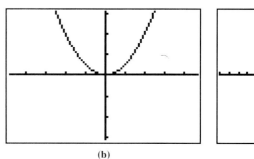
(b)

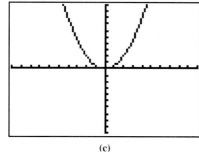
(c)

Range Range Graph .5 X,θ,T SHIFT x² EXE

You should thus obtain a graph like the one in figure 8.4(b). On a *TI-81*, press ZOOM 6 to obtain the default display screen, and then press

Y= .5 X|T x² GRAPH

The result is the graph in figure 8.4(c). ◀

EXAMPLE 8.3

Sketch the graph of $5y^2 + 8x = 0$. Identify its focus, directrix, and length of latus rectum.

Solution

By solving for y^2, we may rewrite this equation in standard form:
$$y^2 = -\tfrac{8}{5}x = 4px.$$

Since $4p = -\tfrac{8}{5}$, we see that $p = -\tfrac{2}{5}$. The graph is a parabola opening to the left, with vertex (0, 0), focus $(-\tfrac{2}{5}, 0)$, directrix $x = \tfrac{2}{5}$, and endpoints of the latus rectum at $(-\tfrac{2}{5}, \tfrac{4}{5})$ and $(-\tfrac{2}{5}, -\tfrac{4}{5})$. The latus rectum has length $4|p| = 4(\tfrac{2}{5}) = \tfrac{8}{5}$. The parabola's graph is in figure 8.5(a).

As in example 8.2, a graphing calculator could be used to sketch the graph of the above parabola. First, you need to understand more mathematics because your calculator can only graph functions. The equation $5y^2 + 8x = 0$ is equivalent to $y = \pm\sqrt{-\tfrac{8}{5}x}$. What does this mean geometrically? Consider the functions f_1 and f_2 given by $f_1(x) = \sqrt{-\tfrac{8}{5}x} = \sqrt{-1.6x}$ and $f_2(x) = -\sqrt{-\tfrac{8}{5}x} = -\sqrt{-1.6x}$. Both f_1 and f_2 have natural domain $(-\infty, 0]$. The graph of f_1 is the upper semiparabola, and the graph of f_2 is the lower semiparabola.

To graph f_1 and f_2 using a *TI-81*, first press ZOOM 6 to obtain the default display screen, and then press

Figure 8.5

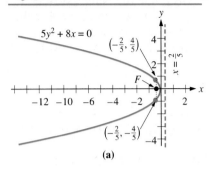
(a)

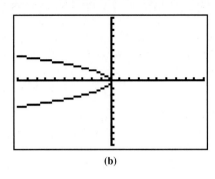
(b)

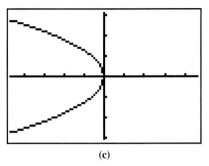
(c)

[Y=] [2nd] [√] [(] [(−)] 1.6 [X|T] [)]
[ENTER] [(−)] [2nd] [√] [(] [(−)] 1.6 [X|T] [)] [GRAPH]

The resulting graph is shown in figure 8.5(b).

To graph f_1 and f_2 on a Casio fx-7700G, first press [Range] [SHIFT] [F1] to set the display to the calculator's default setting, and then press

[Range] [Range] [Graph] [√] [(] [SHIFT] [(−)] 1.6 [X,θ,T] [)] [SHIFT] [↵]
[Graph] [SHIFT] [(−)] [√] [(] [SHIFT] [(−)] 1.6 [X,θ,T] [)] [EXE]

The result is shown in figure 8.5(c).

In your calculator-generated graphs, the upper and lower semiparabolas may not quite meet on the *x*-axis. This technological limitation of your calculator should not affect your understanding of the underlying mathematics.

More realistic graphs of parabolas could be obtained with graphing calculators by proceeding as in the discussion at the end of section 5.3. You should practice this skill in order to obtain a better understanding of the ellipses in section 8.2 and the hyperbolas in section 8.3. This will also enable you to generate "circular" (rather than "oval") graphs of circles! ◀

▶ **EXAMPLE 8.4** Find the standard form equation of the parabola with vertex $V(0, 0)$ and directrix $y = 6$.

Solution The focus F is on the parabola's axis, which is the line through V and perpendicular to $y = 6$. Thus, F is on the *y*-axis. We know that V is midway between F and the directrix, which means F is $(0, −6)$ and $p = −6$. Since a parabola cannot intersect its directrix, this parabola opens downward. It has the standard form equation $x^2 = 4py = −24y$, and its graph is in figure 8.6. ◀

Figure 8.6

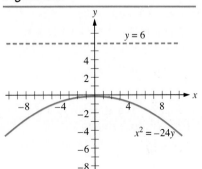

Standard Form Equations for Translated Parabolas

The vertex of a parabola does not have to be at the origin. How can we find such a parabola's equation? In particular, how can we find an equation for a given parabola having horizontal directrix and vertex at (h, k)? The proce-

Parabolas

dure depends on viewing the given parabola as resulting from the translation of a replica having vertex at the origin. Now, we have seen above that the replica has equation $x^2 = 4py$. So, by chapter 2, we can obtain an equation for the given parabola by replacing x by $x - h$ and y by $y - k$. The result is the equation

$$(x - h)^2 = 4p(y - k).$$

The graph of this equation is a parabola with vertex (h, k), focus $(h, k + p)$, and directrix $y = k - p$. This parabola opens upward if $p > 0$ and downward if $p < 0$.

A similar translation of the graph of $y^2 = 4px$ results in the equation

$$(y - k)^2 = 4p(x - h).$$

The graph of this equation is a parabola with vertex (h, k), focus $(h + p, k)$, and directrix $x = h - p$. This parabola opens to the right if $p > 0$, to the left if $p < 0$.

These results are summarized in table 8.1. You should be able to derive such facts quickly in any given case, by reasoning as above. It should not be necessary for you to memorize table 8.1.

Table 8.1

Standard Form Equation	Vertex	Focus	Directrix	Axis	Endpoints of Latus Rectum	Orientation
$(x - h)^2 = 4p(y - k)$	(h, k)	$(h, k + p)$	$y = k - p$	$x = h$	$(h - 2p, k + p)$ $(h + 2p, k + p)$	opens up if $p > 0$ opens down if $p < 0$
$(y - k)^2 = 4p(x - h)$	(h, k)	$(h + p, k)$	$x = h - p$	$y = k$	$(h + p, k - 2p)$ $(h + p, k + 2p)$	opens right if $p > 0$ opens left if $p < 0$

Solving problems involving parabolas often requires you to write the equation in one of the standard forms. To do this, you may need to complete the square.

▶ **EXAMPLE 8.5**

For the parabola $x^2 + 6x - 12y - 15 = 0$, find the vertex, focus, axis, directrix, and length of latus rectum. Sketch its graph.

Solution

We shall complete the square.

$$x^2 + 6x - 12y - 15 = 0$$
$$x^2 + 6x = 12y + 15$$
$$x^2 + 6x + 9 = 12y + 15 + 9$$
$$(x + 3)^2 = 12y + 24$$
$$(x + 3)^2 = 12(y + 2).$$

Figure 8.7

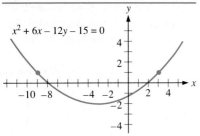

This has the form $(x - h)^2 = 4p(y - k)$, with $h = -3$, $k = -2$, and $p = 3$. Using table 8.1, we have the following results.

Vertex: $(-3, -2)$ Focus: $(-3, 1)$

Directrix: $y = -5$ Axis: $x = -3$

Length of latus rectum: $4|p| = 12$.

The parabola opens upward and passes through the endpoints of the latus rectum, $(-9, 1)$ and $(3, 1)$. The graph of this parabola is in figure 8.7.

▶ **EXAMPLE 8.6**   For the parabola $y^2 - 8y - 8x - 4 = 0$, find the vertex, focus, axis, directrix, and length of latus rectum. Sketch its graph.

Solution   We complete the square.

Figure 8.8

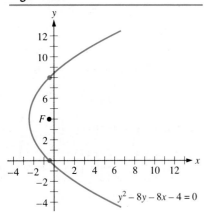

$$y^2 - 8y - 8x - 4 = 0$$
$$y^2 - 8y = 8x + 4$$
$$y^2 - 8y + 16 = 8x + 4 + 16$$
$$(y - 4)^2 = 8x + 20$$
$$(y - 4)^2 = 8(x + 2.5).$$

This has the form $(y - k)^2 = 4p(x - h)$, with $h = -2.5$, $k = 4$, and $p = 2$. Table 8.1 gives the following results.

Vertex: $(-2.5, 4)$ Focus: $(-0.5, 4)$

Directrix: $x = -4.5$ Axis: $y = 4$

Endpoints of latus rectum: $(-0.5, 0)$ and $(-0.5, 8)$

Length of latus rectum: $4|p| = 8$.

The sketch of this parabola is in figure 8.8.

▶ **EXAMPLE 8.7**   Find an equation for the parabola with focus at $(3, -4)$ and directrix the line $x - 5 = 0$.

Solution   A sketch of the given information is in figure 8.9. Since a parabola cannot intersect its directrix, this parabola cannot open up, down, or to the right. Thus, it opens to the left and has an equation of the form $(y - k)^2 = 4p(x - h)$ with $p < 0$. The vertex is midway between the focus and directrix, or at $(4, -4)$. So $h = 4$ and $k = -4$. The distance between the vertex and the directrix is $|p| = 1$. Thus $p = -1$, and we now know enough to write the desired equation: $(y + 4)^2 = -4(x - 4)$. Its graph is in figure 8.9.

Figure 8.9

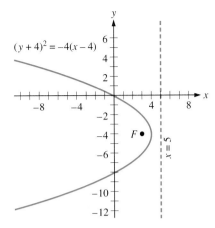

EXAMPLE 8.8

Determine an equation of a parabola with vertex $V(-3, 2)$, one endpoint of its latus rectum at $(5, -4)$, and axis parallel to a coordinate axis.

Solution

The given points are sketched in figure 8.10(a). Two axes satisfy the conditions of this problem: $x = -3$ and $y = 2$. We shall consider each of these in turn. (Recall that V is on the axis, and that the axis is perpendicular to both the latus rectum and the directrix.)

Figure 8.10

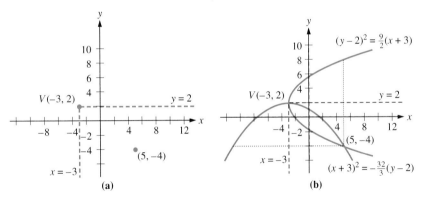

Case I: Axis $x = -3$.

The desired equation has the form $(x - h)^2 = 4p(y - k)$. We are given $h = -3$ and $k = 2$, and so the equation becomes $(x + 3)^2 = 4p(y - 2)$. Since endpoints of the latus rectum lie on the parabola, $(5, -4)$ satisfies the equation. Substituting $x = 5$ and $y = -4$, we get $(5 + 3)^2 = 4p(-4 - 2)$, or $8^2 = 4p(-6)$, and so $p = -\frac{64}{24} = -\frac{8}{3}$. An equation for this parabola is

$$(x + 3)^2 = -\tfrac{32}{3}(y - 2).$$

Case II: Axis $y = 2$.

In this case, the desired equation has the form $(y - k)^2 = 4p(x - h)$, again with $h = -3$ and $k = 2$. This is $(y - 2)^2 = 4p(x + 3)$. Substituting $x = 5$ and $y = -4$, we get $(-4 - 2)^2 = 4p(5 + 3)$, or $36 = 32p$, and so $p = \frac{9}{8}$. Thus, an equation of this parabola is

$$(y-2)^2 = \tfrac{9}{2}(x+3).$$

The graphs of these two parabolas are in figure 8.10(b).

EXAMPLE 8.9

Find an equation for the focal chord of slope 1 for the parabola $y^2 - 4y - 4x = 0$.

Solution

We first need to determine the coordinates of the focus of this parabola. Let's complete the square to get the equation into standard form:

$$y^2 - 4y - 4x = 0$$
$$y^2 - 4y = 4x$$
$$y^2 - 4y + 4 = 4x + 4$$
$$(y-2)^2 = 4(x+1).$$

This has the form $(y-k)^2 = 4p(x-h)$, with $h = -1$, $k = 2$, and $p = 1$. Using table 8.1, or reasoning directly, we see that the focus F is $(h+p, k) = (0, 2)$.

The problem is to find an equation for the line with slope 1 that passes through the focus $F(0, 2)$. Using the point-slope form (section 1.3), we get

$$y - 2 = 1(x - 0),$$

or

$$y = x + 2.$$

The graph of the parabola and the desired chord are shown in figure 8.11.

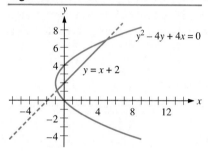

Figure 8.11

Applications

Three-dimensional objects built from parabolas have many uses. These result from the following two geometric/physical facts. You will be able to prove them in a later course, by using calculus and the principle that "the angle of incidence is congruent to the angle of reflection."

(I) Energy rays "entering" a parabola along lines parallel to its axis of symmetry are all reflected through the focus. (This is nearly the case for the sun's rays arriving on Earth. You may have illustrated this fact by focusing solar energy with a magnifying glass in order to burn a pinhole in a piece of paper. Fact I is also used in various radio telescopes and radar antennas, as well as in satellite "dishes" used as a downlink to receive data from a satellite. Such applications explain the origin and suitability of the term "focus.") See figure 8.12(a).

(II) Energy rays emitted from a parabola's focus are all reflected along lines parallel to the parabola's axis. (This fact explains the construction of various headlights, searchlights, and satellite "dishes" used as an uplink—that is, to transmit data to a satellite.) See figure 8.12(b).

Parabolas

Figure 8.12

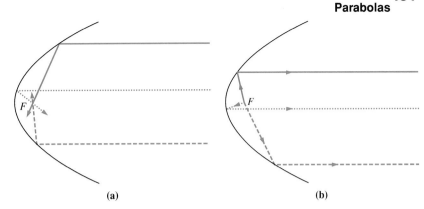

(a) (b)

Figure 8.13

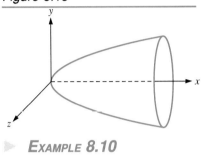

The antennas and headlights referred to above arise as *paraboloids of revolution*. As pictured in figure 8.13, such objects are the result of rigidly revolving a parabola (into three-dimensional space) about its axis of symmetry. Applications involving them will be addressed in example 8.10 and in exercises 51–53. You will study paraboloids of revolution, and the more general topic of solids of revolution, in calculus.

EXAMPLE 8.10

A cable television dish in the shape of a paraboloid of revolution is 1 foot deep and measures 5 feet across at its opening. Where should a technician place the receiver?

Solution

We begin by sketching a cross-section of the television dish on a coordinate system. The vertex is placed at the origin, and the focus is placed on the x-axis, as shown in figure 8.14. Because of the above geometric/physical

Figure 8.14

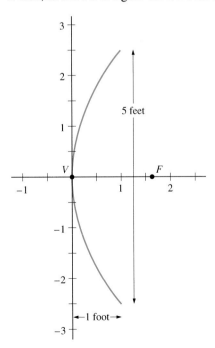

facts, the technician must place the receiver at the focus. We know that the parabola in the figure has an equation of the form $y^2 = 4px$. From the given data, the point (1, 2.5) is on the parabola. Hence $p = \dfrac{y^2}{4x} = \dfrac{(2.5)^2}{4 \cdot 1} = 1.5625$, and so the focus is at (1.5625, 0). We can thus determine that the receiver should be placed 1.5625 feet = $1'\, 6\tfrac{3}{4}''$ from the vertex.

Exercises 8.1

In exercises 1–8, determine the given parabola's vertex, axis, focus, directrix, endpoints of latus rectum, and length of latus rectum. Then sketch the parabola.

1. $x^2 = 9y$
2. $3y^2 - 9x = 0$
3. $y^2 = -3x$
4. $2x^2 + 3y = 0$
5. $y^2 - \pi x + 16y = \pi - 16$
6. $x^2 - 2y + 8x + 10 = 0$
7. $3x^2 + 4y - 6x + 3 = 0$
8. $-4y^2 - 3x - y + 1 = 0$

◆ 9. Prove that the length of the latus rectum of a parabola (written in standard form) is $4|p|$. (Hint: consider $y^2 = 4px$. If $x = p$, then $y = \pm 2p$. What does this mean geometrically?)

10. Prove the assertion made in examples 8.4 and 8.7 that a parabola cannot intersect its directrix. (Hint: this follows from the definition of parabola, using the fact that the focus does not lie on the directrix. An algebraic proof follows by trying to solve $y^2 = 4px$ and $x = -p$ simultaneously for (real) x and y.)

11. Prove that the only point on a parabola that also lies on the parabola's axis is the vertex. (Hint: solve $y^2 = 4px$ and $y = 0$ simultaneously.)

12. Use the techniques of section 2.2 to show that the graph of $y^2 = 4px$ is symmetric with respect to the x-axis, but is not symmetric with respect to any of the following: y-axis, origin, line $y = x$.

In exercises 13–30, find an equation for the parabola with the given properties.

13. Vertex (0, 0), focus (–2, 0)
14. Vertex (0, 0), focus (0, 8)
15. Vertex (–3, 2), focus (–3, –10)
16. Vertex (–2, 3), focus (1, 3)
17. Vertex (0, 0), directrix $y = 2$
18. Vertex (0, 0), directrix $x = \sqrt{2}$
19. Vertex $(\tfrac{1}{2}, -\tfrac{1}{3})$, directrix $x = -1$
20. Vertex (3, –2), directrix $y = -1$
21. Focus $(-\tfrac{1}{3}, \tfrac{1}{2})$, directrix $x = -\tfrac{16}{3}$
22. Focus (–4, –5), directrix $y = -6$
23. Focus (0, 0), directrix $y = 8$
24. Focus (–8, 0), directrix $x = 8$
25. Vertex (1, 2), horizontal axis, (4, 6) on the parabola
26. Vertex (–2, 1), vertical axis, (–9, 6) on the parabola
27. Vertex (5, 4), axis $x = 5$, $(7, \tfrac{5}{3})$ on the parabola
28. Vertex $(\tfrac{1}{3}, -\tfrac{1}{2})$, horizontal axis, $(0, -\tfrac{3}{4})$ on the parabola
29. Vertex (2, 1), axis parallel to a coordinate axis, (4, –3) an endpoint of the latus rectum (Hint: there are two answers.)
30. Vertex (1, –2), axis parallel to a coordinate axis, (–1, 6) an endpoint of the latus rectum

31. Find an equation for the parabola, with vertical axis, passing through the points (–1, 2), (1, –1), and (2, 1). (Hint: it is quickest to write the equation as $y = ax^2 + bx + c$, and then solve three equations for a, b, and c. Be persistent, reviewing section 1.3 or section 3.7 if necessary.)

32. (a) Find an equation for the parabola, with horizontal axis, passing through (–1, 2), (1, –1), and (2, 1). (Hint: modify the hint for Exercise 31.)
    (b) Then find this parabola's focus and length of latus rectum.

33. Prove that a given line and a given parabola can intersect in at most two points. (Hint: you may assume the parabola "is" $y^2 = 4px$ and that the line is either $y = mx + b$ or $x = c$. Reduce matters to solving a quadratic or a linear equation.)

34. Given *any* three distinct points, must there be a parabola passing through all three points? (Hint: apply exercise 33.)

35. Prove that the only nonvertical line through (2, 12) that intersects the parabola $y = 3x^2$ only once is the line $y = 12x - 12$. (Hint: review the method in exercise 33. Recall that the criterion for a quadratic equation to have a double root is that its discriminant be zero.) We call $y = 12x - 12$ the *tangent line to* $y = 3x^2$ *at (2, 12)*.

36. Find an equation of the tangent line to $9y + 4x^2 = 0$ at (–3, –4). (Hint: review exercise 35.)

37. Find an equation for the focal chord of slope –2 for the parabola $y^2 + 8y - 2x = 0$. (Hint: review example 8.9.)

38. Find an equation for the focal chord of the parabola $x^2 - 4x - 4y = 0$ which is parallel to the line $3x + 4y + 5 = 0$.

◆ 39. Prove that $\{(x, y):$ there exists a real number t such that $y = 4t$ and $x = 2t^2\}$ is the parabola $y^2 = 8x$. (Hint: there are two things to do. First, show that $y = 4t$ and $x = 2t^2$ imply $y^2 = 8x$. Next—and more difficult—show that $y^2 = 8x$ implies existence of t such that $y = 4t$ and $x = 2t^2$.)

◆ 40. For any nonzero real number p, prove that $\{(x, y):$ there exists a real number t such that $x = 2pt$ and $y = pt^2\}$ is the parabola $x^2 = 4py$. (Hint: there are two things to do here. Review the hint for exercise 39.)

In exercises 41–44, identify and graph the piece of the parabola $x^2 = 16y$ "parameterized" as $\{(x, y):$ there exists a real number t in S such that $x = 8t$ and $y = 4t^2\}$, in case S is the given interval.

◆ 41. $S = (-\infty, \infty)$ ◆ 42. $S = [0, \infty)$
◆ 43. $S = (-\infty, 0)$ ◆ 44. $S - [0, 1)$

◆ 45. Consider all the lines of slope 3 that intersect the parabola $y^2 = 4x$ in more than one point. For each such line ℓ, consider the midpoint of the line segment connecting the two points of intersection of ℓ and the parabola. Show that all these midpoints lie on the line $y = \frac{2}{3}$. (Hint: as in exercises 39–40, view the parabola as "parameterized" by $y = 2t$, $x = t^2$. If ℓ is $y = 3x + b$, the intersections of ℓ with the parabola correspond to the roots t_1, t_2 of the quadratic equation $3t^2 - 2t + b = 0$. Argue that the y-coordinate of the midpoint arising from ℓ is $t_1 + t_2$. Then use chapter 0 to calculate $t_1 + t_2$.)

◆ 46. (a) (A generalization of exercise 45) Fix a nonzero real number m. Consider all the lines of slope m that intersect a given parabola $y^2 = 4px$ in more than one point. For each such line ℓ, consider the midpoint of the line segment connecting the two points of intersection of ℓ and the parabola. Show that all these midpoints lie on the line $y = \frac{2p}{m}$, which is parallel to the parabola's axis. (Hint: argue as in exercise 45, using $y = 2pt$, $x = pt^2$.)

(b) Notice that as $|m|$ increases without bound, the limiting tendency of $\frac{2}{m}$ is 0. This suggests that if we repeat the process of (a) for *vertical* lines ℓ (and the same parabola), then the midpoints in question all lie on the parabola's axis, $y = 0$. Is this so?

◆ 47. Show that the parabola with focus $(3, 0)$ and directrix $y = 2x$ has equation $x^2 + 4y^2 + 4xy - 30x + 45 = 0$. (Hint: review section 1.3 regarding the distance from a point to a line. Simplify the equation $\frac{|2x - y|}{\sqrt{5}} = \sqrt{(x - 3)^2 + y^2}$.)

48. If a parabola has focus $(-\frac{2}{3}, \frac{4}{7})$ and directrix $21x - 42y + 5 = 0$, prove that its vertex is $(-\frac{107}{210}, \frac{9}{35})$. (Hint: find where the axis meets the directrix. Then use the midpoint formula.)

49. Graph (a) $x = y^2$, (b) $y = \sqrt{x}$, and (c) $y = -\sqrt{x}$. (Hint: the answers are all different.)

50. Graph each of the following, and hence determine which is/are parabolas:
(a) $y^2 = 0$ (b) $y^2 = x$ (c) $y^2 = xy$ (d) $y^2 = x^2$

◆ 51. A parabolic reflector measures 12 feet across at its opening and is 4 feet deep. How far from the vertex is the focus located?

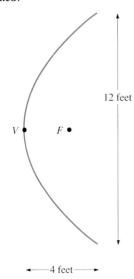

◆ 52. A cable television "dish" is 4 feet deep and measures 10 feet across at its opening. A technician must place the receiver at the focus. How far is this from the vertex?

◆ 53. A flashlight's reflecting surface is part of a paraboloid of revolution, with diameter 12 centimeters and depth 3 centimeters. How far apart are the vertex and the focus?

54. Prove that the point of a parabola that is nearest the parabola's focus is the parabola's vertex. (Hint: you may consider the parabola $y^2 = 4px$. Explain why the problem is equivalent to minimizing $(x - p)^2 + y^2 = (x + p)^2$ for $x \geq 0$.)

55. The Humble Hamburger Hut (of exercise 47 in section 2.3) has grown powerful, and acquires Plentiful Produce as a result of a hostile takeover. HHH decides to change

its subsidiary's name to Parabolic Plentiful Produce (PPP), whose emblem will be a parabolic arch, opening upward, measuring 20 yards across at its opening and 10 yards deep.

(a) How high above the arch's lowest point is its focus?

(b) If PPP prospers and develops a second emblem that uses the old focus but is now 40 yards across at its opening, how deep is PPP's new emblem?

8.2 ELLIPSES

The second type of conic section that we study in this chapter is the ellipse.

Definitions

An **ellipse** is the set of points the sum of whose distances from two given points, the **foci**, is a given constant. Each of the two given points is called a **focus** of the ellipse.

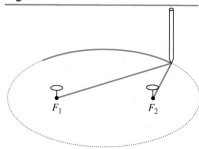

Figure 8.15

One way to draw an ellipse involves placing two tacks in a piece of cardboard. The points at which the tacks pierce the cardboard are the **foci** (plural of "focus"), as shown in figure 8.15. Fasten one end of a piece of string to each tack, and draw the string tight with a pencil. Move the pencil, keeping the string tight. The path traced by the pencil's tip is an ellipse.

The line connecting the foci of an ellipse intersects the ellipse in two points called the **vertices** (plural of "vertex"). The chord joining the vertices is called the **major axis,** and its midpoint is the **center.** It can be shown that the center is also midway between the foci. The **minor axis** is the chord through the center that is perpendicular to the major axis. In exercise 13, you will be asked to show that any ellipse is symmetric with respect to its major axis, its minor axis, and its center. Any chord of the ellipse that passes through a focus is a **focal chord**. A **latus rectum** is a focal chord that is perpendicular to the major axis. Each ellipse has two latera recta (plural of "latus rectum"). A **focal radius** is a segment with one endpoint at a focus and the other endpoint on the ellipse.

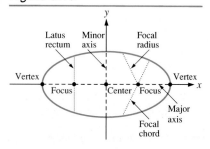

Figure 8.16

Standard Form with Center at the Origin

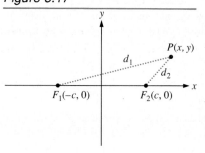

Figure 8.17

To derive a "standard form" for an equation for an ellipse, we shall consider an ellipse centered at the origin, with its major axis along one of the coordinate axes, say the x-axis. Then the foci are $F_1(-c, 0)$ and $F_2(c, 0)$ for some positive number c. Let $2a$ be the constant sum of the lengths of the focal radii for points on the ellipse. Now, consider any point $P(x, y)$. Let $d_1 = d(P, F_1)$, the distance from F_1 to P. Similarly, let $d_2 = d(P, F_2)$, as in figure 8.17. The condition that P lie on the given ellipse is $d_1 + d_2 = 2a$. Using the distance formula, we know that

$$d_1 = \sqrt{(x+c)^2 + y^2} \quad \text{and} \quad d_2 = \sqrt{(x-c)^2 + y^2}.$$

So,
$$\sqrt{(x+c)^2 + y^2} + \sqrt{(x-c)^2 + y^2} = 2a$$
$$\sqrt{(x+c)^2 + y^2} = 2a - \sqrt{(x-c)^2 + y^2}$$
$$[\sqrt{(x+c)^2 + y^2}]^2 = [2a - \sqrt{(x-c)^2 + y^2}]^2$$
$$(x+c)^2 + y^2 = (2a)^2 - 4a\sqrt{(x-c)^2 + y^2} + (x-c)^2 + y^2$$
$$4a\sqrt{(x-c)^2 + y^2} = 4a^2 - 4cx \quad \text{``}-4a\text{'' comes from squaring}$$
$$a\sqrt{(x-c)^2 + y^2} = a^2 - cx$$
$$[a\sqrt{(x-c)^2 + y^2}]^2 = [a^2 - cx]^2$$
$$a^2[(x-c)^2 + y^2] = a^4 - 2a^2cx + c^2x^2$$
$$a^2x^2 - 2a^2cx + a^2c^2 + a^2y^2 = a^4 - 2a^2cx + c^2x^2$$
$$(a^2 - c^2)x^2 + a^2y^2 = a^2(a^2 - c^2).$$

Now, it is important to insist that $a > c$. (Otherwise, certain "ellipses" would degenerate to at most one point.) This follows by applying the geometric triangle inequality to $\triangle PF_1P_2$, for then $d_1 + d_2 > 2c$, that is, $2a > 2c$, or $a > c$. Accordingly, we can define a positive real number b by
$$b = \sqrt{a^2 - c^2}.$$

The above equation for the ellipse becomes $b^2x^2 + a^2y^2 = a^2b^2$. This leads to a standard form equation for an ellipse with a horizontal major axis, as shown in the following box.

Standard Form Equation of an "East-West" Ellipse

$$\frac{x^2}{a^2} + \frac{y^2}{b^2} = 1.$$

↑ "a" larger on this side → E-W

a: major axis
b: minor axis

We have just derived the standard form for an "east-west" ellipse (also called a "horizontal" ellipse, since it has a horizontal major axis), when it is centered at the origin. By setting $x = 0$, we find $\frac{y^2}{b^2} = 1$, or $y = \pm b$. Hence, this ellipse's y-intercepts, which are the endpoints of its minor axis, are $(0, b)$ and $(0, -b)$. By setting $y = 0$ and finding $x = \pm a$, we see that this ellipse's vertices are $(a, 0)$ and $(-a, 0)$, the endpoints of its major axis.

The above reasoning also leads to a standard form equation for a "north-south" (or "vertical") ellipse, with vertical major axis, when it is centered at the origin. In this case, the foci are $F_1(0, -c)$ and $F_2(0, c)$. Then the condition $d_1 + d_2 = 2a$ and the definition $b = \sqrt{a^2 - c^2}$ lead, as above, to another standard form equation for a "north-south" ellipse.

Standard Form Equation of a "North-South" Ellipse

$$\frac{x^2}{b^2} + \frac{y^2}{a^2} = 1.$$

↑ "a" larger here → N-S

The vertices are $(0, a)$ and $(0, -a)$. The endpoints of the minor axis are $(b, 0)$ and $(-b, 0)$.

In <u>both</u> the "east-west" and the "north-south" cases, $a > b$ and $a > c$; $b = \sqrt{a^2 - c^2}$; the length of the major axis is $2a$; and the length of the minor axis is $2b$. Note that a is the length of each **semimajor axis**—that is, each line segment connecting the center and a vertex. Similarly, b is the length of each **semiminor axis**—that is, each segment connecting the center and an endpoint of the minor axis.

▶ **EXAMPLE 8.11**

Sketch the graph of $\dfrac{x^2}{16} + \dfrac{y^2}{9} = 1$. Find the center, foci, vertices, endpoints of the minor axis, length of semimajor axis, and length of semiminor axis.

Solution

As $a^2 > b^2$, we have here that $a^2 = 16$ and $b^2 = 9$. This is an "east-west" ellipse, centered at the origin, with length of semimajor axis $a = 4$, length of semiminor axis $b = 3$, vertices at $(4, 0)$ and $(-4, 0)$, and endpoints of the minor axis at $(0, 3)$ and $(0, -3)$. Using the definition $b = \sqrt{a^2 - c^2}$, we find

$$c = \sqrt{a^2 - b^2}.$$

So, in this example, $c = \sqrt{16 - 9} = \sqrt{7}$, and the foci are $(-\sqrt{7}, 0)$ and $(\sqrt{7}, 0)$. The sketch of this ellipse is in figure 8.18. ◀

Figure 8.18

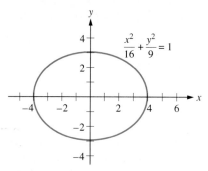

▶ **EXAMPLE 8.12**

Sketch the graph of $25x^2 + 16y^2 = 400$. Find a, b, and c and the graph's center, foci, and vertices.

Solution

To put this in standard form, divide both sides of the equation by 400, to obtain

$$\frac{x^2}{16} + \frac{y^2}{25} = 1.$$

As $a^2 > b^2$, we have $a^2 = 25$ and $b^2 = 16$. Then $c = \sqrt{a^2 - b^2} = \sqrt{9} = 3$. This is a "north-south" ellipse, centered at the origin, with foci $(0, 3)$ and $(0, -3)$, vertices $(0, 5)$ and $(0, -5)$, and endpoints of the minor axis at $(4, 0)$ and $(-4, 0)$. The sketch of the ellipse is in figure 8.19. ◀

Ellipses

Figure 8.19

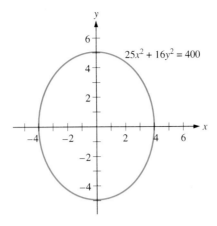

Example 8.13

Find an equation for the ellipse with center at the origin, one focus at $(0, -5)$, and a vertex at $(0, 8)$.

Solution

Vertices and foci lie on the major axis. Hence this is a "north–south" ellipse. Since its center is at the origin, this ellipse has equation $\frac{x^2}{b^2} + \frac{y^2}{a^2} = 1$. We determine from the given information that $c = 5$ and $a = 8$. Using the definition $b = \sqrt{a^2 - c^2}$, we find $b = \sqrt{8^2 - 5^2} = \sqrt{39}$, or $b^2 = 39$. Therefore, the desired equation for this ellipse is

$$\frac{x^2}{39} + \frac{y^2}{64} = 1.$$

It is pictured in figure 8.20.

Figure 8.20

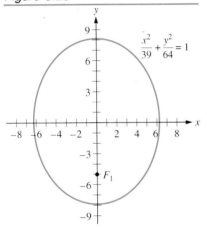

Standard Form Equations for Translated Ellipses

If the center of an ellipse is not at the origin, then, just as in section 8.1, we can use the translation methods of section 2.2 to find a standard form equation for the ellipse. The results of such analyses are summarized in table 8.2.

Table 8.2

Standard Form Equation*	Center	Foci	Endpoints of Major Axis	Endpoints of Minor Axis	Orientation
$\frac{(x-h)^2}{a^2} + \frac{(y-k)^2}{b^2} = 1$	(h, k)	$(h+c, k)$ $(h-c, k)$	$(h-a, k)$ $(h+a, k)$	$(h, k-b)$ $(h, k+b)$	"east–west" or horizontal
$\frac{(x-h)^2}{b^2} + \frac{(y-k)^2}{a^2} = 1$	(h, k)	$(h, k+c)$ $(h, k-c)$	$(h, k-a)$ $(h, k+a)$	$(h-b, k)$ $(h+b, k)$	"north–south" or vertical

*The constants a, b, and c are positive and satisfy $a^2 = b^2 + c^2$ and $a > b$.

Example 8.14

Sketch the graph of $\dfrac{(x-3)^2}{4} + \dfrac{(y+5)^2}{16} = 1$. Find a, b, and c and the graph's center and foci.

Solution

Begin by plotting the center (h, k). In this problem, it is $(3, -5)$. This is the center of the translated coordinate system, so sketch in the translated x- and y-axes, as shown in figure 8.21(a).

From the equation, we see that $a^2 = 16$ and $b^2 = 4$. Hence, $a = 4$ and $b = 2$. Moreover, $c = \sqrt{a^2 - b^2} = \sqrt{12} = 2\sqrt{3}$. As this ellipse is "north-south," table 8.2 tells us that its foci are $(3, -5 + 2\sqrt{3})$ and $(3, -5 - 2\sqrt{3})$. Using the values of h, k, a, and b, we can locate the endpoints of the major and minor axes, as shown in figure 8.21(b). Then the sketch of this ellipse can be drawn, as in figure 8.21(c). ◀

Figure 8.21

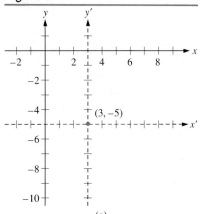
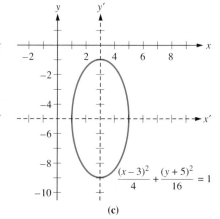

Example 8.15

Sketch the graph of $x^2 + 4y^2 + 2x - 8y + 1 = 0$.

Solution

As in section 8.1, we must first find the equivalent standard form equation, and this is done by completing the square.

$$x^2 + 4y^2 + 2x - 8y + 1 = 0$$
$$x^2 + 2x + 4(y^2 - 2y) = -1$$
$$x^2 + 2x + 1 + 4(y^2 - 2y + 1) = -1 + 1 + 4$$
$$(x + 1)^2 + 4(y - 1)^2 = 4$$
$$\dfrac{(x + 1)^2}{4} + \dfrac{(y - 1)^2}{1} = 1$$

The equation is now in standard form. The ellipse is "east-west," with center at $(-1, 1)$. Plot this new center and draw the translated coordinate axes through it, as shown in figure 8.22(a).

From the equation, we see that $a^2 = 4$ and $b^2 = 1$; that is, $a = 2$ and $b = 1$. We can now proceed to locate the endpoints of the major and minor axes and sketch the ellipse, as shown in figure 8.22(b).

Figure 8.22

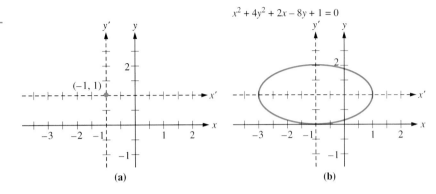

EXAMPLE 8.16

Find the length of a latus rectum of the ellipse $3x^2 + 4y^2 + 6x - 16y + 7 = 0$.

Solution

Recall that a latus rectum is a chord through a focus and perpendicular to the major axis. We shall need to find the coordinates of a focus, so we begin by writing the equation in standard form.

$$3x^2 + 4y^2 + 6x - 16y + 7 = 0$$
$$3(x^2 + 2x) + 4(y^2 - 4y) = -7$$
$$3(x^2 + 2x + 1) + 4(y^2 - 4y + 4) = -7 + 3 + 16$$
$$3(x + 1)^2 + 4(y - 2)^2 = 12$$
$$\frac{(x + 1)^2}{4} + \frac{(y - 2)^2}{3} = 1.$$

This is an "east-west" ellipse, centered at $(-1, 2)$, with $a^2 = 4$, $b^2 = 3$, and $c = \sqrt{a^2 - b^2} = 1$. As $a = 2$ and $b = \sqrt{3}$, we find the foci at $(-2, 2)$ and $(0, 2)$, the vertices at $(1, 2)$ and $(-3, 2)$, and the endpoints of the minor axis at $(-1, 2 + \sqrt{3})$ and $(-1, 2 - \sqrt{3})$. The sketch of this ellipse is in figure 8.23.

Figure 8.23

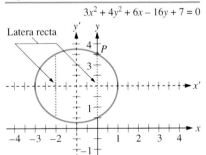

Let P be an endpoint of a latus rectum, as shown in figure 8.23. Then P is on the ellipse and has coordinates either $(-2, y)$ or $(0, y)$. Let's suppose P has coordinates $(0, y)$. Since P's coordinates satisfy the ellipse's equation,

$$4y^2 - 16y + 7 = 0.$$

By the quadratic formula, we see that $y = \frac{4 \pm 3}{2}$, and so y is either $\frac{7}{2}$ or $\frac{1}{2}$. The endpoints of the latus rectum are $(0, \frac{7}{2})$ and $(0, \frac{1}{2})$. The distance between them is 3, and so this is the length of the latus rectum.

460 Chapter 8 / Conic Sections and the General Second-Degree Equation in Two Variables

$$c = \frac{(x, y)(x_1, y_2)}{y + y_2}$$
[wait — handwritten: $c = \frac{(x, y)(x_1, y_2)}{2}$, with $y + y_2$]

You will be asked to show in exercise 16 that if an ellipse has been written in standard form, each of its latera recta has length $\frac{2b^2}{a}$. Notice that this agrees with the results of example 8.16, for there $\frac{2b^2}{a} = \frac{2(3)}{2} = 3$.

EXAMPLE 8.17

Find an equation of the ellipse with foci at $(4, 3)$ and $(4, -5)$ and endpoints of its minor axis at $(4 - \sqrt{7}, -1)$ and $(4 + \sqrt{7}, -1)$.

Solution

Since the foci lie on the major axis, this is a "north-south" ellipse. Its center is at the intersection of the semimajor axis with the semiminor axis, namely $(4, -1)$. You can check that the data yield $c = 4$ and $b = \sqrt{7}$. Hence

$$a = \sqrt{b^2 + c^2} = \sqrt{7 + 16} = \sqrt{23}.$$

Thus, the standard form equation for this ellipse is

$$\frac{(x-4)^2}{7} + \frac{(y+1)^2}{23} = 1.$$

The sketch of the ellipse is in figure 8.24.

Figure 8.24

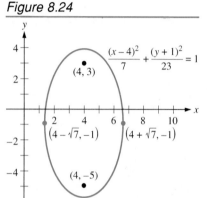

Think back to the definition of a parabola: each of its points is equidistant from a given point (the parabola's focus) and a given line (the directrix). We next indicate an alternative definition of an ellipse that is in the same spirit.

Eccentricity of an Ellipse

[handwritten: between "0" + "1" or $0 \le E < 1$, nonstretched out]

An ellipse is the set of all points the ratio of whose distances to a given point (focus) and to a given line (directrix) is a given constant e such that $0 < e < 1$. The constant e is called the **eccentricity** of the ellipse.

Do not confuse this number e with the constant $e \approx 2.71828$.

In figure 8.25, two foci and their corresponding directrices have been drawn. Notice that each focus must be paired with the directrix corresponding to it. You will essentially show in exercise 48 that

Figure 8.25

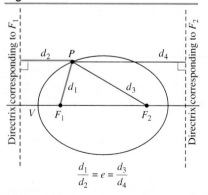

$$e = \frac{c}{a}$$

and the directrices are each $\frac{a}{e}$ units from the ellipse's center.

The condition $0 < e < 1$ guarantees that the directrices are "outside" the ellipse.

Since $c = ae$, **any two of a, b, c, and e determine the other two.** This information is important, especially for an "east-west" or a "north-south" ellipse, for if you know these constants and the center, you can write a standard form equation for the ellipse.

Ellipses

An ellipse's eccentricity e measures how flat (or round) the ellipse is. To see this, first calculate

$$e = \frac{c}{a} = \frac{\sqrt{a^2 - b^2}}{a} = \sqrt{1 - \frac{b^2}{a^2}} = \sqrt{1 - \left(\frac{b}{a}\right)^2}.$$

What does it mean for e to be close to 0? Then $\frac{b}{a}$ is close to 1, the major and minor axes are about the same length, and the ellipse is close to being a circle. At the other extreme, when e is close to 1, the minor axis is much smaller than the major axis, indicating a rather "flat" ellipse.

▶ **EXAMPLE 8.18** Find an equation for the ellipse with a focus at $(2, 1)$, corresponding directrix $x = -3$, and eccentricity $e = \frac{2}{3}$.

Solution A sketch of the given data is in figure 8.26(a).

One way to proceed would be to simplify the equation

$$\frac{\sqrt{(x-2)^2 + (y-1)^2}}{|x+3|} = \frac{2}{3}.$$

This reduces to

$$9(x-2)^2 + 9(y-1)^2 = 4(x+3)^2$$

$$5x^2 - 60x + 9(y-1)^2 = 0$$

$$5(x-6)^2 + 9(y-1)^2 = 180$$

$$\frac{(x-6)^2}{36} + \frac{(y-1)^2}{20} = 1.$$

You could now analyze the ellipse by reasoning as in examples 8.14 and 8.15. The result would be the sketch of this ellipse shown in figure 8.26(b).

We prefer the following alternative, more geometric solution. Since an ellipse's directrix is perpendicular to the major axis, the center is on the line $y = 1$. Denote the center's coordinates by $(h, 1)$. As $c = ae$, we have

$$h - 2 = ae = \tfrac{2}{3}a. \quad (*)$$

Since the directrix is $\dfrac{a}{e}$ units from the center,

$$h + 3 = \frac{a}{e} = \frac{3}{2}a. \quad (**)$$

You should have no trouble in solving $(*)$ and $(**)$ simultaneously, getting $h = 6$ and $a = 6$. Then

$$c = ae = 6(\tfrac{2}{3}) = 4$$

$$b = \sqrt{a^2 - c^2} = \sqrt{36 - 16} = 2\sqrt{5}$$

and we again find the standard form equation for this "east-west" ellipse as

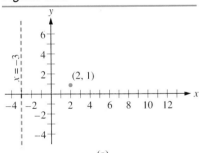

Figure 8.26

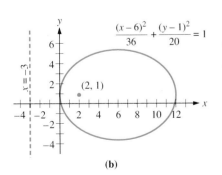

$$\frac{(x-6)^2}{36} + \frac{(y-1)^2}{20} = 1.$$

Applications

You may know from studying physics that, if one ignores air resistance, the path of any projectile is part of a parabola. Ellipses have different, and even more realistic, applications in the heavens. For instance, one of Kepler's Laws states that each planet in our solar system follows the path of an ellipse, with the sun as one of the foci. In a later course, you will be able to prove this fact, using the inverse-square law of universal gravitation, Newton's Second Law of Motion, and, of course, calculus. Exercises 50 and 51 will feature celestial applications of ellipses.

Another type of application for an ellipse, more properly for an **ellipsoid of revolution** generated by rigidly revolving an ellipse about one of its axes, is found in the famous "whispering gallery" in the statuary hall of the Capitol in Washington, D.C. As was the case with the applications of paraboloids mentioned in section 8.1, it follows from a geometric/physical fact. This states that energy rays emanating from (or passing through) one focus of an ellipse are all reflected in such a way that they then pass through the other focus of the ellipse. So, if you position yourself at one focus in a whispering gallery, you will be able to eavesdrop on everything said at the other focus, regardless of the direction in which speakers there address their remarks. Related applications will appear in exercises 52 and 53.

You can probably imagine other applications of the above geometric/physical fact, and indeed many exist. Perhaps the most interesting new application recently suggested involves the crushing of kidney stones: locate the stones at one focus, emit suitable energy from another focus, and let nature take its course.

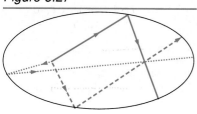

Figure 8.27

Exercises 8.2

In exercises 1–12, determine the given ellipse's center; length of each semimajor axis, a; length of each semiminor axis, b; c; foci; vertices; endpoints of minor axis; eccentricity, e; equations of directrices; and length of each latus rectum. Then sketch the ellipse.

1. $\dfrac{x^2}{64} + \dfrac{y^2}{9} = 1$
2. $10x^2 + 25y^2 = 250$
3. $30x^2 + 25y^2 = 750$
4. $\dfrac{x^2}{9} + \dfrac{y^2}{36} = 1$
5. $\dfrac{(x+3)^2}{64} + \dfrac{(y-4)^2}{81} = 1$
6. $\dfrac{(2x+4)^2}{64} + \dfrac{(3y+9)^2}{81} = 1$
7. $\dfrac{(x-4)^2}{81} + \dfrac{(y+3)^2}{64} = 2$
8. $\dfrac{(x-3)^2}{4} + \dfrac{(y+\sqrt{2})^2}{9} = 1$
9. $x^2 + 9y^2 + 2x - 36y + 28 = 0$
10. $81x^2 + 9y^2 + 54x - 54y - 639 = 0$
11. $16x^2 + 4y^2 - 16x + 12y - 51 = 0$
12. $4x^2 + 16y^2 - 16x + 8y = 47$

◆ 13. Use the techniques of section 2.2 to show that any ellipse is symmetric with respect to its major axis, its minor axis, and its center. (Hint: you may consider an "east-west" ellipse, centered at the origin.)

14. If an ellipse has foci F_1 and F_2, center C, and B an endpoint of the minor axis, prove that $BF_1 = BF_2 = a$, the length of a semimajor axis. (Hint: apply Pythagoras'

Ellipses

Theorem to $\triangle CBF_1$ and $\triangle CBF_2$, and use the definition of b.)

15. It was stated in the text that any two of the constants a, b, c, and e of an ellipse determine the other two. Prove, in particular, that $a = \dfrac{b}{\sqrt{1-e^2}}$.

◆ **16.** Prove the assertion in the text that the latera recta of an ellipse each have length $\dfrac{2b^2}{a}$. (Hint: you may consider $\dfrac{x^2}{a^2} + \dfrac{y^2}{b^2} = 1$. Show that this equation, in conjunction with either $x = c$ or $x = -c$, implies $y = \pm\dfrac{b^2}{a}$.)

In exercises 17–36, find an equation for the ellipse with the given properties.

17. Foci $(8, 9)$ and $(8, 1)$, vertex $(8, -1)$
18. Foci $(8, 9)$ and $(14, 9)$, vertex $(20, 9)$
19. Foci $(8, 9)$ and $(-14, 9)$; $(-3, 0)$ an endpoint of the minor axis
20. Foci $(8, 9)$ and $(8, 21)$; $(12, 15)$ an endpoint of the minor axis
21. Vertices $(-2, 0)$ and $(4, 0)$; $(1, 2)$ an endpoint of the minor axis
22. Vertices $(-2, 9)$ and $(-2, 3)$; $(-1, 6)$ an endpoint of the minor axis
23. Vertices $(-2, -9)$ and $(-2, -3)$, focus $(-2, -8)$
24. Vertices $(-2, -9)$ and $(8, -9)$, focus $(1, -9)$
25. Center $(2, \pi)$, focus $(2, 2\pi)$, vertex $(2, -\pi)$
26. Foci $(8, 9)$ and $(-10, 9)$, $a = 10$
27. Center $(2, -3)$, vertex $(6, -3)$, $b = 2$
28. Center $(2, -3)$, vertex $(2, -8)$, $b = 2$
29. Foci $(8, 9)$ and $(24, 9)$, $e = \dfrac{1}{2}$ (Hint: find $a = \dfrac{c}{e}$.)
30. Vertices $(2, -7)$ and $(2, -1)$, $c = 2$
31. Each point on the ellipse has the sum of its distances from $(-4, 8)$ and $(-4, -12)$ equal to 24.
32. Each point of the ellipse is $\dfrac{5}{3}$ as far from $y = 2$ as it is from $(-4, 1)$.
33. Vertex $(6, -3)$, directrices $x = 18$ and $x = -14$
34. Center $(0, 0)$, vertex $(3, 0)$, passing through $(2, 2)$
35. Focus $(2, 9)$, corresponding directrix $x = 5$, $e = \dfrac{3}{4}$
36. Focus $(-4, -1)$, corresponding directrix $y = -3$, $e = \dfrac{3}{7}$
37. Prove that the only line through $(3, 2)$ that intersects the ellipse $\dfrac{x^2}{18} + \dfrac{y^2}{8} = 1$ only once is the line $y = -\dfrac{2}{3}x + 4$. (Hint: review exercise 35 in section 8.1.) We call $y = -\dfrac{2}{3}x + 4$ the *tangent line* to the given ellipse *at* $(3, 2)$.
38. Find an equation for the tangent line to $18x^2 + 8y^2 - 144 = 0$ at $(2, 3)$. (Hint: review exercise 37.)

◆ **39.** Prove that $\{(x, y)$: there exists a real number θ such that $x = 1 + 3\sin\theta$, $y = 2\cos\theta$, and $0 \le \theta < 2\pi\}$ is the ellipse $\dfrac{(x-1)^2}{9} + \dfrac{y^2}{4} = 1$. (Hint: review the hint for exercise 39 in section 8.1 and apply Pythagoras.)

◆ **40.** (a) Explain why restricting θ in exercise 39 to lie in the interval $[0, \pi]$ converts the given set into only the right-hand half of the stated ellipse.
   (b) If, toward the other extreme, θ is allowed to be any real number, is the converted set then "more" than just the stated ellipse?

41. (Compare with exercise 45 in section 8.1.) Consider all the lines of slope 2 that intersect the ellipse $4x^2 + 9y^2 = 36$ in more than one point. For each such line ℓ, consider the midpoint of the line segment connecting the (two) points of intersection of ℓ and the ellipse. Show that all these midpoints lie on the line $y = -\dfrac{2}{9}x$.

◆ **42.** Consider a standard form "east-west" ellipse, $\dfrac{x^2}{a^2} + \dfrac{y^2}{b^2} = 1$. As a is kept fixed, let b "approach" 0 by assuming smaller and smaller positive values. (Each value of b gives rise to a new ellipse.)
   (a) What is happening to the distance between the foci during this limiting process?
   (b) What is happening to the ellipse's eccentricity e?
   (c) What is happening to the shape of the ellipse? (Experiments of this kind, indeed all graphs involving conic sections, may be aided by the use of graphing calculators or computer graphing programs.)

◆ **43.** Subject the "east-west" ellipse in exercise 42 to a different limiting process, by fixing b and letting a increase without bound (that is, by letting a "approach" infinity). During this limiting process, what is happening to (a) the distance between the foci, (b) the eccentricity e, and (c) the shape of the ellipse? (For help in computing the limiting tendency of y, review the hint for exercise 57 in section 3.6. Your answer explains how a pair of parallel lines arises practically as a degenerate case of an ellipse.)

◆ **44.** (a) In the spirit of exercises 42 and 43, subject the parabola $y^2 = 4px$ to the limiting process in which p approaches zero. Explain how this allows us to view a line as a degenerate case of a parabola.
   (b) If p instead increases without bound, what is the limiting figure?
   (c) Are your answers to (a) and (b) confirmed if you "check" using a graphing calculator or computer graphing program?

◆ **45.** Show that the ellipse with focus $(3, 0)$, corresponding directrix $y = 2x$, and eccentricity $e = \dfrac{1}{2}$ has equation

$16x^2 + 19y^2 + 4xy - 120x + 180 = 0$. (Hint: review the hint for exercise 47 in section 8.1.)

46. Suppose that an ellipse has focus $(0, 0)$, corresponding directrix $y = 2x - 3$, and eccentricity $\frac{2}{7}$. Find the vertex situated between the given focus and directrix. (Hint: this should remind you of exercise 48 in section 3.1, but this is harder. First, observe that the major axis lies along $y = -\frac{x}{2}$, and hence intersects the given directrix at $P(\frac{6}{5}, -\frac{3}{5})$. Next, argue that $\frac{a}{e} - ae = \frac{3}{5}$, yielding $a = \frac{42}{225}$. Finally, use exercise 32 of section 1.2 to locate the point of $y = -\frac{x}{2}$ that is $\frac{a - ae}{\frac{3}{5}}$ of the way from the given focus to P.)

47. Graph (a) $y^2 = \frac{4}{9}(9 - x^2)$, (b) $y = \frac{2}{3}\sqrt{9 - x^2}$, and (c) $y = -\frac{2}{3}\sqrt{9 - x^2}$. (Hint: the answers are all different.)

♦ 48. Verify the alternative definition of ellipse by proving the following. Let a and e be positive numbers, with $0 < e < 1$. Consider the point $F(ae, 0)$ and the line ℓ given by $x = \frac{a}{e}$. Let $P(x, y)$ be a point whose distance from F is d_1 and whose distance from ℓ is d_2. If $\frac{d_1}{d_2} = e$, then $\frac{x^2}{a^2} + \frac{y^2}{a^2(1 - e^2)} = 1$.

49. Let F be a focus of an ellipse.
    (a) Prove that the point on the given ellipse that is closest to F is the vertex that is nearer to F.
    (b) What is the point on the given ellipse that is farthest from F? (Hint: you may consider the ellipse $\frac{x^2}{a^2} + \frac{y^2}{b^2} = 1$. You are seeking to minimize [in (a)] or maximize [in (b)] the expression
    $$(x - c)^2 + y^2 = \frac{c^2}{a^2}x^2 - 2cx + a^2,$$
    considered for $-a \leq x \leq a$. For this, review section 2.3.)

♦ 50. Two amateur scientists, David and John, launch a satellite, SkyMathLab I, which goes into an elliptic orbit, with the Earth at one focus. If the closest that SkyMathLab I comes to the Earth is $P = 80$ miles and the farthest that it is ever from the Earth is $A = 110$ miles, prove that the eccentricity of its orbit is $e = \frac{A - P}{A + P} = \frac{3}{19}$. (Hint: ignore the diameter of the Earth. Use exercise 49 to show that $P = a - ae$ and $A = a + ae$.)

♦ 51. Find the Earth's nearest distance to the Sun if the Earth's orbit has semimajor axis length $a = 92$ million miles and eccentricity $e = 0.017$. (Hint: see P in exercise 50.)

♦ 52. A rogue scientist in Lavir decides to destroy a rival's laboratory 200 miles away by bombarding it with certain energy rays. Since he cannot control the direction of these rays very well, he decides to bounce them off a secret semielliptic surface, of eccentricity $\frac{2}{3}$, with the scientists at the foci. Explain why the greatest height of this surface is approximately 111.8 miles. (Hint: determine $c = 100$ and $a = 150$. Then find b.)

♦ 53. David wishes to spy on John, who is talking to Dwala in a whispering gallery whose highest point is 120 feet above the floor. David positions himself at the center, John and Dwala at one focus 60 feet away, and a small but sensitive tape recorder at the other focus. Show that the eccentricity of the ellipse whose revolution generated the gallery is $e = \frac{1}{\sqrt{5}}$. (Hint: $b = 120$ and $c = 60$.)

54. A bridge spans a river that is 25 meters wide. The bridge is supported by an arch in the shape of the upper half of an "east-west" ellipse. The highest point of the arch is 10 meters above the water.

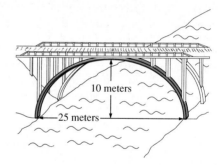

    (a) Determine the eccentricity of the ellipse.
    (b) The Human Fly is walking on the bridge. When she is 5 meters from the middle of the bridge, she falls off, lands on the supporting arch, and somehow manages to hold on. How far did she fall?

    (Hint: for (a), use a and b to find e. For (b), find an equation for the ellipse, calculate y corresponding to $x = 5$, and consider $b - y$.)

55. Two ellipsoids may be formed by rigidly revolving a given ellipse, either about its major axis or about its minor axis. One of the ellipsoids is shaped like an egg or an American football. The other ellipsoid is shaped somewhat like the Earth. Which revolution produced which ellipsoid?

8.3 HYPERBOLAS

The final conic section to be studied is the hyperbola. Its definition is almost identical to that of the ellipse, but the one change in the definition causes quite a difference in the resulting figure.

Definitions

A **hyperbola** is the set of points the absolute value of the difference of whose distances from two given points, the **foci**, is a given constant; this constant is less than the distance between the foci. Each of the two given points is called a **focus** of the hyperbola.

Figure 8.28

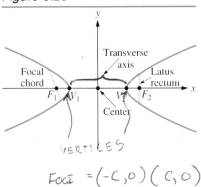

Figure 8.28 shows a hyperbola with foci F_1 and F_2. Its two disconnected pieces are called its **branches**. The line through its foci intersects the hyperbola in two points called the **vertices**. The segment joining the vertices is called the **transverse axis**, and its midpoint is the **center**. It can be shown that the center is also midway between the foci. In exercise 13, you will be asked to show that any hyperbola is symmetric with respect to its transverse axis and its center. Any chord of the hyperbola that contains a focus is a **focal chord** of the hyperbola. A **focal radius** is a segment with one endpoint at a focus and the other endpoint on the hyperbola.

Standard Form with Center at the Origin

A standard form equation for a hyperbola can be derived from the definition. We shall consider a hyperbola with center at the origin, with transverse axis along the x-axis. Then the foci are $F_1(-c, 0)$ and $F_2(c, 0)$ for some positive number c. Let $2a$ be the constant difference of the lengths of the focal radii for points on the hyperbola. From the definition, we have $2a < |c - (-c)| = 2c$, or $a < c$. Now, consider any point $P(x, y)$ and d_1, d_2 the focal radii of P, as shown in figure 8.29. Then

Figure 8.29

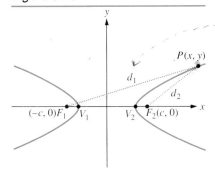

$$d_1 = d(P, F_1) = \sqrt{(x+c)^2 + y^2} \quad \text{and} \quad d_2 = d(P, F_2) = \sqrt{(x-c)^2 + y^2}.$$

The condition that P lie on the given hyperbola is $d_1 - d_2 = \pm 2a$; that is, $|d_1 - d_2| = 2a$, or

$$|\sqrt{(x+c)^2 + y^2} - \sqrt{(x-c)^2 + y^2}| = 2a.$$

Simplifying this in the same way we simplified the corresponding formula for ellipses in section 8.2, we obtain

$$(c^2 - a^2)x^2 - a^2 y^2 = a^2(c^2 - a^2)$$

or

$$\frac{x^2}{a^2} - \frac{y^2}{c^2 - a^2} = 1.$$

Since $0 < a < c$, we can define a positive real number b by

$$b = \sqrt{c^2 - a^2}.$$

Chapter 8 / Conic Sections and the General Second-Degree Equation in Two Variables

✱ A is always under the positive #

(Note: This is *not* the same relation that held for a, b, and c in an ellipse! In particular, $c = \sqrt{a^2 + b^2}$ exceeds both a and b. Moreover, there is no restriction on which of a and b is greater, and they may actually be equal.) Then $c^2 - a^2 = b^2$ and the above equation becomes a standard form equation for a hyperbola.

Standard Form Equation of an "East-West" Hyperbola

The standard form equation of a hyperbola with center at the origin and horizontal transverse axis is

$$\frac{x^2}{a^2} - \frac{y^2}{b^2} = 1.$$

IN OTHER WORDS
1) VERTICES $(a, 0)(-a, 0)$
2) TRAN. AXIS LENGTH $2a$
3) SEMI TRAN. AXIS
4) CONJUGATE AXIS $(0, b), (0, -b)$ (end pts)
5) LENGTH of C.A. $2b$
6) SEMI CON. AXIS

   ✱ $b = \sqrt{c^2 - a^2}$

7) $C = \sqrt{a^2 + b^2}$

8) FOCI $(-c, 0)(c, 0)$

Setting $y = 0$ in the preceding equation leads to $x = \pm a$, and so the vertices of the above horizontal or "east-west" hyperbola are $(a, 0)$ and $(-a, 0)$. The **length of the transverse axis** is thus $2a$. Each segment of length a running from the center to a vertex is a **semitransverse axis**. The segment with endpoints $(0, b)$ and $(0, -b)$ is called the **conjugate axis**; notice that the center is its midpoint. The **length of the conjugate axis** is thus $2b$. Notice that the hyperbola does not intersect the conjugate axis. A **semiconjugate axis** is b units long and runs from the center to an endpoint of the conjugate axis.

The above standard form equation, $\frac{x^2}{a^2} - \frac{y^2}{b^2} = 1$, is equivalent to $y = \pm\frac{b}{a}\sqrt{x^2 - a^2}$. What do the equations mean geometrically? Consider the functions f_1 and f_2 given by

$$f_1(x) = \frac{b}{a}\sqrt{x^2 - a^2} \quad \text{and} \quad f_2(x) = -\frac{b}{a}\sqrt{x^2 - a^2}.$$

Both f_1 and f_2 have natural domain $(-\infty, -a] \cup [a, \infty)$. The graph of f_1 is the union of the upper halves of the branches of the given hyperbola, and the graph of f_2 is the union of the lower halves of these branches.

You may have been disappointed that the graphs of f_1 and f_2 were not the separate branches of the hyperbola. But that would have been unreasonable. For instance, the right-hand branch is not the graph of **any** function of x, since it fails to pass the Vertical Line Test (from section 2.1). However, this branch is the graph of an equation, specifically the graph of $x = \frac{a}{b}\sqrt{b^2 + y^2}$; and the left-hand branch is the graph of the equation $x = -\frac{a}{b}\sqrt{b^2 + y^2}$. Moreover, the right-hand branch consists of the points that satisfy $d_1 - d_2 = 2a$, while the left-hand branch's points satisfy $d_1 - d_2 = -2a$, as in figure 8.30.

What about "north-south" hyperbolas? They, too, have standard form equations, derived by modifying the above reasoning. Consider a "north-south" hyperbola centered at the origin, with vertical transverse axis. In this case, the foci are $(0, c)$ and $(0, -c)$; the vertices are $(0, a)$ and $(0, -a)$; and the endpoints of the conjugate axis are $(b, 0)$ and $(-b, 0)$, where, as before, $b = \sqrt{c^2 - a^2}$.

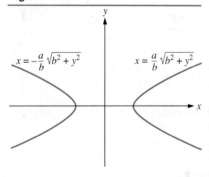

Figure 8.30

Hyperbolas

Standard Form Equation of a "North-South" Hyperbola

(handwritten margin notes:)
1) FOCI $(0,c),(0,-c)$
2) VERTICES $(0,a),(0,-a)$
3) END PTS. of C.A. $(b,0),(-b,0)$
   * $b = \sqrt{c^2-a^2}$

The standard form equation of a hyperbola with center at the origin and vertical transverse axis is

$$\frac{y^2}{a^2} - \frac{x^2}{b^2} = 1.$$

EXAMPLE 8.19

Find the standard form equation of the hyperbola with center at the origin, one focus at (0, 9), and a vertex at (0, –5). Also locate the other focus, the other vertex, and the endpoints of the conjugate axis. Then graph this hyperbola.

Solution

Since foci and vertices lie on the transverse axis, this is a "north-south" hyperbola. As the center is the origin, we see from the given information that $c = 9$ and $a = 5$. Now, $b^2 = c^2 - a^2 = 81 - 25 = 56$ and so $b = \sqrt{56} = 2\sqrt{14}$. We can now conclude that the other focus is at $(0, -c) = (0, -9)$; the other vertex is $(0, a) = (0, 5)$; and the endpoints of the conjugate axis are $(b, 0)$ and $(-b, 0)$, namely $(2\sqrt{14}, 0)$ and $(-2\sqrt{14}, 0)$. The standard form equation for this hyperbola is

$$\frac{y^2}{25} - \frac{x^2}{56} = 1.$$

(handwritten note:) "x" here is negative, therefore ↖↗ was used.

Its graph is in figure 8.31.

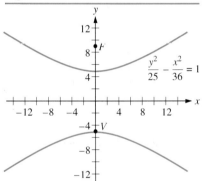

Figure 8.31

We shall next explain how each hyperbola has two asymptotes, which intersect at the hyperbola's center. You may want to review the limiting arguments in exercise 57 of section 3.6. In that discussion, you learned that $y = 3x$ is an asymptote for the function given by $f(x) = 3\sqrt{x^2 - 4}$. The point was that $f(x) - 3x$ has limiting tendency 0 as x increases without bound. Now, by the above comments, the graph of f is the upper half of the hyperbola $\frac{x^2}{4} - \frac{y^2}{36} = 1$, and the lower half is the graph of the function g given by $g(x) = -3\sqrt{x^2 - 4}$. Exactly as in the above cited exercise in section 3.6, one can show that $y = 3x$ is also an asymptote for the graph of g. Indeed, the limiting tendency of $g(x) - 3x$ is 0 as x approaches negative infinity—that is, as x decreases without bound. This means that the lines $y = 3x$ and $y = -3x$ are asymptotes for the hyperbola $\frac{x^2}{4} - \frac{y^2}{36} = 1$. By exactly the same sort of reasoning, one can prove the following general statement.

(handwritten note:)
* remember: a. if "x" is positive the arrows ↘↗ or hyperbola is drawn like what was written.
  b. if "x" is negative then; ↖↗ is used

Asymptotes for an "East-West" Hyperbola

The lines $y = \frac{b}{a}x$ and $y = -\frac{b}{a}x$ are **asymptotes for the "east-west" (horizontal) hyperbola** $\frac{x^2}{a^2} - \frac{y^2}{b^2} = 1.$

(handwritten note:) positive term is x^2

The following is a quick way to obtain the asymptotes' equations. Replace the right-hand side in the above hyperbola's standard form equation with 0. The result is $\frac{x^2}{a^2} - \frac{y^2}{b^2} = 0$. This is algebraically equivalent to $y = \pm\frac{b}{a}x$, the equations of the hyperbola's asymptotes. The same method works in the "north-south" case.

Asymptotes for a "North-South" Hyperbola

The lines $y = \frac{a}{b}x$ and $y = -\frac{a}{b}x$ are **asymptotes for the "north-south" (vertical) hyperbola** $\frac{y^2}{a^2} - \frac{x^2}{b^2} = 1$.

Figure 8.32

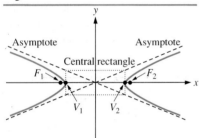

We next introduce a very helpful guide in sketching a hyperbola. Consider a hyperbola, say "east-west," as in figure 8.32. If lines perpendicular to the transverse axis are drawn through the vertices and lines perpendicular to the conjugate axis are drawn through the endpoints of the conjugate axis, a **central rectangle** is formed. The diagonal lines of the central rectangle extend to give the asymptotes of the hyperbola. When you have the asymptotes and the vertices, you can roughly sketch the hyperbola. The extent of "roughness" depends on how quickly the hyperbola approaches its asymptotes as $|x|$ gets large.

▶ **EXAMPLE 8.20**

Sketch the graph of $\frac{x^2}{25} - \frac{y^2}{4} = 1$.

Solution

The given equation takes the form $\frac{x^2}{a^2} - \frac{y^2}{b^2} = 1$. Its graph is therefore an "east-west" hyperbola, with center at the origin. Since $a^2 = 25$ and $b^2 = 4$, we have $a = 5$ and $b = 2$. Thus, the vertices are $(5, 0)$ and $(-5, 0)$, and the endpoints of the conjugate axis are $(0, 2)$ and $(0, -2)$. These four points are plotted in figure 8.33(a). They allow the central rectangle and the asymptotes $y = \pm\frac{2}{5}x$ to be drawn. Using the vertices and these asymptotes as guides, we get the hyperbola sketched in figure 8.33(b). ◀

Figure 8.33

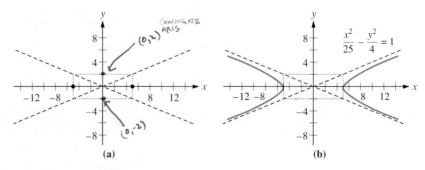

Hyperbolas

EXAMPLE 8.21 Sketch the graph of $64y^2 - 36x^2 = 2304$.

Solution We put this in standard form by dividing by 2304, to obtain

$$\frac{y^2}{36} - \frac{x^2}{64} = 1.$$

This equation takes the form $\frac{y^2}{a^2} - \frac{x^2}{b^2} = 1$. Its graph is therefore a "north-south" hyperbola, with center at the origin. Since $a^2 = 36$ and $b^2 = 64$, we can now conclude that its vertices are $(0, 6)$ and $(0, -6)$ and the endpoints of its conjugate axis are $(8, 0)$ and $(-8, 0)$. Using this information, we sketch the central rectangle and asymptotes $y = \pm\frac{3}{4}x$ in figure 8.34(a). We then get the graph of the hyperbola in figure 8.34(b).

Figure 8.34

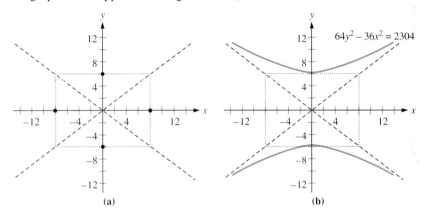

Standard Form Equations for Translated Hyperbolas

If the center of a hyperbola is not at the origin, then, just as in the preceding two sections, we can use the translation methods of section 2.2 to find standard form equations for the hyperbola. This results in table 8.3.

Table 8.3

Standard Form Equation*	Center	Foci	Vertices	Endpoints of Conjugate Axis	Orientation
$\dfrac{(x-h)^2}{a^2} - \dfrac{(y-k)^2}{b^2} = 1$	(h, k)	$(h+c, k)$ $(h-c, k)$	$(h+a, k)$ $(h-a, k)$	$(h, k+b)$ $(h, k-b)$	"east-west" with horizontal transverse axis
$\dfrac{(y-k)^2}{a^2} - \dfrac{(x-h)^2}{b^2} = 1$	(h, k)	$(h, k+c)$ $(h, k-c)$	$(h, k+a)$ $(h, k-a)$	$(h+b, k)$ $(h-b, k)$	"north-south" with vertical transverse axis

*The constants a, b, and c are positive and satisfy $c^2 = a^2 + b^2$.

Chapter 8 / Conic Sections and the General Second-Degree Equation in Two Variables

▶ **EXAMPLE 8.22**

Sketch the graph of $(x - 4)^2 - \dfrac{(y - 2)^2}{16} = 1$. Find its foci and equations for its asymptotes.

Solution

The given equation takes the form $\dfrac{(x - h)^2}{a^2} - \dfrac{(y - k)^2}{b^2} = 1$, with $h = 4$, $k = 2$, $a = 1$, and $b = 4$. Its graph is an "east-west" hyperbola, with center at $(4, 2)$, vertices at $(5, 2)$ and $(3, 2)$, and endpoints of its conjugate axis at $(4, 6)$ and $(4, -2)$. Plot the center, draw the central rectangle, and extend its diagonals to obtain the asymptotes. These are the lines $y - 2 = \pm 4(x - 4)$. Now, use this information to draw the hyperbola in figure 8.35. Finally, to find the foci, rewrite the definition of $b = \sqrt{c^2 - a^2}$ to get

$$c = \sqrt{a^2 + b^2} = \sqrt{1 + 16} = \sqrt{17}.$$

Hence, as in table 8.3, the foci are $(4 + \sqrt{17}, 2)$ and $(4 - \sqrt{17}, 2)$. ◀

Figure 8.35

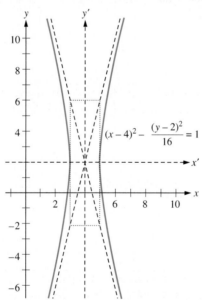

▶ **EXAMPLE 8.23**

Sketch the graph of $16x^2 - 9y^2 - 32x + 36y + 124 = 0$.

Solution

As in the preceding two sections, we must first find the equivalent standard form equation, and this is done by completing the square.

$$16x^2 - 9y^2 - 32x + 36y + 124 = 0$$

$$16(x^2 - 2x) - 9(y^2 - 4y) = -124$$

$$16(x^2 - 2x + 1) - 9(y^2 - 4y + 4) = -124 + 16 - 36$$

$$16(x - 1)^2 - 9(y - 2)^2 = -144$$

Hyperbolas

Figure 8.36

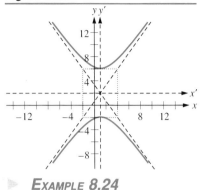

$$-\frac{(x-1)^2}{9} + \frac{(y-2)^2}{16} = 1$$

$$\frac{(y-2)^2}{16} - \frac{(x-1)^2}{9} = 1.$$

The equation is now in standard form for a "north-south" hyperbola, with center at (1, 2), $a = 4$, and $b = 3$. Using this information, we sketch the central rectangle and the asymptotes, $y - 2 = \pm\frac{4}{3}(x - 1)$. This allows us to draw the hyperbola, as shown in figure 8.36. ◀

EXAMPLE 8.24

Find an equation of the hyperbola with center at (1, –3), a focus at (1, –5), and a semiconjugate axis of length $\sqrt{2}$.

Solution

Since the center and focus are both on the line $x = 1$, so is the transverse axis. Hence, this is a "north-south" hyperbola, and so the desired equation has the form

$$\frac{(y+3)^2}{a^2} - \frac{(x-1)^2}{b^2} = 1.$$

Now, we are given $b = \sqrt{2}$. Moreover, c is the distance from the center to a focus, and so $c = 2$. Also, the definition of $b = \sqrt{c^2 - a^2}$ can be rewritten to give

$$a = \sqrt{c^2 - b^2} = \sqrt{4 - 2} = \sqrt{2}.$$

Thus, this hyperbola has the equation

$$\frac{(y+3)^2}{2} - \frac{(x-1)^2}{2} = 1.$$

◀

Example 8.24 presents a situation that we indicated earlier was possible, namely a hyperbola for which $a = b$. We have a name (actually, two names) for such a hyperbola. Any hyperbola in which the transverse axis has the same length as the conjugate axis (that is, for which $a = b$) is called an **equilateral hyperbola**. A hyperbola has perpendicular asymptotes if and only if it is an equilateral hyperbola. Accordingly, an even more popular name for a hyperbola with $a = b$ is a **rectangular hyperbola.**

In section 8.2, we gave an alternative definition of an ellipse in terms of eccentricity. A similar approach to hyperbolas provides the following alternative definition.

Eccentricity of a Hyperbola

A hyperbola is the set of all points the ratio of whose distances from a given point (focus) and from a given line (directrix) is a given constant e (eccentricity) such that $e > 1$.

Figure 8.37

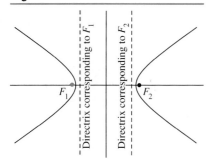

Each hyperbola has two foci, and each focus has a corresponding directrix. You will essentially show in exercise 54 that

$$e = \frac{c}{a}$$

and the directrices are each $\frac{a}{e}$ units from the hyperbola's center.

The condition $e > 1$ guarantees that $a > \frac{a}{e}$. Geometrically, this means that the vertices are farther from the center than are the directrices. In particular, the directrices are located between the branches of the hyperbola, as in figure 8.37.

Since $c = ae$, you can verify that **any two of a, b, c, and e determine the other two.** Thus, you can write down a standard form equation for a hyperbola with either a horizontal or a vertical transverse axis, if you know its center and any two of these constants.

A hyperbola's eccentricity e may be any real number greater than 1. The larger the value of e, the more widely the branches of the hyperbola open (and the "fatter" the hyperbola). For example, figure 8.38 shows two hyperbolas, both centered at the origin, with identical horizontal transverse axes. The "fatter" hyperbola has an eccentricity of 3, and the other hyperbola's eccentricity is 1.1.

The nature of conic sections in terms of eccentricity, e, is summarized in the following chart.

Figure 8.38

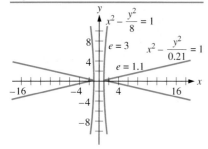

Ellipse	Parabola	Hyperbola
$0 < e < 1$	$e = 1$	$e > 1$

▶ **EXAMPLE 8.25**

Find an equation for the hyperbola with focus $(2, 4)$, corresponding directrix $y = -1$, and eccentricity $e = \frac{3}{2}$.

Solution

A sketch of the data is in figure 8.39(a).

As in example 8.18, one way to proceed would be to simplify the equation

$$\frac{\sqrt{(x-2)^2 + (y-4)^2}}{|y+1|} = \frac{3}{2}.$$

We leave it to you to reduce this to

$$(x-2)^2 + (y-4)^2 = \tfrac{9}{4}(y+1)^2$$

$$4(x-2)^2 + 4(y-4)^2 = 9(y+1)^2$$

$$4(x-2)^2 - 5y^2 - 50y = -55$$

$$4(x-2)^2 - 5(y+5)^2 = -180$$

$$\frac{4(x-2)^2}{-180} - \frac{5(y+5)^2}{-180} = 1$$

$$\frac{(y+5)^2}{36} - \frac{(x-2)^2}{45} = 1.$$

You could now analyze the hyperbola by reasoning as in examples 8.22 and 8.23. The result would be the sketch of this hyperbola shown in figure 8.39(b).

Figure 8.39

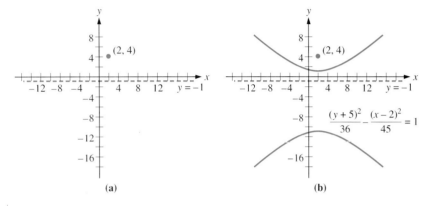

We prefer the following approach. Since the given directrix must be perpendicular to the transverse axis, the center is on the line $x = 2$. Denote the center's coordinates by $(2, k)$. As $c = ae$, we have

$$4 - k = ae = \tfrac{3}{2}a. \qquad (*)$$

Since the directrix is $\dfrac{a}{e}$ units from the center,

$$-k - 1 = \frac{a}{e} = \frac{2}{3}a. \qquad (**)$$

Solving $(*)$ and $(**)$ simultaneously, we get $a = 6$ and $k = -5$. Then

$$c = ae = 6(\tfrac{3}{2}) = 9$$

$$b = \sqrt{c^2 - a^2} = \sqrt{81 - 36} = \sqrt{45}$$

and we again find the above equation for this "north-south" hyperbola. ◀

Applications

Figure 8.40

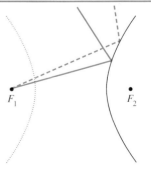

Like the other conic sections, hyperbolas also find application to the real world. Some of these applications arise from a physical/geometric fact which is depicted in figure 8.40. Rays aimed at F_2 from outside a hyperbolic reflector are reflected so as to pass through F_1. On the other hand, rays emitted from F_1 are

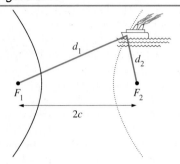

Figure 8.41

reflected so as to **appear** to have originated from F_2. Exercise 57 will feature such an application.

Another type of application for hyperbolas is typified by the navigational system of LORAN C. Suppose that a ship sends a radio signal which is picked up at different times t_1 and t_2 by listening posts F_1 and F_2, respectively. If the ship's distances from F_1 and F_2 are d_1 and d_2, respectively, and $d_1 > d_2$, we can use the known speed v of radio signals and the perceived time lag $t_1 - t_2$ to calculate $d_1 - d_2 = v(t_1 - t_2)$. Then, if $2a$ denotes this constant, the ship is on the branch $d_1 - d_2 = 2a$ of a hyperbola. (See figure 8.41.) Using an additional listening post, we can then locate the ship at the intersection of branches of three different hyperbolas. Exercises 59 and 60 concern a simple situation of this type for which it is feasible to algebraically work out the ship's coordinates. A more realistic situation, and one whose analysis is more algebraically demanding, is given in exercise 58 of section 8.5.

Exercises 8.3

In exercises 1–12, determine the given hyperbola's center; length of each semitransverse axis, a; length of each semiconjugate axis, b; c; foci; vertices; endpoints of the conjugate axis; equations of asymptotes; eccentricity, e; and equations of directrices. Then sketch the hyperbola.

1. $\dfrac{x^2}{64} - \dfrac{y^2}{9} = 1$
2. $25x^2 - 10y^2 = 250$
3. $30y^2 - 25x^2 = 750$
4. $\dfrac{y^2}{36} - \dfrac{x^2}{9} = 1$
5. $\dfrac{(y+4)^2}{81} - \dfrac{(x-3)^2}{64} = 1$
6. $\dfrac{(2x-4)^2}{64} - \dfrac{(3y-9)^2}{81} = 1$
7. $\dfrac{(x-4)^2}{64} - \dfrac{(y+3)^2}{81} = 2$
8. $\dfrac{(x+3)^2}{9} - \dfrac{(y-\sqrt{2})^2}{4} = -1$
9. $4x^2 - y^2 + 16x + 2y + 51 = 0$
10. $x^2 - 9y^2 + 2x + 36y - 71 = 0$
11. $16x^2 - 9y^2 - 16x - 36y - 68 = 0$
12. $4x^2 - 8y^2 + 4x - 24y - 16 = 0$

◆ 13. Use the techniques of section 2.2 to show that any hyperbola is symmetric with respect to its transverse axis, its conjugate axis, and its center. (Hint: review the hint for exercise 13 in section 8.2.)

For exercises 14–18, you will need the following definition. A **latus rectum** of a hyperbola is a chord of the hyperbola that passes through a focus and is perpendicular to the transverse axis.

14. If a hyperbola has a semitransverse axis length a and semiconjugate axis length b, prove that each latus rectum of the hyperbola has length $\dfrac{2b^2}{a}$. (Hint: you may consider $\dfrac{x^2}{a^2} - \dfrac{y^2}{b^2} = 1$. Show that if x is either c or $-c$, then $y = \pm\dfrac{b^2}{a}$.)

In exercises 15–18, use the formula in exercise 14 to find the length of a latus rectum of the given hyperbola.

15. $\dfrac{x^2}{8} - \dfrac{y^2}{16} = 1$
16. $\dfrac{x^2}{16} - \dfrac{y^2}{8} = 1$
17. $9x^2 - 4y^2 + 18x + 16y + 29 = 0$
18. $4x^2 - 9y^2 + 8x + 36y - 31 = 0$

In exercises 19–44, find an equation for the hyperbola with the given properties.

19. Foci (8, 9) and (8, 1), vertex (8, 6)
20. Foci (8, 9) and (14, 9), vertex (12, 9)
21. Foci (8, 9) and (−14, 9); (−3, 19) an endpoint of the conjugate axis
22. Foci (8, 9) and (8, 21); (6, 15) an endpoint of the conjugate axis
23. Foci (8, 9) and (14, 9), asymptote $y - 9 = \tfrac{2}{3}(x - 11)$
    (Hint: use $c^2 = a^2\left(1 + \left(\dfrac{b}{a}\right)^2\right)$ to find a.)
24. Foci (8, 9) and (8, 1), asymptote $y - 5 = -\tfrac{3}{2}(x - 8)$
    (Hint: review the hint for exercise 23.)
25. Vertices (8, 9) and (8, 1), asymptote $y - 5 = -\tfrac{2}{3}(x - 8)$
    (Hint: find b.)
26. Vertices (8, 9) and (14, 9), asymptote $y - 9 = \tfrac{3}{2}(x - 11)$
27. Vertices (−4, 0) and (2, 0); (−1, −4) an endpoint of the conjugate axis

Hyperbolas

28. Vertices $(-4, 9)$ and $(-4, 3)$; $(-6, 6)$ an endpoint of the conjugate axis
29. Vertices $(-4, 9)$ and $(-4, -3)$, focus $(-4, -5)$
30. Vertices $(-4, 9)$ and $(-14, 9)$, focus $(-15, 9)$
31. Center $(2, \pi)$, focus $(2, 2\pi)$, vertex $\left(2, \dfrac{\pi}{3}\right)$
32. Foci $(8, 9)$ and $(-10, 9)$, $a = 8$
33. Center $(2, -3)$, vertex $(6, -3)$, $b = 2$
34. Center $(2, -3)$, vertex $(2, -8)$, $b = 6$
35. Foci $(8, 9)$ and $(24, 9)$, $e = 2$ (Hint: find $a = \dfrac{c}{e}$.)
36. Vertices $(2, -7)$ and $(2, -1)$, $c = 4$
37. Each point on the hyperbola has the difference of its distances from $(-4, 8)$ and $(-4, -12)$ equal to 16.
38. The distance from each point on the hyperbola to $(-4, 1)$ is $\dfrac{5}{3}$ times its distance from $y = 2$. (Hint: review example 8.25.)
39. Center $(0, 0)$, vertex $(4, 0)$; $(-8, 2\sqrt{3})$ a point on the hyperbola. (Hint: find b.)
40. Vertices $(0, 8)$ and $(0, -2)$; $\left(\dfrac{-\sqrt{11}}{15}, 9\right)$ a point on the hyperbola (Hint: find center, equation, and b.)
41. Vertex $(20, -3)$, directrices $x = 18$ and $x = -14$ (Hint: find a, center and $\dfrac{a}{e}$.)
42. Center $(0, 0)$, asymptote $y = \dfrac{3}{5}x$; $\left(\dfrac{50}{3}, -8\right)$ a point on the hyperbola, transverse axis parallel to a coordinate axis
43. Focus $(2, 9)$, corresponding directrix $x = 5$, $e = \dfrac{4}{3}$
44. Focus $(-4, -1)$, corresponding directrix $y = -3$, $e = \dfrac{7}{3}$
45. Show that the only line through $(3\sqrt{6}, 4)$ that intersects the hyperbola $\dfrac{x^2}{18} - \dfrac{y^2}{8} = 1$ only once and is not parallel to an asymptote is the line $y = \dfrac{2x}{\sqrt{6}} - 2$. (Hint: review exercise 35 in section 8.1.) We call $y = \dfrac{2x}{\sqrt{6}} - 2$ the *tangent line to* the given hyperbola *at* $(3\sqrt{6}, 4)$.
46. Find an equation for the tangent line to $x^2 - y^2 - 1 = 0$ at $(3, 2\sqrt{2})$. (Hint: review exercise 45.)
◆ 47. Suppose there exist real numbers c, d, and θ such that $c = 2 \sec \theta$ and $d = -2 + 3 \tan \theta$.
   (a) Prove that (c, d) lies on the hyperbola
   $$\dfrac{x^2}{4} - \dfrac{(y+2)^2}{9} = 1.$$ (Hint: review the trigonometric identities in section 6.1.)
   (b) Suppose that $\dfrac{\pi}{2} < \theta < \pi$. Show that (c, d) is on the lower half of the hyperbola. Which branch of the hyperbola is (c, d) on? (Hint: $\dfrac{S\,|\,A}{T\,|\,C}$)

◆ 48. Subject a standard form "east-west" hyperbola $\dfrac{x^2}{a^2} - \dfrac{y^2}{b^2} = 1$ to a limiting process in which a is kept fixed and b increases without bound. (In other words, let b "approach" infinity.) During this limiting process, what is happening to (a) the distance between the foci; (b) the eccentricity, e; and (c) the shape of the hyperbola? (Your answer explains how a pair of parallel lines arises practically as a degenerate case of a hyperbola.)
49. (Compare with exercise 45 in section 8.1 and exercise 41 in section 8.2.) Consider all the lines of slope 2 that intersect the hyperbola $4x^2 - 9y^2 = 36$ in more than one point. For each such line ℓ, consider the midpoint of the line segment connecting the (two) points of intersection of ℓ and the hyperbola. Show that all these endpoints lie on the line $y = \dfrac{2}{9}x$.
50. (a) Arguing about intuitive "limiting values" as in exercise 57 of section 3.6, explain why the upper branch of the hyperbola $4x^2 - y^2 = -4$ approaches the line $y = 2x$ asymptotically as x approaches infinity, but not as x approaches negative infinity.
   (b) Similarly, discuss the possible asymptotic approach of the lower branch of $4x^2 - y^2 = -4$ to $y = 2x$ under the above two limiting processes.
◆ 51. Show that the hyperbola with focus $(3, 0)$, corresponding directrix $y = 2x$, and eccentricity $e = 2$ has equation $11x^2 - y^2 - 16xy + 30x - 45 = 0$. (Hint: review the hint for exercise 47 in section 8.1.)
52. Suppose that a hyperbola has focus $(0, 0)$, corresponding directrix $y = 2x - 3$, and eccentricity $e = \dfrac{7}{2}$. Find the vertex situated between the given focus and directrix. (Hint: modify the hint for exercise 46 in section 8.2. You will need to consider $ae - \dfrac{a}{e}$.)
53. Graph
   (a) $y^2 = \dfrac{4}{9}(x^2 - 9)$ (b) $y = \dfrac{2}{3}\sqrt{x^2 - 9}$
   (c) $y = -\dfrac{2}{3}\sqrt{x^2 - 9}$ (d) $x = \dfrac{3}{2}\sqrt{y^2 + 4}$
   (e) $x = -\dfrac{3}{2}\sqrt{y^2 + 4}$
   (Hint: the answers are all different.)
◆ 54. Verify the alternative definition of hyperbola by proving the following. Let a and e be positive numbers, with $e > 1$. Consider the point $F(ae, 0)$ and the line ℓ given by $x = \dfrac{a}{e}$. Let $P(x, y)$ be a point whose distance from F is d_1 and whose distance from ℓ is d_2. If $\dfrac{d_1}{d_2} = e$, then $\dfrac{x^2}{a^2} - \dfrac{y^2}{a^2(e^2 - 1)} = 1$.
55. Prove that the point on a given hyperbola that is closest to a given focus F of that hyperbola is the vertex that is

nearer to F. (Hint: use the techniques of section 2.3 to minimize a certain quadratic polynomial function, defined for $x \geq a$. If necessary, review the hint for exercise 49 in section 8.2.)

56. Which of the following hyperbolas is/are rectangular?

    (a) $\dfrac{x^2}{4} - \dfrac{y^2}{4} = 1$ (b) $\dfrac{x^2}{4} - \dfrac{y^2}{4} = -1$ (c) $\dfrac{x^2}{4} - \dfrac{y^2}{4} = 2$

    (d) $\dfrac{(y+1)^2}{8} - \dfrac{(x+2)^2}{8} = 1$ (e) $\dfrac{(y-1)^2}{64} - \dfrac{x^2}{64} = 1$

    (f) $\dfrac{9x^2}{24} - \dfrac{y^2}{429} = 1$

♦ 57. (Return of the rogue from exercise 52 of section 8.2) The rogue decides to humiliate a fast-food restaurant, Hyperbolic Hash (HH), situated 2200 feet away. No fool—having worked the earlier exercise—the owner of HH places herself and the rogue at the foci of a protective hyperbolic reflector with eccentricity $e = 4$. The rogue decides to bounce a ray off the reflector in such a way that (1) the ray will hit a franchise of Humble Hamburger Hut (HHH) situated 1100 feet above the midpoint of the segment connecting the rogue

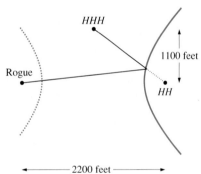

and HH and (2) the ray will *appear* to have emanated from HH. Where does the rogue aim? (Hint: use the first-quadrant intersection point of the hyperbola $\dfrac{x^2}{75{,}625} - \dfrac{y^2}{1{,}134{,}375} = 1$ with the line $y = -x + 1100$.)

58. (HH serves up a just dessert.) Without shooting, the rogue from exercise 57 sees the error of his ways, and moves to a secret location in order to meditate. The hyperbolic reflector is removed. Suddenly, the rogue shouts, "Please forgive me." This cry, moving 1100 feet per second (the speed of sound), reaches HH one second earlier than it arrives at HHH. John and David agree to deliver a tasty peace offering from HH to the rogue. They hope to find him quickly because they know that he is on a branch of the hyperbola $xy - 550x - 550y + 453{,}750 = 0$. How do they know this?

♦ 59. There is a small explosion aboard a ship sailing on a calm sea. View the surface of the sea as a plane, with the ship in the upper half-plane. Suppose individuals at $F_1(-\sqrt{2}, 0)$ and $F_2(\sqrt{2}, 0)$ hear the explosion $\dfrac{2}{1100}$ second apart. If the sound waves generated by the explosion travel 1100 feet per second, explain how it can be determined that the ship is located on the hyperbola $x^2 - y^2 = 1$.

♦ 60. (A continuation of exercise 59) Suppose the explosion was heard at F_2 before it was heard at F_1. Also, suppose that the explosion is heard at $F_3(8 + \sqrt{2}, 0)$ at exactly the same time that it is heard at F_1. Explain how the ship's position can be determined to be $(4, \sqrt{15})$. (Hint: you *could* do this by intersecting hyperbolas. But a quicker way is to intersect $x^2 - y^2 = 1$ with the perpendicular bisector of $\overline{F_1 F_3}$. Why does this method work?)

8.4 ROTATION OF AXES AND THE GENERAL QUADRATIC EQUATION IN x AND y

Parabolas, ellipses, and hyperbolas are alike in at least three ways. First, each of these types of conics can be defined in terms of a focus, a corresponding directrix, and an eccentricity. Second, each arises as the intersection of a double-napped right-circular cone with a plane. Third, each is the graph of a quadratic polynomial equation in x and y,

$$Ax^2 + Bxy + Cy^2 + Dx + Ey + F = 0, \qquad (*)$$

in which not all A, B, C are 0.

In the last three sections, the specific conics studied in the examples had their axes parallel to the coordinate axes and, in their equations, $B = 0$. How-

Rotation of Axes and the General Quadratic Equation in x and y

Figure 8.42

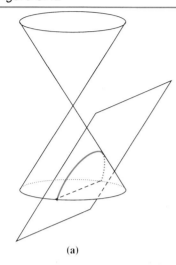

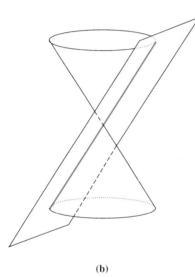

Figure 8.43

ever, when conics' axes have been rotated so they are not parallel to the coordinate axes, the coefficients of the xy-term in their equations need not be 0. Specific equations with $B \neq 0$ have been indicated for a parabola (in exercise 47 of section 8.1), an ellipse (exercise 45 of section 8.2), and a hyperbola (exercise 51 of section 8.3).

In this section, we shall show how to graph the general equation $(*)$. To do so, we must consider the effect of rotating the x- and y-axes through an arbitrary angle α, resulting in x'- and y'-axes. Given a specific $(*)$, our analysis will identify an α such that rotation with respect to α converts $(*)$ into an equation whose coefficient of $x'y'$ is 0. This work will depend on the double- and half-angle trigonometric identities in section 6.3.

Our work will show that some equations of type $(*)$ have graphs that are not parabolas, ellipses, or hyperbolas. In a number of exercises (exercises 42, 43, 44 of section 8.2 and exercise 48 of section 8.3), we have indicated how some of these "degenerate" graphs arise by applying various limiting processes to parabolas, ellipses, or hyperbolas. But many of them actually arise geometrically, as well as special types of intersections of a cone with a plane. So, before considering rotations, we next briefly study just what the degenerate conics are.

Degenerate Conic Sections

Recall that conic sections arise by intersecting a plane with a double-napped right-circular cone. If the plane is parallel to one nappe of the cone, then the resulting intersection is usually a parabola, as in figure 8.42(a). If, however, the plane is tangent to the cone as in figure 8.42(b), then the intersection is a straight line. A line arising in this way is called a **degenerate parabola.** An example of an equation whose graph is a degenerate parabola is $4x^2 - 4xy + y^2 - 4x + 2y + 1 = 0$, since the left-hand side of this equation is just $(y - 2x + 1)^2$.

If a plane cuts a cone as in figure 8.43(a), an ellipse or a circle results. But, if the plane passes through the cone's vertex, as in figure 8.43(b), the intersection is just a point, or **degenerate ellipse.** An example of an equation whose graph is a degenerate ellipse will be given next.

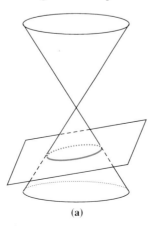
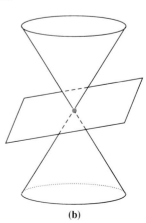

▶ **EXAMPLE 8.26** Sketch the graph of the degenerate ellipse given by

$$\frac{(x+3)^2}{25} + \frac{(y-4)^2}{16} = 0.$$

Solution The only way that a sum of nonnegative real numbers can be zero is for each of the addends to be zero. Hence, only one point satisfies the given equation, and it is determined by

$$\frac{(x+3)^2}{25} = 0 = \frac{(y-4)^2}{16}.$$

That point is $(-3, 4)$; its graph is in figure 8.44.

Notice that the given equation for this degenerate ellipse is similar to the standard form equation for an ellipse, except that the right-hand side has been replaced by 0. If the given equation had been written equivalently as $16x^2 + 25y^2 + 96x - 200y + 544 = 0$, you might not have discovered that its graph was degenerate until after you had completed the squares and were almost finished working. ◀

Figure 8.44

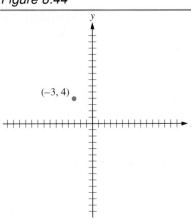

If a plane intersects both nappes of a cone, the intersection is usually a hyperbola, as in figure 8.45(a). However, if this intersection passes through the vertex of the cone as in figure 8.45(b), the result is two intersecting lines, or a **degenerate hyperbola,** as shown in example 8.27.

Figure 8.45

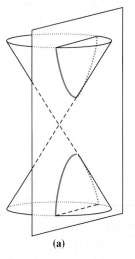
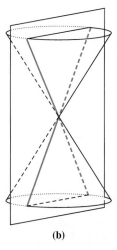

(a)   (b)

▶ **EXAMPLE 8.27** Sketch the graph of the degenerate hyperbola given by

$$\frac{(x+3)^2}{25} - \frac{(y-4)^2}{16} = 0.$$

Solution Except for the fact that its right-hand side is 0 instead of 1, this equation looks like a standard form equation of a hyperbola. Notice that its left-hand

Rotation of Axes and the General Quadratic Equation in x and y

side is the difference of two squares, and so it can be factored. This rewrites the equation as

$$\left(\frac{x+3}{5} + \frac{y-4}{4}\right)\left(\frac{x+3}{5} - \frac{y-4}{4}\right) = 0.$$

So,

either $\quad \dfrac{x+3}{5} + \dfrac{y-4}{4} = 0 \quad$ or $\quad \dfrac{x+3}{5} - \dfrac{y-4}{4} = 0.$

In other words, either $4x + 5y - 8 = 0$ or $4x - 5y + 32 = 0$.

The graphs of these equations are two lines that intersect at $(-3, 4)$, as shown in figure 8.46. ◀

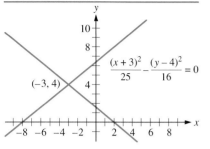

Figure 8.46

The Discriminant

Definition

The quantity $B^2 - 4AC$ is called the **discriminant** of a given quadratic equation

$$Ax^2 + Bxy + Cy^2 + Dx + Ey + F = 0. \qquad (*)$$

The value of the discriminant can be used to classify the graph of an equation. It does not tell you whether that graph is degenerate. What it *does* tell you is summarized next. (You will verify the main assertions in table 8.4 when you work exercises 35, 41, and 42.)

Table 8.4

Discriminant	Type of Curve
$B^2 - 4AC < 0$	Ellipse, point (degenerate ellipse), circle, or empty set
$B^2 - 4AC = 0$	Parabola, line (degenerate parabola), two parallel lines, or empty set
$B^2 - 4AC > 0$	Hyperbola or two intersecting lines (degenerate hyperbola)

The above table mentioned some possibilities that have not yet been illustrated algebraically. For instance, the quadratic equation $x^2 + y^2 + 4 = 0$ has discriminant $0^2 - 4(1)(1) = -4$, and its graph is the empty set. Similarly, the empty set is the graph of the equation $x^2 + 1 = 0$, whose discriminant is $0^2 - 4(1)(0) = 0$. Finally, two parallel lines, $y = \pm 2$, arise when we graph $y^2 - 4 = 0$, which also has discriminant $0^2 - 4(0)1 = 0$.

There is a computational criterion that tells you whether the graph of a given quadratic equation $(*)$ degenerates to either a line or a pair of lines. This is stated in exercise 46, and will be proved in exercise 52 of the section (9.3) on determinants.

Rotation of Axes

We begin the development of the rotation of axes technique with a coordinate system having origin $O(0, 0)$. The x- and y-axes will be rotated about O through an angle α. Since trigonometric functions of α will need to be considered, we measure α in radians, typically $-\frac{\pi}{2} < \alpha < \frac{\pi}{2}$.

For the new coordinate system, the rotated x- and y-axes will be called the x'- and y'-axes. Trigonometric identities will allow us to obtain equations for converting the coordinates of a point in the xy-system to coordinates in the $x'y'$-system, and vice versa. For a rotation in which α is not coterminal with an angle of 0 radians, the origin will be the only point whose coordinates are the same in both systems.

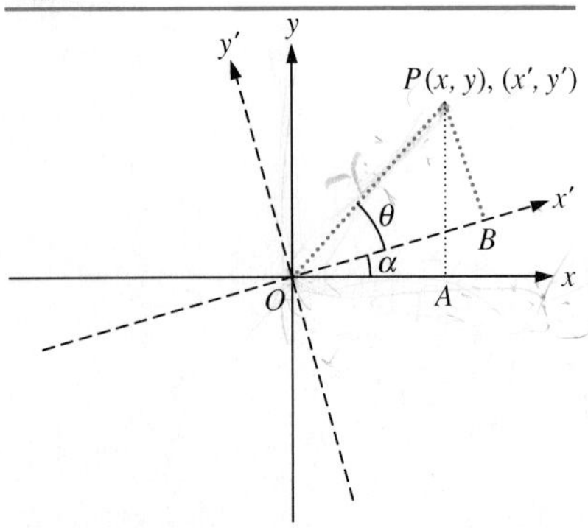

Let $P(x, y)$ be any point in the xy-system. That same point will have coordinates, say (x', y'), in the $x'y'$-system, as shown in figure 8.47. Notice that the angle from the positive x-axis to the positive x'-axis is α.

Let θ denote the angle from the positive x'-axis to the ray $\overrightarrow{OP}$, and let r denote the length OP. Then, by the definition of the trigonometric functions,

$$x' = r \cos \theta \quad \text{and} \quad y' = r \sin \theta. \tag{8.1}$$

The angle from the positive x-axis to $\overrightarrow{OP}$ is $\theta + \alpha$. Using the addition formulas for the sine and cosine functions developed in section 6.2, we have

$$x = r \cos(\theta + \alpha) = r \cos \theta \cos \alpha - r \sin \theta \sin \alpha \tag{8.2}$$

$$y = r \sin(\theta + \alpha) = r \sin \theta \cos \alpha + r \cos \theta \sin \alpha. \tag{8.3}$$

Substituting the values from equation (8.1) into equations (8.2) and (8.3), we obtain

Figure 8.47

Converting from $x'y'$ to xy Coordinates

$$x = x' \cos \alpha - y' \sin \alpha \tag{8.4}$$

$$y = x' \sin \alpha + y' \cos \alpha. \tag{8.5}$$

These equations show how to convert $x'y'$ coordinates into xy coordinates. To handle conversion in the other direction, you will be asked to show in exercise 48 that

Converting from xy to $x'y'$ Coordinates

$$x' = x \cos \alpha + y \sin \alpha$$

$$y' = -x \sin \alpha + y \cos \alpha.$$

▶ **EXAMPLE 8.28** Transform the equation $x^2 + 3xy + y^2 = 2$ by rotating the x- and y-axes through an angle of $\frac{\pi}{4}$ radians. Then graph the equation.

Rotation of Axes and the General Quadratic Equation in x and y

Solution

Let's consider the equivalent equation, $x^2 + 3xy + y^2 - 2 = 0$. By calculating its discriminant, $B^2 - 4AC = 9 - 4 = 5 > 0$, we see that the graph is either a hyperbola or a pair of intersecting lines.

For $\alpha = \dfrac{\pi}{4}$, $\sin\alpha = \cos\alpha = \dfrac{\sqrt{2}}{2}$. From equations (8.4) and (8.5), we obtain

$$x = \frac{\sqrt{2}}{2}(x' - y') \quad \text{and} \quad y = \frac{\sqrt{2}}{2}(x' + y').$$

Substituting for x and y in the given equation, we get

$$\left[\frac{\sqrt{2}}{2}(x'-y')\right]^2 + 3\left[\frac{\sqrt{2}}{2}(x'-y')\right]\left[\frac{\sqrt{2}}{2}(x'+y')\right] + \left[\frac{\sqrt{2}}{2}(x'+y')\right]^2 = 2$$

or

$$\tfrac{1}{2}(x'^2 - 2x'y' + y'^2) + \tfrac{3}{2}(x'^2 - y'^2) + \tfrac{1}{2}(x'^2 + 2x'y' + y'^2) = 2$$

$$\tfrac{5}{2}x'^2 - \tfrac{1}{2}y'^2 = 2$$

$$\frac{x'^2}{\tfrac{4}{5}} - \frac{y'^2}{4} = 1.$$

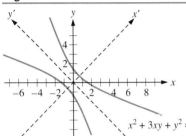

Figure 8.48

Using section 8.3, we see that this is an equation of the hyperbola whose graph is in figure 8.48. ◂

You may have noticed that the equation in example 8.28 that resulted from the specified rotation of axes did not have an $x'y'$ term. We are not so lucky with most rotations! In order to transform an equation which has an xy term into one that does not, it is necessary to rotate the axes through an *appropriate* angle α. To determine the measure of such an appropriate angle α, substitute equations (8.4) and (8.5) into the quadratic equation

$$Ax^2 + Bxy + Cy^2 + Dx + Ey + F = 0,$$

getting

$$A(x'\cos\alpha - y'\sin\alpha)^2 + B(x'\cos\alpha - y'\sin\alpha)(x'\sin\alpha + y'\cos\alpha)$$
$$+ C(x'\sin\alpha + y'\cos\alpha)^2 + D(x'\cos\alpha - y'\sin\alpha)$$
$$+ E(x'\sin\alpha + y'\cos\alpha) + F = 0.$$

After expanding, multiplying, and collecting terms, we find

$$(A\cos^2\alpha + B\sin\alpha\cos\alpha + C\sin^2\alpha)x'^2$$
$$+ [B(\cos^2\alpha - \sin^2\alpha) + 2(C - A)\sin\alpha\cos\alpha]x'y'$$
$$+ (A\sin^2\alpha - B\sin\alpha\cos\alpha + C\cos^2\alpha)y'^2 + (D\cos\alpha + E\sin\alpha)x'$$
$$+ (-D\sin\alpha + E\cos\alpha)y' + F = 0.$$

As bad as this looks, all the terms in parentheses are constant, and so this equation can be written as

$$A'x'^2 + B'x'y' + C'y'^2 + D'x' + E'y' + F = 0.$$

Now, in order to have "no" $x'y'$ term, you want $B' = 0$. Thus, you must select α so that

$$[B(\cos^2 \alpha - \sin^2 \alpha) + 2(C - A) \sin \alpha \cos \alpha] = 0$$

or, by using the double-angle formulas from section 6.3,

$$B \cos 2\alpha + (C - A) \sin 2\alpha = 0.$$

If $A = C$, this may be arranged by taking $\alpha = \dfrac{\pi}{4}$. If $A \neq C$, arrange

$$\frac{\sin 2\alpha}{\cos 2\alpha} = \frac{B}{A - C} \quad \text{or} \quad \tan 2\alpha = \frac{B}{A - C}.$$

In summary, the xy term is eliminated from a quadratic equation

$$Ax^2 + Bxy + Cy^2 + Dx + Ey + F = 0$$

by rotating the axes through an angle α, where

$$\tan 2\alpha = \frac{B}{A - C}.$$

It is always possible to find such an α satisfying $-\dfrac{\pi}{2} < \alpha < \dfrac{\pi}{2}$. In case $A = C$, the above stipulation means that $\tan 2\alpha$ is undefined, and this can be arranged by taking $\alpha = \dfrac{\pi}{4}$.

Guidelines for Graphing Conic Sections When $B \neq 0$

1. Find a suitable angle of rotation α satisfying $\tan 2\alpha = \dfrac{B}{A - C}$. It is always possible to find such an α satisfying $-\dfrac{\pi}{2} < \alpha < \dfrac{\pi}{2}$. If $A = C$, take $\alpha = \dfrac{\pi}{4}$.
2. Substitute
$$x = x' \cos \alpha - y' \sin \alpha$$
$$y = x' \sin \alpha + y' \cos \alpha$$
into the original equation and simplify.
3. Complete squares if necessary. Then sketch the graph of the resulting equation relative to the new $x'y'$ axes.

▶ **EXAMPLE 8.29**

Rotate axes and graph the hyperbola $xy = 8$. Find equations for its asymptotes.

Solution

Here $A = 0$, $B = 1$, and $C = 0$. Since $B^2 - 4AC = 1$, which is positive, we know that the graph is either a hyperbola or a pair of intersecting lines. Since $A = C$, the above guidelines tell us to take $\alpha = \dfrac{\pi}{4}$. Then $\sin \alpha = \cos \alpha = \dfrac{\sqrt{2}}{2}$. The rotation equations are then

Rotation of Axes and the General Quadratic Equation in x and y

$$x = \frac{\sqrt{2}}{2}(x' - y') \quad \text{and} \quad y = \frac{\sqrt{2}}{2}(x' + y').$$

Substituting these values into the original equation, we obtain

$$\left[\frac{\sqrt{2}}{2}(x' - y')\right]\left[\frac{\sqrt{2}}{2}(x' + y')\right] = 8$$

or

$$\tfrac{1}{2}(x'^2 - y'^2) = 8$$

$$\frac{x'^2}{16} - \frac{y'^2}{16} = 1.$$

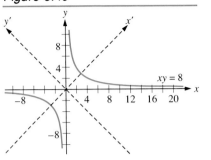

Figure 8.49

This is a hyperbola and, because $a = b\ (= 4)$, it is a rectangular hyperbola. Its graph is in figure 8.49. It is clear that its asymptotes are the original x- and y-axes, namely $y = 0$ and $x = 0$. ◀

In example 8.29, the asymptotes form a degenerate conic, namely the graph of $xy = 0$. This equation could be obtained directly from the hyperbola's equation, by replacing the constant term, 8, with 0. **This method of finding the equations of a hyperbola's asymptotes works whenever $D = E = 0$.**

▶ **EXAMPLE 8.30**

Rotate axes and then graph the hyperbola $4x^2 + 24xy - 3y^2 = 48$. Find equations for its asymptotes.

Solution

Here $A = 4$, $B = 24$, and $C = -3$. Since $B^2 - 4AC = 624$, the graph is either a hyperbola or a pair of intersecting lines. We take $\tan 2\alpha = \dfrac{B}{A - C} = \dfrac{24}{4 + 3} = \dfrac{24}{7}$. We may take $2\alpha = \tan^{-1}\left(\dfrac{24}{7}\right) \approx 73.7398°$, and so $\alpha \approx 36.8699°$.

Calculator round-off errors can seriously mislead us as to the nature of the graph. Although we know α only approximately, we need a way to find $\sin \alpha$ and $\cos \alpha$ exactly. For this, we shall use the half-angle identities from section 6.3,

$$\sin \alpha = \sqrt{\frac{1 - \cos 2\alpha}{2}} \quad \text{and} \quad \cos \alpha = \sqrt{\frac{1 + \cos 2\alpha}{2}}$$

From the right triangle in figure 8.50(a), we see that $\cos 2\alpha = \frac{7}{25}$, and so

$$\sin \alpha = \sqrt{\frac{9}{25}} = \frac{3}{5} \quad \text{and} \quad \cos \alpha = \sqrt{\frac{16}{25}} = \frac{4}{5}.$$

Thus, the rotation equations are

$$x = \tfrac{4}{5}x' - \tfrac{3}{5}y' \text{ and } y = \tfrac{3}{5}x' + \tfrac{4}{5}y'.$$

Substituting these expressions for x and y into the original equation leads to

$$4(\tfrac{4}{5}x' - \tfrac{3}{5}y')^2 + 24(\tfrac{4}{5}x' - \tfrac{3}{5}y')(\tfrac{3}{5}x' + \tfrac{4}{5}y') - 3(\tfrac{3}{5}x' + \tfrac{4}{5}y')^2 = 48$$

or

$$4(\tfrac{16}{25}x'^2 - \tfrac{24}{25}x'y' + \tfrac{9}{25}y'^2) + 24(\tfrac{12}{25}x'^2 + \tfrac{7}{25}x'y' - \tfrac{12}{25}y'^2) - 3(\tfrac{9}{25}x'^2 + \tfrac{24}{25}x'y' + \tfrac{16}{25}y'^2) = 48.$$

Combining terms, we get

$$13x'^2 - \frac{204}{25}y'^2 = 48.$$

Simplifying, we obtain

$$\frac{x'^2}{\frac{48}{13}} - \frac{y'^2}{\frac{300}{51}} = 1,$$

the equation of a hyperbola, with center at the origin and transverse axis along the x'-axis. Its graph is in figure 8.50(b). By the method described following example 8.29, the asymptotes of this hyperbola are given by $4x^2 + 24xy - 3y^2 = 0$. You should check, using the rotation equations, that an equivalent equation for these asymptotes is $\frac{x'^2}{\frac{48}{13}} - \frac{y'^2}{\frac{300}{51}} = 0$, or, in other words, $y' = \pm\frac{5}{2}\sqrt{\frac{13}{51}}\,x'$. ◀

Figure 8.50

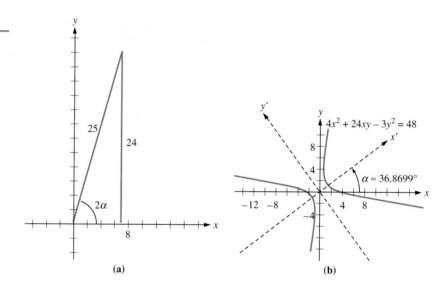

(a)   (b)

EXAMPLE 8.31

Rotate axes and then graph $8x^2 - 12xy + 17y^2 - 56\sqrt{5}\,x + 72\sqrt{5}\,y + 560 = 0$.

Solution

Here $A = 8$, $B = -12$, and $C = 17$. Since $B^2 - 4AC = -400$, the graph is either an ellipse or degenerate. We take $\tan 2\alpha = \dfrac{B}{A - C} = \dfrac{4}{3}$, and hence can draw the right triangle shown in figure 8.51(a). From this, we see that $\cos 2\alpha = \frac{3}{5}$. By using the half-angle formulas, we get $\sin \alpha = \dfrac{1}{\sqrt{5}}$ and $\cos \alpha = \dfrac{2}{\sqrt{5}}$, and so the rotation equations are

$$x = \frac{2}{\sqrt{5}}x' - \frac{1}{\sqrt{5}}y' \quad \text{and} \quad y = \frac{1}{\sqrt{5}}x' + \frac{2}{\sqrt{5}}y'.$$

Rotation of Axes and the General Quadratic Equation in x and y

Substituting these into the original equation, we get

$$8\left(\frac{2}{\sqrt{5}}x' - \frac{1}{\sqrt{5}}y'\right)^2 - 12\left(\frac{2}{\sqrt{5}}x' - \frac{1}{\sqrt{5}}y'\right)\left(\frac{1}{\sqrt{5}}x' + \frac{2}{\sqrt{5}}y'\right) + 17\left(\frac{1}{\sqrt{5}}x' + \frac{2}{\sqrt{5}}y'\right)^2$$
$$- 56\sqrt{5}\left(\frac{2}{\sqrt{5}}x' - \frac{1}{\sqrt{5}}y'\right) + 72\sqrt{5}\left(\frac{1}{\sqrt{5}}x' + \frac{2}{\sqrt{5}}y'\right) + 560 = 0.$$

This simplifies to

$$5x'^2 + 20y'^2 - 40x' + 200y' + 560 = 0$$

or

$$x'^2 + 4y'^2 - 8x' + 40y' + 112 = 0.$$

Completing the square results in

$$(x' - 4)^2 + 4(y' + 5)^2 = 4$$

or

$$\frac{(x' - 4)^2}{4} + (y' + 5)^2 = 1.$$

This is the equation for an ellipse whose center has coordinates $(4, -5)$ in the $x'y'$-system. Its graph is in figure 8.51(b). ◀

Figure 8.51

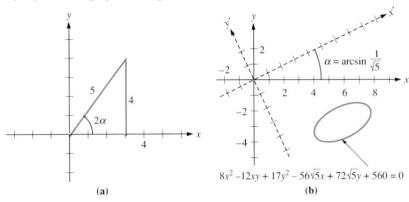

(a)   (b)

▶ **EXAMPLE 8.32** Rotate axes and hence graph $16x^2 - 24xy + 9y^2 + 100x - 200y + 100 = 0$.

Solution The discriminant $B^2 - 4AC = 0$, and so the graph is either a parabola or degenerate. We take $\tan 2\alpha = -\frac{24}{7}$. We may take

$$2\alpha = \pi + \tan^{-1}\left(-\frac{24}{7}\right) \approx 106.2602° \quad \text{and} \quad \alpha \approx 53.1301°.$$

(Notice that $\tan^{-1}\left(-\frac{24}{7}\right)$ is the radian measure of a fourth-quadrant angle. We added π so that 2α would be positive. You should also work this problem using $2\alpha = \tan^{-1}\left(-\frac{24}{7}\right)$. You will get a different rotation, but the same graph!)

From the right triangle in figure 8.52(a), we see that $\cos 2\alpha = \frac{-7}{25}$. Thus, by the half-angle formulas, $\sin \alpha = \frac{4}{5}$ and $\cos \alpha = \frac{3}{5}$. This means that the rotation equations are

$$x = 0.6x' - 0.8y' \quad \text{and} \quad y = 0.8x' + 0.6y'.$$

Substituting these values into the original equation and simplifying, we obtain

$$25y^2 - 100x - 200y + 100 = 0$$

or

$$y^2 - 8y = 4x - 4$$

$$(y - 4)^2 = 4x + 12$$

$$(y - 4)^2 = 4(x + 3).$$

This is a horizontally oriented parabola that opens to the right with its vertex, in the $x'y'$-coordinate system, at $(-3, 4)$ and $p = 1$. The focus is at $(-2, 4)$, the axis is $y' = 4$, and the endpoints of the latus rectum are at $(-2, 2)$ and $(-2, 6)$, where all coordinates are in the $x'y'$ system.

This parabola was graphed in chapter 2 as figure 2.24(d), and is graphed again in figure 8.52(b).

Figure 8.52

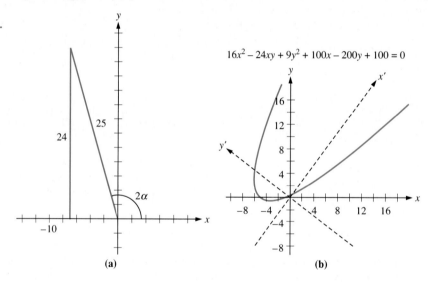

Exercises 8.4

In exercises 1–10, sketch the "degenerate" graph of the given equation.

1. $\dfrac{(x-3)^2}{16} + \dfrac{(y+4)^2}{25} = 0$
2. $\dfrac{(x-3)^2}{16} + \dfrac{(y+4)^2}{25} = -1$
3. $x(x-3) = 0$
4. $\dfrac{(x+1)^2}{9} + \dfrac{(y-3)^2}{4} = 0$
5. $\dfrac{(y+4)^2}{25} + \dfrac{(x-3)^2}{16} = 0$
6. $(y - 2x)(3y + x) = 0$
7. $4x^2 - 4xy + y^2 = 0$. (Hint: factor instead of rotating.)
8. $\dfrac{(x-3)^2}{12} - \dfrac{(y+4)^2}{48} = 0$
9. $4x^2 + 4y^2 + 9 = 0$
10. $4x^2 + y^2 - 4xy - 2x + y = 0$ (Hint: factor or solve for y by the quadratic formula.)

In exercises 11–18, calculate the discriminant and hence classify the graph of the given equation as an ellipse, parabola, or hyperbola. You may assume the graph is not degenerate.

11. $x^2 + 2xy + y^2 - 7x - 2y + 1 = 0$
12. $2x^2 - 2xy + 2y^2 - 7x - 2y + 1 = 0$
13. $3x^2 + 2y^2 - 7x = 0$
14. $9xy + 5y^2 - 2y + 3 = 0$
15. $4x^2 - 9xy + 5y^2 + 7x - 1 = 0$
16. $4x^2 + 7x - 2y + 1 = 0$
17. $x^2 + 4xy + 4y^2 + 2x - 1 = 0$
18. $x^2 + y^2 + 2x - 4y - 9 = 0$

In exercises 19–26, transform the given equation in terms of new coordinates obtained by rotating axes through the given angle α.

19. $3x^2 - 9xy + 4 = 0$; $\alpha = \dfrac{\pi}{6}$
20. $9xy - 3y^2 + 4 = 0$; $\alpha = \dfrac{\pi}{2}$
21. $3x^2 + 2\sqrt{3}xy + y^2 - 8x - \pi = 0$; $\alpha = \dfrac{\pi}{3}$
22. $3x^2 - 2\sqrt{3}xy + y^2 - 8\pi y + 1 = 0$; $\alpha = \pi$
23. $4x^2 + 3xy - 8x - 6y = 0$; $\alpha = 30°$
24. $5x^2 - \sqrt{3}xy + 4y^2 - 2x + 3y - 4 = 0$; $\alpha = 60°$
25. $x^2 - y^2 - \sqrt{2} = 0$; $\alpha = -\dfrac{\pi}{4}$
26. $4x^2 + 4y^2 - 8x + 16y - 80 = 0$; $\alpha = -\dfrac{\pi}{6}$

In exercises 27–34, find an angle α such that rotation through α eliminates the xy term. Identify and sketch the graph, showing both the old axes and the new ones. Give (old) coordinates of all vertices and foci, approximating by calculator if necessary. (You may wish to check if a graphing calculator or computer graphing program confirms the general appearance of the graphs that you find.)

27. $xy = 1$
28. $2x^2 + xy + y^2 + 8x + y + 1 = 0$
29. $2x^2 - 8xy + 8y^2 + 2x - 5 = 0$ (Hint: express α as a value of the inverse sine function.)
30. $x^2 + \sqrt{3}xy - 1 = 0$ 31. $3x^2 - 5xy + 3y^2 - 7 = 0$
32. $2x^2 - y^2 - xy = 0$ (Hint: express α as a value of the inverse sine function.)
33. $x^2 + 2xy + y^2 - 5x - 5y + 6 = 0$
34. $x^2 + 2xy + y^2 - 4x + 8 = 0$

For exercises 35–37, suppose that a rotation of axes through α converts $Ax^2 + Bxy + Cy^2 + Dx + Ey + F$ into $A'x'^2 + B'x'y' + C'y'^2 + D'x' + E'y' + F'$. Explicit formulas for A', B', C', D', E', and F' may be discerned from the text's discussion (but need not be memorized).

◆ 35. Prove that $B^2 - 4AC$ is **invariant under** the **rotation,** in the sense that $B'^2 - 4A'C' = B^2 - 4AC$. (Hint: simplify the left-hand side, using trigonometric identities.)
36. Prove that $A + C$ is invariant under rotation, in the sense that $A' + C' = A + C$.
37. (Invariance of degree) If $A' = B' = C' = 0$, prove that $A = B = C = 0$.
38. (Once a line, always a line.) Use exercise 37 to explain algebraically why no rotation of axes can convert $y^2 - 4x$ into $x' + 4y'$.

For exercises 39–40, suppose that a **translation** of axes from $(0, 0)$ to (h, k) converts $Ax^2 + Bxy + Cy^2 + Dx + Ey + F$ into $A*x'^2 + B*x'y' + C*y'^2 + D*x' + E*y' + F*$.

39. Using $x = x' + h$ and $y = y' + k$, establish that $A* = A$, $B* = B$, and $C* = C$.
40. Use exercise 39 to show that *translation* satisfies the analogues of exercises 35–38. In other words, show that $B^2 - 4AC$ is **invariant under translation,** in the sense that $B*^2 - 4A*C* = B^2 - 4AC$, etc.

For exercises 41–42, let Γ be the graph of a second-degree polynomial equation $Ax^2 + Bxy + Cy^2 + Dx + Ey + F = 0$.

◆ 41. Show the following:
(a) If Γ is an ellipse, then $B^2 - 4AC < 0$. (Hint: by exercises 35 and 40, you may assume the given equation has the form $\dfrac{x^2}{a^2} + \dfrac{y^2}{b^2} = 1$.
(b) If Γ is a parabola, then $B^2 - 4AC = 0$.
(c) If Γ is a hyperbola, then $B^2 - 4AC > 0$.
◆ 42. Show the following:
(a) If Γ is a pair of intersecting lines, then $B^2 - 4AC > 0$. (Hint: as in exercise 41, you may assume the left-hand side of the equation has a simple form, namely one of xy, $(y - mx)x$, $(y - mx)y$ [with $m \neq 0$], and $(y - mx)(y - nx)$ for some $m \neq n$. In the last case, show $B^2 - 4AC = (m - n)^2$.)
(b) If Γ is either a line or a pair of parallel lines, then $B^2 - 4AC = 0$.

For exercises 43–45, write down a second-degree equation for the degenerate conic consisting of the pair of asymptotes of the given hyperbola.

43. $xy = 1$ 44. $4x^2 - 9xy + y^2 + 7 = 0$
45. $9x^2 + xy - y^2 + 4 = 0$
◆ 46. Suppose that a quadratic expression $Ax^2 + Bxy + Cy^2 + Dx + Ey + F$ is converted as a result of either a rotation or a translation of axes. Show that
$$4ACF + BED - CD^2 - FB^2 - AE^2$$
is invariant, in the sense of exercises 35 and 40. (We shall see in exercise 52 of section 9.3 that this invariant is 0 precisely when the nonempty graph of the given quadratic expression is either a line or a pair of lines. A determinantal mnemonic will be given for this invariant there.)
47. (An example in which calculator round-off error makes you fail to recognize a degenerate case) Consider the pair of intersecting lines given by $4x + y - 7 = 0$ and $\sqrt{3}x - y - 2\sqrt{3} - 1 = 0$. Their union is the graph of the degenerate hyperbola given by
$$4\sqrt{3}x^2 + (-4 + \sqrt{3})xy - y^2$$
$$+ (-15\sqrt{3} - 4)x + (-2\sqrt{3} + 6)y + (14\sqrt{3} + 7) = 0.$$

(a) By calculating its discriminant, verify that this quadratic's graph is either a hyperbola or a degenerate hyperbola.
(b) Verify, using the remark in exercise 46, that this quadratic's graph is degenerate.
(c) Use this section's techniques to show that the xy term in the given quadratic can be eliminated by a rotation satisfying $\tan 2\alpha = \dfrac{16 - 17\sqrt{3}}{47}$.
(d) Taking α in $(0, \pi)$ as in (c), use trigonometric identities and your calculator to get $\sin \alpha \approx 0.990312$ and $\cos \alpha \approx 0.1388599$.
(e) Rotate axes using the data in (d). Approximate all the new coefficients via calculator. Did you find $B' = 0$ or $B' = 0.0000004$?
(f) Apply the criterion from exercise 46 to the new coefficients found in (e). Did your new calculations predict a degenerate form or is the expression that you calculated nonzero and only "small" in absolute value?

◆ 48. Show that if rotation through an angle α carries the (x, y) coordinate system into the (x', y') coordinate system, then

$$x' = x \cos \alpha + y \sin \alpha$$
$$y' = -x \sin \alpha + y \cos \alpha.$$

(Hint: one way is to solve (8.4) and (8.5) simultaneously for x' and y'. For another, replace α by $-\alpha$ in the considerations leading to (8.4); remember that cos is an even function and sin is odd.)

8.5 CHAPTER SUMMARY AND REVIEW EXERCISES

Major Concepts

- Discriminant
- Double-napped right-circular cone
- Eccentricity
- Ellipse
  Center
  Degenerate ellipse
  Focal chord
  Focal radius
  Foci
  Latus rectum
  Major axis
  Minor axis

  Vertices
- Hyperbola
  Asymptotes
  Branches
  Center
  Conjugate axis
  Degenerate hyperbola
  Equilateral hyperbola
  Focal chord
  Focal radius
  Foci
  Latus rectum

  Rectangular hyperbola
  Transverse axis
  Vertices
- Parabola
  Axis
  Degenerate parabola
  Directrix
  Focal chord
  Focus
  Latus rectum
  Vertex

Review Exercises

In exercises 1–8, sketch the graph of the given equation. If it is a parabola, determine its vertex, axis, focus, directrix, and length of latus rectum. If it is an ellipse or a hyperbola, determine its center; parameters a, b, and c; foci; vertices; eccentricity e; equations of directrices; and length of each latus rectum. For an ellipse, also find the endpoints of its minor axis. For a hyperbola, also find the endpoints of its conjugate axis and equations of its asymptotes.

1. $y^2 = 16x$
2. $\dfrac{x^2}{25} + \dfrac{y^2}{64} = 1$
3. $\dfrac{(y+5)^2}{4} - \dfrac{(x-3)^2}{25} = 1$
4. $3x^2 - 12x + 4y + 12 = 0$
5. $4x^2 + 9y^2 + 24x - 18y + 9 = 0$
6. $4x^2 - 9y^2 + 4x - 35 = 0$
7. $8y + 3x^2 = 0$
8. $\dfrac{\left(x+\frac{1}{2}\right)^2}{64} + \dfrac{(y-1)^2}{36} = 1$

In exercises 9–30, find an equation for the conic with the given properties.

9. Parabola with vertex $(3, -2)$, focus $(3, 8)$
10. Parabola with vertex $(\frac{1}{2}, -\frac{1}{3})$, directrix $x = \frac{7}{2}$
11. Parabola with focus $(0, 0)$, directrix $x = -2$
12. Parabola passing through the point $(8, 6)$, with vertical axis and vertex $(2, 3)$
13. Parabola with vertical axis, passing through the points $(-9, -7), (-1, 1),$ and $(7, -7)$
14. Parabola with focus $(-1, 1)$, directrix $3y = 4x - 8$

15. Ellipse with center $(-6, -8)$, focus $(-9, -8)$, vertex $(-1, -8)$.
16. Ellipse with foci $(8, 9)$ and $(8, 21)$; $(28, 15)$ an endpoint of the minor axis
17. Ellipse with vertices $(-6, -8)$ and $(0, -8)$; $(-3, -6)$ an endpoint of the minor axis
18. Ellipse with foci $(8, -21)$ and $(8, 9)$, vertex $(8, 14)$
19. Conic with foci $(2, 4)$ and $(2, 14)$, $e = \frac{5}{7}$
20. Conic such that each of its points has the sum of its distances from $(8, -21)$ and $(10, -21)$ equal to 7
21. Ellipse with vertex $(2, 10)$ and directrices $y = -18$ and $y = 14$
22. Ellipse with focus $(4, 5)$, corresponding directrix $2y + x = 0$, $e = \frac{2}{3}$
23. Hyperbola with foci $(8, 9)$ and $(2, 9)$, vertex $(4, 9)$
24. Hyperbola with foci $(8, 9)$ and $(18, 9)$; $(13, 5)$ an endpoint of the conjugate axis
25. Hyperbola with vertices $(8, 6)$ and $(18, 6)$, asymptote $y - 6 = \frac{2}{3}(x - 13)$
26. Hyperbola with center $(-8, -6)$, focus $(5, -6)$, vertex $(-20, -6)$
27. Hyperbola with vertices $(8, 6)$ and $(8, 18)$, focus $(8, 20)$
28. Conic with vertices $(6, \pi)$ and $(-4, \pi)$, $e = 3$
29. Conic with focus $(0, -2)$, corresponding directrix $y = 3x$, $e = \frac{3}{2}$
30. Hyperbola passing through $(0, -3)$, asymptotes $y = 2x$ and $y = 3x$ (Hint: the asymptotes are the graph of the degenerate conic $(y - 2x)(y - 3x) = 0$. The hyperbola has an equation differing from this one only in its constant term.)
31. Determine which of the following hyperbolas is/are rectangular:
    (a) $429y^2 - \frac{x^2}{429} = 1$ (b) $\frac{(y-1)^2}{429} - \frac{(x-3)^2}{429} = 1$
    (c) $\frac{y^2}{429} - \frac{x^2}{429} = 4$ (d) $\frac{x^2}{429} - \frac{y^2}{429} = -1$
32. Write an equation whose graph is the pair of asymptotes of the hyperbola $x^2 - xy = 4$.

In exercises 33–38, sketch the "degenerate" graph of the given equation.

33. $y^2 + 2y = 0$
34. $\frac{(x+3)^2}{25} + \frac{(y-6)^2}{9} = 0$
35. $9x^2 - 12xy + 4y^2 = 0$ (Hint: factor or solve for y by the quadratic formula.)
36. $6x^2 - 5xy + y^2 - 11x + 5y + 4 = 0$
37. $\frac{(y-4)^2}{36} - \frac{(x-5)^2}{9} = 0$ 38. $x^2 + y^2 + 2x - 4y + 5 = 0$

In exercises 39–42, calculate the discriminant and hence classify the graph of the given equation as an ellipse, parabola, or hyperbola. You may assume the graph is not degenerate.

39. $3x^2 + 6xy + 2y^2 - 7 = 0$ 40. $3x^2 - 4xy + 2y^2 + 7 = 0$
41. $2x^2 - 4xy + 2y^2 - \pi x = 0$
42. $2y^2 - \pi x + 7 = 0$

In exercises 43–44, transform the given equation in terms of new coordinates obtained by rotating axes through the given angle α.

43. $xy - 7 = 0$; $\alpha = \frac{\pi}{4}$ 44. $y^2 - 4x = 0$; $\alpha = -\frac{\pi}{6}$

In exercises 45–49, find an angle α such that rotation through α eliminates the xy term. Sketch the conic, giving (old) coordinates of all its vertices and foci, approximating by calculator if necessary. (You may wish to check if a graphing calculator or computer graphing program confirms the general appearance of the graphs that you find.)

45. $3x^2 + 4\sqrt{3}xy - y^2 - 7 = 0$
46. $xy + \sqrt{2}x - 4 = 0$
47. $3x^2 + 6xy + 3y^2 - 4y + 8 = 0$
48. $(1 + \sqrt{3})x^2 + xy + y^2 - 8 = 0$
49. $x^2 + xy - 8 = 0$
50. Graph
    (a) $y = \frac{\sqrt{36 - 4x^2}}{3}$ (b) $x = -\frac{5}{7}\sqrt{49 + y^2}$

In exercises 51–52, identify and graph the given set.

51. $\{(x, y)$: there exists a real number t such that $y = 4t$ and $x = -8t^2\}$
52. $\{(x, y)$: there exists a real number θ in $[0, 2\pi)$ such that $x = 3 \sin \theta$ and $y = -1 + 2 \cos \theta\}$
53. An amateur astronomer builds a telescope containing a parabolic reflector which measures 2 feet across at its opening and is 9 inches deep. How far is the vertex from the focus?
54. John is spying on David in a whispering gallery which has been built using an ellipse with eccentricity $\frac{4}{5}$. They are standing at the foci, 40 meters apart. How far above floor level is the highest point of the gallery's ceiling? (Hint: $c = 20$ and $e = \frac{4}{5}$. Find b.)
55. Find an equation for a conic passing through the five points $P_1(-2, 0)$, $P_2(2, 0)$, $P_3(-2, 2)$, $P_4(2, 2)$, and $P_5(1, \sqrt{2})$. (Hint: the graphs of $y^2 - 2y = 0$ and $x^2 - 4 = 0$ each pass through P_1, P_2, P_3, and P_4. Find a real number k such that P_5 lies on the graph of $k(y^2 - 2y) + (x^2 - 4) = 0$.)

56. (a) Use the method of exercise 55 to find a quadratic equation in x and y whose graph passes through the five points $(-2, 0)$, $(2, 0)$, $(-2, 2)$, $(2, 2)$, and $(0, 1)$.
   (b) Show that the graph obtained in (a) is a degenerate conic.

57. Complete squares, using your calculator, to show that the graph of the equation $6.84x^2 - 0.035y^2 - 12.02x + 0.42y + 4 = 0$ is a hyperbola with eccentricity $e \approx 14.0152$.

◆ 58. An explosion occurs somewhere in the upper half-plane. It is heard later by observers at $F_1(3 - \sqrt{20}, 0)$, $F_2(0, 0)$, and $F_3(3, 0)$. The sound, traveling at 1100 feet per second, reaches F_3 first, F_2 $\frac{1}{1100}$ second later, and F_1 $\frac{1}{1100}$ second after that.
   (a) Show that the x-coordinate where the explosion occurred is a root of the polynomial equation
   $$(20 - 12\sqrt{5})x^2 + (-180 + 92\sqrt{5})x + (360 - 168\sqrt{5}) = 0$$
   and satisfies $x \geq \frac{3}{2}$.
   (b) Explain how the data allow you to conclude that the explosion occurred at $(3, 4)$.

Chapter 9

Systems of Linear Equations and Linear Inequalities

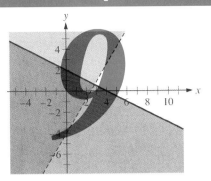

In sections 1.3 and 1.4, you reviewed how to solve certain systems of two or three equations in two variables. Some of these systems were solved in order to find the points of intersection of two lines, three lines, or two circles, while others led to the intersection points of lines, circles, and parabolas. In this chapter, you will continue the study of systems of linear equations. By expanding upon the earlier work on systems of two or three equations in two variables, you will learn systematic ways to solve sets of m linear equations in n variables. Along the way, you will see additional applications for such skills. Many of these involve matrices, a tool introduced in section 9.2 in order to illuminate, and shorten, solution techniques of the type developed in section 9.1. Matrices, and the associated computations of determinants introduced in section 9.3, are of enormous use in many fields outside mathematics. In later courses, many of you will study these topics more deeply.

Not all applications involving optimization can be modeled by equations. In order to partially cope with such situations, we use section 9.4 to introduce the algebra and geometry associated with studying systems of linear inequalities involving two variables.

9.1 SYSTEMS OF LINEAR EQUATIONS

If the numbers a_1, a_2, and b are not all zero, then $a_1x + a_2y = b$ is a linear equation in two variables x and y. The term "linear" is used because this equation's graph in the xy-plane is either empty or a line. Another way to write this equation for a line is $a_1x_1 + a_2x_2 = b$; here, the variables are x_1 and x_2. The advantage of this second method's notation is that it generalizes as follows.

Linear Equation in *n* Variables

> A **linear equation in *n* variables** $x_1, x_2, x_3, \ldots, x_n$ is an equation
> $$a_1x_1 + a_2x_2 + a_3x_3 + \cdots + a_nx_n = b,$$
> where $a_1, a_2, a_3, \ldots, a_n$, and b are complex numbers.

For reasons of simplicity, we shall usually restrict the coefficients to be real numbers. Notice that, to handle degenerate cases, we have permitted $a_1, a_2, \ldots, a_n$ to all be zero.

▶ **EXAMPLE 9.1**

Each of the following is a linear equation:

$$4x - 5y = 9$$

$$\tfrac{17}{2}x + \sqrt{3}\,y - z = \pi$$

$$ex_1 + \sqrt{5}\,x_2 - 9x_3 + \tfrac{3}{5}x_4 = \tfrac{\sqrt{15}}{2}$$

$$7ix_1 + (4 - 3i)x_2 + \pi x_3 + x_4 + x_5 = 0.$$

None of the following is a linear equation:

$$4\sqrt{x_1} + 5x_2 = 9$$

$$7x + \sqrt{3}\,y^2 = 2$$

$$x_1 + 3x_2 + \cos x_3 = \pi.$$ ◀

System of Linear Equations

> A finite set of linear equations in n variables $x_1, x_2, x_3, \ldots, x_n$ is called a **system of linear equations** or a **linear system**.

In section 1.3, you solved linear systems via the method of substitution, primarily in cases where n was 2 or 3. Here, to study linear systems in any number of variables, we shall adopt the following notation for a system of m linear equations in n variables:

$$\begin{cases} a_{11}x_1 + a_{12}x_2 + a_{13}x_3 + \cdots + a_{1n}x_n = b_1 \\ a_{21}x_1 + a_{22}x_2 + a_{23}x_3 + \cdots + a_{2n}x_n = b_2 \\ a_{31}x_1 + a_{32}x_2 + a_{33}x_3 + \cdots + a_{3n}x_n = b_3 \\ \quad\vdots \\ a_{m1}x_1 + a_{m2}x_2 + a_{m3}x_3 + \cdots + a_{mn}x_n = b_m \end{cases}$$

The subscripts are called **double subscripts** and indicate the position of the corresponding constant coefficients. For instance, the subscript "23" indicates that a_{23} is the third constant in the second equation. The number a_{23} should be read "*a* sub two three" and not "*a* sub twenty-three." In general, a_{ij} represents

Systems of Linear Equations

the constant in equation i that is associated with (or multiplies) the jth variable; the coefficient a_{ij} is read "eh sub eye-jay."

A (simultaneous particular) solution for the above linear system consists of an ordered list or "n-tuple" $(s_1, s_2, \ldots, s_n)$ of numbers $s_1, s_2, \ldots, s_n$ such that $a_{i1}s_1 + a_{i2}s_2 + \cdots + a_{in}s_n = b_i$ for all $i = 1, 2, \ldots, m$. **The simultaneous solution** (or **solution set**) for a linear system is the set of all particular solutions for the system. A system of linear equations whose solution set is empty has no particular solutions and is said to be **inconsistent**. A linear system that has one or more particular solutions is said to be **consistent**.

It is also possible for a linear system to have infinitely many particular solutions. In this case the system is said to be **dependent**. Note that any dependent system is also consistent. You will see later that **any consistent linear system either has exactly one particular solution or is dependent**.

Examples of consistent, dependent, and inconsistent systems were given in section 1.3, along with geometric interpretation for this terminology. Although additional examples will be studied below, you may want to pause now and review those earlier examples. The systems in figures 9.1(a), (b), and (c) demonstrate geometrically some ways in which linear systems in two variables can be inconsistent. Analogous situations occur with linear systems of three or more variables, but cannot be graphed in the plane.

Figure 9.1

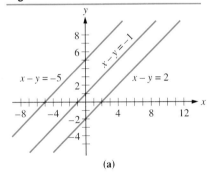 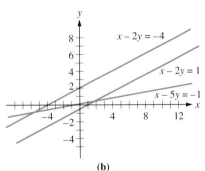 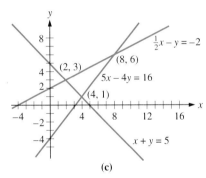

(a)　　　　　　　　　(b)　　　　　　　　　(c)

In section 1.3, you learned to solve systems of linear equations by the substitution method. As you will see here and in the next two sections, there are more efficient methods for solving linear systems. Example 9.2 will set the stage for this work.

EXAMPLE 9.2

Find the solution (set) of the linear system

$$\begin{cases} 4x + 2y - 9z = 7 & (1) \\ 6z = 2 & (2) \\ 5y + 3z = 6 & (3) \end{cases}$$

Solution

This system consists of $m = 3$ equations in $n = 3$ variables.

We begin by solving equation (2) for z, obtaining $z = \frac{2}{6} = \frac{1}{3}$. By substituting this value of z into equations (1) and (3), we have reduced the problem to solving a linear system of 2 ($= m - 1$) equations in 2 ($= n - 1$) variables:

$$\begin{cases} 4x + 2y - 3 = 7 & (4) \\ 5y + 1 = 6 & (5) \end{cases}$$

Solving equation (5) for y, we determine that $y = 1$. Substituting this value for y into equation (4) produces a "system" of 1 ($= m - 2$) linear equation in 1 ($= n - 2$) variable:

$$4x - 1 = 7. \qquad (6)$$

Solving equation (6) for x produces $x = 2$. Thus, we have obtained the only particular solution: $x = 2$, $y = 1$, $z = \frac{1}{3}$. The system's solution set is $\{(2, 1, \frac{1}{3})\}$. ◀

The linear system in example 9.2 was consistent and had a unique (particular) solution. As we saw in figure 9.1, not all linear systems are consistent. We next illustrate in example 9.3 the more complicated situation in which a consistent linear system has infinitely many particular solutions.

▶ **EXAMPLE 9.3**

Find the solution set of the system

$$\begin{cases} 2x + 3y + 5z = 5 & (1) \\ 3x - y - 2z = 2 & (2) \end{cases}$$

Solution

This system consists of $m = 2$ equations in $n = 3$ variables.

We begin by solving equation (2) for y, obtaining

$$y = 3x - 2z - 2.$$

Substituting this value for y into equation (1) produces

$$2x + 3(3x - 2z - 2) + 5z = 5$$

or

$$11x - z = 11. \qquad (3)$$

We have reduced the original linear system of two equations in three variables to a "system" of one linear equation in two variables. To find the solutions to equation (3), we could assign an arbitrary value to x and then solve for z, or we could assign an arbitrary value to z and then solve for x. We shall follow the first approach, and assign x an arbitrary real value t. Then, from equation (3), we determine that

$$x = t \quad \text{and} \quad z = 11t - 11.$$

Next, by substituting these expressions into equation (2), we obtain

$$3t - y - 2(11t - 11) = 2$$

or

$$y = -19t + 20.$$

Thus, we have determined the **general solution** to the given linear system, namely

$$x = t, \quad y = -19t + 20, \quad z = 11t - 11. \qquad (4)$$

(Notice that by broadening the vector algebra developed in section 7.5, we can also write the general solution, $(t, -19t + 20, 11t - 11)$, as the "linear combination" $(0, 20, -11) + t(1, -19, 11)$.)

The formulas in (4) describe the particular solutions in terms of an arbitrary "parameter" t. Since t can be any real number, the linear system has infinitely many particular solutions, one for each value of t, and is thus a dependent system. To obtain some particular solutions for the system, we substitute specific values for t in (4). For example, substituting $t = 1$, we find $x = 1, y = 1, z = 0$, namely the particular solution $(1, 1, 0)$; similarly, substituting $t = 0$ into (4) leads to the particular solution $(0, 20, -11)$. ◀

In example 9.3, we proceeded from equation (3) by assigning an arbitrary value to x and then solving for z. Let's see what would have happened if we had elected to assign an arbitrary value to z and then solved equation (3) for x. For example, if we let $z = s$, then we see that $x = \frac{1}{11}s + 1$. Substituting these expressions into equation (2), we obtain

$$3(\tfrac{1}{11}s + 1) - y - 2s = 2$$
$$y = -\tfrac{19}{11}s + 1.$$

We now have a different-appearing general solution to the given linear system, namely

$$x = \tfrac{1}{11}s + 1, \qquad y = -\tfrac{19}{11}s + 1, \qquad z = s. \tag{5}$$

The solutions given in (4) and (5) yield the same solution set. For example, if $s = 0$, we see that $x = 1, y = 1, z = 0$; this is the solution $(1, 1, 0)$ obtained via $t = 1$. Similarly, if $s = -11$, (5) yields the particular solution $(0, 20, -11)$ which corresponded to $t = 0$. In general, by letting $s = 11t - 11$, you can convert any particular solution resulting from (5) into a solution resulting from (4). Similarly, $t = \dfrac{s}{11} + 1$ converts (4) into (5). (Finally, notice that (5) may be used to write the general solution $\left(\dfrac{s}{11} + 1, -\dfrac{19}{11}s + 1, s\right)$ as the "linear combination" $(1, 1, 0) + s(\tfrac{1}{11}, -\tfrac{19}{11}, 1)$.)

Echelon Form

Examples 9.2 and 9.3 both demonstrated a solution technique known as **back substitution**; this is a version of the substitution method you learned in section 1.3. Our work might have been a little easier if we had written the equations in a different order. For example, consider this system of linear equations:

$$\begin{cases} 4x + 2y - 9z = 7 \\ 5y + 3z = 6 \\ 6z = 2 \end{cases}$$

This is the same set of equations considered in example 9.2, but written in a different order. You can quickly solve this rewritten system, showing that it has the same solution set as in example 9.2, namely $\{(2, 1, \tfrac{1}{3})\}$.

If two systems of linear equations have the same solution set, we say that the systems are **equivalent**. We are going to develop a method for solving a system of linear equations by first rewriting it as an equivalent system that has a "shape" like the one above. This shape is known as echelon form.

Echelon Form of a System of Linear Equations

SORT OF LIKE A PYRAMID
$2x + 3y = 9$
$5y = 8$

A system of linear equations in $x_1, x_2, x_3, \ldots, x_n$ is said to be in **echelon form** provided the following two conditions hold:

1. x_1 appears (with nonzero coefficient) in no equation after the first, x_2 appears in no equation after the second, x_3 appears in no equation after the third, and, in general, x_i appears in no equation after the ith;
2. x_i is not the first variable appearing with nonzero coefficient in two different equations.

It is possible, of course, that x_i does not appear in the ith equation.

▶ **EXAMPLE 9.4**

The following systems are all in echelon form.

$$\begin{cases} 2x + 3y = 9 \\ 2y = 8 \end{cases} \qquad \begin{cases} 4x + 2y - 9z = 7 \\ 5y + 3z = 6 \\ 6z = 2 \end{cases} \qquad \begin{cases} x_1 + 4x_2 - 5x_3 + x_4 = 6 \\ 3x_2 + x_4 = \sqrt{3} \\ -7x_3 + 2x_4 = 9 \\ 5x_4 = \sqrt{2} \end{cases}$$

$$\begin{cases} \sqrt{5}x_1 + 5x_2 - x_3 = 11 \\ 4x_3 = \frac{7}{5} \end{cases} \qquad \begin{cases} 2x_1 - 3x_3 + 4x_4 = 9 \\ 5x_2 - x_4 = 8 \\ 3x_4 = \sqrt{11} \end{cases}$$

None of the following systems is in echelon form.

$$\begin{cases} 3x + 2y = 11 \\ 7x - 4y = 3 \end{cases} \qquad \begin{cases} 6y = 5 \\ 4x - 5y = 2 \end{cases} \qquad \begin{cases} x_2 + x_3 = 11 \\ x_1 - 5x_2 + x_3 = \sqrt{2} \\ x_3 = 9 \end{cases} \qquad \begin{cases} x_1 - x_2 + 5x_3 = 7 \\ x_3 = -4 \\ x_3 = 6 \end{cases}$$

◀

Back Substitution

If a system of linear equations is in echelon form, **back substitution** is a two-step process for finding the system's general solution.

Back Substitution Procedures

Step 1: Solve each equation for its "leading" variable.
Step 2: Substitute the results of step 1 into the other equation(s). Simplify the equations, solving as necessary for the previously designated leading variables.

Example 9.5 illustrates how back substitution can be used on a system of linear equations that is in echelon form.

Systems of Linear Equations

▶ **EXAMPLE 9.5** Use back substitution to solve the (echelon form) linear system

$$\begin{cases} 2x + 5y - 6z = 11 \\ \frac{3}{4}y - 8z = -1 \\ 4z = 2 \end{cases}$$

Solution We begin with step 1, and solve each equation for its leading variable, obtaining

$$\begin{cases} x = -\frac{5}{2}y + 3z + \frac{11}{2} \\ y = \frac{32}{3}z - \frac{4}{3} \\ z = \frac{1}{2} \end{cases}$$

We now proceed to step 2. Substituting $z = \frac{1}{2}$ into the second equation yields

$$y = \frac{32}{3}(\tfrac{1}{2}) - \tfrac{4}{3} = \tfrac{16}{3} - \tfrac{4}{3} = 4.$$

Next, substituting $z = \frac{1}{2}$ and $y = 4$ into the first equation results in

$$x = -\tfrac{5}{2}(4) + 3(\tfrac{1}{2}) + \tfrac{11}{2} = -3.$$

Hence, the only solution to this equation is $x = -3$, $y = 4$, $z = \frac{1}{2}$; the solution set is $\{(-3, 4, \frac{1}{2})\}$. ◀

If a consistent linear system has more variables than equations, then it is a dependent system. If such a system is in echelon form, we solve it by assigning arbitrary values to any nonleading variables, as illustrated in example 9.6.

▶ **EXAMPLE 9.6** Use back substitution to solve the (echelon form) linear system

$$\begin{cases} 4x + 2y - \sqrt{3}\,z = 4\sqrt{3} \\ -3y = 6\sqrt{3} \end{cases}$$

Solution Notice that this system contains three variables but only two equations. We will eventually need to assign an arbitrary value to z, the only nonleading variable.

We begin by solving each equation for its leading variable, with the result

$$\begin{cases} x = -\frac{1}{2}y + \frac{1}{4}\sqrt{3}\,z + \sqrt{3} \\ y = -2\sqrt{3} \end{cases}$$

Substituting the expression for y into the first equation yields

$$x = 2\sqrt{3} + \tfrac{1}{4}\sqrt{3}\,z.$$

If we assign z the arbitrary real value t, then the solution set is given by the formulas

$$x = 2\sqrt{3} + \tfrac{1}{4}\sqrt{3}\,t, \qquad y = -2\sqrt{3}, \qquad z = t.$$

Chapter 9 / Systems of Linear Equations and Linear Inequalities

The general solution is $(x, y, z) = (2\sqrt{3}, -2\sqrt{3}, 0) + t(\frac{1}{4}\sqrt{3}, 0, 1)$ in "vector" notation. ◀

EXAMPLE 9.7 Use back substitution to solve each of the following (echelon form) linear systems:

$$\begin{cases} 2x + 5y - 3z = 1 \\ 4y + 3z = 2 \\ 3z = 6 \end{cases} \qquad \begin{cases} 2.4x + 16.8y + 4.5z = 6.6 \\ 4y + 3z = 2 \\ 3z = 6 \end{cases}$$

Solution Notice that the last two equations in the first system are the same as the last two equations in the second system. Solving each equation for its leading variable, we obtain

$$\begin{cases} x = -\frac{5}{2}y + \frac{3}{2}z + \frac{1}{2} \\ y = -\frac{3}{4}z + \frac{1}{2} \\ z = 2 \end{cases} \qquad \begin{cases} x = -7y - \frac{15}{8}z + \frac{11}{4} \\ y = -\frac{3}{4}z + \frac{1}{2} \\ z = 2 \end{cases}$$

By back substitution, we determine each system has only one particular solution, namely

$$x = 6, \qquad y = -1, \qquad z = 2.$$

In particular, the two given linear systems are equivalent. To motivate the topic of "elementary operations," note that

$$2.4x + 16.8y + 4.5z = 6.6$$

can be obtained by multiplying the equation $2x + 5y - 3z = 1$ through by 1.2, multiplying the equation $4y + 3z = 2$ through by 2.7, and then adding the results. ◀

Your experience with echelon form in examples 9.5 and 9.6 should enable you to prove the following result. It is theoretically important for the following reason. If a linear system can be converted into an equivalent linear system having echelon form, then a mere *glance* at the latter system determines whether the given system is consistent and, if so, whether it is dependent.

Theorem

Suppose that a system of linear equations in the variables $x_1, x_2, x_3, \ldots, x_n$ is given in echelon form. Then

(a) The linear system is consistent if and only if none of its equations has the form $0x_1 + 0x_2 + 0x_3 + \cdots + 0x_n = b_i$, where b_i is a nonzero constant.

(b) If the linear system is consistent and each variable is a leading variable, then the system has exactly one particular solution. If the linear system is consistent and at least one variable is a nonleading variable, then the system is dependent (and hence has infinitely many particular solutions).

Systems of Linear Equations

You should see how the conclusions in (b) apply to the systems studied in examples 9.5 and 9.6. In addition, notice that the situation described in (a) *can* arise, as the system

$$\begin{cases} x_1 + x_2 + x_3 = 1 \\ 0x_1 + 0x_2 + 0x_3 = 1 \end{cases}$$

is in echelon form and is obviously inconsistent. A more substantial application of (a) will be given in example 9.11.

The Elementary Operations

Not every system of linear equations is in echelon form. However the final remark in example 9.7 and our work with example 9.2 indicate some techniques that can be used to transform a given linear system into an equivalent system that is in echelon form. As mentioned earlier, two systems of linear equations are **equivalent systems** if and only if they have the same solution set. The following elementary operations can be used to replace a given system of linear equations with an equivalent system.

Elementary Operations on a System of Linear Equations

OPERATIONS YOU CAN USE (1-3)
SYMBOLS NEXT PAGE:

1. Multiply both sides of an equation by a given nonzero constant.
2. Interchange two equations.
3. Add a multiple of one equation in the system to another equation in the system.

▶ **EXAMPLE 9.8**

Use elementary operations and back substitution to solve the system of linear equations

$$\begin{cases} 2x - 4y - 4z = 6 \\ x - 5y - 3z = 2 \\ -2x + y + z = 3 \end{cases}$$

Solution

We shall use the elementary operations to convert this to a system having echelon form, and then use back substitution to solve the latter system. First, we want to eliminate x from the second and third equations. To eliminate x from the third equation, we shall add the first and third equations. Their sum is a new "third" equation and results in the equivalent system

$$\begin{cases} 2x - 4y - 4z = 6 \\ x - 5y - 3z = 2 \\ - 3y - 3z = 9 \end{cases}$$

Next, to eliminate x from the second equation, add to it $-\frac{1}{2}$ times the first equation. This produces the equivalent system

$$\begin{cases} 2x - 4y - 4z = 6 \\ - 3y - z = -1 \\ - 3y - 3z = 9 \end{cases}$$

Finally, to complete the transformation of the given system to echelon form, we need to eliminate y from the third equation. This is done by adding -1 times the second equation to the third equation, with the result

$$\begin{cases} 2x - 4y - 4z = 6 \\ -3y - z = -1 \\ -2z = 10 \end{cases}$$

Thus, the given system of linear equations has been transformed into an equivalent system in echelon form. Using back substitution on this second system, we obtain the unique particular solution, given by

$$x = -3, \quad y = 2, \quad z = -5. \quad \blacktriangleleft$$

It was necessary to write a sentence or two explaining each elementary operation used in example 9.8. To reduce the amount of writing, we shall adopt some shorthand notation. An arrow will indicate a step involving one elementary operation. Above each arrow will appear a symbol to explain the specific elementary operation used in that step. In all cases, R_i refers to the equation in the ith row, or simply row i.

Elementary Operations

Symbol	Explanation
kR_i	Multiply each element in row i by k, a nonzero real constant.
$R_i \leftrightarrow R_j$	Interchange row i and row j.
$kR_i + R_j$	To row j, add k times row i. (The result is a new row j.)

▶ **EXAMPLE 9.9** Use elementary operations and back substitution to solve the system of linear equations

$$\begin{cases} 9x + 6y - 5z = 11 \\ -3x - 2y + 4z = 8 \\ 6x - 4y + 3z = -4 \end{cases}$$

Solution

$$\begin{cases} 9x + 6y - 5z = 11 \\ -3x - 2y + 4z = 8 \\ 6x - 4y + 3z = -4 \end{cases} \xrightarrow{2R_2 + R_3} \begin{cases} 9x + 6y - 5z = 11 \\ -3x - 2y + 4z = 8 \\ -8y + 11z = 12 \end{cases}$$

$$\xrightarrow{\frac{1}{3}R_1 + R_2} \begin{cases} 9x + 6y - 5z = 11 \\ \frac{7}{3}z = \frac{35}{3} \\ -8y + 11z = 12 \end{cases}$$

$$\xrightarrow{R_2 \leftrightarrow R_3} \begin{cases} 9x + 6y - 5z = 11 \\ -8y + 11z = 12 \\ \frac{7}{3}z = \frac{35}{3} \end{cases}$$

Using back substitution, we obtain the unique particular solution

$$x = \tfrac{5}{12}, \quad y = \tfrac{43}{8}, \quad z = 5.$$

EXAMPLE 9.10

Use the elementary operations and back substitution to solve the system of linear equations

$$\begin{cases} x_1 + 2x_2 + 2x_3 = 3 \\ 2x_1 - 3x_2 - 3x_3 = -1 \\ 3x_1 + 4x_2 + 3x_3 = 2 \\ 4x_1 - 5x_2 - 6x_3 = -6 \end{cases}$$

Solution

$$\begin{cases} x_1 + 2x_2 + 2x_3 = 3 \\ 2x_1 - 3x_2 - 3x_3 = -1 \\ 3x_1 + 4x_2 + 3x_3 = 2 \\ 4x_1 - 5x_2 - 6x_3 = -6 \end{cases} \xrightarrow{-2R_1 + R_2} \begin{cases} x_1 + 2x_2 + 2x_3 = 3 \\ - 7x_2 - 7x_3 = -7 \\ 3x_1 + 4x_2 + 3x_3 = 2 \\ 4x_1 - 5x_2 - 6x_3 = -6 \end{cases}$$

$$\xrightarrow{-3R_1 + R_3} \begin{cases} x_1 + 2x_2 + 2x_3 = 3 \\ - 7x_2 - 7x_3 = -7 \\ - 2x_2 - 3x_3 = -7 \\ 4x_1 - 5x_2 - 6x_3 = -6 \end{cases}$$

$$\xrightarrow{-4R_1 + R_4} \begin{cases} x_1 + 2x_2 + 2x_3 = 3 \\ - 7x_2 - 7x_3 = -7 \\ - 2x_2 - 3x_3 = -7 \\ - 13x_2 - 14x_3 = -18 \end{cases}$$

$$\xrightarrow{-\frac{1}{7}R_2} \begin{cases} x_1 + 2x_2 + 2x_3 = 3 \\   x_2 + x_3 = 1 \\ - 2x_2 - 3x_3 = -7 \\ - 13x_2 - 14x_3 = -18 \end{cases}$$

$$\xrightarrow{2R_2 + R_3} \begin{cases} x_1 + 2x_2 + 2x_3 = 3 \\   x_2 + x_3 = 1 \\   - x_3 = -5 \\ - 13x_2 - 14x_3 = -18 \end{cases}$$

$$\xrightarrow{13R_2 + R_4} \begin{cases} x_1 + 2x_2 + 2x_3 = 3 \\   x_2 + x_3 = 1 \\   - x_3 = -5 \\   - x_3 = -5 \end{cases}$$

At this point, we could solve this system by back substitution, determining that

$$x_1 = 1, \quad x_2 = -4, \quad x_3 = 5.$$

In order to help you understand the echelon form of a system with more equations than variables, we continue applying elementary operations, in order to complete the transformation of the given system into an equivalent system having echelon form. To do this, we apply $(-1)R_3 + R_4$ (subtracting the third equation from the fourth), obtaining a new fourth equation, $0x_1 + 0x_2 + 0x_3 = 0$. The given system has now been transformed to echelon form:

$$\begin{cases} x_1 + 2x_2 + 2x_3 = 3 \\ \phantom{x_1 + {}} x_2 + x_3 = 1 \\ \phantom{x_1 + 2x_2 + {}} -x_3 = -5 \\ 0 = 0 \end{cases}$$

Notice that the earlier theorem predicts correctly that the system has a unique particular solution. ◀

The next example will demonstrate what happens when elementary operations are used to solve an inconsistent system of linear equations. Recall that the earlier theorem tells us to expect a transformed equation $0x_1 + 0x_2 + \cdots + 0x_n = b_i$, with $b_i \neq 0$, in this case.

▶ **EXAMPLE 9.11**

Use elementary operations and back substitution to solve the system of linear equations

$$\begin{cases} 3x_1 + 2x_2 = 5 \\ x_1 - x_2 = 4 \\ 2x_1 - 3x_2 = 4 \end{cases}$$

Solution

$$\begin{cases} 3x_1 + 2x_2 = 5 \\ x_1 - x_2 = 4 \\ 2x_1 - 3x_2 = 4 \end{cases} \xrightarrow{R_1 \leftrightarrow R_2} \begin{cases} x_1 - x_2 = 4 \\ 3x_1 + 2x_2 = 5 \\ 2x_1 - 3x_2 = 4 \end{cases}$$

$$\xrightarrow{-3R_1 + R_2} \begin{cases} x_1 - x_2 = 4 \\ \phantom{x_1 - {}} 5x_2 = -7 \\ 2x_1 - 3x_2 = 4 \end{cases}$$

$$\xrightarrow{-2R_1 + R_3} \begin{cases} x_1 - x_2 = 4 \\ \phantom{x_1 - {}} 5x_2 = -7 \\ {} - x_2 = -4 \end{cases}$$

$$\xrightarrow{\frac{1}{5}R_2 + R_3} \begin{cases} x_1 - x_2 = 4 \\ \phantom{x_1 - {}} 5x_2 = -7 \\ 0 = -\frac{27}{5} \end{cases}$$

The last equation has the form $0x_1 + 0x_2 + 0x_3 = b_3$, with $b_3 \neq 0$. By the criterion noted above, the given linear system is inconsistent (and so has an empty solution set). ◀

Systems of Linear Equations

Any linear system can be converted by elementary operations into an equivalent linear system in echelon form. Thus, we can state the following.

Theorem

> Elementary operations and back substitution enable you to find the solution set of any linear system.

You will learn other methods for solving linear systems in sections 9.2 and 9.3.

As the final example in this section, we consider an application that uses a system of linear equations.

EXAMPLE 9.12

A store owner purchases large quantities of mixed nuts from three distributors, and wants to mix the nuts to form the store's own special mixture. Mixture A (from Distributor A) contains 60 percent peanuts, 10 percent cashews, and 30 percent almonds. Mixture B contains 50 percent peanuts and 50 percent cashews, while Mixture C is 40 percent peanuts, 30 percent cashews, and 30 percent almonds. How much of each mixture must be used in order to get a final 50 kilogram mixture that is 52 percent peanuts, 24 percent cashews, and 24 percent almonds?

Solution

We begin by letting A, B, and C represent the number of kilograms of each mixture that will be used to make the store's special mixture. Thus, $A + B + C = 50$. We know that the store's mixture will contain 52 percent peanuts (or 26 kg), and that this represents 60 percent of A, 50 percent of B, and 40 percent of C. Hence,

$$0.6A + 0.5B + 0.4C = 26.$$

Similarly, the final mixture is 24 percent (or 12 kg) cashews, and this represents 10 percent A, 50 percent B and 30 percent C. Thus,

$$0.1A + 0.5B + 0.3C = 12.$$

Finally, for the almonds, we obtain the equation

$$0.3A + 0.3C = 12.$$

Thus, we have a system of four linear equations in three variables.

$$\begin{cases} A + B + C = 50 \\ 0.6A + 0.5B + 0.4C = 26 \\ 0.1A + 0.5B + 0.3C = 12 \\ 0.3A + 0.3C = 12 \end{cases}$$

Chapter 9 / Systems of Linear Equations and Linear Inequalities

In deriving the solution of this system, we allow most "steps" to combine a few elementary operations, with self-evident notation:

$$\begin{cases} A + B + C = 50 \\ 0.6A + 0.5B + 0.4C = 26 \\ 0.1A + 0.5B + 0.3C = 12 \\ 0.3A \qquad\quad + 0.3C = 12 \end{cases} \xrightarrow{6R_1 - 10R_2} \begin{cases} A + B + C = 50 \\ \quad\quad B + 2C = 40 \\ 0.1A + 0.5B + 0.3C = 12 \\ 0.3A \qquad\quad + 0.3C = 12 \end{cases}$$

$$\xrightarrow{-R_1 + 10R_3} \begin{cases} A + B + C = 50 \\ B + 2C = 40 \\ 4B + 2C = 70 \\ 0.3A \quad\quad + 0.3C = 12 \end{cases}$$

$$\xrightarrow{3R_1 - 10R_4} \begin{cases} A + B + C = 50 \\ B + 2C = 40 \\ 4B + 2C = 70 \\ 3B \quad\quad = 30 \end{cases}$$

$$\xrightarrow{R_2 \leftrightarrow R_4} \begin{cases} A + B + C = 50 \\ 3B \quad\quad = 30 \\ 4B + 2C = 70 \\ B + 2C = 40 \end{cases}$$

$$\xrightarrow{4R_4 - R_3} \begin{cases} A + B + C = 50 \\ 3B \quad\quad = 30 \\ 6C = 90 \\ B + 2C = 40 \end{cases}$$

$$\xrightarrow{\frac{1}{3}(R_2 + R_3) - R_4} \begin{cases} A + B + C = 50 \\ 3B \quad\quad = 30 \\ 6C = 90 \\ 0 = 0 \end{cases}$$

From the first three equations in this last system, we obtain the unique particular solution: $C = 15$, $B = 10$, and $A = 25$. So, the special mixture will use 25 kg from Mixture A, 10 kg from Mixture B, and 15 kg from Mixture C.

◀

Exercises 9.1

1. Determine which of the following equations is/are linear:
(a) $\frac{2}{3}x - y = \sqrt{2}$ (b) $x^{2/3} = 3$ (c) $x - y^2 = 0$
(d) $3x_1 + 2x_2 + \sqrt{2}x_3 = \pi$ (e) $x_3 e^{x_1} - x_2 = 0$
(f) $\sin x - y = i$

2. $\begin{cases} -\frac{2}{3}x + 5z = \sqrt{2} \\ 7y = 1 \end{cases}$

is a system of m linear equations in n variables.
(a) Determine m and n.
(b) Is the given system in echelon form?

3. Prove or disprove: Any two inconsistent linear systems in the same variables are equivalent. (Hint: there is only one empty set.)

4. (a) How many real numbers x satisfy $e^x = e$?
(b) How many real numbers x satisfy $e^{|x|} = e$?
(**Moral:** a **nonlinear** consistent system with more than one particular solution need not have an infinite solution set.)

Systems of Linear Equations

5. Determine which of the following linear systems is/are in echelon form:

   (a) $\begin{cases} 2x - 3y + 4z = 0 \\ - 8y - z = 7\sqrt{2} \end{cases}$
   (b) $\begin{cases} -x + y = 14 \\ 5x = 10 \end{cases}$

   (c) $\begin{cases} 3x_1 + 2x_2 = 6 \\ 3x_2 - 4x_3 = 1 \\ -8x_1 + x_3 = 0 \end{cases}$
   (d) $\begin{cases} 2x_1 - x_3 = 1 \\ 3x_2 - 4x_3 = 1 \\ \sqrt{2}x_3 = 4 \end{cases}$

6. (a) *Without finding solution sets*, determine which of the linear systems having echelon form in exercise 5 is/are consistent.
   (b) Of the consistent systems identified in (a), determine which is dependent and which has a unique particular solution. (Hint: it is *not* necessary to find the solution sets here either.)

7. Without referring to echelon form, give a direct proof that a consistent linear system with at least two particular solutions has infinitely many particular solutions. (Hint: if $v = (r_1, r_2, \ldots, r_n)$ and $w = (s_1, s_2, \ldots, s_n)$ are distinct particular solutions and t is any real number, consider $v + t(w - v) = ((1 - t)r_1 + ts_1, (1 - t)r_2 + ts_2, \ldots, (1 - t)r_n + ts_n).$)

8. Show that each elementary operation converts a linear system into an equivalent linear system. (Hint: this is evident except possibly for the third type of operation. For this, you need to verify the following. Let $a, b, c, d,$ and k be numbers. Then $a = b$ and $c = d$ if (and only if) $a = b$ and $ka + c = kb + d$.)

In exercises 9–10, apply the elementary operations (a) $2R_3$, (b) $R_1 \leftrightarrow R_2$, and (c) $-4R_2 + R_3$ separately (not in sequence) to the given linear system.

9. $\begin{cases} \sqrt{2}x \phantom{- \frac{y}{2}} - 5z = 0 \\ 4x - \frac{y}{2} + z = 6 \\ - 3y + 2z = \frac{3}{4} \end{cases}$

10. $\begin{cases} \phantom{\frac{2}{3}x_1 -} 3x_2 - 2x_3 + x_4 = 1 \\ \frac{2}{3}x_1 - x_2 + 5x_3 = 6 \\ \phantom{\frac{2}{3}x_1 - x_2 +} 6x_3 - x_4 = \pi \\ \phantom{\frac{2}{3}x_1 -} 3x_2 = 7 \\ 5x_1 - 6x_4 = 5 \end{cases}$

11. (a) What is the effect of following an elementary operation $R_i \leftrightarrow R_j$ by $R_j \leftrightarrow R_i$?
    (b) What is the effect of following an elementary operation kR_i by $k^{-1}R_i$?
    (c) What is the effect of following an elementary operation $kR_i + R_j$ by $-kR_i + R_j$?
    (**Moral**: the "inverse" of each elementary operation is also an elementary operation.)

12. Show how the elementary operation $R_i \leftrightarrow R_j$ results by applying four suitable elementary operations of type $kR_p + R_q$ or kR_p. (Hint: consider $-R_j + R_i$, $(-1)R_i$, $-R_i + R_j$, and $R_j + R_i$.)

13. Consider the linear system
$$\begin{cases} 2x + 3y + 4z = 48 \\ 3x = -12 \end{cases}$$
    (a) Show that this system has the general solution
    $$(x, y, z) = \left(-4, t, 14 - \frac{3t}{4}\right).$$
    (b) Show that this system's general solution can also be expressed as $(x, y, z) = \left(-4, \frac{56 - 4s}{3}, s\right)$.
    (c) Given any particular solution expressed as in (a) via t, show which s (defined in terms of t) recovers this particular solution in the form expressed in (b).

14. Show that the linear system
$$\begin{cases} 2x_1 + 3x_2 - 4x_3 + x_4 = 0 \\ 2x_2 + x_3 - x_4 = 1 \\ 2x_1 + 5x_2 - 3x_3 = 1 \\ -4x_1 - 4x_2 + 9x_3 - 3x_4 = 1 \end{cases}$$
has the general solution
$$(x_1, x_2, x_3, x_4) = \left(\frac{-3 + 11s - 5t}{4}, \frac{1 - s + t}{2}, s, t\right)$$
in terms of the "independent" parameters s and t.

In exercises 15–38, find the solution set of the given linear system by first using elementary operations to transform it to an equivalent system in echelon form and then using back substitution.

15. $\begin{cases} x - 2y - z = 0 \\ 4y + 3z = 4 \\ \frac{z}{2} = 1 \end{cases}$
16. $\begin{cases} 2x - 3y + 5z = 4 \\ 3z = -15 \end{cases}$

17. $\begin{cases} 2x_1 + x_2 = 5 \\ x_1 - 2x_2 = -4 \\ 0x_1 + 0x_2 = 2 \end{cases}$
18. $\begin{cases} 2x_1 - x_2 - x_3 = -\frac{1}{\pi} \\ 3x_2 - \pi x_3 = 2 \\ 4\pi x_3 = 4 \end{cases}$

19. $\begin{cases} 2x - 3y + 6z = 1 \\ 5y - 7z = 0 \\ 0x + 0y + 0z = 0 \end{cases}$
20. $\begin{cases} \sqrt{2}x - 2z = 0 \\ 0x + 0y + 0z = \sqrt{2} \end{cases}$

21. $\begin{cases} 2x_1 - 3x_2 = 7 \\ -x_1 + 2x_2 + x_3 = 1 \\ 8x_1 - 13x_2 - 2x_3 = 19 \end{cases}$
22. $\begin{cases} x_1 + 2x_2 - 4x_3 = 0 \\ -3x_1 + 6x_2 + x_3 = 2 \\ 5x_1 - 2x_2 - 9x_3 = 1 \end{cases}$

23. $\begin{cases} 3x - 5y + 6z = 60 \\ \dfrac{x}{5} - \dfrac{y}{3} + \dfrac{2}{5}z = 4 \end{cases}$

24. $\begin{cases} 12x - 36y + 48z = 24 \\ x - 2y + 5z = 2 \end{cases}$

25. $\begin{cases} 12x_1 - 36x_2 + 48x_3 = 24 \\ x_1 - 2x_2 + 5x_3 = 2 \\ x_1 - 3x_2 + 4x_3 = 2 \\ 2x_1 - x_2 + 13x_3 = 5 \end{cases}$

26. $\begin{cases} 5x - y = 2 \\ 2x + z = 0 \\ 2y + 7z = 0 \\ 3x + 3y + 13z = 2 \end{cases}$

27. $\begin{cases} x_1 + 2x_3 = 1 \\ 3x_2 - x_3 = 0 \\ 5x_1 - 12x_2 + 14x_3 = 6 \end{cases}$

28. $\begin{cases} x + 2y - 4z = 0 \\ -3x - 2y + 9z = -12 \\ -\dfrac{x}{2} + \dfrac{5}{4}z = -3 \end{cases}$

29. $\begin{cases} 4x - 8y + 12z = 8 \\ -5x + 10y - 15z = 10 \end{cases}$

30. $\begin{cases} 4x - 8y + 12z = 8 \\ -5x + 10y - 15z = -10 \end{cases}$

31. $\begin{cases} x_1 + \dfrac{1}{2}x_2 = \dfrac{1}{2} \\ -2x_1 - x_2 = -1 \\ 2x_2 - x_3 = 6 \\ x_2 - \pi x_3 = 3 \end{cases}$

32. $\begin{cases} x_2 + 5x_3 = 0 \\ x_1 + 2x_2 = -1 \\ 4x_1 + 11x_2 + 15x_3 = -4 \\ -x_1 - 3x_2 - 5x_3 = 1 \end{cases}$

33. $\begin{cases} \dfrac{1}{2}x_1 - x_3 = 0 \\ 3x_1 + 2x_2 = 2 \\ -x_1 + x_2 = -9 \end{cases}$

34. $\begin{cases} x_1 - x_2 = 1 \\ x_2 + x_4 = 0 \\ x_2 - x_3 - x_4 = 2 \\ x_1 + x_3 = 0 \end{cases}$

35. $\begin{cases} 12x - 36y + 48z = 24 \\ x - 3y + 4z = 2 \end{cases}$

36. $\begin{cases} 12x - 36y + 48z = 24 \\ x - 3y + 4z = 3 \end{cases}$

37. $\begin{cases} x_1 - x_2 + x_3 = 4 \\ -2x_1 + 2x_2 - 3x_3 = 8 \\ 4x_1 - 4x_2 + 5x_3 = 0 \\ x_1 - x_2 = 20 \end{cases}$

38. $\begin{cases} x_1 - x_2 + x_3 = 4 \\ 2x_1 + 2x_2 = 1 \\ x_1 - 5x_2 + 3x_3 = 10 \\ -x_1 + x_2 - x_3 = -4 \end{cases}$

39. Tohoot University's groundskeeper needs to mix 200 kilograms of fertilizer that will be 24.95 percent nitrogen (N). To do this, she uses Mixture A (which is 20 percent N), Mixture B (which is 15 percent N), and Mixture C (which is 30 percent N). Assuming that she uses 4 kg more of Mixture B than of Mixture A, how many kilograms of each fertilizer mixture does she use?

40. The only road connecting points A and B is 75 miles long. As one travels from A to B, this road is level for x miles, rises for y miles, and falls for z miles. (So, $x + y + z = 75$.) It is known that $x = y + 30$. In his old car, David chooses to travel 40 mph on level roads, 20 mph when the road is rising, and 30 mph when descending. If David takes 144 minutes to drive this car from A to B and 135 minutes to drive it from B to A, determine x, y, and z. (Hint: for how many miles does the road from B to A rise?)

41. Find an equation for the circle passing through the points $(-2, 3)$, $(4, 0)$, and $(-1, 5)$. (Hint: the equation may be taken in the form $x^2 + y^2 + ax + by + c = 0$. Construct a linear system of three equations in the three "variables" a, b, and c.)

42. (a) Show that if you adapt the method suggested in exercise 41 to find an equation for a circle passing through the points $(-2, 3)$, $(4, 0)$, and $(2, 1)$, the resulting linear system is inconsistent.
    (b) What does the result of (a) imply about the three points mentioned in (a)?

43. Louise, Lewis, and Lois are house painters. As a team, Louise and Lewis could paint a certain house in $\dfrac{180}{11}$ hours. Working together, Louise and Lois could paint it in $\dfrac{360}{19}$ hours, while the team of Lewis and Lois would take $\dfrac{120}{7}$ hours. Determine how many hours each of the painters would need to paint this house if he or she worked alone. (Hint: let A, B, and C denote these times. Construct a linear system of three equations in the "variables" $\dfrac{1}{A}$, $\dfrac{1}{B}$, and $\dfrac{1}{C}$. The first equation is $\dfrac{1}{A} + \dfrac{1}{B} = \dfrac{11}{180}$.)

44. An industrial vat has two inlet tubes and one drain. One tube can fill at twice the rate of the other tube. When both tubes and the drain are operating, the vat fills at the rate of 65 ml/min. When only the weaker tube and the drain are operating, the vat drains at 27 ml/min. Find the rates at which the tubes and the drain can operate. (Hint: if those rates are x, y, and z ml/min, take $x = 2y$, $x + y - z = 65$, and $y - z = -27$.)

In exercises 45–49, determine which (if any) value(s) of k make the given linear system (a) inconsistent, (b) dependent, (c) consistent, with a unique particular solution.

45. $\begin{cases} 2x_1 - 3x_2 + x_3 = 5 \\ \dfrac{1}{2}x_2 + \pi x_3 = 0 \\ kx_3 = k \end{cases}$

46. $\begin{cases} 2x_1 - 3x_2 + x_3 = 5 \\ \dfrac{1}{2}x_2 + \pi x_3 = 0 \\ kx_3 = 2 \end{cases}$

47. $\begin{cases} 2x_1 - 3x_2 + x_3 = 5 \\ \dfrac{1}{2}x_2 + \pi x_3 = 0 \\ 2x_3 = k \end{cases}$

48. $\begin{cases} 2x_1 - 3x_2 + x_3 = 5 \\ kx_2 + \pi x_3 = 0 \\ 2x_3 = 2 \end{cases}$

49. $\begin{cases} 2x_1 - 3x_2 + x_3 = 5 \\ 2x_1 - 3x_2 + kx_3 = k \end{cases}$

50. By generalizing the reasoning in example 9.8, give an intuitive argument showing that any linear system can be transformed into an equivalent system in echelon form by successively applying suitable elementary operations. (Hint: if $a_{i1} \neq 0$, begin with $R_1 \leftrightarrow R_i$ and then follow with $-a_{11}^{-1} a_{j1} R_1 + R_j$ for $j = 2, \ldots, m$.)

*9.2 MATRICES AND THEIR USE IN SOLVING SYSTEMS OF LINEAR EQUATIONS

This section introduces certain mathematical tools, called matrices, that are used in studying many problems, both within and outside mathematics. The introduction of matrices will lead to a streamlined method for solving systems of linear equations. In addition, the text and exercises will give you a glimpse into the extensive theory of matrices.

Matrix

A **matrix** (plural: **matrices**) is a rectangular arrangement of numbers, enclosed in parentheses or brackets.

For instance,

$$\begin{pmatrix} 9 & 7 \\ -5 & \sqrt{3} \end{pmatrix} \quad \text{and} \quad \begin{bmatrix} 8 & \frac{1}{2} & 4 \\ -2 & \sqrt{5} & \pi \end{bmatrix}$$

are examples of matrices. In general, we have the following definition of a matrix.

$m \times n$ Matrix

Let m and n be positive integers. An **$m \times n$ matrix** is an array of numbers arranged in a form like

$$\begin{bmatrix} a_{11} & a_{12} & a_{13} & \cdots & a_{1n} \\ a_{21} & a_{22} & a_{23} & \cdots & a_{2n} \\ a_{31} & a_{32} & a_{33} & \cdots & a_{3n} \\ \vdots & & & & \vdots \\ a_{m1} & a_{m2} & a_{m3} & \cdots & a_{mn} \end{bmatrix}$$

Notice that this definition employs the same double subscript notation used in section 9.1 to describe systems of linear equations. Specifically, a_{ij} denotes the number that appears in row i and column j. Each a_{ij} is called an **entry** (or **element**) **of the matrix**. (While the entries a_{ij} can be complex numbers, we shall restrict them to real numbers.) The elements $a_{11}, a_{22}, a_{33}, \ldots$ are called the **main diagonal entries**. It is customary to denote a matrix by a capital letter, such as A. The above matrix may also be denoted by $[a_{ij}]$.

The notation $m \times n$ in the above definition is read "m by n" and indicates the **size** of the matrix. Here, m denotes the number of rows and n the number of columns of the matrix. If $m = n$, the matrix is called a **square matrix of order n**.

In a sense, the ith row $[a_{i1} \quad a_{i2} \quad \cdots \quad a_{in}]$ is a "disembodied" form of the linear expression $a_{i1}x_1 + a_{i2}x_2 + \cdots + a_{in}x_n$. The row notation is more com-

---

*This section is optional. However, the material in it is needed for the next optional section, section 9.3.

Chapter 9 / Systems of Linear Equations and Linear Inequalities

pact, in the same manner as the "detached coefficients" were when we studied synthetic division in section 3.2.

Algebra of Matrices

Being able to calculate with matrices is important in many areas. Much of the algebra of matrices is just as one would expect, and will be included in the exercises. As you will see, however, multiplication of matrices requires additional attention.

Equal Matrices

Two matrices are said to be **equal** if and only if they have the same size and their corresponding entries are equal.

▶ **EXAMPLE 9.13**

$$\begin{bmatrix} 5 & \sqrt{3} \\ 2 & 0 \end{bmatrix} = \begin{bmatrix} 5 & \sqrt{3} \\ 2 & 0 \end{bmatrix}, \quad \begin{bmatrix} \sqrt{2} & 3 \\ \pi & 4 \end{bmatrix} = \begin{bmatrix} \frac{2}{\sqrt{2}} & 2+1 \\ \pi & 7-3 \end{bmatrix},$$

$$\begin{bmatrix} 1 & 2 & 3 \\ 4 & 5 & 6 \end{bmatrix} \neq \begin{bmatrix} 1 & 2 \\ 4 & 5 \end{bmatrix}, \quad \begin{bmatrix} 3 & 1 \\ 4 & 2 \end{bmatrix} \neq \begin{bmatrix} 4 & 1 \\ 3 & 2 \end{bmatrix}, \quad \begin{bmatrix} 1 & 2 \\ 0 & 0 \end{bmatrix} \neq \begin{bmatrix} 1 & 2 \end{bmatrix} \quad ◀$$

Next we see that, just as when we studied vectors, scalar multiplication means multiplying each entry by a real number constant.

Scalar Multiple

If k is a real number and A is an $m \times n$ matrix, then kA denotes the **scalar multiple** (of A by k), the $m \times n$ matrix formed by multiplying each entry of A by k.

▶ **EXAMPLE 9.14**

If $A = \begin{bmatrix} 3 & 5 \\ 2 & -1 \end{bmatrix}$, determine $5A$.

Solution

$$5A = 5\begin{bmatrix} 3 & 5 \\ 2 & -1 \end{bmatrix} = \begin{bmatrix} 5(3) & 5(5) \\ 5(2) & 5(-1) \end{bmatrix} = \begin{bmatrix} 15 & 25 \\ 10 & -5 \end{bmatrix}. \quad ◀$$

Row Vector

A matrix that consists of exactly one row is called a **row vector**.

The 1×2 matrix $A = [a_{11} \quad a_{12}]$ is a row vector, which may be identified with the vector notation $\mathbf{v} = \langle a_{11}, a_{12} \rangle$ introduced in section 7.4. Notice that if k is a real number, the scalar multiple $kA = [ka_{11} \quad ka_{12}]$ is then essentially the same as the scalar multiplication $k\mathbf{v} = \langle ka_{11}, ka_{12} \rangle$ in the sense of section 7.4.

Matrices and Their Use in Solving Systems of Linear Equations

Column Vector

A matrix that consists of exactly one column is called a **column vector**.

Now, consider a row vector R and a column vector C, each having n entries:

$$R = [r_1 \quad r_2 \quad r_3 \quad \cdots \quad r_n], \qquad C = \begin{bmatrix} c_1 \\ c_2 \\ c_3 \\ \vdots \\ c_n \end{bmatrix}.$$

In this case, we have the following definition.

Inner Product

The **inner product** $R \cdot C$ is defined to be the following number:

$$R \cdot C = [r_1 \quad r_2 \quad r_3 \quad \cdots \quad r_n] \cdot \begin{bmatrix} c_1 \\ c_2 \\ c_3 \\ \vdots \\ c_n \end{bmatrix} = r_1 c_1 + r_2 c_2 + r_3 c_3 + \cdots + r_n c_n.$$

If $n = 2$ and we consider the vectors $\mathbf{v} = \langle r_1, r_2 \rangle$ and $\mathbf{w} = \langle c_1, c_2 \rangle$, notice that $R \cdot C$ is the same as the dot product $\mathbf{v} \cdot \mathbf{w}$, as defined in section 7.5. Thus, the notion of inner product generalizes that of dot product. Exercise 7 will confirm that the earlier properties of dot product carry over as properties of inner product.

You should note the following two points about the inner product notation $R \cdot C$:

(1) R and C must have the same number of entries.
(2) The row vector R is written to the left of the column vector C.

EXAMPLE 9.15

Calculate the inner product of $[3 \quad 0 \quad 2 \quad -7]$ and $\begin{bmatrix} 4 \\ 5 \\ -1 \\ 6 \end{bmatrix}$.

Solution

$$[3 \quad 0 \quad 2 \quad -7] \cdot \begin{bmatrix} 4 \\ 5 \\ -1 \\ 6 \end{bmatrix} = 3(4) + 0(5) + 2(-1) + (-7)6 = 12 + 0 - 2 - 42 = -32.$$

◀

We next use the concept of inner product to define matrix multiplication.

Chapter 9 / Systems of Linear Equations and Linear Inequalities

Matrix Multiplication

If A is an $m \times r$ matrix and B is an $r \times n$ matrix, then the **product** AB is the $m \times n$ matrix $[p_{ij}]$, where the entry p_{ij} is the inner product of the ith row of A and the jth column of B.

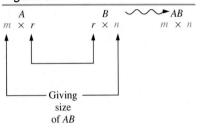

Figure 9.2

Notice that p_{ij} is formed by taking the product of each entry in row i of A with the corresponding entry in column j of B and then adding these r products together. For a matrix product AB to make sense, the definition **requires** that the number of columns in A be the same as the number of rows in B: see figure 9.2.

▶ **EXAMPLE 9.16**

If $C = [3 \ -2]$ and $D = \begin{bmatrix} 1 & 5 & 6 \\ 4 & 0 & -3 \end{bmatrix}$, calculate CD.

Solution

Here C is a 1×2 matrix and D is a 2×3 matrix. The product $P = CD$ is therefore defined and is a 1×3 matrix. If $P = [p_{11} \ p_{12} \ p_{13}]$, then

$$p_{11} = 3(1) + (-2)4 = 3 - 8 = -5$$
$$p_{12} = 3(5) + (-2)0 = 15 - 0 = 15$$
$$p_{13} = 3(6) + (-2)(-3) = 18 + 6 = 24.$$

Thus, we have $[3 \ -2] \begin{bmatrix} 1 & 5 & 6 \\ 4 & 0 & -3 \end{bmatrix} = [-5 \ 15 \ 24]$. ◀

▶ **EXAMPLE 9.17**

If $E = \begin{bmatrix} 1 & 2 & 3 \\ 4 & -2 & -1 \end{bmatrix}$ and $F = \begin{bmatrix} 2 & -7 & 5 & -3 \\ 0 & 1 & 4 & 1 \\ -1 & 6 & -2 & 0 \end{bmatrix}$, calculate EF.

Solution

Here, E is 2×3 and F is 3×4. The product EF thus has size 2×4; that is,

$$EF = \begin{bmatrix} p_{11} & p_{12} & p_{13} & p_{14} \\ p_{21} & p_{22} & p_{23} & p_{24} \end{bmatrix}.$$

To determine p_{11}, we single out row 1 from E and column 1 from F, and then calculate their inner product. Schematically, the process looks like this:

$$\begin{bmatrix} 1 & 2 & 3 \\ 4 & -2 & -1 \end{bmatrix} \cdot \begin{bmatrix} 2 & -7 & 5 & -3 \\ 0 & 1 & 4 & 1 \\ -1 & 6 & -2 & 0 \end{bmatrix} = \begin{bmatrix} -1 & p_{12} & p_{13} & p_{14} \\ p_{21} & p_{22} & p_{23} & p_{24} \end{bmatrix}$$

$$p_{11} = 1(2) + 2(0) + 3(-1) = 2 + 0 - 3 = -1.$$

Similarly, p_{23} is the inner product of the second row of E and the third column of F. Schematically, it looks like this:

$$\begin{bmatrix} 1 & 2 & 3 \\ 4 & -2 & -1 \end{bmatrix} \cdot \begin{bmatrix} 2 & -7 & 5 & -3 \\ 0 & 1 & 4 & 1 \\ -1 & 6 & -2 & 0 \end{bmatrix} = \begin{bmatrix} -1 & p_{12} & p_{13} & p_{14} \\ p_{21} & p_{22} & 14 & p_{24} \end{bmatrix}$$

$$p_{23} = 4(5) + (-2)4 + (-1)(-2) = 20 - 8 + 2 = 14.$$

Matrices and Their Use in Solving Systems of Linear Equations

The remaining elements in the product are found in a similar manner. The result is

$$EF = \begin{bmatrix} -1 & 13 & 7 & -1 \\ 9 & -36 & 14 & -14 \end{bmatrix}.$$

It is tedious, but possible, to show that **matrix multiplication is associative**. This means that if A is $m \times r$, B is $r \times p$, and C is $p \times n$, then $(AB)C = A(BC)$. You will be asked to verify this in some special cases in exercise 15.

Is matrix multiplication commutative? Obviously not, since AB and BA need not have the same size even if both products are defined: just consider the case of a 2×3 matrix A and a 3×2 matrix B. So, let's refine our question: must AB and BA be equal if A and B are square matrices of the same size? Example 9.18 will provide a negative answer to *this* question.

▶ **EXAMPLE 9.18**   If $A = \begin{bmatrix} 1 & 2 \\ 0 & 4 \end{bmatrix}$ and $B = \begin{bmatrix} 2 & 1 \\ 3 & 0 \end{bmatrix}$, is $AB = BA$?

Solution

$$AB = \begin{bmatrix} 1 \cdot 2 + 2 \cdot 3 & 1 \cdot 1 + 2 \cdot 0 \\ 0 \cdot 2 + 4 \cdot 3 & 0 \cdot 1 + 4 \cdot 0 \end{bmatrix} = \begin{bmatrix} 8 & 1 \\ 12 & 0 \end{bmatrix}$$

$$BA = \begin{bmatrix} 2 \cdot 1 + 1 \cdot 0 & 2 \cdot 2 + 1 \cdot 4 \\ 3 \cdot 1 + 0 \cdot 0 & 3 \cdot 2 + 0 \cdot 4 \end{bmatrix} = \begin{bmatrix} 2 & 8 \\ 3 & 6 \end{bmatrix}.$$

We thus see that $AB \neq BA$. ◀

Identity Matrix of Order n

> The (multiplicative) **identity matrix of order n** is the $n \times n$ matrix I_n, whose main diagonal entries are each 1 and whose other entries are all 0. If there is no confusion about n, we simply write I instead of I_n.

Thus,

$$I_1 = [1], \quad I_2 = \begin{bmatrix} 1 & 0 \\ 0 & 1 \end{bmatrix}, \quad I_3 = \begin{bmatrix} 1 & 0 & 0 \\ 0 & 1 & 0 \\ 0 & 0 & 1 \end{bmatrix}, \quad I_4 = \begin{bmatrix} 1 & 0 & 0 & 0 \\ 0 & 1 & 0 & 0 \\ 0 & 0 & 1 & 0 \\ 0 & 0 & 0 & 1 \end{bmatrix}, \text{ and so on.}$$

Identity matrices are important for the following reason. If A is $m \times n$, then $AI_n = A$; and if B is $r \times n$, then $I_r B = B$. In particular, if A is square of order n, then $I = I_n$ satisfies $AI = A = IA$.

Invertible Matrix

> If A is an $n \times n$ matrix and if a matrix B can be found such that $AB = BA = I_n$, then A is said to be **invertible** (or **nonsingular**).

As you will show in exercise 17, **if** a matrix A is invertible, only one matrix B satisfies $AB = BA = I$. This B is called the **(multiplicative) inverse** of A and is denoted by A^{-1}. Thus, we have the following.

Chapter 9 / Systems of Linear Equations and Linear Inequalities

Multiplicative Inverse

> If A is invertible, then
> $$AA^{-1} = A^{-1}A = I.$$

A preferred method for determining whether a given square matrix A is invertible and, if so, for calculating A^{-1} will be illustrated in example 9.23. Another method will be described in section 9.3.

▶ **EXAMPLE 9.19** The matrices $A = \begin{bmatrix} 5 & 2 & -3 \\ -1 & -1 & 1 \\ -3 & -1 & 2 \end{bmatrix}$ and $B = \begin{bmatrix} 1 & 1 & 1 \\ 1 & -1 & 2 \\ 2 & 1 & 3 \end{bmatrix}$ are inverses of each other, since you can verify that $AB = I = BA$. ◀

Using Matrices to Solve Systems of Linear Equations

Matrices provide a natural way to describe and then solve a system of linear equations.

Coefficient Matrix

> The **coefficient matrix** of a linear system is the matrix whose elements are the coefficients of the variables written in the same position as they appear in the system.

For example, consider the system of linear equations

$$\begin{cases} 2x - 4y - 4z = 6 \\ x - 5y - 3z = 2 \\ -2x + z = 1 \end{cases}$$

For this system, the coefficient matrix is

$$[a_{ij}] = \begin{bmatrix} 2 & -4 & -4 \\ 1 & -5 & -3 \\ -2 & 0 & 1 \end{bmatrix}.$$

Notice that $a_{32} = 0$ because the coefficient of y in the third equation is 0.

A principal motivation for the definition of matrix multiplication is that this definition permits us to write any linear system in a compact form. For instance, the above linear system amounts to the matrix equation

$$\begin{bmatrix} 2 & -4 & -4 \\ 1 & -5 & -3 \\ -2 & 0 & 1 \end{bmatrix} \begin{bmatrix} x \\ y \\ z \end{bmatrix} = \begin{bmatrix} 6 \\ 2 \\ 1 \end{bmatrix}.$$

Matrices and Their Use in Solving Systems of Linear Equations

In general, let's consider a linear system

$$\begin{cases} a_{11}x_1 + \cdots + a_{1n}x_n = b_1 \\ a_{21}x_1 + \cdots + a_{2n}x_n = b_2 \\ \vdots \\ a_{m1}x_1 + \cdots + a_{mn}x_n = b_m \end{cases}$$

If we let $A = [a_{ij}]$ be the coefficient matrix for this system, $X = \begin{bmatrix} x_1 \\ x_2 \\ \vdots \\ x_n \end{bmatrix}$ be the column vector formed by the variables, and $B = \begin{bmatrix} b_1 \\ b_2 \\ \vdots \\ b_m \end{bmatrix}$ be the column vector formed by the right-hand-side constants, then this linear system can be written equivalently as

$$AX = B.$$

In particular, we have the following conclusion.

Theorem

> If A is invertible, the system $AX = B$ is consistent and has the unique particular solution $X = A^{-1}B$.

This observation, which you will implement in exercise 26, provides interest for developing methods to calculate inverses of (nonsingular) matrices. Linear systems of the kind described in the above theorem may be solved by using a graphing calculator. To see how to input matrices and how to implement the $X = A^{-1}B$ formula, consult the owner's manual for your calculator. There, you can also find instructions on implementing other material about matrices and determinants from this section and section 9.3.

Augmented Matrix

> The **augmented matrix** of a linear system is formed by "augmenting" the system's coefficient matrix with the column of constant terms from the right-hand sides of the system's equations.

For the above example, the augmented matrix is

$$\begin{bmatrix} 2 & -4 & -4 & 6 \\ 1 & -5 & -3 & 2 \\ -2 & 0 & 1 & 1 \end{bmatrix}.$$

Chapter 9 / Systems of Linear Equations and Linear Inequalities

▶ **EXAMPLE 9.20** Write the system

$$\begin{cases} 5x_1 - 3x_2 + 5x_4 = 7 \\ x_2 - 4x_3 + 2x_4 = 1 \\ 9x_1 + 4x_2 - 7x_3 + 6 = 0 \\ 4x_1 - 3x_2 + 2x_3 - x_4 = 0 \end{cases}$$

in matrix form as $AX = B$. Find the system's coefficient matrix and augmented matrix.

Solution The system is written as $AX = B$, where

$$X = \begin{bmatrix} x_1 \\ x_2 \\ x_3 \\ x_4 \end{bmatrix}, \quad B = \begin{bmatrix} 7 \\ 1 \\ -6 \\ 0 \end{bmatrix},$$

and A is the coefficient matrix

$$\begin{bmatrix} 5 & -3 & 0 & 5 \\ 0 & 1 & -4 & 2 \\ 9 & 4 & -7 & 0 \\ 4 & -3 & 2 & -1 \end{bmatrix}.$$

This system's augmented matrix, formed by "augmenting" A with B, is

$$\begin{bmatrix} 5 & -3 & 0 & 5 & 7 \\ 0 & 1 & -4 & 2 & 1 \\ 9 & 4 & -7 & 0 & -6 \\ 4 & -3 & 2 & -1 & 0 \end{bmatrix}.$$

◀

Reduced Row-Echelon Form

It is convenient to say that a matrix is in echelon form if it is the coefficient matrix of a linear system in echelon form. When working with a matrix, it is often desirable to transform it to a special type of echelon form called reduced row-echelon form. We next indicate what this is, how to recognize it, how to achieve (reduction to) it, and how to use it.

Reduced Row-Echelon Form of a Matrix

A matrix is said to be in **reduced row-echelon form** if and only if it satisfies the following conditions:

1. If not all the entries in a given row are 0, then the first nonzero entry in that row (read from left to right) is 1. (This entry is that row's "leading 1.")
2. If two consecutive rows each have leading 1's, the column containing the "top" row's leading 1 is to the left of the column containing the next row's leading 1.
3. If there are any rows consisting entirely of zeros, these rows are grouped together at the "bottom" of the matrix.
4. If a column contains a leading 1, then all its other entries are 0.

Matrices and Their Use in Solving Systems of Linear Equations

It is evident from conditions 1, 2, and 4 that any matrix in reduced row-echelon form is also in echelon form. However, as you will see via matrices A and D in example 9.21, a matrix in echelon form need not be in reduced row-echelon form.

EXAMPLE 9.21

Each of the following matrices is in reduced row-echelon form:

$$\begin{bmatrix} 1 & 0 \\ 0 & 1 \end{bmatrix}, \quad \begin{bmatrix} 1 & 0 & 3 \\ 0 & 1 & 4 \end{bmatrix}, \quad \begin{bmatrix} 1 & 0 & 2 & 0 & -9 \\ 0 & 1 & 7 & 0 & 2 \\ 0 & 0 & 0 & 1 & 4 \end{bmatrix}, \quad \begin{bmatrix} 1 & 2 & 0 & 4 & 0 \\ 0 & 0 & 1 & 6 & 0 \\ 0 & 0 & 0 & 0 & 1 \\ 0 & 0 & 0 & 0 & 0 \end{bmatrix}$$

None of the following matrices is in reduced row-echelon form:

$$A = \begin{bmatrix} 1 & 0 & 3 \\ 0 & 2 & 1 \\ 0 & 0 & 1 \end{bmatrix}, \quad B = \begin{bmatrix} 1 & 0 & 5 \\ 0 & 0 & 1 \\ 0 & 1 & 2 \end{bmatrix}, \quad C = \begin{bmatrix} 1 & 0 & 3 & 1 \\ 0 & 0 & 0 & 0 \\ 0 & 1 & 2 & 3 \end{bmatrix}, \quad D = \begin{bmatrix} 1 & 2 & 3 & 4 \\ 0 & 1 & 5 & 6 \\ 0 & 0 & 1 & 7 \\ 0 & 0 & 0 & 1 \end{bmatrix}$$

Matrix A violates condition 1 because the first nonzero number in its second row is 2, *not* 1.

Matrix B violates condition 2. Indeed, the leading 1 in its second row is in the third column, which is *not* to the left of the (second) column containing the leading 1 in the third row.

Matrix C violates condition 3 because its second row, which consists entirely of zeros, is followed by a row which is *not* all-zero.

Matrix D violates condition 4 in several ways. For instance, notice that the second column contains the second row's leading 1, but the second column's other entries are *not* all zero, for $d_{12} = 2$. ◀

Gauss-Jordan Elimination

In section 9.1, you saw how to use the elementary operations and back substitution to solve any linear system. Next, you will see a method called Gauss-Jordan elimination. It, too, is based on "elementary operations," but does not require back substitution. This new method gains efficiency, at least when we work longhand, by using matrix notation.

Gauss-Jordan Elimination

> **Gauss-Jordan elimination** involves transforming a system's augmented matrix by "elementary operations" until its "coefficient matrix" part has been transformed into reduced row-echelon form.

Gauss-Jordan elimination is advantageous since it leads to a unique description of the given system's general solution. (Contrast with example 9.3 in section 9.1.) The basic reason (which we state without proof here) is that each matrix can be transformed to only one matrix in reduced row-echelon form.

In order to use Gauss-Jordan elimination, we need to introduce the following elementary row operations on a matrix. These operations are analogous to

Chapter 9 / Systems of Linear Equations and Linear Inequalities

the elementary operations on a system of linear equations that were listed in section 9.1. We shall denote them with the same notation developed for the corresponding elementary operations on systems, with R_i now denoting row i of a matrix.

Elementary Row Operations on a Matrix

1. Multiply each entry in a given row by a given nonzero constant.
2. Interchange two rows.
3. Add a scalar multiple of one row to another row.

▶ **EXAMPLE 9.22**

Use Gauss-Jordan elimination to solve the linear systems

(a) $\begin{cases} 9x + 6y - 5z = 11 \\ -3x - 2y + 4z = 8 \\ 6x - 4y + 3z = -4 \end{cases}$ (b) $\begin{cases} -2x_1 - 6x_2 = 4 \\ x_1 + 3x_2 - 3x_3 - 12x_4 = -17 \\ x_3 + 4x_4 = 5 \\ x_1 + 3x_2 - x_3 - 4x_4 = -7 \end{cases}$

and (c) $\begin{cases} 3x + 6y + 15z = 21 \\ -x - 2y - 5z = 6 \end{cases}$

Solutions

(a) $\begin{bmatrix} 9 & 6 & -5 & 11 \\ -3 & -2 & 4 & 8 \\ 6 & -4 & 3 & -4 \end{bmatrix} \xrightarrow{2R_2 + R_3} \begin{bmatrix} 9 & 6 & -5 & 11 \\ -3 & -2 & 4 & 8 \\ 0 & -8 & 11 & 12 \end{bmatrix}$

$\xrightarrow{\frac{1}{3}R_1 + R_2} \begin{bmatrix} 9 & 6 & -5 & 11 \\ 0 & 0 & \frac{7}{3} & \frac{35}{3} \\ 0 & -8 & 11 & 12 \end{bmatrix}$

$\xrightarrow{3R_2} \begin{bmatrix} 9 & 6 & -5 & 11 \\ 0 & 0 & 7 & 35 \\ 0 & -8 & 11 & 12 \end{bmatrix}$

$\xrightarrow{R_2 \leftrightarrow R_3} \begin{bmatrix} 9 & 6 & -5 & 11 \\ 0 & -8 & 11 & 12 \\ 0 & 0 & 7 & 35 \end{bmatrix}$

$\xrightarrow{\frac{1}{7}R_3} \begin{bmatrix} 9 & 6 & -5 & 11 \\ 0 & -8 & 11 & 12 \\ 0 & 0 & 1 & 5 \end{bmatrix}$

$\xrightarrow{\frac{3}{4}R_2 + R_1} \begin{bmatrix} 9 & 0 & \frac{13}{4} & 20 \\ 0 & -8 & 11 & 12 \\ 0 & 0 & 1 & 5 \end{bmatrix}$

$\xrightarrow{-11R_3 + R_2} \begin{bmatrix} 9 & 0 & \frac{13}{4} & 20 \\ 0 & -8 & 0 & -43 \\ 0 & 0 & 1 & 5 \end{bmatrix}$

Matrices and Their Use in Solving Systems of Linear Equations

$$\xrightarrow{-\frac{13}{4}R_3 + R_1} \begin{bmatrix} 9 & 0 & 0 & \frac{15}{4} \\ 0 & -8 & 0 & -43 \\ 0 & 0 & 1 & 5 \end{bmatrix}$$

$$\xrightarrow{\frac{1}{9}R_1} \begin{bmatrix} 1 & 0 & 0 & \frac{5}{12} \\ 0 & -8 & 0 & -43 \\ 0 & 0 & 1 & 5 \end{bmatrix}$$

$$\xrightarrow{-\frac{1}{8}R_2} \begin{bmatrix} 1 & 0 & 0 & \frac{5}{12} \\ 0 & 1 & 0 & \frac{43}{8} \\ 0 & 0 & 1 & 5 \end{bmatrix}.$$

Therefore (and this is why Gauss-Jordan elimination used the *augmented* matrix) the corresponding system of elementary operations on the given linear system would transform it to the following equivalent system:

$$\begin{cases} 1x + 0y + 0z = \frac{5}{12} \\ 0x + 1y + 0z = \frac{43}{8} \\ 0x + 0y + 1z = 5 \end{cases}$$

We see immediately that the unique particular solution of the given system is

$$x = \frac{5}{12}, \quad y = \frac{43}{8}, \quad z = 5.$$

This is the same system that we solved in section 9.1 in example 9.9. Compare how the two examples were worked. It may look as if Gauss-Jordan elimination involved more steps, but remember, in example 9.9 you still had to back-substitute after transforming the matrix into echelon form. (**Moral:** Gauss-Jordan "eliminates" back substitution.)

There are several sequences of elementary row operations that could be used to work any example of Gauss-Jordan elimination. We have chosen to show one way to produce such a sequence. It is not necessarily the most "efficient" method, if you wish to delay the use of fractions.

(b) You can (and should) verify that an appropriate "sequence" of elementary row operations transforms the augmented matrix as follows:

$$\begin{bmatrix} -2 & -6 & 0 & 0 & 4 \\ 1 & 3 & -3 & -12 & -17 \\ 0 & 0 & 1 & 4 & 5 \\ 1 & 3 & -1 & -4 & -7 \end{bmatrix} \xrightarrow{\cdots} \begin{bmatrix} 1 & 3 & 0 & 0 & -2 \\ 0 & 0 & 0 & 0 & 0 \\ 0 & 0 & 1 & 4 & 5 \\ 0 & 0 & 0 & 0 & 0 \end{bmatrix}.$$

The corresponding sequence of elementary operations converts the given linear system to

$$\begin{cases} 1x_1 + 3x_2 & = -2 \\ 0x_1 + 0x_2 + 0x_3 + 0x_4 = 0 \\ 1x_3 + 4x_4 = 5 \\ 0x_1 + 0x_2 + 0x_3 + 0x_4 = 0 \end{cases}$$

In such a (dependent) situation, each variable that does not "receive" a leading 1 is to be assigned an arbitrary parametric value. So, in this example, we set $x_2 = s$ and $x_4 = t$. The general solution for the given linear system is thus

$$(x_1, x_2, x_3, x_4) = (-2 - 3s, s, 5 - 4t, t).$$

For example, you can check that $s = 1$ and $t = -2$ lead to the particular solution $(-5, 1, 13, -2)$. With practice, you should be able to write a general solution immediately after transforming the augmented matrix.

(c) A suitable sequence of elementary row operations transforms the augmented matrix to $\begin{bmatrix} 1 & 2 & 5 & 7 \\ 0 & 0 & 0 & 1 \end{bmatrix}$. The second equation of the resulting linear system is $0x + 0y + 0z = 1$, which is inconsistent. Thus, the given linear system is inconsistent. In Gauss-Jordan elimination, inconsistency is always indicated by finding that some row of the augmented matrix transforms to $[0 \ \cdots \ 0 \ d]$, where d is a nonzero number. ◀

To close this section, we give the promised method for computing inverses. To do so, we need to state some theoretical results without proof. (Anyone wanting to see or develop proofs of these assertions needs first to notice the following. An elementary row operation on a matrix A is achieved by multiplying A on the left by the matrix resulting when that same elementary row operation is applied to I.)

Let A be a square matrix of order n. It can be shown that A is invertible if and only if the reduced row-echelon form of A is I_n—that is, if and only if it is possible to transform A to I_n by successively applying finitely many elementary row operations to A. As a consequence, one can show that *if* A is invertible, then A^{-1} may be computed as follows. Record, in order, any "sequence" of elementary row operations that transforms A to I. Then the "same" sequence of elementary row operations transforms the "partitioned" matrix $[A \,|\, I]$ into $[I \,|\, A^{-1}]$.

The above results can be summarized as follows.

Theorem

Let A be a square matrix of order n, with reduced row-echelon form F. Suppose that a sequence of elementary row operations transforms the partitioned matrix $[A \,|\, I_n]$ into $[F \,|\, B]$. Then A is invertible if and only if $F = I_n$; and if A is invertible, then $B = A^{-1}$.

Matrices and Their Use in Solving Systems of Linear Equations

EXAMPLE 9.23 Determine whether $A = \begin{bmatrix} 2 & -3 \\ 7 & 9 \end{bmatrix}$ is invertible, and if it is, compute A^{-1}.

Solution

$[A \mid I] = \begin{bmatrix} 2 & -3 \\ 7 & 9 \end{bmatrix} \begin{bmatrix} 1 & 0 \\ 0 & 1 \end{bmatrix}$

$\xrightarrow{-3R_1 + R_2} \begin{bmatrix} 2 & -3 \\ 1 & 18 \end{bmatrix} \begin{bmatrix} 1 & 0 \\ -3 & 1 \end{bmatrix} \xrightarrow{R_1 \leftrightarrow R_2} \begin{bmatrix} 1 & 18 \\ 2 & -3 \end{bmatrix} \begin{bmatrix} -3 & 1 \\ 1 & 0 \end{bmatrix}$

$\xrightarrow{-2R_1 + R_2} \begin{bmatrix} 1 & 18 \\ 0 & -39 \end{bmatrix} \begin{bmatrix} -3 & 1 \\ 7 & -2 \end{bmatrix} \xrightarrow{-\frac{1}{39}R_2} \begin{bmatrix} 1 & 18 \\ 0 & 1 \end{bmatrix} \begin{bmatrix} -3 & 1 \\ -\frac{7}{39} & \frac{2}{39} \end{bmatrix}$

$\xrightarrow{-18R_2 + R_1} \begin{bmatrix} 1 & 0 \\ 0 & 1 \end{bmatrix} \begin{bmatrix} \frac{3}{13} & \frac{1}{13} \\ -\frac{7}{39} & \frac{2}{39} \end{bmatrix} = [F \mid B].$

Since $F = I$, A is invertible; and $A^{-1} = B = \begin{bmatrix} \frac{3}{13} & \frac{1}{13} \\ -\frac{7}{39} & \frac{2}{39} \end{bmatrix}$

Exercises 9.2

1. Determine the size ($m \times n$) of each of the following matrices:

   (a) $\begin{bmatrix} -2 & 1 & 5 \\ 3 & 4 & 0 \end{bmatrix}$ (b) $\begin{bmatrix} 4 & 0 & -6 \\ 2 & -1 & \pi \\ 0 & 5 & 0 \end{bmatrix}$

   (c) $\begin{bmatrix} \frac{1}{2} & 0 & -1 \\ e & 0 & 4 \\ \sqrt{2} & 0 & -2 \\ \pi & 0 & 5 \end{bmatrix}$ (d) $\begin{bmatrix} 6 \\ -4 \end{bmatrix}$ (e) $[1 \ \ \pi]$ (f) $[5]$

2. Which of the matrices in exercise 1 is square? Which are row vectors? Which are column vectors?
3. What can you conclude about m and n if an $m \times n$ matrix is both a row vector and a column vector?
4. Determine which of the following matrices is/are equal to $\begin{bmatrix} -2 & 0 & 1 \\ 4 & \pi & 0 \end{bmatrix}$:

   (a) $\begin{bmatrix} 4 & \pi & 0 \\ -2 & 0 & 1 \end{bmatrix}$ (b) $\begin{bmatrix} \sqrt{4} & 0 & 1 \\ 4 & \pi & 0 \end{bmatrix}$

   (c) $\begin{bmatrix} -2 & 4 \\ 0 & \pi \\ 1 & 0 \end{bmatrix}$ (d) $\begin{bmatrix} -2 & 4-4 & 1 \\ 3+1 & \pi & 0 \end{bmatrix}$

   (e) $\begin{bmatrix} -2 & 0 \\ 4 & \pi \end{bmatrix}$ (f) $\begin{bmatrix} -\sqrt{4} & 0 & 1 \\ 4 & \pi & 0 \end{bmatrix}$

5. Perform the indicated scalar multiplications:

   (a) $-\frac{3}{2} \begin{bmatrix} 8 & -6 \\ -10 & -3 \end{bmatrix}$ (b) $4[-2 \ 1 \ 0 \ 5]$

   (c) $-1 \begin{bmatrix} 1 \\ 2 \\ 0 \\ \pi \end{bmatrix}$ (d) $0 \begin{bmatrix} -2 & 3 & \pi \\ e & 4 & 0 \end{bmatrix}$

6. Calculate the following inner products:

   (a) $[1 \ -2 \ 3]\begin{bmatrix} -4 \\ -6 \\ 8 \end{bmatrix}$ (b) $[-1 \ 0]\begin{bmatrix} 0 \\ 5 \end{bmatrix}$

   (c) $[-6 \ 24]\begin{bmatrix} 12 \\ 3 \end{bmatrix}$ (d) $[0 \ \frac{1}{2} \ \pi]\begin{bmatrix} 5 \\ -2\pi \\ 1 \end{bmatrix}$

   (e) $[\frac{1}{2}][7]$ (f) $[-4 \ 0 \ 5]\begin{bmatrix} -4 \\ \pi \\ 0 \end{bmatrix}$

◆ 7. Prove that inner product satisfies the analogues of properties 4 and 5 stated for dot product in section 7.5. (For the latter, you will need to decide which row vectors are denoted by 0.)
8. Find the column vector C such that $[-1 \ 3 \ 2] \cdot C = 5$, $[0 \ -1 \ 1] \cdot C = -1$, and $[0 \ 0 \ -4] \cdot C = -4$. (Hint: review exercise 7 in section 7.5.)
9. Which of the following "products" of matrices is/are defined? (Do not calculate these products in this exercise.)

   (a) $[-2 \ 4]\begin{bmatrix} -5 \\ -6 \end{bmatrix}$ (b) $\begin{bmatrix} 0 & 1 \\ \pi & 0 \end{bmatrix}\begin{bmatrix} \pi & 0 & e \\ \sqrt{2} & 5 & 1 \\ 3 & 4 & -1 \end{bmatrix}$

(c) $\begin{bmatrix} 0 & 1 \\ \pi & 0 \end{bmatrix} \begin{bmatrix} \pi & 0 & e \\ \sqrt{2} & 5 & 1 \end{bmatrix}$ (d) $\begin{bmatrix} 0 & 1 & 2 \\ \pi & 0 & 5 \end{bmatrix} \begin{bmatrix} -2 \\ 4 \\ e \end{bmatrix}$

(e) $\begin{bmatrix} 0 & 1 & 2 \\ \pi & 0 & 5 \end{bmatrix} \begin{bmatrix} \pi & 0 & e \\ \sqrt{2} & 5 & 1 \end{bmatrix}$

(f) $\begin{bmatrix} 0 & 1 & 2 \\ \pi & 0 & 5 \end{bmatrix} \begin{bmatrix} -2 & 0 & -1 & \frac{2}{3} \\ 4 & 0 & 3 & 2 \\ e & 0 & -2 & 1 \end{bmatrix}$

10. Prove that if A and B are matrices such that both AB and BA are defined products, then there exist positive integers m and n such that A has size $m \times n$ and B has size $n \times m$.

In exercises 11–14, calculate the given product of matrices.

11. $\begin{bmatrix} 0 & 3 & 2 \\ 4 & 0 & -1 \end{bmatrix} \begin{bmatrix} -2 & 0 & -1 & 1 \\ 4 & 1 & 0 & -1 \\ -1 & 5 & 2 & 3 \end{bmatrix}$

12. $\begin{bmatrix} 0 & 3 & 2 \\ 4 & 0 & -1 \end{bmatrix} \begin{bmatrix} 9 \\ -24 \\ 36 \end{bmatrix}$

13. $\begin{bmatrix} 1 & -2 & 3 \\ 4 & -5 & -6 \\ 0 & -3 & 18 \end{bmatrix} \begin{bmatrix} -108 & 36 & 27 \\ -72 & 24 & 18 \\ -12 & 4 & 3 \end{bmatrix}$

14. $\begin{bmatrix} 5 & 6 \\ -7 & 8 \end{bmatrix} \begin{bmatrix} \frac{1}{2} & 0 \\ -\frac{3}{2} & 4 \end{bmatrix}$

♦ 15. Verify that $(AB)C = A(BC)$ in the following cases:

(a) $A = \begin{bmatrix} 4 & -3 & 0 \\ 1 & 0 & 2 \end{bmatrix}$, $B = \begin{bmatrix} 3 \\ 4 \\ 5 \end{bmatrix}$, $C = [6 \ 1 \ 5 \ 3]$

(b) $A = \begin{bmatrix} 6 & 9 \\ 4 & 6 \end{bmatrix}$, $B = \begin{bmatrix} 6 & -9 \\ -4 & 6 \end{bmatrix}$, $C = \begin{bmatrix} -2 & -7 & 0 & 4 \\ 0 & 1 & 3 & 0 \end{bmatrix}$

16. Consider the matrices

$A = \begin{bmatrix} \pi & \sqrt{2} & e \\ 1 & 0 & -1 \end{bmatrix}$, $B = \begin{bmatrix} \frac{1}{2} & 4 \\ 9 & 5 \\ -6 & 0 \end{bmatrix}$, $C = \begin{bmatrix} -2 & 3 & 1 \\ 0 & 0 & 5 \\ 4 & -1 & 0 \end{bmatrix}$

Verify that
(a) $AI = A$ (b) $IB = B$ (c) $CI = C = IC$

♦ 17. (Uniqueness of inverse) Show that if A, B, and C are matrices such that $AB = BA = I$ and $AC = CA = I$, then $B = C$. (Hint: $(CA)B = C(AB)$.)

18. Give an example of matrices A, B, and C such that $AB = BA = AC = CA$ and $B \neq C$. (Hint: take A, B as in exercise 15(b).)

In exercises 19–22, determine whether B is the inverse of A.

19. $A = \begin{bmatrix} 4 & -5 \\ 5 & -6 \end{bmatrix}$, $B = \begin{bmatrix} -6 & 5 \\ -5 & 4 \end{bmatrix}$

20. $A = \begin{bmatrix} \frac{1}{2} & -\frac{2}{3} & 0 \\ 0 & \frac{1}{3} & 0 \\ 0 & 0 & 1 \end{bmatrix}$, $B = \begin{bmatrix} 2 & 4 & 0 \\ 0 & 3 & 0 \\ 0 & 0 & 1 \end{bmatrix}$

21. $A = \begin{bmatrix} 2 & 4 & 0 \\ 0 & 3 & 0 \\ 0 & 0 & 1 \end{bmatrix}$, $B = \begin{bmatrix} 3 & -4 & 0 \\ 0 & 2 & 0 \\ 0 & 0 & 6 \end{bmatrix}$

22. $A = \begin{bmatrix} 6 & 9 \\ 4 & 6 \end{bmatrix}$, $B = \begin{bmatrix} \frac{1}{6} & 3 \\ 0 & -2 \end{bmatrix}$

23. (a) If A is an invertible matrix, show that A^{-1} is invertible and that $(A^{-1})^{-1} = A$. (Hint: what properties should $(A^{-1})^{-1}$ have? Use exercise 17.)

   (b) If A and B are invertible matrices of the same size, show that AB is invertible and that $(AB)^{-1} = B^{-1}A^{-1}$. (Hint: adapt the hint given in (a).)

24. (a) Find matrices A and B such that AB is invertible and A is not invertible. (Hint: take A to be 1×2 and B to be 2×1.)

   (b) Find invertible 2×2 matrices A and B such that $(AB)^{-1} \neq A^{-1}B^{-1}$. (Hint: find invertible 2×2 matrices C and D such that $CD \neq DC$. Then take $A = D^{-1}$ and $B = C^{-1}$. Use exercise 23(b).)

25. Write the linear system

$$\begin{cases} -3x + 4y = 2 \\ \frac{1}{2}x - y + 6z = 0 \end{cases}$$

in matrix form as $AX = B$. Find this system's coefficient matrix and augmented matrix.

♦ 26. Solve the linear system

$$\begin{cases} x - y + z = 4 \\ 5x + 2y - z = 0 \\ 2x - 3y + z = -2 \end{cases}$$

given that

$\begin{bmatrix} 1 & -1 & 1 \\ 5 & 2 & -1 \\ 2 & -3 & 1 \end{bmatrix}^{-1} = \begin{bmatrix} \frac{1}{13} & \frac{2}{13} & \frac{1}{13} \\ \frac{7}{13} & \frac{1}{13} & -\frac{6}{13} \\ \frac{19}{13} & -\frac{1}{13} & -\frac{7}{13} \end{bmatrix}$.

(Hint: given $AX = B$ and the value of A^{-1}, what is X?)

27. Determine which of the following matrices is/are in reduced row-echelon form:

(a) $\begin{bmatrix} 2 & 0 & 4 \\ 0 & 1 & -2 \\ 0 & 0 & 0 \end{bmatrix}$ (b) $\begin{bmatrix} 1 & 0 & 2 \\ 0 & 1 & 0 \\ 0 & 0 & 0 \end{bmatrix}$

Matrices and Their Use in Solving Systems of Linear Equations

(c) $\begin{bmatrix} 1 & -3 & 5 \\ 0 & 0 & 0 \\ 0 & 0 & 0 \end{bmatrix}$ (d) $\begin{bmatrix} 1 & 0 & 0 \\ 0 & 0 & 1 \\ 0 & 1 & 0 \end{bmatrix}$

(e) $\begin{bmatrix} 1 & 0 & -2 \\ 0 & 0 & 0 \\ 0 & 1 & 3 \end{bmatrix}$ (f) $\begin{bmatrix} 1 & 0 & 2 & 0 \\ 0 & 0 & 0 & 0 \end{bmatrix}$

(g) $\begin{bmatrix} 1 \\ 0 \end{bmatrix}$ (h) $\begin{bmatrix} 1 & -2 & 0 \\ 0 & 1 & 0 \\ 0 & 0 & 1 \end{bmatrix}$

28. Explain how any interchange of a pair of rows can be obtained by successively applying four suitable elementary row operations of the first and third types. (Hint: review exercise 12 in section 9.1.)

In exercises 29–30, apply the elementary row operations (a) $-\frac{1}{2}R_3$, (b) $R_3 \leftrightarrow R_2$, and (c) $2R_3 + R_1$ separately (not successively) to the given matrix.

29. $\begin{bmatrix} -3 & 12 & 8 & -9 \\ -1 & 0 & 3 & -5 \\ 2 & -8 & 0 & 4 \end{bmatrix}$ **30.** $\begin{bmatrix} -10 & 6 \\ \sqrt{2} & 0 \\ 16 & 1 \end{bmatrix}$

In exercises 31–34, use elementary row operations to transform the given matrix to a matrix in reduced row-echelon form.

31. $\begin{bmatrix} 3 & -8 & -19 \\ -3 & 4 & 11 \end{bmatrix}$ **32.** $\begin{bmatrix} -4 & 2 & 6 \\ 0 & 0 & 1 \\ -8 & 3 & 5 \end{bmatrix}$

33. $\begin{bmatrix} \frac{1}{2} & \frac{1}{3} \\ 1 & 1 \\ \frac{1}{4} & \frac{1}{5} \end{bmatrix}$ **34.** $\begin{bmatrix} -2 & 0 \\ 0 & 3 \\ -8 & 9 \end{bmatrix}$

In exercises 35–40, use Gauss-Jordan elimination to solve the given linear system.

35. $\begin{cases} -8x + 3y + 5z = 0 \\ 4z = -12 \\ -4x + 2y + 6z = 3 \end{cases}$ **36.** $\begin{cases} -4x + 2y + 6z = 2 \\ 3x - 5z = 1 \\ -4y + z = -6 \\ 5x + y - 7z = 0 \end{cases}$

37. $\begin{cases} 7x + 0y + 0z = -14 \\ 17x + 0y + 0z = -34 \end{cases}$ **38.** $\begin{cases} 4x + 0z = 8 \\ - 8y + 0z = 88 \\ 8x - 4y + 0z = 888 \end{cases}$

39. $\begin{cases} -2x_1 = 5 \\ 3x_2 = -7 \\ -8x_1 + 9x_2 = 1 \end{cases}$

40. $\begin{cases} 3x_1 + 4x_2 - 5x_3 + 6x_4 = 2 \\ 2x_1 + 3x_2 + 4x_3 - 5x_4 = -6 \\ x_1 + 4x_2 + 57x_3 - 70x_4 = -58 \\ x_1 + 2x_2 + 13x_3 - 16x_4 = -14 \end{cases}$

In exercises 41–42, argue as in example 9.23 in order to determine whether the given matrix is invertible, and if it is, compute its inverse.

41. $\begin{bmatrix} 2 & -1 & -3 \\ 7 & 1 & 9 \\ -2 & 0 & 5 \end{bmatrix}$ **42.** $\begin{bmatrix} 9 & -12 \\ 12 & -16 \end{bmatrix}$

◆ **43.** Let A be a square matrix of order n. Show that A is invertible if and only if the only column vector X such that $AX = \begin{bmatrix} 0 \\ 0 \\ 0 \\ \vdots \\ 0 \end{bmatrix}$ is $X = \begin{bmatrix} 0 \\ 0 \\ 0 \\ \vdots \\ 0 \end{bmatrix}$. (Hint: for the "if" assertion, notice that Gauss-Jordan elimination transforms $\begin{bmatrix} A & \begin{matrix} 0 \\ 0 \\ \vdots \\ 0 \end{matrix} \end{bmatrix}$ to $\begin{bmatrix} I & \begin{matrix} 0 \\ 0 \\ \vdots \\ 0 \end{matrix} \end{bmatrix}$. Hence, determine the reduced row-echelon form matrix "equivalent" to A. Next, use a result asserted prior to example 9.23.)

In view of exercise 43, it is clearly useful to have a notation for the $m \times n$ matrix whose entries are all zeros. We denote this matrix by $\mathbf{0}_{m \times n}$ or, if no confusion will result, by $\mathbf{0}$. It plays the expected role in **matrix addition**, which is defined below.

If $A = [a_{ij}]$ and $B = [b_{ij}]$ are matrices of the same size, say $m \times n$, then their sum $A + B$ is defined to be the $m \times n$ matrix $[c_{ij}]$, where $c_{ij} = a_{ij} + b_{ij}$ for all $i = 1, 2, \ldots, m; j = 1, 2, \ldots, n$.

Evidently, matrix addition is commutative and associative. (In other words, $A + B = B + A$ and $(A + B) + C = A + (B + C)$ if A, B, and C all have the same size.) As suggested above, $A + \mathbf{0} = A$; and if the **additive inverse** $-A = [d_{ij}]$ is defined by $d_{ij} = -a_{ij}$, we infer that $A + (-A) = \mathbf{0}$ and $-(-A) = A$. As expected, **matrix subtraction** is defined by $B - A = B + (-A)$.

Matrix addition and matrix multiplication are the basic operations in **matrix algebra**. Their "compatibility" is enshrined in the distributive properties:

$$(A + B)C = AC + BC$$

if A and B are each $m \times r$ matrices and C is $r \times p$, and

$$C(A + B) = CA + CB$$

if A and B are each $m \times r$ matrices and C is $p \times m$.

Exercises 44–50 will make use of the material just presented.

44. (Matrix addition generalizes vector addition.) As in the text, identify a row vector $A = [a_{11} \quad a_{12}]$ with a vector

$\mathbf{v} = \langle a_{11}, a_{12} \rangle$ in the plane; and, similarly, identify a row vector $B = [b_{11} \ b_{12}]$ with the planar vector $\mathbf{w} = \langle b_{11}, b_{12} \rangle$. Show that $\mathbf{v} + \mathbf{w}$ is the planar vector with which the matrix sum $A + B$ is identified.

45. Given $A = \begin{bmatrix} 8 & 0 & -4 \\ 16 & -6 & 2 \end{bmatrix}$ and $B = \begin{bmatrix} -1 & 5 & -3 \\ 3 & 0 & 4 \end{bmatrix}$, determine the entries of each of the following matrices:
    (a) $A + B$ (b) $A - B$ (c) $-\frac{1}{2}A + B$ (d) $3A - 2B$

46. (Uniqueness of additive inverse) If A, B, and C are matrices such that $A + B = A + C$, prove that $B = C$. (Hint: add $-A$ to both sides.)

47. (a) Using the above definition of matrix addition (and identifying parentheses with brackets), verify the assertion made in example 9.3 in section 9.1 that
    $$(t, -19t + 20, 11t - 11) = (0, 20, -11) + t(1, -19, 11)$$
    for any real number t.
    (b) Find row vectors u, v, and w so that the general solution found in example 9.22(b) can be rewritten as
    $$(-2 - 3s, s, 5 - 4t, t) = u + sv + tw.$$

48. (a) Find invertible 2×2 matrices A and B such that $A + B$ is not invertible. (Hint: **0** is not invertible.)
    (b) Find invertible 2×2 matrices A and B such that $A + B$ is invertible, but $(A + B)^{-1} \neq A^{-1} + B^{-1}$.
    (Hint: what is $\begin{bmatrix} 3 & 0 \\ 0 & 3 \end{bmatrix}^{-1}$? Is $\frac{1}{3} = 1 + \frac{1}{2}$?)

49. (Compare with exercise 7 in section 9.1.) Let A be an $m \times n$ matrix and B a column vector with m entries.
    (a) If X and Y are column vectors such that $AX = B$ and $AY = \mathbf{0}$, prove that $A(X + Y) = B$. (Hint: use a distributive property.)
    (b) If X_1 and X_2 are column vectors such that $A(X_1) = B = A(X_2)$, show that $A(X_1 - X_2) = \mathbf{0}$. (Hint: first determine $A(-X_2)$.)
    (c) Let X_1 be a particular column vector such that $A(X_1) = B$. Show that a column vector X satisfies $AX = B$ if and only if there exists a column vector Y such that $X = X_1 + Y$ and $AY = \mathbf{0}$. (Hint: $X_2 = X_1 + (X_2 - X_1)$.)

50. (Compare with exercise 52 in section 7.4 and exercise 52 in section 7.5.) One may "identify" a complex number $z = a + bi$ (with a, b real numbers) with the 2×2 "real" matrix $\begin{bmatrix} a & b \\ -b & a \end{bmatrix}$. Suppose that $z_1 = a_1 + b_1 i$ and $z_2 = a_2 + b_2 i$ are thus identified with the matrices
    $$A = \begin{bmatrix} a_1 & b_1 \\ -b_1 & a_1 \end{bmatrix} \text{ and } B = \begin{bmatrix} a_2 & b_2 \\ -b_2 & a_2 \end{bmatrix},$$
    respectively. Show that the complex number $z_1 + z_2$ is identified with the matrix sum $A + B$, and the complex number $z_1 z_2$ is identified with the matrix product AB.

*9.3 DETERMINANTS AND CRAMER'S RULE

Each square matrix A has associated with it a number called the **determinant of A**, denoted $|A|$ or **det** A. (Do not confuse the notation $|A|$ with the symbol for the absolute value of a real number.) Determinants play many important roles in mathematics and in a variety of applications outside mathematics. In a later course on calculus, you will study a special type of determinant, called a Jacobian, which explains geometric phenomena like those in exercises 49–50 of section 6.6. The exercises at the end of this section will introduce you to several more accessible applications, as well as other particular kinds of determinants. The text of this section will emphasize how to compute determinants and how to use them to solve linear systems. In order to avoid awkward notation (and because we know you'll study determinants again), we shall omit a number of formal proofs in this section.

A determinant of an $n \times n$ matrix is called a **determinant of order n**. We proceed to show how to calculate such a determinant, emphasizing the cases $n \leq 4$ and beginning with the definition of determinants of order 1 or 2.

---

*This section is optional because the material in it is not essential for later material in the book.

Evaluating Determinants of Order 1 or Order 2

Determinants of Order 1

If $A = [a_{11}]$ is a square matrix of order 1, we define det $A = a_{11}$.

For instance, det $[-7] = -7 = |[-7]|$.

Determinants of Order 2

If $A = \begin{bmatrix} a_{11} & a_{12} \\ a_{21} & a_{22} \end{bmatrix}$ is a square matrix of order 2, then its determinant is defined by

$$\begin{vmatrix} a_{11} & a_{12} \\ a_{21} & a_{22} \end{vmatrix} = a_{11}a_{22} - a_{21}a_{12}.$$

Notice that in the notation for the determinant of A, the brackets of the matrix A are replaced with vertical bars.

EXAMPLE 9.24

If $A = [\sqrt{3}]$, $B = \begin{bmatrix} 4 & 6 \\ -2 & 7 \end{bmatrix}$, and $C = \begin{bmatrix} \sqrt{3} & \frac{1}{2} \\ \frac{2}{3} & 4\sqrt{3} \end{bmatrix}$, find $|A|$, $|B|$, and $|C|$.

Solution

By definition, $|A| = \sqrt{3}$.
Also by definition,

$$|B| = \begin{vmatrix} 4 & 6 \\ -2 & 7 \end{vmatrix} = 4(7) - (-2)6 = 28 + 12 = 40,$$

and

$$|C| = \begin{vmatrix} \sqrt{3} & \frac{1}{2} \\ \frac{2}{3} & 4\sqrt{3} \end{vmatrix} = (\sqrt{3})(4\sqrt{3}) - (\tfrac{2}{3})(\tfrac{1}{2}) = 12 - \tfrac{1}{3} = \tfrac{35}{3}.$$ ◀

We emphasize that, even though a matrix is an *array* of numbers, **the determinant of a (square) matrix is a number.**

Before defining determinants of order greater than 2, we need to consider some terminology. For $n \geq 2$, it will eventually help us to reduce the calculation of a determinant of order n to calculations of related determinants of order $n - 1$.

Minors and Cofactors

Minors

If A is a square matrix of order n, then the **minor of the entry** a_{ij}, denoted M_{ij}, is the determinant of the square matrix of order $n - 1$ formed by deleting row i and column j from A.

Chapter 9 / Systems of Linear Equations and Linear Inequalities

For example, if A is a 3×3 matrix

$$A = \begin{bmatrix} a_{11} & a_{12} & a_{13} \\ a_{21} & a_{22} & a_{23} \\ a_{31} & a_{32} & a_{33} \end{bmatrix}$$

then M_{21} is the determinant of the matrix formed by deleting row 2 and column 1 from A. In this case,

$$M_{21} = \begin{vmatrix} a_{11} & a_{12} & a_{13} \\ a_{21} & a_{22} & a_{23} \\ a_{31} & a_{32} & a_{33} \end{vmatrix} = \begin{vmatrix} a_{12} & a_{13} \\ a_{32} & a_{33} \end{vmatrix} = a_{12}a_{33} - a_{32}a_{13}.$$

▶ **EXAMPLE 9.25**

If $A = \begin{bmatrix} 4 & 3 & -1 \\ 2 & 0 & 4 \\ 6 & -2 & 5 \end{bmatrix}$, find the minors M_{12} and M_{31}.

Solution

To determine M_{12}, we "cross out" row 1 and column 2 and then evaluate the determinant of the remaining matrix:

$$M_{12} = \begin{vmatrix} 4 & 3 & -1 \\ 2 & 0 & 4 \\ 6 & -2 & 5 \end{vmatrix} = \begin{vmatrix} 2 & 4 \\ 6 & 5 \end{vmatrix} = 2(5) - 6(4) = 10 - 24 = -14.$$

Similarly,

$$M_{31} = \begin{vmatrix} 4 & 3 & -1 \\ 2 & 0 & 4 \\ 6 & -2 & 5 \end{vmatrix} = \begin{vmatrix} 3 & -1 \\ 0 & 4 \end{vmatrix} = 3(4) - 0(-1) = 12 - 0 = 12.$$

◀

Cofactors

The number $(-1)^{i+j}M_{ij}$ is called the **cofactor of the entry a_{ij}** and is denoted C_{ij}.

▶ **EXAMPLE 9.26**

If $A = \begin{bmatrix} 4 & 3 & -1 \\ 2 & 0 & 4 \\ 6 & -2 & 5 \end{bmatrix}$, calculate C_{12} and C_{31}.

Solution

In example 9.25, we found that $M_{12} = -14$ and $M_{31} = 12$. Hence,

$$C_{12} = (-1)^{1+2}M_{12} = (-1)^3(-14) = 14$$

and

$$C_{31} = (-1)^{3+1}M_{31} = (-1)^4(12) = 12.$$

◀

Evaluating Determinants of Order n

We are now ready to evaluate determinants of order n.

Cofactor Expansion Method for Evaluating a Determinant

> The **cofactor expansion method** for evaluating a determinant states that a determinant of order n is computed by adding the products formed by multiplying each element in a given row or column by its cofactor.

A more careful theoretical treatment might take the preceding statement, with "a given row or column" replaced by "the first row," as the **definition** of a determinant of order n. Then, the given statement would become a theorem. The natural proofs of this theorem would take us too far afield. Some related properties will be given without proof later in this section, but for now we proceed to see how to implement the cofactor expansion method.

EXAMPLE 9.27

Evaluate $|A| = \begin{vmatrix} 4 & 3 & -1 \\ 2 & 0 & 4 \\ 6 & -2 & 5 \end{vmatrix}$.

Solution

We shall evaluate first by "expanding about" the first row and then by "expanding about" the second column. According to the above theorem, the resulting answers should be the same.

Expanding about the first row leads to

$$|A| = a_{11}C_{11} + a_{12}C_{12} + a_{13}C_{13}$$

$$= 4(-1)^{1+1}\begin{vmatrix} 0 & 4 \\ -2 & 5 \end{vmatrix} + 3(-1)^{1+2}\begin{vmatrix} 2 & 4 \\ 6 & 5 \end{vmatrix} + (-1)(-1)^{1+3}\begin{vmatrix} 2 & 0 \\ 6 & -2 \end{vmatrix}$$

$$= 4(0+8) - 3(10-24) - 1(-4-0) = 32 + 42 + 4 = 78.$$

Expanding about the second column, we have

$$|A| = a_{12}C_{12} + a_{22}C_{22} + a_{32}C_{32}$$

$$= (-3)\begin{vmatrix} 2 & 4 \\ 6 & 5 \end{vmatrix} + 0\begin{vmatrix} 4 & -1 \\ 6 & 5 \end{vmatrix} + 2\begin{vmatrix} 4 & -1 \\ 2 & 4 \end{vmatrix}$$

$$= -3(10-24) + 0 + 2(16+2) = 42 + 0 + 36 = 78,$$

agreeing with the earlier answer. ◀

Elementary Operations on Determinants

In example 9.27, we expanded the determinant of A about row 1 and also about column 2. Less computation was required in the latter case because one of the elements in column 2 was 0. Computation may be further reduced if we transform the matrix A in order to increase the number of zeros in one of its rows or columns.

Chapter 9 / Systems of Linear Equations and Linear Inequalities

Now, we saw in sections 9.1 and 9.2 that an appropriate sequence of elementary row operations transforms any given matrix A to an echelon matrix. This is particularly useful, and you will be asked to verify the following statement in exercise 8.

Proposition

> The determinant of any square matrix in echelon form is the product of the matrix's main diagonal entries. (More generally, it is enough to assume that the matrix is **upper triangular** in the sense of condition (1) in the definition of echelon form, namely that each entry below the main diagonal is zero.)

To relate this product to det A, we need to examine the effect on the determinant of a matrix when the matrix is changed by an elementary operation. The basic facts are summarized next without proof.

The Effect of Elementary Operations on a Determinant

> Let A be a square matrix, and let A' be the matrix that results when the given operation is performed.
>
> *Operation*
>
> 1. Multiply each element in a given row or column of A by a nonzero constant k.
> 2. Interchange two rows or two columns.
> 3. Add a scalar multiple of one row (or column) to another row (or column) in the matrix.
>
> *Effect*
>
> $|A'| = k|A|$
> $|A'| = -|A|$
>
> $|A'| = |A|$

▶ **EXAMPLE 9.28**

Use elementary operations to show that

(a) $\begin{vmatrix} 1 & 1 & 2 \\ 3 & 4 & 5 \\ 6 & 7 & 8 \end{vmatrix} = -3$ and (b) $\begin{vmatrix} 2 & 8 & 10 \\ 1 & 5 & 9 \\ 0 & 0 & 1 \end{vmatrix} = 2.$

Solutions

(a) We transform the given matrix to echelon form by elementary row operations and record at each step the effect on the determinant:

$\begin{vmatrix} 1 & 1 & 2 \\ 3 & 4 & 5 \\ 6 & 7 & 8 \end{vmatrix} \stackrel{(\text{use } -3R_1 + R_2)}{=} \begin{vmatrix} 1 & 1 & 2 \\ 0 & 1 & -1 \\ 6 & 7 & 8 \end{vmatrix} \stackrel{(\text{use } -6R_1 + R_3)}{=} \begin{vmatrix} 1 & 1 & 2 \\ 0 & 1 & -1 \\ 0 & 1 & -4 \end{vmatrix}$

$\stackrel{(\text{use } -R_2 + R_3)}{=} \begin{vmatrix} 1 & 1 & 2 \\ 0 & 1 & -1 \\ 0 & 0 & -3 \end{vmatrix} \stackrel{\substack{(\text{use product of main} \\ \text{diagonal entries} \\ \text{in echelon form})}}{=} 1(1)(-3) = -3.$

(b) Similarly,

$\begin{vmatrix} 2 & 8 & 10 \\ 1 & 5 & 9 \\ 0 & 0 & 1 \end{vmatrix} \stackrel{(\frac{1}{2}R_1)}{=} 2\begin{vmatrix} 1 & 4 & 5 \\ 1 & 5 & 9 \\ 0 & 0 & 1 \end{vmatrix} \stackrel{(-R_1 + R_2)}{=} 2\begin{vmatrix} 1 & 4 & 5 \\ 0 & 1 & 4 \\ 0 & 0 & 1 \end{vmatrix}$

$= 2(1)(1)(1) = 2.$

Determinants and Cramer's Rule

Notice, in the first step, that applying $\tfrac{1}{2}R_1$ had the effect of "factoring 2 out of the determinant." ◀

To compute a square matrix's determinant, it is usually not necessary to transform the matrix all the way to echelon form. As the next example shows, it is often faster to transform until reaching a matrix with some row or column having at most one nonzero entry. We then expand about that row or column.

▶ **EXAMPLE 9.29**

Use elementary row and column operations to evaluate

$$\begin{vmatrix} 1 & 2 & -3 & 4 \\ 3 & -1 & 2 & 4 \\ -2 & -4 & -6 & -8 \\ 0 & 3 & -1 & 2 \end{vmatrix}$$

Solution

$$\begin{vmatrix} 1 & 2 & -3 & 4 \\ 3 & -1 & 2 & 4 \\ -2 & -4 & -6 & -8 \\ 0 & 3 & -1 & 2 \end{vmatrix} = -2 \begin{vmatrix} 1 & 2 & -3 & 4 \\ 3 & -1 & 2 & 4 \\ 1 & 2 & 3 & 4 \\ 0 & 3 & -1 & 2 \end{vmatrix} = -2(2) \begin{vmatrix} 1 & 2 & -3 & 2 \\ 3 & -1 & 2 & 2 \\ 1 & 2 & 3 & 2 \\ 0 & 3 & -1 & 1 \end{vmatrix}$$

$$= -4 \begin{vmatrix} 1 & 2 & -3 & 2 \\ 0 & -7 & 11 & -4 \\ 1 & 2 & 3 & 2 \\ 0 & 3 & -1 & 1 \end{vmatrix} = -4 \begin{vmatrix} 1 & 2 & -3 & 2 \\ 0 & -7 & 11 & -4 \\ 0 & 0 & 6 & 0 \\ 0 & 3 & -1 & 1 \end{vmatrix}$$

$$= -4(6) \begin{vmatrix} 1 & 2 & 2 \\ 0 & -7 & -4 \\ 0 & 3 & 1 \end{vmatrix} = (-24)(1) \begin{vmatrix} -7 & -4 \\ 3 & 1 \end{vmatrix}$$

$$= -24(-7 + 12) = -120.$$

You should verify that the same answer results (more tediously) if you directly apply the cofactor expansion method to the given determinant. ◀

The Transpose and Adjoint of a Matrix

Transpose of a Matrix

If A is an $m \times n$ matrix, the **transpose of A**, denoted A^T, is the $n \times m$ matrix whose ith row is the ith column of A.

Thus, the first column of A is the first row of A^T, the second column of A is the second row of A^T, and so on.

▶ **EXAMPLE 9.30**

If $A = \begin{bmatrix} 2 & 3 & -1 & 4 \\ 0 & 2 & 5 & -9 \end{bmatrix}$, then $A^T = \begin{bmatrix} 2 & 0 \\ 3 & 2 \\ -1 & 5 \\ 4 & -9 \end{bmatrix}$. ◀

One benefit of the notation for transposes is that any column vector C may be written compactly as R^T, where R is a suitable row vector. (In fact, $R = C^T$.) More generally, $(A^T)^T = A$ for any matrix A. Some other properties of the transpose operation will be developed in exercises 22, 30, and 31. One result worthy of highlighting here is the following.

Theorem

For any square matrix A,
$$|A^T| = |A|.$$

If A is any $n \times n$ matrix and C_{ij} is the cofactor of a_{ij}, then the matrix

$$\begin{bmatrix} C_{11} & C_{12} & \cdots & C_{1n} \\ C_{21} & C_{22} & \cdots & C_{2n} \\ \vdots & & & \vdots \\ C_{n1} & C_{n2} & \cdots & C_{nn} \end{bmatrix}$$

is the matrix of cofactors of A. The transpose of this matrix is called the **adjoint of A**, denoted **adj(A)**. Thus,

$$\text{adj}(A) = \begin{bmatrix} C_{11} & C_{21} & \cdots & C_{n1} \\ C_{12} & C_{22} & \cdots & C_{n2} \\ \vdots & & & \vdots \\ C_{1n} & C_{2n} & \cdots & C_{nn} \end{bmatrix}.$$

▶ **EXAMPLE 9.31**

If $A = \begin{bmatrix} 2 & 1 & 1 \\ 4 & -1 & -3 \\ 3 & 5 & 1 \end{bmatrix}$, determine adj($A$).

Solution

As in example 9.26, you can verify that the matrix of cofactors of A is

$$\begin{bmatrix} 14 & -13 & 23 \\ 4 & -1 & -7 \\ -2 & 10 & -6 \end{bmatrix}.$$

Transposing this matrix produces

$$\text{adj}(A) = \begin{bmatrix} 14 & 4 & -2 \\ -13 & -1 & 10 \\ 23 & -7 & -6 \end{bmatrix}.$$

◀

A theoretical result of considerable importance is the following.

Theorem

For any square matrix A,
$$A \cdot \text{adj}(A) = \text{adj}(A) \cdot A = |A|I.$$

Determinants and Cramer's Rule

One consequence is that if $|A| \neq 0$, then A is invertible and its inverse is given by the following formula.

Theorem

$$A^{-1} = \frac{1}{|A|} \text{adj}(A).$$

The next example illustrates how to use this formula, but you should be warned that it is generally quicker to find inverses by using the method in example 9.23 in section 9.2.

EXAMPLE 9.32   Consider $A = \begin{bmatrix} 2 & 1 & 1 \\ 4 & -1 & -3 \\ 3 & 5 & 1 \end{bmatrix}$, the matrix from the last example. Determine A^{-1} using the method of this section.

Solution   We begin by finding the determinant of A.

$$|A| = 2\begin{vmatrix} -1 & -3 \\ 5 & 1 \end{vmatrix} - \begin{vmatrix} 4 & -3 \\ 3 & 1 \end{vmatrix} + \begin{vmatrix} 4 & -1 \\ 3 & 5 \end{vmatrix} = 2(14) - 13 + 23 = 38.$$

From example 9.31, we know that

$$\text{adj}(A) = \begin{bmatrix} 14 & 4 & -2 \\ -13 & -1 & 10 \\ 23 & -7 & -6 \end{bmatrix}.$$

Hence,

$$A^{-1} = \frac{1}{|A|}\text{adj}(A) = \frac{1}{38}\begin{bmatrix} 14 & 4 & -2 \\ -13 & -1 & 10 \\ 23 & -7 & -6 \end{bmatrix} = \begin{bmatrix} \frac{7}{19} & \frac{2}{19} & -\frac{1}{19} \\ -\frac{13}{38} & -\frac{1}{38} & \frac{5}{19} \\ \frac{23}{38} & -\frac{7}{38} & -\frac{3}{19} \end{bmatrix}.$$ ◀

We have seen that if $|A| \neq 0$, then A is invertible. The converse of this statement is also true. In other words, if A is invertible, then $|A| \neq 0$. The proof of this depends on the following rather difficult result.

Theorem

If A and B are square matrices of the same size, then

$$|AB| = |A|\,|B|.$$

Indeed, given this result, we see that if A were invertible, then $|A|\,|A^{-1}| = |AA^{-1}| = |I| = 1$, from which we could conclude that $|A| \neq 0$. To sum up, we now know the following.

Chapter 9 / Systems of Linear Equations and Linear Inequalities

Criterion for Invertibility

A square matrix A is invertible if and only if $\det A \neq 0$.

In view of the characterization of invertible matrices given in exercise 43 of section 9.2, it is now apparent that determinants have a role to play in solving linear systems. In fact, **we shall show later in this section how to use determinants to solve any system of m linear equations in n unknowns**. First, we consider some special situations.

Solving Linear Systems with Invertible Coefficient Matrices

In exercise 26 of section 9.2, we illustrated the observation that if A is an invertible matrix, then the unique particular solution to a linear system $AX = B$ is given by $X = A^{-1}B$. The next example reinforces this method.

▶ **EXAMPLE 9.33**

Solve the linear system

$$\begin{cases} 2x_1 + x_2 + x_3 = 3 \\ 4x_1 - x_2 - 3x_3 = 9 \\ 3x_1 + 5x_2 + x_3 = 10 \end{cases}$$

Solution

We have $A = \begin{bmatrix} 2 & 1 & 1 \\ 4 & -1 & -3 \\ 3 & 5 & 1 \end{bmatrix}$, $X = \begin{bmatrix} x_1 \\ x_2 \\ x_3 \end{bmatrix}$, and $B = \begin{bmatrix} 3 \\ 9 \\ 10 \end{bmatrix}$. In example 9.32, we

determined that $A^{-1} = \begin{bmatrix} \frac{7}{19} & \frac{2}{19} & -\frac{1}{19} \\ -\frac{13}{38} & -\frac{1}{38} & \frac{5}{19} \\ \frac{23}{38} & -\frac{7}{38} & -\frac{3}{19} \end{bmatrix}$, and so

$$X = A^{-1}B = \begin{bmatrix} \frac{7}{19} & \frac{2}{19} & -\frac{1}{19} \\ -\frac{13}{38} & -\frac{1}{38} & \frac{5}{19} \\ \frac{23}{38} & -\frac{7}{38} & -\frac{3}{19} \end{bmatrix} \begin{bmatrix} 3 \\ 9 \\ 10 \end{bmatrix} = \begin{bmatrix} \frac{29}{19} \\ \frac{26}{19} \\ -\frac{27}{19} \end{bmatrix}$$

Thus, the unique particular solution is $x_1 = \frac{29}{19}$, $x_2 = \frac{26}{19}$, and $x_3 = -\frac{27}{19}$. ◀

Cramer's Rule

Consider a linear system, $AX = B$, consisting of n linear equations in n variables, such that the system's coefficient matrix A is invertible. We know that this system has a unique particular solution, namely $X = A^{-1}B$. It would be helpful to find this solution without having to resort to either of the methods that we have seen for computing A^{-1}. A theoretical result that accomplishes this is Cramer's Rule.

Cramer's Rule

Consider

$$\begin{cases} a_{11}x_1 + a_{12}x_2 + a_{13}x_3 + \cdots + a_{1n}x_n = b_1 \\ a_{21}x_1 + a_{22}x_2 + a_{23}x_3 + \cdots + a_{2n}x_n = b_2 \\ \vdots \\ a_{n1}x_1 + a_{n2}x_2 + a_{n3}x_3 + \cdots + a_{nn}x_n = b_n \end{cases}$$

where $|A| \neq 0$. Let $|A_i|$ denote the determinant of the matrix obtained by replacing the ith column of A by $[b_1 \ \cdots \ b_n]^T$. Then, Cramer's Rule states that (if $|A| \neq 0$, then)

$$x_i = \frac{|A_i|}{|A|}.$$

Proof: For simplicity of notation, we consider the case $n = 3$ and $i = 2$, after which the general argument will be apparent. We proceed via elementary operations:

$$x_2|A| = x_2 \begin{vmatrix} a_{11} & a_{12} & a_{13} \\ a_{21} & a_{22} & a_{23} \\ a_{31} & a_{32} & a_{33} \end{vmatrix} = \begin{vmatrix} a_{11} & a_{12}x_2 & a_{13} \\ a_{21} & a_{22}x_2 & a_{23} \\ a_{31} & a_{32}x_2 & a_{33} \end{vmatrix} = \begin{vmatrix} a_{11} & a_{12}x_2 + a_{11}x_1 & a_{13} \\ a_{21} & a_{22}x_2 + a_{21}x_1 & a_{23} \\ a_{31} & a_{32}x_2 + a_{31}x_1 & a_{33} \end{vmatrix}$$

$$= \begin{vmatrix} a_{11} & a_{12}x_2 + a_{11}x_1 + a_{13}x_3 & a_{13} \\ a_{21} & a_{22}x_2 + a_{21}x_1 + a_{23}x_3 & a_{23} \\ a_{31} & a_{32}x_2 + a_{31}x_1 + a_{33}x_3 & a_{33} \end{vmatrix} = \begin{vmatrix} a_{11} & b_1 & a_{13} \\ a_{21} & b_2 & a_{23} \\ a_{31} & b_3 & a_{33} \end{vmatrix} = |A_2|.$$

Dividing by $|A|$, we have $x_2 = \dfrac{|A_2|}{|A|}$, completing the proof.

Another proof of Cramer's rule, using the formula $A^{-1} = \dfrac{1}{|A|}\text{adj}(A)$, will be requested in exercise 51.

We now give two examples indicating how to implement Cramer's Rule.

▶ **EXAMPLE 9.34**

Use Cramer's Rule to solve the system of linear equations

$$\begin{cases} 2x - 5y = 16 \\ 4x + 3y = 6 \end{cases}$$

Solution

Here, $|A| = \begin{vmatrix} 2 & -5 \\ 4 & 3 \end{vmatrix} = 26$, $|A_x| = \begin{vmatrix} 16 & -5 \\ 6 & 3 \end{vmatrix} = 78$, and $|A_y| = \begin{vmatrix} 2 & 16 \\ 4 & 6 \end{vmatrix} = -52$.

Hence, since $|A| \neq 0$, Cramer's Rule may be applied, yielding

$$x = \frac{|A_x|}{|A|} = \frac{78}{26} = 3 \quad \text{and} \quad y = \frac{|A_y|}{|A|} = \frac{-52}{26} = -2.$$

◀

Example 9.35

Use Cramer's Rule to solve the linear system

$$\begin{cases} 2x_1 + 5x_2 - 3x_3 = 0 \\ 4x_1 - 6x_3 = 1 \\ 4x_2 - x_3 = 5 \end{cases}$$

Solution

Here,

$$|A| = \begin{vmatrix} 2 & 5 & -3 \\ 4 & 0 & -6 \\ 0 & 4 & -1 \end{vmatrix} = \begin{vmatrix} 2 & 5 & -3 \\ 0 & -10 & 0 \\ 0 & 4 & -1 \end{vmatrix} = 2\begin{vmatrix} -10 & 0 \\ 4 & -1 \end{vmatrix} = 20 \neq 0$$

$$|A_1| = \begin{vmatrix} 0 & 5 & -3 \\ 1 & 0 & -6 \\ 5 & 4 & -1 \end{vmatrix} = \begin{vmatrix} 0 & 5 & -3 \\ 1 & 0 & -6 \\ 0 & 4 & 29 \end{vmatrix} = (-1)\begin{vmatrix} 5 & -3 \\ 4 & 29 \end{vmatrix} = -157$$

$$|A_2| = \begin{vmatrix} 2 & 0 & -3 \\ 4 & 1 & -6 \\ 0 & 5 & -1 \end{vmatrix} = \begin{vmatrix} 2 & 0 & -3 \\ 0 & 1 & 0 \\ 0 & 5 & -1 \end{vmatrix} = 2\begin{vmatrix} 1 & 0 \\ 5 & -1 \end{vmatrix} = -2$$

$$|A_3| = \begin{vmatrix} 2 & 5 & 0 \\ 4 & 0 & 1 \\ 0 & 4 & 5 \end{vmatrix} = \begin{vmatrix} 2 & 5 & 0 \\ 0 & -10 & 1 \\ 0 & 4 & 5 \end{vmatrix} = 2\begin{vmatrix} -10 & 1 \\ 4 & 5 \end{vmatrix} = -108.$$

Hence, by Cramer's Rule, $x_1 = \dfrac{|A_1|}{|A|} = \dfrac{-157}{20} = -7.85$, $x_2 = \dfrac{|A_2|}{|A|} = \dfrac{-2}{20} = -0.1$, and $x_3 = \dfrac{|A_3|}{|A|} = \dfrac{-108}{20} = -5.4$. ◀

Note: the following is enrichment material. It should be of help in your later studies on the rank of matrices. However, students or instructors who pursue this material should be warned to allocate extra time.

Frankly, Cramer's Rule is not a very efficient method, especially for $n > 3$. Indeed, the amount of arithmetic computation involved in finding determinants of order n increases very rapidly as n increases. Nevertheless, Cramer's Rule will be important for many theoretical purposes in several of your later courses. To preview one such topic, we close this section with yet another approach to solving linear systems. As you will see, it is markedly different from the methods in sections 9.1 and 9.2.

Consider a matrix A. By definition, a **submatrix** of A is obtained by "crossing out" various rows and columns of A. If s is the largest positive integer for which A has an invertible $s \times s$ submatrix, then s is called the **rank** of A. In view of the cofactor expansion method, this definition can be restated as follows.

Rank of a Matrix

If s is a positive integer such that $|D| \neq 0$ for some $s \times s$ submatrix D of A and $|C| = 0$ for all $(s + 1) \times (s + 1)$ submatrices C of A, then s is the rank of A. By convention, the rank of $\mathbf{0}$ is taken to be 0.

Determinants and Cramer's Rule

The promised (third, and slowest) method to solve linear systems is described in the following.

Theorem

Let $A = [a_{ij}]$ be an $m \times n$ matrix of rank r, and let B be a column vector with m entries. Then

(a) The linear system $AX = B$ is consistent if and only if the rank of the partitioned matrix $[A \mid B]$ is r (that is, if and only if each $(r+1) \times (r+1)$ submatrix of $[A \mid B]$ has determinant equal to 0).

(b) Suppose that the linear system $AX = B$ is consistent. Choose an invertible $r \times r$ submatrix

$$D = \begin{bmatrix} a_{i_1 j_1} & a_{i_1 j_2} & \cdots & a_{i_1 j_r} \\ a_{i_2 j_1} & a_{i_2 j_2} & \cdots & a_{i_2 j_r} \\ \vdots & & & \vdots \\ a_{i_r j_1} & a_{i_r j_2} & \cdots & a_{i_r j_r} \end{bmatrix}$$

of A. Rewrite the equations labeled $i_1, i_2, \ldots, i_r$ in the original linear system as a new linear system

$$\begin{cases} a_{i_1 j_1} x_{j_1} + a_{i_1 j_2} x_{j_2} + \cdots + a_{i_1 j_r} x_{j_r} = c_1 \\ \vdots \\ a_{i_r j_1} x_{j_1} + a_{i_r j_2} x_{j_2} + \cdots + a_{i_r j_r} x_{j_r} = c_r \end{cases}$$

having coefficient matrix D. Solving this new linear system (for instance, by Cramer's Rule) produces the general solution of the linear system $AX = B$.

It should be clear that if $m = n = r$, the recipe in part (b) of the above theorem amounts to just a restatement of Cramer's Rule. The next two examples illustrate this recipe in other cases.

EXAMPLE 9.36

Use determinants to solve the linear system

$$\begin{cases} 2x - 3y = 12 \\ -\frac{1}{3}x + \frac{1}{2}y = -4 \end{cases}$$

Solution

The graphs of the two given equations are distinct parallel lines. Since these lines do not intersect, the given linear system has an empty solution set, and is therefore inconsistent. Let's see how part (a) of the above theorem would lead to the same conclusion.

Write the given system as $AX = B$ in the usual way. As $|[2]| = 2 \neq 0$ and you can check that $|A| = 0$, it follows that the rank of A is $r = 1$. However,

$$[A \mid B] = \begin{bmatrix} 2 & -3 & 12 \\ -\frac{1}{3} & \frac{1}{2} & -4 \end{bmatrix}$$

has an invertible $(r+1) \times (r+1) = 2 \times 2$ submatrix since, for instance,

$$\begin{vmatrix} -3 & 12 \\ \frac{1}{2} & -4 \end{vmatrix} = 6 \neq 0.$$

Thus, the rank of $[A \mid B]$ is $2 \neq r$, and so part (a) of the theorem yields that the given system is inconsistent. ◀

Notice that if we had ignored the fact that $|A| = 0$ in example 9.36, a careless attempt to apply Cramer's Rule would have led to

$$x_1 = \frac{|A_1|}{|A|} = \frac{\begin{vmatrix} 12 & -3 \\ -4 & \frac{1}{2} \end{vmatrix}}{0} = \frac{-6}{0},$$

which is undefined. Any time some $|A_i| \neq 0$ and $|A| = 0$, you can be sure that the relevant linear system is inconsistent. However—and this is why a careful theory must consider rank—some inconsistent (and all dependent) "square" linear systems satisfy $|A| = 0$ and $|A_i| = 0$ for **all** i. Exercise 50 will reinforce this point.

▶ **EXAMPLE 9.37**

Use determinants to solve the linear system

$$\begin{cases} 2x - y + z = 4 \\ x + 3y + 2z = 12 \\ 3x + 2y + 3z = 16 \end{cases}$$

Solution

Write the system as $AX = B$ as usual. Since $\begin{vmatrix} 2 & 1 \\ 3 & 3 \end{vmatrix} = 3 \neq 0$ and $|A| = 0$, the rank of A is 2. You can check that all four 3×3 submatrices of $[A \mid B]$ have determinants equal to 0. By part (a) of the theorem, the given system is consistent.

Following the recipe in part (b), we proceed to solve the new system

$$\begin{cases} 2x + z = 4 + y \\ 3x + 3z = 16 - 2y \end{cases}$$

Cramer's Rule gives

$$x = \frac{\begin{vmatrix} 4+y & 1 \\ 16-2y & 3 \end{vmatrix}}{\begin{vmatrix} 2 & 1 \\ 3 & 3 \end{vmatrix}} = \frac{(4+y)3 - (16-2y)1}{3} = \frac{-4+5y}{3}$$

and

$$z = \frac{\begin{vmatrix} 2 & 4+y \\ 3 & 16-2y \end{vmatrix}}{\begin{vmatrix} 2 & 1 \\ 3 & 3 \end{vmatrix}} = \frac{2(16-2y) - 3(4+y)}{3} = \frac{20-7y}{3}.$$

Determinants and Cramer's Rule

The general solution of the given system is therefore

$$(x, y, z) = \left(\frac{-4 + 5y}{3}, y, \frac{20 - 7y}{3}\right),$$

with y playing the role of the $n - r = 3 - 2 = 1$ parameter.

You can check that if we had instead used the invertible $r \times r$ submatrix $\begin{bmatrix} 2 & -1 \\ 1 & 3 \end{bmatrix}$ of A, the theorem's recipe would have produced the required general solution in the form

$$(x, y, z) = \left(\frac{24 - 5z}{7}, \frac{20 - 3z}{7}, z\right),$$

with parameter z. Thus, this section's method shares a "defect" with the method of section 9.1 (but *not* of section 9.2): it may lead to several (equivalent, but different-looking) forms of a system's general solution. ◀

Exercises 9.3

1. Calculate the determinant of each given matrix:

   (a) $[-2]$ (b) $\begin{bmatrix} -12 & 2 \\ 3 & \frac{3}{4} \end{bmatrix}$ (c) $\begin{bmatrix} \sqrt{\pi} & 0 \\ -2 & \sqrt{\pi} \end{bmatrix}$ (d) $[0]$

2. Calculate the following determinants:

   (a) $\begin{vmatrix} -12 & -16 \\ 3 & 4 \end{vmatrix}$ (b) $|[-4]|$

   (c) $\det([9])$ (d) $\det\begin{pmatrix} \frac{1}{2} & 1 \\ 6 & 8 \end{pmatrix}$

3. Calculate the minors (i) M_{12} and (ii) M_{24} for each of the following matrices:

   (a) $\begin{bmatrix} 2 & -1 & 0 & 9 \\ 3 & 6 & -2 & 0 \\ 4 & 0 & 0 & -3 \\ 5 & 1 & -5 & 0 \end{bmatrix}$ (b) $\begin{bmatrix} 4 & 5 & 0 & 1 \\ -2 & -1 & 0 & 0 \\ 3 & 4 & 0 & 0 \\ -5 & \pi & 4 & 3 \end{bmatrix}$

4. Calculate the cofactors (i) C_{22} and (ii) C_{12} for each of the following matrices:

   (a) $\begin{bmatrix} 0 & 9 & 0 \\ -2 & -3 & 4 \\ 3 & 4 & 6 \end{bmatrix}$ (b) $\begin{bmatrix} -11 & \sqrt{2} & 6 \\ 0 & 0 & 0 \\ 2 & e & -1 \end{bmatrix}$

In exercises 5–7, evaluate each of the given determinants by the cofactor expansion method, expanding about the indicated row or column.

5. (a) ▸ $\begin{vmatrix} 1 & -2 & -3 \\ 0 & 4 & 0 \\ 5 & 0 & 6 \end{vmatrix}$ (b) ▸ $\begin{vmatrix} -1 & 4 & 0 \\ -2 & -3 & 0 \\ -4 & 1 & 5 \end{vmatrix}$

   (c) ▸ $\begin{vmatrix} -3 & -4 \\ 2 & -1 \end{vmatrix}$

6. (a) $\begin{vmatrix} -2 & e & \pi \\ 3 & 5 & 0 \\ 1 & -4 & 2 \end{vmatrix}$ ▼ (b) $\begin{vmatrix} -2 & -3 \\ -5 & -4 \end{vmatrix}$ ▼

   (c) $\begin{vmatrix} \frac{1}{2} & \pi & 2 \\ -1 & 0 & -4 \\ \frac{1}{2} & -5 & 6 \end{vmatrix}$ ▼

7. (a) ▸ $\begin{vmatrix} \sqrt{2} & 0 & 0 \\ 0 & \pi & 0 \\ 0 & 0 & e \end{vmatrix}$ (b) $\begin{vmatrix} -2 & e & \frac{1}{2} \\ 0 & \pi & 4 \\ 0 & 0 & e \end{vmatrix}$ ▼

8. ◆ (A generalization of exercise 7) Prove the assertion made in the text that if $A = [a_{ij}]$ is an upper triangular matrix, then $\det(A) = a_{11}a_{22} \cdots a_{nn}$, the product of the main diagonal entries. **Comment:** you may assume $n = 4$ if you want a concrete case. A fully precise way to handle such a "repetitive" argument for general n involves mathematical induction, a topic discussed later in section 10.5.

9. (a) If A is a square matrix with a row consisting entirely of zeros, show that $|A| = 0$. (Hint: use the cofactor expansion method.)

   (b) Do (a) with "row" replaced by "column."

10. If A is an $n \times n$ matrix and k is a real number, prove that $\det(kA) = k^n \cdot \det(A)$. **Comment:** as in exercise 8, you may consider the case $n = 4$. (Hint: review the last remark in the solution of example 9.28(b).)

11. (a) If A is a square matrix with (at least) two equal rows, show that $|A| = 0$. (Hint: you should show

that $|A| = -|A|$. What is the effect of interchanging two equal rows?)
(b) Do (a) with "rows" replaced by "columns."

12. (a) Let A be a square matrix with two proportional rows. (This means $R_j = kR_i$ for some nonzero real number k and unequal indices i and j.) Prove that $|A| = 0$. (Hint: show that $|A| = k \cdot 0$. Exercises 9 and 11 may help.)
(b) Redo (a) with "rows" replaced by "columns."

In exercises 13–14, use elementary operations to evaluate the given determinants:

13. (a) $\begin{vmatrix} -4 & 3 & 1 \\ 2 & 6 & -1 \\ 6 & -3 & -4 \end{vmatrix}$ (b) $\begin{vmatrix} -2 & 4 \\ 6 & 13 \end{vmatrix}$

(c) $\begin{vmatrix} -2 & 0 & -1 & 0 \\ 6 & 1 & 0 & -4 \\ 0 & 0 & -1 & 3 \\ 0 & 2 & 4 & 5 \end{vmatrix}$

14. (a) $\begin{vmatrix} -1 & -3 & 7 \\ 2 & 4 & -20 \\ 5 & 17 & -27 \end{vmatrix}$ (b) $\begin{vmatrix} 4 & 3 & 2 & 1 \\ 1 & 2 & 3 & 4 \\ -4 & 1 & -2 & 3 \\ 1 & 6 & 3 & 8 \end{vmatrix}$

(c) $\begin{vmatrix} -\frac{1}{4} & \frac{3}{2} \\ \frac{7}{12} & \frac{7}{2} \end{vmatrix}$

15. (A Vandermonde determinant) If r, s, t are real numbers, show that

$$\begin{vmatrix} 1 & r & r^2 \\ 1 & s & s^2 \\ 1 & t & t^2 \end{vmatrix} = (s-r)(t-r)(t-s).$$

(Hint: here is an approach using a result from section 3.4. By exercise 11, you may assume that r, s, and t are pairwise distinct. Hence, $f(x) = \begin{vmatrix} 1 & x & x^2 \\ 1 & s & s^2 \\ 1 & t & t^2 \end{vmatrix}$ is a quadratic polynomial, with leading coefficient $t - s$ and roots s, t. Thus $f(x) = (t-s)(x-s)(x-t)$.)

16. (A generalization of exercise 15) If r, s, t, w are real numbers, show that

$$\begin{vmatrix} 1 & r & r^2 & r^3 \\ 1 & s & s^2 & s^3 \\ 1 & t & t^2 & t^3 \\ 1 & w & w^2 & w^3 \end{vmatrix} = (s-r)(t-r)(w-r)(t-s)(w-s)(w-t).$$

(Hint: argue as in, and use, exercise 15.)

17. (a) Verify that

$$\begin{vmatrix} 22 & 16 & 11 \\ 22-6 \cdot 5 & 16-6 \cdot 7 & 11+6 \cdot 4 \\ 5 & 7 & -4 \end{vmatrix} = 0.$$

(Hint: a quick way uses elementary row operations and exercise 9.)

(b) Prove that

$$\begin{vmatrix} a+r & b+s & c+t \\ d & e & f \\ g & h & k \end{vmatrix} = \begin{vmatrix} a & b & c \\ d & e & f \\ g & h & k \end{vmatrix} + \begin{vmatrix} r & s & t \\ d & e & f \\ g & h & k \end{vmatrix}$$

(Hint: expand about the first row and use the distributive property of real numbers.)

18. Verify that

$$\begin{vmatrix} x & y & 1 \\ 2 & -3 & 1 \\ -4 & 7 & 1 \end{vmatrix} = 0$$

is an equation for the line passing through the points $(2, -3)$ and $(-4, 7)$. (Hint: is this a linear equation? Do the given points satisfy it? Exercise 11 is relevant.)

19. Let $P_1(x_1, y_1)$ and $P_2(x_2, y_2)$ be distinct points. Prove that $\begin{vmatrix} x & y & 1 \\ x_1 & y_1 & 1 \\ x_2 & y_2 & 1 \end{vmatrix} = 0$ is an equation for the line passing through P_1 and P_2. (Hint: see the hint for exercise 18.)

20. Using the result in exercise 19, write down and simplify an equation for the line through the two given points:
(a) $(-3, 2)$ and $(-4, 7)$ (b) $(-3, 2)$ and $(-3, 7)$
(c) $(-3, 2)$ and $(-4, 2)$

21. Calculate the transpose of the given matrix:

(a) $\begin{bmatrix} -2 & \pi \\ \sqrt{2} & 0 \end{bmatrix}$ (b) $[-1 \ 4 \ 0 \ 7]$

(c) $\begin{bmatrix} e & 4 & \frac{1}{2} \\ \pi & -1 & 0 \end{bmatrix}$ (d) $\begin{bmatrix} -1 & 0 \\ 2 & 1 \\ e & 4 \end{bmatrix}$

◆ 22. Show that if $A = [a_{ij}]$ is an $m \times r$ matrix and $B = [b_{ks}]$ is an $r \times p$ matrix, then $(AB)^T = (B^T)(A^T)$. (Hint: the entry in row s and column i of $(AB)^T$ is $a_{i1}b_{1s} + a_{i2}b_{2s} + \cdots + a_{ir}b_{rs}$. Calculate the corresponding entry of $(B^T)(A^T)$.)

In exercises 23–24, if A denotes the given matrix, determine adj(A).

23. (a) $\begin{bmatrix} 0 & 2 & 1 \\ 6 & 3 & 2 \\ 9 & 5 & 3 \end{bmatrix}$ (b) $\begin{bmatrix} 2 & 4 \\ 8 & 0 \end{bmatrix}$

24. (a) $\begin{bmatrix} 4 & 8 \\ 0 & 2 \end{bmatrix}$ (b) $\begin{bmatrix} 0 & 9 & 5 \\ 4 & 16 & 9 \\ 2 & 8 & -1 \end{bmatrix}$

Determinants and Cramer's Rule

In exercises 25–26, verify that the given matrix A satisfies $A \cdot \text{adj}(A) = \text{adj}(A) \cdot A = |A|I$.

25. $\begin{bmatrix} \frac{1}{2} & 0 & \frac{1}{5} \\ \frac{1}{4} & \frac{1}{3} & 0 \\ 0 & 120 & 0 \end{bmatrix}$

26. $\begin{bmatrix} \frac{3}{4} & -\frac{2}{7} \\ -42 & 16 \end{bmatrix}$

In exercises 27–28, use the methods of this section to determine if the given matrix is invertible, and if it is, calculate its inverse.

27. $\begin{bmatrix} -20 & 3 \\ -7 & 2 \end{bmatrix}$

28. $\begin{bmatrix} \frac{1}{2} & \frac{1}{4} & \frac{2}{3} \\ -6 & -3 & -8 \\ 2 & 4 & \frac{3}{2} \end{bmatrix}$

29. Solve the linear system
$$\begin{cases} 2x + 5y - z = 6 \\ 3x + 2z = 5 \\ \frac{x}{5} + \frac{z}{5} = 4 \end{cases}$$
given that
$$\begin{bmatrix} 2 & 5 & -1 \\ 3 & 0 & 2 \\ \frac{1}{5} & 0 & \frac{1}{5} \end{bmatrix}^{-1} = \begin{bmatrix} 0 & 1 & -10 \\ \frac{1}{5} & -\frac{3}{5} & 7 \\ 0 & -1 & 15 \end{bmatrix}$$
(Hint: review example 9.33 or exercise 26 in section 9.2.)

◆ 30. Prove that if A is an invertible matrix, then so is A^T and $(A^T)^{-1} = (A^{-1})^T$. (Hint: use exercise 22 to simplify $A^T(A^{-1})^T$ and $(A^{-1})^T A^T$. A harder proof is available using this section's formula for A^{-1}, provided you first establish that $\text{adj}(A^T) = (\text{adj}(A))^T$.)

◆ 31. Let A and B be invertible $n \times n$ matrices. Prove that A^T, B^T, AB, and BA are all invertible. (Hint: show that each of these matrices has a nonzero determinant.)

32. If A and B are $n \times n$ matrices such that $BA = I$, show that A and B are invertible. (Hint: it **is** possible to show that $AB = I$, but it's much faster to show that $|A|$ and $|B|$ are each nonzero. What is $|B| \cdot |A|$?)

33. Let z be a complex number. As in exercise 50 of section 9.2, "identify" z with a certain 2×2 matrix A (with real entries). Similarly, identify the complex conjugate $\bar{z}$ with a matrix B.
    (a) Show that $\det A = |z|^2$, where $|z|$ is the absolute value of z, in the sense of section 1.5.
    (b) Show that $B = A^T$.

34. Let $P_1(x_1, y_1)$, $P_2(x_2, y_2)$, and $P_3(x_3, y_3)$ be noncollinear points. Prove that
$$\begin{vmatrix} x^2 + y^2 & x & y & 1 \\ x_1^2 + y_1^2 & x_1 & y_1 & 1 \\ x_2^2 + y_2^2 & x_2 & y_2 & 1 \\ x_3^2 + y_3^2 & x_3 & y_3 & 1 \end{vmatrix} = 0$$
is an equation of the circle passing through P_1, P_2, and P_3. (Hint: reason as in exercises 18 and 19.)

35. Let $P_1(x_1, y_1)$, $P_2(x_2, y_2)$, and $P_3(x_3, y_3)$ be noncollinear points. Prove that the area of the triangle $\triangle P_1 P_2 P_3$ is given by
$$\frac{1}{2} \left| \det \begin{bmatrix} x_1 & y_1 & 1 \\ x_2 & y_2 & 1 \\ x_3 & y_3 & 1 \end{bmatrix} \right|$$
(The vertical bars denote absolute value here. These are needed here since the determinant may be negative.)
(Hint: let ℓ denote the line $\overleftrightarrow{P_2 P_3}$. By exercise 19 or otherwise, show that $(y_2 - y_3)x - (x_2 - x_3)y + (x_2 y_3 - x_3 y_2) = 0$ is an equation for ℓ. Express d, the distance from P_1 to ℓ. (In this book, we have already derived the formula for the distance from a point to a line four ways.) Use the fact that the required area is $\frac{d(P_2 P_3)}{2}$. The distance formula from section 1.2 is relevant.)

36. Use the formula developed in exercise 35 in order to find the area of $\triangle P_1 P_2 P_3$:
    (a) $P_1(-2, 1)$, $P_2(1, \sqrt{2})$, $P_3(0, -1)$
    (b) $P_1(8, 0)$, $P_2(0, 0)$, $P_3(0, 4)$

37. If $P_1(x_1, y_1)$, $P_2(x_2, y_2)$, and $P_3(x_3, y_3)$ are three vertices of a parallelogram, show that the parallelogram's area is given by
$$\left| \det \begin{bmatrix} x_1 & y_1 & 1 \\ x_2 & y_2 & 1 \\ x_3 & y_3 & 1 \end{bmatrix} \right|$$
(Hint: by elementary geometry, the area in question equals twice the area of $\triangle P_1 P_2 P_3$. Apply exercise 35.)

38. Use the formula developed in exercise 37 in order to find the area of a parallelogram three of whose vertices are P_1, P_2, and P_3:
    (a) $P_1(0, 0)$, $P_2(-4, 0)$, $P_3(-4, 12)$
    (b) $P_1(-1, 0)$, $P_2(2, 6)$, $P_3(5, 14)$

Chapter 9 / Systems of Linear Equations and Linear Inequalities

In exercises 39–42, use Cramer's Rule to solve the given linear system:

39. $\begin{cases} x - 4z = 1 \\ -2x + 6y + z = 0 \\ 3y - 2z = 4 \end{cases}$

40. $\begin{cases} 5x + 6y = 7 \\ 3x + 4y = 1 \end{cases}$

41. $\begin{cases} \sqrt{2}x_1 + x_2 = 0 \\ x_1 - x_2 = \sqrt{2} \end{cases}$

42. $\begin{cases} 4x_1 + x_2 = 0 \\ 2x_2 - x_3 = 4 \\ x_1 - x_2 = 2 \end{cases}$

In exercises 43–49, use determinants to solve the given linear system.

43. $\begin{cases} -2x + 3y = 4 \\ 10x - 15y = 20 \end{cases}$

44. $\begin{cases} x - y + z = 1 \\ y - 2z = 0 \\ 2x + y = -1 \\ 3x - y + 3z = 2 \end{cases}$

45. $\begin{cases} 2x - 3y = -4 \\ 10x - 15y = -20 \\ -4x + 9y = 13 \end{cases}$

46. $\begin{cases} 2x - 3y = -4 \\ 20x - 15y = 20 \\ 20x - 30y = -40 \end{cases}$

47. $\begin{cases} -y + z = -5 \\ 3x + y - z = 2 \\ 3x + 3y - 3z = 12 \end{cases}$

48. $\begin{cases} 3x - y + z = 24 \\ -6x + 2y - 2z = -48 \\ x - 2y = 19 \end{cases}$

49. $\begin{cases} -2x_1 + x_2 - x_4 = -13 \\ x_1 + 3x_2 - x_3 = -7 \\ 5x_1 + 8x_2 - 3x_3 + x_4 = -8 \\ -9x_1 + x_2 + x_3 - 4x_4 = -45 \end{cases}$

◆ 50. Show that the linear system

$$\begin{cases} x_1 + x_2 + x_3 = 1 \\ 2x_1 + 2x_2 + 2x_3 = 2 \\ 3x_1 + 3x_2 + 3x_3 = 4 \end{cases}$$

is inconsistent, although, using the notation of Cramer's Rule, it satisfies $0 = |A| = |A_1| = |A_2| = |A_3|$.

◆ 51. Give an alternative proof of Cramer's Rule, using the formula $A^{-1} = \frac{1}{|A|}\text{adj}(A)$. (Hint: since $X = A^{-1}B$, conclude that

$$|A|x_i = C_{1i}b_1 + C_{2i}b_2 + \cdots + C_{ni}b_n,$$

where $\text{adj}(A) = [C_{ij}]^T$. Next, expand $|A_i|$ about column i.)

◆ 52. (This exercise provides a criterion promised in exercise 46 of section 8.4.) Let Γ be the graph of a quadratic expression

$$Ax^2 + Bxy + Cy^2 + Dx + Ey + F = 0.$$

Let $\delta = 4 \begin{vmatrix} A & \frac{B}{2} & \frac{D}{2} \\ \frac{B}{2} & C & \frac{E}{2} \\ \frac{D}{2} & \frac{E}{2} & F \end{vmatrix}$.

(a) Prove that if Γ "degenerates" to either one line or a pair of lines, then $\delta = 0$. (Hint: see exercise 42 of section 8.4.)

(b) If $\delta = 0$ and Γ is nonempty, prove that Γ degenerates to either one line or a pair of lines. **Comment:** you may assume that $C \neq 0$ and $B = 0 = E$. (Hint: solve for $y = \pm\sqrt{\frac{Ax^2 + Dx + F}{-C}}$. Use the condition $\delta = 0$ to show $4AF = D^2$. If $A = 0$, show the graph is a line or two parallel lines. If $A \neq 0$, explain how to arrange $D = 0$, and show the graph is a pair of intersecting lines.)

(c) Show that $x^2 - 2xy + y^2 + 1 = 0$ is an example for which $\delta = 0$ and Γ is empty. (Hint: solving for $x - y$ is "complex.")

9.4 LINEAR INEQUALITIES AND LINEAR PROGRAMMING

In chapter 1, we introduced inequalities and discussed how to solve linear and quadratic inequalities. Since then, each time a new type of function was introduced, we have studied inequalities involving that function. Following that tradition, we shall now devote a section to studying systems of linear inequalities. While it is possible to develop mathematical techniques for studying systems of linear inequalities in more than two variables, the background available to us at this point confines us to planar domains. For this reason, we shall restrict attention to systems containing two variables. Then, we shall introduce an area called linear programming designed to solve certain applications involving these systems.

Linear Inequalities and Linear Programming

To visualize the planar regions that occur throughout this section, especially to approximate the "corners" or vertices of these regions, you may find it useful to have a graphing calculator or computer graphing software handy.

Linear Inequalities

Let a, b, and c be real numbers such that a and b are not both zero. Then the following are **linear inequalities** in x and y:

$$ax + by + c < 0 \qquad ax + by + c > 0$$
$$ax + by + c \leq 0 \qquad ax + by + c \geq 0.$$

The first two inequalities are called **strict inequalities** because they do not contain an equal sign. If substituting x_0 for x and y_0 for y in an inequality produces a true statement, then (x_0, y_0) is called a (**particular**) **solution** for the inequality. To **solve an inequality** means to find the set of all of its particular solutions; this is usually an infinite set. The **graph of an inequality** is the graph of all the particular solutions of the inequality. Two inequalities are **equivalent** if and only if they have exactly the same solutions, and hence if and only if they have the same graphs.

Just as the graphs of linear inequalities in x were intervals, the graphs of linear inequalities in x and y are subsets of the plane. A line separates the plane into three mutually disjoint sets: the line itself and two **open half-planes**. In figure 9.3(a), the regions marked A and B are each open half-planes. The line separating these two half-planes, called the **boundary**, is dashed to indicate that it is not part of either open half-plane. The union of an open half-plane and its boundary is called a **closed half-plane**. A solid line indicates that the boundary is to be included; thus, figure 9.3(b) is the graph of a closed half-plane.

Figure 9.3

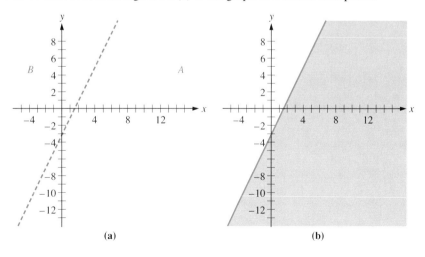

Every linear inequality determines an associated linear equation whose graph is the boundary of the solution set of the inequality. As in chapter 1,

Chapter 9 / Systems of Linear Equations and Linear Inequalities

we shall find it easiest to solve a linear inequality by first solving the associated linear equation. Next, we shall use a "key point" to decide which half-plane is included in the inequality's graph. The next two examples illustrate the method.

▶ **EXAMPLE 9.40**

Graph the inequality $4x - 2y < 7$.
Comment: this inequality is equivalent to the linear inequality $4x - 2y - 7 < 0$ and so it, too, is considered to be a linear inequality.

Solution

We begin by graphing the boundary and then consider one of the regions on either side of the boundary.

The boundary line is the graph of the associated linear equation $4x - 2y = 7$, or $y = 2x - \frac{7}{2}$. Because the given inequality is a strict inequality, the boundary is drawn as a dashed line, as in figure 9.4(a).

Now, select some convenient key point not on the boundary, say $(0, 0)$. If the point $(0, 0)$ is a solution of the given inequality, then the points in the open half-plane containing $(0, 0)$ form the solution set of this inequality. If $(0, 0)$ is not a solution, then the solution set is the open half-plane on the other side of the boundary.

Substituting $(0, 0)$ for (x, y) in the given inequality, we obtain

$$4(0) - 2(0) = 0 < 7,$$

which is a true statement. Thus, $(0, 0)$ is a solution, and hence the half-plane containing $(0, 0)$ is the solution set. This is shown in figure 9.4(b). ◀

Figure 9.4

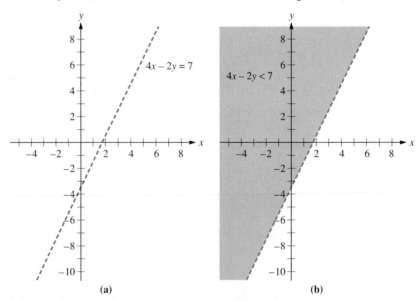

We shall use a different approach to solving the inequality in example 9.41. Either of the methods demonstrated in example 9.40 and example 9.41 can be used to solve any linear inequality.

Linear Inequalities and Linear Programming

EXAMPLE 9.41   Graph the inequality $4x - 5y + 8 \geq 0$.

Solution   We begin by solving this inequality for y, obtaining

$$-5y \geq -4x - 8$$
$$y \leq \tfrac{4}{5}x + \tfrac{8}{5}.$$

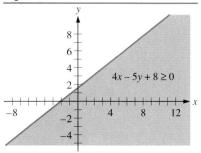

Figure 9.5

From this, we can see that the boundary is the line $y = \tfrac{4}{5}x + \tfrac{8}{5}$. Because the inequality indicates that y is *less than* (or equal to) $\tfrac{4}{5}x + \tfrac{8}{5}$, the solution set is the (closed) half-plane (containing and) *below* the line $y = \tfrac{4}{5}x + \tfrac{8}{5}$. You can verify that the solution includes the open half-plane below this line (as shown in figure 9.5) by substituting a key point from this half-plane into the given inequality.

The procedure used in example 9.41 can be generalized to solve any inequality of the form $y < f(x)$ for any function f. In particular, we have the following.

Theorem

If f is a function, then

(i) the graph of the inequality $y < f(x)$ is the open region that lies *below* the graph of $y = f(x)$;
(ii) the graph of the inequality $y > f(x)$ is the open region that lies *above* the graph of $y = f(x)$.

Systems of Linear Inequalities

The (general) solution to a system of linear inequalities is the set of points whose coordinates satisfy **every** inequality of the system. This general solution may be graphed by intersecting the solution sets of the individual inequalities in the system (when these individual solution sets are graphed on the same coordinate system).

EXAMPLE 9.42   Graph the solution set of the system of linear inequalities

$$\begin{cases} 2x + 4y \leq 8 & (1) \\ 5x - 3y < 12 & (2) \end{cases}$$

Solution   Solving inequality (1) for y, we obtain

$$y \leq -\tfrac{1}{2}x + 2.$$

The graph for this inequality is in figure 9.6(a).
Solving inequality (2) for y produces

$$y > \tfrac{5}{3}x - 4,$$

and its solution set is shown in figure 9.6(b).

542 Chapter 9 / Systems of Linear Equations and Linear Inequalities

Figure 9.6

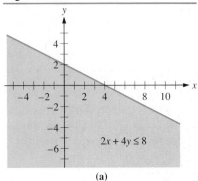
(a)

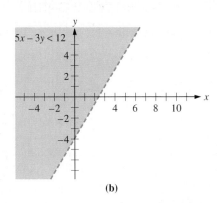
(b)

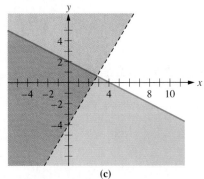
(c)

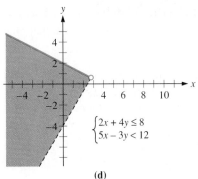
$\begin{cases} 2x + 4y \le 8 \\ 5x - 3y < 12 \end{cases}$
(d)

The graphs of these two inequalities are combined in figure 9.6(c). The given system's solution set is the intersection of these graphs. This is the region where the shading overlaps; it is in figure 9.6(d). As usual, the open circle at the vertex $(\frac{36}{13}, \frac{8}{13})$ indicates that this point is not in the solution set.

Graphing calculators may often be used to help solve systems of inequalities. To do this, begin by rewriting the inequalities, if necessary, so that only "y" appears on the left-hand side of the inequality symbol. For the above system, we have already done this and obtained

$$\begin{cases} y \le -\frac{1}{2}x + 2 \\ y > \frac{5}{3}x - 4 \end{cases}$$

TI-81 Version

Figure 9.6 (continued)

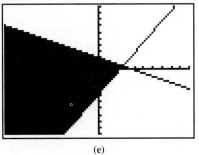
(e)

The *TI-81* can shade only the region between two graphs. It shades above the graph of the first-entered function and below the graph of the second-entered function. To graph the above system using a *TI-81*, first clear any equations from the "equation screen" by pressing the Y= key and using the ▼ and CLEAR keys as needed. Next, press

2nd DRAW 7 ((5 ÷ 3) XIT − 4 ALPHA ,
(−) ((1 ÷ 2) XIT + 2) ENTER

You should obtain a graph like the one in figure 9.6(e).

The shaded region in figure 9.6(e) indicates the solution set of the system of inequalities. By using the TRACE key and the cursor keys, you can approximate the coordinates of the vertex $(\frac{36}{13}, \frac{8}{13})$. Notice that the graphing calculator does not permit us to indicate an open circle at this vertex even though it is not in the solution set.

Casio *fx-7700G* Version

To graph the above system using a Casio *fx-7700G*, first clear the graphing screen by pressing SHIFT Cls EXE . Next, put the calculator in the inequality graphing mode by pressing MODE SHIFT ÷ . Pressing Graph gives you the display shown in figure 9.6(f). As you can see, this display has a menu of four inequality indicators along the bottom of the screen. So, to graph the above system, press

Linear Inequalities and Linear Programming

[F4] [SHIFT] [(−)] .5 [X,θ,T] [+] 2 [SHIFT] [↵] [Graph]

[F1] [(] 5 [÷] 3 [)] [X,θ,T] [−] 4 [EXE]

The result is the graph in figure 9.6(g).

Figure 9.6 (continued)

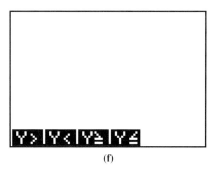
(f)

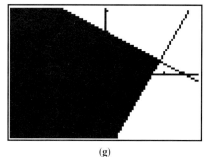
(g)

The shaded region in figure 9.6(g) indicates the solution set of the system of inequalities. By using the [SHIFT] [Trace] keys and the cursor keys, you can approximate the coordinates of the vertex $(\frac{36}{13}, \frac{8}{13})$. Notice that the graphing calculator does not permit us to indicate an open circle at this vertex even though it is not in the solution set.   ◀

> **EXAMPLE 9.43**

Graph the solution set of the system

$$\begin{cases} 2x + 3y < 9 & (1) \\ 4x - 5y \geq 10 & (2) \\ x \geq 0 & (3) \\ y > -3 & (4) \end{cases}$$

Solution

This system has four inequalities. Drawing each boundary and shading the appropriate half-plane would require a lot of shading. Such work produces a picture from which it is difficult to discern the intersection of the shaded regions. As an alternative method, we prefer to use little arrows on the boundaries to show which side of each boundary to consider in the solution set.

We begin by solving the first two inequalities for y. The given system then may be rewritten as the equivalent system

$$\begin{cases} y < -\frac{2}{3}x + 3 & (5) \\ y \leq \frac{4}{5}x - 2 & (6) \\ x \geq 0 & (7) \\ y > -3 & (8) \end{cases}$$

Inequality (1) is thus satisfied by the points below the line $y = -\frac{2}{3}x + 3$. The second inequality describes the set of points on or below the line $y = \frac{4}{5}x - 2$.

The third inequality describes the points on or to the right of the y-axis, and inequality (4) is satisfied by the points above the line $y = -3$. These conclusions are summarized graphically in figure 9.7(a). Notice the use of the little arrows in this figure. The given system's solution set is graphed in figure 9.7(b). The points in the shaded region are the points that satisfy all four given inequalities. As before, open circles are used to indicate the vertices that are not in the solution set.

Figure 9.7

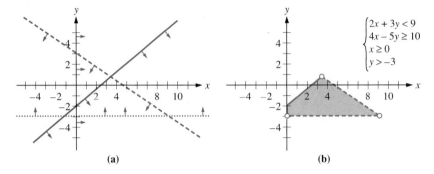

(a)   (b)

Casio *fx-7700G* Version

Figure 9.7 (continued)

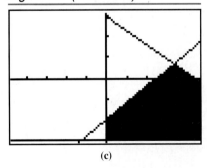

(c)

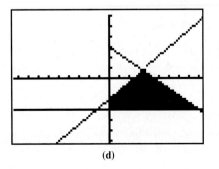

(d)

The Casio *fx-7700G* allows you to graph a system of more than two inequalities. You can use the instructions given in example 9.42 to determine how to input the inequalities (5), (6), and (8). (Each of these is an inequality with "y" followed by an inequality symbol.) But, how can you handle inequality (7)? Since the Casio *fx-7700G* permits you to specify a domain of interest and (7) leads to the domain $[0, \infty)$, you need to specify $[0, \infty)$ in this example. This can be done after one of the inequalities has been entered into the calculator. (We will do this after entering (5).) To graph the above system, first clear any previous graphs from the "equation screen," put your calculator in inequality mode, press Graph , and respond to the displayed menu by entering the following keystrokes:

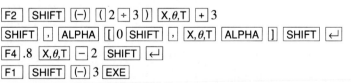

The result is shown in figure 9.7(c). Since this does not include the entire solution set of the system of inequalities, "zoom out" by pressing SHIFT Zoom F4 . You can now see the solution set, determined by the shaded region in figure 9.7(d).

If the graph of a system of linear inequalities extends infinitely far in at least one direction, we say the graph is **unbounded**. Systems of linear inequalities whose graphs are **bounded** are "cut off" in every direction by line segments. (Technically, a region is said to be bounded if it can be wholly contained

Linear Inequalities and Linear Programming

within a disk.) The solution set in figure 9.6(d) is an example of an unbounded graph, while the one in figure 9.7(b) is bounded.

Other Systems of Inequalities

This chapter focuses on systems of linear equations and linear inequalities. Some systems of nonlinear equations were considered briefly in section 1.8. While it is beyond the scope of this book to study systems of nonlinear inequalities, we shall now provide an example of a nonlinear system and its graphical solution.

EXAMPLE 9.44

Graph the solution of the system

$$\begin{cases} y \geq \cos x + \frac{3}{2} \\ x \geq (y-2)^2 - 5 \\ y \leq -2x + 6 \end{cases}$$

Solution

This system has three inequalities. The equation associated with each inequality is graphed in figure 9.8(a). Arrows have been used to indicate which side of each associated graph to include in the solution set.

The graph of the solution set for the given system of inequalities is shown in figure 9.8(b).

Figure 9.8

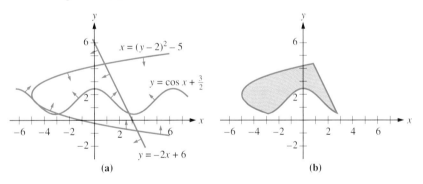

(a)   (b)

Linear Programming

One application of systems of linear inequalities that developed rapidly during the last half-century is called **linear programming**. Here, "programming" is used in the sense of "planning." Linear programming has its most important uses in helping managers decide how to allocate resources in order to minimize costs or maximize profits. A typical business application realistically involves linear inequalities in many variables. However, we can still indicate the flavor of linear programming by restricting our work to inequalities in x and y. First, we shall need to introduce some terminology associated with planar regions.

Convex Set

A set of points S is called a **convex set** if and only if, for any two points P and Q in S, the entire segment $\overline{PQ}$ is in S. The **vertices** of a set are the corners or points where adjacent boundary sides meet.

A convex set is shown in figure 9.9(a). An example of a set that is not convex is in figure 9.9(b). Glancing back at figure 9.7(b), you can see that the vertices are $(9, -3)$, $(0, -3)$, $(0, -2)$, and $(\frac{75}{22}, \frac{8}{11})$.

Figure 9.9

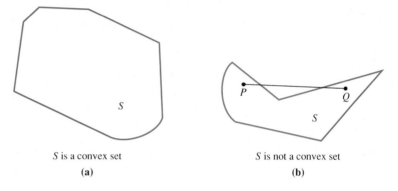

S is a convex set
(a)

S is not a convex set
(b)

One reason for the importance of convex sets in linear programming is the following result.

Theorem

The solution set of any system of linear inequalities is (either empty or) a convex set.

An **objective function** is an expression in terms of x and y whose maximum or minimum value is desired. The "domain" of the objective function is subject to certain restrictions, called **constraints**. These constraints are specified by a system of linear inequalities whose solution set S is convex. Each point of S is called a **feasible point**, and S is called the **feasible region**. The following example will illustrate this terminology.

▶ **EXAMPLE 9.45**

Let $P = 5x + 8y$. Find the maximum and minimum values of P subject to the following constraints:

$$\begin{cases} 4x - y \geq 0 \\ x - 2y \geq -14 \\ x - 4y \leq -1 \\ x + y \leq 20 \\ 2x + 8y \geq 41 \end{cases}$$

Solution

The constraints determine the feasible region, shown by the graph of this system of linear inequalities in figure 9.10(a). The objective function is $P = 5x + 8y$. We want to determine the points in the feasible region that yield the largest (maximum) and the smallest (minimum) possible values for P.

Linear Inequalities and Linear Programming

What are some possible values for *P*? Can $P = 160$? The graph of $5x + 8y = 160$ is shown in figure 9.10(b). As you can see, it does not intersect the feasible region. Hence, given the constraints, it is not possible for *P* to have the value of 160.

Can $P = 120$? The graph of $5x + 8y = 120$ is in figure 9.10(c). As you can see, it passes through the feasible region. Thus, there are infinitely many values for *x* and *y* that will allow $P = 120$. Notice that the line $5x + 8y = 120$ is parallel to the line $5x + 8y = 160$.

Figure 9.10

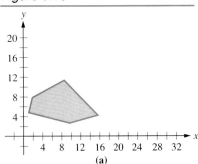
(a)
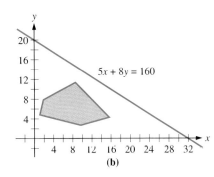
(b)
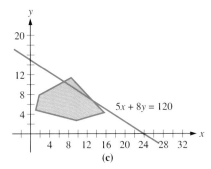
(c)

In fact, if we solve the objective function for *y*, we obtain

$$y = -\tfrac{5}{8}x + \tfrac{1}{8}P.$$

This describes the family of parallel lines with slope $-\tfrac{5}{8}$. If you continue to draw lines with slope $-\tfrac{5}{8}$, you will "see" that the highest such line that will intersect the feasible region meets the feasible region at the vertex $(\tfrac{26}{3}, \tfrac{34}{3})$, and the lowest such line meets at the vertex $(\tfrac{41}{34}, \tfrac{82}{17})$. Thus, the desired maximum for *P* is $5(\tfrac{26}{3}) + 8(\tfrac{34}{3}) = 134$; and the desired minimum for *P* is $5(\tfrac{41}{34}) + 8(\tfrac{82}{17}) = \tfrac{1517}{34}$. ◀

Example 9.45 illustrated the following linear programming theorem, which is stated without proof.

Theorem

> Let $P = ax + by + c$ be an objective function in two variables, considered over a convex feasible set *S* determined by the constraints imposed by a system of linear inequalities. If *S* is bounded and contains its vertices, then *P* assumes an optimum value (maximum or minimum) at some vertex of *S*. If *S* is not bounded, then *P* need not assume an optimum value, but if it does, this occurs at a vertex.

The next two examples illustrate how to implement the above linear programming theorem. In exercise 42, you will encounter an objective function *P* without an optimum value on an unbounded feasible set *S*.

EXAMPLE 9.46

Dobberson Company, Inc. manufactures two models of industrial robots. One is a cylindrical coordinate robot (Model C), and the other is a spherical coordinate robot (Model S). There are two assembly stations, and each robot needs to be worked on at both of these stations. The table below shows the number of hours needed at each station to manufacture a robot and the profit derived from each manufactured robot.

Robot	Station A (hours)	Station B (hours)	Profit
C	2	7	$1250
S	4	2	$1620

If each station can run 24 hours a day, how many of each robot should be manufactured each day in order to maximize the profit?

Solution

We shall let C represent the number of Model C robots that are to be manufactured, and S the number of Model S robots that are manufactured. We know that $C \geq 0$ and $S \geq 0$ since even Dobberson cannot make fewer than zero robots. Because the maximum number of hours for Station A is 24, we know that the number of hours Station A operates must satisfy $2C + 4S \leq 24$. Similarly, the 24-hour maximum for Station B results in the inequality $7C + 2S \leq 24$. Thus, the constraints are given by the system of inequalities

$$\begin{cases} C \geq 0 \\ S \geq 0 \\ 2C + 4S \leq 24 \\ 7C + 2S \leq 24 \end{cases}$$

The objective function for the profit P is $P = 1250C + 1620S$.

The feasible region for this system of inequalities is shown in figure 9.11, with the horizontal axis representing C and the vertical axis S. We wish to maximize P over this feasible region.

According to the above theorem, we need only check the value of P at the vertices of this bounded region in order to find the optimum value. The vertices of the feasible region are (0, 0), (0, 6), (2, 5), and (3.5, 0). Substituting each of these points into the objective function, we obtain the following:

Point (C, S)	P
(0, 0)	$0
(0, 6)	$9,720
(2, 5)	$10,600
(3.5, 0)	$4,375

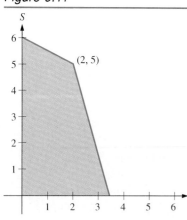

Figure 9.11

Linear Inequalities and Linear Programming

From this table, you can see that the largest daily profit is $10,600, which occurs when two Model C and five Model S robots are manufactured each day. ◀

EXAMPLE 9.47

Tohoot University wants to build some new air-conditioned dormitories. Each dormitory will contain at least 552 students and will have two types of rooms: doubles, which each house two students, and quads, which each house four students. A double room requires 10 units of concrete and 6 units of wood and costs $6,000 to construct. A quad requires 24 units of concrete and 8 units of wood, at a cost of $10,500. Certain contracts of the university require that at least 3020 units of concrete and at least 1148 units of wood be used. Find the number of each type of room the university should build in order to minimize construction costs.

Solution

Suppose D double rooms and Q quads are built. Since a negative number of rooms cannot be built, we know $D \geq 0$ and $Q \geq 0$. The doubles hold a total of $2D$ students and the quads a total of $4Q$ students. Since the total capacity of the rooms must be at least 552 students, we have $2D + 4Q \geq 552$. Similarly, the amount of concrete used must satisfy the inequality $10D + 24Q \geq 3020$, and the amount of wood used must satisfy the inequality $6D + 8Q \geq 1148$. The dollar amount of the construction cost is $C = 6000D + 10500Q$.

The problem requires C to be minimized over the feasible region S determined by the system of inequalities

$$\begin{cases} D \geq 0 \\ Q \geq 0 \\ 2D + 4Q \geq 552 \\ 10D + 24Q \geq 3020 \\ 6D + 8Q \geq 1148 \end{cases}$$

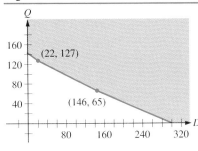

Figure 9.12

The graph of S is in figure 9.12, with the horizontal axis representing the number of double rooms and the vertical axis the number of quads.

The coordinates of the vertices are (0, 143.5), (22, 127), (146, 65), and (302, 0). The value of C at each of these vertices is tabulated in the following table.

Point (D, Q)	C
(0, 143.5)	$1,506,750
(22, 127)	$1,465,500
(146, 65)	$1,558,500
(302, 0)	$1,812,000

Although S is unbounded, it is possible to show that C achieves a minimum at a vertex. (Essentially, the reason is that $D \geq 0$, $Q \geq 0$ in S and the coefficients, 6000 and 10,500, of C are positive. If follows that the value of

C at any given feasible point is at least as great as the value of C at a suitable point on the boundary.) Thus, from this table, you can see that the least cost is \$1,465,500 and that this will occur when 22 double rooms and 127 quads are built. ◂

Exercises 9.4

1. Determine which of the following is/are (equivalent to) a linear inequality:
   (a) $-2x + 3y > 1$ (b) $-x^2 + 3y - 1 > 0$
   (c) $x + \sqrt{y} \leq -1$ (d) $\pi x - ey + 1 \leq 0$
2. Determine which, if any, of the following inequalities has $(2, -3)$ as a particular solution:
   (a) $3x + 2y < 0$ (b) $3x + 2y \geq 0$
   (c) $5x^2 + 8y + 6 \leq 1$ (d) $-5x^2 + 8y + 40 > 2$

In exercises 3–14, graph the given inequality.

3. $2x - 4y < 7$
4. $-x - 4y + 5 > 0$
5. $-x + 4y + 5 \geq 0$
6. $-8x - 2y \leq 8$
7. $2x + y \leq 3$
8. $y < x^2$
9. $-x - 2y + 5 > 0$
10. $\sqrt{y} < x$ (Hint: this is not equivalent to the inequality in exercise 8.)
11. $3x < 2$
12. $2x + 5 \geq 0$
13. $-5y + 6 > 0$
14. $\sqrt{x - 1} > y + 2$
15. Determine which of the following are open half-planes and which are closed half-planes:

    (a)

    (b)
    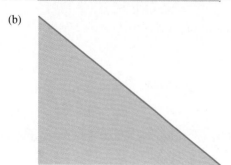

    (c)

    (d)
    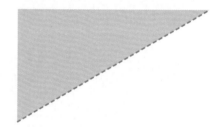

16. Determine which of the following are convex sets:

    (a)

(b)

(c)

(d)

(e)

(f)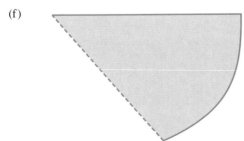

Linear Inequalities and Linear Programming

In exercises 17–26, graph the solution set of the given system of inequalities. State whether or not this graph is bounded. Find the coordinates of all the vertices.

17. $\begin{cases} 4x + 2y \le 8 \\ 5x + 3y > 12 \end{cases}$

18. $\begin{cases} -4x + 2y \ge 8 \\ 2x - 4y < -8 \end{cases}$

19. $\begin{cases} -4x + 2y < 8 \\ -4x + 4y \ge 3 \\ 2x + y \le 0 \end{cases}$

20. $\begin{cases} -4x + 2y < 8 \\ -4x + 4y \ge 3 \\ 2x + 8 \le 0 \end{cases}$

21. $\begin{cases} (x+1)^2 < x^2 + 1 \\ (x-1)^2 < x^2 + 1 \\ y > 3 \end{cases}$

22. $\begin{cases} 3x + 2y > 6 \\ x \ge 0 \\ y \ge 0 \\ x < 3 \end{cases}$

23. $\begin{cases} x^2 + y^2 \le 25 \\ (x-1)^2 + y^2 \ge 25 \end{cases}$

24. $\begin{cases} 2y + x^2 > 0 \\ y - 2x \ge 0 \end{cases}$

25. $\begin{cases} x \ge 0 \\ y \ge 0 \\ 2x + 3y \ge 6 \\ y - 2x < 0 \end{cases}$

26. $\begin{cases} x \ge 0 \\ y \ge 0 \\ 2x + 3y < 6 \end{cases}$

27. Graph the solution of the nonlinear system $\begin{cases} y < \sin x + 3 \\ x \le y^2 \\ y > 2x - 6 \end{cases}$

28. Determine which, if any, of the following systems of linear inequalities has $(2, -3)$ as a feasible point:

(a) $\begin{cases} 9x + 4y > 3 \\ 5x + 2y \le 5 \\ x - y \ge 5 \end{cases}$ (b) $\begin{cases} 3x + 2y < 2 \\ 2x + 3y \ge 1 \\ y - x + 6 > 0 \end{cases}$

In exercises 29–32, find the maximum and minimum values of the given objective function P, subject to the given constraints. (If the feasible region is unbounded, you may assume that only one of the desired optimum values exists and that it is attained at a vertex.)

29. $P = 5x - 8y$, with constraints $\begin{cases} 2x - y \ge 0 \\ x - y \ge -5 \\ x - 2y \le 10 \\ x + y \le 10 \end{cases}$

30. $P = -5x + 8y$, with constraints $\begin{cases} 3x - y \ge 0 \\ x - y \le 0 \\ x + y \le 8 \end{cases}$

31. $P = 8x + 5y$, with constraints $\begin{cases} x \ge 0 \\ y \ge 0 \\ x + y \le 10 \\ 2x + y \ge 15 \end{cases}$

32. $P = 8x + 5y$, with constraints $\begin{cases} x \le 0 \\ y \le 0 \\ x + y + 10 \le 0 \\ x - y \le 0 \end{cases}$

33. A manufacturer produces two brands of stereos. A profit of $60 is made from each brand A stereo that is produced (and eventually sold), but only $50 profit from each brand B stereo. On a daily basis, the factory can produce at most 100 units of brand A and must produce at least 400 units of brand B. If the factory cannot produce more than 550 stereos daily, how many stereos of each brand should be produced daily in order to maximize profit?

34. Two department stores (called I and II) order 35 refrigerators and 45 refrigerators, respectively, from a manufacturer. The 80 refrigerators will be delivered from two depots (called A and B). There are 60 refrigerators in A and 70 refrigerators in B. Delivery costs are summarized as follows:

Cost of Shipping Each Refrigerator to Store

Depot	I	II
A	$20	$30
B	$30	$35

How many refrigerators should each depot deliver to each store in order to minimize the delivery costs?

35. The manager of a hockey club must order at least 12 hockey sticks and at most 20 pucks, but cannot spend more than $450. At least 34 items must be ordered, but the manager cannot order more pucks than sticks. If hockey sticks cost $18 each and pucks cost $4.50 each, how many of each should be ordered if the manager intends to minimize cost?

36. A gourmet coffee store finds that it has a surplus of 640 ounces of Brazilian coffee and 800 ounces of Colombian coffee. The manager intends to use this coffee to make two types of 16 ounce packages. The first type of package will contain 4 ounces of Brazilian and 12 ounces of Colombian coffee; such a package will be sold at a $0.25 profit. The second type of package will contain 10 ounces of Brazilian coffee and 6 ounces of Colombian coffee; such a package will be sold at a $0.50 profit. How many packages of each type should the manager mix in order to maximize her profit?

37. A farmer intends to plant crops of corn and wheat on (at most) 100 acres. Planting, harvesting, and other expenses amount to $110 per acre of corn and $200 per acre of wheat. The farmer has (at most) $12,000 available to meet these expenses. The expected yields are 108 bushels per acre of corn and 25 bushels per acre of wheat. After harvesting, the crops must be stored, and the farmer's silos can hold at most 3500 bushels. If the net profit (after expenses have been met) is $1.25 per bushel of corn and $2.00 per bushel of wheat, how many acres of each crop should the farmer plant in order to maximize profit? (Comment: the answers need not be integers.)

38. John and David need at least 1000 pages typed within five hours. Their secretaries are willing to work overtime, typing 80 pages per hour at a cost of $15 per hour. A company of freelance typists can type 160 pages per hour, at a cost of $25 per hour. If John and David intend to minimize the cost of typing, calculate how long their secretaries type and how long the freelancers type. (Comment: the answers need not be whole numbers of hours.)

39. The following information about nutrition is given on cereal boxes:

Brand of Cereal	One Ounce of Cereal, with Milk, Provides This Number of Grams of	
	Protein	Carbohydrate
A	7	25
B	5	32

A dietician in charge of a dozen patients intends to mix portions of brand A and brand B so as to produce at least 75 grams of protein and at least 310 grams of carbohydrate. Brand A costs $0.08 per ounce, and brand B costs $0.16 per ounce. Ignoring the cost of milk, how should the dietician mix the cereals in order to minimize costs? (Comment: the numbers of ounces of brand A and brand B need not be integers.)

40. The Human Fly plays a carnival game repeatedly. Each time she wins, she gains $3; but each time she loses, she must pay $2. She plays at least twice and at most eighteen times. If she cannot win more often than she loses, what is the greatest net profit that she can hope to gain? (Comment: one way to work this problem does not require this section's techniques!)

41. If an objective function $P = ax + by + c$ takes the same value at two distinct points $P_1(x_1, y_1)$ and $P_2(x_2, y_2)$, show that P takes this same value at every point of the line $\overleftrightarrow{P_1P_2}$. (Hint: if $\overleftrightarrow{P_1P_2}$ is vertical, show that $b = 0$. If $\overleftrightarrow{P_1P_2}$ has slope m, show that $m = -\dfrac{a}{b}$ and that the coordinates of any point on $\overleftrightarrow{P_1P_2}$ satisfy $y - y_1 = -\dfrac{a}{b}(x - x_1)$. Hence infer that $ax + by = ax_1 + by_1$.)

♦ **42.** (A generalization of exercise 41, containing the main idea needed for a proof of the linear programming theorem) Let Q be a point between $P_1(x_1, y_1)$ and $P_2(x_2, y_2)$. Suppose that the value of an objective function $P = ax + by + c$ is less at P_1 than it is at P_2. Show that the value of P at Q is between the values of P at P_1 and at P_2. (Hint: one way is to adapt the argument suggested for exercise 41. Here's another way. First, take care of the cases in which $\overleftrightarrow{P_1P_2}$ is horizontal or vertical. Next, the coordinates of Q satisfy
$$\frac{y - y_1}{y_2 - y_1} = \frac{x - x_1}{x_2 - x_1} = t, \textbf{ parametric equations of } \overleftrightarrow{P_1P_2},$$
in a sense to be developed fully in section 10.4. Solve for x and y in terms of t. Show that
$$ax + by + c = t(ax_2 + by_2 + c) + (1 - t)(ax_1 + by_1 + c)$$
and notice that $0 < t < 1$.)

43. Show that the objective function $P = x - y$ attains neither a maximum value nor a minimum value when subjected to the constraints

$$\begin{cases} x - 2y \leq -1 \\ 2x + y \geq 2 \\ x \geq 0 \\ y \geq 0 \end{cases}$$

(Hint: the value of P at a point (x, y) on the line $x - 2y = -1$ is $y - 1$, and this increases without bound (has "limit infinity") as y increases without bound. Also, the value of P at a point $(0, b)$ is $-b$, and this has limit "negative infinity" as b increases without bound. By graphing the solution set of the given system of inequalities, you can verify that both these limiting processes are available within the feasible region.)

44. Give an example of a linear programming problem in which the feasible region is bounded but the objective function attains neither a maximum nor a minimum. (Hint: according to the linear programming theorem, the feasible region should not contain all the vertices. Notice that $x + y$ is neither maximized nor minimized when subjected to $0 < x < 1, 0 < y < 1$.)

9.5 CHAPTER SUMMARY AND REVIEW EXERCISES

Major Concepts

- $|A|$
- A^T
- Additive identity
- Additive inverse
- adj(A)
- Adjoint of A
- Augmented matrix
- Back substitution
- Boundary
- C_{ij}
- Closed half-plane
- Coefficient matrix
- Cofactor expansion method
- Cofactor of the entry a_{ij}
- Column vector
- Constraints
- Convex set
- Cramer's Rule
- det A
- Determinant of a matrix
- Determinant of order n
- Double subscripts
- Echelon form
- Effect of elementary operations on a determinant
- Elementary operations on systems of linear equations
- Elementary row operations on a matrix
- Equal matrices
- Feasible point
- Feasible region
- Gauss-Jordan elimination
- Graph
  - Bounded
  - of an inequality
  - Unbounded
- I_n
- Identity matrix of order n
- Inner product
- Linear equation in n variables
- Linear inequalities
  - Equivalent
  - Strict
- Linear programming
- Linear system
  - Consistent
  - Dependent
  - Equivalent
  - Inconsistent
- M_{ij}
- Matrix
  - Addition
  - Entry (or element)
- Inverse
- Invertible (or nonsingular)
- $m \times n$
- Main diagonal entry
- Multiplication
- Rank
- Scalar multiple
- Size
- Square, of order n
- Submatrix
- Subtraction
- Transpose
- Upper triangular
- Minor of the entry a_{ij}
- Objective function
- Open half-plane
- Reduced row-echelon form
- Row vector
- Simultaneous particular solution
- Simultaneous solution (or solution set) for a linear system
- System of linear equations
- Vertex
- $0_{m \times n}$

Chapter 9 / Systems of Linear Equations and Linear Inequalities

Review Exercises

1. Determine which of the following equations is/are linear:
   (a) $3x^2 - y = 4$ (b) $3x + 2y = 4$
   (c) $\pi x_1 - 4x_2 = \sqrt{2}$ (d) $4x_1 + \ln x_2 = 0$
   (e) $-3x = 4$ (f) $2xy - 3z = 4$

2. Determine which of the following linear systems is/are in echelon form:

   (a) $\begin{cases} -4x \quad - \quad z = 6 \\ 3y + 2z = 7 \\ y + \sqrt{2}z = 0 \end{cases}$ (b) $\begin{cases} -4x \quad - \quad z = 6 \\ 3y + 2z = 7 \\ \sqrt{2}z = 1 \end{cases}$

   (c) $\begin{cases} -4x_1 + x_2 - x_3 = 0 \\ 3x_2 + 2x_3 = 1 \end{cases}$ (d) $\begin{cases} -4x_1 \quad - \quad x_3 = 6 \\ 2x_1 + 3x_2 \quad = 7 \\ \sqrt{2}x_3 = 1 \end{cases}$

3. *Without finding solution sets*, determine which (if any) of the linear systems having echelon form in exercise 2 is inconsistent, which is dependent, and which has a unique particular solution.

4. Apply the elementary operations (i) $3R_2$, (ii) $R_2 \leftrightarrow R_3$, and (iii) $-2R_3 + R_1$ separately (not in sequence) to the given linear system:

   (a) $\begin{cases} 4x_2 + x_3 = 0 \\ \frac{2}{3}x_1 + \frac{5}{3}x_2 \quad - x_4 = 1 \\ -4x_1 \quad - x_3 + x_4 = \pi \\ ex_1 - x_2 + 2x_3 \quad = \sqrt{2} \end{cases}$

   (b) $\begin{cases} \frac{x}{2} + 4y - 5z = -2 \\ -x + \frac{4}{3}y + z = 0 \\ \frac{x}{4} + 2y + z = -3 \end{cases}$

5. Consider the linear system

   $$\begin{cases} 4x - 4y + 2z = -8 \\ 2x + 4y = -14 \end{cases}$$

   (a) Show that this system has the general solution
   $$(x, y, z) = (-7 - 2t, t, 10 + 6t).$$

   (b) Show that this system's general solution can also be expressed as
   $$(x, y, z) = \left(\frac{-11 - s}{3}, \frac{-10 + s}{6}, s \right).$$

6. Consider the linear system
   $$\begin{cases} a_{11}x_1 + a_{12}x_2 + \cdots + a_{1n}x_n = b_1 \\ \vdots \\ a_{m1}x_1 + a_{m2}x_2 + \cdots + a_{mn}x_n = b_m \end{cases}$$

   (a) If $m < n$, prove that the given system is either inconsistent or dependent. (Hint: transform to echelon form. Can every x_i be a leading variable?)
   (b) If $m < n$ and each $b_i = 0$, prove that the given system has infinitely many particular solutions. (Hint: find *one* particular solution, and then apply (a).)

In exercises 7–12, find the solution set of the given linear system by first using elementary operations to transform it into an equivalent system in echelon form and then using back substitution.

7. $\begin{cases} 2x - y + 3z = -6 \\ -2x + 5y + 2z = 7 \\ -12x + 13y + 11z = 35 \end{cases}$ 8. $\begin{cases} 2x_1 - x_2 + 3x_3 = -6 \\ -2x_1 + 5x_2 + 2x_3 = 7 \\ 6x_1 + x_2 + 14x_3 = -15 \end{cases}$

9. $\begin{cases} 2x_1 - x_2 + 3x_3 = -6 \\ -2x_1 + x_2 - 3x_3 = 7 \end{cases}$ 10. $\begin{cases} 2x - y + 3z = -6 \\ -2x + 5y + 2z = 7 \end{cases}$

11. $\begin{cases} 2x - y = -6 \\ -x + \frac{y}{2} = 3 \\ -4x + 2y = 12 \end{cases}$ 12. $\begin{cases} 2x_1 - x_2 = -6 \\ -x_1 + 2x_2 = 6 \\ -3x_1 + x_2 = 8 \end{cases}$

13. An industrial vat has two inlet tubes and one drain. One tube can fill at four times the rate of the other tube. When both tubes and the drain are operating, the vat fills at the rate of 60 ml/min. When only the weaker tube and drain are operating, the vat drains at 52 ml/min. Find the rates at which the tubes and the drain can operate. (Hint: see the hint for exercise 44 in section 9.1.)

14. An industrial chemist hopes to create a new soup for possible marketing. To do so, he uses three mixtures that had been prepared earlier. By volume, Mixture A contains 60 percent mushroom soup, 10 percent celery soup, and 30 percent chicken noodle soup. Mixture B contains 40 percent mushroom soup, 30 percent celery soup, and 30 percent chicken noodle soup. The corresponding percentages for Mixture C are 0 percent, 40 percent, and 60 percent. Is it possible for the chemist to get a final 680 ml mixture that is 42 percent mushroom soup, 24 percent celery soup, and 34 percent chicken noodle soup? (Hint: model your solution after example 9.12, going from soups to nuts.)

15. Determine which (if any) value(s) of k make the given linear system (i) inconsistent, (ii) dependent, or (iii) consistent, with a unique particular solution:

Chapter Summary and Review Exercises

(a) $\begin{cases} 3x - 2y + 4z = 0 \\ 4y + \pi z = 5 \\ kz = \sqrt{2} \end{cases}$ (b) $\begin{cases} 3x - 2y + 4z = 5 \\ 3x - 2y + 2kz = 4 \end{cases}$

16. By solving a system of linear equations or by using a determinant, write down an equation for the circle passing through the points $(1, 8)$, $(0, \sqrt{2}\,)$, and $(-1, 0)$.

17. Determine the size ($m \times n$) of each of the following matrices:

(a) $\begin{bmatrix} -2 & 3 \\ 1 & 4 \\ 5 & 0 \end{bmatrix}$ (b) $\begin{bmatrix} 0 & -1 & 0 \\ -1 & 0 & 0 \\ 2 & \pi & 0 \end{bmatrix}$

(c) $\begin{bmatrix} -4 \\ 6 \end{bmatrix}$ (d) $[-\frac{1}{2}]$

18. Determine which of the following matrices is/are equal to $\begin{bmatrix} 0 & -1 & 0 \\ 2 & \pi & 0 \end{bmatrix}$:

(a) $\begin{bmatrix} 2-2 & -1 & 0 \\ 2 & \pi & 1-1 \end{bmatrix}$ (b) $\begin{bmatrix} 2 & \pi & 0 \\ 0 & -1 & 0 \end{bmatrix}$

(c) $\begin{bmatrix} 0 & -\sqrt{1} & 0 \\ 1+1 & \pi & 0 \end{bmatrix}$ (d) $\begin{bmatrix} 0 & 2 \\ -1 & \pi \\ 0 & 0 \end{bmatrix}$

19. Perform the indicated scalar multiplications and inner products.

(a) $-\frac{2}{3}\begin{bmatrix} -3 & 6 \\ -1 & 0 \end{bmatrix}$ (b) $3[0 \ -1 \ 4 \ \frac{7}{3}]$

(c) $[4 \ \sqrt{2} \ -3]\begin{bmatrix} -1 \\ 0 \\ -2 \end{bmatrix}$ (d) $[-4 \ 0 \ 6]\begin{bmatrix} 9 \\ \pi \\ 6 \end{bmatrix}$

20. What can you conclude if R is a row vector with n (real) entries such that $R \cdot R^T = 0$? Explain. (Hint: let $R = [r_1 \ r_2 \ \cdots \ r_n]$.)

In exercises 21–23, perform the indicated matrix multiplication.

21. $\begin{bmatrix} 2 & -3 & 1 \\ 4 & -6 & -1 \end{bmatrix}\begin{bmatrix} \frac{1}{2} & 0 & 2 & -3 \\ 0 & -\frac{1}{3} & 3 & -2 \\ -1 & 1 & 4 & 0 \end{bmatrix}$

22. $\begin{bmatrix} 2 & -3 \\ 4 & 6 \end{bmatrix}\begin{bmatrix} -3 & \frac{3}{2} \\ -2 & 1 \end{bmatrix}$ 23. $\begin{bmatrix} 0 & 0 \\ 1 & 0 \end{bmatrix}^2$

24. Let r and s be real numbers, A and B be $m \times n$ matrices, and C be an $n \times p$ matrix. Prove each of the following assertions.

(a) $(r + s)A = rA + sA$ (Hint: let $A = [a_{ij}]$ and compare corresponding entries.)

(b) $r(A + B) = rA + rB$ (Hint: adapt the preceding hint.)

(c) $(rs)A = r(sA)$ (d) $(rA)C = r(AC) = A(rC)$

25. (a) If r is a real number and A is an $n \times n$ matrix, verify that $(rI)A = A(rI)$.

(b) If B is an $n \times n$ matrix such that $BA = AB$ for all $n \times n$ matrices A, show that there exists a real number r such that $B = rI$. (Hint: what does $BA = AB$ imply if A has only one nonzero entry?)

26. Write the linear system

$$\begin{cases} -3x + 4y = 2 \\ 2x - y + 6z = 3 \\ 2y - z = 0 \end{cases}$$

in matrix form as $AX = B$. Find this system's coefficient matrix and augmented matrix.

27. Solve the linear system

$$\begin{cases} -x - y + z = -3 \\ 3y = 2 \\ 2y - z = -1 \end{cases}$$

given that

$$\begin{bmatrix} -1 & -1 & 1 \\ 0 & 3 & 0 \\ 0 & 2 & -1 \end{bmatrix}^{-1} = \begin{bmatrix} -1 & \frac{1}{3} & -1 \\ 0 & \frac{1}{3} & 0 \\ 0 & \frac{2}{3} & -1 \end{bmatrix}.$$

28. Determine which of the following matrices is/are in reduced row-echelon form:

(a) $\begin{bmatrix} 1 & 0 & 0 & 3 \\ 0 & 0 & 1 & 0 \\ 0 & 0 & 0 & 0 \end{bmatrix}$ (b) $\begin{bmatrix} 0 & 1 & 0 \\ 0 & 0 & 1 \\ 0 & 0 & 0 \\ 0 & 0 & 0 \end{bmatrix}$

(c) $\begin{bmatrix} 1 & 2 & 0 \\ 0 & 1 & 0 \\ 0 & 0 & 1 \end{bmatrix}$ (d) $\begin{bmatrix} 1 & 0 & 0 & 0 \\ 0 & 0 & 0 & 0 \\ 0 & 1 & 0 & 1 \end{bmatrix}$

29. Apply the elementary row operations (i) $4R_2$, (ii) $R_1 \leftrightarrow R_3$, and (iii) $3R_1 + R_2$ separately (not successively) to the given matrix:

(a) $\begin{bmatrix} -1 & 0 & 2 \\ 0 & \sqrt{2} & -1 \\ -3 & \pi & 5 \\ 0 & 1 & 0 \end{bmatrix}$ (b) $\begin{bmatrix} \frac{1}{12} & 0 & \frac{1}{6} & 1 \\ \frac{3}{4} & -1 & -\frac{1}{2} & 0 \\ 0 & 2 & -4 & 5 \end{bmatrix}$

In exercises 30–32, use elementary row operations to transform the given matrix to an equivalent matrix in reduced row-echelon form.

30. $\begin{bmatrix} \frac{2}{7} & 9 \\ 1 & 63 \end{bmatrix}$ 31. $\begin{bmatrix} \frac{2}{7} & 9 \\ 1 & 63 \\ 1 & 0 \end{bmatrix}$

Chapter 9 / Systems of Linear Equations and Linear Inequalities

32. $\begin{bmatrix} -7 & 0 & -14 \\ -5 & 5 & -7 \end{bmatrix}$

In exercises 33–36, use Gauss-Jordan elimination to solve the given linear system.

33. $\begin{cases} \frac{2}{7}x + 9y = 4 \\ 2x + 63y = 28 \end{cases}$

34. $\begin{cases} \frac{2}{7}x + 9y = -27 \\ x + 63y = -126 \\ x = -63 \end{cases}$

35. $\begin{cases} 2x_1 - x_2 + 2x_3 = -1 \\ -3x_1 \quad - x_3 = 0 \\ x_1 + x_2 + x_3 = 1 \\ 5x_1 - 4x_2 + 7x_3 = -6 \end{cases}$

36. $\begin{cases} -7x_1 \quad - 14x_3 = 2 \\ -5x_1 + 5x_2 - 7x_3 = 35 \end{cases}$

37. Argue as in example 9.23 (transforming $[A \mid I]$) to determine whether the given matrix is invertible and, if it is, compute its inverse:

(a) $\begin{bmatrix} \frac{1}{3} & -\frac{1}{4} & \frac{1}{5} \\ -20 & 15 & -12 \\ -3 & 0 & 1 \end{bmatrix}$ (b) $\begin{bmatrix} \frac{3}{2} & -7 \\ \frac{1}{7} & \frac{2}{3} \end{bmatrix}$

38. Given $A = \begin{bmatrix} 1 & -1 \\ 2 & 4 \\ 0 & -5 \end{bmatrix}$ and $B = \begin{bmatrix} -1 & 1 \\ -3 & -2 \\ 2 & 0 \end{bmatrix}$, determine the entries of each of the following matrices:
(a) $A + B$ (b) $A - B$ (c) $-A + 3B$ (d) $-2A + 5B$

39. Calculate the following determinants:

(a) $|[\pi]|$ (b) $\begin{vmatrix} -\frac{2}{7} & 7 \\ -2 & \frac{7}{2} \end{vmatrix}$

(c) $\begin{vmatrix} -2 & 4 & 1 \\ 0 & \sqrt{2} & 0 \\ 0 & 0 & 3 \end{vmatrix}$ (d) $\begin{vmatrix} -2 & 0 & 0 \\ 4 & \sqrt{2} & 0 \\ 1 & 0 & 3 \end{vmatrix}$

(e) $\begin{vmatrix} -1 & 0 & 2 & 16 \\ 4 & 0 & 9 & -5 \\ 0 & 0 & 6 & 3 \\ 1 & 0 & -2 & 9 \end{vmatrix}$ (f) $\begin{vmatrix} -1 & 2 & 0 \\ 0 & 3 & 8 \\ -2 & -4 & 5 \end{vmatrix}$

40. Evaluate each of the given determinants by the cofactor expansion method, expanding about the indicated row or column.

(a) $\begin{vmatrix} -2 & -3 \\ -5 & -4 \end{vmatrix}$ (b)→ $\begin{vmatrix} 8 & 1 & -7 \\ 0 & -2 & -3 \\ -4 & -5 & 6 \end{vmatrix}$

(c) $\begin{vmatrix} 0 & -1 & 2 & 3 \\ 1 & 2 & -3 & -1 \\ 0 & -1 & 0 & -1 \\ -2 & 0 & 0 & 4 \end{vmatrix}$

41. Use elementary operations to evaluate the given determinants:

(a) $\begin{vmatrix} 1 & 0 & 1 \\ 7 & 4 & -1 \\ -7 & -4 & 1 \end{vmatrix}$ (b) $\begin{vmatrix} 1 & 2 & 3 & 4 \\ 2 & 3 & 4 & 5 \\ 3 & 4 & 5 & 6 \\ 4 & 5 & 6 & 9 \end{vmatrix}$

(c) $\begin{vmatrix} 1 & 2 & 3 \\ 2 & 3 & 4 \\ 3 & 4 & 8 \end{vmatrix}$ (d) $\begin{vmatrix} 1 & 3 & 9 \\ 1 & 2 & 4 \\ 1 & 1 & 1 \end{vmatrix}$

42. Consider the points $P_1(-2, 3)$, $P_2(4, 0)$, and $P_3(3, 6)$. Use determinants to find each of the following:
(a) an equation for the line through P_1 and P_2
(b) the area of the triangle $\triangle P_1 P_2 P_3$
(c) the area of a parallelogram three of whose vertices are P_1, P_2, and P_3

43. Calculate the transpose of the given matrix:

(a) $\begin{bmatrix} 6 & 9 \\ -5 & 8 \\ 4 & 7 \end{bmatrix}$ (b) $\begin{bmatrix} \frac{1}{2} & 0 & -1 \\ -5 & 1 & 3 \\ 4 & 6 & -\frac{1}{2} \end{bmatrix}$

(c) $\begin{bmatrix} \pi \\ 0 \\ -4 \\ i \end{bmatrix}$ (d) $\begin{bmatrix} 0 & 1 & -4 \\ \sqrt{3} & 0 & 6 \end{bmatrix}$

44. Let r be a real number and let A and B be $m \times n$ matrices. Prove the following assertions:
(a) $(rA)^T = r(A^T)$ (Hint: write A as $[a_{ij}]$.)
(b) $(A + B)^T = A^T + B^T$

45. If A denotes the given matrix, determine adj(A):

(a) $\begin{bmatrix} 5 & 1 & 0 \\ 32 & -2 & -8 \\ 4 & 0 & -1 \end{bmatrix}$ (b) $\begin{bmatrix} -\frac{2}{3} & \frac{4}{5} \\ 20 & -27 \end{bmatrix}$

(c) $\begin{bmatrix} -\frac{2}{3} & \frac{4}{5} \\ 20 & -24 \end{bmatrix}$ (d) $\begin{bmatrix} -16 & -24 & 64 \\ 10 & 15 & -40 \\ 2 & 3 & -8 \end{bmatrix}$

46. Use the methods of section 9.3 to determine if the given matrix is invertible and, if it is, calculate its inverse:

(a) $\begin{bmatrix} 4 & 0 & -1 \\ 1 & -5 & 3 \\ 0 & 0 & -2 \end{bmatrix}$ (b) $\begin{bmatrix} -\frac{3}{2} & 2 \\ -12 & 16 \end{bmatrix}$

In exercises 47–48, use Cramer's Rule to solve the given linear system.

47. $\begin{cases} 6x - 2y - z = 1 \\ 3y + 2z = 0 \\ x \quad + 3z = 4 \end{cases}$

48. $\begin{cases} 5x - 6y = 7 \\ 3x - 4y = 1 \end{cases}$

In exercises 49–52, use determinants to solve the given linear system.

49. $\begin{cases} -4x + y - z = -4 \\ 2x + 3z = 13 \\ x + 5y = -16 \\ -x + y - z = -7 \end{cases}$

50. $\begin{cases} -4x_1 + x_2 - x_3 - 2x_4 = 0 \\ -x_2 + x_4 = 1 \\ 8x_3 - x_4 = -1 \end{cases}$

51. $\begin{cases} 2x_1 - x_2 + x_3 - 5x_4 = 4 \\ -6x_1 + 3x_2 - 3x_3 + 15x_4 = -12 \\ 4x_1 - 2x_2 + 2x_3 - 10x_4 = 8 \\ x_1 - x_2 + x_3 - 5x_4 = 1 \end{cases}$

52. $\begin{cases} -4x + y - z = -4 \\ 2x + 3z = 13 \\ x + 5y = -16 \\ x - 4y + 8z = 50 \end{cases}$

53. If A is an invertible matrix, prove that $\det(A^{-1}) = \dfrac{1}{\det A}$.
(Hint: consider $\det(A^{-1}A)$.)

54. Determine which, if any, of the following systems of linear inequalities has $(-3, 2)$ as a feasible point:

(a) $\begin{cases} 3x + 2y > -4 \\ 2x - 3y \le -12 \end{cases}$ (b) $\begin{cases} 2x - 3y \le -10 \\ 2x + 6y > 5 \\ y \ge 0 \end{cases}$

In exercises 55–60, graph the given inequality.

55. $-2x + 4y \ge 8$
56. $8x - 2y > 4$
57. $x + 2y - 3 < 0$
58. $-x + 3y + 6 \le 0$
59. $-y \le \sqrt{x}$
60. $y \ge e^x$

In exercises 61–64, graph the solution set of the given system of inequalities. State whether or not this graph is bounded. Find the coordinates of all the vertices.

61. $\begin{cases} y \ge 4x \\ 0 < x \\ x + y \le 10 \end{cases}$

62. $\begin{cases} y < 4x \\ x \ge 0 \\ x - y < -10 \end{cases}$

63. $\begin{cases} x < 0 \\ y \ge 0 \\ x < y^2 - 1 \end{cases}$

64. $\begin{cases} x \ge 0 \\ y > 0 \\ x \le 1 - y^2 \end{cases}$

In exercises 65–66, find the maximum and minimum values of the given objective function P, subject to the given constraints. (If the feasible region is unbounded, you may assume that only one of the desired optimum values exists and that it is attained at a vertex.)

65. $P = 8x - 5y$, with constraints
$\begin{cases} x - 2y \ge 0 \\ x - y \ge 4 \\ x + y \le -6 \\ y + 8 \ge 0 \end{cases}$

66. $P = 3x + 4y$, with constraints
$\begin{cases} x \ge 0 \\ y \ge 0 \\ x - 2y \le 8 \\ x + y \ge 3 \end{cases}$

67. A manufacturer produces two brands of stereos. A profit of $60 is made from each brand A stereo that is produced, and a $50 profit is made from each brand B stereo. The factory must produce at least 100 units of brand A and at least 400 units of brand B daily, but cannot produce more than 550 stereos daily. How many stereos of each brand should be produced daily in order to maximize profit?

68. Tohoston University intends to construct an academic office building for at least 180 faculty. It will have two types of rooms: singles, each housing only one professor, and doubles, each housing two professors. A single room requires 3 units of concrete and 4 units of wood, at a cost of $3,500. A double requires 8 units of concrete and 5 units of wood and costs $5,000 to construct. Certain contracts require that at least 600 units of concrete and at least 600 units of wood be used. Find the number of each type of office that Tohoston should build in order to minimize construction costs.

Chapter 10

Additional Topics in Geometry and Algebra

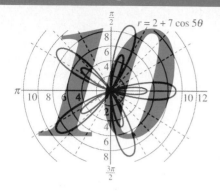

Our earlier discussions of analytic geometry, in chapters 1 and 8, have involved graphing relative to a rectangular or Cartesian coordinate system. In the first three sections of this chapter, we shall consider another type of coordinate system called polar coordinates. Rather than being based on a grid consisting of horizontal and vertical lines, polar coordinates are taken relative to a "grid" consisting of concentric circles and lines through the origin. As you will see, polar coordinates can be used effectively in a number of situations, including some that involve the conic sections. To prepare for sections 10.1–10.3, you might want to review section 6.6 on the polar form of a complex number.

Our study of polar graphs, as well as some of the exercises in chapter 8, motivates the work in section 10.4 on parametric equations. Additional applications will be given to certain types of motion in the plane.

Sections 10.5 and 10.6 discuss two algebraic topics, mathematical induction and the binomial theorem, which will be helpful in your later study of calculus. The exercise sets of these sections include optional enrichment material on limits of sequences and combinatorial symbols.

10.1 POLAR COORDINATES

Consider a point P in the plane, and suppose that P has Cartesian coordinates (x, y). The material in section 6.6 motivates us to make the following definition.

Polar Coordinates

Polar Coordinates

An ordered pair of real numbers (r, θ) is a **pair of polar coordinates** for $P(x, y)$ if and only if

$$x = r \cos \theta \quad \text{and} \quad y = r \sin \theta. \tag{1}$$

EXAMPLE 10.1

$(r,\theta) \qquad (x,y)$

$\left(2, \dfrac{\pi}{3}\right)$ is a pair of polar coordinates for $(1, \sqrt{3})$ since

"$r \cos \theta = x = x$"
$$2 \cos \dfrac{\pi}{3} = 2\left(\dfrac{1}{2}\right) = 1 \quad \text{and} \quad$$

"$r \sin \theta = y = y$"
$$2 \sin \dfrac{\pi}{3} = 2\left(\dfrac{\sqrt{3}}{2}\right) = \sqrt{3}.$$

More generally, *if* the complex number $x + iy$ has the polar (or "trigonometric") form r cis θ, *then* (r, θ) is a pair of polar coordinates for (x, y).

However—and this is perhaps the most important difference between polar and rectangular coordinates—**each point P has infinitely many different pairs of polar coordinates.**

EXAMPLE 10.2

I. $\left(2, \dfrac{7\pi}{3}\right)$ is another pair of polar coordinates for $(1, \sqrt{3})$. This follows because

$$\cos \dfrac{7\pi}{3} = \cos \dfrac{\pi}{3} \quad \text{and} \quad \sin \dfrac{7\pi}{3} = \sin \dfrac{\pi}{3}.$$

II. For a similar reason, $\left(2, -\dfrac{5\pi}{3}\right)$ is also a pair of polar coordinates for $(1, \sqrt{3})$.

BY USING THIS, All EQUATIONS FOR X,Y WILL BE THE SAME. FOR EX. I, II BOTH HAVE $(1, \sqrt{3})$ FOR AN EQUATION.

More generally, if (r, θ) is a pair of polar coordinates for (x, y), then so is $(r, \theta + k2\pi)$ for each integer k.

Moreover, if (r, θ) is a pair of polar coordinates for (x, y), then

$$(-r, \theta + \pi + k2\pi)$$

is another pair of polar coordinates for (x, y). [To verify this, just note via the trigonometric identities in sections 6.1 and 6.2 that

$$-r \cos (\theta + \pi + k2\pi) = -r \cos (\theta + \pi) = -r (- \cos \theta) = r \cos \theta$$

$$-r \sin (\theta + \pi + k2\pi) = -r \sin (\theta + \pi) = -r (- \sin \theta) = r \sin \theta.]$$

In particular, **the first coordinate in a pair of polar coordinates can be negative**. Thus, given one pair of polar coordinates (r, θ) for a point P, we have an "essentially different" pair of polar coordinates for P, namely $(-r, \theta + \pi)$.

For instance, $\left(-2, \dfrac{\pi}{3} + \pi\right) = \left(-2, \dfrac{4\pi}{3}\right)$ is another pair of polar coordinates for $P(1, \sqrt{3})$. So is $\left(-2, -\dfrac{2\pi}{3}\right)$.

In discussing polar coordinates, it is customary to refer to the origin O as the **pole** and to the ray starting at O and running along the positive x-axis as the

Chapter 10 / Additional Topics in Geometry and Algebra

Figure 10.1

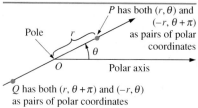

P has both (r, θ) and $(-r, \theta + \pi)$ as pairs of polar coordinates

Q has both $(r, \theta + \pi)$ and $(-r, \theta)$ as pairs of polar coordinates

polar axis. These appear in figure 10.1, which summarizes the above discussion. It is also customary to refer to the y-axis as the $\frac{\pi}{2}$-line.

For reasons that will become obvious, polar coordinate graphing paper has a "grid" consisting of concentric circles centered at O and rays having endpoint O. If we need to plot a point P having a given (r, θ) as a pair of polar coordinates, it is most convenient to locate the angle θ in standard position and find a point R on the terminal side of θ such that R is $|r|$ units from O. If $r \geq 0$, then $P = R$; and if $r < 0$, then P is located by reflecting R through O. This is illustrated in the following example.

▶ **EXAMPLE 10.3**

Plot the points having the indicated pairs of polar coordinates:

$A\left(3, \frac{\pi}{4}\right)$, $B\left(-3, \frac{\pi}{4}\right)$, $C\left(-7, \frac{2\pi}{3}\right)$, $D\left(5, -\frac{\pi}{3}\right)$, $E(6, \pi)$, and $F\left(5, \frac{13\pi}{6}\right)$.

Solution The points are plotted in figure 10.2.

Figure 10.2

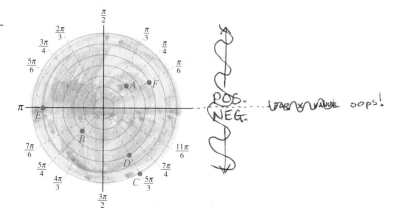

Converting Between Polar and Rectangular Coordinates

Some equations are easier to deal with when they are written in polar, rather than rectangular, coordinates. For other equations, the reverse is true. Thus, it is often useful to convert from one system to the other. To convert from polar to rectangular coordinates, you use the equations in (1), as the next example illustrates.

▶ **EXAMPLE 10.4**   CONVERTING: POLAR ⟶ RECTANGULAR

Find the rectangular coordinates for the points with the following pairs of polar coordinates: **(a)** $\left(6, \frac{\pi}{6}\right)$, **(b)** $\left(2, \frac{3\pi}{4}\right)$, and **(c)** $(-2, \pi)$.

Solutions **(a)** Here, $r = 6$ and $\theta = \frac{\pi}{6}$. So

$$x = r \cos \theta = 6 \cos \frac{\pi}{6} = 3\sqrt{3} \quad \text{and} \quad y = r \sin \theta = 6 \sin \frac{\pi}{6} = 3.$$

Polar Coordinates

As a result, the rectangular coordinates of this point are $(3\sqrt{3}, 3)$.

(b) In this problem, $r = 2$ and $\theta = \dfrac{3\pi}{4}$. So

$$x = 2 \cos \dfrac{3\pi}{4} = -\sqrt{2} \quad \text{and} \quad y = 2 \sin \dfrac{3\pi}{4} = \sqrt{2}.$$

Thus, the rectangular coordinates of this point are $(-\sqrt{2}, \sqrt{2})$.

(c) We have $r = -2$ and $\theta = \pi$. So

$$x = -2 \cos \pi = 2 \quad \text{and} \quad y = -2 \sin \pi = 0.$$

The requested rectangular coordinates are $(2, 0)$.

Converting from Rectangular Coordinates to Polar Coordinates

To convert from rectangular coordinates (x, y) to polar coordinates (r, θ), it is helpful to recall (as in section 6.6) that

$$r^2 = x^2 + y^2 \quad \text{and} \quad \tan \theta = \dfrac{y}{x} \text{ if } x \neq 0. \qquad (2)$$

As noted above, $(-r, \theta + \pi)$ is then an "essentially different" pair of polar coordinates for the same point. The next example illustrates the use of (2).

EXAMPLE 10.5 *IN OTHER WORDS THEY WANT YOU TO FIND 2 OTHER POLAR COORDINATES.*

Find two "essentially different" pairs of polar coordinates for each of the following points: (a) $(5, -5\sqrt{3})$, (b) $(-\sqrt{2}, \sqrt{2})$, and (c) $(-7, -24)$.

Solutions

(a) Here we have $x = 5$ and $y = -5\sqrt{3}$. So $r^2 = 5^2 + (-5\sqrt{3})^2 = 25 + 75 = 100$, which means that $r = \pm 10$. Also, $\tan \theta = \dfrac{y}{x} = \dfrac{-5\sqrt{3}}{5} = -\sqrt{3}$. The two numbers in the interval $[0, 2\pi)$ that satisfy this are $\theta = \dfrac{2\pi}{3}$ and $\theta = \dfrac{5\pi}{3}$. Since the point $(5, -5\sqrt{3})$ is in the fourth quadrant, we take $\theta = \dfrac{5\pi}{3}$ and $r = 10$. In other words, one pair of polar coordinates for $(5, -5\sqrt{3})$ is $\left(10, \dfrac{5\pi}{3}\right)$. An "essentially different" pair is $\left(-10, \dfrac{5\pi}{3} + \pi\right) = \left(-10, \dfrac{8\pi}{3}\right)$ ⒶA; another is $\left(-10, \dfrac{2\pi}{3}\right)$ ⒷB.

FROM THIS, WE CAN GET A + B

(b) Here $x = -\sqrt{2}$ and $y = \sqrt{2}$. So $r^2 = (-\sqrt{2})^2 + (\sqrt{2})^2 = 4$ and $r = \pm 2$. We also have $\tan \theta = -1$, which means $\dfrac{3\pi}{4}$ and $\dfrac{7\pi}{4}$ are candidates for θ. Since $(-\sqrt{2}, \sqrt{2})$ is in quadrant II, we conclude that $\theta = \dfrac{3\pi}{4}$ may be used with $r = 2$. In other words, $\left(2, \dfrac{3\pi}{4}\right)$ is one pair of polar coordinates for $(-\sqrt{2}, \sqrt{2})$.

(Compare this to example 10.4(b).) An "essentially different" pair is $\left(-2, \frac{3\pi}{4} + \pi\right) = \left(-2, \frac{7\pi}{4}\right)$.

(c) For $(-7, -24)$, $x = -7$ and $y = -24$. So $r^2 = 625$ and $r = \pm 25$. From $\tan\theta = \frac{24}{7}$, we see that $r = 25$ may be coupled with $\theta = \pi + \tan^{-1}\left(\frac{24}{7}\right) \approx 4.429$. An "essentially different" pair of polar coordinates for the same point is $(-25, 2\pi + \tan^{-1}\frac{24}{7}) \approx (-25, 7.570)$ or, for that matter, $(-25, \tan^{-1}\frac{24}{7}) \approx (-25, 1.287)$.

Conversions between rectangular coordinates and polar coordinates can also be carried out on graphing calculators such as a *TI-81* and a Casio *fx-7700G*. Consult the owner's manual of your calculator for details.

Symmetry

If (r, θ) is a pair of polar coordinates for (x, y), then $(r, -\theta)$ is a pair of polar coordinates for $(x, -y)$. (To verify this, recall that cosine is an even function and sine is odd.) However, you will recall that the two points (x, y) and $(x, -y)$ are symmetric with respect to the x–axis. Hence, we can conclude the following.

Symmetry with Respect to the Polar Axis

The points with polar coordinates (r, θ) and $(r, -\theta)$ are **symmetric with respect to the polar axis.**

By similar reasoning (which you will be asked to supply in exercises 33 and 34), the following can be shown.

Symmetry with Respect to the $\frac{\pi}{2}$-line and the Pole

The points with polar coordinates (r, θ) and $(r, \pi - \theta)$ are **symmetric about the $\frac{\pi}{2}$-line.**

The points with polar coordinates (r, θ) and $(-r, \theta)$ are **symmetric about the pole**.

These observations about symmetry will be useful in carrying out the graphing in sections 10.2 and 10.3. We now proceed to a brief introduction to polar graphing, emphasizing familiar curves.

Polar Equations and Their Graphs

An equation involving one or both of the letters r and θ representing polar coordinates (and no other "unknowns") is a **polar equation**. The **(polar) graph of a polar equation** is the set of all points having at least one pair of polar coordinates that satisfy the given equation.

Polar Coordinates

For instance, the point with polar coordinates $\left(2, \frac{\pi}{3}\right)$ is on the polar graph of $r = 2$. So is the point with polar coordinates $\left(-2, \frac{\pi}{3}\right)$, since it has the polar coordinates $\left(2, \frac{4\pi}{3}\right)$. On the other hand, the point with polar coordinates $\left(3, \frac{\pi}{3}\right)$ is not on the polar graph of $r = 2$ since *none* of this point's pairs of polar coordinates has the form $(2, \theta)$.

The above type of reasoning leads us to the conclusion that the polar graph of $r = 2$ is the circle, with radius 2, centered at the pole (origin). This same circle is the polar graph of $r = -2$. A similar conclusion holds for the polar graph of any equation of the form $r = $ constant. Later in this section, you will see a polar equation of a "general" circle, not necessarily centered at the pole.

▶ **EXAMPLE 10.6**

Graph the polar equation $r = 6 \sin \theta$.

Solution

Just as for graphs involving rectangular coordinates, we begin by constructing a table of values. To do this, we shall assign various values of θ and then calculate the value of r corresponding to each of them. For instance, if $\theta = \frac{\pi}{6}$, then $r = 6 \sin \frac{\pi}{6} = 6(\frac{1}{2}) = 3$. Since sine has period 2π, the entire graph is swept out as θ goes from 0 to 2π, and so the table below only covers the interval $[0, 2\pi]$.

θ	0	$\frac{\pi}{6}$	$\frac{\pi}{4}$	$\frac{\pi}{3}$	$\frac{\pi}{2}$	$\frac{2\pi}{3}$	$\frac{3\pi}{4}$	$\frac{5\pi}{6}$	π	$\frac{7\pi}{6}$	$\frac{5\pi}{4}$	$\frac{4\pi}{3}$	$\frac{3\pi}{2}$	$\frac{5\pi}{3}$	$\frac{7\pi}{4}$	$\frac{11\pi}{6}$	2π
r	0	3	$3\sqrt{2} \approx 4.24$	$3\sqrt{3} \approx 5.20$	6	$3\sqrt{3} \approx 5.20$	$3\sqrt{2} \approx 4.24$	3	0	-3	$-3\sqrt{2} \approx -4.24$	$-3\sqrt{3} \approx -5.20$	-6	$-3\sqrt{3} \approx -5.20$	$-3\sqrt{2} \approx -4.24$	-3	0

Figure 10.3

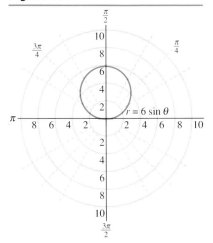

The graph of this curve is in figure 10.3. Notice that, even though sine has a period of 2π, the entire graph is actually swept out as θ goes from 0 to π, and the graph is **retraced** as θ goes from π to 2π. ◀

The above work suggests that the polar graph of $r = 6 \sin \theta$ is a circle with radius 3 and center having Cartesian coordinates (0, 3). Here is an analytic way to prove this. Multiply $r = 6 \sin \theta$ through by r, getting $r^2 = 6r \sin \theta$. (This has the same polar graph since $r = 0$, giving the origin, satisfies $r = 6 \sin \theta$ with θ equal, for instance, to π.) Equivalently, $x^2 + y^2 = 6y$, or $x^2 + (y - 3)^2 = 9$. The graph of this equation is a circle centered at (0, 3), with radius 3.

Let's determine the polar equation for a general circle. In rectangular coordinates, a circle with center $C(h, k)$ and radius a has the equation

$$(x - h)^2 + (y - k)^2 = a^2.$$

To convert this to polar coordinates, we let polar coordinates of the center C be (ρ, ϕ) and let $P(x, y)$, with polar coordinates (r, θ), be any point on the circle. Converting, we see that

$$h = \rho \cos \phi, \quad k = \rho \sin \phi, \quad x = r \cos \theta, \quad y = r \sin \theta.$$

Substituting these into the Cartesian equation for the circle yields

$$(r \cos \theta - \rho \cos \phi)^2 + (r \sin \theta - \rho \sin \phi)^2 = a^2.$$

Expanding and then simplifying via the Pythagorean identity $\sin^2 \alpha + \cos^2 \alpha = 1$, we have

$$\rho^2 + r^2 - 2\rho r(\cos \theta \cos \phi + \sin \theta \sin \phi) = a^2.$$

By the subtraction formula of section 6.2, this can be simplified to the following equation.

Polar Equation of a Circle

The polar equation of the circle with radius a and center having polar coordinates (ρ, ϕ) is

$$\rho^2 + r^2 - 2\rho r \cos(\theta - \phi) = a^2. \tag{3}$$

Figure 10.4

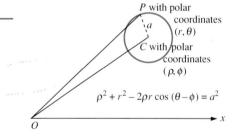

Equation (3) is awkward, but simplifies in the following special cases.

Special Cases of the Polar Equation of a Circle

If the center C is at the pole, then $\rho = 0$ and equation (3) becomes $r^2 = a^2$, which has the same polar graph as

$$r = a. \tag{4}$$

If C lies on the polar axis and has polar coordinates $(a, 0)$, we may take $\rho = a$ and $\phi = 0$. Then equation (3) has the same polar graph as

$$r = 2a \cos \theta. \tag{5}$$

If C has polar coordinates $\left(a, \dfrac{\pi}{2}\right)$, equation (3) has the same graph as

$$r = 2a \sin \theta. \tag{6}$$

Polar Coordinates

If you look again at $r = 6 \sin \theta$, the equation in example 10.6, you will see that it has the form of equation (6). This proves again that its graph is a circle of radius $a = 3$, with center having polar coordinates $\left(3, \frac{\pi}{2}\right)$.

▶ **EXAMPLE 10.7** Find a polar equation of, and graph, the circle with radius 6 and center C having polar coordinates $\left(2, \frac{5\pi}{3}\right)$.

Solution Here we have $a = 6$, $\rho = 2$, and $\phi = \frac{5\pi}{3}$. So equation (3) yields

$$6^2 = r^2 + 2^2 - 2 \cdot 2r \cos\left(\theta - \frac{5\pi}{3}\right)$$

or $$r^2 - 4r \cos\left(\theta - \frac{5\pi}{3}\right) = 32.$$

Using the subtraction formula for $\cos(\alpha - \beta)$, we can write this as

$$r^2 - 4r\left(\cos \frac{5\pi}{3} \cos \theta + \sin \frac{5\pi}{3} \sin \theta\right) = 32$$

$$r^2 - 4r\left(\frac{1}{2} \cos \theta - \frac{\sqrt{3}}{2} \sin \theta\right) = 32$$

or $$r^2 - 2r(\cos \theta - \sqrt{3} \sin \theta) = 32.$$

The graph of this equation is in figure 10.5.

therefore $R = \sqrt{32}$ or 5.656

Figure 10.5

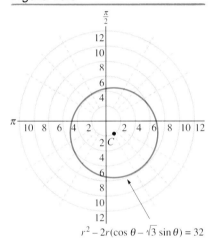

$r^2 - 2r(\cos \theta - \sqrt{3} \sin \theta) = 32$

▶ **EXAMPLE 10.8** Find the center and radius of the circle given by

$$r^2 - 4\sqrt{3}\ r \cos \theta - 4r \sin \theta + 15 = 0.$$

Solution One way to work this example is to convert it to rectangular coordinates, and then analyze the resulting equation by the methods of section 1.4. We would find that

$$x^2 + y^2 - 4\sqrt{3}x - 4y + 15 = 0$$

and, after completing the square,

$$(x - 2\sqrt{3})^2 + (y - 2)^2 = -15 + 12 + 4 = 1.$$

The graph is a circle with center $(2\sqrt{3}, 2)$ and radius 1. Notice that $\left(4, \frac{\pi}{6}\right)$ is a pair of polar coordinates for the center, since $4 \cos \frac{\pi}{6} = 2\sqrt{3}$ and $4 \sin \frac{\pi}{6} = 2$.

Chapter 10 / Additional Topics in Geometry and Algebra

Next, we shall show how to rework this example solely in terms of polar coordinates. Let's begin by writing the equation as

$$r^2 - (4\sqrt{3} \cos \theta + 4 \sin \theta)r + 15 = 0.$$

To get this into standard form, we need to write $4\sqrt{3} \cos \theta + 4 \sin \theta$ as $2\rho \cos (\theta - \phi)$, where ϕ and ρ are constants to be determined. Recall from section 6.2 that $\cos (\theta - \phi) = \cos \theta \cos \phi + \sin \theta \sin \phi$. If we choose $2\rho = 8$, then ϕ is to satisfy

$$2\rho \cos (\theta - \phi) = 8\left(\frac{\sqrt{3}}{2} \cos \theta + \frac{1}{2} \sin \theta\right),$$

which means that $\cos \phi = \frac{\sqrt{3}}{2}$ and $\sin \phi = \frac{1}{2}$. Thus, with $\rho = 4$, we may choose $\phi = \frac{\pi}{6}$. It follows (again) that $\left(4, \frac{\pi}{6}\right)$ is a pair of polar coordinates for the circle's center. Finally, to find the circle's radius, use equation (3) and the given equation to conclude that $\rho^2 - a^2 = 15$. Since $\rho = 4$, it must be that $a = 1$ is the radius. ◀

Polar Equation of a Line

Suppose ℓ is a line in a polar coordinate system, as in figure 10.6, and suppose that ℓ does not contain the pole. Let N be the projection of O on ℓ; let (ρ, ω), with $\rho > 0$, be a pair of polar coordinates for N. (Thus, $\overline{ON} \perp \ell$.) Let P be any other point on ℓ; let (r, θ), with $r > 0$, be a pair of polar coordinates for P. By considering the right triangle $\triangle ONP$, we can reach the following conclusions about the polar equation of the line.

Polar Equation of a Line Not Containing the Pole

Any line ℓ not containing the pole has a polar equation

$$r \cos (\theta - \omega) = \rho. \qquad (7)$$

If ℓ is parallel to the polar axis, ω can be taken as either $\frac{\pi}{2}$ or $\frac{3\pi}{2}$, and equation (7) becomes either

$$r \sin \theta = \rho \quad \text{or} \quad r \sin \theta = -\rho. \qquad (8)$$

If ℓ is perpendicular to the polar axis, ω can be taken as either 0 or π, and equation (7) becomes either

$$r \cos \theta = \rho \quad \text{or} \quad r \cos \theta = -\rho, \qquad (9)$$

where ρ is a nonnegative constant.

There is one "degenerate" case to consider.

Polar Equation of a Line Through the Pole

If a line ℓ contains the pole O, then a polar equation of ℓ is

$$\theta = \theta_1, \qquad (10)$$

where θ_1 is a constant.

Figure 10.6

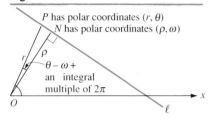

P has polar coordinates (r, θ)
N has polar coordinates (ρ, ω)
$\theta - \omega +$ an integral multiple of 2π

If α is the (radian measure of the) angle of inclination of ℓ, then θ_1 can be $\alpha + n\pi$, for any integer n.

In short, each line has a polar equation having one of the forms given by (7) (with ρ possibly negative) and (10).

EXAMPLE 10.9

Determine a polar equation of, and graph, the line ℓ passing through N with polar coordinates $\left(4, \dfrac{\pi}{3}\right)$ and perpendicular to the polar axis.

Solution

N has Cartesian coordinates $\left(4 \cos \dfrac{\pi}{3}, 4 \sin \dfrac{\pi}{3}\right) = (2, 2\sqrt{3})$. Moreover, the line is vertical, and so has Cartesian equation $x = $ constant, namely $x = 2$. Converting, we have the polar equation $r \cos \theta = 2$, which has the form (9). Its graph is shown in figure 10.7.

Figure 10.7

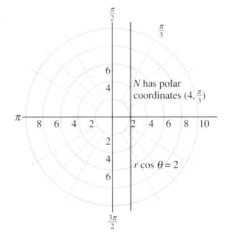

EXAMPLE 10.10

Determine the polar equation of, and graph, the line ℓ through the point N with polar coordinates $\left(5, \dfrac{7\pi}{6}\right)$ and perpendicular to $\overleftrightarrow{ON}$.

Solution

Here, we have $\rho = 5$ and $\omega = \dfrac{7\pi}{6}$. Thus, via equation (7), the required equation is

$$r \cos\left(\theta - \dfrac{7\pi}{6}\right) = 5.$$

Its graph is in figure 10.8.

Figure 10.8

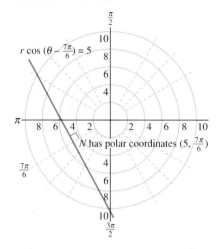

Graphing Calculator Activity

In this graphing calculator activity, we demonstrate how to graph a polar equation.

▶ **EXAMPLE 10.11**

Use a graphing calculator to graph the function $r(\theta) = a + \cos \theta$ for each of the following values of a: (a) 1.05, (b) 1.25, (c) 1.75, and (d) 1.95.

Solutions

Casio fx-7700G Version

(a) First, change the calculator from the rectangular (REC) graphing mode to the polar (POL) graphing mode. To do this, press [SHIFT] [DRG] [F2] [EXE] [MODE] [+] [MODE] [SHIFT] [−]. Also make sure that your calculator is in radian mode.

Next, set the display parameters for the screen to Xmin = −3, Xmax = 3, Xscl = 1, Ymin = −2, Ymax = 2, and Yscl = 1. As soon as you press [EXE], you are shown another screen with the heading Range and then T, θ. Because this example involves graphing an equation in terms of θ, you will need to input values of θ. Since $\cos \theta$ has period 2π, we consider $0 \le \theta \le 2\pi$; set min = 0, max = 2π, and ptch = $\dfrac{\pi}{36}$. When you have finished, you should have the screen in figure 10.9(a). Press the [EXE] key to exit from this display.

Now, graph $r(\theta) = 1.05 + \cos \theta$ by pressing [Graph] 1.05 [+] [cos] [X,θ,T]. (Notice that when you pressed the [Graph] key, the display read "$r =$". Also, notice that the [X,θ,T] key is now used to input the variable θ.) The screen should now look like the one in figure 10.9(b).

Press [EXE]. You should obtain the screen in figure 10.9(c).

Figure 10.9

(a)

Polar Coordinates

Figure 10.9 (continued)

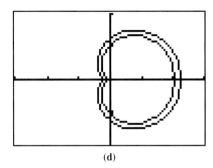
(b)

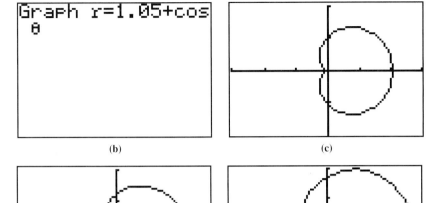
(c)   (d)

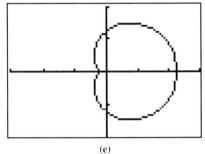
(e)

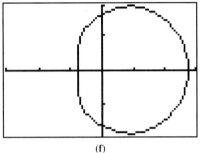
(f)

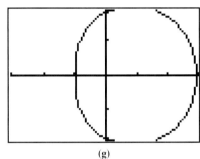
(g)

(b) Graph $r(\theta) = 1.25 + \cos\theta$ by pressing [Graph] 1.25 [+] [cos] [X,θ,T] [EXE].
You should get the display in figure 10.9(d). If you do not want this second graph to appear on the same screen as the first, press [SHIFT] [Cls] [EXE] before you press the [Graph] key. You will then get the graph in figure 10.9(e).

(c) Graph $r(\theta) = 1.75 + \cos\theta$ by pressing [Graph] 1.75 [+] [cos] [X,θ,T] [EXE].
You should get the display in figure 10.9(f).

(d) Graph $r(\theta) = 1.95 + \cos\theta$ by pressing [Graph] 1.95 [+] [cos] [X,θ,T] [EXE].
You should get the display in figure 10.9(g).

Compare your graphs in (f) and (g) with the ones from (d). What changes can you see? What has happened to the size of the figure? What changes have occurred in the "dimple?"

TI-81 Version (a) The following instructions will allow you to graph these polar equations on your *TI-81* calculator. The reason they work will be explained in section 10.4.

First, change the calculator from the rectangular graphing mode to the parametric graphing mode and make sure that your calculator is in radian mode. To do this, press [MODE] and change the settings to those in figure 10.9(h).

Next, press the [RANGE] key and set the display parameters to Tmin = 0, Tmax = 6.30 ($\approx 2\pi$), Tstep = 0.08 ($\approx \pi \div 36$), Xmin = –3, Xmax = 3, Xscl = 1, Ymin = –2, Ymax = 2, and Yscl = 1. When you press the [RANGE] key, you will see an arrow pointing downward in the position of the equal sign that had followed Ymin. This arrow indicates that there are additional lines

Chapter 10 / Additional Topics in Geometry and Algebra

"below" the screen. When you press the ENTER key at Ymin, the top line will scroll off the top of the screen and the next line, in this case Ymax, will appear at the bottom.

We will actually input the formula $r(\theta) = a + \cos \theta$ by using the variable a, and then change the values of a in the calculator's memory to obtain each of the four graphs. Press the Y= key to see the "equation screen." We will need the first two lines, giving the equations for X1T and Y1T, in order to graph a polar equation. To enter the equation for X1T, press (ALPHA A + COS X|T) COS X|T . Press ENTER to move to the next line. To enter the equation for Y1T, press (ALPHA A + COS X|T) SIN X|T . (Notice that the X|T key was used to input the variable T.) Your screen should now look like the one in figure 10.9(i).

Because we wrote the above equations in terms of a, we need to tell the calculator what value we want a to assume. For the first graph, $a = 1.05$; so type 2nd QUIT to exit from the equation screen, and then type 1.05 STO▶ A ENTER . (You should not press the ALPHA key before typing A because pressing the STO▶ key automatically places the calculator in alphabetic mode.) Now, press GRAPH and you should see the display in figure 10.9(j).

(b) To graph $r = 1.25 + \cos \theta$, press 1.25 STO▶ A ENTER GRAPH . You should get the display in figure 10.9(k).

(c) To graph $r = 1.75 + \cos \theta$, press 1.75 STO▶ A ENTER GRAPH . You should get the display in figure 10.9(l).

(d) To graph $r = 1.95 + \cos \theta$, press 1.95 STO▶ A ENTER GRAPH . You should get the display in figure 10.9(m).

Figure 10.9 (continued)

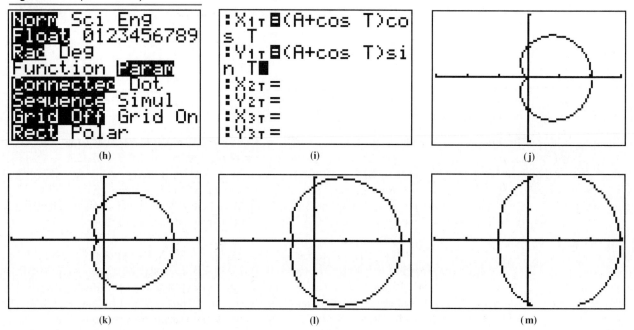

Compare your graphs in (l) and (m) with the ones from (j) and (k). What changes can you see? What has happened to the size of the figure? What changes have occurred in the "dimple?"

We will look more closely in the next section at the so-called "limaçon" graphs generated by equations of the type $r(\theta) = a + b \cos \theta$. For the moment, you may wish to experiment further with your calculator to see how the graphs change as you vary a and b.

Exercises 10.1

In exercises 1–14, plot the point with the given pair of polar coordinates, and then find the Cartesian coordinates of this point.

1. $\left(3, \dfrac{\pi}{6}\right)$
2. $\left(2, \dfrac{\pi}{4}\right)$
3. $\left(-2, \dfrac{5\pi}{3}\right)$
4. $\left(\sqrt{3}, -\dfrac{16\pi}{3}\right)$
5. $\left(\sqrt{2}, \dfrac{13\pi}{4}\right)$
6. $\left(-3, \dfrac{17\pi}{6}\right)$
7. $(\tfrac{1}{3}, 6)$ (Hint: use your calculator to approximate the coordinates.)
8. $\left(-2, \dfrac{13\pi}{6}\right)$
9. $(-\tfrac{1}{2}, \pi)$
10. $(\tfrac{1}{2}, 4\pi)$
11. $\left(\dfrac{1}{2}, -\dfrac{5\pi}{2}\right)$
12. $\left(2, \dfrac{\pi}{2}\right)$
13. $\left(0, \dfrac{\pi}{4}\right)$
14. $\left(0, -\dfrac{\pi}{4}\right)$

15. (a) If θ is any real number, prove that $(0, \theta)$ is a pair of polar coordinates for the pole (that is, for the origin).
    (b) If (r, θ) is a pair of polar coordinates for the pole, show that $r = 0$.

16. Let r and s be nonzero (possibly equal) real numbers, and also let θ and ϕ be real numbers. Prove that (r, θ) and (s, ϕ) are polar coordinates for the same point if and only if exactly one of the following two conditions holds:
    (i) $r = s$ and $\theta - \phi = k2\pi$ for some integer k;
    (ii) $r = -s$ and $\theta - \phi = \pi + k2\pi$ for some integer k.
    (Hint: review exercise 11(b) in section 5.2.)

In exercises 17–24, find two "essentially different" pairs of polar coordinates (r, θ) [one with $r > 0$, the other with $r < 0$] for the point with the given Cartesian coordinates. (If necessary, use a calculator to approximate r and/or θ.)

17. $(2, 2\sqrt{3}\,)$
18. $(-2\sqrt{3}, 2)$
19. $(-4, -4)$
20. $(2, -2\sqrt{3}\,)$
21. $(2, -3)$
22. $(\tfrac{1}{2}, \tfrac{2}{3})$
23. $(-\sqrt{2}, 5)$
24. $(-\sqrt{2}, 0)$

In exercises 25–26, prove that the points with the given pairs of polar coordinates are symmetric with respect to the polar axis. (Hint: use the criterion for symmetry given in the text.)

25. $\left(-2, \dfrac{\pi}{3}\right)$ and $\left(-2, -\dfrac{\pi}{3}\right)$
26. $\left(\sqrt{3}, -\dfrac{\pi}{6}\right)$ and $\left(\sqrt{3}, \dfrac{\pi}{6}\right)$

In exercises 27–28, prove that the points with the given pairs of polar coordinates are symmetric with respect to the $\dfrac{\pi}{2}$-line.

27. $\left(\sqrt{2}, \dfrac{\pi}{6}\right)$ and $\left(\sqrt{2}, \dfrac{5\pi}{6}\right)$
28. $\left(-3, -\dfrac{\pi}{3}\right)$ and $\left(-3, \dfrac{4\pi}{3}\right)$

In exercises 29–30, prove that the points with the given pairs of polar coordinates are symmetric with respect to the pole.

29. $\left(\sqrt{3}, \dfrac{11\pi}{5}\right)$ and $\left(-\sqrt{3}, \dfrac{11\pi}{5}\right)$
30. $(-e, -2)$ and $(e, -2)$

31. (a) Prove that the points with $\left(2, \dfrac{\pi}{3}\right)$ and $\left(2, \dfrac{5\pi}{3}\right)$ as pairs of polar coordinates are symmetric about the polar axis.
    (b) Can the conclusion in (a) be reached by reasoning as in exercises 25–26?
    (c) Based on (a), conjecture whether points having polar coordinates (r, θ) and $(r, 2\pi - \theta)$ are necessarily symmetric with respect to the polar axis.
    (d) Prove the conjecture made in (c).

32. Prove that the points with $\left(2, \frac{3\pi}{4}\right)$ and $\left(2, \frac{7\pi}{4}\right)$ as pairs of polar coordinates are *not* symmetric with respect to the polar axis. (Hint: check quadrants or convert to rectangular coordinates.)

◆ 33. Prove that points with polar coordinates (r, θ) and $(r, \pi - \theta)$ are symmetric with respect to the $\frac{\pi}{2}$-line.

◆ 34. Prove that points with polar coordinates (r, θ) and $(-r, \theta)$ are symmetric about the pole.

In exercises 35–38, verify that the point with the given polar coordinates is on the polar graph of the given polar equation.

35. $\left(-2\sqrt[4]{75}, \frac{\pi}{3}\right)$; $r^2 - 4\sqrt{3} = 32 \sin \theta$

36. $(2, -2)$; $r - \theta = 4$

37. $(2, \pi - 2)$; $r = \theta$ (Hint: a point has more than one pair of polar coordinates!)

38. $\left(-1, \frac{\pi}{2}\right)$; $r = |\sin \theta|$ (Hint: see the preceding hint.)

39. Prove that the point with polar coordinates $\left(\frac{9}{2}, \frac{77\pi}{4}\right)$ is **not** on the polar graph of $r - 1 = 3 \cos \theta$. (Hint: what is the range of the cosine function? What are the possible values of $3 \cos \theta$?)

40. (a) Let α be a real number and k an integer. Explain why the polar graphs of $\theta = \alpha$ and $\theta = \alpha + k\pi$ are the same.
(b) Identify the graph(s) discussed in (a).

In exercises 41–50, identify and sketch the polar graph of the given polar equation.

41. $r = 2$
42. $\theta = \frac{\pi}{3}$
43. $\theta = \frac{7\pi}{6}$
44. $r = -2$
45. $r = -3$
46. $\theta = -\frac{\pi}{3}$
47. $r = 8 \sin \theta$
48. $r = 12 \cos \theta$
49. $r = \frac{1}{2} \cos \theta$
50. $r = \frac{1}{3} \sin \theta$

In exercises 51–58, find a polar equation for the circle whose radius is the given number a and whose center has the given polar coordinates.

51. $a = 5$; $(0, \pi)$
52. $a = 3$; $(3, 0)$
53. $a = 9$; $(9, 0)$
54. $a = 16$; $(0, -\pi)$
55. $a = \sqrt{2}$; $\left(\sqrt{2}, \frac{\pi}{2}\right)$
56. $a = \sqrt{3}$; $\left(\sqrt{3}, \frac{\pi}{2}\right)$
57. $a = 6$; $\left(-2, \frac{5\pi}{3}\right)$
58. $a = 7$; $\left(3, \frac{11\pi}{6}\right)$

In exercises 59–62, find the radius and (either Cartesian or polar) coordinates of the center of the circle which is the polar graph of the given polar equation. (Approximate by calculator only if necessary.)

59. $r^2 - (2 \cos \theta + 2 \sin \theta)r + \frac{7}{4} = 0$
60. $r^2 - (4\sqrt{3} \cos \theta + 4 \sin \theta)r - 20 = 0$
61. $r^2 + 6r \cos \theta - 8r \sin \theta = 0$ (Hint: try $2\rho = 10$.)
62. $r^2 - 4r \cos \theta + 6r \sin \theta - 12 = 0$ (Hint: find 2ρ as in example 10.8; or consider $(r \cos \theta - 2)^2 + (r \sin \theta + 3)^2$.)

In exercises 63–70, find a polar equation for the line ℓ with the given properties, and then graph ℓ.

63. ℓ is perpendicular to the polar axis and passes through the point with polar coordinates $\left(2, -\frac{\pi}{6}\right)$.

64. ℓ is perpendicular to the polar axis and passes through the point with polar coordinates $\left(-4, \frac{\pi}{3}\right)$.

65. ℓ is parallel to the polar axis and passes through the point with polar coordinates $\left(-2, \frac{\pi}{3}\right)$.

66. ℓ is parallel to the polar axis and passes through the point with polar coordinates $\left(4, \frac{\pi}{4}\right)$.

67. ℓ passes through the point N with polar coordinates $\left(-5, \frac{5\pi}{3}\right)$ and ℓ is perpendicular to $\overleftrightarrow{ON}$.

68. ℓ passes through the point N with polar coordinates $\left(2, \frac{4\pi}{3}\right)$ and ℓ is perpendicular to $\overleftrightarrow{ON}$.

In exercises 69–76, convert the given polar equation into an equation involving rectangular coordinates and identify the graph of the equation.

◆ 69. $r \cos \theta = -2$
◆ 70. $r \sin \theta = 2$
◆ 71. $\sqrt{2} r \cos \theta - 3r \sin \theta = 4$
◆ 72. $4r \cos \theta - \sqrt{2} r \sin \theta = -3$
◆ 73. $r^2 - 2r \cos \theta = 10$
◆ 74. $r^2 + 2r \cos \theta - 4r \sin \theta = 31$
◆ 75. $r = \dfrac{2}{\cos \theta}$
◆ 76. $3r = \dfrac{-4}{\sin \theta}$
◆ 77. Use your graphing calculator to sketch the graph of $r(\theta) = \cos n\theta$ for the following values of n: 2, 3, 4, 5, 6, and 7. From these results, conjecture about the effect n has on the number of "loops" in these graphs. Try to prove your conjecture.

10.2 CURVES IN POLAR COORDINATES

In section 10.1, we graphed some simple polar equations. This section deals more deeply with graphing in polar coordinates. We shall examine some interesting new curves and obtain new analytic descriptions of the familiar conic sections.

In this section and the next, you will find it particularly helpful to have access to a graphing calculator or computer graphing software. In the graphing calculator activity in the preceding section, we used a graphing calculator to show how to sketch some unusual polar graphs. The shapes of such curves would be very time-consuming to discover without today's technology.

Curve Sketching

In example 10.6, we sketched the circle $r = 6 \sin \theta$ by making a table of values and plotting the corresponding points. The table was somewhat redundant; for example, the polar coordinates $\left(3, \frac{\pi}{6}\right)$ and $\left(-3, \frac{7\pi}{6}\right)$ represented the same point, namely $(\frac{3}{2}\sqrt{3}, \frac{3}{2})$. In order to reduce the time needed to sketch a polar graph, one should be alert for other types of redundancy, such as those arising from possible symmetry of the polar graph. The following guidelines for curve sketching include a number of time-saving criteria for symmetry. They are in the spirit of the algebraic criteria in section 2.2, and may be verified by the sort of observations about symmetry made in section 10.1. It will be convenient to say that an equation is **unchanged** if a pertinent change of variables converts the equation to an equivalent equation (that is, an equation with the same solution set).

Guidelines for Graphing Polar Equations

1. Find all intercepts on the lines $\theta = 0, \frac{\pi}{2}, \pi,$ and $\frac{3\pi}{2}$.
2. Test the curve for symmetry by using the following criteria:
    (i) A curve is symmetric with respect to the polar axis if some polar equation for the curve is unchanged when (r, θ) is replaced by $(r, -\theta)$ or when (r, θ) is replaced by $(-r, \pi - \theta)$.
    (ii) A curve is symmetric with respect to the $\frac{\pi}{2}$-line if one of its polar equations is unchanged when (r, θ) is replaced by $(-r, -\theta)$ or when (r, θ) is replaced by $(r, \pi - \theta)$.
    (iii) A curve is symmetric with respect to the pole if one of its polar equations is unchanged when (r, θ) is replaced by $(-r, \theta)$ or when (r, θ) is replaced by $(r, \pi + \theta)$.
3. Plot points arising from the **"crucial"** values of θ—that is, the values of θ for which r is maximized or minimized. Determine the values of θ for which r is undefined.
4. Determine the extent of "variation" of r as θ increases or decreases.

Hearts and Roses

EXAMPLE 10.12 Sketch the polar graph of $r = 2 - 2\sin\theta$.

Solution We shall use the above guidelines, after first rewriting the given equation as $r = 2(1 - \sin\theta)$.

(1) *Intercepts:* Setting $\theta = 0, \frac{\pi}{2}, \pi,$ and $\frac{3\pi}{2}$, we find that the corresponding values of r are 2, 0, 2, and 4. We thus find four points on the polar graph, with polar coordinates $(2, 0), \left(0, \frac{\pi}{2}\right), (2, \pi),$ and $\left(4, \frac{3\pi}{2}\right)$.

(2) *Symmetry:* Since $\sin\theta = \sin(\pi - \theta)$, the criterion in 2(i) shows that this curve is symmetric with respect to the $\frac{\pi}{2}$-line. The other symmetry criteria do not give additional information in this example.

(3) *Crucial values of θ:* Observe that each real number θ is in the domain of r, and each resulting value of r is nonnegative. The maximum and minimum values of $\sin\theta$ are 1 and -1, and these lead to the minimum and maximum values of r, namely 0 and 4. We saw in step 2 that these extreme values of r were attained at various intercepts.

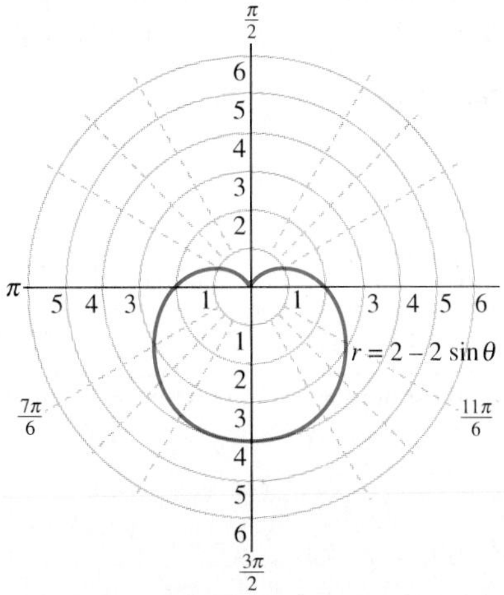

(4) *Extent of variation of r:* The periodicity of the sine function shows that all the points on this curve are obtained as θ increases from 0 to 2π.

As $r \geq 0$ and the curve is symmetric about the $\frac{\pi}{2}$-line, we only need to plot points corresponding to $0 \leq \theta \leq \frac{\pi}{2}$ and $\frac{3\pi}{2} \leq \theta \leq 2\pi$ (that is, points in the "right half-plane"), and then reflect them through the $\frac{\pi}{2}$-line to get the "left half" of the graph. Notice that as θ goes from 0 to $\frac{\pi}{2}$, r goes from 2 to 0; and as θ goes from $\frac{3\pi}{2}$ to 2π, r goes from 4 to 2.

The result is the heart-shaped curve called a **cardioid** shown in figure 10.10.

Figure 10.10

EXAMPLE 10.13 Sketch the polar graph of $r = 2\cos 3\theta$.

Solution
(1) *Intercepts:* It will clearly be useful to set $3\theta = 0, \frac{\pi}{2}, \pi, \frac{3\pi}{2}, \ldots, \frac{11\pi}{2},$ and 6π. We find that the corresponding pairs of polar coordinates (r, θ) are $(2, 0), \left(0, \frac{\pi}{6}\right), \left(-2, \frac{\pi}{3}\right), \left(0, \frac{\pi}{2}\right), \ldots, \left(0, \frac{11\pi}{6}\right),$ and $(2, 2\pi)$.

Curves in Polar Coordinates

(2) *Symmetry:* Since cos is an even function, $\cos 3\theta = \cos 3(-\theta)$, and so the criterion in 2(i) shows that this curve is symmetric with respect to the polar axis.

(3) *Crucial values of θ:* The maximum value of r is 2; this arises from $\cos 3\theta = 1$, for instance at $\theta = 0, \frac{2\pi}{3}, \frac{4\pi}{3}, 2\pi$. The minimum value of r is -2, which arises at $\theta = \frac{\pi}{3}, \pi, \frac{5\pi}{3}, \frac{7\pi}{3}, \ldots$. Each real value of θ gives rise to a well-defined real number r.

(4) *Extent of variation of r:* As θ varies from 0 to $\frac{\pi}{6}$, r decreases from 2 to 0; as θ varies from $\frac{\pi}{6}$ to $\frac{\pi}{3}$, r decreases from 0 to -2; as θ varies from $\frac{\pi}{3}$ to $\frac{\pi}{2}$, r increases from -2 to 0; as θ varies from $\frac{\pi}{2}$ to $\frac{2\pi}{3}$, r increases from 0 to 2; and so on.

Figure 10.11

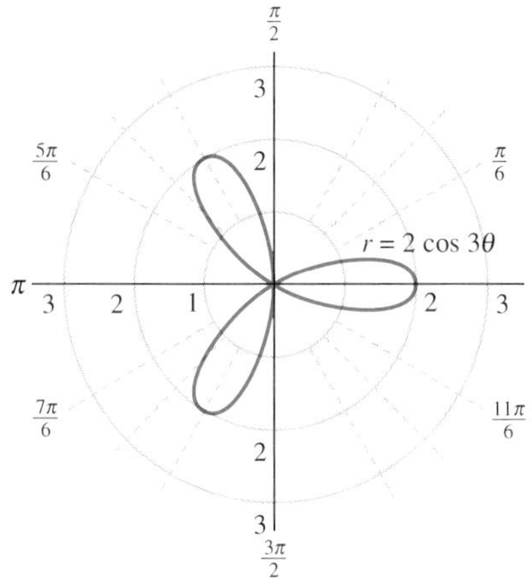

A table of values for θ over the interval $\left[0, \frac{2\pi}{3}\right]$ is given below. The resulting points are plotted and smoothly connected; then symmetry over the polar axis is used to obtain the graph in figure 10.11. This curve is an example of a class of curves called **roses**. For obvious reasons, it is called a **3-leaved rose** or a **rose with 3 petals**. (You first graphed roses in exercise 77, section 10.1.) Notice that the values of θ in $\left[0, \frac{2\pi}{3}\right]$ only give rise to one petal and two half-petals; the other two half-petals were found by symmetry (or by considering $\frac{2\pi}{3} \leq \theta \leq 2\pi$).

θ	0	$\frac{\pi}{12}$	$\frac{\pi}{6}$	$\frac{\pi}{4}$	$\frac{\pi}{3}$	$\frac{5\pi}{12}$	$\frac{\pi}{2}$	$\frac{7\pi}{12}$	$\frac{2\pi}{3}$
r	2	$\sqrt{2}$	0	$-\sqrt{2}$	-2	$-\sqrt{2}$	0	$\sqrt{2}$	2

▶ **EXAMPLE 10.14**

Sketch the polar graph of $r = 1 + 3\cos\theta$.

Solution

The period of $1 + 3\cos\theta$ is 2π, and so the values of θ in $[0, 2\pi]$ generate all the points on this curve.

(1) *Intercepts:* $(4, 0), \left(1, \frac{\pi}{2}\right), (-2, \pi)$, and $\left(1, \frac{3\pi}{2}\right)$ are pairs of polar coordinates for intercepts.

(2) *Symmetry:* Polar axis, by the criterion in 2(i).

Chapter 10 / Additional Topics in Geometry and Algebra

(3) *Crucial values of θ*: Each real θ leads to a real r. The maximum value of r, namely 4, arises when $\cos \theta = 1$; the minimum r, -2, arises from $\cos \theta = -1$.

(4) *Extent of variation of r*: r decreases from 4 to -2 over $[0, \pi]$ and increases from -2 to 4 over $[\pi, 2\pi]$.

By plotting points via the table below and using symmetry with respect to the polar axis, we find a curve called a **limaçon** which is shown in figure 10.12. (You have already seen limaçons in example 10.11.) Notice that its "inner loop" was swept out exactly once as θ varied from $\pi - \cos^{-1}(\frac{1}{3})$ to $\pi + \cos^{-1}(\frac{1}{3})$. It will be necessary in your later study of integral calculus to be able to determine intervals of θ values which sweep out a given loop, petal, etc.

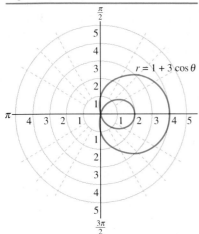

Figure 10.12

θ	0	$\frac{\pi}{6}$	$\frac{\pi}{3}$	$\frac{\pi}{2}$	$\frac{2\pi}{3}$	$\frac{5\pi}{6}$	π
r	4	$1 + \frac{3}{2}\sqrt{3}$	2.50	1	-0.50	$1 - \frac{3}{2}\sqrt{3}$	-2

Each of the curves discussed in the last three examples can be classified as either a limaçon or a rose. (The cardioid is a special type of limaçon in which the "inner loop" degenerates to a point.) In general, we have the following conclusions.

Roses

The polar graphs of equations of the form

$$r = a \cos n\theta \quad \text{or} \quad r = a \sin n\theta \quad \text{for an integer } n \geq 2$$

are roses. If n is odd, the rose has n petals; and if n is even, the rose has $2n$ petals.

Figure 10.11 shows a rose with $n = 3$ petals.

Limaçons

The polar graphs of equations of the form

$$r = a \pm b \cos \theta \quad \text{or} \quad r = a \pm b \sin \theta$$

are called limaçons.

- If $|a| < |b|$, the limaçon has two loops, as in figure 10.12;
- If $|a| = |b|$, the limaçon has one "heart-shaped" loop, as in figure 10.10, and is called a cardioid;
- If $2|b| > |a| > |b|$, the limaçon has one "dimpled" loop, as in figures 10.9(a) and (b);
- If $|a| \geq 2|b|$, the limaçon consists of one loop and encloses a convex region.

Curves in Polar Coordinates

Generalized Roses and Butterflies

Roses and limaçons are attractive and add a certain amount of beauty to mathematics. So do figures known as generalized roses, which are generated by combining the equations for the rose and the limaçon. For example, the generalized rose in figure 10.13(a) is the polar graph of $r = 2 + 7 \cos 5\theta$. Notice that this "rose" has 5 petal-pairs. The outer petal of each pair is 9 units long, and the inner petal is 5 units long. Three other generalized roses are shown in figures 10.13(b), (c), and (d).

Figure 10.13

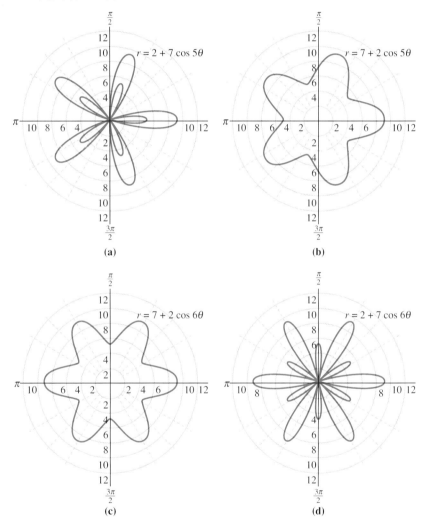

Roses, limaçons, generalized roses, and all the other graphs of polar equations we have studied so far have had their extent of variation of r limited by the periodicity of the trigonometric functions. The graphs in figures 10.14(a) and (b) are examples of **butterfly curves**. Both resulted from graphing

$$r = e^{\cos \theta} - 2\cos(4\theta) + \sin^5\left(\frac{\theta}{12}\right).$$

In figure 10.14(a) the curve is graphed for values of θ in the interval $[0, 4\pi]$, and in figure 10.14(b) for θ in the interval $[0, 16\pi]$. As θ increases, this butterfly curve does not retrace itself until θ reaches 24π.

Figure 10.14

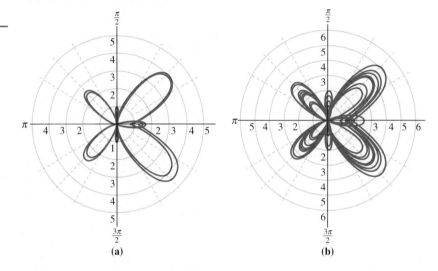

Polar Equations of Conic Sections

Each type of "nondegenerate" conic section studied in chapter 8 could be defined in terms of a focus and a directrix. These definitions are recalled below.

Conic Section

> A conic section can be defined as the set of points the ratio of whose distances to a given point (focus) and to a given line (directrix) is a given positive constant e (eccentricity). If $0 < e < 1$, the conic is an ellipse; if $e = 1$, a parabola; and if $e > 1$, a hyperbola.

Figure 10.15

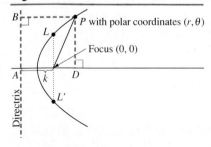

The polar equation of a conic takes a fairly simple and useful form as follows. Let the pole of the polar coordinate system be placed at the focus F, and let the polar axis fall along the conic's axis of symmetry going through F (and perpendicular to the directrix). Let the distance from F to the directrix be k, and let (r, θ) be a pair of polar coordinates for a point P on the given conic. The situation is depicted in figure 10.15, where B is the projection of P on the directrix, and D is the projection of P on the polar axis. Then, by definition of a conic section with eccentricity e, we have

$$FP = e \cdot PB. \qquad (*)$$

Assuming $r \geq 0$ and the (vertical) directrix is to the left of F, we have $FP = r$ and $PB = AF + FD = k + r\cos\theta$.

Curves in Polar Coordinates

Substituting these values into equation (∗), we obtain

$$r = e(k + r \cos \theta)$$

which, when solved for r, becomes

$$r = \frac{ek}{1 - e \cos \theta}. \qquad (1)$$

Equation (1) is thus a polar equation of the conic with eccentricity e, focus at the pole, and corresponding directrix k units to the left of the pole.

It turns out that if $\theta = \frac{\pi}{2}$, $FP = FL = ek$, from which we see that the length of the latus rectum is $LL' = 2ek$.

By similar reasoning, we have the following conclusions.

Polar Equations of Conic Sections

Let a conic with eccentricity e have a focus F at the pole. Let k be the distance from F to the corresponding directrix.

If the directrix is *perpendicular* to the polar axis, then a polar equation for the conic is

$$r = \frac{ek}{1 \mp e \cos \theta}, \qquad (2)$$

where the minus sign would indicate that the directrix is to the left of the pole and the plus sign that it is to the right.

If the directrix is *parallel* to the polar axis, then a polar equation for the conic is

$$r = \frac{ek}{1 \mp e \sin \theta}, \qquad (3)$$

where the minus sign would indicate that the directrix is below the polar axis and the plus sign that it is above.

▶ **EXAMPLE 10.15**

Describe and sketch the polar graph of $r = \dfrac{4}{2 - \cos \theta}$.

Solution

This is almost in the form of equation (1), but the constant term of the denominator is not 1. To get it to that form, both the numerator and denominator are divided by 2, with the result

$$r = \frac{2}{1 - \frac{1}{2} \cos \theta} = \frac{\frac{1}{2}(4)}{1 - \frac{1}{2} \cos \theta}.$$

From this, we can readily see that $e = \frac{1}{2}$ and $k = 4$. Since $e < 1$, the curve is an ellipse; it has one focus at the pole and corresponding directrix 4 units to the left of the pole. A rough sketch of the ellipse can be made by plotting the four intercepts having polar coordinates $(4, 0)$, $\left(2, \frac{\pi}{2}\right)$, $\left(\frac{4}{3}, \pi\right)$, and $\left(2, \frac{3\pi}{2}\right)$. When these are smoothly connected, we obtain the ellipse in figure 10.16.

Figure 10.16

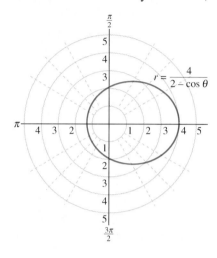

EXAMPLE 10.16

Describe and sketch the polar graph of $r = \dfrac{8}{1 + 3 \sin \theta}$.

Solution

Figure 10.17

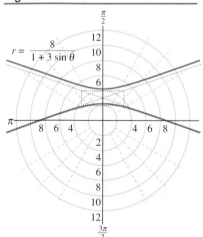

This is already in the form of equation (3), with $e = 3$ and $(8 = ek = 3k$, so) $k = \frac{8}{3}$. Since $e > 1$, this is a polar equation of a hyperbola; it has one focus at the pole, with the corresponding directrix $\frac{8}{3}$ units above the pole. The transverse axis of the hyperbola therefore lies on the $\frac{\pi}{2}$-line. Thus we can determine the vertices as certain intercepts:

$$\theta = \frac{\pi}{2}, r = 2 \qquad \theta = \frac{3\pi}{2}, r = -4.$$

From figure 10.17, we see that the distance $2a$ between these vertices is 2, and so $a = 1$. Since $e = \dfrac{c}{a} = 3$, we see that $c = 3$, and so $b = \sqrt{c^2 - a^2} = 2\sqrt{2}$. Thus, we find the "central rectangle" leading to the sketch in figure 10.17.

Another way you might have graphed the last two examples would have involved converting to Cartesian coordinates. You had some exposure to this method in section 10.1 in example 10.8 and a number of exercises, including exercises 69–76. We next illustrate the method for a conic. You could work this example also by applying equation (2).

EXAMPLE 10.17

Convert the polar equation $r = \dfrac{1}{2 + 2 \cos \theta}$ to rectangular form.

Curves in Polar Coordinates

Solution

We need to rewrite the given equation in terms of expressions that involve r^2, $r \cos \theta$, and $r \sin \theta$.

$$r = \frac{1}{2 + 2\cos\theta} = \frac{1}{2(1 + \cos\theta)}$$

$$r + r\cos\theta = \tfrac{1}{2}$$

$$r = \tfrac{1}{2} - r\cos\theta$$

$$r^2 = (\tfrac{1}{2} - r\cos\theta)^2.$$

We next substitute $r^2 = x^2 + y^2$ and $x = r\cos\theta$, with the result

$$x^2 + y^2 = (\tfrac{1}{2} - x)^2$$
$$= \tfrac{1}{4} - x + x^2.$$

Thus

$$y^2 = \tfrac{1}{4} - x = -(x - \tfrac{1}{4}).$$

As shown in figure 10.18, this is the equation of a parabola that opens to the left and has vertex at $(\tfrac{1}{4}, 0)$. ◀

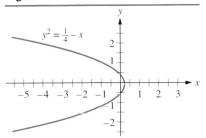

Figure 10.18

Rotation of Axes

In chapter 8, we discussed how to rotate the axes of a rectangular coordinate system. The corresponding problem for polar coordinates is easily answered. One may show the following.

Rotation of Axes

If α is a given constant, then the polar graph of $r = f(\theta - \alpha)$ can be obtained by revolving the polar graph of $r = f(\theta)$ through an angle α (radians). In particular,

$$r = \frac{ek}{1 \pm e\cos(\theta - \alpha)} \quad \text{or} \quad r = \frac{ek}{1 \pm e\sin(\theta - \alpha)}$$

is the polar equation of a conic with eccentricity e and one focus at the pole.

For instance, consider the ellipse in figure 10.19. One of its foci is at the pole, and this ellipse is the result of revolving the polar graph of an equation of type (1) or (2) about the pole through a positive angle α.

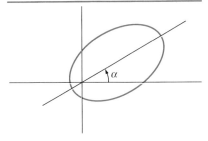

Figure 10.19

▷ **EXAMPLE 10.18**

Discuss and sketch the graph of the polar equation

$$r = \frac{3}{2 + 2\cos\left(\theta + \dfrac{\pi}{3}\right)}.$$

Solution

As in the solution of example 10.15, we must begin by dealing with the fact that the constant term of the denominator is not 1. Dividing both the numerator and denominator by 2, we rewrite the equation as

$$r = \frac{\frac{3}{2}}{1 + \cos\left(\theta + \frac{\pi}{3}\right)}.$$

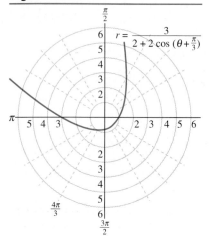

Figure 10.20

This equation has the above form, with $e = 1$, $k = \frac{3}{2}$, and $\alpha = -\frac{\pi}{3}$. Since the eccentricity is $e = 1$, the graph is a parabola. Its focus is at the pole, and its axis (of symmetry) is the polar graph of $\theta = -\frac{\pi}{3}$. To facilitate graphing, we shall locate the parabola's vertex. Putting $\theta = -\frac{\pi}{3}$ in the given equation leads to $r = \frac{3}{4}$, and so $\left(\frac{3}{4}, -\frac{\pi}{3}\right)$ is a pair of polar coordinates for the vertex. The parabola is sketched in figure 10.20. ◂

Other Polar Graphs

A number of other interesting graphs are described by means of polar equations. These include lemniscates and a variety of spirals. Exercises 21–26 will introduce you to these additional equations.

Exercises 10.2

In exercises 1–20, identify and sketch the polar graph of the given polar equation. In the case of a rose, find an interval of θ values that sweeps out exactly one petal. In the case of a limaçon with two loops, find an interval of θ values that exactly sweeps out the smaller loop. In the case of a conic with focus at the pole, find the corresponding directrix and the eccentricity.

1. $r = 3(1 - \sin \theta)$
2. $r = 4 \sin 2\theta$
3. $r = \frac{1}{2} \cos 5\theta$
4. $r = 5 + 5 \cos \theta$
5. $r = 2 \sin 4\theta$
6. $r = 5(1 - \sin \theta)$
7. $r = 1 - 2 \cos \theta$
8. $r = 1 + 2 \sin \theta$
9. $r = 2 + \sqrt{3} \sin \theta$
10. $r = 2 - \sqrt{3} \cos \theta$
11. $r = 6 + 6 \cos\left(\theta - \frac{\pi}{3}\right)$
12. $r = 2 \sin 3\left(\theta - \frac{\pi}{4}\right)$
13. $r = \frac{6}{4 + 3 \cos \theta}$
14. $r = \frac{2}{3 + 3 \sin \theta}$
15. $r = \frac{12}{-1 + 4 \cos \theta}$
16. $r = \frac{2}{2 - \sin \theta}$
17. $r = \frac{3 \csc \theta}{2 \csc \theta + 2}$
18. $r = \frac{3 \sec \theta}{2 \sec \theta - 4}$
19. $r = \frac{\sqrt{2}}{1 - \sin\left(\theta - \frac{\pi}{4}\right)}$
20. $r = \frac{4}{\sqrt{2} + 2 \cos\left(\theta - \frac{\pi}{6}\right)}$

In exercises 21–26, sketch the polar graph of the given polar equation. (You may find that a graphing calculator or computer graphing program helps you discover these answers.)

◆ 21. $r = 2\theta$ (Hint: this is an **Archimedean spiral**. Observe that this graph is symmetric about the $\frac{\pi}{2}$-line. So, it is enough to plot the points corresponding to $\theta \geq 0$, and then reflect about the $\frac{\pi}{2}$-line to get the "other half" of the graph.)

◆ 22. $r = \frac{\theta}{2}$ (Hint: review the hint for exercise 21.)

◆ 23. $r\theta = 2$ (Hint: this is a **hyperbolic spiral**. Observe that it is symmetric about the $\frac{\pi}{2}$-line. Consider the limiting tendencies of r as θ approaches either 0 or infinity.)

◆ 24. $r = 2^\theta$ (Hint: this is a **logarithmic spiral**. Notice how it differs from an Archimedean spiral.)

◆ 25. $r^2 = 9 \cos 2\theta$ (Hint: this is a **lemniscate**. Observe that this graph is symmetric about the polar axis and the $\frac{\pi}{2}$-line (and the pole). So, it is enough to graph the portion in the "first quadrant" and then reflect about both axes. As θ goes from 0 to $\frac{\pi}{4}$, r goes from ± 3 to 0, with the first quadrant portion of the curve approaching the origin "tangentially" to the line $\theta = \frac{\pi}{4}$. Why are there no points on the graph corresponding to values of θ in $\left(\frac{\pi}{4}, \frac{\pi}{2}\right]$?)

◆ 26. $r^2 = 4 \sin 2\theta$ (Hint: review the hint for exercise 25.)

In exercises 27–29, convert the given equation in rectangular coordinates into a polar equation having the same graph.

27. $xy = 4$
28. $y = 4x$
29. $x^2 + y^2 = 9$

In exercises 30–33, convert the given (possibly unfamiliar) polar equation into an equation in rectangular coordinates and identify its (familiar) graph.

30. $r = 8 \sin\left(\theta - \frac{\pi}{6}\right)$ (Hint: since the origin is on the graph, there is no harm in multiplying both sides by r. Next, use an identity from section 6.2.)

31. $r^2 = \sec 2\theta$ (Hint: first rewrite $\sec 2\theta$ in terms of $\cos^2 \theta$ by using various trigonometric identities.)

32. $r^2 \sin 2\theta = 1$ (Hint: apply section 6.3.)

33. $r \cos^2 \frac{\theta}{2} = 1$

34. Convert the polar equation $r = 2 \tan \theta$ into rectangular coordinates. (Hint: rewrite $\pm x \sqrt{x^2 + y^2} = 2y$ as a "simpler" equation.)

In exercises 35–38, determine which conclusions about symmetry of the polar graph of the given polar equation follow from the symmetry criteria in the text.

35. $r^2 \cos \theta = 6$
36. $r^2 \sin \theta = 6$
37. $r^3 \sin \theta = 6$
38. $r = \sin \theta + \cos \theta$

39. Prove that the polar graph of $r^2 - 4r \sin \theta + 3 = 0$ is not symmetric about the polar axis. (Hint: it is *not* enough to notice that the criteria in 2(i) are not satisfied. You must find a specific *counterexample*, as in section 6.1. For this, notice that the point with polar coordinates $\left(1, \frac{\pi}{2}\right)$ is on the given graph, but none of the polar coordinates $\left(1, \frac{3\pi}{2} + k2\pi\right)$ or $\left(-1, \frac{3\pi}{2} + \pi + k2\pi\right)$ gives a point on the graph.)

40. There are two criteria in the text for symmetry with respect to the $\frac{\pi}{2}$-line. Show that exactly one of these criteria can be used to show that the polar graph of $r^2 = \sin \theta$ is symmetric with respect to the $\frac{\pi}{2}$-line.

41. Identify the polar graph of the inequality $r < 2$. (This consists of the points having at least one pair of polar coordinates (r, θ) such that $r < 2$.)

42. Identify the polar graph of $\theta^2 = \pi^2$.

In exercises 43–48, find a polar equation for the conic having the given eccentricity e, focus at the pole, and corresponding directrix the given line ℓ.

43. $e = \sqrt{2}$; $\ell: r \cos \theta = 5$ (Hint: apply equation (2), with $k = 5$.)

44. $e = 1$; $\ell: r \sin \theta = -4$
45. $e = \frac{1}{2}$; $\ell: r \sin \theta = -3$
46. $e = 3$; $\ell: r \cos \theta = -2$
47. $e = 1$; $\ell: r \sin \theta = \frac{1}{4}$
48. $e = \frac{1}{\sqrt{3}}$; $\ell: r \cos \theta = 5$

49. Find a polar equation for the ellipse with focus at the pole and vertices having polar coordinates $\left(2, \frac{\pi}{2}\right)$ and $\left(8, \frac{3\pi}{2}\right)$. (Hint: find in turn the constants a, c, e, $\frac{a}{e}$, and k.)

50. (The limiting case of a "flattened-loop" limaçon is a circle.)

(a) Sketch the polar graph of $r = 1 + \frac{\cos \theta}{n}$ for $n = 2$, 3, 4. (If you use a graphing calculator or computer graphing program, you may wish to develop additional examples, say for $n = 5$, 10, and 100.)

(b) On the basis of the graphs in (a), make a conjecture about the "limiting tendency" of the polar graph of $r = 1 + \frac{\cos \theta}{n}$ as n approaches infinity.

(c) Give an intuitive argument to support the conjecture made in (b). (Hint: explain why, if θ is fixed and n approaches infinity, the limiting tendency of $1 + \frac{\cos \theta}{n}$ is 1.)

*10.3 INTERSECTION OF POLAR GRAPHS

To find the points of intersection of the graphs of two functions f and g, we only need to solve the equations $f(x) = y = g(x)$. This technique is successful because each point in the plane has a unique pair of Cartesian coordinates (x, y). Such is not the case with polar coordinates. Indeed, to locate all the points of intersection of the polar graphs of polar equations $r = f(\theta)$ and $r = g(\theta)$, it may *not* be enough to solve the equations $f(\theta) = r = g(\theta)$. As the following examples illustrate, one must take into account all the different pairs of polar coordinates representing possible points of intersection.

▶ **EXAMPLE 10.19** Find (all) the points of intersection of the polar graphs of $r = 2 \sin \theta$ and $r = 2 \cos \theta$.

Solution To find at least *some* of the points of intersection, we shall begin by solving the equations simultaneously. Eliminating r results in

$$2 \sin \theta = 2 \cos \theta$$

or

$$\tan \theta = 1.$$

The values of θ in the interval $[0, 2\pi)$ that satisfy this last equation are $\theta = \dfrac{\pi}{4}$ and $\theta = \dfrac{5\pi}{4}$. Thus, some points of intersection are given by the pairs of polar coordinates $\left(\sqrt{2}, \dfrac{\pi}{4}\right)$ and $\left(-\sqrt{2}, \dfrac{5\pi}{4}\right)$. However, these polar coordinate pairs describe the same point in the plane, namely $(1, 1)$. Thus, it would seem from a casual algebraic approach that this is the only point of intersection.

Using the methods of section 10.1, we see that the polar graphs of $r = 2 \sin \theta$ and $r = 2 \cos \theta$ are circles of radius 1, with centers having polar coordinates $\left(1, \dfrac{\pi}{2}\right)$ and $(1, 0)$, respectively. Examination of these graphs, as shown in figure 10.21, shows that they intersect in two points. One of these points is $(1, 1)$, which we had already found algebraically. The other point of intersection is the pole (or origin). Notice that the pole is on the polar graph of $r = 2 \sin \theta$ by virtue of $(r, \theta) = (0, 0)$; and it is on the polar graph of $r = 2 \cos \theta$ because of $(r, \theta) = \left(0, \dfrac{\pi}{2}\right)$. ◀

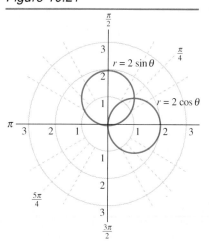

Figure 10.21

The above example suggests a first hint for finding all the points of intersection arising from two polar equations: **sketch their polar graphs.**

▶ **EXAMPLE 10.20** Find all points of intersection of the polar graphs of $r = 3$ and $r = 3 \cos 2\theta$.

---

*This section is optional because the material in it is not essential for later material in the book.

Intersection of Polar Graphs

Solution

Solving simultaneously, we have $3 = 3\cos 2\theta$ or $\cos 2\theta = 1$. This means that $2\theta = 2n\pi$, and so $\theta = n\pi$, where n is an integer. Using this result for θ in the interval $[0, 2\pi)$, we see that $(3, 0)$ and $(3, \pi)$ are pairs of polar coordinates for two points of intersection, namely $(3, 0)$ and $(-3, 0)$. A sketch of the polar graphs is in figure 10.22, and from it we can see that there are two additional points of intersection, namely $(0, 3)$ and $(0, -3)$. Thus, the four-petal rose given by $r = 3\cos 2\theta$ intersects the circle given by $r = 3$ in exactly four points. Notice that it is possible to verify, after the fact, that $(0, 3)$ and $(0, -3)$ are indeed in the intersection. For instance, you can see that $(0, -3)$ is on the polar graphs of $r = 3$ and $r = 3\cos 2\theta$ by using $(r, \theta) = \left(3, \dfrac{3\pi}{2}\right)$ and $(r, \theta) = \left(-3, \dfrac{\pi}{2}\right)$, respectively.

Figure 10.22

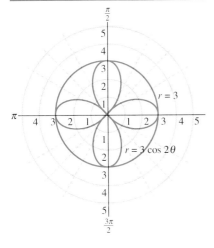

The above two examples suggest a second hint for finding additional points of intersection: **simultaneously solve $r = f(\theta)$ and $-r = g(\theta + \pi + 2k\pi)$**. Actually, even more possibilities need to be considered, such as $r = g(\theta + k2\pi)$, not to mention the possibilities arising when the roles of f and g are interchanged. In practice, the boldface rule is often enough, even with $k = 0$. If you can guess from the graphs what the additional points of intersections are (as we did in the last two examples), you can verify them, after the fact, by checking the various polar coordinate pairs for these points.

EXAMPLE 10.21

Find all points of intersection of the ellipse $r = \dfrac{2}{2 + \sin \theta}$ and the limaçon $r = 1 - 3\sin\theta$.

Solution

Solving simultaneously, we have

$$1 - 3\sin\theta = \dfrac{2}{2 + \sin\theta}$$

$$2 - 5\sin\theta - 3\sin^2\theta = 2$$

$$\sin\theta(5 + 3\sin\theta) = 0.$$

So, either $\sin\theta = 0$ or $\sin\theta = -\dfrac{5}{3}$.

Now, $\sin\theta = 0$ when θ is 0 or π, and $\sin\theta$ is never $-\dfrac{5}{3}$. Thus, we have found two points of intersection, having polar coordinates $(1, 0)$ and $(1, \pi)$.

To test for other possible points of intersection, we use the above hint. Let's replace r by $-r$ and θ by $\theta + \pi$ in the equation $r = 1 - 3\sin\theta$. The result is

$$-r = 1 - 3\sin(\theta + \pi)$$

or

$$r = 3\sin(\theta + \pi) - 1.$$

Since $\sin(\theta + \pi) = -\sin\theta$, this can be simplified to

$$r = -3\sin\theta - 1.$$

Now, simultaneously solving the equations

$$r = -3 \sin \theta - 1 \quad \text{and} \quad r = \frac{2}{2 + \sin \theta}$$

we obtain

$$-3 \sin \theta - 1 = \frac{2}{2 + \sin \theta}$$
$$-2 - 7 \sin \theta - 3 \sin^2 \theta = 2$$
$$3 \sin^2 \theta + 7 \sin \theta + 4 = 0$$

or
$$(3 \sin \theta + 4)(\sin \theta + 1) = 0.$$

The factor $3 \sin \theta + 4$ is never zero, and so the only possible solutions arise when $\sin \theta = -1$, for instance when $\theta = \frac{3\pi}{2}$. Hence, $\left(2, \frac{3\pi}{2}\right)$ is a pair of polar coordinates for a (third) point of intersection.

The graphs in figure 10.23 show that we have found all the (three) points of intersection. To sum up, they have polar coordinates $(1, 0)$, $(1, \pi)$, and $\left(2, \frac{3\pi}{2}\right)$. ◂

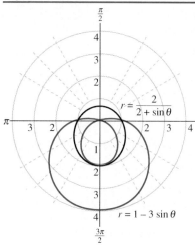

Figure 10.23

▶ **EXAMPLE 10.22** Find all points of intersection of the polar graphs of $r = 1 + \cos \theta$ and $r^2 = 4 \cos \theta$.

Solution The graph of $r = 1 + \cos \theta$ is a cardioid. The graph of $r^2 = 4 \cos \theta$ is a figure-eight curve, reminiscent of the lemniscates studied in exercises 25 and 26 of section 10.2. This figure-eight curve is shown in figure 10.24, which indicates four points of intersection with the cardioid. We shall algebraically determine all four of these points.

Given the two equations, we begin by eliminating r. Since

$$(1 + \cos \theta)^2 = r^2 = 4 \cos \theta$$

we can see that

$$\cos^2 \theta - 2 \cos \theta + 1 = 0$$

or
$$(\cos \theta - 1)^2 = 0.$$

We conclude that $\cos \theta = 1$, and so θ must be of the form $2k\pi$, where k is an integer. The cardioid and the figure-eight curve are each completely swept out as θ varies from 0 to 2π, and so it appears that this approach finds only one of the points of intersection, namely the one with polar coordinates $(2, 0)$. We still need to identify the other three points of intersection.

Let's replace r by $-r$ and θ by $\theta + \pi$ in the equation $r = 1 + \cos \theta$, with the result that $-r = 1 + \cos (\theta + \pi) = 1 - \cos \theta$. Once again, we proceed to eliminate r, obtaining

Intersection of Polar Graphs

and so
$$4\cos\theta = r^2 = (1 - \cos\theta)^2$$
$$\cos^2\theta - 6\cos\theta + 1 = 0.$$

Using the quadratic formula, we see that
$$\cos\theta = 3 \pm 2\sqrt{2}.$$

Since $|\cos\theta| \leq 1$, the value $3 + 2\sqrt{2}$ is discarded. Thus,
$$\cos\theta = 3 - 2\sqrt{2} \approx 0.1715729.$$

The possible values of θ in $[0, 2\pi)$ are $\arccos(3 - 2\sqrt{2}) \approx 1.3983703$ and $2\pi - \arccos(3 - 2\sqrt{2}) \approx 4.8848150$. Each of these values of θ leads to
$$r = -(1 - \cos\theta) = -(1 - (3 - 2\sqrt{2})) = 2 - 2\sqrt{2} \approx -0.8284271.$$

Thus, two more points of intersection have approximate polar coordinates $(-0.8284, 1.3984)$ and $(-0.8284, 4.8848)$.

The last point of intersection is found by examining the graph in figure 10.24. It looks as if the pole is this fourth point of intersection. We can check this algebraically as follows. The pole is on the polar graph of $r^2 = 4\cos\theta$ by virtue of $(r, \theta) = \left(0, \dfrac{\pi}{2}\right)$, and it is on the polar graph of $r = 1 + \cos\theta$ because of $(r, \theta) = (0, \pi)$. ◀

Summarizing this section, we have the following.

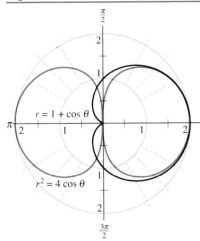

Figure 10.24

Hints for Finding All the Points of Intersection of Two Polar Graphs

If $r = f(\theta)$ and $r = g(\theta)$ are polar graphs, then to find all the points of intersection of these graphs,

1. Sketch the polar graphs of the given polar equations.
2. Simultaneously solve $r = f(\theta)$ and $r = g(\theta)$.
3. Simultaneously solve $r = f(\theta)$ and $-r = g(\theta + \pi + 2k\pi)$, k an integer.

Exercises 10.3

In exercises 1–46, sketch the polar graphs of the two given polar equations, and then find all their points of intersection. (On particularly difficult examples, a graphing calculator or computer graphing program can help you estimate the number, and location, of these points.)

1. $r = 3\sin\theta$, $r = 3\cos\theta$
2. $r = 4\sin\theta$, $r = 4\cos\theta$
3. $r = 3\sin\theta$, $r = 4\sin\theta$
4. $r = 4\sin\theta$, $r = 2$
5. $r = 4$, $r = 2$
6. $r = 2\sin\theta$, $r = 4$
7. $r = 2\sin\theta$, $\theta = \dfrac{\pi}{4}$
8. $r = 2$, $\theta = -\dfrac{5\pi}{6}$
9. $\theta = \dfrac{\pi}{6}$, $\theta = \dfrac{2\pi}{3}$
10. $\theta = \pi$, $\theta = -\pi$
11. $r = 4$, $r = 4\cos 2\theta$
12. $r = 2$, $r = 2\sin 4\theta$
13. $r = 3$, $r = 3\sin 5\theta$
14. $r = 2$, $r = 2\cos 3\theta$
15. $r = \sqrt{3}$, $r = 2\cos 2\theta$ (Hint: consider figure 10.22 for exercises 15–18.)
16. $\theta = \dfrac{\pi}{6}$, $r = 3\cos 2\theta$
17. $\theta = \dfrac{\pi}{4}$, $r = 3\cos 2\theta$
18. $r = \cos 2\theta$, $r = 3\cos 2\theta$
19. $\theta = \dfrac{\pi}{4}$, $r = 2\theta$
20. $\theta = \dfrac{\pi}{4}$, $r = 2^\theta$
21. $r = 1$, $r = 1 - 3\sin\theta$ (Hint: consider figure 10.23. Some of the θ values may be expressed via inverse trigonometric functions.)

22. $r = 3, r = 1 + 3\cos\theta$ 23. $r = 4, r = 1 + 3\cos\theta$
24. $r = 2, r = 1 - 3\sin\theta$
25. $r = \dfrac{4}{1 - \sin\theta}, r = 1 - 3\sin\theta$
26. $r = \dfrac{2}{2 - \cos\theta}, r = 1 + 3\cos\theta$ (Hint: "rotate" example 10.21.)
27. $r = \sin\theta, r = 2 - 3\sin\theta$
28. $\theta = \pi, r = 1 + 3\cos\theta$
29. $\theta = \dfrac{\pi}{3}, r = 1 + 2\cos\theta$
30. $r = \dfrac{4}{1 - \sin\theta}, \theta = \dfrac{\pi}{6}$
31. $r = 1 + \cos\theta, r = 1 - \cos\theta$
32. $r = 2 - 2\sin\theta, r = 2 + 2\sin\theta$
33. $r = 1 + \cos\theta, r = 1 + \sin\theta$
34. $r = \dfrac{3}{2}(1 + \sin\theta), r^2 = 9\sin\theta$
35. $r^2 = 4\cos\theta, r = 2$ 36. $r^2 = 4\cos\theta, r = 1$
37. $\theta = \dfrac{\pi}{2}, r^2 = 9\cos 2\theta$ 38. $\theta = \dfrac{\pi}{3}, r^2 = 4\cos\theta$
39. $r = 3, r = 3 - 3\sin\theta$
40. $r = \sin\theta, r = 1 + \cos\theta$ (Hint: consider r^2 and use Pythagoras' Theorem.)
41. $r = \cos\theta, r = 1 + \cos\theta$ 42. $r = 2, r = 2 + 2\cos\theta$
43. $\theta = \dfrac{\pi}{3}, r = 4 - 4\sin\theta$ 44. $\theta = 0, r = 1 + \cos\theta$
45. $r = \dfrac{1}{2}, r = 2\theta$ 46. $r = 2, r = -2$
47. From a figure, how many points of intersection do the polar graphs of $r = 1 + \sin\theta$ and $r^2 = 4\cos\theta$ appear to have?
48. Consider the polar graphs of $r = 3$, $\theta = \dfrac{\pi}{6}$, and $\theta = -\dfrac{\pi}{6}$.
    (a) Show that each pair of these graphs intersects in at least one point.
    (b) Show that the intersection of all three of these graphs is empty.

10.4 PARAMETRIC EQUATIONS

In chapters 1 and 2, you learned how to graph various equations and relations. This skill is important because not all interesting curves arise as graphs of functions. For instance, the parabola $y^2 = 4x$ consists of two pieces: the graphs of the functions given by $y = f_1(x) = 2\sqrt{x}$ and $y = f_2(x) = -2\sqrt{x}$. You may, however, point out that this parabola is the graph of $x = g(y) = \dfrac{1}{4}y^2$, for which "$x$ is a function of y." This is correct, but no such point of view accommodates an example like the ellipse $4x^2 + 9y^2 = 36$. In this example, *neither x nor y is a function of the other*. Such curves can, however, be described by introducing a third variable, often denoted t and called the **parameter**, such that both x and y are functions of t. One way to do this for the above ellipse is by means of $x = 3\sin t$ and $y = 2\cos t$.

The most natural application of such **parametric equations** concerns a particle moving in the plane. If the particle's position at time t is the point $(f(t), g(t))$, then $x = f(t), y = g(t)$ are parametric equations for the **parameterized curve**, $\{(x, y): \text{there exists } t \in I \text{ such that } x = f(t) \text{ and } y = g(t)\}$. Here, the set I is the interval of t-values during which the motion is considered.

In general, a parameterized curve is given by equations $x = f(t)$ and $y = g(t)$, where f and g are continuous functions and t varies over some interval I. (If I is not specified, you may suppose that I is the intersection of the natural domains of f and g.) There is no need, in general, to interpret the parameter t as time. In fact, we have already previewed parametric equations for the various conic sections, and some uses for them, with parameters other than time, in sections 8.1 (exercises 39–46), 8.2 (exercises 39–40), and 8.3 (exercise 47).

We turn now to examples of parameterized curves. The most trivial arise from the graph of a continuous function $y = g(x)$, $a \le x \le b$. Parametrically, this may be viewed by means of $x = t$, $y = g(t)$ and $I = [a, b]$. In the same way, the graph of an equation $x = f(y)$ is parameterized via $x = f(t)$ and $y = t$.

Parametric Equations

More interestingly, any polar graph $r = f(\theta)$ is a special kind of parameterized curve (assuming that f is continuous): use $t = \theta$, $x = f(\theta) \cos \theta$, and $y = f(\theta) \sin \theta$. (This explains the keystrokes used in the TI-81 version of example 10.11.) The next example pursues this idea.

EXAMPLE 10.23

Sketch the parameterized curve determined by the parametric equations $x = 2 \cos 3\theta \cos \theta$, $y = 2 \cos 3\theta \sin \theta$ if (a) θ varies over $I = \mathbf{R}$ and (b) θ varies over $I = \left[0, \dfrac{2\pi}{3}\right]$.

Solutions

(a) This is just the polar graph of the polar equation $r = 2 \cos 3\theta$. This 3-petal rose appeared in section 10.2 in figure 10.11 and is reproduced below in figure 10.25(a).

(b) As θ varies from 0 to $\dfrac{2\pi}{3}$, only a *portion* of the 3-petal rose in (a) is swept out. This portion appears in figure 10.25(b). You should check carefully which of its petals or half-petals are swept out as θ varies over $\left[0, \dfrac{\pi}{6}\right]$, $\left[\dfrac{\pi}{6}, \dfrac{\pi}{2}\right]$, and $\left[\dfrac{\pi}{2}, \dfrac{2\pi}{3}\right]$.

Figure 10.25

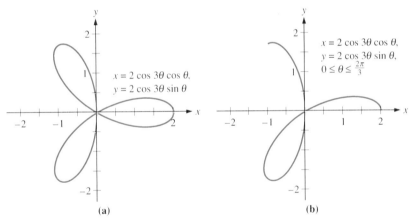

The next example illustrates the flexibility of the parametric equation approach by describing a graph of a type that we have not been able to study earlier.

EXAMPLE 10.24

Sketch the graph determined by the parametric equations

$$x = \cos t, \quad y = \sin 3t.$$

Solution

We shall build a table with three rows—with labels t, x, and y. Since the periods of $\cos t$ and $\sin 3t$ are 2π and $\dfrac{2\pi}{3}$, respectively, we expect the entire curve to be generated by the values of t in the interval $[0, 2\pi]$.

Chapter 10 / Additional Topics in Geometry and Algebra

t	0	$\frac{\pi}{6}$	$\frac{\pi}{3}$	$\frac{\pi}{2}$	$\frac{2\pi}{3}$	$\frac{5\pi}{6}$	π	$\frac{7\pi}{6}$	$\frac{4\pi}{3}$	$\frac{3\pi}{2}$	$\frac{5\pi}{3}$	$\frac{11\pi}{6}$	2π
$x = \cos t$	1	$\frac{\sqrt{3}}{2}$	0.5	0	-0.5	$-\frac{\sqrt{3}}{2}$	-1	$-\frac{\sqrt{3}}{2}$	-0.5	0	0.5	$\frac{\sqrt{3}}{2}$	1
$y = \sin 3t$	0	1	0	-1	0	1	0	-1	0	1	0	-1	0

The graph, as shown in figure 10.26, is a smooth curve that begins at $(1, 0)$ when $t = 0$ and traces the path indicated by the arrows, returning to $(1, 0)$ when $t = 2\pi$. Notice that the values of x and y (not of t) are used to plot the points. ◀

Figure 10.26

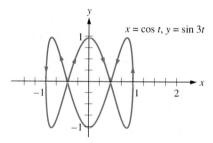

EXAMPLE 10.25

Determine the parameterized curve given by

$$x = t + 2, \quad y = t^2 - 4$$

for $-3 \leq t \leq 3$.

Solution

Once again we shall use a table to find some of the points on the curve.

Figure 10.27

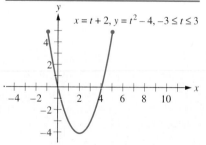

t	-3	-2	-1	0	1	2	3
x	-1	0	1	2	3	4	5
y	5	0	-3	-4	-3	0	5

The graph of the curve described by these parametric equations is shown in figure 10.27. You will probably notice that the curve appears to be a portion of a parabola. Soon, we shall see how to verify this assertion algebraically. ◀

Elimination of Parameters

How can one *prove* that the curve in example 10.25 is part of a parabola? One way is to find a **Cartesian equation** for the parameterized curve. (Another name for the Cartesian equation is the **rectangular equation**.) This is done by **eliminating the parameter**. The particular method by which the parameter is

EXAMPLE 10.26

Find a Cartesian equation for the curve parameterized by $x = t + 2$ and $y = t^2 - 4$.

Solution

These are the parametric equations examined in example 10.25. To eliminate the parameter here, we shall solve for t in terms of x and substitute the result into the equation defining y. In this way, we shall find y in terms of x, or, in other words, a Cartesian equation for the curve.

Since $x = t + 2$, we know that $t = x - 2$. Substituting this into $y = t^2 - 4$, we obtain

$$y = (x - 2)^2 - 4.$$

This is indeed the equation of a parabola. Its equivalent standard form is $y + 4 = (x - 2)^2$, indicating a parabola that opens upward with vertex at $(2, -4)$ and (vertical) axis $x = 2$. Notice, however, that the parameterized curve in example 10.25 is only a *portion* of this parabola. To know *just which portion it is*, one needs to consider a table of values and a sketch based on it, as in example 10.25. ◀

EXAMPLE 10.27

Find a Cartesian equation for, and sketch the graph of, the curve with parametric equations $x = 3 + 5 \sin t$ and $y = 6 - 4 \cos t$.

Solution

One cannot eliminate t in this example in a completely algebraic way. What algebra *does* tell us, via the given parametric equations, is that

$$\sin t = \frac{x - 3}{5} \quad \text{and} \quad \cos t = \frac{6 - y}{4}.$$

Now, to eliminate t, we use the trigonometric (Pythagorean) identity $\sin^2 t + \cos^2 t = 1$. Thus

$$\left(\frac{x-3}{5}\right)^2 + \left(\frac{6-y}{4}\right)^2 = 1$$

or

$$\frac{(x-3)^2}{25} + \frac{(y-6)^2}{16} = 1.$$

This is an equation of an "east-west" ellipse centered at $(3, 6)$, with semimajor axis $a = 5$ and semiminor axis $b = 4$. The ellipse is sketched in figure 10.28.

It is important to notice that the *entire* ellipse is the parameterized curve in question. You should verify that each point (x, y) on the ellipse arises from a real number t such that $x = 3 + 5 \sin t$ and $y = 6 - 4 \cos t$. In fact, for a given (x, y), there is only one such t in $[0, 2\pi)$. ◀

Figure 10.28

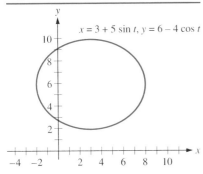

▷ **EXAMPLE 10.28** Find a Cartesian equation for, and sketch the graph of, the curve parameterized by $x = \sin t$ and $y = \cos 2t$.

Solution Here, the key to eliminating t is the trigonometric identity $\sin^2 t = \frac{1}{2}(1 - \cos 2t)$ from section 6.3. By substituting $x = \sin t$ and $y = \cos 2t$, we obtain

$$x^2 = \tfrac{1}{2}(1 - y)$$

or

$$x^2 = -\tfrac{1}{2}(y - 1).$$

This is the equation of a parabola that opens downward, with vertex at $(0, 1)$ and (vertical) axis $x = 0$. The graph of this parabola is shown in figure 10.29(a).

Returning to the parametric equations, we see that x and y each extend over at most $[-1, 1]$. In fact, the parameterized curve in question is only the solid part of this parabola shown in figure 10.29(b). Moreover, each point on this solid portion arises from a unique t in $\left[\dfrac{\pi}{2}, \dfrac{3\pi}{2}\right]$. ◀

Figure 10.29

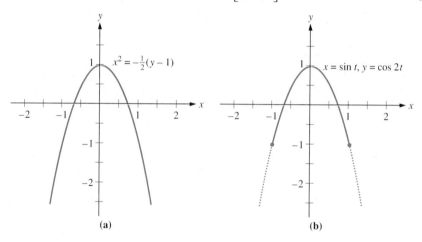

(a)   (b)

In the spirit of exercise 40 of section 8.1, we remark that the above parameterized curve has another parametrization different from the one studied in example 10.28: consider $x = -\dfrac{t}{4}$, $y = 1 - \dfrac{t^2}{8}$, $-4 \le t \le 4$.

Deriving Parametric Equations

As with converting parametric equations to a Cartesian equation, there is no general method for reversing the process. Indeed, the reverse process is sure to be hard since, as the above parabola indicated, each curve has many parametrizations. In the next three examples, we shall indicate parametrizations for some standard kinds of curves.

Parametric Equations

EXAMPLE 10.29 Derive a pair of parametric equations for the line $y = \frac{4}{3}x - 2$.

Solution One way to proceed is to recognize that this line is the graph of the function $y = g(x) = \frac{4}{3}x - 2$. Then, by the opening comments of this section, $x = t$ and $y = \frac{4}{3}t - 2$ suffice. We shall next give another parametrization for this line. You will find in a later course that this new method generalizes naturally to describe lines in space.

The given line's equation can be rewritten as $\dfrac{x}{3} = \dfrac{y+2}{4}$. If this common value is chosen as the parameter t, we find $x = 3t$ and $y = 4t - 2$ as the desired new parametrization of the line. In the same way, any line has a parametrization of the form $x = at + b$, $y = ct + d$ for suitable constants a, b, c, and d; this is particularly useful if one is interested in points "near" (b, d), for the point (b, d) corresponds to the parameter value $t = 0$. ◀

EXAMPLE 10.30 Find a pair of parametric equations for the circle, with radius r, centered at the point (h, k).

Solution This has already been done, in case $r = 1$ and $(h, k) = (0, 0)$, in (section 5.3) the "circular function" approach to trigonometry. In the general case, the same ideas carry over, as follows.

Let $P(x, y)$ be any point on the circle, as shown in figure 10.30. If Q denotes (h, k) and $r = QP$, then we know from trigonometry that

$$x - h = r \cos \theta \quad \text{and} \quad y - k = r \sin \theta.$$

It follows that

$$x = r \cos \theta + h \quad \text{and} \quad y = r \sin \theta + k$$

are the desired parametric equations. ◀

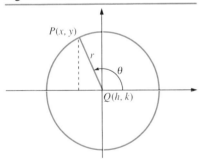

Figure 10.30

To be concrete, let's record one application of the method developed in example 10.30. One pair of parametric equations for the circle, with radius 2, centered at $(3, 4)$ is $x = 2 \cos \theta + 3$, $y = 2 \sin \theta + 4$.

EXAMPLE 10.31 Give a pair of parametric equations for an ellipse

$$\frac{(x-h)^2}{a^2} + \frac{(y-k)^2}{b^2} = 1.$$

Solution One answer, in the spirit of exercise 39 of section 8.2, is given by $x = a \sin t + h$, $y = b \cos t + k$. Another parametrization, which is more geometrically inspired, is reminiscent of the preceding example:

$$x = a \cos \theta + h, \qquad y = b \sin \theta + k.$$

In this pair of parametric equations, it is traditional to refer to the parameter θ as the **eccentric angle** of the point (x, y) on the ellipse. Notice that if we restrict θ to lie in $[0, \pi]$, the resulting parameterized curve is only the top half of the ellipse. ◀

Finally, we next give an example to illustrate how one may develop a set of parametric equations for a curve that results from a specific type of planar motion. This curve, called a **cycloid,** has several "relatives," two of which will be introduced in exercises 51–52.

▶ **EXAMPLE 10.32**

Obtain a pair of parametric equations for the path traced by a point P on the edge of a circle, as the circle rolls without slipping along a straight line.

Solution

We shall assume that the circle has radius r and center C and that the circle rolls to the right along the x-axis, as shown in figure 10.31. For convenience, we also assume that P touched the x-axis at the origin. Let θ represent the radian measure of the angle through which the circle has turned since P was at the origin. This means in particular that P is at its highest point when $\theta = \pi, 3\pi, 5\pi, \ldots$.

Figure 10.31

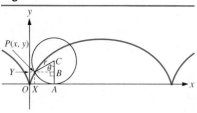

In figure 10.31, the circle has rolled to the point A on the x-axis. Since there was no slippage, the length of arc PA is OA. However, we know from section 5.1 that the length of arc PA is $r\theta$. Thus, we have

$$OA = r\theta.$$

Also, by the very definition of the trigonometric functions, as interpreted in $\triangle PCB$, we have

$$PB = r \sin \theta \quad \text{and} \quad BC = r \cos \theta.$$

Solving for the coordinates of P leads to the desired parametric equations, as follows:

$$x = OX = OA - XA = OA - PB$$
$$= r\theta - r \sin \theta = r(\theta - \sin \theta)$$

and
$$y = OY = AC - BC = r - r \cos \theta = r(1 - \cos \theta). \quad \blacktriangleleft$$

Exercises 10.4

In exercises 1–4, write a pair of parametric equations for each graph.

1. The graph of the function given by $y = x^3 - x + 7$
2. The graph of the equation $x = y^3 + y - 8$
3. The polar graph of the polar equation $r = 2 + 3 \sin \theta$
4. The polar graph of the polar equation $r = 2\theta$

In exercises 5–6, sketch the given parameterized curve.

5. $x = \cos \theta + \cos^2 \theta,\ y = \sin \theta + \cos \theta \sin \theta,\ 0 \leq \theta \leq \dfrac{3\pi}{2}$
6. $x = 2 \sin \theta \cos \theta,\ y = 2 \cos^2 \theta,\ 0 \leq \theta < \dfrac{\pi}{2}$

In exercises 7–36, eliminate the parameter in order to find a Cartesian equation of a curve which contains the given parameterized curve. Then sketch the parameterized curve.

7. $x = 3t,\ y = t^2 + 1$
8. $x = t - 1,\ y = t^2$
9. $x = t^2 - 1,\ y = t + 1,\ t \geq -\dfrac{1}{2}$
10. $x = e^{2t},\ y = e^t$
11. $x = 2t^{1/3},\ y = 8t^{1/3}$
12. $x = 2t^{1/4},\ y = 8t^{1/4},\ t \geq 0$
13. $x = t^4,\ y = t + 1$
14. $x = t^3,\ y = t - 1$
15. $x = \cos \theta,\ y = \cos^2 \theta$
16. $x = \cos^2 \theta,\ y = \sin \theta$ (Hint: Pythagoras.)
17. $x = 2 \cos \theta,\ y = 4 \sin \theta,\ \dfrac{\pi}{2} \leq \theta \leq 2\pi$
18. $x = 2 \cos \theta,\ y = 4 \cos \theta$

19. $x = 5 + 3 \cos \theta, y = 1 + \sin \theta, 0 \le \theta \le \frac{3\pi}{2}$
20. $x = 4 \cos \theta, y = 2 \sin \theta$
21. $x = 5 + 5 \cos \theta, y = -3 + 5 \sin \theta, \frac{\pi}{2} < \theta \le \frac{3\pi}{2}$
22. $x = -4 + 4 \sin \theta, y = 4 + 4 \cos \theta, 0 \le \theta \le \pi$
23. $x = \cos 2\theta, y = 2 \cos^2 \theta$
24. $x = 5 + 3 \sin \theta, y = 1 + \cos \theta, 0 \le \theta \le \frac{3\pi}{2}$
25. $x = \cos 2\theta, y = \sin \theta$ 26. $x = 2 \sin^2 \theta, y = \cos 2\theta$
27. $x = \sin \theta, y = 2$
28. $x = \tan \theta, y = \tan 2\theta$ (Hint: sections 6.3 and 3.6 are relevant.)
29. $x = -1 + 3 \sec \theta, y = 2 \tan \theta$
30. $x = 1 + \sec \theta, y = -3 + 2 \tan \theta$
31. $x = 1 + \frac{1}{t}, y = t + 1$ 32. $x = t^3, y = t^2$
33. $x = 7 + 2t, y = 3 + 4t$
34. $x = -4 - 2t, y = -3 - 4t, -2 \le t \le 3$
35. $x = -4 + 2t, y = 8 - 4t, -3 \le t \le 2$
36. $x = 6 + 2t, y = -3 - 2t$

In exercises 37–50, give at least one pair of parametric equations whose parameterized curve is the given graph. If possible, identify a domain I so that each point on the graph arises from just one parameter value in I.

37. Circle, center $(3, -2)$, radius π
38. Circle, center $(-2, 0)$, radius $\sqrt{2}$
39. "North-south" ellipse, center at origin, $a = 2, b = \sqrt{2}$
40. "East-west" ellipse, center $(3, -2)$, $a = \sqrt{2}, b = 1$
41. "East-west" hyperbola, center $(-2, 3)$, $a = \sqrt{2}, b = \sqrt{3}$ (Hint: review exercise 29 above or exercise 47 in section 8.3.)
42. "North-south" hyperbola, center $(-3, -2)$, $a = \sqrt{2}, b = \sqrt{2}$
43. Parabola, vertex $(3, 2)$, opening downward, passing through $(-1, -5)$ (Hint: motivated by exercise 40 in section 8.1 or the above exercises 7–9, you should determine a negative constant p such that the answer is given by $x = 3 + 2pt, y = 2 + pt^2$.)
44. Parabola, vertex $(-2, 3)$, opening upward, passing through $(2, 5)$ (Hint: review the hint for exercise 43.)
45. Line passing through $(5, 2)$ and $(-1, 18)$ (Hint: the answer should be of the form $x = at + 5, y = ct + 2$. To find a suitable pair (a, c), observe that $\frac{-1 - 5}{a} = \frac{18 - 2}{c}$.)
46. Line passing through $(2, -5)$ and $(-18, 3)$ (Hint: review the hint for exercise 45.)
47. Line passing through $(5, 2)$ and $(5, 5)$ (Hint: adjust the method suggested for exercise 45, using $a = 0$.)
48. Line passing through $(5, 2)$ and $(12, 2)$
49. The line $y = \frac{3}{4}x + 2$ 50. The line $2x + 3y + 4 = 0$

◆ 51. (**Curtate cycloid**)
  (a) In the context of example 10.32, suppose that the point Q is on the moving disc at a distance of s units from the disc's center. Suppose also that Q is directly above P when P is at the origin. Show that a pair of parametric equations for the path traced by Q is given by $x = r\theta - s \sin \theta, y = r - s \cos \theta$.
  (b) Give a physical interpretation for the answer in (a) in case $s = 0$.

◆ 52. (**Epicycloid**) Show that $x = 3 \cos \theta + \cos 3\theta, y = 3 \sin \theta + \sin 3\theta$ describes the path of a point on a circle of radius 1 that rolls without slipping on the outside of a circle of radius 2.

10.5 MATHEMATICAL INDUCTION, SUMMATION NOTATION, AND SEQUENCES

In this section, we shall examine an important proof procedure that is often used in calculus and more advanced mathematics. Along the way, we shall introduce some concepts and formulas that will be helpful when you study calculus or discrete mathematics. We shall also be able to prove a couple of results that were only stated in earlier chapters. First, we shall look at a few examples and introduce some notation.

▷ **EXAMPLE 10.33**

What is the sum of the first n odd positive integers?

Solution

We begin by making a table and looking for a pattern in the sequence of numbers that results.

n	Sum of First n Terms		
1	1	=	1
2	1 + 3	=	4
3	1 + 3 + 5	=	9
4	1 + 3 + 5 + 7	=	16
5	1 + 3 + 5 + 7 + 9	=	25
6	1 + 3 + 5 + 7 + 9 + 11	=	36

Two patterns are evident from this table. First, it appears from the sequence of sums 1, 4, 9, 16, 25, 36, ... that the sum of the first n odd positive integers is n^2. Secondly, the last addend in each sum is $2n - 1$. Thus, we assert that

$$1 + 3 + 5 + 7 + \cdots + (2n - 1) = n^2 \quad \text{for all positive integers } n.$$

Is this assertion true? Soon, we shall prove that it is, by using a principle of proof called mathematical induction. ◀

Summation Notation

Before looking at the next example, we shall introduce some notation that is particularly useful in developing scientific applications of integral calculus. Example 10.33 involved a sum. A symbol frequently used for summation is the Greek letter Σ (capital sigma). The fifth line of the above table was $1 + 3 + 5 + 7 + 9 = 25$ and could be written

$$\sum_{i=1}^{5} (2i - 1) = 25.$$

The **index of summation** i takes on successive integral values which are used to form the different terms, $2i - 1$, of the sum. The equation below the Σ symbol, namely $i = 1$, indicates the index of summation and the first integral value of the index. The numeral above the Σ symbol indicates the last integral value of the index. The index of summation is an example of a "dummy variable," having no intrinsic meaning other than indicating the "extent" of the summation process. So, for example, the fifth line of the above table could be written as

$$\sum_{j=1}^{5} (2j - 1) = 25 \quad \text{or} \quad \sum_{k=1}^{5} (2k - 1) = 25 \quad \text{or} \quad \sum_{t=1}^{5} (2t - 1) = 25.$$

▶ **EXAMPLE 10.34**   Determine $\sum_{i=1}^{7} (3i + 1)$.

Mathematical Induction, Summation Notation, and Sequences

Solution

To evaluate this sum, successively substitute the values $i = 1, 2, 3, 4, 5, 6$, and 7 into the expression $3i + 1$ and add the resulting terms. This produces

$$\sum_{i=1}^{7}(3i + 1) = (3 \cdot 1 + 1) + (3 \cdot 2 + 1) + (3 \cdot 3 + 1) + (3 \cdot 4 + 1)$$
$$+ (3 \cdot 5 + 1) + (3 \cdot 6 + 1) + (3 \cdot 7 + 1)$$
$$= 4 + 7 + 10 + 13 + 16 + 19 + 22 = 91.$$

▶ **EXAMPLE 10.35**

Evaluate the following sums:

(a) $\sum_{j=2}^{5} j(j-1)^2$ (b) $\sum_{k=0}^{3} \frac{k+1}{k+2}$.

Solutions (a)

$$\sum_{j=2}^{5} j(j-1)^2 = 2(2-1)^2 + 3(3-1)^2 + 4(4-1)^2 + 5(5-1)^2$$
$$= 2 \cdot 1^2 + 3 \cdot 2^2 + 4 \cdot 3^2 + 5 \cdot 4^2$$
$$= 2 + 12 + 36 + 80 = 130.$$

(b)

$$\sum_{k=0}^{3} \frac{k+1}{k+2} = \frac{0+1}{0+2} + \frac{1+1}{1+2} + \frac{2+1}{2+2} + \frac{3+1}{3+2}$$
$$= \tfrac{1}{2} + \tfrac{2}{3} + \tfrac{3}{4} + \tfrac{4}{5}$$
$$= \tfrac{163}{60}.$$

Another Example

▶ **EXAMPLE 10.36**

What is $\sum_{i=1}^{n} i$?

Solution

As in example 10.33, we begin by making a table and looking for a pattern in the resulting sequence of answers. In order to help find the pattern, we have added a third column that expresses the sum of the first n terms as a multiple of n.

n	Sum of First n Terms			Sum as a Multiple of n
1	1	=	1	$1 \cdot 1$
2	$1 + 2$	=	3	$2 \cdot \frac{3}{2}$
3	$1 + 2 + 3$	=	6	$3 \cdot 2$
4	$1 + 2 + 3 + 4$	=	10	$4 \cdot \frac{5}{2}$
5	$1 + 2 + 3 + 4 + 5$	=	15	$5 \cdot 3$
6	$1 + 2 + 3 + 4 + 5 + 6$	=	21	$6 \cdot \frac{7}{2}$

If the right-hand factors in the last column are rewritten as $1 = \frac{2}{2}$, $2 = \frac{4}{2}$, and $3 = \frac{6}{2}$, a discernible pattern emerges. It appears that the sequence of answers is described by $n\left(\frac{n+1}{2}\right)$, and so we conjecture that $\sum_{i=1}^{n} i = n\left(\frac{n+1}{2}\right)$. This, along with the assertion made in example 10.33, will soon be proved. It is now time to introduce the appropriate proof procedure. ◀

The Principle of Mathematical Induction

How could we hope to prove conjectures of the kind made in examples 10.33 and 10.36? One method of proof, called the **principle of mathematical induction**, can often be used effectively to prove assertions about the positive integers.

The Principle of Mathematical Induction

Let $P(n)$ be a statement, for instance an equation or an inequality, that involves a positive integer n. Then $P(n)$ is true for every positive integer n if

(i) $P(1)$ is true and
(ii) for each positive integer k, if $P(k)$ is true then $P(k + 1)$ is also true.

This principle suggests that the following steps should be used when using mathematical induction to prove a proposition.

Steps in a Mathematical Induction Proof

The steps to follow when trying to prove a proposition by mathematical induction are

(i) (The "induction basis" step) Verify directly that $P(1)$ is true.
(ii) (The "induction step") Assuming that $P(k)$ is true, prove that $P(k + 1)$ is also true.
(iii) Having carried out (i) and (ii), conclude that $P(n)$ is true for all positive integers n.

We shall now return to the conjectures that were made in examples 10.33 and 10.36 and prove each of them by mathematical induction. Then we shall do a more difficult example of this type that will be useful when you study calculus.

▶ **EXAMPLE 10.37**

Prove that $\sum_{i=1}^{n} (2i - 1) = n^2$ for all positive integers n.

Solution

We shall let $P(n)$ be the assertion $\sum_{i=1}^{n}(2i-1) = n^2$. According to the above plan, we begin the proof with the "induction basis" step. This involves showing that $P(1)$ is true, namely that

$$\sum_{i=1}^{1}(2i-1) = 1^2.$$

This is equivalent to $\quad\quad 2(1) - 1 = 1^2$

or $\quad\quad\quad\quad\quad\quad\quad\quad 1 = 1$

which is clearly true.

Next, we carry out the "induction step": assuming $P(k)$, prove $P(k+1)$. In other words, assume

$$\sum_{i=1}^{k}(2i-1) = k^2 \quad\quad (*)$$

and prove

$$\sum_{i=1}^{k+1}(2i-1) = (k+1)^2. \quad\quad (**)$$

(The strategy is similar to that of proving trigonometric identities, as in section 6.1.) Using (∗) to rewrite the left-hand side of (∗∗), we get

$$\sum_{i=1}^{k+1}(2i-1) = \left(\sum_{i=1}^{k}(2i-1)\right) + \{2(k+1) - 1\}$$
$$= k^2 + \{2(k+1) - 1\}$$
$$= k^2 + 2k + 1.$$

Is this $(k+1)^2$, the right-hand side of (∗∗)? Yes, by the elementary algebra in chapter 0! Thus, the induction step has been established. So, by the principle of mathematical induction, $\sum_{i=1}^{n}(2i-1) = n^2$ for all positive integers n. ◀

▷ **EXAMPLE 10.38**

Use mathematical induction to prove that $\sum_{i=1}^{n} i = n\left(\dfrac{n+1}{2}\right)$ for all positive integers n.

Solution

This time, we shall let $P(n)$ be the statement $\sum_{i=1}^{n} i = n\left(\frac{n+1}{2}\right)$. The "induction basis" amounts to showing that

$$1 = 1\left(\frac{1+1}{2}\right)$$

which *is* true.

For the "induction step," assume that $\sum_{i=1}^{k} i = 1 + 2 + 3 + \cdots + k = k\left(\frac{k+1}{2}\right)$ is true. We must show

$$\sum_{i=1}^{k+1} i = 1 + 2 + 3 + \cdots + k + (k+1) = \frac{(k+1)(k+2)}{2}.$$

To do this, notice that

$$\sum_{i=1}^{k+1} i = \sum_{i=1}^{k} i + (k+1)$$

$$= k\left(\frac{k+1}{2}\right) + (k+1)$$

$$= (k+1)\left(\frac{k}{2} + 1\right) = (k+1)\left(\frac{k+2}{2}\right).$$

This establishes the "induction step." Thus, by the principle of mathematical induction, we have proved $P(n)$ for every positive integer n. ◀

Three Useful Formulas

$$\sum_{i=1}^{n} i = \frac{n(n+1)}{2}$$

$$\sum_{i=1}^{n} i^2 = \frac{n(n+1)(2n+1)}{6}$$

and

$$\sum_{i=1}^{n} i^3 = \frac{n^2(n+1)^2}{4} = \left(\frac{n(n+1)}{2}\right)^2.$$

We proved the first of these in example 10.38. We shall establish the second in example 10.39 by means of mathematical induction. You will be asked to establish the third in exercise 11.

Mathematical Induction, Summation Notation, and Sequences

with satisfactory $q = xq_1 + q_3$ and $r = r_3$. This establishes the induction step, and so completes the proof.

In the next section, you will see how to use mathematical induction to prove another valuable result, the binomial theorem.

It is important to emphasize that a valid proof by means of mathematical induction must carefully establish both the induction basis and the induction step. In general, neither of these steps *by itself* can guarantee the other step. Since students often assume otherwise, we have included exercises 43–46, to help make the point.

In closing, we note an important extension of the principle of mathematical induction. How can we prove an inequality like $n^2 < 2^n$ which holds for all positive integers $n \geq 5$ (but is false for $n = 1, 2, 3, 4$)? In this case, we take $P(n)$ as follows:

$$P(1): 1 + 6 = 6 + 1$$
$$P(2): 2 + 6 = 6 + 2$$
$$P(3): 3 + 6 = 6 + 3$$
$$P(4): 4 + 6 = 6 + 4$$
$$P(n): n^2 < 2^n, \quad \text{for } n \geq 5.$$

(Any valid $P(1), P(2), P(3), P(4)$ could have been used instead of the ones given above.) Step (i) of the modified mathematical induction for this problem would involve directly verifying $P(5)$, namely that $5^2 < 2^5$. Step (ii) would involve assuming $k^2 < 2^k$ and $k \geq 5$, and deducing from these conditions that $(k + 1)^2 < 2^{k+1}$. You will be asked to carry this out in exercise 48. Additional practice with this modified type of mathematical induction will be given in exercise 47.

Exercises 10.5

In exercises 1–6, determine the given sum.

1. $\sum_{i=1}^{4} (3i + 4)$

2. $\sum_{j=2}^{6} j^2(j - \frac{1}{2})$

3. $\sum_{k=3}^{4} (3k - 5)^2$

4. $\sum_{i=1}^{5} \left(1 + \frac{1}{i}\right)$

5. $\sum_{i=1}^{4} -3(\frac{2}{5})^{i-1}$

6. $1 + \frac{1}{2} + \frac{1}{4} + \cdots + \frac{1}{2^{n-1}}$

In exercises 7–26, use mathematical induction to prove the given assertion.

7. $\sum_{i=1}^{n} (3i - 2) = \frac{n(3n - 1)}{2}$ for all positive integers n

8. $\sum_{i=1}^{n} (-2i + 4) = n(3 - n)$ for all positive integers n

9. $\sum_{j=1}^{n} (-6j^2 + j) = \frac{-n(n + 1)(4n + 1)}{2}$ for all positive integers n

10. $\sum_{j=1}^{n} (3j - 1)^2 = \frac{n(6n^2 + 3n - 1)}{2}$ for all positive integers n

◆ 11. $\sum_{k=1}^{n} k^3 = \frac{n^2(n + 1)^2}{4}$ for all positive integers n

12. $\sum_{k=1}^{n} \left(\frac{1}{k} - \frac{1}{k + 1}\right) = \frac{n}{n + 1}$ for all positive integers n

13. $\sum_{i=1}^{n} \frac{i}{2}(i+3) = \frac{n(n+1)(n+5)}{6}$ for all positive integers n

14. $\sum_{i=1}^{n} i(3i-2) = \frac{n(n+1)(2n-1)}{2}$ for all positive integers n

15. $1 \cdot 2 + 3 \cdot 4 + \cdots + (2n-1)2n = \frac{n(n+1)(4n-1)}{3}$ for all positive integers n

16. If $a_1, a_2, \ldots$ is a geometric sequence with ratio r, then $a_n = a_1 r^{n-1}$ for all positive integers n. (Hint: $a_{k+1} = ra_k$.)

17. $2n < 3^n$ for all positive integers n (Hint: $3 \le 3^n$.)

18. $5n + 3 \le (n+2)^2$ for all positive integers n

19. $n + 5 \le 3 \cdot 2^n$ for all positive integers n

20. $2^n + \frac{1}{2} < e^n$ for all positive integers n (Hint: $\frac{5}{2} < e$.)

21. $(15n+9)^2 - 1$ is an integral multiple of 5 for all positive integers n.

22. $(18n+5)^2 - 1$ is an integral multiple of 6 for all positive integers n.

23. $n^3 + 2n$ is an integral multiple of 3 for all positive integers n.

24. $2^{4n} - 1$ is an integral multiple of 3 for all positive integers n. (Hint: $2^{4k+4} = 16(2^{4k} - 1) + 16$.)

25. $3^{2n} - 1$ is an integral multiple of 4 for all positive integers n.

26. $3^{2n+1} + 2^{n+2}$ is an integral multiple of 7 for all positive integers n. (Hint: at some point, you should consider $9(3^{2k+1} + 2^{k+2}) - 7 \cdot 2^{k+2}$.)

27. Suppose that you invest \$100 and interest is 7 percent compounded annually. Then your investment is worth $\$100(1.07)$ after one year, $\$100(1.07)^2$ after two years, $\$100(1.07)^3$ after three years, and so on.
    (a) What is your investment worth after n years?
    (b) Using the formula surmised in (a), show that the sequence of values for your investment after n years is a geometric sequence. Find its ratio r.

28. It is known that if a tennis ball is dropped onto a clay court from a height h, then it rebounds to a height $0.55h$. Suppose that such a ball is dropped onto such a court from a height of 5 feet and you watch it bounce repeatedly.
    (a) How far does the ball rise after its first bounce before it starts to fall again?
    (b) How far does the ball rise after its second bounce before it starts to fall again?
    (c) At the moment of its twentieth bounce, how far has the ball traveled? (Hint: use (a) and (b) to surmise a formula for the height to which the ball rises after its nth bounce. Then example 10.40 becomes relevant.)

29. A certain bacterial colony, of original size N, is observed to increase in size by 15 percent per hour. Its size after n hours is Ne^{kn}, where k is an experimentally determined constant.
    (a) Show that the sequence of sizes Ne^{kn} is a geometric sequence. Find its ratio.
    (b) Prove that $k = \ln 1.15$.

30. (Examples of some geometric sequences) Suppose that C_1, a circle of radius r_1, is given.
    (a) Show that it is possible to inscribe in C_1 a square S_1 of side $s_1 = \sqrt{2}r_1$. (Hint: see the figure and consider C_1 to be the graph of $x^2 + y^2 = r_1^2$.)

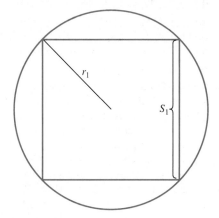

(b) Show that it is possible to inscribe in S_1 a circle C_2 of radius $r_2 = \frac{s_1}{2}$. (Hint: see the figure.)

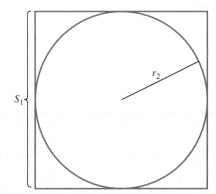

(c) Repeat the procedures in (a) and (b) to get a sequence of circles $C_1, C_2, C_3, \ldots, C_n, \ldots$ and a sequence of squares $S_1, S_2, S_3, \ldots, S_n, \ldots$. Let A_n denote the area of C_n, and let a_n denote the area of S_n. Show that $A_1, A_2, A_3, \ldots, A_n, \ldots$ is a geometric sequence with ratio $r = \frac{1}{2}$.

(d) Is $a_1, a_2, a_3, \ldots, a_n, \ldots$ a geometric sequence? If so, determine its ratio.

You will need the following definition in exercises 31–42. A sequence $a_1, a_2, a_3, \ldots, a_n, \ldots$ of numbers is called an **arithmetic sequence** if each difference of consecutive terms is the same. If this holds, then $d = a_{n+1} - a_n$ is called the **difference** of the arithmetic sequence.

31. Determine whether the following sequences are arithmetic, geometric, or neither. If arithmetic, calculate d; if geometric, calculate r.
    (a) $1, 3, 5, 7, \ldots, 2n-1, \ldots$
    (b) $3, 12, 27, 48, \ldots, 3n^2, \ldots$
    (c) $2, \frac{3}{2}, \frac{9}{8}, \frac{27}{32}, \ldots, 2(\frac{3}{4})^{n-1}, \ldots$
    (d) $0, 0, 0, 0, \ldots, 0, \ldots$
    (e) $1, -\frac{1}{2}, -2, -\frac{7}{2}, \ldots, 1-\frac{3}{2}(n-1), \ldots$
    (f) $1, 3, 7, 15, \ldots, 2^n - 1, \ldots$

32. Suppose that a new professor is hired at an annual salary of $\$S$. She is told that she will receive a raise of $\$2,000$ at the end of each year of service.
    (a) Explain intuitively why the raise after her nth year of service produces a salary of $\$(S + 2000n)$.
    (b) Is the sequence given by $S + 2000n$ an arithmetic sequence? If so, calculate its difference d.

33. Prove by mathematical induction that if $a_1, a_2, a_3, \ldots, a_n, \ldots$ is an arithmetic sequence with difference d, then $a_n = a_1 + (n-1)d$ for all positive integers n. (Hint: $a_{k+1} = a_k + d$.)

34. Let $a_1, a_2, a_3, \ldots$ be an arithmetic sequence with difference d. Prove by mathematical induction that the sum of the first n terms of this sequence is $\dfrac{2na_1 + (n-1)nd}{2}$. (Hint: you will need exercise 33 to describe a_{k+1}.)

35. Compute $1 + 2 + 3 + 4 + \cdots + 86$. (Hint: use example 10.36 or exercise 34.)

36. The professor in exercise 32 retires after twenty-five years of service. If she saved 10 percent of her salary during those twenty-five years, how much has she saved? (Ignore any interest that her savings may have earned. Your answer will involve S. Your first step should be to find her total earnings. For this, combine the results of exercise 32(a) and exercise 34.)

37. There are eighteen rows in a theater. The first row seats six people, the second row seats eight people, the third row seats ten people, and so on. What is the seating capacity of the theater? (Hint: use exercise 34.)

38. At n o'clock, a chime clock strikes n times and does not chime again during the next hour. Knowing this, John and David seek to determine how many times the clock will chime during a 24-hour period. A weary David sits beside the clock at one minute past midnight and vows to stay up for the next 24 hours, counting all the chimes that he hears. An equally weary John quickly calculates that David will count 156 chimes in all. John then has a good night's sleep while a wary David grows wearier.
    (a) Is John right? (Hint: exercise 34 is relevant.)
    (b) Can you suggest a way for skeptical David to avoid chime-lag and get up to 12 hours of sleep while conducting his experiment? (Hint: the clock is not on military or European time. One hour after noon, the clock chimes only once, not thirteen times. Example 10.33 is relevant.)

39. Prove that if $a_1, a_2, a_3, \ldots$ is a geometric sequence of positive real numbers, then $\ln(a_1), \ln(a_2), \ln(a_3), \ldots$ is an arithmetic sequence. (Hint: show $\ln(a_{n+1}) - \ln(a_n) = \ln r$.)

40. (A converse of exercise 33) If f is a linear function, show that $f(1), f(2), f(3), \ldots, f(n), \ldots$ constitutes an arithmetic sequence. (Hint: $f(x) = ax + b$. Calculate $f(n+1) - f(n)$.)

41. (Another way to discover example 10.36 and exercise 34.)
    (a) Let $S = \sum_{i=1}^{n} i$. In the display

1	2	3	$\cdots$	i	$\cdots$	$n-2$	$n-1$	n
n	$n-1$	$n-2$	$\cdots$	$n-i+1$	$\cdots$	3	2	1

    what is the sum of the entries in each column? How many columns are there? Compute the sum of all the entries in two ways. (Hint: this matrix has rows too.) Hence, show that $S = \dfrac{n(n+1)}{2}$.
    (b) Use a display similar to that in (a) to show that
    $$\sum_{i=1}^{n}(a_1 + (i-1)d) = \dfrac{2na_1 + (n-1)nd}{2}.$$

42. (a) If b is a nonzero real number, prove that the "constant" sequence $b, b, b, \ldots, b, \ldots$ is both a geometric sequence and an arithmetic sequence.
    (b) If $a_1, a_2, a_3, \ldots, a_n, \ldots$ is both a geometric sequence and an arithmetic sequence, prove that $a_1 = a_2 = a_3 = \cdots = a_n = \cdots$. (Hint: by exercises 16 and 33, $a_1 + (n-1)d = a_1 r^{n-1}$ for each positive integer n. Use this fact for $n = 2, 3, 4$ to conclude that $a_1(r^2 - r) = a_1(r^3 - r^2)$. Hence show $r = 1$.)

Chapter 10 / Additional Topics in Geometry and Algebra

In exercises 43–46, each of the assertions is false. For each assertion, show that one step (induction basis or induction step) of an attempted mathematical induction "proof" *is* valid. Determine the smallest positive integer n for which the given assertion is false.

♦ 43. $\sum_{i=1}^{n} 2i = n^2 + n + 4$

♦ 44. $\sum_{i=1}^{n} (3i - 1) = \left[\frac{n^2 + 3}{2}\right]$, where $[\cdots]$ is the greatest integer function introduced in section 2.1

♦ 45. $n^4 - 10n^3 + 35n^2 - 50n + 24 = 0$

♦ 46. $2n < n^2 - 1$

In exercises 47–48, use the modified method of mathematical induction to prove the given assertion.

♦ 47. $n + 2 < n^2$ for all positive integers $n \geq 3$

♦ 48. $n^2 < 2^n$ for all positive integers $n \geq 5$ (Hint: see the remarks at the end of the section.)

In exercises 49–50, you will need the following definition of **product notation**. Given a sequence $a_1, a_2, a_3, \ldots$, then

$$\prod_{i=1}^{n} a_i = a_1 a_2 a_3 \cdots a_n.$$

49. Determine the following products:

(a) $\prod_{i=1}^{2} (3i + 4)$ (b) $\prod_{j=2}^{5} j$ (c) $\prod_{k=4}^{6} (k^2 - 1)$

50. Use mathematical induction to prove that $\prod_{k=1}^{n} \left(1 + \frac{1}{k}\right) = n + 1$ for all positive integers n.

Note: exercises 51–60 are enrichment material on limits of sequences, from both the theoretical and the computational points of view. This additional material should be of help in studying calculus later. However, students or instructors who pursue exercises 51–60 should be warned to allocate extra time for these exercises beyond the usual demands of one section in this book.

51. Prove by mathematical induction that if h is any real number greater than -1, then $(1 + h)^n \geq 1 + nh$ for all positive integers n. (Hint: you will need to show that $(1 + kh)(1 + h) \geq 1 + kh + h$.)

52. (a) Using a calculator, find $(\frac{1}{2})^n$ for $n = 2, 4, 8, 16, 32$; $(\frac{3}{4})^n$ for $n = 2, 4, 8, 16, 32, 54$; and $(\frac{7}{4})^n$ for $n = 2, 4, 8, 16, 32, 64, 65$.

(b) What conclusions do the data in (a) suggest to you concerning the limiting tendency of r^n as n increases without bound—that is, as n "approaches infinity"? (Proof of these conclusions will be requested in exercise 53.)

53. (a) Let $r > 1$. Give an intuitive argument that as n increases without bound, the limiting tendency of r^n is infinity. (Hint: let $h = r - 1$. By exercise 51, $r^n \geq 1 + nh$. What is the limiting tendency of $1 + nh$?)

(b) Let r be a real number such that $|r| < 1$. Give an intuitive argument that as n increases without bound, the limiting tendency of r^n is 0. (Hint: according to (a), what is the limiting tendency of

$$\left(\frac{1}{|r|}\right)^n?$$ What about $|r|^n$?)

(c) Use properties of the natural logarithm function to give another argument for the assertion in (a). (Hint: what is the limiting tendency of $n \ln r$? Of $\ln r^n$?)

54. (a) Using a calculator, find $2^{1/n}$ for $n = 32, 64, 128, 139$; $3^{1/n}$ for $n = 32, 64, 128, 221$; and $(\frac{2}{5})^{1/n}$ for $n = 32, 64, 128, 181$.

(b) What conclusions do the data in (a) suggest to you concerning the limiting tendency of $r^{1/n}$ as n increases without bound? (Proofs of these conclusions will be requested in exercise 55.)

55. (a) Let $r > 0$. Give an intuitive argument that as n increases without bound, the limiting tendency of $r^{1/n}$ is 1. (Hint: reduce to the case $r > 1$. Let $h_n = r^{1/n} - 1$. Use exercise 51 to conclude that $r \geq 1 + nh_n$. Observe that $0 < h_n \leq \frac{r-1}{n}$. What is the limiting tendency of $\frac{r-1}{n}$? Finally, use the "sandwiching" idea seen in exercise 19 of section 5.4.)

(b) Use properties of the natural logarithm function to give another argument for the assertion in (a). (Hint: what is the limiting tendency of $\frac{\ln r}{n}$? Of $\ln (r^{1/n})$?)

56. (a) Using a calculator, find $n^{1/n}$ for $n = 2, 64, 128, 256, 999, 1462$.

(b) What conclusion do the data in (a) suggest to you concerning the limiting tendency of $n^{1/n}$ as n increases without bound? (A proof of this conclusion will be requested in exercise 57.)

57. (a) Give an intuitive argument that as n increases without bound, the limiting tendency of $n^{1/n}$ is 1.

(Hint: let $h_n = \sqrt{n^{\sqrt{n}}} - 1$. Use exercise 51 to conclude that $\sqrt{n} \geq 1 + nh_n$. Observe that $0 < h_n < \frac{1}{\sqrt{n}}$. What is the limiting tendency of $\frac{1}{\sqrt{n}}$? "Sandwich," as in exercise 55(a), to show that the limiting tendency of h_n is 0. But $n^{1/n} = 1 + 2h_n + (h_n)^2$.)

(b) Use the result in (a) and the "sandwiching" technique to obtain yet another proof of the result in exercise 55.

58. (a) Using a calculator, find $\frac{\ln n}{n}$ for $n = 64, 128, 256, 999, 1457$; and $\frac{n}{\ln n}$ for $n = 64, 128, 256, 999, 9119$.

(b) What conclusions do the data in (a) suggest to you concerning the limiting tendencies of $\frac{\ln n}{n}$ and $\frac{n}{\ln n}$ as n increases without bound? (Proofs of these conclusions will be requested in exercise 59.)

59. Use properties of the natural logarithm function to give intuitive arguments that as n increases without bound, the limiting tendency of $\frac{\ln n}{n}$ is 0 and the limiting tendency of $\frac{n}{\ln n}$ is infinity. (Hint: for the first assertion, recall exercise 57(a) and use continuity of ln.)

60. Let's again consider the bouncing ball in exercise 28.

(a) Show that if $n > 1$, then at the moment of its nth bounce, the ball has traveled $5 + 10(0.55) + 10(0.55)^2 + \cdots + 10(0.55)^{n-1} = 5 + \frac{110}{9}(1 - (0.55)^{n-1})$ feet. (Hint: example 10.40.)

(b) Show that the ball travels $\frac{155}{9}$ feet. (Hint: using exercise 53(b), find the limiting tendency, as n increases without bound, for the expression in (a).) **Comment:** Congratulations! You have just summed your first **infinite series**. In studying calculus, you will often reencounter such ideas.

10.6 BINOMIAL THEOREM

Many problems refer to an integral power of a sum. In this section, we shall explore a method for easily expanding such an expression $(a + b)^n$. By time-consuming applications of the technique of repeated multiplication, we can verify that

$$(a + b)^1 = a + b$$
$$(a + b)^2 = a^2 + 2ab + b^2$$
$$(a + b)^3 = a^3 + 3a^2b + 3ab^2 + b^3$$
$$(a + b)^4 = a^4 + 4a^3b + 6a^2b^2 + 4ab^3 + b^4$$
$$(a + b)^5 = a^5 + 5a^4b + 10a^3b^2 + 10a^2b^3 + 5ab^4 + b^5$$
$$(a + b)^6 = a^6 + 6a^5b + 15a^4b^2 + 20a^3b^3 + 15a^2b^4 + 6ab^5 + b^6.$$

The expression on the right-hand side of each equation is called the **expansion** of the left-hand side. Examination of these six expansions leads to some conjectures about the expansion of $(a + b)^n$ for any positive integer n.

(1) The expansion has $n + 1$ terms.
(2) Its first term is a^n, and its last term is b^n.
(3) In each term, the sum of the exponent of a and the exponent of b is n.
(4) In each successive term, the exponent of a decreases by 1 and the exponent of b increases by 1.
(5) The numerical coefficients of the terms may be obtained from the following array, known as **Pascal's Triangle**. Each number, except

those at the ends of the rows, is the sum of the two nearest numbers in the row directly above it. The numbers at the ends of the rows are always 1. The first row is obtained from $(a + b)^0$.

$$
\begin{array}{ccccccccccccc}
 & & & & & & 1 & & & & & & \\
 & & & & & 1 & & 1 & & & & & \\
 & & & & 1 & & 2 & & 1 & & & & \\
 & & & 1 & & 3 & & 3 & & 1 & & & \\
 & & 1 & & 4 & & 6 & & 4 & & 1 & & \\
 & 1 & & 5 & & 10 & & 10 & & 5 & & 1 & \\
1 & & 6 & & 15 & & 20 & & 15 & & 6 & & 1 \\
 & & & & & & \cdot & & & & & & \\
 & & & & & & \cdot & & & & & & \\
 & & & & & & \cdot & & & & & & \\
\end{array}
$$

Pascal's Triangle is **not** a very convenient method for determining the numerical coefficients in the expansion when n is large. An alternative method is

(6)   The coefficient of any term after the first may be found from the preceding term as follows. Denote the coefficients of the terms by $c_0, c_1, c_2, \ldots, c_n$, where the subscript of each coefficient corresponds to the exponent of b in the corresponding term of the expansion. Then for $r \geq 1$, the coefficient of the $a^{n-r}b^r$ term is

$$c_r = c_{r-1}\left(\frac{n - r + 1}{r}\right).$$

The next example will illustrate how to implement the above method.

▶ **EXAMPLE 10.43**   Write the expansion of $(a + b)^7$.

Solution   By property 2, the first term is a^7. (Note that $c_0 = 1$ and the exponent of b in the term $a^7 = a^7 b^0$ is 0.) Thus, $(a + b)^7 = a^7 + \cdots$.

By property 4, the next term is $c_1 a^6 b^1$. By applying property 6, with $n = 7$ and $r = 1$, we find $c_1 = c_0 \left(\frac{7 - 1 + 1}{1}\right) = 1 \cdot 7 = 7$. Thus, $(a + b)^7 = a^7 + 7a^6 b + \cdots$.

To obtain the coefficient of the next term, $c_2 a^5 b^2$, we apply property 6, with $n = 7$ and $r = 2$, to get $c_2 = c_1 \left(\frac{7 - 2 + 1}{2}\right) = 7 \cdot 3 = 21$. Thus, we see that the third term is $21 a^5 b^2$.

For the fourth term, $n = 7$ and $r = 3$, and so the coefficient is $c_3 = c_2 \left(\frac{7 - 3 + 1}{3}\right) = \frac{(21)5}{3} = 35$. The fourth term is thus $35 a^4 b^3$. At this stage, we have $(a + b)^7 = a^7 + 7a^6 b + 21 a^5 b^2 + 35 a^4 b^3 + \cdots$.

Continuing in the above manner, we obtain the remaining terms, yielding the expansion

$$(a + b)^7 = a^7 + 7a^6 b + 21a^5 b^2 + 35a^4 b^3 + 35a^3 b^4 + 21a^2 b^5 + 7ab^6 + b^7. \quad (1)$$

◀

Binomial Theorem

If we examine the above coefficients more closely, a new pattern begins to emerge. The coefficient of the a^4b^3 term is 35, which was obtained as $\frac{(21)5}{3}$. However, the factor 21 is the coefficient of the previous term, and was itself obtained as $\frac{7 \cdot 6}{2}$. Substituting this into the work that produced 35, we obtain $35 = \frac{7 \cdot 6 \cdot 5}{2 \cdot 3}$. One may well speculate the following.

$$c_r = \frac{n(n-1)(n-2)(n-3)\cdots(n-r+1)}{1 \cdot 2 \cdot 3 \cdots r} \tag{2}$$

The next example will illustrate how to use this formula effectively.

▷ **EXAMPLE 10.44**

Determine the numerical coefficient of the a^2b^5 term in the expansion of $(a + b)^7$.

Solution

Here $n = 7$, $r = 5$, and so $n - r + 1 = 3$. Thus,

$$c_5 = \frac{(7)(6)(5)(4)(3)}{(1)(2)(3)(4)(5)} = 21.$$

The term is thus $21a^2b^5$, just as we found in (1) by the more tedious iterative process. ◁

The above conjectures and suggested formulas are facts. Indeed, the expansion of $(a + b)^n$ that we have been describing is established in general in the **binomial theorem**.

Binomial Theorem

For any numbers a and b and any positive integer n,

$$(a+b)^n = a^n + na^{n-1}b + \frac{n(n-1)}{1 \cdot 2}a^{n-2}b^2 + \frac{n(n-1)(n-2)}{1 \cdot 2 \cdot 3}a^{n-3}b^3$$
$$+ \cdots + \frac{n(n-1)(n-2)\cdots(n-r+1)}{1 \cdot 2 \cdot 3 \cdots r}a^{n-r}b^r + \cdots + b^n.$$

The binomial theorem will be proved by using mathematical induction later in this section. A more conceptual proof of it will be sketched in the enrichment exercises on combinatorics.

▷ **EXAMPLE 10.45**

Use the binomial theorem to expand $(3x - 2)^5$. Algebraically simplify the terms of the expansion.

Solution

We are considering $(a + b)^5$, where $a = 3x$ and $b = -2$. So, the binomial theorem yields

$$(3x-2)^5 = (3x)^5 + \frac{5}{1}(3x)^4(-2) + \frac{5 \cdot 4}{1 \cdot 2}(3x)^3(-2)^2 + \frac{5 \cdot 4 \cdot 3}{1 \cdot 2 \cdot 3}(3x)^2(-2)^3$$
$$+ \frac{5 \cdot 4 \cdot 3 \cdot 2}{1 \cdot 2 \cdot 3 \cdot 4}(3x)(-2)^4 + (-2)^5.$$

Using laws of exponents and elementary algebra, we simplify this expansion to

$$3^5x^5 + 5(3^4)(-2)x^4 + 10(3^3)(-2)^2x^3 + 10(3^2)(-2)^3x^2 + 5 \cdot 3(-2)^4x + (-2)^5$$
$$= 243x^5 - 81x^4 + 1080x^3 - 720x^2 + 240x^4 - 32.$$ ◀

Factorials

If you examine formula (2) you will notice that the denominator of the right-hand side is the product of the first r positive integers. In general, the symbol $n!$, which is read **n factorial**, is used to denote the product of the first n positive integers. Thus,

$$5! = 5 \cdot 4 \cdot 3 \cdot 2 \cdot 1 = 120$$
$$8! = 8 \cdot 7 \cdot 6 \cdot 5 \cdot 4 \cdot 3 \cdot 2 \cdot 1 = 40{,}320$$

and, in general, we have the following.

n factorial

$$n! = n(n-1)(n-2) \cdots 3 \cdot 2 \cdot 1.$$

Since $(n-1)! = (n-1)(n-2)(n-3) \cdots 3 \cdot 2 \cdot 1$, you can see the following is true for each positive integer $n \geq 2$.

$$n! = n(n-1)!$$

It is convenient to have this formula also for $n = 1$. This means $1! = 1 \cdot 0!$, or $1! = 0!$. For this reason, we define $0!$ as follows.

$$0! = 1.$$

Many calculators have a key that can be used to calculate factorials. This key, usually designated $\boxed{x!}$, is frequently used in conjunction with the $\boxed{\text{2nd}}$ key. Thus, on many algebraic calculators, 9! would be calculated by pressing 9 $\boxed{\text{2nd}}$ $\boxed{x!}$, with the result 362880. On an RPN calculator, the keystroke sequence would be 9 $\boxed{\text{ENTER}}$ $\boxed{\text{2nd}}$ $\boxed{x!}$, with the same result. On a TI-81, you would press 9 $\boxed{\text{MATH}}$ 5 $\boxed{\text{ENTER}}$. On a Casio fx-7700G, you would press 9 $\boxed{\text{SHIFT}}$ $\boxed{\text{MATH}}$ $\boxed{\text{F2}}$ $\boxed{\text{F1}}$ $\boxed{\text{EXE}}$.

▶ **EXAMPLE 10.46**

Evaluate each of the following:

(a) $\dfrac{6!}{4!}$ (b) $\dfrac{10!}{9!}$ (c) $\dfrac{12!5!}{8!6!}$ (d) $\dfrac{15!}{12!3!}$

Solutions

(a) Since $6! = 6 \cdot 5 \cdot 4!$, we can simplify this as $\dfrac{6!}{4!} = \dfrac{6 \cdot 5 \cdot 4!}{4!} = 6 \cdot 5 = 30$.

(b) $\dfrac{10!}{9!} = \dfrac{10 \cdot 9!}{9!} = 10$.

Binomial Theorem

(c) $\dfrac{12!5!}{8!6!} = \dfrac{12 \cdot 11 \cdot 10 \cdot 9 \cdot 8! \cdot 5!}{8! \cdot 6 \cdot 5!} = \dfrac{12 \cdot 11 \cdot 10 \cdot 9}{6} = \dfrac{11880}{6} = 1980.$

(d) $\dfrac{15!}{12!3!} = \dfrac{15 \cdot 14 \cdot 13 \cdot 12!}{12! \cdot 3 \cdot 2 \cdot 1} = \dfrac{15 \cdot 14 \cdot 13}{3 \cdot 2 \cdot 1} = 455.$

Factorial notation can be used to express the binomial theorem as follows:

$$(a+b)^n = a^n + \dfrac{n!}{(n-1)!1!}a^{n-1}b + \dfrac{n!}{(n-2)!2!}a^{n-2}b^2 + \dfrac{n!}{(n-3)!3!}a^{n-3}b^3$$
$$+ \cdots + \dfrac{n!}{(n-r)!r!}a^{n-r}b^r + \cdots + b^n.$$

Thus, using factorials and the summation notation from section 10.5, we have the following.

Compact Statement of the Binomial Theorem

$$(a+b)^n = \sum_{r=0}^{n} \dfrac{n!}{(n-r)!r!} a^{n-r} b^r.$$

If you study the terms in this expansion, you will notice in each case that the numerator of the coefficient is $n!$ and the denominator is the product of the factorials of the exponents in that term.

Sometimes you may need to find a certain term, say the mth term, in the binomial expansion of $(a+b)^n$. Now, b occurs in the second term, b^2 in the third term, b^3 in the fourth term, and so on. In general, b^{m-1} occurs in the mth term. In fact, we can state the following.

mth Term of Binomial Expansion

The mth term of $(a+b)^n$ is $\dfrac{n!}{(n-m+1)!(m-1)!} a^{n-m+1} b^{m-1}$.

The next two examples illustrate this.

▶ **EXAMPLE 10.47**

Find the fourth term in the expansion of $(a+b)^{12}$.

Solution

The exponent of b in the fourth term is 3, and so the exponent of a in this term is $12 - 3 = 9$. Thus, the fourth term of $(a+b)^{12}$ is

$$\dfrac{12!}{9!3!} a^9 b^3 = 220 a^9 b^3. \qquad ◀$$

▶ **EXAMPLE 10.48**

Determine the sixth term in the expansion of $(x - 2y)^9$.

Solution

Here, $a = x$ and $b = -2y$. The exponent of b in the sixth term is 5, and so the exponent of a in this term is $9 - 5 = 4$. Thus, we see that the sixth term is

$$\dfrac{9!}{4!5!} a^4 b^5 = \dfrac{9!}{4!5!} x^4 (-2y)^5$$

$$= 126x^4(-32y^5)$$
$$= -4032x^4y^5.$$

EXAMPLE 10.49

Find the term in the expansion of $(x^3 + \frac{1}{2}y^2)^{12}$ that involves x^{12}.

Solution

We are considering $(a + b)^{12}$, with $a = x^3$ and $b = \frac{1}{2}y^2$.

Since $x^{12} = (x^3)^4$, we seek the term in the expansion of $(a + b)^{12}$ that involves a^4. This term involves b^8, and so is the ninth term. This term is

$$\frac{12!}{4!8!}a^4b^8 = \frac{12!}{4!8!}(x^3)^4\left(\frac{1}{2}y^2\right)^8$$

$$= 495x^{12}(\tfrac{1}{256}y^{16})$$

$$= \tfrac{495}{256}x^{12}y^{16}.$$

Many scientific calculators have a key to provide rapid calculations of $\frac{n!}{(n-r)!r!}$. This key is usually designated by $\boxed{\text{nCr}}$ or $\boxed{\text{Cy,x}}$. If your calculator has a $\boxed{\text{Cy,x}}$ key, then consider $y = n$ and $x = r$. On some algebraic calculators, you should use the keystroke combination $n.rrr$ $\boxed{\text{nCr}}$. On an RPN calculator you would press n $\boxed{\text{ENTER}}$ r $\boxed{\text{Cy,x}}$.

For example, suppose that a calculator is used to evaluate $\frac{12!}{8!4!}$. The keystroke combination on an algebraic calculator would be 12.004 $\boxed{\text{nCr}}$. (As you probably noticed, the two numbers are entered as a single decimal number. The value of r must be entered as a three-digit number to the right of the decimal point. Be careful, because all these decimal places are used to determine r. Thus, if you enter 12.04, the calculator will consider $r = 40$; if you enter 12.4, r will be considered as 400; and if you enter 12.004, r will be 4.) On an RPN calculator, you would press 12 $\boxed{\text{ENTER}}$ 4 $\boxed{\text{Cy,x}}$. In both cases, the result is 495.

As you might expect, some mathematicians use the notation $_nC_r$ to represent $\frac{n!}{(n-r)!r!}$. Others use the notation $\binom{n}{r}$.

Some graphing calculators use the $_nC_r$ notation in certain menus. For example, to calculate $_{12}C_4$ on a Casio fx-7700G, press 12 $\boxed{\text{SHIFT}}$ $\boxed{\text{MATH}}$ $\boxed{\text{F2}}$ $\boxed{\text{F3}}$ 4 $\boxed{\text{EXE}}$, and the result, 495, appears on the screen. Similarly, to calculate $_{12}C_4$ on a TI-81, press 12 $\boxed{\text{MATH}}$ $\boxed{\blacktriangleright}$ $\boxed{\blacktriangleright}$ $\boxed{\blacktriangleright}$ 3 4 $\boxed{\text{ENTER}}$.

Exercise 50 will indicate a combinatorial interpretation for the symbol $_nC_r$.

The rest of this section is devoted to some theoretical matters. The first of these is a proof of the binomial theorem. It will be convenient to separate the following preliminary result. Essentially, it validates the construction of Pascal's Triangle. A combinatorial proof of this result is indicated in enrichment exercise 56.

Binomial Theorem

Proposition: If r and k are positive integers and $r \leq k$, then

$$\binom{k}{r} + \binom{k}{r-1} = \binom{k+1}{r}.$$

Proof: We simplify the left-hand side to the right-hand side, by repeated use of the formula $n! = n(n-1)!$ as follows:

$$\binom{k}{r} + \binom{k}{r-1} = \frac{k!}{(k-r)!r!} + \frac{k!}{(k-r+1)!(r-1)!}$$

$$= \frac{(k+1-r) \cdot k!}{(k+1-r) \cdot (k-r)!r!} + \frac{r \cdot k!}{(k-r+1)!r \cdot (r-1)!}$$

$$= \frac{(k+1-r) \cdot k!}{(k+1-r)!r!} + \frac{r \cdot k!}{(k+1-r)!r!}$$

$$= \frac{[(k+1-r)+r] \cdot k!}{(k+1-r)!r!} = \frac{(k+1) \cdot k!}{(k+1-r)!r!}$$

$$= \frac{(k+1)!}{(k+1-r)!r!} = \binom{k+1}{r}.$$

This completes the proof of the proposition.

Proof of the Binomial Theorem: We shall proceed by mathematical induction. Since $(a+b)^1 = a+b$, the induction basis is evident.

For the induction step, we assume

$$(a+b)^k = \sum_{r=0}^{k} \binom{k}{r} a^{k-r} b^r$$

and must show that

$$(a+b)^{k+1} = \sum_{r=0}^{k+1} \binom{k+1}{r} a^{k+1-r} b^r. \tag{*}$$

Now,

$$(a+b)^{k+1} = (a+b)^k (a+b) = \left(\sum_{r=0}^{k} \binom{k}{r} a^{k-r} b^r \right)(a+b)$$

which, by the generalized distributive law, is

$$\sum_{r=0}^{k} \binom{k}{r} a^{k+1-r} b^r + \sum_{r=0}^{k} \binom{k}{r} a^{k-r} b^{r+1}. \tag{**}$$

By taking a new index of summation defined by $s = r+1$, we may rewrite the second summation in (**) as

$$\sum_{s=1}^{k+1} \frac{k!}{(k+1-s)!(s-1)!} a^{k+1-s} b^s = \left(\sum_{t=1}^{k} \binom{k}{t-1} a^{k+1-t} b^t \right) + b^{k+1}.$$

Rewriting the first summation in (**) as $a^{k+1} + \sum_{t=1}^{k} \binom{k}{t} a^{k+1-t} b^t$, we may now rewrite (**) as

$$a^{k+1} + \sum_{t=1}^{k} \binom{k}{t} a^{k+1-t} b^t + \sum_{t=1}^{k} \binom{k}{t-1} a^{k+1-t} b^t + b^{k+1}.$$

By combining like terms, we reduce this to

$$a^{k+1} + \sum_{t=1}^{k} \left(\binom{k}{t} + \binom{k}{t-1} \right) a^{k+1-t} b^t + b^{k+1}. \qquad (***)$$

However, the above proposition gives

$$\binom{k}{t} + \binom{k}{t-1} = \binom{k+1}{t}.$$

Substituting this into (***) yields

$$(a+b)^{k+1} = a^{k+1} + \sum_{t=1}^{k} \binom{k+1}{t} a^{k+1-t} b^t + b^{k+1}$$

$$= \sum_{t=0}^{k+1} \binom{k+1}{t} a^{k+1-t} b^t$$

which, after changing the "dummy variable" t to r, is just (*). This establishes the induction step, and by the principle of mathematical induction, this completes the proof of the binomial theorem.

The section's last topic concerns the natural constant e. It was stated in section 4.1 that as n increases without bound, the limiting tendency of the sequence $\left(1 + \frac{1}{n}\right)^n$ is e. A proof of this would take us too far afield, and so it is deferred to a course on calculus. However, as our last example, we shall show how to use this fact, in conjunction with the binomial theorem, to prove that $e \leq 3$.

▶ **EXAMPLE 10.50**

Prove that $e \leq 3$.

Solution

Using the binomial theorem, we obtain

$$\left(1 + \frac{1}{n}\right)^n$$

$$= 1 + n\left(\frac{1}{n}\right) + \frac{n(n-1)}{2}\left(\frac{1}{n}\right)^2 + \frac{n(n-1)(n-2)}{2 \cdot 3}\left(\frac{1}{n}\right)^3 + \cdots + \frac{n!}{n!}\left(\frac{1}{n}\right)^n$$

$$= 1 + 1 + \frac{n(n-1)}{n \cdot n}\left(\frac{1}{2}\right) + \frac{n(n-1)(n-2)}{n \cdot n \cdot n}\left(\frac{1}{2 \cdot 3}\right) + \cdots + \frac{n(n-1)\cdots 1}{n \cdot n \cdots n}\left(\frac{1}{n!}\right)$$

$$< 1 + 1 + \frac{1}{2} + \frac{1}{2 \cdot 3} + \cdots + \frac{1}{2 \cdot 3 \cdots n}$$

$$< 1 + 1 + \frac{1}{2} + \frac{1}{2 \cdot 2} + \cdots + \frac{1}{2 \cdot 2 \cdots 2}$$

$$= 1 + \sum_{i=1}^{n} 1\left(\frac{1}{2}\right)^{i-1}$$

By the formula developed in example 10.40 for the sum of the first n terms of a geometric sequence, this summation is

$$\sum_{i=1}^{n} 1\left(\frac{1}{2}\right)^{i-1} = \frac{1\left(1 - \left(\frac{1}{2}\right)^n\right)}{1 - \frac{1}{2}} = 2\left(1 - \left(\frac{1}{2}\right)^n\right) < 2 \cdot 1 = 2.$$

Hence, $\left(1 + \frac{1}{n}\right)^n < 1 + 2 = 3$ for all positive integers n. By considering the limiting tendency as n increases without bound, we may conclude $e \leq 3$, as asserted. ◄

Any student who has reached this point (and worked the exercises for this section!) has seen, with ample opportunity to use and explore, all the concepts and tools that are needed to begin the study of calculus. Thus ends

PRECALCULUS

Exercises 10.6

1. Write the line of Pascal's Triangle corresponding to $(a + b)^8$.
2. Write the line of Pascal's Triangle corresponding to $(a + b)^9$.
3. Write the expansion of $(a + b)^8$.
4. Write the expansion of $(a + b)^9$.
5. Determine the numerical coefficient of the $a^3 b^9$ term in the expansion of $(a + b)^{12}$.
6. Determine the numerical coefficient of the $a^6 b^5$ term in the expansion of $(a + b)^{11}$.

In exercises 7–14, use the binomial theorem to expand the given expression, and then algebraically simplify the terms of the expansion.

7. $(x + 2)^4$
8. $(x - 3)^5$
9. $(2x - 5)^7$
10. $(\sqrt{2}x + 1)^6$
11. $(\pi x - 3)^5$
12. $\left(\frac{2}{x} + x^2\right)^3$ (Hint: $a = 2x^{-1}$ and $b = x^2$.)
13. $\left(x^5 - \frac{4}{\sqrt{y}}\right)^3$ (Hint: $a = x^5$ and $b = 4y^{-1/2}$.)

14. $\left(3y - \dfrac{2}{x}\right)^4$

In exercises 15–16, simplify the given expression.

15. $(x+h)^3 - x^3$

16. $\dfrac{(x+h)^5 - x^5}{h}$ for $h \neq 0$

In exercises 17–26, calculate the given expression without using a calculator.

17. $5!$

18. $\dfrac{18!}{16!}$

19. $\dfrac{7!}{3!}$

20. $\dfrac{0!}{2!3!}$

21. $\dfrac{14!}{11!3!}$

22. $\dfrac{11!}{5!6!}$

23. $_{18}C_{16}$

24. $_9C_3$

25. $\binom{8}{6}$

26. $\binom{12}{5}$

27. Determine the fifth term in the expansion of $(a+b)^{15}$.

28. Determine the sixth term in the expansion of $(3x-2y)^9$.

29. Determine the fourth term in the expansion of
$$\left(\dfrac{3}{x} - 5x\right)^{12}$$

30. Determine the seventh term in the expansion of $(a+b)^{12}$

In exercises 31–36, find the term in the expansion of the given expression that involves the given power of x.

31. $(2+x)^{90}$; x^4

32. $(2x-3)^{60}$; x^{57}

33. $(3x-2)^{27}$; x^{24}

34. $\left(4\sqrt{x} - \dfrac{5}{\sqrt{x}}\right)^{16}$; x^{-4} (Hint: use laws of exponents to reduce the problem to finding m such that $\dfrac{16-m}{2} + \dfrac{m}{-2} = -4$.)

35. $(-x^2 + 3y)^{16}$; x^{24}

36. $\left(\dfrac{4}{x} + 3x\right)^8$; x^0

37. Find, but do not simplify algebraically, the term in the expansion of $(x^{2/3}y - y^2x^{-1/4})^{22}$ that does not involve x.

38. If r and n are positive integers and $r \leq n$, prove that $_nC_{n-r} = {_nC_r}$.

39. (Compare with example 6.56 of section 6.7.)
(a) Use the binomial theorem to simplify $(\cos\theta + i\sin\theta)^9$.
(b) Use (a) and De Moivre's Theorem to prove the trigonometric identity
$$\cos 9\theta = \cos^9\theta - 36\cos^7\theta\sin^2\theta + 126\cos^5\theta\sin^4\theta - 84\cos^3\theta\sin^6\theta + 9\cos\theta\sin^8\theta.$$

40. Use the binomial theorem and De Moivre's Theorem to prove the trigonometric identity
$$\sin 4\theta = 4\cos^3\theta\sin\theta - 4\cos\theta\sin^3\theta.$$

41. (a) Write down and add up the first four terms in the expansion of $(1 + 0.01)^{10}$.
(b) Use your calculator to approximate $(1.01)^{10}$, and hence show that the sum in (a) is an approximation accurate to the hundred-thousandths place.

42. (a) Write down and add up the first four terms in the expansion of $(1 + 0.01)^{100}$.
(b) Using your calculator, show that the sum in (a) is an approximation accurate to the tenths, but *not* to the hundredths, place.

43. (a) Show that the matrices $A = \begin{bmatrix} 1 & 0 \\ 0 & 0 \end{bmatrix}$, $B = \begin{bmatrix} 0 & 1 \\ 0 & 0 \end{bmatrix}$ do not satisfy $(A+B)^2 = A^2 + 2AB + B^2$.
(b) Prove that 2×2 matrices C and D satisfy $(C+D)^2 = C^2 + 2CD + D^2$ if and only if $CD = DC$. (Hint: what *is* $(C+D)^2$ equal to if you carefully expand using the distributive law?)

Comment: by the proof of the binomial theorem, if A and B are 2×2 matrices such that $AB = BA$, then for *all* positive integers n, $(A+B)^n$ is given by $\sum_{r=0}^{n} \binom{n}{r} A^{n-r} B^r$.

44. There is only one place in the text's proof of the binomial theorem that (implicitly) used the assumption that $ab = ba$. Find it.

Note: exercises 45–56 provide enrichment material on combinatorics that illuminates the meaning of the $_nC_r$ symbol and leads to a natural proof of the binomial theorem. This additional material should be of help in studying discrete mathematics later. However, students or instructors who pursue exercises 45–54 should be warned to allocate extra time for this material.

45. Let $A = [a_{ij}]$ be an $m \times n$ matrix with distinct entries. How many entries does A have in the following situations?
(a) $m = 4, n = 3$ (b) $n = 2, m$ unspecified
(c) $m = 3, n$ unspecified (d) m, n both unspecified

46. (A fundamental counting principle) Natalie Attired owns 5 sweaters, 4 pairs of jeans, and 7 pairs of earrings.
(a) Ignoring questions of style, how many outfits can Natalie select consisting of one sweater and one pair of jeans? (Hint: one way to proceed is to let the outfits be the entries of a 5×4 matrix, with each row corresponding to one of the sweaters that

Natalie owns and each column corresponding to one of her pairs of jeans. Then reason as in exercise 45.)

(b) How many outfits can Natalie create if, in addition to choosing a sweater and a pair of jeans, she must also select a pair of earrings? (Hint: now, the outfits from (a) correspond to the rows of a new matrix and the pairs of earrings correspond to the columns of this new matrix.)

47. Natalie nattily visits the local Humble Hamburger Hut. She must choose which, if any, of the following her hamburger will have: pickle, lettuce, cheese, mayonnaise. (The hamburger may have all, some, or none of these items.) In how many different ways can Natalie order her hamburger at HHH? (Hint: reason as in exercise 46. First she decides whether or not she'll have any pickles; then she decides whether or not to have lettuce; and so on.)

48. (a) Marla, Elaine, Dwala, John, and David run a 100-meter dash. In how many different ways might the first-, second-, and third-place ribbons be given out? (Hint: first you determine the winner. After you do so, there are four individuals who might still finish second.)

(b) (A generalization) Suppose you have a set of n objects and wish to select an ordered subset of r objects, designating one of the selected objects as the "first," another as the "second," and so on. Show that the number of ways in which you may choose the ordered subset is

$$n(n-1)(n-2)\cdots(n-r+1) = \frac{n!}{(n-r)!}.$$

(c) Interpret the formula in (b) in case $r = n$.

49. From among fifty faculty members at Tohoot University, the Dean must form a four-person committee consisting of a Chairperson, a Parliamentarian, a Recording Secretary, and a Treasurer. In how many ways may this committee be chosen? (Hint: exercise 48(b).)

◆ **50.** (Why $_nC_r$ is a positive integer) Suppose you have a set of n objects and wish to select an unordered subset of r objects. Show that the number of different such subsets that you may choose is $_nC_r$. (Hint: according to exercise 48(b) and (c), $\frac{n!}{(n-r)!}$ is the result if each of the subsets is counted $r!$ times.)

51. The Egalitarian Society of Tohoot University convinces the Dean that the four-person committee that was selected in exercise 49 is to be replaced with a five-person committee in which none of the members has a designated role or title. Assuming that all fifty faculty members are eligible for selection, in how many ways may the Dean choose the new committee? (Hint: apply exercise 50, after determining n and r.)

52. Sean wishes to make a bouquet consisting of six flowers. He visits a florist near closing time, and finds only nine flowers available. In how many ways may Sean choose a suitable bouquet from this florist's flowers? (Hint: see the hint for exercise 51.)

53. (a) Morag wishes to make a bouquet consisting of four roses and two snapdragons. She visits a florist who has only five roses and four snapdragons. In how many ways may Morag choose a suitable bouquet from this florist's flowers? (Hint: there are three aspects to this problem. To count the number of ways to choose the roses in the bouquet, reason as in exercises 50–52; proceed similarly for the choice of the snapdragons. To put all this together and answer the problem, reason as in exercises 45–49.)

(b) Generalize (a) to the case in which Morag's bouquet is to consist of r roses and s snapdragons, while the florist's stock consists of n roses and m snapdragons.

54. Use exercise 50, in conjunction with the generalized distributive law, to obtain a new proof of the binomial theorem. (Hint: before combining like terms, we find an expansion of $(a + b)^n = (a + b)(a + b)\cdots(a + b)$ consisting of 2^n terms. For each r between 1 and n, exactly $_nC_r$ of these 2^n terms equal $a^{n-r}b^r$.)

55. The Textbook Committee decides to consider fifty candidates as possible texts for the Precalculus course at Tohoot University. One of these fifty texts is Dobbs-Peterson. The committee's first task is to determine a set of six finalists.

(a) How many sets of six finalists can be chosen from the fifty texts?

(b) How many sets of six finalists can be chosen if Dobbs-Peterson is definitely one of the finalists?

(c) How many sets of six finalists can be chosen if Dobbs-Peterson is definitely out of the running?

(d) Explain intuitively why the sum of your answers in (b) and (c) is your answer in (a).

◆ **56.** (Dobbs-Peterson leads to another generalization.) Use exercise 50 to obtain a new proof that if r and k are positive integers such that $r \leq k$, then

$$\binom{k}{r-1} + \binom{k}{r} = \binom{k+1}{r}.$$

(Hint: argue as in exercise 55, with $r = 6$ and $k + 1 = 50$.)

10.7 CHAPTER SUMMARY AND REVIEW EXERCISES

Major Concepts

- Archimedean spiral
- Arithmetic sequence
- Binomial theorem
- Cardioid
- $_nC_r$
- Compact statement of the binomial theorem
- Convert from polar to rectangular coordinates
- Convert from rectangular to polar coordinates
- Crucial values of θ
- Cycloid
- Elimination of parameters
- Expansion of $(a+b)^n$
- Geometric sequence
- Graph of a polar equation
- Guidelines for graphing polar equations
- Hyperbolic spiral
- Index of summation
- Lemniscate
- Limaçon
- Logarithmic spiral
- $n!$
- $\binom{n}{r}$
- n factorial
- Pair of polar coordinates
- Parameter
- Parameterized curve
  - Circle
  - Ellipse
  - Hyperbola
  - Line
  - Parabola
- Parametric equations
- Pascal's Triangle
- Polar axis
- Polar equation
  - of a circle
  - of a line not containing the pole
  - of a line through the pole
- Polar equations of conic sections
- Pole
- Principle of mathematical induction
- Product notation
- (r, θ)
- Rose
- Steps to follow when trying to prove a proposition by mathematical induction
- Summation notation
- Symmetry with respect to the $\frac{\pi}{2}$-line
- Symmetry with respect to the polar axis
- Symmetry with respect to the pole

Review Exercises

Exercises 35–40 refer to section 10.3, which is an optional section. Exercise 72(c) assumes familiarity with the enrichment material on limits of sequences in exercises 51–60 of section 10.5. Exercise 80 relates to the enrichment material on combinatorics discussed in exercises 45–56 of section 10.6.

In exercises 1–2, plot the point with the given pair of polar coordinates and then find the Cartesian coordinates of this point.

1. $(-1, \pi)$

2. $\left(\frac{1}{2}, \frac{10\pi}{3}\right)$

In exercises 3–4, find two "essentially different" pairs of polar coordinates (r, θ) [one with $r > 0$, the other with $r < 0$] for the point with the given Cartesian coordinates.

3. $(-4, 4\sqrt{3})$

4. $(\sqrt{2}, 0)$

In exercises 5–6, determine which conclusions about symmetry of the polar graph of the given polar equation follow from the symmetry criteria in the text.

5. $r^3 \cos\theta = 6$

6. $r^2 \theta^4 = 2\pi$

In exercises 7–8, verify that the point with the given polar coordinates is on the polar graph of the given polar equation.

7. $\left(-2, \frac{\pi}{3}\right)$; $\sqrt{3}\, r = 2\tan\theta$

8. $\left(-\frac{\sqrt{3}}{2}, \frac{\pi}{3}\right)$; $r = \cos\frac{\theta}{2}$

In exercises 9–27, identify and sketch the polar graph of the given polar equation.

9. $r = 3$

10. $r = -4$

11. $\theta = -\frac{4\pi}{3}$

12. $r = 4\sin\theta$

13. $r = 6\cos\theta$

14. $r^2 - 8r\cos\theta + 6r\sin\theta = 0$

15. $r\cos\left(\theta - \frac{\pi}{4}\right) = 2$

16. $2r = 3\sec\theta$

17. $r = 4(1 + \cos\theta)$

18. $r = 4\sin 5\theta$

19. $r = \frac{1}{2}\cos 2\theta$

20. $r = 2 - \sqrt{3}\sin\left(\theta + \frac{\pi}{6}\right)$

21. $r = \sqrt{3} - 2\sin\theta$

22. $r = \frac{3}{1 - \cos\theta}$

23. $r = \frac{2}{2 + 3\sin\theta}$

24. $r = \frac{3\csc\theta}{3\csc\theta - 2}$

25. $2r\theta = 1$

26. $r^2 = 16\cos 4\theta$

27. $r = \sin\frac{\theta}{2}$ (Hint: the answer is *not* a cardioid.)

Chapter Summary and Review Exercises

In exercises 28–30, convert the given equation in rectangular coordinates into a polar equation having the same graph.

28. $3x - 4y + 7 = 0$
29. $4x^2 + 9y^2 = 36$
30. $x^2 + y^2 = 4$

In exercises 31–34, find a polar equation for the given graph.

31. The circle with radius $\sqrt{5}$ and center having polar coordinates $\left(-2, \dfrac{5\pi}{6}\right)$
32. The line ℓ which passes through the point N with polar coordinates $\left(2, \dfrac{5\pi}{4}\right)$ such that ℓ is perpendicular to $\overleftrightarrow{ON}$
33. The ellipse with eccentricity $\dfrac{3}{5}$, focus at the pole, and corresponding directrix $r\cos\theta = -8$
34. The hyperbola with focus at the pole and vertices having polar coordinates $(3, \pi)$ and $(-7, 0)$

In exercises 35–40, sketch the polar graphs of the two given polar equations, and then find all their points of intersection.

35. $r = 2\sin\theta$, $r = -2\cos\theta$
36. $r = -2$, $r = 2\sin 3\theta$
37. $r = 2 - 2\sin\theta$, $r = 2$
38. $r = \dfrac{1}{1 - \cos\theta}$, $\theta = \dfrac{\pi}{3}$
39. $r = 1 + 2\sin\theta$, $r\cos\theta = 2$
40. $r = 2\cos\theta$, $r\sin\theta = -1$

41. Write a pair of parametric equations for the graph of the equation $x = 2y^3 + \dfrac{1}{3}y^2 + \sqrt{2}y - \pi$.
42. Sketch the parameterized curve given by $x = 3\cos 2\theta \cos\theta$, $y = 3\cos 2\theta \sin\theta$, $0 \le \theta < \dfrac{\pi}{2}$.

In exercises 43–50, eliminate the parameter in order to find a Cartesian equation of a curve which contains the given parameterized curve. Then sketch the parameterized curve.

43. $x = 3t^2 - 2$, $y = 5t$, $-1 \le t \le 2$
44. $x = \sin^2\theta$, $y = \cos^2\theta$
45. $x = \sqrt{t}$, $y = 2t$
46. $x = -3 + 5\cos\theta$, $y = -5 + 3\sin\theta$, $0 < \theta \le \pi$
47. $x = e^t + 1$, $y = e^t - 1$
48. $x = -5 + 3\sec\theta$, $y = -3 + 5\tan\theta$
49. $x = \cos\dfrac{t}{2}$, $y = \cos t$
50. $x = 1 + 2\theta$, $y = 3 + 4\theta$, $-2 \le \theta \le 1$

In exercises 51–56, give a pair of parametric equations which parameterizes the given graph. Find a domain I so that each point on the graph arises from only one parameter value in I.

51. Circle, center (e, π), radius $\sqrt{3}$
52. North-south ellipse, center $(4, -5)$, $a = 3$, $b = 2$
53. East-west hyperbola, center $(4, -5)$, $a = 3$, $b = 2$
54. Parabola, vertex $(-5, 4)$, opening upward, passing through $(0, 10)$
55. Line passing through $(-3, 2)$ and $(\sqrt{2} - 3, 2 - 3\sqrt{2})$
56. The line $\sqrt{7}x + \sqrt{6}y + \sqrt{5} = 0$

In exercises 57–59, determine the given sum.

57. $\displaystyle\sum_{i=1}^{4}\left(2 - \dfrac{3}{i}\right)$
58. $\displaystyle\sum_{j=2}^{5} 3j^2$
59. $3 - 6 + 12 - 24 + \cdots + 3(-2)^{n-1}$

In exercises 60–65, use mathematical induction to prove the given assertion for all positive integers n.

60. $\displaystyle\sum_{i=1}^{n}(-3i + 4) = \dfrac{n(5 - 3n)}{2}$
61. $\displaystyle\sum_{j=1}^{n}(12j^2 - j) = \dfrac{n(n+1)(8n+3)}{2}$
62. $10n + 10 < (n+4)^2$
63. $2^{3n} - 1$ is an integral multiple of 7.
64. $\sin n\pi = 0$ (Hint: $\sin(k+1)\pi = \sin(k\pi + \pi)$ can be expanded via a formula in section 6.2.)
65. $\cos n\pi = \begin{cases} 1 & \text{if } n \text{ is even} \\ -1 & \text{if } n \text{ is odd}\end{cases}$ (Hint: see exercise 64.)

66. Prove that $3n < 2^n$ for each positive integer $n \ge 4$. (Hint: $2^{k+1} - 2^k = 2^k$.)
67. Determine whether the following sequences are arithmetic, geometric, or neither. If arithmetic, calculate d; if geometric, calculate r.
    (a) $4, 2, 1, \dfrac{1}{2}, \ldots, 4\left(\dfrac{1}{2}\right)^{n-1}, \ldots$
    (b) $2, 16, 162, \ldots, 2n^3, \ldots$
    (c) $3, 1, -1, \ldots, 3 - 2(n - 1), \ldots$
    (d) $-2, 3, 8, \ldots, -2 + 5(n - 1), \ldots$
    (e) $-4, 6, -9, \ldots, -4\left(-\dfrac{3}{2}\right)^{n-1}, \ldots$
    (f) $2, 3, 5, \ldots, 2^{n-1} + 1, \ldots$

68. A new professor is hired and offered a choice. Either (I) her initial salary will be $30,000 per year, with annual raises of $1,500, or (II) her initial salary will be $25,000, with annual raises of 10 percent. Determine which option she should choose if she wishes to have
    (a) the higher of the possible salaries at the start of her fifth year of employment
    (b) earned the most money possible during her first four years of employment

69. The seating capacity of a Precalculus classroom at Tohoot University is not sufficient to meet the demand. Two seats are therefore added to the first row, four seats are added to the second row, six seats are added to the third row, and so on.

(a) If there are seven rows of seats, how many seats have been added?
(b) How many seats have been added if the classroom is really one side of a stadium with forty rows of seats?

70. Suppose that an arithmetic sequence has first term 4 and the tenth term –2. What is the sum of the first twelve terms of this arithmetic sequence?

71. Suppose that a geometric sequence has the second term $\frac{3}{16}$ and the fifth term $\frac{1}{18}$. Find the ninth term of this geometric sequence.

72. It is known that if a certain ball is dropped onto a certain court from a height h, then it rebounds to a height $0.43h$. Suppose that this ball is dropped onto this court from a height of 6 feet and you watch it bounce repeatedly.
    (a) How far does the ball rise after its first bounce before it starts to fall again?
    (b) At the moment of its nth bounce ($n \geq 2$), how far has the ball traveled?
    (c) Show that the ball travels $\frac{286}{19}$ feet in all.

73. (a) Write down the line of Pascal's Triangle corresponding to $(a + b)^7$.
    (b) Using (a), write the expansion of $(a + b)^7$.
    (c) Using the binomial theorem, expand $(a + b)^7$.

74. Determine the numerical coefficient of
    (a) the $a^4 b^7$ term in the expansion of $(a + b)^{11}$
    (b) the x^{24} term in the expansion of $(x^{3/2} - x^{-1/2})^{40}$

    (Express it as a symbol of the form $\binom{n}{r}$.)

In exercises 75–76, use the binomial theorem to expand the given expression, and then algebraically simplify the terms of the expansion.

75. $(x + 2)^5$
76. $(\sqrt{2}x - 4)^3$

77. Simplify $\dfrac{(x + h)^4 - x^4}{h}$ for $h \neq 0$.

78. Calculate the following expressions without using a calculator:

    (a) $6!$ (b) $\dfrac{14!}{3!12!}$ (c) $_8C_3$ (d) $\binom{11}{4}$

79. (a) Write down and add up the first three terms in the expansion of $(1 + .02)^5$.
    (b) Use your calculator to approximate $(1.02)^5$, and hence show that the sum in (a) is an approximation accurate to the thousandths place.
    (c) How many terms of the expansion are needed to give a sum that approximates correctly to at least the hundred-thousandths place?

80. Dobbs and Peterson wish to send a bouquet consisting of six roses and six forget-me-nots to their publisher. They visit a florist who has ten roses and fourteen forget-me-nots. In how many ways may the authors choose a suitable bouquet from this florist's flowers? (To the reader: if you wish a bouquet of thanks, please let us know your reactions to this book. Good luck with calculus!)

Appendix I / Powers of e

x	e^x
0.00	1.0000
0.05	1.0513
0.10	1.1052
0.15	1.1618
0.20	1.2214
0.25	1.2840
0.30	1.3499
0.35	1.4191
0.40	1.4918
0.45	1.5683
0.50	1.6487
0.55	1.7333
0.60	1.8221
0.65	1.9155
0.70	2.0138
0.75	2.1170
0.80	2.2255
0.85	2.3396
0.90	2.4596
0.95	2.5857
1.00	2.7183
1.05	2.8577
1.10	3.0042
1.15	3.1582
1.20	3.3201
1.25	3.4903
1.30	3.6693
1.35	3.8574
1.40	4.0552
1.45	4.2631
1.50	4.4817
1.55	4.7115
1.60	4.9530
1.65	5.2070
1.70	5.4739
1.75	5.7546
1.80	6.0496
1.85	6.3598
1.90	6.6859
1.95	7.0287

x	e^x
2.00	7.3891
2.05	7.7679
2.10	8.1662
2.15	8.5849
2.20	9.0250
2.25	9.4877
2.30	9.9742
2.35	10.4856
2.40	11.0232
2.45	11.5883
2.50	12.1825
2.55	12.8071
2.60	13.4637
2.65	14.1540
2.70	14.8797
2.75	15.6426
2.80	16.4446
2.85	17.2878
2.90	18.1741
2.95	19.1060
3.00	20.0855
3.05	21.1153
3.10	22.1980
3.15	23.3361
3.20	24.5325
3.25	25.7903
3.30	27.1126
3.35	28.5027
3.40	29.9641
3.45	31.5004
3.50	33.1155
3.55	34.8133
3.60	36.5982
3.65	38.4747
3.70	40.4473
3.75	42.5211
3.80	44.7012
3.85	46.9931
3.90	49.4024
3.95	51.9354

x	e^x
4.00	54.5982
4.05	57.3975
4.10	60.3403
4.15	63.4340
4.20	66.6863
4.25	70.1054
4.30	73.6998
4.35	77.4785
4.40	81.4509
4.45	85.6269
4.50	90.0171
4.55	94.6324
4.60	99.4843
4.65	104.5850
4.70	109.9472
4.75	115.5843
4.80	121.5104
4.85	127.7404
4.90	134.2898
4.95	141.1750
5.00	148.4132
5.50	244.6919
6.00	403.4288
6.50	665.1416
7.00	1,096.6332
7.50	1,808.0424
8.00	2,980.9580
8.50	4,914.7688
9.00	8,103.0839
9.50	13,359.7268
10.00	22,026.4658

Appendix II / Common Logarithms

N	0	1	2	3	4	5	6	7	8	9
1.0	.0000	.0043	.0086	.0128	.0170	.0212	.0253	.0294	.0334	.0374
1.1	.0414	.0453	.0492	.0531	.0569	.0607	.0645	.0682	.0719	.0755
1.2	.0792	.0828	.0864	.0899	.0934	.0969	.1004	.1038	.1072	.1106
1.3	.1139	.1173	.1206	.1239	.1271	.1303	.1335	.1367	.1399	.1430
1.4	.1461	.1492	.1523	.1553	.1584	.1614	.1644	.1673	.1703	.1732
1.5	.1761	.1790	.1818	.1847	.1875	.1903	.1931	.1959	.1987	.2014
1.6	.2041	.2068	.2095	.2122	.2148	.2175	.2201	.2227	.2253	.2279
1.7	.2304	.2330	.2355	.2380	.2405	.2430	.2455	.2480	.2504	.2529
1.8	.2553	.2577	.2601	.2625	.2648	.2672	.2695	.2718	.2742	.2765
1.9	.2788	.2810	.2833	.2856	.2878	.2900	.2923	.2945	.2967	.2989
2.0	.3010	.3032	.3054	.3075	.3096	.3118	.3139	.3160	.3181	.3201
2.1	.3222	.3243	.3263	.3284	.3304	.3324	.3345	.3365	.3385	.3404
2.2	.3424	.3444	.3464	.3483	.3502	.3522	.3541	.3560	.3579	.3598
2.3	.3617	.3636	.3655	.3674	.3692	.3711	.3729	.3747	.3766	.3784
2.4	.3802	.3820	.3838	.3856	.3874	.3892	.3909	.3927	.3945	.3962
2.5	.3979	.3997	.4014	.4031	.4048	.4065	.4082	.4099	.4116	.4133
2.6	.4150	.4166	.4183	.4200	.4216	.4232	.4249	.4265	.4281	.4298
2.7	.4314	.4330	.4346	.4362	.4378	.4393	.4409	.4425	.4440	.4456
2.8	.4472	.4487	.4502	.4518	.4533	.4548	.4564	.4579	.4594	.4609
2.9	.4624	.4639	.4654	.4669	.4683	.4698	.4713	.4728	.4742	.4757
3.0	.4771	.4786	.4800	.4814	.4829	.4843	.4857	.4871	.4886	.4900
3.1	.4914	.4928	.4942	.4955	.4969	.4983	.4997	.5011	.5024	.5038
3.2	.5051	.5065	.5079	.5092	.5105	.5119	.5132	.5145	.5159	.5172
3.3	.5185	.5198	.5211	.5224	.5237	.5250	.5263	.5276	.5289	.5302
3.4	.5315	.5328	.5340	.5353	.5366	.5378	.5391	.5403	.5416	.5428
3.5	.5441	.5453	.5465	.5478	.5490	.5502	.5514	.5527	.5539	.5551
3.6	.5563	.5575	.5587	.5599	.5611	.5623	.5635	.5647	.5658	.5670
3.7	.5682	.5694	.5705	.5717	.5729	.5740	.5752	.5763	.5775	.5786
3.8	.5798	.5809	.5821	.5832	.5843	.5855	.5866	.5877	.5888	.5899
3.9	.5911	.5922	.5933	.5944	.5955	.5966	.5977	.5988	.5999	.6010
4.0	.6021	.6031	.6042	.6053	.6064	.6075	.6085	.6096	.6107	.6117
4.1	.6128	.6138	.6149	.6160	.6170	.6180	.6191	.6201	.6212	.6222
4.2	.6232	.6243	.6253	.6263	.6274	.6284	.6294	.6304	.6314	.6325
4.3	.6335	.6345	.6355	.6365	.6375	.6385	.6395	.6405	.6415	.6425
4.4	.6435	.6444	.6454	.6464	.6474	.6484	.6493	.6503	.6513	.6522
4.5	.6532	.6542	.6551	.6561	.6571	.6580	.6590	.6599	.6609	.6618
4.6	.6628	.6637	.6646	.6656	.6665	.6675	.6684	.6693	.6702	.6712
4.7	.6721	.6730	.6739	.6749	.6758	.6767	.6776	.6785	.6794	.6803
4.8	.6812	.6821	.6830	.6839	.6848	.6857	.6866	.6875	.6884	.6893
4.9	.6902	.6911	.6920	.6928	.6937	.6946	.6955	.6964	.6972	.6981
5.0	.6990	.6998	.7007	.7016	.7024	.7033	.7042	.7050	.7059	.7067
5.1	.7076	.7084	.7093	.7101	.7110	.7118	.7126	.7135	.7143	.7152
5.2	.7160	.7168	.7177	.7185	.7193	.7202	.7210	.7218	.7226	.7235
5.3	.7243	.7251	.7259	.7267	.7275	.7284	.7292	.7300	.7308	.7316
5.4	.7324	.7332	.7340	.7348	.7356	.7364	.7372	.7380	.7388	.7396

Appendix II (continued)

N	0	1	2	3	4	5	6	7	8	9
5.5	.7404	.7412	.7419	.7427	.7435	.7443	.7451	.7459	.7466	.7474
5.6	.7482	.7490	.7497	.7505	.7513	.7520	.7528	.7536	.7543	.7551
5.7	.7559	.7566	.7574	.7582	.7589	.7597	.7604	.7612	.7619	.7627
5.8	.7634	.7642	.7649	.7657	.7664	.7672	.7679	.7686	.7694	.7701
5.9	.7709	.7716	.7723	.7731	.7738	.7745	.7752	.7760	.7767	.7774
6.0	.7782	.7789	.7796	.7803	.7810	.7818	.7825	.7832	.7839	.7846
6.1	.7853	.7860	.7868	.7875	.7882	.7889	.7896	.7903	.7910	.7917
6.2	.7924	.7931	.7938	.7945	.7952	.7959	.7966	.7973	.7980	.7987
6.3	.7993	.8000	.8007	.8014	.8021	.8028	.8035	.8041	.8048	.8055
6.4	.8062	.8069	.8075	.8082	.8089	.8096	.8102	.8109	.8116	.8122
6.5	.8129	.8136	.8142	.8149	.8156	.8162	.8169	.8176	.8182	.8189
6.6	.8195	.8202	.8209	.8215	.8222	.8228	.8235	.8241	.8248	.8254
6.7	.8261	.8267	.8274	.8280	.8287	.8293	.8299	.8306	.8312	.8319
6.8	.8325	.8331	.8338	.8344	.8351	.8357	.8363	.8370	.8376	.8382
6.9	.8388	.8395	.8401	.8407	.8414	.8420	.8426	.8432	.8439	.8445
7.0	.8451	.8457	.8463	.8470	.8476	.8482	.8488	.8494	.8500	.8506
7.1	.8513	.8519	.8525	.8531	.8537	.8543	.8549	.8555	.8561	.8567
7.2	.8573	.8579	.8585	.8591	.8597	.8603	.8609	.8615	.8621	.8627
7.3	.8633	.8639	.8645	.8651	.8657	.8663	.8669	.8675	.8681	.8686
7.4	.8692	.8698	.8704	.8710	.8716	.8722	.8727	.8733	.8739	.8745
7.5	.8751	.8756	.8762	.8768	.8774	.8779	.8785	.8791	.8797	.8802
7.6	.8808	.8814	.8820	.8825	.8831	.8837	.8842	.8848	.8854	.8859
7.7	.8865	.8871	.8876	.8882	.8887	.8893	.8899	.8904	.8910	.8915
7.8	.8921	.8927	.8932	.8938	.8943	.8949	.8954	.8960	.8965	.8971
7.9	.8976	.8982	.8987	.8993	.8998	.9004	.9009	.9015	.9020	.9025
8.0	.9031	.9036	.9042	.9047	.9053	.9058	.9063	.9069	.9074	.9079
8.1	.9085	.9090	.9096	.9101	.9106	.9112	.9117	.9122	.9128	.9133
8.2	.9138	.9143	.9149	.9154	.9159	.9165	.9170	.9175	.9180	.9186
8.3	.9191	.9196	.9201	.9206	.9212	.9217	.9222	.9227	.9232	.9238
8.4	.9243	.9248	.9253	.9258	.9263	.9269	.9274	.9279	.9284	.9289
8.5	.9294	.9299	.9304	.9309	.9315	.9320	.9325	.9330	.9335	.9340
8.6	.9345	.9350	.9355	.9360	.9365	.9370	.9375	.9380	.9385	.9390
8.7	.9395	.9400	.9405	.9410	.9415	.9420	.9425	.9430	.9435	.9440
8.8	.9445	.9450	.9455	.9460	.9465	.9469	.9474	.9479	.9484	.9489
8.9	.9494	.9499	.9504	.9509	.9513	.9518	.9523	.9528	.9533	.9538
9.0	.9542	.9547	.9552	.9557	.9562	.9566	.9571	.9576	.9581	.9586
9.1	.9590	.9595	.9600	.9605	.9609	.9614	.9619	.9624	.9628	.9633
9.2	.9638	.9643	.9647	.9652	.9657	.9661	.9666	.9671	.9675	.9680
9.3	.9685	.9689	.9694	.9699	.9703	.9708	.9713	.9717	.9722	.9727
9.4	.9731	.9736	.9741	.9745	.9750	.9754	.9759	.9763	.9768	.9773
9.5	.9777	.9782	.9786	.9791	.9795	.9800	.9805	.9809	.9814	.9818
9.6	.9823	.9827	.9832	.9836	.9841	.9845	.9850	.9854	.9859	.9863
9.7	.9868	.9872	.9877	.9881	.9886	.9890	.9894	.9899	.9903	.9908
9.8	.9912	.9917	.9921	.9926	.9930	.9934	.9939	.9943	.9948	.9952
9.9	.9956	.9961	.9965	.9969	.9974	.9978	.9983	.9987	.9991	.9996

Use properties of logarithms for other values. For instance, $\log 0.256 = \log 2.56 - \log 10$
$\approx 0.4082 - 1 = -0.5918$.

Appendix III / Natural Logarithms

n	0.0	0.1	0.2	0.3	0.4	0.5	0.6	0.7	0.8	0.9
1	0.0000	0.0953	0.1823	0.2624	0.3365	0.4055	0.4700	0.5306	0.5878	0.6419
2	0.6931	0.7419	0.7885	0.8329	0.8755	0.9163	0.9555	0.9933	1.0296	1.0647
3	1.0986	1.1314	1.1632	1.1939	1.2238	1.2528	1.2809	1.3083	1.3350	1.3610
4	1.3863	1.4110	1.4351	1.4586	1.4816	1.5041	1.5261	1.5476	1.5686	1.5892
5	1.6094	1.6292	1.6487	1.6677	1.6864	1.7047	1.7228	1.7405	1.7579	1.7750
6	1.7918	1.8083	1.8245	1.8405	1.8563	1.8718	1.8871	1.9021	1.9169	1.9315
7	1.9459	1.9601	1.9741	1.9879	2.0015	2.0149	2.0281	2.0412	2.0541	2.0669
8	2.0794	2.0919	2.1041	2.1163	2.1282	2.1401	2.1518	2.1633	2.1748	2.1861
9	2.1972	2.2083	2.2192	2.2300	2.2407	2.2513	2.2618	2.2721	2.2824	2.2925
10	2.3026	2.3125	2.3224	2.3321	2.3418	2.3514	2.3609	2.3702	2.3795	2.3888

Use properties of logarithms for other values. For instance, $\ln 0.25 = \ln 2.5 - \ln 10$
$\approx 0.9163 - 2.3026 = -1.3863$.

Appendix IV Values of Trigonometric Functions—Degree Measure

t degrees	sin t	cos t	tan t	cot t	sec t	csc t	
0.0	0.0000	1.0000	0.0000	—	1.0000	—	90.0
0.1	0.0017	1.0000	0.0017	572.9572	1.0000	572.9581	89.9
0.2	0.0035	1.0000	0.0035	286.4777	1.0000	286.4795	89.8
0.3	0.0052	1.0000	0.0052	190.9842	1.0000	190.9868	89.7
0.4	0.0070	1.0000	0.0070	143.2371	1.0000	143.2406	89.6
0.5	0.0087	1.0000	0.0087	114.5887	1.0000	114.5930	89.5
0.6	0.0105	0.9999	0.0105	95.4895	1.0001	95.4947	89.4
0.7	0.0122	0.9999	0.0122	81.8470	1.0001	81.8531	89.3
0.8	0.0140	0.9999	0.0140	71.6151	1.0001	71.6221	89.2
0.9	0.0157	0.9999	0.0157	63.6567	1.0001	63.6646	89.1
1.0	0.0175	0.9998	0.0175	57.2900	1.0002	57.2987	89.0
1.1	0.0192	0.9998	0.0192	52.0807	1.0002	52.0903	88.9
1.2	0.0209	0.9998	0.0209	47.7395	1.0002	47.7500	88.8
1.3	0.0227	0.9997	0.0227	44.0661	1.0003	44.0775	88.7
1.4	0.0244	0.9997	0.0244	40.9174	1.0003	40.9296	88.6
1.5	0.0262	0.9997	0.0262	38.1885	1.0003	38.2016	88.5
1.6	0.0279	0.9996	0.0279	35.8006	1.0004	35.8145	88.4
1.7	0.0297	0.9996	0.0297	33.6935	1.0004	33.7083	88.3
1.8	0.0314	0.9995	0.0314	31.8205	1.0005	31.8362	88.2
1.9	0.0332	0.9995	0.0332	30.1446	1.0006	30.1612	88.1
2.0	0.0349	0.9994	0.0349	28.6363	1.0006	28.6537	88.0
2.1	0.0366	0.9993	0.0367	27.2715	1.0007	27.2898	87.9
2.2	0.0384	0.9993	0.0384	26.0307	1.0007	26.0499	87.8
2.3	0.0401	0.9992	0.0402	24.8978	1.0008	24.9179	87.7
2.4	0.0419	0.9991	0.0419	23.8593	1.0009	23.8802	87.6
2.5	0.0436	0.9990	0.0437	22.9038	1.0010	22.9256	87.5
2.6	0.0454	0.9990	0.0454	22.0217	1.0010	22.0444	87.4
2.7	0.0471	0.9989	0.0472	21.2049	1.0011	21.2285	87.3
2.8	0.0488	0.9988	0.0489	20.4465	1.0012	20.4709	87.2
2.9	0.0506	0.9987	0.0507	19.7403	1.0013	19.7656	87.1
3.0	0.0523	0.9986	0.0524	19.0811	1.0014	19.1073	87.0
3.1	0.0541	0.9985	0.0542	18.4645	1.0015	18.4915	86.9
3.2	0.0558	0.9984	0.0559	17.8863	1.0016	17.9142	86.8
3.3	0.0576	0.9983	0.0577	17.3432	1.0017	17.3720	86.7
3.4	0.0593	0.9982	0.0594	16.8319	1.0018	16.8616	86.6
3.5	0.0610	0.9981	0.0612	16.3499	1.0019	16.3804	86.5
3.6	0.0628	0.9980	0.0629	15.8945	1.0020	15.9260	86.4
3.7	0.0645	0.9979	0.0647	15.4638	1.0021	15.4961	86.3
3.8	0.0663	0.9978	0.0664	15.0557	1.0022	15.0889	86.2
3.9	0.0680	0.9977	0.0682	14.6685	1.0023	14.7026	86.1
	cos t	sin t	cot t	tan t	csc t	sec t	t degrees

Appendix IV (continued)

t degrees	sin t	cos t	tan t	cot t	sec t	csc t	
4.0	0.0698	0.9976	0.0699	14.3007	1.0024	14.3356	86.0
4.1	0.0715	0.9974	0.0717	13.9507	1.0026	13.9865	85.9
4.2	0.0732	0.9973	0.0734	13.6174	1.0027	13.6541	85.8
4.3	0.0750	0.9972	0.0752	13.2996	1.0028	13.3371	85.7
4.4	0.0767	0.9971	0.0769	12.9962	1.0030	13.0346	85.6
4.5	0.0785	0.9969	0.0787	12.7062	1.0031	12.7455	85.5
4.6	0.0802	0.9968	0.0805	12.4288	1.0032	12.4690	85.4
4.7	0.0819	0.9966	0.0822	12.1632	1.0034	12.2043	85.3
4.8	0.0837	0.9965	0.0840	11.9087	1.0035	11.9506	85.2
4.9	0.0854	0.9963	0.0857	11.6645	1.0037	11.7073	85.1
5.0	0.0872	0.9962	0.0875	11.4301	1.0038	11.4737	85.0
5.1	0.0889	0.9960	0.0892	11.2048	1.0040	11.2493	84.9
5.2	0.0906	0.9959	0.0910	10.9882	1.0041	11.0336	84.8
5.3	0.0924	0.9957	0.0928	10.7797	1.0043	10.8260	84.7
5.4	0.0941	0.9956	0.0945	10.5789	1.0045	10.6261	84.6
5.5	0.0958	0.9954	0.0963	10.3854	1.0046	10.4334	84.5
5.6	0.0976	0.9952	0.0981	10.1988	1.0048	10.2477	84.4
5.7	0.0993	0.9951	0.0998	10.0187	1.0050	10.0685	84.3
5.8	0.1011	0.9949	0.1016	9.8448	1.0051	9.8955	84.2
5.9	0.1028	0.9947	0.1033	9.6768	1.0053	9.7283	84.1
6.0	0.1045	0.9945	0.1051	9.5144	1.0055	9.5668	84.0
6.1	0.1063	0.9943	0.1069	9.3572	1.0057	9.4105	83.9
6.2	0.1080	0.9942	0.1086	9.2052	1.0059	9.2593	83.8
6.3	0.1097	0.9940	0.1104	9.0579	1.0061	9.1129	83.7
6.4	0.1115	0.9938	0.1122	8.9152	1.0063	8.9711	83.6
6.5	0.1132	0.9936	0.1139	8.7769	1.0065	8.8337	83.5
6.6	0.1149	0.9934	0.1157	8.6427	1.0067	8.7004	83.4
6.7	0.1167	0.9932	0.1175	8.5126	1.0069	8.5711	83.3
6.8	0.1184	0.9930	0.1192	8.3863	1.0071	8.4457	83.2
6.9	0.1201	0.9928	0.1210	8.2636	1.0073	8.3238	83.1
7.0	0.1219	0.9925	0.1228	8.1443	1.0075	8.2055	83.0
7.1	0.1236	0.9923	0.1246	8.0285	1.0077	8.0905	82.9
7.2	0.1253	0.9921	0.1263	7.9158	1.0079	7.9787	82.8
7.3	0.1271	0.9919	0.1281	7.8062	1.0082	7.8700	82.7
7.4	0.1288	0.9917	0.1299	7.6996	1.0084	7.7642	82.6
7.5	0.1305	0.9914	0.1317	7.5958	1.0086	7.6613	82.5
7.6	0.1323	0.9912	0.1334	7.4947	1.0089	7.5611	82.4
7.7	0.1340	0.9910	0.1352	7.3962	1.0091	7.4635	82.3
7.8	0.1357	0.9907	0.1370	7.3002	1.0093	7.3684	82.2
7.9	0.1374	0.9905	0.1388	7.2066	1.0096	7.2757	82.1
	cos t	sin t	cot t	tan t	csc t	sec t	t degrees

Appendix IV (continued)

t degrees	sin t	cos t	tan t	cot t	sec t	csc t	
8.0	0.1392	0.9903	0.1405	7.1154	1.0098	7.1853	82.0
8.1	0.1409	0.9900	0.1423	7.0264	1.0101	7.0972	81.9
8.2	0.1426	0.9898	0.1441	6.9395	1.0103	7.0112	81.8
8.3	0.1444	0.9895	0.1459	6.8548	1.0106	6.9273	81.7
8.4	0.1461	0.9893	0.1477	6.7720	1.0108	6.8454	81.6
8.5	0.1478	0.9890	0.1495	6.6912	1.0111	6.7655	81.5
8.6	0.1495	0.9888	0.1512	6.6122	1.0114	6.6874	81.4
8.7	0.1513	0.9885	0.1530	6.5350	1.0116	6.6111	81.3
8.8	0.1530	0.9882	0.1548	6.4596	1.0119	6.5366	81.2
8.9	0.1547	0.9880	0.1566	6.3859	1.0122	6.4637	81.1
9.0	0.1564	0.9877	0.1584	6.3138	1.0125	6.3925	81.0
9.1	0.1582	0.9874	0.1602	6.2432	1.0127	6.3228	80.9
9.2	0.1599	0.9871	0.1620	6.1742	1.0130	6.2546	80.8
9.3	0.1616	0.9869	0.1638	6.1066	1.0133	6.1880	80.7
9.4	0.1633	0.9866	0.1655	6.0405	1.0136	6.1227	80.6
9.5	0.1650	0.9863	0.1673	5.9758	1.0139	6.0589	80.5
9.6	0.1668	0.9860	0.1691	5.9124	1.0142	5.9963	80.4
9.7	0.1685	0.9857	0.1709	5.8502	1.0145	5.9351	80.3
9.8	0.1702	0.9854	0.1727	5.7894	1.0148	5.8751	80.2
9.9	0.1719	0.9851	0.1745	5.7297	1.0151	5.8164	80.1
10.0	0.1736	0.9848	0.1763	5.6713	1.0154	5.7588	80.0
10.1	0.1754	0.9845	0.1781	5.6140	1.0157	5.7023	79.9
10.2	0.1771	0.9842	0.1799	5.5578	1.0161	5.6470	79.8
10.3	0.1788	0.9839	0.1817	5.5026	1.0164	5.5928	79.7
10.4	0.1805	0.9836	0.1835	5.4486	1.0167	5.5396	79.6
10.5	0.1822	0.9833	0.1853	5.3955	1.0170	5.4874	79.5
10.6	0.1840	0.9829	0.1871	5.3435	1.0174	5.4362	79.4
10.7	0.1857	0.9826	0.1890	5.2924	1.0177	5.3860	79.3
10.8	0.1874	0.9823	0.1908	5.2422	1.0180	5.3367	79.2
10.9	0.1891	0.9820	0.1926	5.1929	1.0184	5.2883	79.1
11.0	0.1908	0.9816	0.1944	5.1446	1.0187	5.2408	79.0
11.1	0.1925	0.9813	0.1962	5.0970	1.0191	5.1942	78.9
11.2	0.1942	0.9810	0.1980	5.0504	1.0194	5.1484	78.8
11.3	0.1959	0.9806	0.1998	5.0045	1.0198	5.1034	78.7
11.4	0.1977	0.9803	0.2016	4.9594	1.0201	5.0593	78.6
11.5	0.1994	0.9799	0.2035	4.9152	1.0205	5.0159	78.5
11.6	0.2011	0.9796	0.2053	4.8716	1.0209	4.9732	78.4
11.7	0.2028	0.9792	0.2071	4.8288	1.0212	4.9313	78.3
11.8	0.2045	0.9789	0.2089	4.7867	1.0216	4.8901	78.2
11.9	0.2062	0.9785	0.2107	4.7453	1.0220	4.8496	78.1
	cos t	sin t	cot t	tan t	csc t	sec t	t degrees

Appendix IV / Values of Trigonometric Functions—Degree Measure

Appendix IV (continued)

t degrees	sin t	cos t	tan t	cot t	sec t	csc t	
12.0	0.2079	0.9781	0.2126	4.7046	1.0223	4.8097	78.0
12.1	0.2096	0.9778	0.2144	4.6646	1.0227	4.7706	77.9
12.2	0.2113	0.9774	0.2162	4.6252	1.0231	4.7321	77.8
12.3	0.2130	0.9770	0.2180	4.5864	1.0235	4.6942	77.7
12.4	0.2147	0.9767	0.2199	4.5483	1.0239	4.6569	77.6
12.5	0.2164	0.9763	0.2217	4.5107	1.0243	4.6202	77.5
12.6	0.2181	0.9759	0.2235	4.4737	1.0247	4.5841	77.4
12.7	0.2198	0.9755	0.2254	4.4373	1.0251	4.5486	77.3
12.8	0.2215	0.9751	0.2272	4.4015	1.0255	4.5137	77.2
12.9	0.2233	0.9748	0.2290	4.3662	1.0259	4.4793	77.1
13.0	0.2250	0.9744	0.2309	4.3315	1.0263	4.4454	77.0
13.1	0.2267	0.9740	0.2327	4.2972	1.0267	4.4121	76.9
13.2	0.2284	0.9736	0.2345	4.2635	1.0271	4.3792	76.8
13.3	0.2300	0.9732	0.2364	4.2303	1.0276	4.3469	76.7
13.4	0.2317	0.9728	0.2382	4.1976	1.0280	4.3150	76.6
13.5	0.2334	0.9724	0.2401	4.1653	1.0284	4.2837	76.5
13.6	0.2351	0.9720	0.2419	4.1335	1.0288	4.2527	76.4
13.7	0.2368	0.9715	0.2438	4.1022	1.0293	4.2223	76.3
13.8	0.2385	0.9711	0.2456	4.0713	1.0297	4.1923	76.2
13.9	0.2402	0.9707	0.2475	4.0408	1.0302	4.1627	76.1
14.0	0.2419	0.9703	0.2493	4.0108	1.0306	4.1336	76.0
14.1	0.2436	0.9699	0.2512	3.9812	1.0311	4.1048	75.9
14.2	0.2453	0.9694	0.2530	3.9520	1.0315	4.0765	75.8
14.3	0.2470	0.9690	0.2549	3.9232	1.0320	4.0486	75.7
14.4	0.2487	0.9686	0.2568	3.8947	1.0324	4.0211	75.6
14.5	0.2504	0.9681	0.2586	3.8667	1.0329	3.9939	75.5
14.6	0.2521	0.9677	0.2605	3.8391	1.0334	3.9672	75.4
14.7	0.2538	0.9673	0.2623	3.8118	1.0338	3.9408	75.3
14.8	0.2554	0.9668	0.2642	3.7848	1.0343	3.9147	75.2
14.9	0.2571	0.9664	0.2661	3.7583	1.0348	3.8890	75.1
15.0	0.2588	0.9659	0.2679	3.7321	1.0353	3.8637	75.0
15.1	0.2605	0.9655	0.2698	3.7062	1.0358	3.8387	74.9
15.2	0.2622	0.9650	0.2717	3.6806	1.0363	3.8140	74.8
15.3	0.2639	0.9646	0.2736	3.6554	1.0367	3.7897	74.7
15.4	0.2656	0.9641	0.2754	3.6305	1.0372	3.7657	74.6
15.5	0.2672	0.9636	0.2773	3.6059	1.0377	3.7420	74.5
15.6	0.2689	0.9632	0.2792	3.5816	1.0382	3.7186	74.4
15.7	0.2706	0.9627	0.2811	3.5576	1.0388	3.6955	74.3
15.8	0.2723	0.9622	0.2830	3.5339	1.0393	3.6727	74.2
15.9	0.2740	0.9617	0.2849	3.5105	1.0398	3.6502	74.1
	cos t	sin t	cot t	tan t	csc t	sec t	t degrees

Appendix IV (continued)

t degrees	sin t	cos t	tan t	cot t	sec t	csc t	
16.0	0.2756	0.9613	0.2867	3.4874	1.0403	3.6280	74.0
16.1	0.2773	0.9608	0.2886	3.4646	1.0408	3.6060	73.9
16.2	0.2790	0.9603	0.2905	3.4420	1.0413	3.5843	73.8
16.3	0.2807	0.9598	0.2924	3.4197	1.0419	3.5629	73.7
16.4	0.2823	0.9593	0.2943	3.3977	1.0424	3.5418	73.6
16.5	0.2840	0.9588	0.2962	3.3759	1.0429	3.5209	73.5
16.6	0.2857	0.9583	0.2981	3.3544	1.0435	3.5003	73.4
16.7	0.2874	0.9578	0.3000	3.3332	1.0440	3.4799	73.3
16.8	0.2890	0.9573	0.3019	3.3122	1.0446	3.4598	73.2
16.9	0.2907	0.9568	0.3038	3.2914	1.0451	3.4399	73.1
17.0	0.2924	0.9563	0.3057	3.2709	1.0457	3.4203	73.0
17.1	0.2940	0.9558	0.3076	3.2506	1.0463	3.4009	72.9
17.2	0.2957	0.9553	0.3096	3.2305	1.0468	3.3817	72.8
17.3	0.2974	0.9548	0.3115	3.2106	1.0474	3.3628	72.7
17.4	0.2990	0.9542	0.3134	3.1910	1.0480	3.3440	72.6
17.5	0.3007	0.9537	0.3153	3.1716	1.0485	3.3255	72.5
17.6	0.3024	0.9532	0.3172	3.1524	1.0491	3.3072	72.4
17.7	0.3040	0.9527	0.3191	3.1334	1.0497	3.2891	72.3
17.8	0.3057	0.9521	0.3211	3.1146	1.0503	3.2712	72.2
17.9	0.3074	0.9516	0.3230	3.0961	1.0509	3.2535	72.1
18.0	0.3090	0.9511	0.3249	3.0777	1.0515	3.2361	72.0
18.1	0.3107	0.9505	0.3269	3.0595	1.0521	3.2188	71.9
18.2	0.3123	0.9500	0.3288	3.0415	1.0527	3.2017	71.8
18.3	0.3140	0.9494	0.3307	3.0237	1.0533	3.1848	71.7
18.4	0.3156	0.9489	0.3327	3.0061	1.0539	3.1681	71.6
18.5	0.3173	0.9483	0.3346	2.9887	1.0545	3.1515	71.5
18.6	0.3190	0.9478	0.3365	2.9714	1.0551	3.1352	71.4
18.7	0.3206	0.9472	0.3385	2.9544	1.0557	3.1190	71.3
18.8	0.3223	0.9466	0.3404	2.9375	1.0564	3.1030	71.2
18.9	0.3239	0.9461	0.3424	2.9208	1.0570	3.0872	71.1
19.0	0.3256	0.9455	0.3443	2.9042	1.0576	3.0716	71.0
19.1	0.3272	0.9449	0.3463	2.8878	1.0583	3.0561	70.9
19.2	0.3289	0.9444	0.3482	2.8716	1.0589	3.0407	70.8
19.3	0.3305	0.9438	0.3502	2.8556	1.0595	3.0256	70.7
19.4	0.3322	0.9432	0.3522	2.8397	1.0602	3.0106	70.6
19.5	0.3338	0.9426	0.3541	2.8239	1.0608	2.9957	70.5
19.6	0.3355	0.9421	0.3561	2.8083	1.0615	2.9811	70.4
19.7	0.3371	0.9415	0.3581	2.7929	1.0622	2.9665	70.3
19.8	0.3387	0.9409	0.3600	2.7776	1.0628	2.9521	70.2
19.9	0.3404	0.9403	0.3620	2.7625	1.0635	2.9379	70.1
	cos t	sin t	cot t	tan t	csc t	sec t	t degrees

Appendix IV / Values of Trigonometric Functions—Degree Measure

Appendix IV (continued)

t degrees	sin t	cos t	tan t	cot t	sec t	csc t	
20.0	0.3420	0.9397	0.3640	2.7475	1.0642	2.9238	70.0
20.1	0.3437	0.9391	0.3659	2.7326	1.0649	2.9099	69.9
20.2	0.3453	0.9385	0.3679	2.7179	1.0655	2.8960	69.8
20.3	0.3469	0.9379	0.3699	2.7034	1.0662	2.8824	69.7
20.4	0.3486	0.9373	0.3719	2.6889	1.0669	2.8688	69.6
20.5	0.3502	0.9367	0.3739	2.6746	1.0676	2.8555	69.5
20.6	0.3518	0.9361	0.3759	2.6605	1.0683	2.8422	69.4
20.7	0.3535	0.9354	0.3779	2.6464	1.0690	2.8291	69.3
20.8	0.3551	0.9348	0.3799	2.6325	1.0697	2.8161	69.2
20.9	0.3567	0.9342	0.3819	2.6187	1.0704	2.8032	69.1
21.0	0.3584	0.9336	0.3839	2.6051	1.0711	2.7904	69.0
21.1	0.3600	0.9330	0.3859	2.5916	1.0719	2.7778	68.9
21.2	0.3616	0.9323	0.3879	2.5782	1.0726	2.7653	68.8
21.3	0.3633	0.9317	0.3899	2.5649	1.0733	2.7529	68.7
21.4	0.3649	0.9311	0.3919	2.5517	1.0740	2.7407	68.6
21.5	0.3665	0.9304	0.3939	2.5386	1.0748	2.7285	68.5
21.6	0.3681	0.9298	0.3959	2.5257	1.0755	2.7165	68.4
21.7	0.3697	0.9291	0.3979	2.5129	1.0763	2.7046	68.3
21.8	0.3714	0.9285	0.4000	2.5002	1.0770	2.6927	68.2
21.9	0.3730	0.9278	0.4020	2.4876	1.0778	2.6811	68.1
22.0	0.3746	0.9272	0.4040	2.4751	1.0785	2.6695	68.0
22.1	0.3762	0.9265	0.4061	2.4627	1.0793	2.6580	67.9
22.2	0.3778	0.9259	0.4081	2.4504	1.0801	2.6466	67.8
22.3	0.3795	0.9252	0.4101	2.4383	1.0808	2.6354	67.7
22.4	0.3811	0.9245	0.4122	2.4262	1.0816	2.6242	67.6
22.5	0.3827	0.9239	0.4142	2.4142	1.0824	2.6131	67.5
22.6	0.3843	0.9232	0.4163	2.4023	1.0832	2.6022	67.4
22.7	0.3859	0.9225	0.4183	2.3906	1.0840	2.5913	67.3
22.8	0.3875	0.9219	0.4204	2.3789	1.0848	2.5805	67.2
22.9	0.3891	0.9212	0.4224	2.3673	1.0856	2.5699	67.1
23.0	0.3907	0.9205	0.4245	2.3559	1.0864	2.5593	67.0
23.1	0.3923	0.9198	0.4265	2.3445	1.0872	2.5488	66.9
23.2	0.3939	0.9191	0.4286	2.3332	1.0880	2.5384	66.8
23.3	0.3955	0.9184	0.4307	2.3220	1.0888	2.5282	66.7
23.4	0.3971	0.9178	0.4327	2.3109	1.0896	2.5180	66.6
23.5	0.3987	0.9171	0.4348	2.2998	1.0904	2.5078	66.5
23.6	0.4003	0.9164	0.4369	2.2889	1.0913	2.4978	66.4
23.7	0.4019	0.9157	0.4390	2.2781	1.0921	2.4879	66.3
23.8	0.4035	0.9150	0.4411	2.2673	1.0929	2.4780	66.2
23.9	0.4051	0.9143	0.4431	2.2566	1.0938	2.4683	66.1
	cos t	sin t	cot t	tan t	csc t	sec t	t degrees

Appendix IV (continued)

t degrees	sin t	cos t	tan t	cot t	sec t	csc t	
24.0	0.4067	0.9135	0.4452	2.2460	1.0946	2.4586	66.0
24.1	0.4083	0.9128	0.4473	2.2355	1.0955	2.4490	65.9
24.2	0.4099	0.9121	0.4494	2.2251	1.0963	2.4395	65.8
24.3	0.4115	0.9114	0.4515	2.2148	1.0972	2.4300	65.7
24.4	0.4131	0.9107	0.4536	2.2045	1.0981	2.4207	65.6
24.5	0.4147	0.9100	0.4557	2.1943	1.0989	2.4114	65.5
24.6	0.4163	0.9092	0.4578	2.1842	1.0998	2.4022	65.4
24.7	0.4179	0.9085	0.4599	2.1742	1.1007	2.3931	65.3
24.8	0.4195	0.9078	0.4621	2.1642	1.1016	2.3841	65.2
24.9	0.4210	0.9070	0.4642	2.1543	1.1025	2.3751	65.1
25.0	0.4226	0.9063	0.4663	2.1445	1.1034	2.3662	65.0
25.1	0.4242	0.9056	0.4684	2.1348	1.1043	2.3574	64.9
25.2	0.4258	0.9048	0.4706	2.1251	1.1052	2.3486	64.8
25.3	0.4274	0.9041	0.4727	2.1155	1.1061	2.3400	64.7
25.4	0.4289	0.9033	0.4748	2.1060	1.1070	2.3314	64.6
25.5	0.4305	0.9026	0.4770	2.0965	1.1079	2.3228	64.5
25.6	0.4321	0.9018	0.4791	2.0872	1.1089	2.3144	64.4
25.7	0.4337	0.9011	0.4813	2.0778	1.1098	2.3060	64.3
25.8	0.4352	0.9003	0.4834	2.0686	1.1107	2.2976	64.2
25.9	0.4368	0.8996	0.4856	2.0594	1.1117	2.2894	64.1
26.0	0.4384	0.8988	0.4877	2.0503	1.1126	2.2812	64.0
26.1	0.4399	0.8980	0.4899	2.0413	1.1136	2.2730	63.9
26.2	0.4415	0.8973	0.4921	2.0323	1.1145	2.2650	63.8
26.3	0.4431	0.8965	0.4942	2.0233	1.1155	2.2570	63.7
26.4	0.4446	0.8957	0.4964	2.0145	1.1164	2.2490	63.6
26.5	0.4462	0.8949	0.4986	2.0057	1.1174	2.2412	63.5
26.6	0.4478	0.8942	0.5008	1.9970	1.1184	2.2333	63.4
26.7	0.4493	0.8934	0.5029	1.9883	1.1194	2.2256	63.3
26.8	0.4509	0.8926	0.5051	1.9797	1.1203	2.2179	63.2
26.9	0.4524	0.8918	0.5073	1.9711	1.1213	2.2103	63.1
27.0	0.4540	0.8910	0.5095	1.9626	1.1223	2.2027	63.0
27.1	0.4555	0.8902	0.5117	1.9542	1.1233	2.1952	62.9
27.2	0.4571	0.8894	0.5139	1.9458	1.1243	2.1877	62.8
27.3	0.4586	0.8886	0.5161	1.9375	1.1253	2.1803	62.7
27.4	0.4602	0.8878	0.5184	1.9292	1.1264	2.1730	62.6
27.5	0.4617	0.8870	0.5206	1.9210	1.1274	2.1657	62.5
27.6	0.4633	0.8862	0.5228	1.9128	1.1284	2.1584	62.4
27.7	0.4648	0.8854	0.5250	1.9047	1.1294	2.1513	62.3
27.8	0.4664	0.8846	0.5272	1.8967	1.1305	2.1441	62.2
27.9	0.4679	0.8838	0.5295	1.8887	1.1315	2.1371	62.1
	cos t	sin t	cot t	tan t	csc t	sec t	t degrees

Appendix IV (continued)

t degrees	sin t	cos t	tan t	cot t	sec t	csc t	
28.0	0.4695	0.8829	0.5317	1.8807	1.1326	2.1301	62.0
28.1	0.4710	0.8821	0.5340	1.8728	1.1336	2.1231	61.9
28.2	0.4726	0.8813	0.5362	1.8650	1.1347	2.1162	61.8
28.3	0.4741	0.8805	0.5384	1.8572	1.1357	2.1093	61.7
28.4	0.4756	0.8796	0.5407	1.8495	1.1368	2.1025	61.6
28.5	0.4772	0.8788	0.5430	1.8418	1.1379	2.0957	61.5
28.6	0.4787	0.8780	0.5452	1.8341	1.1390	2.0890	61.4
28.7	0.4802	0.8771	0.5475	1.8265	1.1401	2.0824	61.3
28.8	0.4818	0.8763	0.5498	1.8190	1.1412	2.0757	61.2
28.9	0.4833	0.8755	0.5520	1.8115	1.1423	2.0692	61.1
29.0	0.4848	0.8746	0.5543	1.8040	1.1434	2.0627	61.0
29.1	0.4863	0.8738	0.5566	1.7966	1.1445	2.0562	60.9
29.2	0.4879	0.8729	0.5589	1.7893	1.1456	2.0498	60.8
29.3	0.4894	0.8721	0.5612	1.7820	1.1467	2.0434	60.7
29.4	0.4909	0.8712	0.5635	1.7747	1.1478	2.0371	60.6
29.5	0.4924	0.8704	0.5658	1.7675	1.1490	2.0308	60.5
29.6	0.4939	0.8695	0.5681	1.7603	1.1501	2.0245	60.4
29.7	0.4955	0.8686	0.5704	1.7532	1.1512	2.0183	60.3
29.8	0.4970	0.8678	0.5727	1.7461	1.1524	2.0122	60.2
29.9	0.4985	0.8669	0.5750	1.7391	1.1535	2.0061	60.1
30.0	0.5000	0.8660	0.5774	1.7321	1.1547	2.0000	60.0
30.1	0.5015	0.8652	0.5797	1.7251	1.1559	1.9940	59.9
30.2	0.5030	0.8643	0.5820	1.7182	1.1570	1.9880	59.8
30.3	0.5045	0.8634	0.5844	1.7113	1.1582	1.9821	59.7
30.4	0.5060	0.8625	0.5867	1.7045	1.1594	1.9762	59.6
30.5	0.5075	0.8616	0.5890	1.6977	1.1606	1.9703	59.5
30.6	0.5090	0.8607	0.5914	1.6909	1.1618	1.9645	59.4
30.7	0.5105	0.8599	0.5938	1.6842	1.1630	1.9587	59.3
30.8	0.5120	0.8590	0.5961	1.6775	1.1642	1.9530	59.2
30.9	0.5135	0.8581	0.5985	1.6709	1.1654	1.9473	59.1
31.0	0.5150	0.8572	0.6009	1.6643	1.1666	1.9416	59.0
31.1	0.5165	0.8563	0.6032	1.6577	1.1679	1.9360	58.9
31.2	0.5180	0.8554	0.6056	1.6512	1.1691	1.9304	58.8
31.3	0.5195	0.8545	0.6080	1.6447	1.1703	1.9249	58.7
31.4	0.5210	0.8536	0.6104	1.6383	1.1716	1.9194	58.6
31.5	0.5225	0.8526	0.6128	1.6319	1.1728	1.9139	58.5
31.6	0.5240	0.8517	0.6152	1.6255	1.1741	1.9084	58.4
31.7	0.5255	0.8508	0.6176	1.6191	1.1753	1.9031	58.3
31.8	0.5270	0.8499	0.6200	1.6128	1.1766	1.8977	58.2
31.9	0.5284	0.8490	0.6224	1.6066	1.1779	1.8924	58.1
	cos t	sin t	cot t	tan t	csc t	sec t	t degrees

Appendix IV / Values of Trigonometric Functions—Degree Measure

Appendix IV (continued)

t degrees	sin t	cos t	tan t	cot t	sec t	csc t	
32.0	0.5299	0.8480	0.6249	1.6003	1.1792	1.8871	58.0
32.1	0.5314	0.8471	0.6273	1.5941	1.1805	1.8818	57.9
32.2	0.5329	0.8462	0.6297	1.5880	1.1818	1.8766	57.8
32.3	0.5344	0.8453	0.6322	1.5818	1.1831	1.8714	57.7
32.4	0.5358	0.8443	0.6346	1.5757	1.1844	1.8663	57.6
32.5	0.5373	0.8434	0.6371	1.5697	1.1857	1.8612	57.5
32.6	0.5388	0.8425	0.6395	1.5637	1.1870	1.8561	57.4
32.7	0.5402	0.8415	0.6420	1.5577	1.1883	1.8510	57.3
32.8	0.5417	0.8406	0.6445	1.5517	1.1897	1.8460	57.2
32.9	0.5432	0.8396	0.6469	1.5458	1.1910	1.8410	57.1
33.0	0.5446	0.8387	0.6494	1.5399	1.1924	1.8361	57.0
33.1	0.5461	0.8377	0.6519	1.5340	1.1937	1.8312	56.9
33.2	0.5476	0.8368	0.6544	1.5282	1.1951	1.8263	56.8
33.3	0.5490	0.8358	0.6569	1.5224	1.1964	1.8214	56.7
33.4	0.5505	0.8348	0.6594	1.5166	1.1978	1.8166	56.6
33.5	0.5519	0.8339	0.6619	1.5108	1.1992	1.8118	56.5
33.6	0.5534	0.8329	0.6644	1.5051	1.2006	1.8070	56.4
33.7	0.5548	0.8320	0.6669	1.4994	1.2020	1.8023	56.3
33.8	0.5563	0.8310	0.6694	1.4938	1.2034	1.7976	56.2
33.9	0.5577	0.8300	0.6720	1.4882	1.2048	1.7929	56.1
34.0	0.5592	0.8290	0.6745	1.4826	1.2062	1.7883	56.0
34.1	0.5606	0.8281	0.6771	1.4770	1.2076	1.7837	55.9
34.2	0.5621	0.8271	0.6796	1.4715	1.2091	1.7791	55.8
34.3	0.5635	0.8261	0.6822	1.4659	1.2105	1.7745	55.7
34.4	0.5650	0.8251	0.6847	1.4605	1.2120	1.7700	55.6
34.5	0.5664	0.8241	0.6873	1.4550	1.2134	1.7655	55.5
34.6	0.5678	0.8231	0.6899	1.4496	1.2149	1.7610	55.4
34.7	0.5693	0.8221	0.6924	1.4442	1.2163	1.7566	55.3
34.8	0.5707	0.8211	0.6950	1.4388	1.2178	1.7522	55.2
34.9	0.5721	0.8202	0.6976	1.4335	1.2193	1.7478	55.1
35.0	0.5736	0.8192	0.7002	1.4281	1.2208	1.7434	55.0
35.1	0.5750	0.8181	0.7028	1.4229	1.2223	1.7391	54.9
35.2	0.5764	0.8171	0.7054	1.4176	1.2238	1.7348	54.8
35.3	0.5779	0.8161	0.7080	1.4124	1.2253	1.7305	54.7
35.4	0.5793	0.8151	0.7107	1.4071	1.2268	1.7263	54.6
35.5	0.5807	0.8141	0.7133	1.4019	1.2283	1.7221	54.5
35.6	0.5821	0.8131	0.7159	1.3968	1.2299	1.7179	54.4
35.7	0.5835	0.8121	0.7186	1.3916	1.2314	1.7137	54.3
35.8	0.5850	0.8111	0.7212	1.3865	1.2329	1.7095	54.2
35.9	0.5864	0.8100	0.7239	1.3814	1.2345	1.7054	54.1
	cos t	sin t	cot t	tan t	csc t	sec t	t degrees

Appendix IV / Values of Trigonometric Functions—Degree Measure

Appendix IV *(continued)*

t degrees	sin t	cos t	tan t	cot t	sec t	csc t	
36.0	0.5878	0.8090	0.7265	1.3764	1.2361	1.7013	54.0
36.1	0.5892	0.8080	0.7292	1.3713	1.2376	1.6972	53.9
36.2	0.5906	0.8070	0.7319	1.3663	1.2392	1.6932	53.8
36.3	0.5920	0.8059	0.7346	1.3613	1.2408	1.6892	53.7
36.4	0.5934	0.8049	0.7373	1.3564	1.2424	1.6852	53.6
36.5	0.5948	0.8039	0.7400	1.3514	1.2440	1.6812	53.5
36.6	0.5962	0.8028	0.7427	1.3465	1.2456	1.6772	53.4
36.7	0.5976	0.8018	0.7454	1.3416	1.2472	1.6733	53.3
36.8	0.5990	0.8007	0.7481	1.3367	1.2489	1.6694	53.2
36.9	0.6004	0.7997	0.7508	1.3319	1.2505	1.6655	53.1
37.0	0.6018	0.7986	0.7536	1.3270	1.2521	1.6616	53.0
37.1	0.6032	0.7976	0.7563	1.3222	1.2538	1.6578	52.9
37.2	0.6046	0.7965	0.7590	1.3175	1.2554	1.6540	52.8
37.3	0.6060	0.7955	0.7618	1.3127	1.2571	1.6502	52.7
37.4	0.6074	0.7944	0.7646	1.3079	1.2588	1.6464	52.6
37.5	0.6088	0.7934	0.7673	1.3032	1.2605	1.6427	52.5
37.6	0.6101	0.7923	0.7701	1.2985	1.2622	1.6390	52.4
37.7	0.6115	0.7912	0.7729	1.2938	1.2639	1.6353	52.3
37.8	0.6129	0.7902	0.7757	1.2892	1.2656	1.6316	52.2
37.9	0.6143	0.7891	0.7785	1.2846	1.2673	1.6279	52.1
38.0	0.6157	0.7880	0.7813	1.2799	1.2690	1.6243	52.0
38.1	0.6170	0.7869	0.7841	1.2753	1.2708	1.6207	51.9
38.2	0.6184	0.7859	0.7869	1.2708	1.2725	1.6171	51.8
38.3	0.6198	0.7848	0.7898	1.2662	1.2742	1.6135	51.7
38.4	0.6211	0.7837	0.7926	1.2617	1.2760	1.6099	51.6
38.5	0.6225	0.7826	0.7954	1.2572	1.2778	1.6064	51.5
38.6	0.6239	0.7815	0.7983	1.2527	1.2796	1.6029	51.4
38.7	0.6252	0.7804	0.8012	1.2482	1.2813	1.5994	51.3
38.8	0.6266	0.7793	0.8040	1.2437	1.2831	1.5959	51.2
38.9	0.6280	0.7782	0.8069	1.2393	1.2849	1.5925	51.1
39.0	0.6293	0.7771	0.8098	1.2349	1.2868	1.5890	51.0
39.1	0.6307	0.7760	0.8127	1.2305	1.2886	1.5856	50.9
39.2	0.6320	0.7749	0.8156	1.2261	1.2904	1.5822	50.8
39.3	0.6334	0.7738	0.8185	1.2218	1.2923	1.5788	50.7
39.4	0.6347	0.7727	0.8214	1.2174	1.2941	1.5755	50.6
39.5	0.6361	0.7716	0.8243	1.2131	1.2960	1.5721	50.5
39.6	0.6374	0.7705	0.8273	1.2088	1.2978	1.5688	50.4
39.7	0.6388	0.7694	0.8302	1.2045	1.2997	1.5655	50.3
39.8	0.6401	0.7683	0.8332	1.2002	1.3016	1.5622	50.2
39.9	0.6414	0.7672	0.8361	1.1960	1.3035	1.5590	50.1
	cos t	sin t	cot t	tan t	csc t	sec t	t degrees

Appendix IV (continued)

t degrees	sin t	cos t	tan t	cot t	sec t	csc t	
40.0	0.6428	0.7660	0.8391	1.1918	1.3054	1.5557	50.0
40.1	0.6441	0.7649	0.8421	1.1875	1.3073	1.5525	49.9
40.2	0.6455	0.7638	0.8451	1.1833	1.3093	1.5493	49.8
40.3	0.6468	0.7627	0.8481	1.1792	1.3112	1.5461	49.7
40.4	0.6481	0.7615	0.8511	1.1750	1.3131	1.5429	49.6
40.5	0.6494	0.7604	0.8541	1.1708	1.3151	1.5398	49.5
40.6	0.6508	0.7593	0.8571	1.1667	1.3171	1.5366	49.4
40.7	0.6521	0.7581	0.8601	1.1626	1.3190	1.5335	49.3
40.8	0.6534	0.7570	0.8632	1.1585	1.3210	1.5304	49.2
40.9	0.6547	0.7559	0.8662	1.1544	1.3230	1.5273	49.1
41.0	0.6561	0.7547	0.8693	1.1504	1.3250	1.5243	49.0
41.1	0.6574	0.7536	0.8724	1.1463	1.3270	1.5212	48.9
41.2	0.6587	0.7524	0.8754	1.1423	1.3291	1.5182	48.8
41.3	0.6600	0.7513	0.8785	1.1383	1.3311	1.5151	48.7
41.4	0.6613	0.7501	0.8816	1.1343	1.3331	1.5121	48.6
41.5	0.6626	0.7490	0.8847	1.1303	1.3352	1.5092	48.5
41.6	0.6639	0.7478	0.8878	1.1263	1.3373	1.5062	48.4
41.7	0.6652	0.7466	0.8910	1.1224	1.3393	1.5032	48.3
41.8	0.6665	0.7455	0.8941	1.1184	1.3414	1.5003	48.2
41.9	0.6678	0.7443	0.8972	1.1145	1.3435	1.4974	48.1
42.0	0.6691	0.7431	0.9004	1.1106	1.3456	1.4945	48.0
42.1	0.6704	0.7420	0.9036	1.1067	1.3478	1.4916	47.9
42.2	0.6717	0.7408	0.9067	1.1028	1.3499	1.4887	47.8
42.3	0.6730	0.7396	0.9099	1.0990	1.3520	1.4859	47.7
42.4	0.6743	0.7385	0.9131	1.0951	1.3542	1.4830	47.6
42.5	0.6756	0.7373	0.9163	1.0913	1.3563	1.4802	47.5
42.6	0.6769	0.7361	0.9195	1.0875	1.3585	1.4774	47.4
42.7	0.6782	0.7349	0.9228	1.0837	1.3607	1.4746	47.3
42.8	0.6794	0.7337	0.9260	1.0799	1.3629	1.4718	47.2
42.9	0.6807	0.7325	0.9293	1.0761	1.3651	1.4690	47.1
43.0	0.6820	0.7314	0.9325	1.0724	1.3673	1.4663	47.0
43.1	0.6833	0.7302	0.9358	1.0686	1.3696	1.4635	46.9
43.2	0.6845	0.7290	0.9391	1.0649	1.3718	1.4608	46.8
43.3	0.6858	0.7278	0.9424	1.0612	1.3741	1.4581	46.7
43.4	0.6871	0.7266	0.9457	1.0575	1.3763	1.4554	46.6
43.5	0.6884	0.7254	0.9490	1.0538	1.3786	1.4527	46.5
43.6	0.6896	0.7242	0.9523	1.0501	1.3809	1.4501	46.4
43.7	0.6909	0.7230	0.9556	1.0464	1.3832	1.4474	46.3
43.8	0.6921	0.7218	0.9590	1.0428	1.3855	1.4448	46.2
43.9	0.6934	0.7206	0.9623	1.0392	1.3878	1.4422	46.1
	cos t	sin t	cot t	tan t	csc t	sec t	t degrees

Appendix IV (continued)

t degrees	sin t	cos t	tan t	cot t	sec t	csc t	
44.0	0.6947	0.7193	0.9657	1.0355	1.3902	1.4396	46.0
44.1	0.6959	0.7181	0.9691	1.0319	1.3925	1.4370	45.9
44.2	0.6972	0.7169	0.9725	1.0283	1.3949	1.4344	45.8
44.3	0.6984	0.7157	0.9759	1.0247	1.3972	1.4318	45.7
44.4	0.6997	0.7145	0.9793	1.0212	1.3996	1.4293	45.6
44.5	0.7009	0.7133	0.9827	1.0176	1.4020	1.4267	45.5
44.6	0.7022	0.7120	0.9861	1.0141	1.4044	1.4242	45.4
44.7	0.7034	0.7108	0.9896	1.0105	1.4069	1.4217	45.3
44.8	0.7046	0.7096	0.9930	1.0070	1.4093	1.4192	45.2
44.9	0.7059	0.7083	0.9965	1.0035	1.4118	1.4167	45.1
45.0	0.7071	0.7071	1.0000	1.0000	1.4142	1.4142	45.0
	cos t	sin t	cot t	tan t	csc t	sec t	t degrees

Appendix V / Values of Trigonometric Functions—Radian Measure

t	sin t	cos t	tan t	cot t	sec t	csc t
0.00	0.0000	1.0000	0.0000	—	1.0000	—
0.01	0.0100	1.0000	0.0100	99.9967	1.0001	100.0017
0.02	0.0200	0.9998	0.0200	49.9933	1.0002	50.0033
0.03	0.0300	0.9996	0.0300	33.3233	1.0005	33.3383
0.04	0.0400	0.9992	0.0400	24.9867	1.0008	25.0067
0.05	0.0500	0.9988	0.0500	19.9833	1.0013	20.0083
0.06	0.0600	0.9982	0.0601	16.6467	1.0018	16.6767
0.07	0.0699	0.9976	0.0701	14.2624	1.0025	14.2974
0.08	0.0799	0.9968	0.0802	12.4733	1.0032	12.5133
0.09	0.0899	0.9960	0.0902	11.0811	1.0041	11.1261
0.10	0.0998	0.9950	0.1003	9.9666	1.0050	10.0167
0.11	0.1098	0.9940	0.1104	9.0542	1.0061	9.1093
0.12	0.1197	0.9928	0.1206	8.2933	1.0072	8.3534
0.13	0.1296	0.9916	0.1307	7.6489	1.0085	7.7140
0.14	0.1395	0.9902	0.1409	7.0961	1.0099	7.1662
0.15	0.1494	0.9888	0.1511	6.6166	1.0114	6.6917
0.16	0.1593	0.9872	0.1614	6.1966	1.0129	6.2767
0.17	0.1692	0.9856	0.1717	5.8256	1.0146	5.9108
0.18	0.1790	0.9838	0.1820	5.4954	1.0164	5.5857
0.19	0.1889	0.9820	0.1923	5.1997	1.0183	5.2950
0.20	0.1987	0.9801	0.2027	4.9332	1.0203	5.0335
0.21	0.2085	0.9780	0.2131	4.6917	1.0225	4.7971
0.22	0.2182	0.9759	0.2236	4.4719	1.0247	4.5823
0.23	0.2280	0.9737	0.2341	4.2709	1.0270	4.3864
0.24	0.2377	0.9713	0.2447	4.0864	1.0295	4.2069
0.25	0.2474	0.9689	0.2553	3.9163	1.0321	4.0420
0.26	0.2571	0.9664	0.2660	3.7591	1.0348	3.8898
0.27	0.2667	0.9638	0.2768	3.6133	1.0376	3.7491
0.28	0.2764	0.9611	0.2876	3.4776	1.0405	3.6185
0.29	0.2860	0.9582	0.2984	3.3511	1.0436	3.4971
0.30	0.2955	0.9553	0.3093	3.2327	1.0468	3.3839
0.31	0.3051	0.9523	0.3203	3.1218	1.0501	3.2781
0.32	0.3146	0.9492	0.3314	3.0176	1.0535	3.1790
0.33	0.3240	0.9460	0.3425	2.9195	1.0570	3.0860
0.34	0.3335	0.9428	0.3537	2.8270	1.0607	2.9986
0.35	0.3429	0.9394	0.3650	2.7395	1.0645	2.9163
0.36	0.3523	0.9359	0.3764	2.6567	1.0685	2.8387
0.37	0.3616	0.9323	0.3879	2.5782	1.0726	2.7654
0.38	0.3709	0.9287	0.3994	2.5037	1.0768	2.6960
0.39	0.3802	0.9249	0.4111	2.4328	1.0812	2.6303

Appendix V (continued)

t	sin t	cos t	tan t	cot t	sec t	csc t
0.40	0.3894	0.9211	0.4228	2.3652	1.0857	2.5679
0.41	0.3986	0.9171	0.4346	2.3008	1.0904	2.5087
0.42	0.4078	0.9131	0.4466	2.2393	1.0952	2.4524
0.43	0.4169	0.9090	0.4586	2.1804	1.1002	2.3988
0.44	0.4259	0.9048	0.4708	2.1241	1.1053	2.3478
0.45	0.4350	0.9004	0.4831	2.0702	1.1106	2.2990
0.46	0.4439	0.8961	0.4954	2.0184	1.1160	2.2525
0.47	0.4529	0.8916	0.5080	1.9686	1.1216	2.2081
0.48	0.4618	0.8870	0.5206	1.9208	1.1274	2.1655
0.49	0.4706	0.8823	0.5334	1.8748	1.1334	2.1248
0.50	0.4794	0.8776	0.5463	1.8305	1.1395	2.0858
0.51	0.4882	0.8727	0.5594	1.7878	1.1458	2.0484
0.52	0.4969	0.8678	0.5726	1.7465	1.1523	2.0126
0.53	0.5055	0.8628	0.5859	1.7067	1.1590	1.9781
0.54	0.5141	0.8577	0.5994	1.6683	1.1659	1.9450
0.55	0.5227	0.8525	0.6131	1.6310	1.1730	1.9132
0.56	0.5312	0.8473	0.6269	1.5950	1.1803	1.8826
0.57	0.5396	0.8419	0.6410	1.5601	1.1878	1.8531
0.58	0.5480	0.8365	0.6552	1.5263	1.1955	1.8247
0.59	0.5564	0.8309	0.6696	1.4935	1.2035	1.7974
0.60	0.5646	0.8253	0.6841	1.4617	1.2116	1.7710
0.61	0.5729	0.8196	0.6989	1.4308	1.2200	1.7456
0.62	0.5810	0.8139	0.7139	1.4007	1.2287	1.7211
0.63	0.5891	0.8080	0.7291	1.3715	1.2376	1.6974
0.64	0.5972	0.8021	0.7445	1.3431	1.2467	1.6745
0.65	0.6052	0.7961	0.7602	1.3154	1.2561	1.6524
0.66	0.6131	0.7900	0.7761	1.2885	1.2658	1.6310
0.67	0.6210	0.7838	0.7923	1.2622	1.2758	1.6103
0.68	0.6288	0.7776	0.8087	1.2366	1.2861	1.5903
0.69	0.6365	0.7712	0.8253	1.2116	1.2966	1.5710
0.70	0.6442	0.7648	0.8423	1.1872	1.3075	1.5523
0.71	0.6518	0.7584	0.8595	1.1634	1.3186	1.5341
0.72	0.6594	0.7518	0.8771	1.1402	1.3301	1.5166
0.73	0.6669	0.7452	0.8949	1.1174	1.3420	1.4995
0.74	0.6743	0.7385	0.9131	1.0952	1.3542	1.4830
0.75	0.6816	0.7317	0.9316	1.0734	1.3667	1.4671
0.76	0.6889	0.7248	0.9505	1.0521	1.3796	1.4515
0.77	0.6961	0.7179	0.9697	1.0313	1.3929	1.4365
0.78	0.7033	0.7109	0.9893	1.0109	1.4066	1.4219
0.79	0.7104	0.7038	1.0092	0.9908	1.4208	1.4078

Appendix V *(continued)*

t	sin t	cos t	tan t	cot t	sec t	csc t
0.80	0.7174	0.6967	1.0296	0.9712	1.4353	1.3940
0.81	0.7243	0.6895	1.0505	0.9520	1.4503	1.3807
0.82	0.7311	0.6822	1.0717	0.9331	1.4658	1.3677
0.83	0.7379	0.6749	1.0934	0.9146	1.4818	1.3551
0.84	0.7446	0.6675	1.1156	0.8964	1.4982	1.3429
0.85	0.7513	0.6600	1.1383	0.8785	1.5152	1.3311
0.86	0.7578	0.6524	1.1616	0.8609	1.5327	1.3195
0.87	0.7643	0.6448	1.1853	0.8437	1.5508	1.3083
0.88	0.7707	0.6372	1.2097	0.8267	1.5695	1.2975
0.89	0.7771	0.6294	1.2346	0.8100	1.5888	1.2869
0.90	0.7833	0.6216	1.2602	0.7936	1.6087	1.2766
0.91	0.7895	0.6137	1.2864	0.7774	1.6293	1.2666
0.92	0.7956	0.6058	1.3133	0.7615	1.6507	1.2569
0.93	0.8016	0.5978	1.3409	0.7458	1.6727	1.2475
0.94	0.8076	0.5898	1.3692	0.7303	1.6955	1.2383
0.95	0.8134	0.5817	1.3984	0.7151	1.7191	1.2294
0.96	0.8192	0.5735	1.4284	0.7001	1.7436	1.2207
0.97	0.8249	0.5653	1.4592	0.6853	1.7690	1.2123
0.98	0.8305	0.5570	1.4910	0.6707	1.7953	1.2041
0.99	0.8360	0.5487	1.5237	0.6563	1.8225	1.1961
1.00	0.8415	0.5403	1.5574	0.6421	1.8508	1.1884
1.01	0.8468	0.5319	1.5922	0.6281	1.8802	1.1809
1.02	0.8521	0.5234	1.6281	0.6142	1.9107	1.1736
1.03	0.8573	0.5148	1.6652	0.6005	1.9424	1.1665
1.04	0.8624	0.5062	1.7036	0.5870	1.9754	1.1595
1.05	0.8674	0.4976	1.7433	0.5736	2.0098	1.1528
1.06	0.8724	0.4889	1.7844	0.5604	2.0455	1.1463
1.07	0.8772	0.4801	1.8270	0.5473	2.0828	1.1400
1.08	0.8820	0.4713	1.8712	0.5344	2.1217	1.1338
1.09	0.8866	0.4625	1.9171	0.5216	2.1622	1.1279
1.10	0.8912	0.4536	1.9648	0.5090	2.2046	1.1221
1.11	0.8957	0.4447	2.0143	0.4964	2.2489	1.1164
1.12	0.9001	0.4357	2.0660	0.4840	2.2952	1.1110
1.13	0.9044	0.4267	2.1198	0.4718	2.3438	1.1057
1.14	0.9086	0.4176	2.1759	0.4596	2.3947	1.1006
1.15	0.9128	0.4085	2.2345	0.4475	2.4481	1.0956
1.16	0.9168	0.3993	2.2958	0.4356	2.5041	1.0907
1.17	0.9208	0.3902	2.3600	0.4237	2.5631	1.0861
1.18	0.9246	0.3809	2.4273	0.4120	2.6252	1.0815
1.19	0.9284	0.3717	2.4979	0.4003	2.6906	1.0772

Appendix V / Values of Trigonometric Functions—Radian Measure

Appendix V / (continued)

t	sin t	cos t	tan t	cot t	sec t	csc t
1.20	0.9320	0.3624	2.5722	0.3888	2.7597	1.0729
1.21	0.9356	0.3530	2.6503	0.3773	2.8327	1.0688
1.22	0.9391	0.3436	2.7328	0.3659	2.9100	1.0649
1.23	0.9425	0.3342	2.8198	0.3546	2.9919	1.0610
1.24	0.9458	0.3248	2.9119	0.3434	3.0789	1.0573
1.25	0.9490	0.3153	3.0096	0.3323	3.1714	1.0538
1.26	0.9521	0.3058	3.1133	0.3212	3.2699	1.0503
1.27	0.9551	0.2963	3.2236	0.3102	3.3752	1.0470
1.28	0.9580	0.2867	3.3413	0.2993	3.4878	1.0438
1.29	0.9608	0.2771	3.4672	0.2884	3.6085	1.0408
1.30	0.9636	0.2675	3.6021	0.2776	3.7383	1.0378
1.31	0.9662	0.2579	3.7471	0.2669	3.8782	1.0350
1.32	0.9687	0.2482	3.9033	0.2562	4.0294	1.0323
1.33	0.9711	0.2385	4.0723	0.2456	4.1933	1.0297
1.34	0.9735	0.2288	4.2556	0.2350	4.3715	1.0272
1.35	0.9757	0.2190	4.4552	0.2245	4.5661	1.0249
1.36	0.9779	0.2092	4.6734	0.2140	4.7792	1.0226
1.37	0.9799	0.1994	4.9131	0.2035	5.0138	1.0205
1.38	0.9819	0.1896	5.1774	0.1931	5.2731	1.0185
1.39	0.9837	0.1798	5.4707	0.1828	5.5613	1.0166
1.40	0.9854	0.1700	5.7979	0.1725	5.8835	1.0148
1.41	0.9871	0.1601	6.1654	0.1622	6.2459	1.0131
1.42	0.9887	0.1502	6.5811	0.1519	6.6567	1.0115
1.43	0.9901	0.1403	7.0555	0.1417	7.1260	1.0100
1.44	0.9915	0.1304	7.6018	0.1315	7.6673	1.0086
1.45	0.9927	0.1205	8.2381	0.1214	8.2986	1.0073
1.46	0.9939	0.1106	8.9886	0.1113	9.0441	1.0062
1.47	0.9949	0.1006	9.8874	0.1011	9.9378	1.0051
1.48	0.9959	0.0907	10.9834	0.0910	11.0288	1.0041
1.49	0.9967	0.0807	12.3499	0.0810	12.3903	1.0033
1.50	0.9975	0.0707	14.1014	0.0709	14.1368	1.0025
1.51	0.9982	0.0608	16.4281	0.0609	16.4585	1.0019
1.52	0.9987	0.0508	19.6695	0.0508	19.6949	1.0013
1.53	0.9992	0.0408	24.4984	0.0408	24.5188	1.0008
1.54	0.9995	0.0308	32.4611	0.0308	32.4765	1.0005
1.55	0.9998	0.0208	48.0785	0.0208	48.0889	1.0002
1.56	0.9999	0.0108	92.6205	0.0108	92.6259	1.0001
1.57	1.0000	0.0008	1,255.7656	0.0008	1,255.7660	1.0000

Answers to All Exercises in Chapter 0

1. (a), (d) 2. (b), (d)
3. (a), (b), (e), and (f)
4.
```
     -3/5 1/3
  -7-6-5-4-3-2-1 0 1 2 3 4 5 6 7
```
5. (a) Closure (b) Commutative
   (c) Associative and Commutative
   (d) Commutative and Distributive
6. (a) -2 (b) $\frac{2}{3}$ (c) $-\pi$ (d) 0
7. (a) $\frac{1}{2}$ (b) $-\frac{3}{2}$ (c) $\frac{1}{\sqrt{2}} = \frac{\sqrt{2}}{2}$ (d) -1
8. (a) $ac + ad + bc + bd$ (b) $x^2 + 6x + 9$
   (c) $x^2 + y^2 + 1 + 2xy + 2x + 2y$
   (d) $ax + ay + az + a + bx + by + bz + b + cx + cy + cz + c$
9. (a) $>, \geq$ (b) $<, \leq$ (c) $\leq, \geq$
   (d) $>, \geq$ (e) $>, \geq$ (f) $<, \leq$
10. (a) $(-2, 1)$
    (b) $[3, \infty)$
    (c) $[0, 2]$
    (d) $(-\infty, 5)$
    (e) $(-\infty, 3)$
    (f) $[2, 3)$
11. If $a < b$ fails, then $b - a$ is not positive; as $b \neq a$, $b - a \neq 0$, and it would follow that $b - a$ is negative. If $b < a$ also fails, we see similarly that $a - b$ is negative. Since the sum of negative numbers is negative, we would have $(b - a) + (a - b)$ being negative. But this is a contradiction since 0 is not negative.
12. (a) -5 (b) 13 (c) 5 (d) 5 (e) $\pi - 3$ (f) $3\sqrt{2} - 4$
13. (a) 2 (b) $\frac{5}{3}$ (c) 20 (d) $\frac{7}{3}$ (e) 16 (f) 14
14. If $a > 0$, then $|a| = a$, a is positive, $-a$ is negative, $-a < 0$, and $|-a| = -(-a) = a = |a|$. If $a < 0$, then $|a| = -a$, a is negative, $-a$ is positive, $-a > 0$, and $|-a| = -a = |a|$. If $a = 0$, then $|a| = 0$, $-a = 0$, and $|-a| = 0 = |a|$.
15. (a) inconsistent (b) identity (c) neither (d) identity
16. (a) $\{\frac{8}{3}\}$ (b) $\{-2\}$
17. (a) $(x - \frac{25}{2})^2 - \frac{625}{4}$ (b) $(t + \frac{3}{14})^2 - \frac{9}{196}$
    (c) $5(x + 3)^2 - 45$
18. (a) $\{-4, \frac{3}{2}\}$ (b) $\{-3\sqrt{2}, \sqrt{2}\}$
19. (a) $\{\frac{1}{2}, -\frac{3}{2}\}$ (b) $\left\{\frac{-1 + 2\sqrt{3}}{2}, \frac{-1 - 2\sqrt{3}}{2}\right\}$
20. (a) $\{3, -2\}$ (b) $\{1, -\frac{1}{3}\}$
21. $4(\sqrt{10} + \sqrt{6})$ ft by $4(\sqrt{10} - \sqrt{6})$ ft

22. 23.

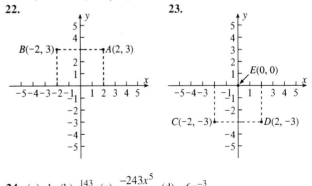

24. (a) 1 (b) $\frac{143}{9}$ (c) $\frac{-243x^5}{32}$ (d) $-6x^{-3}$
25. (a) $\frac{-9x^{13}}{y^8}$ (b) $\frac{2u^3}{v^2}$ (c) $\frac{u^9}{27v^3}$ (d) $\frac{-32x^5}{y^6}$
26. (a) (i) 5 (ii) 3 (b) (i) 3 (ii) -1 (c) (i) 0 (ii) 12
    (d) They do not exist.
27. (a) $3x^4 - 3x^3 + x^2 + 5x - 4$ (b) $x^5 + 5x^3 - 4x^2 + 3x + 2$
    (c) $-15x^5 + 18x^4 - 9x^3 + x^2 + x - 2$
28. (a) $18x^2 + 23x - 6$ (b) $9x^4 + 6x^2y + y^2$
    (c) $64u^9 - 96u^6v + 48u^3v^2 - 8v^3$

Answers to Exercises

(d) $u^4 + 12u^3v + 54u^2v^2 + 108uv^3 + 81v^4$

29. (a) $16x^4 - 96x^3y + 216x^2y^2 - 216xy^3 + 81y^4$
(b) $a^5 + 5a^4b + 10a^3b^2 + 10a^2b^3 + 5ab^4 + b^5$

30. (a) $(3xy - 5)(x + 2y)$ (b) $(\frac{3}{4}x - 2)^2$
(c) $(6xy^3 - 1)(6xy^3 + 1)$ (d) $(v + 4)(v^2 - 4v + 16)$
(e) $4(x - 5)(x + 2)$ (f) $(2x - 1)(x + 3)$

31. (a), (c), and (d)

32. (a) quotient $= 8x + 11$, remainder $= 44$
(b) quotient $= 2x - \frac{3}{2}$, remainder $= -\frac{119}{2}$
(c) quotient $= \frac{x}{3} - 2$, remainder $= -\frac{2}{3}x + 6$

33. (a) $\dfrac{x(x + 7)}{x + 6}$ (b) $\dfrac{x + 2y}{3xy + 5}$ (c) $\dfrac{x^2 + 4x + 16}{x - 4}$

34. (a) $\dfrac{x(x + 3)}{2y(x + 5)}$ (b) $\dfrac{(1 + 2x + 4x^2)(x + 1)}{(1 - x)(2x + 1)}$
(c) $\dfrac{2}{x^2 - 16}$ (d) $\dfrac{2(x^3 + x^2 - 2x - 1)}{x(x - 1)(x + 1)}$

35. $\dfrac{-(3x^2 + 3xh + h^2)}{x^3(x + h)^3}$

36. (a) no, $\dfrac{x^2 - x + 1}{x - 1}$ (b) yes (c) no, $\dfrac{x + 6}{x - 3}$ (d) no, $\dfrac{5}{x - 7x^2}$

37. (a) no solution (b) $\{-1, 2\}$ (c) $\{\frac{5}{3}\}$

38. (a) 4 (b) $-\frac{2}{3}$ (c) $\frac{4}{3}$ (d) $\frac{1}{2}$ (e) 0

39. (a) $(x^2 + y^2)^{1/2}$ (b) $4^{1/3}$ (c) $-z^{2/3}$ (d) $x^{8/15}$

40. (a) $\frac{1}{3}$ (b) 14 (c) $4x^6 + x^{-1/6}$ (d) $x^{8/21} + \dfrac{1}{x^3y^{3/2}} - 1$

41. (a) $-\pi$ (b) π (c) $\dfrac{\sqrt{7}}{3}$ (d) $2\sqrt[3]{x}$
(e) $2y\sqrt[5]{\dfrac{4}{x^2}}$ (f) $(2 + 4|x|)\sqrt{2}$

42. (a) $\dfrac{\sqrt{21}}{3}$ (b) $\dfrac{10\sqrt{3} + 5\sqrt{6}}{6}$ (c) $\dfrac{2\sqrt[3]{xy}}{y}$

43. (a) $\dfrac{6}{5(2\sqrt{3} + \sqrt{6})}$ (b) $\dfrac{3x}{\sqrt{6xy}}$ (c) $\dfrac{-7}{\sqrt{-7(x + h)} + \sqrt{-7x}}$

44. (a) $\dfrac{(x^3 - 1)(5x^3 + 1)}{x^2}$ (b) $\dfrac{3x}{2\sqrt{x + 1}}$
(c) $\dfrac{1 + x^2}{x^2(1 - x^2)^{1/2}}$ (d) $\dfrac{1 - 5x^3}{2(x^3 + 1)^2 x^{1/2}}$

45. $\{-62\}$ **46.** $\{5\}$ **47.** $\{\frac{5}{4}\}$

48. (a) $6x^2 + 17$ (b) $2|x^5 + 5|$ (c) $5y^2|x|$ (d) $\dfrac{x^2}{9|y|}$

49. $\{\frac{2}{3}, 4\}$ **50.** $\{-\frac{1}{5}, 9\}$
51. $\{-1, 0, 1\}$ **52.** $\{0\}$ **53.** $\{9\}$

Answers to Odd-Numbered Exercises in Chapters 1–10

Section 1.1

1. $-6 > -12$
3. $-8 = 2(-4)$ and $-12 = 3(-4)$, but $c = -4$ is not positive.
7. $(-\infty, \frac{1}{2}]$
9. $(\frac{21}{8}(-\pi - 2), \infty)$ $\quad -\frac{21}{8}(2+\pi) \approx -13.497$
11. $(-\infty, \frac{46}{45})$
13. $[-8, \infty)$
15. $[-3, 4)$
17. $(0, 10)$
19. $(-\infty, -2) \cup (\frac{14}{3}, \infty)$
21. $(-\infty, -3] \cup [\sqrt{2}, \infty)$
23. $(-4, 3)$
25. $[-\sqrt{2}, \sqrt{2}\,]$
27. $\left(\frac{-1-\sqrt{5}}{2}, \frac{-1+\sqrt{5}}{2}\right)$
29. $[1, 2] \cup [3, 4]$
31. $(0, \frac{3}{2}) \cup (\frac{3}{2}, 4)$
33. $(\frac{3}{2}, 3) \cup (4, \infty)$
35. $(-\infty, -3) \cup [8, \infty)$
37. $(-3, -2)$
39. $(-1, 4)$
41. $[2.3, 2.7]$
43. The solution set is empty.
45. $(-\infty, -\frac{4}{3}] \cup [\frac{4}{3}, \infty)$

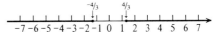

645

Answers to Exercises

47. $(-\infty, \pi - 1] \cup [\pi + 1, \infty)$

49. $[-\frac{1}{2}, 1) \cup (1, \frac{5}{2}]$

55. $(-5, 5)$ **57.** $[17, 33]$ **59.** The solution set is empty.

61. $[5, \infty)$ **63.** $(-3, 2)$ **65.** $[20, \frac{260}{9}]$

67. $61\frac{1}{7}$ miles **69.** $\left[\frac{2-\sqrt{3}}{2}, \frac{2+\sqrt{3}}{2}\right]$ **71.** $0 < R_2 < 4$

Section 1.2

1. **3.**

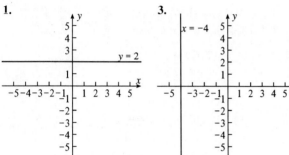

5. **7.**

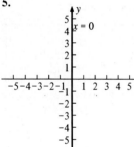 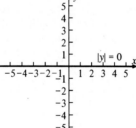

9. **11.**

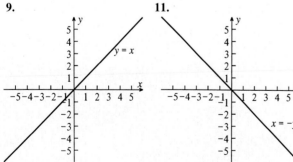

13.

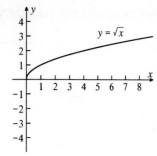

15. (a) 3 (b) $(-2, -\frac{9}{2})$ **17.** (a) 10 (b) (6, 11)

19. (a) $\sqrt{11}$ (b) $\left(-2, \frac{\sqrt{2}}{2}\right)$ **21.** 2 and 14 **25.** yes **27.** no

29. $x^2 + y^2 = 25$ **31.** $3x^2 + 4y^2 = 12$ **33.** $(\frac{1}{16}, \frac{1}{24})$ **35.** $\frac{1}{3}$

37. -1 **39.** (a) ℓ_2 (b) ℓ_4 (c) ℓ_3 (d) ℓ_1 **41.** ℓ_1

43. **45.**

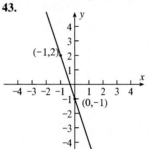 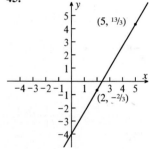

47. yes **49.** no **51.** no, since $m_{AB} = \frac{7}{12} \neq \frac{72}{19} = m_{AC}$

53. yes **55.** no

Section 1.3

1. (a) $x = -4$ (b) $x = -2$ (c) $x = 0$ (d) $x = 3$

3. $2x + y = -6$ **5.** $x - 3y = 12$ **7.** $y = 1$

9. $y = 2x - 8$ **11.** $y = \frac{1}{3}$ **13.** $x + 2y = 2\sqrt{14}$

15. (a) 3 (b) $(0, -4)$ **17.** (a) 0 (b) $(0, \frac{7}{2})$

19. (a) (b)

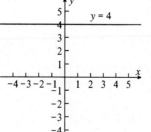 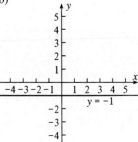

Answers to Exercises

(c)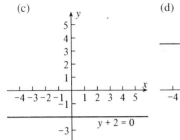

(d)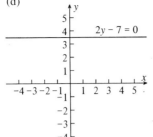

21. (a) $8x + 10y = -51$ (b) $5x - 4y = -37$
23. (a) $y = \pi$ (b) $x = \sqrt{2}$ **25.** $y - 3x - 7 = 0$
27. 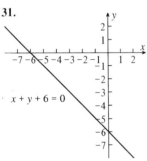 **29.**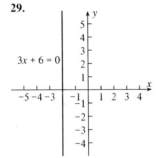

31.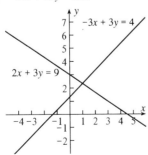

33. 1 **35.** $\frac{1}{\sqrt{5}}$ **37.** 0
39. $(-3x + 4y + 7)^2 = 100$ or
 $9x^2 + 16y^2 - 24xy - 42x + 56y = 51$.
41. consistent with a unique particular solution

43. dependent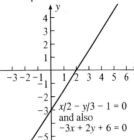

45. inconsistent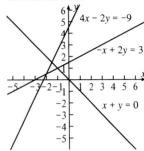

47. (a) -12 (b) none (c) any value of $k \neq -12$
49. (a) none (b) all values of k (c) none
51. (b) **55.** $(x, y) = (2, -1)$ **57.** $(x, y) = (-2, -13)$
59. $(x, y, z) = (4, 1, 6)$ **61.** $y = 0.45x$
63. (a) 650 miles (b) $242.50
65. 3080 **67.** $y = 0.52x - 85$ **69.** $a = -\frac{2}{19}$, $b = -\frac{3}{19}$

Section 1.4

1. $x^2 + y^2 + 4x - 8y + 11 = 0$ **3.** $x^2 + y^2 - 2\pi y + \pi^2 - 2 = 0$
5. $x^2 + y^2 - 169 = 0$ **7.** $4x^2 + 4y^2 - 4x + 8y - 27 = 0$
9. center $(-3, 4)$, radius 4 **11.** center $(-\frac{2}{3}, \frac{1}{3})$, radius 3
13. $\{(-3, 4)\}$ **15.** empty set
17. vertex $(0, 0)$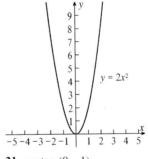

19. vertex $(0, 0)$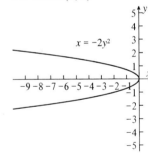

21. vertex $(0, -1)$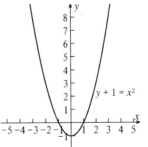

23. vertex $(0, -1)$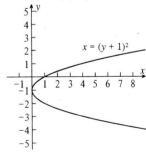

25. $x^2 + y^2 + 2x - 4y + 1 = 0$ **27.** $x^2 + y^2 - 1 = 0$
29. $y = \frac{2}{5}\sqrt{6}x + 7$ and $y = -\frac{2}{5}\sqrt{6}x + 7$
31. $y = 4$ and $45y - 28x - 68 = 0$ **33.** 38.5 miles
35. $y = -3x + 2\sqrt{10}$ and $y = -3x - 2\sqrt{10}$

Answers to Exercises

37. $y = -2x + 5 + 3\sqrt{5}$ and $y = -2x + 5 - 3\sqrt{5}$
39. The line intersects the circle at $(-4, -3)$ and $(3, 4)$.
41. The solution set is empty; the line does not intersect the circle.
43. The solution set is $\{(0, 5), (0, -5), (5, 0), (-5, 0)\}$; the axes intersect the circle four times.
45. The solution set is $\{(-2, 0)\}$. The circle does not intersect the y–axis; and the x-axis is tangent to the circle at $(-2, 0)$.
47. The solution set is $\{(-4, -1), (2, 5)\}$. The circles intersect twice.
49. The solution set is $\{(0, 0)\}$. The circles "touch" at the origin.
51. The solution set is $\{(\sqrt{2}, -2), (-\sqrt{2}, -2)\}$. The circle and the parabola intersect twice.
53. The solution set is empty. The line does not intersect the parabola.
55. $\{(3, 2), (-3, -2), (2, 3), (-2, -3)\}$ **57.** $\{(-8, 10), (1, 4)\}$

Section 1.5

1. (a) $4i$ (b) $\sqrt{15}i$ (c) $\sqrt{\pi}i$ (d) $\frac{2}{3}i$
3. (a) $3\sqrt{5}i$ (b) $-5i$ (c) -21 (d) $\sqrt{3}$
5. (a) (i) $\frac{3}{2}$ (ii) -5 (b) (i) $-\sqrt{2}$ (ii) 0 (c) (i) 0 (ii) 7
 (d) (i) $-\frac{3}{2}$ (ii) 5 (e) (i) 7 (ii) 0 (f) (i) 0 (ii) 0
7. (a) $\frac{11}{4} + (-2)i$ (b) $-2 + \sqrt{2} + \frac{1}{3}i$
 (c) $1 + (-2)i$ (d) $-6 + 10i$
9. (a) $-14 + 22i$ (b) $14 + 22i$ (c) $\frac{31}{9} + (-\frac{11}{3})i$ (d) $7 + 7i$
11. (a) $\frac{3}{5} + \frac{4}{5}i$ (b) $-1 + \frac{\sqrt{2}}{3}i$ (c) $-\frac{20}{41} + (-\frac{16}{41})i$ (d) $-4 + 3i$
13. (a) $-4 + 2i$ (b) $-46 + 9i$ (c) $-22 + (-23)i$ (d) $0 + (-1)i$
15. (a) $\sqrt{3} - \frac{1}{2}i$ (b) $2 + i$ (c) $-2 - i$ (d) $-3i$ (e) -5 (f) 0
19. (a) $\frac{5}{2}$ (b) $\sqrt{3}$ (c) $\frac{\sqrt{265}}{13}$ (d) 7 **25.** $-\frac{11}{3} - 3i$
27. $x = -\frac{5}{4}i$ **29.** $-\frac{12}{5} + \frac{24}{125}i$ **31.** $x = 8 + 6i$
33. discriminant $= -23$; two nonreal complex roots
35. discriminant $= 0$; a real double root
37. discriminant $= 16$; two unequal real roots **39.** $x = \dfrac{1 \pm \sqrt{13}}{3}$
41. $x = \pm 2\sqrt{\frac{2}{3}}i$ **43.** $x = \frac{2}{3}$ is a double root. **45.** $x = 1 \pm \frac{\pi}{2}$
47. $\{-2, 1 + \sqrt{3}i, 1 - \sqrt{3}i\}$ **49.** $\left\{\frac{1}{\sqrt{2}} + \frac{1}{\sqrt{2}}i, -\frac{1}{\sqrt{2}} - \frac{1}{\sqrt{2}}i\right\}$
51. $x = \pm v$, where $v = \sqrt{\dfrac{1 + \sqrt{2}}{2}} + \dfrac{1}{\sqrt{2 + 2\sqrt{2}}}i$
53. $\left\{-\frac{1}{2} + \left(\frac{1}{\sqrt{2}} - \frac{1}{2}\right)i, -\frac{1}{2} - \left(\frac{1}{\sqrt{2}} + \frac{1}{2}\right)i\right\}$
55. $\{1, -1, i, -i\}$ **57.** $\{3, -3, 2i, -2i\}$

Section 1.6 (Review Exercises)

1. $(-\infty, \frac{51}{29}]$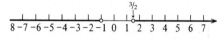
3. $(-\infty, -\frac{9}{5}) \cup (3, \infty)$
5. $[-3, 2]$
7. $(-\infty, -1) \cup (\frac{3}{2}, \infty)$
9. $(-6.01, -5.99)$
11. $(4, \infty)$
13. (a) (b) (c) (d)

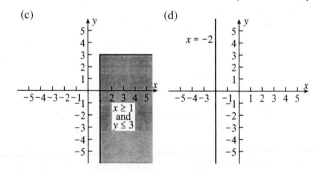

15. (a) yes (b) no **17.** (a) -3 (b) 0 (c) $\frac{5}{2}$

Answers to Exercises

19.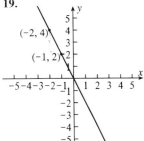

21. (a)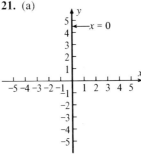

(b) inconsistent (c) dependent

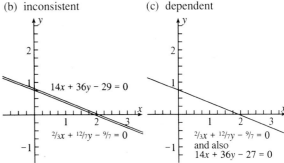

(b)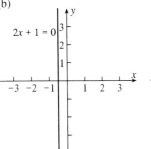

(c)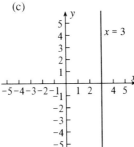

23. (a) $-\frac{1}{2}$ (b) $(0, -2)$
(c)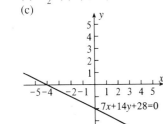

25. (a) consistent with a unique particular solution
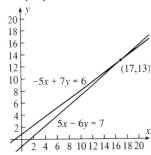

29. $(x, y) = (-2, 8)$ **31.** $y = 2.5x + 121.5$
33. (a) $x^2 + y^2 - 4x + 8y + 17 = 0$
(b) $x^2 + y^2 + 2x - 10y + 1 = 0$
(c) $25x^2 + 25y^2 - 300x - 100y + 999 = 0$
(d) $3x^2 + 3y^2 - 3x - 11y = 0$
35. empty set
37. $y = \left(\dfrac{-5 + \sqrt{153}}{16}\right)x + \dfrac{3 + \sqrt{153}}{8}$ and
$y = \left(\dfrac{-5 - \sqrt{153}}{16}\right)x + \dfrac{3 - \sqrt{153}}{8}$
39. The solution set is $\{(0, 1), (-1, 0)\}$. The line intersects the circle at $(0, 1)$ and $(-1, 0)$.
41. The solution set is empty. The circles do not intersect.
43. $\{(4, 2), (-4, -2), (2, 4), (-2, -4)\}$
45. (a) $4\sqrt{6}i$ (b) $-4i$ (c) $-\dfrac{2\sqrt{2}}{15}$ (d) 4
47. (a) $2 + (-2)i$ (b) $-2 + 3i$ (c) $-18 + 24i$ (d) $-5 + (-10)i$
(e) $28 + 23i$
49. (a) (i) $\dfrac{1}{\sqrt{2}} - \dfrac{1}{\sqrt{2}}i$ (ii) 1 (b) (i) $3 + 4i$ (ii) 5
(c) (i) $-\dfrac{\pi}{2}$ (ii) $\dfrac{\pi}{2}$ (d) (i) $-\sqrt{2}i$ (ii) $\sqrt{2}$
51. discriminant $= -44$; two nonreal complex roots;
$\left\{\dfrac{1}{3} + \dfrac{\sqrt{11}}{3}i, \dfrac{1}{3} - \dfrac{\sqrt{11}}{3}i\right\}$
53. discriminant $= 52$; two unequal real roots;
$\left\{\dfrac{-1 + \sqrt{13}}{3}, \dfrac{-1 - \sqrt{13}}{3}\right\}$
55. $\{-1 + i, 1 - i\}$

Section 2.1

1. yes **3.** no **5.** yes **7.** yes **9.** yes **11.** no
13. (a) At any given time (in a monogamous society), no woman could have more than one husband.
(b) the set of women who were married at the given time
(c) the set of men who were married at the given time
(d) no

Answers to Exercises

15. yes. If two pairs of elements of the type (x, y_1) and (x, y_2) are in the subset, they would also be in the given function, a contradiction.

17. $\{x: x \leq 4\}$ or $(-\infty, 4]$ **19.** $\{x: -2 \leq x \leq 2\}$ or $[-2, 2]$
21. R or $(-\infty, \infty)$ **23.** R **25.** R
27. $\{t: t \neq -\frac{3}{2}\}$ or $(-\infty, -\frac{3}{2}) \cup (-\frac{3}{2}, \infty)$
29. $\{t: t \neq 1\}$ **31.** $\{t: t > 2\}$
33. range is $\{y: y \geq 0\}$ or $[0, \infty)$
35. range is $\{y: 0 \leq y \leq 2\pi\}$ or $[0, 2\pi]$

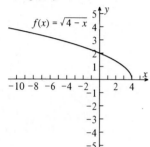
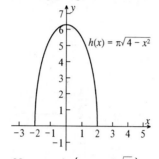

37. range is $\{y: y \geq 0\}$
39. range is $\{y: y \geq 8\sqrt{5}\}$ or $[8\sqrt{5}, \infty)$

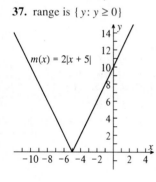
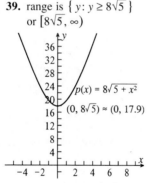

41. (a) $D = (-\infty, \infty)$, $R = (0, \infty)$ (b) $D = (0, \infty)$, $R = (-\infty, \infty)$
(c) $D = (-\infty, \infty)$, $R = [-1, 1]$ (d) $D = \left(-\frac{\pi}{2}, \frac{\pi}{2}\right)$, $R = (-\infty, \infty)$
(e) $D = [-1, 1]$, $R = \left[-\frac{\pi}{2}, \frac{\pi}{2}\right]$

43. (a) In Lavir, the cost of mailing less than 1 gram is 9 spendies, of mailing an envelope weighing at least 1 gram but less than 2 grams is 15 spendies, of mailing an envelope weighing at least 2 grams but less than 3 grams is 21 spendies, and so on.

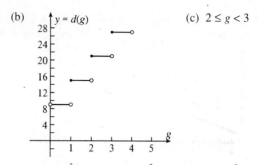

(b) $y = d(g)$ (c) $2 \leq g < 3$

45. (a) $f(0) = \pi\sqrt[3]{5}$ (b) $f(1) = \pi\sqrt[3]{4}$ (c) $f(-1) = \pi\sqrt[3]{6}$
(d) $f(2.7) = \pi\sqrt[3]{2.3}$ (e) $f(3 + h) = \pi\sqrt[3]{2 - h}$
(f) $f(x + h) = \pi\sqrt[3]{5 - x - h}$ (g) $f(-x) = \pi\sqrt[3]{5 + x}$
(h) $f\left(\frac{1}{x}\right) = \pi\sqrt[3]{5 - \frac{1}{x}}$

47. (a) $f(0) = 2$ (b) $f(1) = -2$ (c) $f(-1) = -2$
(d) $f(2.7) = -27.16$ (e) $f(3 + h) = -34 - 24h - 4h^2$
(f) $f(x + h) = -4x^2 + 2 - 8hx - 4h^2$ (g) $f(-x) = -4x^2 + 2$
(h) $f\left(\frac{1}{x}\right) = -\frac{4}{x^2} + 2$

49. (a) $f(0) = -6$ (b) $f(1) = -6$ (c) $f(-1) = -6$
(d) $f(2.7) = -6$ (e) $f(3 + h) = -6$ (f) $f(x + h) = -6$
(g) $f(-x) = -6$ (h) $f\left(\frac{1}{x}\right) = -6$

51. (a) $f(0) = \frac{5}{2}$ (b) $f(1) = 1$ (c) $f(-1) = 9$
(d) $f(2.7) = \frac{419}{470}$ (e) $f(3 + h) = \frac{5 + 3h + h^2}{5 + h}$
(f) $f(x + h) = \frac{x^2 + (2h - 3)x + h^2 - 3h + 5}{x + h + 2}$
(g) $f(-x) = \frac{x^2 + 3x + 5}{-x + 2}$ (h) $f\left(\frac{1}{x}\right) = \frac{1 - 3x + 5x^2}{x + 2x^2}$

53. (a) $f(0) = 6$ (b) $f(1) = 4$ (c) $f(-1) = 8$ (d) $f(2.7) = 0.6$
(e) $f(3 + h) = 2|h|$ (f) $f(x + h) = 2|x + h - 3|$
(g) $f(-x) = 2|x + 3|$ (h) $f\left(\frac{1}{x}\right) = 2\left|\frac{1}{x} - 3\right|$

55. (a) $f(0) = 0$ (b) $f(1) = 2$ (c) $f(-1) = 0$ (d) $f(2.7) = 5.4$
(e) $f(3 + h) = 3 + h + |3 + h|$ (f) $f(x + h) = x + h + |x + h|$
(g) $f(-x) = -x + |x|$ (h) $f\left(\frac{1}{x}\right) = \frac{1}{x} + \frac{1}{|x|}$

57. 5; 5 **59.** 0 if $h \neq 0$; 0 if $h \neq 0$ **61.** 2; $-\frac{1}{h}$ **63.** $-1; -1$
65. -1 **67.** $-14x - 7h + 3$ **73.** $A = A(d) = \frac{\pi d^2}{4}$ **75.** $25t$ miles

Answers to Exercises

77. (a) $f(x) = \begin{cases} \sqrt{x} & \text{if } 0 \le x < 4 \\ x - 1 & \text{if } 4 \le x \end{cases}$

(b) $f(4) = 4 - 1 = 3$
(c) $(0, 4)$ and $(4, \infty)$
(d) none

51.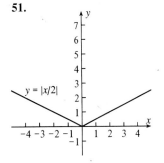

53.

Section 2.2

1. $y = -3x + 12$ **3.** $x = -\frac{1}{2}$ **5.** $(10, 9)$
7. 1 **9.** infinitely many
13. (a) $f(-x) = \frac{3}{7}(-x)^4 - 9(-x)^2 + \pi = \frac{3}{7}x^4 - 9x^2 + \pi = f(x)$
(b) $g(-x) = \frac{3}{7}(-x)^5 - \frac{2}{3}(-x)^3 + 4(-x) = -\frac{3}{7}x^5 + \frac{2}{3}x^3 - 4x = -g(x)$
15. symmetry with respect to origin **17.** none of these
19. symmetry with respect to y-axis
21. symmetry with respect to x-axis **23.** none of these
25. symmetry with respect to y-axis, x-axis, and origin
27. none of these **29.** symmetry with respect to x-axis
31. symmetry with respect to x-axis **33.** none of these
35. symmetry with respect to y-axis **37.** none of these
39. symmetry with respect to x-axis
41. symmetry with respect to y-axis

43.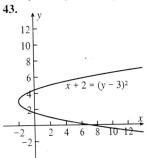

45.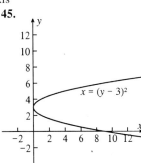

47.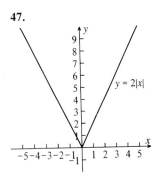

49.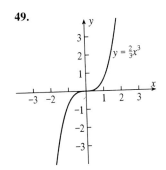

55.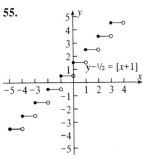

57.

Function	Domain	Range
g	$[-3, 9)$	same as the range of f
h	$[-7, 5)$	$\{-4y : y \text{ is in the range of } f\}$
k	$[-\frac{7}{4}, \frac{5}{4})$	same as the range of f

59. (a) $a = -2, b = 0, c = 1, d = 1$ (b) $a = 0, b = 3, c = 1, d = 1$
(c) $a = 3, b = 0, c = -1, d = 1$
(d) $a = 3, b = 0, c = -2, d = 1$ (e) $a = 2, b = -1, c = \frac{1}{2}, d = 1$

Section 2.3

1. increasing; y-intercept $(0, -4)$

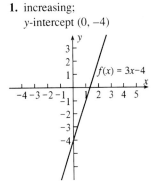

3. decreasing; y-intercept $(0, 0)$

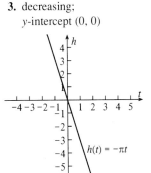

Answers to Exercises

5. neither; y-intercept $(0, -\sqrt{7})$

7. 4511

11. Vertically stretch by the factor $\sqrt{2}$. The domain is $\mathbf{R}$; the range is $[0, \infty)$.

13. Move to new origin at $(-\sqrt{3}, -5)$. The domain is $\mathbf{R}$; the range is $[-5, \infty)$.

15. Move to new origin at $(-\sqrt{3}, 0)$, then vertically stretch (relative to the new axes) by the factor 2, and then reflect through the x-axis. The domain is $\mathbf{R}$; the range is $(-\infty, 0]$.

17. Move to new origin at $(7, -2\pi)$ and vertically stretch (relative to the new axes) by the factor 2. The domain is $\mathbf{R}$; the range is $[-2\pi, \infty)$.

19.

Exercise	Vertex	Axis
11	$(0, 0)$	$x = 0$
12	$(0, 0)$	$x = 0$
13	$(-\sqrt{3}, -5)$	$x = -\sqrt{3}$
14	$(7, -\pi)$	$x = 7$
15	$(-\sqrt{3}, 0)$	$x = -\sqrt{3}$
16	$(7, 0)$	$x = 7$
17	$(7, -2\pi)$	$x = 7$
18	$(-\sqrt{3}, -5)$	$x = -\sqrt{3}$

23. (a) $b = 0$ (b) none
(c) The quadratics with graph symmetric to the y-axis are given by $ax^2 + c$ (with $a \neq 0$). No quadratic has its graph symmetric to the x-axis.

25. $-\frac{3}{8}x^2 + \frac{9}{2}x - \frac{37}{2}$

27. $2(x - \frac{7}{4})^2 - \frac{89}{8}$; opens up; low point $(\frac{7}{4}, -\frac{89}{8})$

29. $-2(x + \frac{7}{4})^2 + \frac{89}{8}$; opens down; high point $(-\frac{7}{4}, \frac{89}{8})$

31. $-\pi(x - 0)^2 + \pi$; opens down; high point $(0, \pi)$

33. (a) y-intercept $(0, -4)$; x-intercepts $\left(\frac{3 + \sqrt{41}}{4}, 0\right)$ and $\left(\frac{3 - \sqrt{41}}{4}, 0\right)$

(b) y-intercept $(0, 4)$; no x-intercepts

(c) y-intercept $(0, 1)$; x-intercept $\left(-\frac{1}{\sqrt{\pi}}, 0\right)$

(d) y-intercept $(0, 2)$; x-intercepts $\left(\sqrt{\frac{2}{3}}, 0\right)$ and $\left(-\sqrt{\frac{2}{3}}, 0\right)$

(e) y-intercept $(0, 0)$; x-intercept $(0, 0)$

35. In succession, you obtain the graphs of $y = 2x^2$; $y + 4 = 2(x - 3)^2$; and $-y + 4 = 2(x - 3)^2$. The last-obtained equation is equivalent to $y = -2(x - 3)^2 + 4$.

39. (a) F is magnified by a factor of k.
(b) If r is magnified by a factor of k, then F is diminished by a factor of $\frac{1}{k^2}$, shrinking to $\frac{F}{k^2}$.
(c) If two bodies are the same distance from a third body, the ratio of the gravitational forces that they exert on the third body is the same as the ratio of their masses. If two bodies have the same mass, the ratio of the gravitational forces that they exert on a third body is the reciprocal of the square of the ratio of their distances from the third body.

41. The "two" numbers are each equal to 2. (There is no solution if you insist that the numbers be distinct.)

45. $\left(\sqrt{\frac{3}{8}}, \frac{1}{4}\right)$ and $\left(-\sqrt{\frac{3}{8}}, \frac{1}{4}\right)$

Section 2.4

	New Function	Domain	Range
1.	$(f + g)(x) = -x^2 + x + 12$	$(-\infty, \infty)$	$(-\infty, \frac{49}{4}]$
	$(f - g)(x) = x^2 + x - 6$	$(-\infty, \infty)$	$[-\frac{25}{4}, \infty)$
	$(f \cdot g)(x) = -x^3 - 3x^2 + 9x + 27$	$(-\infty, \infty)$	$(-\infty, \infty)$
	$(f/g)(x) = \frac{x+3}{9-x^2}$	$\{x : x \neq 3$ and $x \neq -3\}$	$\{y : y \neq 0$ and $y \neq \frac{1}{6}\}$
3.	$(f + g)(x) = \sqrt{x - 3} + 2\sqrt{x}$	$[3, \infty)$	$[2\sqrt{3}, \infty)$
	$(f - g)(x) = \sqrt{x - 3} - 2\sqrt{x}$	$[3, \infty)$	$(-\infty, -3]$
	$(f \cdot g)(x) = 2\sqrt{x^2 - 3x}$	$[3, \infty)$	$[0, \infty)$
	$(f/g)(x) = \frac{1}{2}\sqrt{\frac{x-3}{x}}$	$[3, \infty)$	$[0, \frac{1}{2})$
5.	$f + g = \{(-1, 0), (3, 0)\}$	$\{-1, 3\}$	$\{0\}$
	$f - g = \{(-1, 4), (3, 8)\}$	$\{-1, 3\}$	$\{4, 8\}$
	$f \cdot g = \{(-1, -4), (3, -16)\}$	$\{-1, 3\}$	$\{-4, -16\}$
	$f/g = \{(-1, -1), (3, -1)\}$	$\{-1, 3\}$	$\{-1\}$

	$f \circ g$	Domain of $f \circ g$	Range of $f \circ g$	$g \circ f$	Domain of $g \circ f$	Range of $g \circ f$
11.	$(2x - 1)^4$	$(-\infty, \infty)$	$[0, \infty)$	$2x^4 - 1$	$(-\infty, \infty)$	$[-1, \infty)$
13.	$\sqrt{x} + 1$	$[0, \infty)$	$[1, \infty)$	$\sqrt{x+2} - 1$	$[-2, \infty)$	$[-1, \infty)$
15.	$\frac{1}{x^2} + 2$	$\{x : x \neq 0\}$	$(2, \infty)$	$\frac{1}{x^2 + 2}$	$(-\infty, \infty)$	$(0, \frac{1}{2}]$
17.	$\|x\|$	$(-\infty, \infty)$	$[0, \infty)$	$\|x + 2\| - 2$	$(-\infty, \infty)$	$[-2, \infty)$
19.	$2(x - 3)^{4/5}$	$(-\infty, \infty)$	$[0, \infty)$	$2x^{4/5} - 3$	$(-\infty, \infty)$	$[-3, \infty)$

Answers to Exercises

$f \circ g$	Domain of $f \circ g$	Range of $f \circ g$	$g \circ f$	Domain of $g \circ f$	Range of $g \circ f$
21. if n is odd, x	$(-\infty, \infty)$	$(-\infty, \infty)$	x	$(-\infty, \infty)$	$(-\infty, \infty)$
if n is even, $\lvert x \rvert$	$(-\infty, \infty)$	$[0, \infty)$	x	$[0, \infty)$	$[0, \infty)$

23. $\{(-1, 3), (3, 1)\}$ $\{-1, 3\}$ $\{3, 1\}$ $\{(2, -1)\}$ $\{2\}$ $\{-1\}$
25. $f(x) = 2x^{4/5}$, $g(x) = x - 3$ **27.** $f(x) = -3x^2 - \pi$, $g(x) = x + 4$
29. $f = \{(1, 2), (5, -1), (4, \sqrt{2})\}$, $g = \{(0, 1), (1, 5), (\pi, 4)\}$
31. $A = 9\pi t^{4/5} = (f \circ r)(t)$, where $f(x) = \pi x^2$ **47.** yes
49. no **51.** no **53.** yes **55.** no
57. $\left\{(0, 0), (7, 2), \left(\dfrac{7\pi}{2}, \pi\right), \left(-\dfrac{7}{2}, -1\right)\right\}$
59. $yx^2 = 7$ **61.** $(y + 1)(x - 2) = 3$ **63.** $3\sqrt{y-1} = x$
67. Call the given function f. Its domain is $\{0, 2, \pi, -1\}$; its range is $\left\{0, 7, \dfrac{7\pi}{2}, -\dfrac{7}{2}\right\}$. As f is one-to-one, f^{-1} is a function, namely $\left\{(0, 0), (7, 2), \left(\dfrac{7\pi}{2}, \pi\right), \left(-\dfrac{7}{2}, -1\right)\right\}$, with domain $\left\{0, 7, \dfrac{7\pi}{2}, -\dfrac{7}{2}\right\}$ and range $\{0, 2, \pi, -1\}$.

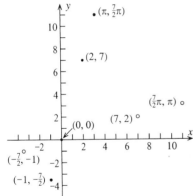

69. no
71. h has range $[\tfrac{40}{3}, \infty)$. There is an inverse function, given by the rule $h^{-1}(x) = -\tfrac{2}{3}x + \tfrac{8}{9}$, with domain $[\tfrac{40}{3}, \infty)$ and range $(-\infty, -8]$.

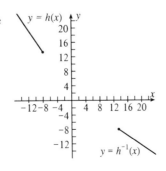

73. k has domain and range equal to $(-\infty, \infty)$. There is an inverse function, given by the rule $k^{-1}(x) = \sqrt[3]{x+7}$, with domain and range each equal to $(-\infty, \infty)$.

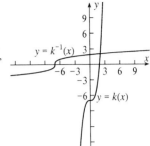

75. no
77. q has range $(-\infty, 0)$. There is an inverse function given by the rule $q^{-1}(x) = \dfrac{-2}{x}$, with domain $(-\infty, 0)$ and range $(0, \infty)$. [Note: $q \neq q^{-1}$.]

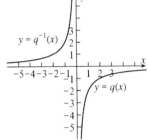

79. s has domain and range each equal to $(-\infty, \infty)$. So does the inverse function, which is given by the rule $s^{-1}(x) = \dfrac{x^3 + 7}{2}$.

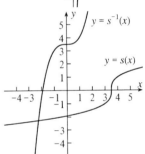

81. v has domain $[3, \infty)$ and range $[0, \infty)$. The inverse function is given by the rule $v^{-1}(x) = x^2 + 3$, with domain $[0, \infty)$ and range $[3, \infty)$.

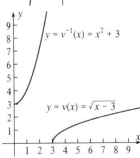

83. F has domain $\{x: x \neq 2\}$ and range $\{y: y \neq \tfrac{6}{5}\}$. The inverse function is given by the rule $F^{-1}(x) = \dfrac{-10x + 24}{-5x + 6}$, with domain $\{x: x \neq \tfrac{6}{5}\}$ and range $\{y: y \neq 2\}$.
85. no

Answers to Exercises

Section 2.5 (Review Exercises)

1. yes **3.** yes
5. domain = {−1, 1, π}; range = {0, 1}
7. domain = (−∞, ∞); range = {√2}

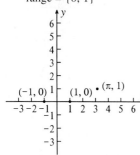
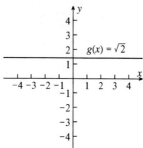

9. domain = [−3, 3]; range = [−9, 0]

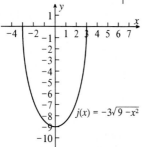

11. no, since, for instance, $f(-1) = -1 \neq 1 = g(-1)$

$f(0)$	$f(-1)$	$f(3.5)$	$f\left(\dfrac{2}{x}\right)$	$\dfrac{f(x+h) - f(x)}{h}$
13. $\dfrac{1}{2}$	$\dfrac{9}{2} + \pi$	$\dfrac{99}{2} - 3.5\pi$	$\dfrac{x^2 - 2\pi x + 16}{x^2}$	$8x + 4h - \pi$
15. $\sqrt{5}$	$\sqrt{6}$	$\sqrt{1.5}$	$\sqrt{5 - \dfrac{2}{x}}$	$\dfrac{-1}{\sqrt{5-x-h} + \sqrt{5-x}}$

17. (a) decreasing (b) increasing (c) neither (d) neither
19. (−9, −10)
21. symmetry with respect to origin and the line $y = x$
23. symmetry with respect to x-axis
25. symmetry with respect to y-axis, x-axis, origin, and the line $y = x$.

27.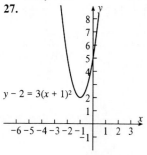
29.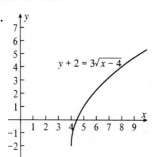

31. vertex (0, 0); axis $x = 0$; high point (0, 0)
33. vertex (3, −7); axis $x = 3$; low point (3, −7)
35. $\dfrac{7 + \sqrt{113}}{4}$ and $\dfrac{7 - \sqrt{113}}{4}$
37. Translate to new origin at (2, 120), stretch vertically (relative to the new axes) by the factor 24, and reflect through the new horizontal axis.
39. (0, 0) **41.** 360 watts
43. $f + g = \{(1, 2\pi), (2, 1), (3, 2 + \sqrt{2})\}$,
domain = {1, 2, 3}, range = {2π, 1, 2 + √2}
$f - g = \{(1, 0), (2, -1), (3, \sqrt{2} - 2)\}$,
domain = {1, 2, 3}, range = {0, −1, √2 − 2}
$f \cdot g = \{(1, \pi^2), (2, 0), (3, 2\sqrt{2})\}$,
domain = {1, 2, 3}, range = {π², 0, 2√2}
$f/g = \left\{(1, 1), (2, 0), \left(3, \dfrac{1}{\sqrt{2}}\right)\right\}$,
domain = {1, 2, 3}, range = $\left\{1, 0, \dfrac{1}{\sqrt{2}}\right\}$
$f \circ g = \{(3, 0), (2, \pi), (-3, 0)\}$,
domain = {3, 2, −3}, range = {0, π}
45. $(f + g)(x) = \sqrt{x} + x^2 + 4$, domain = [0, ∞), range = [4, ∞)
$(f - g)(x) = \sqrt{x} - x^2 - 4$, domain = [0, ∞),
range = $(-\infty, 4^{-1/3} - 4^{-4/3} - 4]$
$(f \cdot g)(x) = x^{5/2} + 4\sqrt{x}$, domain = [0, ∞), range = [0, ∞)
$(f/g)(x) = \dfrac{\sqrt{x}}{x^2 + 4}$, domain = [0, ∞), range = $\left[0, \dfrac{3\sqrt{2}}{16\sqrt[4]{3}}\right]$
$(f \circ g)(x) = \sqrt{x^2 + 4}$, domain = (−∞, ∞), range = [2, ∞)
47. $(f \circ g)(x) = 3\sqrt{x - 2} - 4$, domain = [2, ∞), range = [−4, ∞)
49. $(f \circ g)(x) = 3(2x - 5)^{2/7}$, domain = (−∞, ∞), range = [0, ∞)
51. (a) $f \circ g$, where $f(x) = 3x^{7/2} + 6$ and $g(x) = 5x + 2$
(b) $f \circ g$, where $f(x) = x^3$ and $g(x) = x^{1/3}$
(c) $f \circ g$, where $f(x) = \dfrac{x^2 - 2x + 1}{x - 2}$ and $g(x) = x + 1$ for $x \neq 1$
53. (a) yes (b) no (c) no (d) yes
55. f has range $\left(\dfrac{1}{3}, \infty\right)$. The inverse function is given by $f^{-1}(x) = \dfrac{3}{2}(x + 1)$, with domain $\left(\dfrac{1}{3}, \infty\right)$ and range (2, ∞).

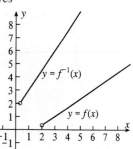

57. h has both domain and range equal to $(-\infty, \infty)$, and hence so does h^{-1}, which is given by

$$h^{-1}(x) = \sqrt[3]{\frac{x-7}{2}}.$$

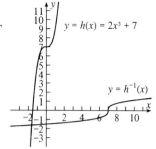

59. m has domain $(-\infty, -5) \cup [4, \infty)$ and range $[0, 1) \cup (1, \infty)$. The inverse function is given by

$$m^{-1}(x) = \frac{-9}{x^2 - 1} - 5,$$

with domain $[0, 1) \cup (1, \infty)$ and range $(-\infty, -5) \cup [4, \infty)$.

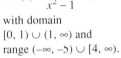
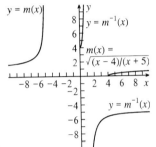

Section 3.1

1. (a) f, g, h, and m (b) all five
3. (a) a polynomial each of whose coefficients is an integer   (b) f and m
5. $(g+h)(x) = \frac{3}{2}x^7 + \sqrt{2}x^3 + 4x^2 - 7x + \frac{1}{2}$
   $(g-h)(x) = \frac{-3}{2}x^7 + \sqrt{2}x^3 - 4x^2 - 7x + \frac{1}{2}$
   $(g \cdot h)(x) = \frac{3}{\sqrt{2}}x^{10} - \frac{21}{2}x^8 + \frac{3}{4}x^7 + 4\sqrt{2}x^5 - 28x^3 + 2x^2$
   $(3g - 2h)(x) = -3x^7 + 3\sqrt{2}x^3 - 8x^2 - 21x + \frac{3}{2}$
9. Because of the closure property (see chapter 0), $P(k)$ is a real number for each real polynomial function P and real number k.
11. Any real quadratic, for instance x^2, will do.
13. f: even function, graph symmetric with respect to y-axis
    g: odd function, graph symmetric with respect to origin
    h: odd function, graph symmetric with respect to origin
    k: neither even nor odd, none of these symmetries

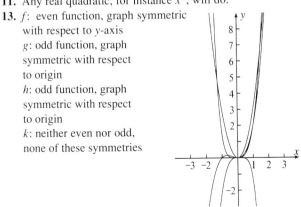

15. f: odd function, graph symmetric with respect to origin
    g, h, and k: neither even nor odd, none of these symmetries

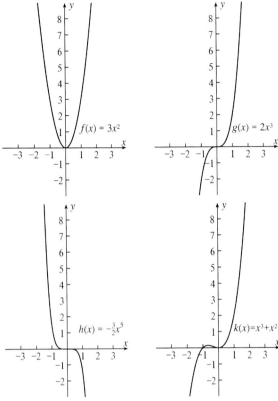

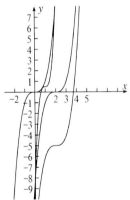

(continued)

Answers to Exercises

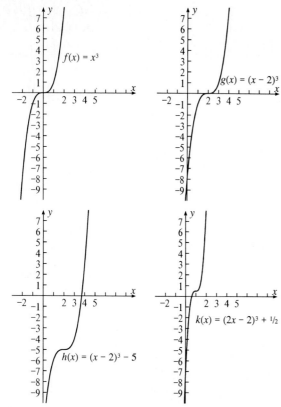

17. $(-\infty, \infty)$ in each case
19. Consider $a(x - h)^2 + r$ with $a > 0$; for instance, $x^2 + r$.
21. $0, \sqrt{2}$, and i 23. none
25. $q(x) = -2x^2 - 18x - 165, r(x) = -1478, k = 9$
27. $q(x) = -\frac{2}{3}ix^3 - \frac{11}{9}x, r(x) = -\frac{44}{9}x + 5$ 29. $q(x) = 0, r(x) = D(x)$
31. $P(-2) = 32 + 16 - 52 - 2 + 6 = 0$ and
    $P(3) = 162 - 54 - 117 + 3 + 6 = 0$.
33. When you divide $P(x)$ by $x - 7i$, (the quotient is $3x^2$ and) the remainder is 0.
35. When you divide $P(x)$ by $x - \frac{2}{5}$, (the quotient is
    $\frac{1}{2}x^2 + \frac{1}{5}x - \frac{23}{25}$ and) the remainder is $\frac{204}{125}$. Hence, $P(\frac{2}{5}) = \frac{204}{125}$.
37. $4, -\frac{2}{3}, \sqrt{2}$, and $-\sqrt{2}$ 39. $\frac{7 + \sqrt{23}i}{6}, \frac{7 - \sqrt{23}i}{6}, 0$, and 5
41. all complex numbers
47. $P(x) = (x - \frac{1}{2})(x + 3)x = x^3 + \frac{5}{2}x^2 - \frac{3}{2}x$ works; so does $cP(x)$ for any nonzero number c.
51. 6, 7, and 8 units

Section 3.2

1. $q(x) = 2x^2 + 18x + 159, r = 1424$
3. $q(x) = \frac{-3}{2}x^3 + 3x^2 - 7x + 14, r = -3$
5. $q(x) = \frac{3}{2}x^4 + \left(\frac{1}{2} - \frac{3}{2\sqrt{2}}\right)x^3 + \left(\frac{1}{2\sqrt{2}} + \frac{3}{4}\right)x^2 + \left(\frac{-1}{4} - \frac{3}{4\sqrt{2}}\right)x$
   $+ \left(\frac{1}{4\sqrt{2}} - \frac{1}{8}\right), r = \frac{1 + 19\sqrt{2}}{4\sqrt{2}}$
7. $q(x) = \sqrt{3}x + 5, r = 0$
9. $q(x) = x^2 + (8 - 2i)x + (46 - 20i), r = 263 - 166i$
15. $\frac{17}{3}$ 17. $-149 - \sqrt{2} - 191i$ 19. $-\frac{2}{3}$ and $2 - i$
21. 1 and $2 - i$ 23. 0, 1 25. $-1, 1$
27. $P(x) = ([(8x + 32)x + 0]x - 2)x + 5; P(-\frac{1}{2}) = \frac{5}{2}$;
    $P(2.7) = 1054.6088$
29. $P(x) = [(2ix + 16i)x + 4]x - 1; P(-\frac{1}{2}) = -3 + \frac{15}{4}i$;
    $P(2.7) = 9.8 + 156.006i$
31. $q(x) = x^3 + x^2 + 4x + 5, r = 21$
33. $q(x) = 2x^2 - 5x - 3, r = 0$
35. $P(2.3) = -5.661, P(2.4) = -1.352, P(2.5) = 3.625$,
    $P(2.6) = 9.312, P(2.7) = 15.751$, and so one root is approximately 2.45.

Section 3.3

1. continuous, not smooth 3. not continuous, not smooth
5. continuous, smooth 7. not continuous, not smooth
9. $P(x) = 14x^3 - 7x^2; P(a) = 84 < 147 < 315 = P(b)$; one such c, approximately 2.35
11. $P(x) = -3x^4; P(a) = 0 > -3 > -48 = P(b); c = 1$
13. The theorem does *not* say that if $a < c < b$ then $P(c)$ is between $P(a)$ and $P(b)$.
15. right: infinity; left: negative infinity
17. right: negative infinity; left: negative infinity
19. right: infinity; left: infinity
21. right: negative infinity; left: infinity
25. (a) For instance, consider x^2 or $x^2 - 2x + 1$ or $x^2 - 3x + 1$.
    (b) Consider $x^2 + a^2$ with $a \neq 0$; for instance, $x^2 + 1$.
29. Consider $x^2 - 2x + 1$ or $x^2 - 3x + 1$, for instance.
31. Limiting tendencies:
    to the right: infinity;
    to the left: negative
    infinity; symmetric about
    the point (1, 1); intercepts
    $(0, -1), \left(1 - \frac{1}{\sqrt[3]{2}}, 0\right)$

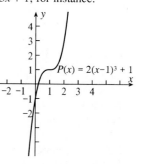

Answers to Exercises

33. Limiting tendencies: to the right: negative infinity; to the left: infinity; symmetric about the origin; intercept (0, 0)

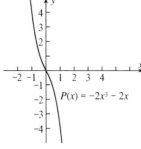

35. Limiting tendencies: to the right: infinity; to the left: negative infinity; symmetric about the point (0, −1); intercepts (0, −1) and three x-intercepts $(a, 0)$, $(b, 0)$, and $(c, 0)$, with $-2 < a < -1$, $-1 < b < 0$, and $1 < c < 2$

37. Limiting tendencies: to the right: infinity; to the left: infinity; symmetric about the y-axis; intercepts $(0, -1)$ and two x-intercepts $(-a, 0)$ and $(a, 0)$, with $1 < a < 2$

39. Limiting tendencies: to the right: negative infinity; to the left: negative infinity; symmetric about the y-axis; intercepts (0, 5) and two x-intercepts $(-a, 0)$ and $(a, 0)$, with $1 < a < 2$

41. 3.715
43. (a) [0, 24], possibly excluding an endpoint

Section 3.4

1. 7 3. 4; 3 5. 10; 2 7. 1; 1
9. $2x^2 + 10 = 0$ 11. $5x^3 + (-5\sqrt{2} - 5i)x^2 - x + (\sqrt{2} + i) = 0$

Root	Multiplicity	Behavior of Graph near Corresponding x-intercept
13. 0, 4, and −4	5 (each)	tangent to, and crosses, the x-axis
15. 2	1	crosses, but is not tangent to, the x-axis
$-\frac{1}{3}$	5	tangent to, and crosses, the x-axis
$\sqrt{\pi}i, -\sqrt{\pi}i$	1 (each)	no corresponding x-intercept
17. 2	1	crosses, but is not tangent to, the x-axis
1, −1	3 (each)	tangent to, and crosses, the x-axis
$-1 + \sqrt{3}i, -1 - \sqrt{3}i$	1 (each)	no corresponding x-intercept
$i, -i$	1 (each)	no corresponding x-intercept

19. $P(-2) = -8 + 12 - 6 + 2 = 0$; $-2, \dfrac{-1 + \sqrt{3}i}{2}$, and $\dfrac{-1 - \sqrt{3}i}{2}$

21. $\frac{1}{28}x^3 - \frac{1}{14}x^2 - \frac{5}{28}x + \frac{3}{14}$ 25. $x^3 - \frac{11}{2}x^2 + x + 2$

27. $x^4 + (-7 + \sqrt{3})x^3 + (11 - 2\sqrt{3})x^2 + (-7 + \sqrt{3})x + 10 - 2\sqrt{3}$

29. $x^4 + 6x^3 - 4x^2 - 64x - 64$

31. $x^6 - \frac{11}{6}x^5 - \frac{44}{9}x^4 + \frac{109}{9}x^3 - \frac{133}{9}x^2 + \frac{91}{18}x$

37. (a) Let $P(x) = x^2 - 6x + 13$. Then
$P(x)P(-x) = (x^2 - 6x + 13)(x^2 + 6x + 13)$
$= x^4 - 10x^2 + 169 = Q(x^2)$,
where $Q(x) = x^2 - 10x + 169$. By exercise 35, the roots of Q are r_1^2, r_2^2.

(b) $r_1, r_2 = \dfrac{6 \pm \sqrt{36 - 52}}{2} = \dfrac{6 \pm 4i}{2} = 3 \pm 2i$, and the roots of $x^2 - 10x + 169$ are $s_1, s_2 = \dfrac{10 \pm \sqrt{100 - 676}}{2} = \dfrac{10 \pm 24i}{2} = 5 \pm 12i$. Calculate $r_1^2 = (3 + 2i)^2 = 5 + 12i = s_1$ and $r_2^2 = (3 - 2i)^2 = 5 - 12i = s_2$.

41. $\pm 1, \pm 2, \pm 3, \pm 6, \pm 7, \pm 14, \pm 21, \pm 42, \pm\frac{1}{2}, \pm\frac{3}{2}, \pm\frac{7}{2}, \pm\frac{21}{2}, \pm\frac{1}{3}, \pm\frac{2}{3}, \pm\frac{7}{3}, \pm\frac{14}{3}, \pm\frac{1}{6}, \pm\frac{7}{6}$

43. $0, \pm 1, \pm\frac{1}{3}$ 45. $\pm 1, \pm 2, \pm 3, \pm 5, \pm 6, \pm 10, \pm 15, \pm 30$

658
Answers to Exercises

47. candidates: $\pm 1, \pm 2, \pm\frac{1}{3}, \pm\frac{2}{3}$
    actual roots: $2, \frac{1}{3}, i,$ and $-i$

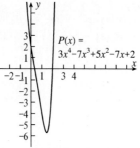
$P(x) = 3x^4 - 7x^3 + 5x^2 - 7x + 2$

49. candidates:
    $\pm 1, \pm 3, \pm 9, \pm\frac{1}{7}, \pm\frac{3}{7}, \pm\frac{9}{7}$
    actual roots:
    $-1, -1, -1, \frac{3}{7}, \sqrt{3}, -\sqrt{3}$

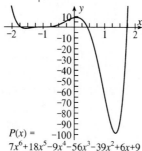
$P(x) = 7x^6 + 18x^5 - 9x^4 - 56x^3 - 39x^2 + 6x + 9$

51. Consider, for instance, $2\sqrt{2}x - 3\sqrt{2}$.

Section 3.5

1. (a) 1 (b) one (c) $-2x^5 + x^4 - \frac{3}{2}x^2 - 7$; 2
   (d) two or zero (e) yes
3. (a) 2 (b) two or zero (c) $2x^6 - \frac{3}{2}x^5 + x^4 + 4x^3 - 5x$; 3
   (d) three or one (e) do not know
5. (a) 3 (b) three or one (c) $-\frac{2}{7}x^3 - \frac{3}{5}x^2 - 4x - \frac{2}{3}$; 0
   (d) none (e) do not know
7. (a) 3 (b) three or one (c) $P(-x) = P(x)$; 3
   (d) three or one (e) yes
9. (a) 0 (b) none (c) $x^6 - x^5 + 2x^4 - 6x^3$; 3
   (d) three or one (e) do not know
15. $U = 1, L = -1$ 17. $U = 2, L = -2$
19. $U = 2, L = -3$; roots $2, -\frac{1}{2}, -2$
21. $U = 1, L = -1$; roots $1, 2i, -2i$
25. $(-\infty, -3) \cup (0, \frac{1}{2}) \cup (\frac{7}{3}, \infty)$ 27. $[-1, 2]$

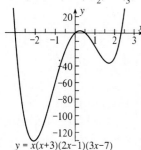
$y = x(x+3)(2x-1)(3x-7)$

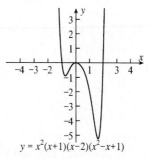
$y = x^2(x+1)(x-2)(x^2-x+1)$

29. $(-\infty, -3) \cup (\frac{4}{3}, \infty)$

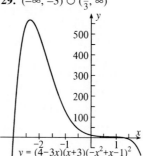
$y = (4-3x)(x+3)(-x^2+x-1)^2$

33. $(-\infty, \frac{1}{2}) \cup (6, 7) \cup (7, \infty)$

31. $[1, 2] \cup \{3\} \cup [4, \infty)$

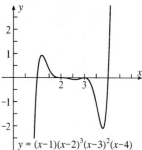
$y = (x-1)(x-2)^3(x-3)^2(x-4)$

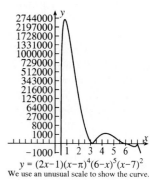
$y = (2x-1)(x-\pi)^4(6-x)^5(x-7)^2$
We use an unusual scale to show the curve.

35. $(-\infty, -\frac{2}{7}) \cup (1, \infty)$ 37. $(-\infty, -\frac{5}{2}] \cup [1, \frac{3}{2}]$
41. (a) at the initial midnight and at all times from noon until the midnight that ends the 24-hour period

Section 3.6

1. (a) $\{x: x \neq -2\}$ (b) 2 (c) yes, namely $\frac{x-2}{2+x}$
3. (a) $\{x: x \neq 2, -1\}$ (b) no zeros (c) no
5. (a) $\{x: x \neq 1\}$ (b) $2, -1$ (c) yes, namely $\frac{x^2-x-2}{x^2-2x+1}$

7.
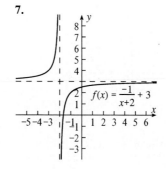
$f(x) = \frac{-1}{x+2} + 3$

9.
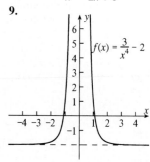
$f(x) = \frac{3}{x^4} - 2$

Answers to Exercises

11.
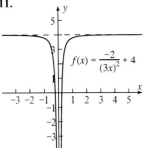

13.
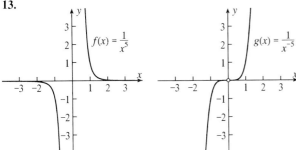

15. (a) $\dfrac{4x+3}{x^2+x-6}$, domain $\{x: x \neq -3, 2\}$

(b) $\dfrac{-2x+3}{x^2+x-6}$, domain $\{x: x \neq -3, 2\}$

(c) $\dfrac{3x}{(x-2)(x^2+x-6)}$, domain $\{x: x \neq -3, 2\}$

(d) $\dfrac{x^2+x-6}{3x^2-6x}$, domain $\{x: x \neq -3, 2, 0\}$

(e) $\dfrac{1}{\dfrac{-2x^2+x+12}{x^2+x-6}}$, domain $\left\{x: x \neq -3, 2, \dfrac{1+\sqrt{97}}{4}, \dfrac{1-\sqrt{97}}{4}\right\}$

21. Consider, for instance, x or $3x-7$ or x^3. (Any answer to exercise 14(b) works here, too.)

23. $x=-2, x=5$ **25.** $x=-1, x=0$ **27.** $x=-3$ **29.** none
31. $y=0$ **33.** does not exist **35.** $y=\dfrac{2}{9}$
37. $y=9x+1$ **39.** does not exist **41.** $y=-3x+4$
43.
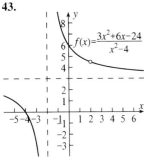

45.
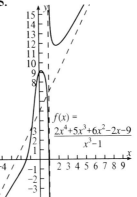

47.
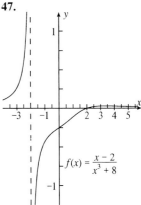

49.
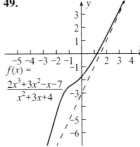

51. $(-\infty, -4) \cup (-3, -2) \cup (1, \infty)$ **53.** $(-1, 1)$
55. Consider $P(x) = -x^3 + 1$ and $D(x) = x+1$, with $n=3$ and $m=1$. As $|x|$ increases without bound, the limiting tendency of $\dfrac{P(x)}{D(x)}$ is negative infinity, not $-x^2$ (which is not a number or a "type of infinity" and hence cannot be a limiting tendency).
59. at least 25 ml

Section 3.7

1. $\dfrac{-3}{x+5} + \dfrac{4}{x-6}$ **3.** $\dfrac{2}{x+3} + \dfrac{-17}{(x+3)^2}$

5. $-x+1+\dfrac{1}{x}+\dfrac{-1}{x^2}+\dfrac{2}{x+2}$ **7.** $\dfrac{\frac{2}{3}}{x}+\dfrac{-\sqrt{2}}{x^2}+\dfrac{9}{3x-3\sqrt{2}}$

9. $1+\dfrac{-\frac{1}{3}}{x}+\dfrac{-\frac{5}{3}}{x+3}+\dfrac{1}{x-2}$ **11.** $\dfrac{\frac{5}{6}}{(x-6)^2}$

13. $\dfrac{-\frac{5}{4}}{x+3}+\dfrac{-\frac{3}{4}}{(x+3)^2}+\dfrac{\frac{5}{4}}{x+1}+\dfrac{-\frac{3}{4}}{(x+1)^2}$ **15.** $2x-3+\dfrac{3x-7}{2x^2+8}$

17. $\dfrac{x}{2}-1+\dfrac{1}{2x}+\dfrac{\frac{1}{2}x-\frac{1}{2}}{x^2+x+1}$ **19.** $\dfrac{3x-5}{x^2+9}+\dfrac{-2}{(x^2+9)^2}$

21. $\dfrac{-3}{x+1}+\dfrac{4}{x-1}+\dfrac{2}{x^2+1}$ **23.** $\dfrac{-x+2}{x^2+1}+\dfrac{-1}{x+1}+\dfrac{8}{(x+1)^2}$

Answers to Exercises

25. $\dfrac{-5}{x^2 + \sqrt{2}x + 1} + \dfrac{9}{x^2 - \sqrt{2}x + 1}$ 27. $\dfrac{\frac{3}{2}x + 2\pi}{x^2 + 6} + \dfrac{\frac{-1-2\pi}{2}}{x}$

29. $\dfrac{x - \frac{2}{5}}{x^2 + 8} + \dfrac{2x + \frac{2}{5}}{x^2 + 3}$ 31. $-1 + \dfrac{3}{(x-2)^2} + \dfrac{8}{x+2} + \dfrac{2}{(x+2)^2}$

33. $\dfrac{-7}{2x - 2} + \dfrac{\frac{1}{2}}{x + 4}$ 35. $\dfrac{x-1}{(3x^2 + 1)^4} + \dfrac{2}{x} + \dfrac{-3}{x^2}$

37. $\dfrac{\frac{232}{243}}{x - 2} + \dfrac{-\frac{155}{81}}{(x - 2)^2} + \dfrac{-\frac{124}{81}}{3x + 3} + \dfrac{\frac{58}{9}}{(3x + 3)^2}$

39. $\dfrac{\frac{6}{5}x - \frac{4}{5}}{x^2 + 4} + \dfrac{-6x - 36}{(x^2 + 4)^2} + \dfrac{-92x - 240}{(x^2 + 4)^3} + \dfrac{-\frac{6}{5}}{x - 1} + \dfrac{3}{(x - 1)^2}$

41.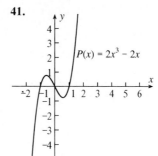

43.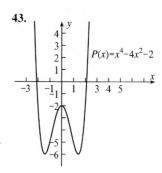

45.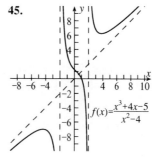

47. $\dfrac{4}{x - 3} + \dfrac{2}{(x - 3)^2} + \dfrac{-1}{x + 1}$ 49. $\dfrac{2x + 1}{x^2 + x + 1}$

51. $\dfrac{1}{x - 1} + \dfrac{1}{x + 1} + \dfrac{-1}{x^2 + 1}$

Section 4.1

1. 12.6029 3. 3020.2932 5. 12.5297 7. 200.3368

9. (a)

x	-3	-2	-1	0	1	1.5	2
$f(x) = 4^x$	0.0156	0.0625	0.25	1	4	8	16

(b) domain $(-\infty, \infty)$, range $= (0, \infty)$

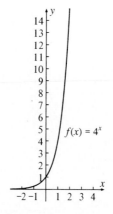

Section 3.8 (Review Exercises)

1. (a) f (b) f and g (c) all three
3. (a) domain $= (-\infty, \infty)$ = range
   (b) domain $= (-\infty, \infty)$, range $= [-3, \infty)$
   (c) domain $= \{x: x \neq 0\}$, range $= (3, \infty)$
   (d) domain $= \{x: x \neq 5\}$, range $= \{y: y \neq 4\}$
5. $P(2) = 48 - 8 - 68 + 58 - 30 = 0$, and so $x - 2$ is a factor of $P(x)$. Similarly, $P(-3) = 0$, and so $x + 3$ is a factor. Then $P(x) = (x - 2)(x + 3)(3x^2 - 4x + 5)$.
7. $q(x) = 2x^2 - 18x + 159$, $r(x) = -1424$
9. $q(x) = ix + 2$, $r(x) = 0$
11. $P(x) = [(-4x - 16)x + 4]x + 5$; $P(1.2) = -20.152$
13. $x^6 - 16x^5 + 105x^4 - 388x^3 + 819x^2 - 820x + 299$
15. $x^4 - (\sqrt{3} + 2)x^3 + (2\sqrt{3} + 1)x^2 - (\sqrt{3} + 2)x + 2\sqrt{3}$
17. (a) continuous, not smooth (b) continuous, smooth
    (c) not continuous, not smooth
    (d) not continuous, not smooth

Root	Multiplicity
21. $-2 + 2\sqrt{3}i, -2 - 2\sqrt{3}i, 4i, -4i, \frac{1}{3}$	1 (each)
-4	2
4	3
23. 0	2
$-1, -\frac{5}{2}, \frac{3}{2}, 1$	1 (each)
25. -1	2
$\frac{1}{5}, -2 + i, -2 - i$	1 (each)

27. (a) two or zero (b) one (c) yes
29. (a) five or three or one (b) three or one (c) yes
31. $(-\infty, -3] \cup \{0\} \cup [3, \infty)$ 33. $(-\frac{3}{2}, 2)$
35. (a) $\{x: x \neq 1, 2, 3\}$ (b) $\{x: x \neq -3 + \sqrt{2}, -3 - \sqrt{2}\}$
    (c) $\{x: x \neq 4\}$
37. $x = 3, y = 1$ 39. $x = -1, x = 3, y = 0$

11. (a)

x	−1	−0.5	−0.1	0	0.1	0.2	0.25
$h(x) = -2 \cdot 5^{3x}$	−0.016	−0.1789	−1.234	−2	−3.2413	−5.2531	−6.6874

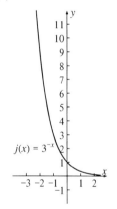

(b) domain = $(-\infty, \infty)$, range = $(-\infty, 0)$

15. (a)

x	−1	−0.5	−0.25	0	0.5	1	1.25
$g(x) = 3 \cdot 4^{-x}$	12	6	4.2426	3	1.5	0.75	0.5303

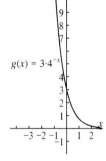

(b) domain = $(-\infty, \infty)$, range = $(0, \infty)$

13. (a)

x	−2	−1.5	−1	0	1	2	3
$j(x) = 3^{-x}$	9	5.1962	3	1	0.3333	0.1111	0.0370

(b) domain = $(-\infty, \infty)$, range = $(0, \infty)$

17. (a)

x	−3	−2	−1	0	0.5	1	1.5
$k(x) = 3^{x+1/2}$	0.0642	0.1925	0.5774	1.7321	3	5.1962	9

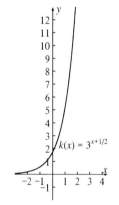

(b) domain = $(-\infty, \infty)$, range = $(0, \infty)$

662 Answers to Exercises

19. (a)

x	-2	-1.5	-1	0	1	1.5	2
$f(x) = 3^{|x|}$	9	5.1962	3	1	3	5.1962	9

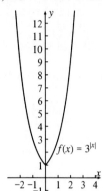

(b) domain $= (-\infty, \infty)$, range $= [1, \infty)$

21. (a)

x	-1.5	-1	-0.5	0	0.5	1	1.5
$h(x) = 4^x + 4^{-x}$	8.125	4.25	2.5	2	2.5	4.25	8.125

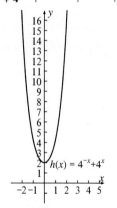

(b) domain $= (-\infty, \infty)$, range $= [2, \infty)$

23.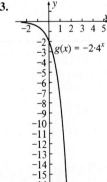

25.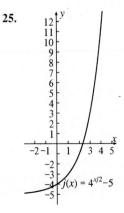

27.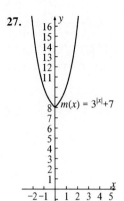

29. (a) $(0, 1)$ (b) $(0, 5)$
31. (a) f is increasing; g, h, and k are each decreasing.
(b) $y = 0$
33. (a) $\$4110.26$ (b) $\$4130.68$ (c) $\$4141.26$ (d) $\$4148.45$
(e) $\$4151.97$ (f) $\$4152.09$
35. $\$13.29$ **37.** $\$8468$ (of which $\$4468$ is interest)
39. 999.8784 grams after 1 year; 999.75682 grams after 2 years; 999.63525 grams after 3 years; 500 grams; 250 grams
41. $e^\pi \approx 23.140693$;
$\pi^e \approx 22.459158$;
$e^\pi > \pi^e$

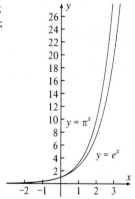

Section 4.2

1. $6^3 = 216$ **3.** $4^2 = 16$ **5.** $(\frac{1}{7})^2 = \frac{1}{49}$ **7.** $2^{-5} = \frac{1}{32}$
9. $9^{7/2} = 2187$ **11.** $\log_5 125 = 3$ **13.** $\log_2 256 = 8$
15. $\log_{7.2} (2687.3856) = 4$ **17.** $\log_5 (\frac{1}{125}) = -3$
19. $\log_4 512 = \frac{9}{2}$ **21.** $\log_{1/2} (\frac{1}{8}) = 3$ **23.** $\log_{1/5} (\frac{1}{125}) = 3$
25. -7 **27.** 0.5 **29.** 343 **31.** 1.3739687 **33.** 1.1325798
35. 2.9184628 **37.** -2.0061911 **39.** -0.964774

663
Answers to Exercises

41.

x	0.5	1	2	3	4	5	6
$g(x) = \log x$	−0.3010	0	0.3010	0.4771	0.6021	0.6990	0.7782

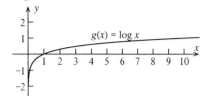

43.

x	0.5	1	2	3	4	5	6
$k(x) = \log_{15} x$	−0.2560	0	0.2560	0.4057	0.5119	0.5943	0.6616

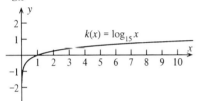

45.

x	0.5	1	2	3	4	5	6
$m(x) = \log_{1/4} x$	0.5	0	−0.5	−0.7925	−1	−1.1610	−1.2925

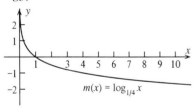

47. vertical asymptote
$x = -2$
x-intercept
$(-\frac{3}{2}, 0)$
increasing
function

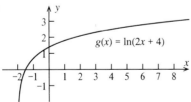

49. vertical asymptote
$x = 0$
x-intercept
$(500, 0)$
increasing
function

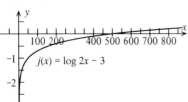

51. vertical asymptote
$x = 3$
x-intercepts
$(2, 0)$ and $(4, 0)$
neither
increasing
nor decreasing

53. vertical asymptote
$x = -4$
x-intercept
$(e^{-5/3} - 4, 0) \approx$
$(-3.811, 0)$
decreasing
function

55. vertical asymptote
$x = 0$
x-intercept
$(1, 0)$
decreasing
function

57. vertical asymptote
$x = 0$
x-intercept
$(\frac{1}{2}, 0)$
increasing
function

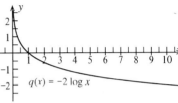

59. $10^{0.05} \approx 1.1220185$ **61.** $e^{0.75} \approx 2.117$
63. $12^{-2.5} \approx 0.0020047$ **65.** $e^{\ln 5/6.2} \approx 1.2963943$
67. $\frac{1}{8}$ **69.** $e^{2/3} \approx 1.947734$
71. $I = 10^R$; R increases by 1 if I is multiplied by 10; R increases by 2; R increases by 3
73. approximately 7.6484 years; 7.6253 years
75. approximately 7534.9901 years to decay down to 400 grams; 9900.7039 years; 13234.99 years; 18934.99 years
77. (a) x (b) x (c) $3x$ (d) $\frac{1}{x}$ (e) x (f) x (g) $\frac{1}{x}$ (h) $\frac{1}{x}$
79. $\dfrac{\ln\left(\frac{x}{2}\right)}{\ln 3}$; $\dfrac{\ln\left(\frac{x}{3}\right)}{\ln 2}$; $\dfrac{\ln\left(\frac{x}{c}\right)}{\ln b}$

Section 4.3

1. $\log 2 - \log 3$ **3.** $\ln 5 + \ln e^2 = \ln 5 + 2$
5. $\log_2 46 - \log_2 15 - \log_2 y = 1 + \log_2 23 - \log_2 15 - \log_2 y$
7. $\ln 15 + \ln x$
9. $2 + 2 \log_2 |a| + 3 \log_2 x - \log_2 3 - 4 \log_2 |b|$
11. $\frac{1}{7}(3 \log x + 3 \log y - \log(z + 1))$ **13.** $\log 28$ **15.** $\ln 12$

Answers to Exercises

17. $\log \frac{(x+1)^4}{(y-2)^3}$ **19.** $\ln 40$ **21.** $\log(3^{6/5} \cdot 2^{4/7})$ **23.** $\log \frac{5}{2}$
25. 0.9030 **27.** 1.1761 **29.** 1.5562 **31.** -1.1303
33. 2.3010 **35.** 0.58805 **41.** $x = -2 + \sqrt{6}$ **43.** $x = 2$
45. $x = 4$ **47.** $x = -2$ **49.** $x = \frac{21}{124}$
51. $10^{(x^4+1)\log(2\sqrt{x})}$ (Use $f^g = 10^{g \log f}$.)

Section 4.4

1. 90590
5. (a) $k = \frac{1}{20} \ln 2 \approx 0.0346574$ (b) approximately 51.7 years
7. (a) 112500
  (b) approximately 20.7 years after 1985, namely 2006
9. $\frac{1}{3} \ln(\frac{4}{3}) \approx 0.095894$
11. (b) only $\frac{1}{4}$ of the radioactive material remains at $t = \frac{\ln 4}{k}$; $\frac{1}{8}$ remains; $\frac{1}{12}$ remains
13. $k = \frac{1}{3} \ln(\frac{4}{3}) \approx 0.095894$; approximately 16.4565 hours after the initial observation
15. (a) $|T - A| = |T_0 - A|e^{-kt}$ since $|e^{-kt}| = e^{-kt}$
  (b) 20 minutes after the bath is drawn
  (c) $\frac{1}{10} \ln 2 \approx 0.0693147$
17. (a) 14.7 psi (b) approximately 10.727896 psi
  (c) $\frac{\ln 2}{0.21} \approx 3.3007009$ miles
  (d) approximately 14.865802 psi
19. (a) $\$1136.62$
  (b) $\frac{\ln 2}{0.075} \approx 9.242$ years after the investment
  (c) 18.484 (d) $\frac{\ln 5}{0.075} \approx 21.459$
21. (a) 2.5939942 A (b) 0.0202138 A (c) 0.0001362 A

Section 4.5

1. The solution set is $\{4\}$. **3.** $\{-\frac{5}{4}\}$
5. $x = \frac{\ln 129}{\ln 8} \approx 2.3370758$ **7.** $\{3\}$
9. $x = \frac{1}{2}\left\{\frac{\ln \frac{3}{12345}}{\ln \frac{2}{5}} - 1\right\} \approx 4.0413502$ **11.** $\{-\frac{5}{3}\}$
13. $x = \frac{6 \ln 3 + \ln 5}{2 \ln 5 - 4 \ln 3} \approx -6.9762655$
15. $x = \frac{1}{4}\left(\frac{\ln 127}{\ln 10} - 1\right) \approx 0.2759509$
17. $x = \frac{1}{2}\left(\frac{\ln 0.167}{\ln 10} - 3\right) \approx -1.8886418$
19. $x = \frac{1}{3}(\ln 4.472 - 1) \approx 0.1659452$
21. $\{0\}$ **23.** $x = \ln \frac{3 \pm \sqrt{5}}{2} \approx 0.9624236, -0.9624236$
25. $x = \frac{1}{2} \ln \frac{13}{7} \approx 0.3095196$ **27.** $\{32\}$ **29.** $\{4, -1\}$

31. $x = \frac{3 \pm \sqrt{37}}{2} \approx 4.54138126, -1.54138126$
33. The solution set is empty. **35.** $\{-1\}$
37. The solution set is empty.
39. $(-\infty, a)$, where $a = \frac{\ln 5 - \ln 2}{2 \ln 2 - 5 \ln 5} \approx -0.1375627$
41. $(-\infty, \infty)$ **43.** $(-\frac{1}{2}, \infty)$ **45.** $\{\frac{5}{2}\}$
47. (a) 123 (b) $t = \frac{\ln \frac{17}{8}}{0.63} \approx 1.1964632$
  (c) $k = \frac{1}{3} \ln \frac{17}{8} \approx 0.2512573$ (d) 17
53. (a) At $x = e$, the graph of f lies above the graph of g. As in example 4.37, $e^\pi - \pi^e \approx 0.68$....
  (b) The value is approximately 3.14. (It is exactly π.)
  (c) It appears that one such x exists.

Section 4.6 (Review Exercises)

1. 33.1155
3.

x	-0.75	-0.5	-0.25	0	1	2	3
$f(x) = 3^{-x+2}$	20.5156	15.5885	11.8447	9	3	1	0.3333

domain = $(-\infty, \infty)$
range = $(0, \infty)$
y-intercept $(0, 9)$
horizontal asymptote $y = 0$
decreasing function

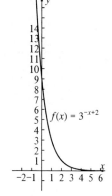

5. $\$938.86$ **7.** (a) $\log_{1/2} 16 = -4$ (b) $\log_5 \frac{1}{25} = -2$
9. $\frac{\ln \frac{x}{5}}{\ln 2}; \frac{e^x}{2}$

Answers to Exercises

11.

x	0.75	1	2	3	6	9
$f(x) = \log_4 (2x - 1)$	-0.5	0	0.7925	1.4037	1.7297	2.0437

domain = $(\frac{1}{2}, \infty)$
range = $(-\infty, \infty)$
x-intercept $(1, 0)$
vertical asymptote $x = \frac{1}{2}$
increasing function

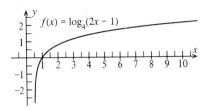

13. $\frac{1}{9}$ **15.** (a) $\log \frac{27}{4}$ (b) $\ln \frac{36x^8}{y^{3/7}}$

17. (a) 2.1972 (b) 7.6495 (c) 1.354
(d) -4.8444 (e) -8.6711

19. $k = \frac{1}{1600} \ln 2 \approx 0.0004332$ **21.** $\frac{10 \ln 2}{\ln (\frac{25}{16})} \approx 15.531419$ years

23. approximately $\frac{\ln(\frac{302}{382})}{\ln(\frac{342}{382})} - 1 \approx 1.1245264$ minutes

25. $\{-\frac{3}{2}\}$ **27.** $x = \frac{\ln 3}{4 \ln 3 - 2 \ln 5} \approx 0.9345332$

29. $\{\frac{8}{3}\}$ **31.** $\{4, -4\}$ **33.** $\{2\}$

35. $(-\infty, a]$, where $a = 4 + \ln 4 \approx 5.3862944$

37. $(0, a]$, where $a = \frac{-2 + \sqrt{10}}{6} \approx 0.1937129$

39. $y^x = b^{\alpha/\beta} = b^{(\log a)/(\log b)} = b^{\log_b a} = a$; it did, presumably with negligible round-off errors and the intermediate value $x = 1.0685057$.

Section 5.1

1. (a) (b)

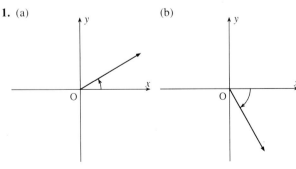

3. (a) (b)

(c) (d)

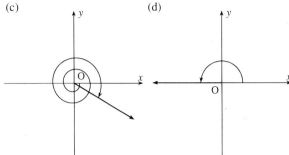

(c) (d)

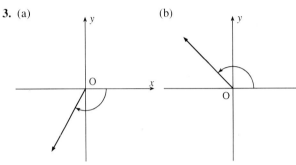

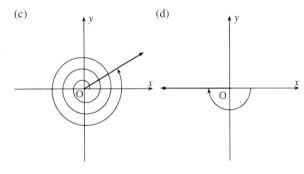

5. (a) 2.36 radians (b) -0.53 radian (c) 0 radians
(d) 15.78 radians

7. (a) $180°$ (b) $57.30°$ (c) $-401.07°$ (d) $-30°$

9. (a) $2°30'$ (b) $-30°7'30''$ (c) $65°27'16.36''$
(d) $458°21'58.45''$

11. (a) $270°$ (b) $135°$ (c) $10°$ (d) $180°$ (e) $184°$ (f) $320°$

13. $\frac{10\pi}{3}$ cm **15.** $\frac{9}{2\pi}$ feet **17.** $\frac{3\pi}{4}$ radians **19.** $\frac{40\pi}{9}$ cm

21. $\theta = 1' \approx 0.0002909$ radian, and so 1 nautical mile is $s = r\theta \approx 3960(0.0002909) \approx 1.15$ miles

23. $\frac{16}{25}$ radian

25. the (second) slice whose radius is 8 inches; 8π square inches

27. (b) $\frac{x}{60}$ **29.** (a) $\frac{25}{\pi}$ (b) 150 feet per second

Answers to Exercises

31. $\frac{42}{\pi}$ cm **33.** 30 cm
35. (a) no
(b) Then, once again, ignoring units, the area is the product of the angular speed and the linear speed.

Section 5.2

1.

	sin θ	cos θ	tan θ	cot θ	sec θ	csc θ
(a)	$\frac{4}{5}$	$\frac{3}{5}$	$\frac{4}{3}$	$\frac{3}{4}$	$\frac{5}{3}$	$\frac{5}{4}$
(b)	$\frac{2\sqrt{2}}{3}$	$-\frac{1}{3}$	$-2\sqrt{2}$	$\frac{-1}{2\sqrt{2}}$	-3	$\frac{3}{2\sqrt{2}}$
(c)	0	-1	0	undefined	-1	undefined
(d)	1	0	undefined	0	undefined	1
(e)	$\frac{3}{\sqrt{13}}$	$\frac{2}{\sqrt{13}}$	$\frac{3}{2}$	$\frac{2}{3}$	$\frac{\sqrt{13}}{2}$	$\frac{\sqrt{13}}{3}$

3. (a) positive (b) positive (c) positive (d) negative
5. $\sin\theta = \frac{3}{\sqrt{10}}$, $\cos\theta = \frac{-1}{\sqrt{10}}$, $\tan\theta = -3$, $\cot\theta = \frac{-1}{3}$, $\sec\theta = -\sqrt{10}$, $\csc\theta = \frac{\sqrt{10}}{3}$

7. Take $P(1, -1)$ on the terminal side of θ in standard position. Then $r = \sqrt{1^2 + (-1)^2} = \sqrt{2}$ and $\csc\theta = \frac{r}{y} = \frac{\sqrt{2}}{-1} = -\sqrt{2}$.

9. (a) sin θ (b) cos θ (c) tan θ **13.** (b), (c), (e), (f), and (g)
21. $\sin\theta = \frac{12}{13}$, $\cos\theta = \frac{5}{13}$, $\tan\theta = \frac{12}{5}$, $\cot\theta = \frac{5}{12}$, $\sec\theta = \frac{13}{5}$, $\csc\theta = \frac{13}{12}$; and $\sin\phi = \frac{5}{13}$, $\cos\phi = \frac{12}{13}$, $\tan\phi = \frac{5}{12}$, $\cot\phi = \frac{12}{5}$, $\sec\phi = \frac{13}{12}$, $\csc\phi = \frac{13}{5}$

23. (a) $\sin\theta = \frac{1}{2}$, $\cos\theta = \frac{\sqrt{3}}{2}$, $\tan\theta = \frac{1}{\sqrt{3}}$, $\cot\theta = \sqrt{3}$, $\csc\theta = 2$
(b) $\sin\theta = \frac{\sqrt{3}}{2}$, $\cos\theta = \frac{1}{2}$, $\tan\theta = \sqrt{3}$, $\cot\theta = \frac{1}{\sqrt{3}}$, $\sec\theta = 2$

31. (a) 2 (b) $\frac{7}{5}$ **33.** $\frac{6}{5}$ **35.** 75 tan 30° feet
37. 40 sin 70° miles **39.** $\frac{25000}{29}$ tan 30° feet per second
41. 6 miles

Section 5.3

1. (a) (1, 0)
(b) sin 0 = 0, cos 0 = 1, tan 0 = 0, cot 0 is undefined, sec 0 = 1, csc 0 is undefined
3. (a) (0, 1)
(b) $\sin\frac{\pi}{2} = 1$, $\cos\frac{\pi}{2} = 0$, $\tan\frac{\pi}{2}$ is undefined, $\cot\frac{\pi}{2} = 0$, $\sec\frac{\pi}{2}$ is undefined, $\csc\frac{\pi}{2} = 1$
5. (a) $\left(-\frac{1}{\sqrt{2}}, -\frac{1}{\sqrt{2}}\right)$

(b) $\sin\frac{5\pi}{4} = \frac{-1}{\sqrt{2}}$, $\cos\frac{5\pi}{4} = \frac{-1}{\sqrt{2}}$, $\tan\frac{5\pi}{4} = 1$, $\cot\frac{5\pi}{4} = 1$, $\sec\frac{5\pi}{4} = -\sqrt{2}$, $\csc\frac{5\pi}{4} = -\sqrt{2}$

7. (a) (−1, 0)
(b) sin 3π = 0, cos 3π = −1, tan 3π = 0, cot 3π is undefined, sec 3π = −1, csc 3π is undefined
9. (a) (−1, 0)
(b) sin (−π) = 0, cos (−π) = −1, tan (−π) = 0, cot (−π) is undefined, sec (−π) = −1, csc (−π) is undefined

11. (a) $\left(-\frac{1}{\sqrt{2}}, \frac{1}{\sqrt{2}}\right)$

(b) $\sin\left(-\frac{5\pi}{4}\right) = \frac{1}{\sqrt{2}}$, $\cos\left(-\frac{5\pi}{4}\right) = \frac{-1}{\sqrt{2}}$, $\tan\left(-\frac{5\pi}{4}\right) = -1$, $\cot\left(-\frac{5\pi}{4}\right) = -1$, $\sec\left(-\frac{5\pi}{4}\right) = -\sqrt{2}$, $\csc\left(-\frac{5\pi}{4}\right) = \sqrt{2}$

13. (a) −cos α (b) cot α (c) −sin α
15. (a) −sin α (b) sin α **17.** (a) $-\frac{1}{2}$ (b) $\cos\alpha = \pm\frac{\sqrt{3}}{2}$
19. Since $P(\theta + 2\pi) = P(\theta)$, the answer to (a) is sin θ. Similarly, since $P(\theta - 2\pi) = P(\theta)$, the answer to (b) is cos θ; and since $P(\theta + 6\pi) = P(\theta)$, the answer to (c) is tan θ.

21.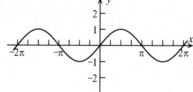

23.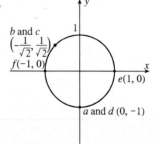

25. the upper unit semicircle, including endpoints
27. the left unit semicircle, including endpoints
29. the unit circle

Section 5.4

1. (a) 58° (b) 36° (c) 14° (d) 41°
3. (a) $\frac{7\pi}{15}$ (b) $\frac{\pi}{3}$ (c) $\frac{\pi}{5}$ (d) $\frac{\pi}{3}$ (e) $\frac{3\pi}{8}$ (f) $\pi - 2.9$
5. (a) $-\frac{1}{2}$ (b) $-\frac{1}{2}$ (c) -2 (d) 1 **7.** (a) $-\frac{2}{\sqrt{13}}$ (b) $\frac{\sqrt{119}}{5}$

9. (a) IV (b) II (c) III (d) III (e) II (f) II

13.

	sin	cos	tan
(a)	−0.6047381	0.7964244	−0.7593164
(b)	−0.8190964	0.5736559	−1.4278533
(c)	0.0174524	−0.9998477	−0.0174551
(d)	−0.0610485	0.9981348	−0.0611626

	cot	sec	csc
(a)	−1.316974	1.255612	−1.6536083
(b)	−0.7003521	1.7432054	−1.2208575
(c)	−57.289965	−1.0001523	57.298685
(d)	−16.349856	1.0018687	−16.380408

15.

	sin	cos	tan
(a)	−0.6427876	−0.7660444	0.8390996
(b)	0.8476778	−0.5305113	−1.5978506
(c)	−0.6569866	0.7539023	−0.8714480
(d)	−0.9880316	0.1542514	−6.4053312

	cot	sec	csc
(a)	1.1917536	−1.3054073	−1.5557238
(b)	−0.6258408	−1.8849739	1.1796935
(c)	−1.1475154	1.3264319	−1.5221011
(d)	−0.15612	6.4829213	−1.0121134

17. Values of sin and cos are to four decimal places:

θ	1	0.1	0.01	0.001	0.0001	negative values of θ
$\sin\theta$	0.8415	0.0998	0.0100	0.0010	0.0001	recall sin is odd
$\cos\theta$	0.5403	0.9950	1	1	1	recall cos is even

This suggests that as θ approaches 0, the limiting tendency of $\sin\theta$ is 0 and the limiting tendency of $\cos\theta$ is 1.

21. $\dfrac{\pi}{4}$

23. yes, because Sinus swims $700\sin 40° \approx 449.951$ yards
25. 60 feet (ignoring Petra's height) 27. 29.119776 meters

Section 5.5

1. All except (f) and (g) are periodic. 3. (a) yes; no

7. (a) $\left(0, \dfrac{\pi}{2}\right)$ and $\left(\dfrac{3\pi}{2}, 2\pi\right)$

(b) $\left(\dfrac{\pi}{2}, \pi\right), \left(\pi, \dfrac{3\pi}{2}\right)$, and $\left(\dfrac{\pi}{2}, \dfrac{3\pi}{2}\right)$

9. (a) $-y = \sin 2(-x) = \sin(-2x) = -\sin 2x$ is equivalent to $y = \sin 2x$.
(b) $4(-x) = \sin 3(-y)$ is equivalent to $4x = \sin 3y$ since sin is an odd function.

11. Consider, for instance, $\left(\dfrac{\pi}{2}, 1\right)$ which is in sin, although $\left(-\dfrac{\pi}{2}, 1\right)$ is not in sin.

13. (a) $x = \cos(-y) = \cos y$ is (equivalent to) $x = \cos y$.
(b) $2(-y)^2 = \cos 3x$ is (equivalent to) $2y^2 = \cos 3x$.

15.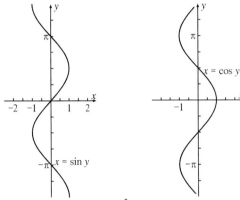

19. (a) $2\pi, 3$ (b) $\pi, 1$ (c) $\dfrac{2\pi}{3}, 1$
(d) $\dfrac{2\pi}{3}, 2$ (e) $\dfrac{2\pi}{3}, 2$ (f) $\dfrac{2\pi}{3}, 2$

21. (a) −3 (b) 0 (c) 3

23. (a) $f(x) = A\sin\left(\dfrac{\pi}{2} + (Bx - C)\right) = A\sin\left(Bx - \left(-\dfrac{\pi}{2} + C\right)\right)$,

with phase shift $\dfrac{-\dfrac{\pi}{2} + C}{B} = \dfrac{C - \dfrac{\pi}{2}}{B}$

(b) $\dfrac{2}{3}\sin\left(-9x + \dfrac{3\pi}{4}\right)$

25.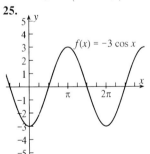

27.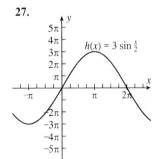

29.

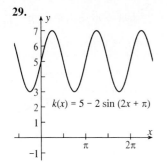

31.

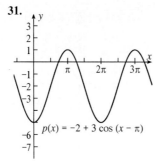

33.

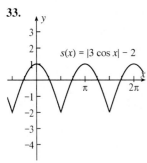

(c) $\sec\left(\left(-\frac{\pi}{4} - \frac{\sqrt{\pi}}{2} - \frac{\pi}{4}\right)^2\right) = \sec\left(\frac{\pi^2 + 2\pi^{3/2} + \pi}{4}\right) \approx$

1.0310977, but $1.4142136 \approx \sqrt{2} = \sec\frac{\pi}{4} = \sec\left(\left(\frac{\sqrt{\pi}}{2}\right)^2\right) =$

$\sec\left(\left(\frac{\pi}{4} + \frac{\sqrt{\pi}}{2} - \frac{\pi}{4}\right)^2\right)$

(d) $2\left(-\frac{\pi}{4}\right) - \sec\left(-\frac{\pi}{4}\right) = -\frac{\pi}{2} - \sqrt{2} \neq \frac{\pi}{2} - \sqrt{2} = 2\left(\frac{\pi}{4}\right) - \sec\left(\frac{\pi}{4}\right)$

7. (a) $\left\{x: x \neq \frac{\pi}{4} + n\frac{\pi}{2} \text{ for } n \text{ an integer}\right\}$

(b) $\left\{x: x \neq \frac{\pi}{2} + n\pi \text{ for } n \text{ an integer}\right\}$

(c) $\left\{x: x \neq n\pi - \frac{\pi}{4} \text{ for } n \text{ an integer}\right\}$

9. (a) $\mathbf{R}$ (b) $\mathbf{R}$ (c) $(-\infty, -3] \cup [1, \infty)$

11. (a) $\left(\frac{\pi}{2}, \pi\right)$ and $\left(\pi, \frac{3\pi}{2}\right)$ (b) $\left(0, \frac{\pi}{2}\right)$

13. (a) for instance, $x = \frac{\pi}{6}$ (b) 0

35. $A = 2, B = \frac{1}{2}, C = 0 = D$ **37.** $A = 2, B = 3, C = 0, D = -1$
39. $A = \frac{1}{2}, B = 1, C = \frac{\pi}{2}, D = 0$ **41.** $A = 1, B = \frac{1}{2}, C = 0, D = \frac{3}{2}$
43. $A = 16.73, B = \frac{\pi}{0.26} \approx 12.083049$

15. Values of $\frac{\tan x}{x}$ are rounded to seven decimal places:

x	1	0.1	0.01	0.001	0.0001
$\frac{\tan x}{x}$	1.5574077	1.0033467	1.0000332	1	1

For $x < 0$, notice that $\frac{\tan x}{x}$ is an even function. The above information suggests that as x approaches 0, the limiting tendency of $\frac{\tan x}{x}$ is 1.

Section 5.6

1. (a) $\frac{\pi}{2}$ (b) π (c) π (d) $\frac{\pi}{2}$
(e) $\frac{\pi}{2}$ (f) 2π (g) π (h) π (i) $\frac{\pi}{4}$

3. (a) $\sec(-x) = \frac{1}{\cos(-x)} = \frac{1}{\cos x} = \sec x$
(b) $5 - 3\sec(2(-x)) = 5 - 3\sec(-2x) = 5 - 3\sec(2x)$, by (a)
(c) $-1 + 4\sec(-2(-x)) = -1 + 4\sec(2x) = -1 + 4\sec(-2x)$, by (a)
(d) $2\sec((-x)^2 + 4) = 2\sec(x^2 + 4)$
(e) $\csc((-x)^2 + 4) = \csc(x^2 + 4)$ (f) $\csc(|-x|) = \csc(|x|)$

5. (a) $\tan\left(-\frac{\pi}{4}\right) = -1 \neq 1 = \tan\frac{\pi}{4}$

(b) $\sec\left(-\frac{\pi}{2} - \frac{\pi}{4}\right) = \sec\left(-\frac{3\pi}{4}\right) = -\sqrt{2} \neq \sqrt{2} = \sec\left(\frac{\pi}{4}\right) = \sec\left(\frac{\pi}{2} - \frac{\pi}{4}\right)$

17. As x approaches 0 (from either side), the limiting tendency of $\frac{\sin x}{x}$ is 1, not infinity or negative infinity.

19. (a) 5 (b) $\frac{1}{3}$ (c) 0 **21.** $5 - 4\cot\left(-2x + 3 + \frac{\pi}{2}\right)$

23.

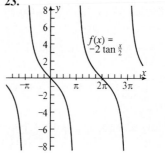

25.

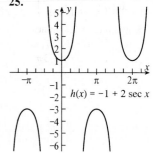

27.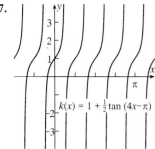

29. $A = 1, B = 1, C = \dfrac{\pi}{4}, D = -1$

31. $A = \dfrac{1}{2}, B = 2, C = \dfrac{\pi}{2}, D = 0$

Section 5.7

1. (a) $-\dfrac{\pi}{3}$ (b) $\dfrac{\pi}{2}$ (c) $\dfrac{\pi}{6}$ (d) 0 (e) $\dfrac{5\pi}{6}$

3. (a) -0.961411 (b) 1.4056476 (c) $\dfrac{\pi}{3}$

5. (a) $\dfrac{2}{3}$ (b) -0.8 (c) π (d) $\dfrac{2}{\pi}$ (e) $\dfrac{\pi}{2}$ (f) -2

7. (a) $\dfrac{2\pi}{7}$ (b) $\dfrac{\pi}{4}$ (c) $\dfrac{\pi}{6}$ **9.** (a) $\dfrac{-1}{2\sqrt{2}}$ (b) $\dfrac{-1}{\sqrt{10001}}$

11. It's false. For instance, consider $x = -\dfrac{\sqrt{3}}{2}$. Then
$\sec^{-1}\left(\dfrac{1}{x}\right) = \sec^{-1}\left(\dfrac{-2}{\sqrt{3}}\right) = \dfrac{7\pi}{6}$, but
$\cos^{-1} x = \cos^{-1}\left(\dfrac{-\sqrt{3}}{2}\right) = \dfrac{5\pi}{6}$.

13. (a) 0.2942268 (b) 2.7388768 (c) 3.2847343 (d) 4.9573676

15. (a) $[-2, 2]$ (b) $\mathbf{R}$ (c) $(-\infty, 0] \cup [2, \infty)$

17. for 15(a), $\left[-\dfrac{\pi}{2}, \dfrac{\pi}{2}\right]$; for 15(b), $(-2\pi, 0)$;
for 15(c), $\left[-\dfrac{3\pi}{4} - 2, -\dfrac{\pi}{2} - 2\right) \cup \left[-\dfrac{\pi}{4} - 2, -2\right]$; for 16(a), $[0, 2\pi]$;
for 16(b), $\left(\dfrac{3\pi}{4}, \dfrac{5\pi}{4}\right)$; for 16(c), $\left[1, \dfrac{\pi}{2} + 1\right] \cup \left[\pi + 1, \dfrac{3\pi}{2} + 1\right)$

19. (a) none (b) none (c) $(-\infty, \infty)$ (d) none (e) $[1, \infty)$ (f) $[1, \infty)$

21. (a) $\dfrac{2\pi}{3}$ (b) $\dfrac{\pi}{4}, \dfrac{3\pi}{4}$ (c) none

23. $\arccos(-1) = \pi \neq 0 = -0 = -\arccos(1)$; so arccos is not odd.

29.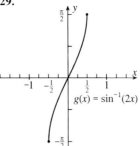

31.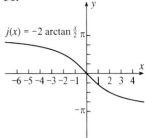

33. $\dfrac{169}{2}(\sin^{-1}(\tfrac{12}{13}) - \sin^{-1}(\tfrac{5}{13})) \approx 66.01259$

35. yes, since the maximum reach of the ladder is $25 \sin \dfrac{\pi}{3} \approx 21.65$ meters

Section 5.8

1. $\left\{\dfrac{\pi}{3}, \dfrac{2\pi}{3}\right\}$ **3.** $\{0.5880026, 3.7295953\}$

5. The solution set is empty. **7.** $\left\{\dfrac{\pi}{6}, \dfrac{\pi}{3}, \dfrac{5\pi}{3}, \dfrac{11\pi}{6}\right\}$

9. $\left\{0, \dfrac{\pi}{2}, \pi\right\}$ **11.** $\{0, 1.6814535, \pi, 4.8230462\}$

13. $\left\{\dfrac{\pi}{4}, 2.0344439, \dfrac{5\pi}{4}, 5.1760366\right\}$ **15.** $\left\{\dfrac{\pi}{3}, \dfrac{2\pi}{3}, \dfrac{4\pi}{3}, \dfrac{5\pi}{3}\right\}$

17. $\left\{\dfrac{\pi}{3}, \dfrac{2\pi}{3}, \dfrac{4\pi}{3}, \dfrac{5\pi}{3}\right\}$ **19.** $\left\{\dfrac{5\pi}{12}, \dfrac{13\pi}{12}\right\}$

21. $\left\{\dfrac{\pi}{12}, \dfrac{5\pi}{12}, \dfrac{13\pi}{12}, \dfrac{17\pi}{12}\right\}$ **23.** $\left\{\dfrac{\pi}{2}, \pi, \dfrac{3\pi}{2}\right\}$

25. $\{1.791815, 4.4913703\}$ **27.** $\left\{\dfrac{\pi}{6}, \dfrac{5\pi}{6}\right\}$ **29.** $\left\{\dfrac{\pi}{3}, \dfrac{5\pi}{3}\right\}$

31. $\left\{\dfrac{\pi}{12}, \dfrac{5\pi}{12}, \dfrac{13\pi}{12}, \dfrac{17\pi}{12}\right\}$

33. $\{0.3690496, 1.4162471, 2.4634447, 3.5106422, 4.5578398, 5.6050373\}$

35. The solution set is empty. **37.** $\left\{\dfrac{\pi}{4}, \dfrac{3\pi}{4}, \dfrac{5\pi}{4}, \dfrac{7\pi}{4}\right\}$

39. (b) $\left\{\dfrac{\pi}{6}, \dfrac{\pi}{2}, \dfrac{5\pi}{6}\right\}$ **41.** $\left[0, \dfrac{\pi}{6}\right) \cup \left(\dfrac{5\pi}{6}, 2\pi\right)$ **43.** $\left[\dfrac{\pi^2}{2}, 2\pi\right)$

45. $\left(\dfrac{\pi}{4}, \dfrac{3\pi}{4}\right) \cup \left(\dfrac{5\pi}{4}, \dfrac{7\pi}{4}\right)$ **47.** $\{0\} \cup [\pi, 2\pi)$

49. $\left[0, \dfrac{\pi}{4}\right] \cup \left[\dfrac{3\pi}{4}, \dfrac{5\pi}{4}\right] \cup \left[\dfrac{7\pi}{4}, 2\pi\right)$

Answers to Exercises

Section 5.9 (Review Exercises)

1. (a) $60° = \dfrac{\pi}{3}$ radians (b) $-30° = -\dfrac{\pi}{6}$ radians
   $\theta_{\text{ref}} = 60°$ $\theta_{\text{ref}} = 30°$

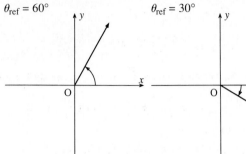

(c) $-210° = -\dfrac{7\pi}{6}$ radians   (d) $930° = \dfrac{31\pi}{6}$ radians
$\theta_{\text{ref}} = 30°$ $\theta_{\text{ref}} = 30°$

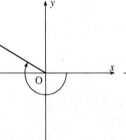 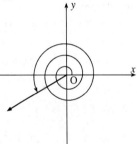

3. 0.93
5. (a) 54π cm^2; 81π cm^2
   (b) 6° per second, or $\dfrac{\pi}{30}$ radians per second
   (c) $\dfrac{3\pi}{5}$ cm per second
7. In each case, the graph is symmetric about the y-axis.
   (a) $\cos(-x) = \cos x$ (b) $\sin((-x)^2) = \sin(x^2)$
   (c) $3 - 2\sec(5(-x)) = 3 - \dfrac{2}{\cos(-5x)} =$
   $3 - \dfrac{2}{\cos 5x} = 3 - 2\sec 5x$
   (d) $3\tan|-x| = 3\tan|x|$
9. $\sin \alpha = \tfrac{3}{5}$, $\cos \alpha = \tfrac{4}{5}$, $\tan \alpha = \tfrac{3}{4}$, $\cot \alpha = \tfrac{4}{3}$, $\sec \alpha = \tfrac{5}{4}$,
   $\csc \alpha = \tfrac{5}{3}$
13. $\dfrac{\sqrt{3}}{2}$
15. $\left(-\dfrac{1}{\sqrt{2}}, -\dfrac{1}{\sqrt{2}}\right)$; $\sin \alpha = -\dfrac{1}{\sqrt{2}} = \cos \alpha$, $\tan \alpha = 1 = \cot \alpha$,
    $\sec \alpha = -\sqrt{2} = \csc \alpha$
17. $50\sqrt{3}$ yards; 50 yards

19. (a) $-\dfrac{1}{2}$ (b) $\dfrac{\pi}{3}$ (c) -1.1071487 (d) $-\sqrt{2}$
    (e) $\dfrac{7\pi}{6}$ (f) 2.1874468
21. (a) tan, sec, and csc (b) sec^{-1} and tan^{-1} (c) sin^{-1}
23.

	Domain	Range	Maximum	Minimum
(a)	$(-\infty, \infty)$	$[5, 9]$	9	5
(b)	$(-\infty, \infty)$	$\left(-\dfrac{3\pi}{2}, \dfrac{3\pi}{2}\right)$	does not exist	does not exist
(c)	$[-3, -1]$	$[-\pi, \pi]$	π	$-\pi$
(d)	$\left\{x: x \neq n\dfrac{\pi}{2} + \dfrac{\pi}{4} + 2\right\}$	$(-\infty, \infty)$	does not exist	does not exist
(e)	$\left\{x: x \neq n\pi + \dfrac{\pi}{2}\right\}$	$(-\infty, \infty)$	does not exist	does not exist
(f)	$(-\infty, \infty)$	$(0, \pi)$	does not exist	does not exist

25. $4 + 2\cos\left(-2x + \dfrac{\pi}{2} - 3\right)$
27. Since $\sin \theta = \dfrac{5x}{6}$, we have $\sqrt{36 - 25x^2} =$
    $\sqrt{36 - 25\left(\dfrac{6\sin\theta}{5}\right)^2} = \sqrt{36 - 36\sin^2\theta} =$
    $\sqrt{36}\sqrt{1 - \sin^2\theta} = 6|\cos\theta| = 6\cos\theta$, as $-\dfrac{\pi}{2} \leq \theta \leq \dfrac{\pi}{2}$.

29.

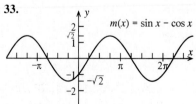

$f(x) = \cos(2x - \pi)$

31.

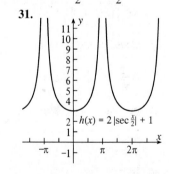

$h(x) = 2|\sec\tfrac{x}{2}| + 1$

33.

$m(x) = \sin x - \cos x$

35. $A = \tfrac{1}{2}$, $B = 2$, $C = -\dfrac{\pi}{2}$, $D = -1$
37. $\left[0, \dfrac{\pi}{2}\right) \cup \left(\dfrac{\pi}{2}, \dfrac{3\pi}{2}\right) \cup \left(\dfrac{3\pi}{2}, 2\pi\right)$

39. $\left\{\dfrac{\pi}{2}, \pi, \dfrac{3\pi}{2}\right\}$ **41.** $\left\{\dfrac{\pi}{6}, \dfrac{5\pi}{6}\right\}$ **43.** $\left\{0, \dfrac{\pi}{2}\right\}$

45. $\left[0, \dfrac{\pi - \alpha}{2}\right) \cup \left(\dfrac{\pi + \alpha}{2}, \dfrac{3\pi - \alpha}{2}\right) \cup \left(\dfrac{3\pi + \alpha}{2}, 2\pi\right)$, where $\alpha = \cos^{-1}\dfrac{2}{3} \approx 0.8410687$.

49. yes. To see this, use $\sec x = \dfrac{1}{\cos x}$, expand the left-hand side, write both sides with common denominator $\cos x$, reduce the problem to showing that $\sin^2 x - \cos^2 x = 1 - 2\cos^2 x$, and invoke the Pythagorean identity in the form $\sin^2 x = 1 - \cos^2 x$.

Section 6.1

1. $2 \sin x$ **3.** $\dfrac{1 + \sin^2 x}{\cos^2 x}$

41. (a) Since $-\dfrac{\pi}{2} < \theta < \dfrac{\pi}{2}$, $\sec \theta > 0$, and so $|\sec \theta| = \sec \theta$.

Notice also that $\tan \theta = \dfrac{u}{\sqrt{7}}$, and so
$\sqrt{7 + u^2} = \sqrt{7 + (\sqrt{7} \tan \theta)^2} = \sqrt{7 + 7\tan^2 \theta} = \sqrt{7}\sqrt{1 + \tan^2 \theta} = \sqrt{7}\sqrt{\sec^2 \theta} = \sqrt{7}\,|\sec \theta| = \sqrt{7}\sec \theta$.

(b) As in (a), $|\sec \theta| = \sec \theta$ and $\tan \theta = \dfrac{Bu}{A}$. Also $|A| = A$.

Then $\sqrt{A^2 + B^2 u^2} = \sqrt{A^2 + B^2 \left(\dfrac{A\tan\theta}{B}\right)^2} = \sqrt{A^2(1 + \tan^2\theta)} = \sqrt{A^2}\sqrt{\sec^2\theta} = |A||\sec\theta| = A\sec\theta$.

45. Consider, for instance, $x = \dfrac{3\pi}{4}$: $\sqrt{1 - \sin^2 x} = \sqrt{1 - \left(\dfrac{1}{\sqrt{2}}\right)^2} = \dfrac{1}{\sqrt{2}} \neq -\dfrac{1}{\sqrt{2}} = \cos x$.

47. Consider, for instance, $\theta = \dfrac{\pi}{6}$: $\tan \theta = \dfrac{1}{\sqrt{3}} \neq \sqrt{3} = \dfrac{\sqrt{3}/2}{1/2} = \dfrac{\cos \theta}{\sin \theta}$.

49. $\left\{0, \dfrac{2\pi}{3}, \dfrac{4\pi}{3}\right\}$

51. $(2\alpha, \pi)$, where $\alpha = \arctan\left(\dfrac{1}{2}\right) \approx 0.4636476$; or approximately $(0.9273, \pi)$.

53. $\left[0, \dfrac{\pi}{3}\right] \cup \{\pi\} \cup \left[\dfrac{5\pi}{3}, 2\pi\right)$

55. Consider, for instance, $\theta = \dfrac{3\pi}{4}$: $\dfrac{\sqrt{1 + \tan^2\theta}}{\tan\theta} = \dfrac{\sqrt{1 + (-1)^2}}{-1} = -\sqrt{2} \neq \sqrt{2} = \sec\theta$.

Section 6.2

1. $\dfrac{1 + \sqrt{3}}{2\sqrt{2}}$ **3.** $\dfrac{-1 - \sqrt{3}}{2\sqrt{2}}$ **5.** $\dfrac{\sqrt{3} + 1}{\sqrt{3} - 1} = 2 + \sqrt{3}$

7. $\sin\left(x - \dfrac{\pi}{3}\right)$ **9.** $2\sin\left(\dfrac{\pi}{3} + x\right)$ **11.** $\cos(x - \alpha)$

13. $\sin\left(x + \dfrac{\pi}{5}\right)$ **15.** $\dfrac{-12 + 5\sqrt{3}}{26}$ **17.** $\dfrac{3}{4}$

43. Consider, for instance, $x = \dfrac{\pi}{3}$ and $y = \dfrac{\pi}{6}$: then $\sin(x - y) = \sin\left(\dfrac{\pi}{3} - \dfrac{\pi}{6}\right) = \sin\dfrac{\pi}{6} = \dfrac{1}{2} \neq \dfrac{\sqrt{3}}{2} - \dfrac{1}{2} = \sin\dfrac{\pi}{3} - \sin\dfrac{\pi}{6} = \sin x - \sin y$.

47. $\sqrt{2}\sin\left(x - \dfrac{\pi}{4}\right) = \sqrt{2}\left(\sin x \cos\dfrac{\pi}{4} - \cos x \sin\dfrac{\pi}{4}\right) = \sqrt{2}\left[(\sin x)\dfrac{1}{\sqrt{2}} - (\cos x)\dfrac{1}{\sqrt{2}}\right] = \dfrac{\sqrt{2}\sin x}{\sqrt{2}} - \dfrac{\sqrt{2}\cos x}{\sqrt{2}} = \sin x - \cos x$

49. $[0, 2\pi)$

51. $\left(\dfrac{\alpha}{2}, \dfrac{\pi}{4}\right) \cup \left(\dfrac{\pi + \alpha}{2}, \dfrac{3\pi}{4}\right) \cup \left(\pi + \dfrac{\alpha}{2}, \dfrac{5\pi}{4}\right) \cup \left(\dfrac{3\pi + \alpha}{2}, \dfrac{7\pi}{4}\right)$, where $\alpha = \tan^{-1} 2 \approx 1.10714872$

Section 6.3

1. $\sin 2\theta = 1.4\sqrt{0.51} \approx 0.9998$; $\cos 2\theta = 0.02$; $\tan 2\theta = 70\sqrt{0.51} \approx 49.9900$; 2θ in quadrant I

3. $\sin 2\theta = \dfrac{3}{5}$; $\cos 2\theta = -\dfrac{4}{5}$; $\tan 2\theta = -\dfrac{3}{4}$; 2θ in quadrant II

5. $\sin \theta = \pm\sqrt{0.2}$; $\cos \theta = \pm\sqrt{0.8}$; $\tan \theta = \dfrac{1}{2}$

7. $\sin \theta = \pm\sqrt{\dfrac{1 - 3/\sqrt{10}}{2}}$; $\cos \theta = \pm\sqrt{\dfrac{1 + 3/\sqrt{10}}{2}}$; $\tan \theta = \sqrt{\dfrac{\sqrt{10} - 3}{\sqrt{10} + 3}} = \sqrt{10} - 3$

9. $\dfrac{1}{\sqrt{3}}$ **11.** $\sqrt{\dfrac{\sqrt{5} - 1}{2\sqrt{5}}}$ **13.** $-\dfrac{1}{\sqrt{2}}$ **15.** $-\sqrt{\dfrac{\sqrt{10} + 1}{\sqrt{10} - 1}}$

17. $\dfrac{\sqrt{3}}{2}$ **19.** $\sqrt{\dfrac{2 - \sqrt{2}}{4}}$ **21.** $\sqrt{\dfrac{2 - \sqrt{3}}{2 + \sqrt{3}}} = 2 - \sqrt{3}$

23. $\dfrac{3 - 4\cos 2\theta + \cos 4\theta}{8}$ **41.** $\left\{\dfrac{\pi}{3}, \dfrac{5\pi}{3}\right\}$

43. $\left\{\dfrac{\pi}{2}, \dfrac{2\pi}{3}, \dfrac{4\pi}{3}, \dfrac{3\pi}{2}\right\}$ **45.** $\left\{0, \dfrac{\pi}{3}, \dfrac{2\pi}{3}, \pi, \dfrac{4\pi}{3}, \dfrac{5\pi}{3}\right\}$

47. $\{0, \pi\}$ **49.** $\left[0, \dfrac{2\pi}{3}\right)$ **51.** $\left[0, \dfrac{2\pi}{3}\right] \cup \left[\dfrac{4\pi}{3}, 2\pi\right)$

53. $\left(\dfrac{\pi}{2}, \dfrac{2\pi}{3}\right) \cup \left(\pi, \dfrac{4\pi}{3}\right) \cup \left(\dfrac{3\pi}{2}, 2\pi\right)$

Answers to Exercises

55.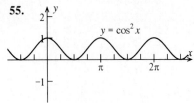

57. She will charge a fee since $\cos 2\theta = \frac{23}{49}$ and so $2\theta = \arccos\left(\frac{23}{49}\right) \approx 1.0822 > 1.0815 \approx \arccos 0.47$.

Section 6.4

1.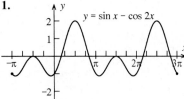

3.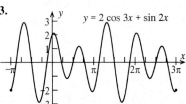

5.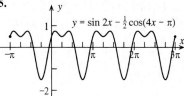

7.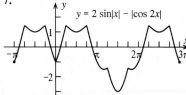

9.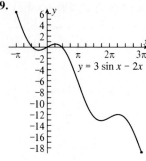

11.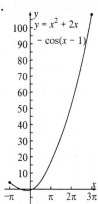

13.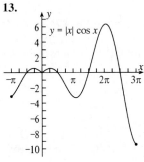

15.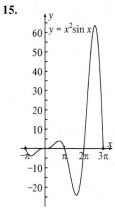

17.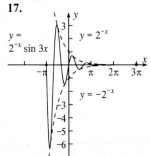

19.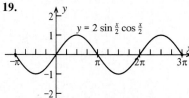

21. 2π **27.** 2   **29.** $\frac{3}{4}$ **31.** 0

Answers to Exercises

Section 6.5

1. $\frac{1}{2}(\sin 8\theta - \sin 2\theta)$ 3. $\frac{1}{8}(\cos 5x + \cos x)$
5. $\frac{5}{2}(\cos 4\alpha - \cos 10\alpha)$ 7. $2 \cos 4\theta \cos 2\theta$
9. $2 \cos(\frac{9}{2}x) \sin(-\frac{1}{2})x = -2 \cos(\frac{9}{2}x) \sin(\frac{1}{2}x)$
11. $-2 \sin \alpha \sin(\frac{1}{5}\alpha)$
13. $8 \sin(\frac{9}{4}\theta) \cos(-\frac{7}{4}\theta) = 8 \sin(\frac{9}{4}\theta) \cos(\frac{7}{4}\theta)$
15. $\frac{\sqrt{3}-1}{4}$ 17. $\frac{\sqrt{3}-1}{2}$ 27. $\{0, \frac{\pi}{2}, \pi, \frac{3\pi}{2}\}$
29. $\{\frac{\pi}{4}, \frac{\pi}{2}, \frac{3\pi}{4}, \frac{5\pi}{4}, \frac{3\pi}{2}, \frac{7\pi}{4}\}$ 31. $\{0, \frac{\pi}{3}, \frac{2\pi}{3}, \pi, \frac{4\pi}{3}, \frac{5\pi}{3}\}$
33. $(\frac{\pi}{4}, \frac{3\pi}{4}) \cup (\pi, \frac{5\pi}{4}) \cup (\frac{7\pi}{4}, 2\pi)$
35. $\sqrt{34} \sin(\theta + \arctan \frac{3}{5})$
37. $-\sqrt{41} \sin(\theta + \arctan(-\frac{4}{5}))$ or $\sqrt{41} \sin(\theta + \pi - \arctan(\frac{4}{5}))$
39. $-\sqrt{41} \sin(\theta + \arctan(-\frac{5}{4}))$ or $\sqrt{41} \sin(\theta + \pi - \arctan(\frac{5}{4}))$
41. $\sqrt{5} \cos(\theta - \arctan 2)$ 43. $\sqrt{5} \cos(\theta - \arctan(\frac{1}{2}))$
45. $\sqrt{5} \cos(\theta - \arctan(-\frac{1}{2}))$ or $-\sqrt{5} \cos(\theta - (\pi - \arctan(\frac{1}{2})))$

Section 6.6

1. [Plot on complex plane showing points labeled (a) through (e) at approximately: (a) $-4i$, (b) on real axis around 9, (c) $2i$, (d) near -2, (e) $-3i$]

3. (a) $\sqrt{3}$ (b) 1 (c) 2 (d) 0 (e) 5 (f) 2
9. $2\sqrt{5}$ cis $\left(\arcsin\left(\frac{2}{\sqrt{5}}\right)\right)$ 11. $2\sqrt{5}$ cis $\left(\arcsin\left(\frac{2}{\sqrt{5}}\right) + \pi\right)$
13. r cis 0 15. cis $\frac{3\pi}{2}$ 17. 5 cis $(\arcsin(-\frac{4}{5}) + 2\pi)$
19. 5 cis $(\pi - \arcsin(\frac{4}{5}))$ 21. 13
23. (b) the circle with center $(2, -3) = 2 - 3i$ and radius 4
    (c) the circle with center $(0, \frac{1}{2}) = \frac{1}{2}i$ and radius $\frac{3}{2}$
29. (a) $\frac{1}{2}$ (b) $-2 - 2\sqrt{3}i$ (c) $1 - i$ (d) $-4i$
31. $\sqrt{2}$ cis $\frac{19\pi}{12}$; $\frac{\sqrt{3}-1}{2} - \frac{\sqrt{3}+1}{2}i$
33. 65 cis $(\pi + \arcsin \frac{63}{65}) \approx 65$ cis 4.4637; $-16 - 63i$
35. $\sqrt{3}$ cis $\frac{\pi}{12}$; $\frac{\sqrt{6}+3\sqrt{3}}{2} + \frac{\sqrt{6}-3\sqrt{3}}{2}i$
37. 12 cis $\frac{\pi}{12}$; $6\sqrt{2+\sqrt{3}} + 6\sqrt{2-\sqrt{3}}\,i = 3(\sqrt{6}+\sqrt{2}) + 3(\sqrt{6}-\sqrt{2})i$

47. yes

Section 6.7

1. 3^5 cis $\frac{5\pi}{4} = 3^5 \left(-\frac{\sqrt{2}}{2} - \frac{\sqrt{2}}{2}i\right) = \frac{-243}{\sqrt{2}} - \frac{243}{\sqrt{2}}i$
3. 32 cis $\frac{4\pi}{3} = -16 - 16\sqrt{3}\,i$ 5. 2^9 cis $\frac{3\pi}{2} = -512i$
7. $\frac{81}{5^5}$ cis $(2\pi - 5 \arctan(\frac{3}{4})) \approx \frac{81}{3125}$ cis $3.0657 \approx -0.02584535 + 0.00196577i$
9. cis α, where $0 \leq \alpha < 2\pi$ and $\alpha + 4\theta = n2\pi$ for some integer n
13. $\frac{1}{2}$ cis $\frac{\pi}{6}$ 15. $\frac{1}{3}$ cis $\frac{5\pi}{6}$
17. $\sqrt[3]{2}$ cis $0 = \sqrt[3]{2}$; $\sqrt[3]{2}$ cis $\frac{2\pi}{3} = -\frac{\sqrt[3]{2}}{2} + \frac{\sqrt[3]{2}\sqrt{3}}{2}i$;
    $\sqrt[3]{2}$ cis $\frac{4\pi}{3} = -\frac{\sqrt[3]{2}}{2} - \frac{\sqrt[3]{2}\sqrt{3}}{2}i$
19. cis $\frac{\pi}{4} = \frac{\sqrt{2}}{2} + \frac{\sqrt{2}}{2}i$; cis $\frac{5\pi}{4} = -\frac{\sqrt{2}}{2} - \frac{\sqrt{2}}{2}i$
21. $\sqrt[8]{2}$ cis $\frac{7\pi}{16} \approx 0.2127 + 1.0696i$; $\sqrt[8]{2}$ cis $\frac{15\pi}{16} \approx -1.0696 + 0.2127i$;
    $\sqrt[8]{2}$ cis $\frac{23\pi}{16} \approx -0.2127 - 1.0696i$; $\sqrt[8]{2}$ cis $\frac{31\pi}{16} \approx 1.0696 - 0.2127i$
23. 2 cis $\frac{\pi}{6} = \sqrt{3} + i$; 2 cis $\frac{5\pi}{12} \approx 0.5176 + 1.9319i$;
    2 cis $\frac{2\pi}{3} = -1 + \sqrt{3}i$; 2 cis $\frac{11\pi}{12} \approx -1.9319 + 0.5176i$;
    2 cis $\frac{7\pi}{6} = -\sqrt{3} - i$; 2 cis $\frac{17\pi}{12} \approx -0.5176 - 1.9319i$;
    2 cis $\frac{5\pi}{3} = 1 - \sqrt{3}i$; 2 cis $\frac{23\pi}{12} \approx 1.9319 - 0.5176i$
25. $\sqrt[4]{2}$ cis $\frac{\pi}{4} = \frac{1}{\sqrt[4]{2}} + \frac{1}{\sqrt[4]{2}}i$; $\sqrt[4]{2}$ cis $\frac{3\pi}{4} = \frac{-1}{\sqrt[4]{2}} + \frac{1}{\sqrt[4]{2}}i$;
    $\sqrt[4]{2}$ cis $\frac{5\pi}{4} = \frac{-1}{\sqrt[4]{2}} - \frac{1}{\sqrt[4]{2}}i$; $\sqrt[4]{2}$ cis $\frac{7\pi}{4} = \frac{1}{\sqrt[4]{2}} - \frac{1}{\sqrt[4]{2}}i$
31. (a) $\left(-\frac{1}{2} + \frac{\sqrt{3}}{2}i\right)^3 = \left(\text{cis}\,\frac{2\pi}{3}\right)^3 = $ cis $2\pi = 1$
    (b) cis $\frac{4\pi}{3} = -\frac{1}{2} - \frac{\sqrt{3}}{2}i$ and cis $2\pi = 1$
35. cis 0, cis $\frac{\pi}{3}$, cis $\frac{2\pi}{3}$, cis π, cis $\frac{4\pi}{3}$, cis $\frac{5\pi}{3}$ 37. $1, i, -1, -i$
39. 2 cis $0 = 2$; 2 cis $\frac{2\pi}{5} \approx 0.6180 + 1.9021i$;
    2 cis $\frac{4\pi}{5} \approx -1.6180 + 1.1756i$; 2 cis $\frac{6\pi}{5} \approx -1.6180 - 1.1756i$;
    2 cis $\frac{8\pi}{5} \approx 0.6180 - 1.9021i$

41. $2 \text{ cis } \frac{\pi}{4} = \sqrt{2} + \sqrt{2}i$; $2 \text{ cis } \frac{7\pi}{12} \approx -0.5176 + 1.9319i$;

$2 \text{ cis } \frac{11\pi}{12} \approx -1.9319 + 0.5176i$;

$2 \text{ cis } \frac{5\pi}{4} = -\sqrt{2} - \sqrt{2}i$; $2 \text{ cis } \frac{19\pi}{12} \approx 0.5176 - 1.9319i$;

$2 \text{ cis } \frac{23\pi}{12} \approx 1.9319 - 0.5176i$

43. $2 \text{ cis } \frac{\pi}{6} = \sqrt{3} + i$; $2 \text{ cis } \frac{5\pi}{6} = -\sqrt{3} + i$; $2 \text{ cis } \frac{3\pi}{2} = -2i$

45. $\sqrt[3]{16} \text{ cis } \frac{\pi}{9} \approx 2.3679 + 0.8618i$; $\sqrt[3]{16} \text{ cis } \frac{7\pi}{9} \approx$

$-1.9303 + 1.6197i$; $\sqrt[3]{16} \text{ cis } \frac{13\pi}{9} \approx -0.4376 - 2.4816i$

47. approximately $0.0493421 + 0.2275449i \approx$
$0.2328333 \text{ cis } 1.3572568$; and $-1.0493421 - 0.2275449i \approx$
$1.0737297 \text{ cis } 6.069646$

49. $\frac{1}{\sqrt[4]{2}} \text{ cis } \frac{\pi}{8} \approx 0.776887 + 0.3217971i$;

$\frac{1}{\sqrt[4]{2}} \text{ cis } \frac{9\pi}{8} \approx -0.776887 - 0.3217971i$;

$\frac{1}{\sqrt[4]{2}} \text{ cis } \frac{7\pi}{8} \approx -0.776887 + 0.3217971i$; and

$\frac{1}{\sqrt[4]{2}} \text{ cis } \frac{15\pi}{8} \approx 0.776887 - 0.3217971i$

Section 6.8 (Review Exercises)

1. $\dfrac{\cos x + \sin^2 x}{\cos^2 x}$

3. (a) $\dfrac{\sqrt{2 + \sqrt{2}}}{2} = \sqrt{\dfrac{\sqrt{2} + 1}{2\sqrt{2}}}$ (b) $-\dfrac{\sqrt{2 - \sqrt{3}}}{2}$

(c) $-\sqrt{\dfrac{2 + \sqrt{2}}{2 - \sqrt{2}}} = -\sqrt{3 + 2\sqrt{2}} = \dfrac{1}{1 - \sqrt{2}}$

5. $\dfrac{3\sqrt{3} + 4}{4\sqrt{3} - 3}$

7. $\sin \theta = \pm \sqrt{\dfrac{1 - \sqrt{0.6975}}{2}}$, $\cos \theta = \pm \sqrt{\dfrac{1 + \sqrt{0.6975}}{2}}$,

$\tan \theta = \sqrt{\dfrac{1 - \sqrt{0.6975}}{1 + \sqrt{0.6975}}}$

9. (a) $\frac{1}{2}(\sin 7x - \sin 3x)$ (b) $\frac{1}{2}(\sin 7x + \sin 3x)$

11. (a) $\sqrt{41} \sin(\theta + \pi - \arctan(\frac{5}{4})) =$
$\sqrt{41} \cos(\theta - (\pi - \arctan(-\frac{4}{5})))$

(b) $-\dfrac{\sqrt{17}}{4} \sin(\theta + \arctan(\frac{1}{4})) = -\dfrac{\sqrt{17}}{4} \cos(\theta - \arctan 4)$

31. $[0, 2\pi)$ **33.** $\left\{0, \dfrac{2\pi}{3}, \pi, \dfrac{4\pi}{3}\right\}$ **35.** $\left\{\dfrac{\pi}{4}, \dfrac{\pi}{2}, \dfrac{3\pi}{4}, \dfrac{5\pi}{4}, \dfrac{3\pi}{2}, \dfrac{7\pi}{4}\right\}$

37. $\left(0, \dfrac{\pi}{4}\right) \cup \left(\dfrac{\pi}{4}, \dfrac{\pi}{2}\right) \cup \left(\dfrac{\pi}{2}, \dfrac{3\pi}{4}\right) \cup \left(\dfrac{3\pi}{4}, \pi\right) \cup \left(\pi, \dfrac{5\pi}{4}\right) \cup$

$\left(\dfrac{5\pi}{4}, \dfrac{3\pi}{2}\right) \cup \left(\dfrac{3\pi}{2}, \dfrac{7\pi}{4}\right) \cup \left(\dfrac{7\pi}{4}, 2\pi\right)$

39. $(2 \arcsin(\frac{1}{4}), \pi) \cup (2\pi - 2 \arcsin(\frac{1}{4}), 2\pi)$

41.

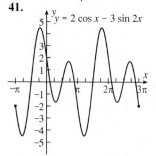

43.

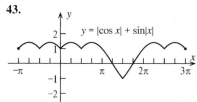

45.

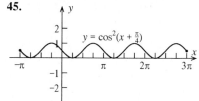

47. π **49.** (a) $1 - \sqrt{3}i$ (b) -3 (c) $-32i$

51. absolute value = 16; principal argument = $\dfrac{2\pi}{3}$; polar form = $16 \text{ cis } \dfrac{2\pi}{3}$

53. absolute value = 2; principal argument = $\dfrac{\pi}{2}$; polar form = $2 \text{ cis } \dfrac{\pi}{2}$

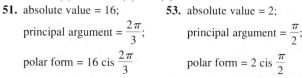

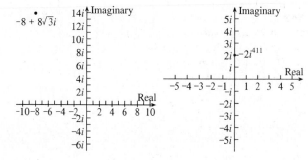

Answers to Exercises

55. absolute value $= \frac{1}{18}$;

principal argument $= \frac{\pi}{6}$;

polar form $= \frac{1}{18}$ cis $\frac{\pi}{6}$

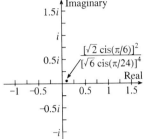

57. absolute value $= 2$;

principal argument $= \frac{\pi}{4}$;

polar form $= 2$ cis $\frac{\pi}{4}$

59. absolute value ≈ 0.7283511;

principal argument $2\pi - 5 \arctan \frac{4}{\sqrt{3}}$;

polar form ≈ 0.7283511 cis $\left(2\pi - 5 \arctan \frac{4}{\sqrt{3}}\right)$

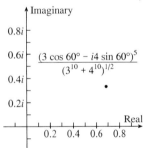

61. $\sqrt{2}$ cis $\frac{5\pi}{12}$; $\sqrt{2}$ cis $\frac{11\pi}{12}$; $\sqrt{2}$ cis $\frac{17\pi}{12}$; $\sqrt{2}$ cis $\frac{23\pi}{12}$

63. $\sqrt[3]{2}$ cis $\frac{5\pi}{9}$; $\sqrt[3]{2}$ cis $\frac{11\pi}{9}$; $\sqrt[3]{2}$ cis $\frac{17\pi}{9}$

65. $i =$ cis $\frac{\pi}{2}$ is a root with multiplicity 2

Section 7.1

Note: many of the answers to exercises in Chapter 7 are given as rounded-off decimal approximations.

1. $\beta = 53°$, $\gamma = 90°$, $a \approx 9.0426$, $c \approx 15.0256$
3. $\beta = 22°30'$, $\gamma = 90°$, $a \approx 1.1087$, $b \approx 0.4592$
5. $\alpha = \frac{5\pi}{12}$, $\gamma = \frac{\pi}{2}$, $a \approx 57.1003$, $c \approx 59.1147$
7. $\alpha \approx 0.6435$ radian ($36.87°$), $\beta \approx 0.9273$ radian ($53.13°$), $\gamma = \frac{\pi}{2}$ ($90°$), $c = 5$
9. $\alpha \approx 1.0911$ ($62.51°$), $\beta \approx 0.4797$ ($27.49°$), $\gamma = \frac{\pi}{2}$ ($90°$), $a \approx 11.5326$
11. $\beta = 41.76°$, $\gamma = 90°$, $a \approx 26.5443$, $c \approx 35.5850$
13. $\alpha \approx 1.1760$ ($67.38°$), $\beta \approx 0.3948$ ($22.62°$), $\gamma = \frac{\pi}{2}$ ($90°$), $b = 8.5$
15. false. Consider similar right triangles. For instance, $\alpha = \frac{\pi}{6}$, $\beta = \frac{\pi}{3}$, and $\gamma = \frac{\pi}{2}$ could correspond to $a = 1$, $b = \sqrt{3}$, and $c = 2$; or to $a = 2$, $b = 2\sqrt{3}$, and $c = 4$, among others.
17. true. In view of exercise 16, we can assume given at least one side and at least one non-right angle. Then all angles are known (since their measures add to $180°$) and the conclusion follows by the angle-angle-side criterion for congruence of triangles.
19. $43.1736°$ **21.** 35.9704 feet **23.** 78.5717 feet
25. 1254.1002 feet **27.** 326.5716 mph, bearing $95.6233°$
29. 5218.52 yards

Section 7.2

1. $\gamma = 73.7°$, $a \approx 11.8280$ cm, $c \approx 12.8221$ cm
3. $\alpha = \frac{\pi}{6}$, $a = b - \sqrt{6} \approx 2.4495$ feet
5. $\beta = 90.3°$, $a \approx 2.8936$ m, $c \approx 3.6884$ m
7. two: $\beta_1 \approx 48.5904°$, $\gamma_1 \approx 101.4096°$, $c_1 \approx 7.8419$ cm
$\beta_2 \approx 131.4096°$, $\gamma_2 \approx 18.5904°$, $c_2 \approx 2.5504$ cm
9. one; $\alpha \approx 0.5590$, $\beta \approx 1.7972$, $b \approx 11.0250$ feet **11.** none
13. one; $\beta \approx 0.6155$, $\gamma \approx 0.4317$, $c \approx 2.8989$ inches
15. one; $\alpha = 106°$, $\gamma = 37°$, $a \approx 4.7918$ m
17. none **19.** none
21. 8.6113 miles from Tom, 6.7945 miles from Dick
23. 4.1818 feet **25.** 61.2836 feet **27.** 338.6334 feet
29. 23.3638 feet or 34.0818 feet

Section 7.3

1. $\alpha \approx 147.1114°$, $\beta \approx 8.8886°$, $c \approx 8.2919$ cm
3. $a \approx 3.4071$ feet, $\beta \approx 0.3142$, $\gamma \approx 1.0322$
5. $b \approx 28.2522$, $\alpha \approx 137.7385°$, $\gamma \approx 12.2615°$
7. $\alpha \approx 0.6846$ ($39.2247°$), $\beta \approx 1.1196$ ($64.1484°$), $\gamma \approx 1.3374$ ($76.6274°$)
9. $\alpha \approx 0.4460$ ($25.5547°$), $\beta \approx 1.1243$ ($64.4169°$), $\gamma \approx 1.5713$ ($90.0289°$)
11. 790.2658 miles **13.** 255.7285 mph
15. 104.8872 feet downhill and 89.6698 feet uphill
17. (a) 63.7170 feet (b) $92.8244°$
19. (a) 144.8889 m (b) $\frac{75}{\sqrt{2}}$ m ≈ 53.0330 m **25.** 3.9937 cm^2
27. $s = \frac{\sqrt{8} + \sqrt{13} + \sqrt{41}}{2} \approx 6.418\,551\,320$. The exact answer is $\sqrt{s(s-\sqrt{8})(s-\sqrt{13})(s-\sqrt{41})} = 1$. Calculator approximations range from $0.999\,999\,987$ to $1.000\,000\,053$
29. 11.6190 **31.** 35

Answers to Exercises

Section 7.4

1. (a) scalar (b) vector (c) vector (d) scalar
3. (a)

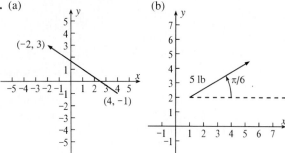

5. If $v = PQ$ then $v + 0 = PQ + QQ = PQ = v$ **7.** E
9.

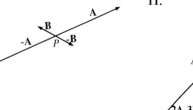

11.

13. a southeast wind blowing at 15 mph
15. $\langle -4, -2 \rangle$ **17.** $(e - 4)\mathbf{i} + 4\mathbf{j}$
19. $\langle 1, \sqrt{3} \rangle$ **21.** $\left\langle \frac{3\sqrt{2}}{8}, \frac{3\sqrt{2}}{8} \right\rangle$

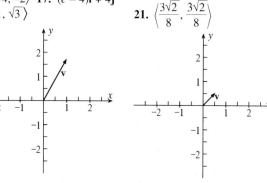

23. $\left\langle \frac{-3\sqrt{3}}{4}, \frac{3}{4} \right\rangle$ **25.** $\langle -1, 0 \rangle$

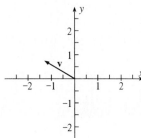

27. $\mathbf{v} \approx \langle -1.4295, -2.7975 \rangle$

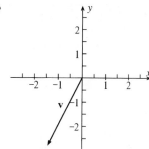

29. $|\mathbf{v}| = 20; \theta = \frac{\pi}{3}$ **31.** $|\mathbf{v}| = 5\sqrt{5}; \theta = 2\pi - \arccos \frac{1}{\sqrt{5}} \approx 5.1760$
33. $|\mathbf{v}| = 5; \theta = \frac{3\pi}{2}$ **35.** $|\mathbf{v}| = 4; \theta = 0$ **37.** $(3, 4)$ **39.** (a) and (b)
41. (a) $\langle 5, -6 \rangle$ (b) $\langle -9, -10 \rangle$ (c) $\langle -10, -40 \rangle$
(d) $\langle -\frac{14}{3}, -\frac{4}{3} \rangle$ (e) $\langle \frac{25}{2}, -2 \rangle$
43. (a) $\frac{7}{2}\mathbf{i} + 3\mathbf{j}$ (b) $-\frac{1}{2}\mathbf{i} + 9\mathbf{j}$ (c) $\frac{15}{2}\mathbf{i} + 30\mathbf{j}$
(d) $-\frac{4}{3}\mathbf{i} + 2\mathbf{j}$ (e) $\frac{41}{8}\mathbf{i} - \frac{3}{2}\mathbf{j}$
45. $\frac{3}{\sqrt{34}}\mathbf{i} - \frac{5}{\sqrt{34}}\mathbf{j}$ **47.** $-\frac{1}{\sqrt{2}}\mathbf{i} + \frac{1}{\sqrt{2}}\mathbf{j}$
51. (a) $\langle 45 \cos 5.323, 45 \sin 5.323 \rangle \approx \langle 25.8016, -36.8684 \rangle$
(b) $\langle 60 \cos 40°, 60 \sin 40° \rangle \approx \langle 45.9627, 38.5673 \rangle$
53. 289.2199 mph with direction $\approx 284.2497°$
55. (a) 479.9967 miles (b) 152.2475 miles
(c) 455.2115 miles
57. 78.3536 pounds on the cable at the right and 41.5097 pounds on the cable at the left
59. (a) $44.9229\mathbf{i} + 0.0003\mathbf{j}$ (b) $-44.9229\mathbf{i} - 0.0003\mathbf{j}$

Section 7.5

1. $-\frac{51}{5}$ **3.** 3 **7.** $\mathbf{u} = -\frac{53}{2}\mathbf{i} - 20\mathbf{j}$ **9.** 1.8034 **11.** π **13.** $\frac{4}{5}$
15. $2.7526°$ **19.** $-36\mathbf{i} + 48\mathbf{j} = -12(3\mathbf{i} - 4\mathbf{j})$ **21.** (a) no (b) yes
23. (a) $\frac{7}{2}$ (b) -14 **25.** $\frac{405}{109}\mathbf{i} - \frac{1350}{109}\mathbf{j}$ **27.** $36\mathbf{i} - 24\mathbf{j}$
29. $\mathbf{a}_1 = -0.2\mathbf{i} + 0.4\mathbf{j}; \mathbf{a}_2 = 5.2\mathbf{i} + 2.6\mathbf{j}$
31. $\mathbf{a}_1 = \mathbf{0}$ and $\mathbf{a}_2 = 10\mathbf{i} - 9\mathbf{j}$
33. $30 + 24\sqrt{3} \approx 71.5692$ N·m **35.** 2
37. 140 ft-lb **39.** $\frac{2}{\sqrt{365}}$ **41.** 0

Section 7.6

1. $s(t) = 10 \sin \left(\frac{40}{13} \pi t - \frac{\pi}{2} \right)$ inches
3. $s(t) = 25 \sin \left(\frac{5}{2} \pi t - \frac{\pi}{2} \right)$ cm
5. $s(t) = 12 \sin \left(\frac{20}{13} \pi t + \frac{\pi}{2} \right)$ cm
7. $s(t) = 6.2 \sin \left(\frac{25}{9} \pi t + \frac{\pi}{2} \right)$ inches
9. (a) 5 cm (b) $\frac{1}{7}$ (c) 7

(d) $\frac{5\sqrt{2}}{2}$ cm above the equilibrium position

(e) $\frac{5}{\sqrt{2}}$ cm below the equilibrium position (f) $\frac{7}{8}$

11. $\cos(5\pi t - 3\pi) = \cos(3\pi - 5\pi t) = \sin\left(\frac{\pi}{2} - (3\pi - 5\pi t)\right) = \sin\left(5\pi t - \frac{\pi}{2} - 2\pi\right) = \sin\left(5\pi t - \frac{\pi}{2}\right) = \sin\left(5\pi\left(t - \frac{1}{10}\right)\right).$

13. (a) $y = 0.01 \sin(800\pi t) \cos(400\pi t)$
    (b) frequency = 200, period = $\frac{1}{200}$
    (c)

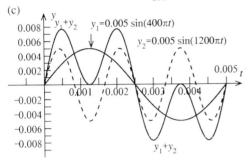

15. (a) 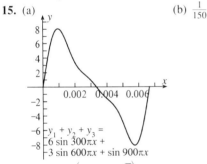 (b) $\frac{1}{150}$

17. $I = 6.8 \sin\left(160\pi t + \frac{\pi}{3}\right)$ A

19. (a) $\frac{3}{2}$ (b) $\frac{2}{3}$
    (c) $\sqrt{3}e^{-4} \approx 0.0317$ ft above the "equilibrium" position
    (d) about 0.8893 ft above the "equilibrium" position

21. (a) $\frac{2}{\pi}$ (b) $\frac{\pi}{2}$
    (c) $-2e^{-6} \sin 17 \approx 0.0048$ m above the "equilibrium" position
    (d) about 0.1595 m below the "equilibrium" position

23. $m \approx 0.0429$ 25. $m \approx 0.2231$ 31. Rowena

Section 7.7 (Review Exercises)

1. one; $\alpha = 17°$, $a \approx 1.4619$ cm, $b \approx 4.7815$ cm
3. one; $\beta = 117°$, $b \approx 12.1270$ feet, $c \approx 6.1790$ feet 5. none
7. one; $\alpha \approx 0.8054$ (46.1462°), $\beta = \frac{\pi}{2}$ (90°), $\gamma \approx 0.7654$ (43.8538°)
9. one; $\alpha \approx 0.5236$ ($\alpha = 30°$), $\beta \approx 2.3754$ (136.1021°), $\gamma \approx 0.2426$ (13.8979°)
11. one; $\alpha = \frac{\pi}{12}$, $a = c \approx 1.5529$
13. one; $\alpha \approx 0.5528$, $b \approx 5.5215$ yards, $\gamma \approx 0.2326$ 15. none
17. two: $\alpha_1 \approx 63.5680°$, $\beta_1 \approx 90.9320°$, $b_1 \approx 5.8063$ cm; and $\alpha_2 \approx 116.4320°$, $\beta_2 \approx 38.0680°$, $b_2 \approx 3.5806$ cm
19. 214.2145 feet 21. 40.9459 feet
23. 25.7099 mph, in the direction 13.4957°
25. 279.4884 miles 27. 84.5100 m
29. 72.9327° and 107.0673° 31. 153.8187° 33. 1.8385
35. 0.8204 39. 7.6111 square inches 41. 10.0794 m²
43. 105.8977 feet 45. $\alpha = \arccos\sqrt{\frac{13}{29}} \approx 0.8372$
47. (a) (b)

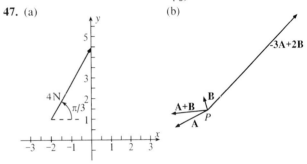

49. $\langle -5, 9 \rangle = -5\mathbf{i} + 9\mathbf{j}$ 51. magnitude 4, direction $\frac{5\pi}{6}$
53. $-0.5\mathbf{i} + 0.5\sqrt{3}\mathbf{j}$
55. 130.1205 pounds on the cable at the right and 242.6216 pounds on the cable on the left
57. (a) 23 (b) 5 (c) -2 (d) 0
59. (a) 1.3834 (b) 2.7611 61. (a) $\frac{21}{8}$ (b) $-\frac{56}{3}$
63. $\mathbf{a}_1 = \frac{35}{41}\mathbf{i} + \frac{28}{41}\mathbf{j}$ and $\mathbf{a}_2 = \frac{88}{41}\mathbf{i} - \frac{110}{41}\mathbf{j}$
67. a, b, c are *not* determined by α, β, γ since corresponding angles in similar triangles are congruent.
69. $s(t) = 5 \sin\left(\frac{200\pi}{37}t - \frac{\pi}{2}\right)$ cm
71. (a) 4 inches (b) $\frac{1}{5}$ (c) 5
    (d) $2\sqrt{3} \approx 3.4641$ inches above the equilibrium position
    (e) about 2.4846 inches above the equilibrium position
    (f) $\frac{5}{6}$
73. (a) 3 m above the "equilibrium" (b) 2 (c) $\frac{1}{2}$
    (d) $3e^{-1}$ m ≈ 1.1036 m above the "equilibrium"
75. 17.9181 m

Answers to Exercises

Section 8.1

1. vertex $(0, 0)$; axis $x = 0$;
focus $(0, \frac{9}{4})$; directrix $y = -\frac{9}{4}$;
endpoints of latus rectum
$(-\frac{9}{2}, \frac{9}{4})$, $(\frac{9}{2}, \frac{9}{4})$;
length of latus rectum 9

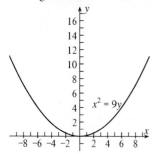

3. vertex $(0, 0)$; axis $y = 0$;
focus $(-\frac{3}{4}, 0)$; directrix
$x = \frac{3}{4}$; endpoints of latus
rectum $(-\frac{3}{4}, -\frac{3}{2})$, $(-\frac{3}{4}, \frac{3}{2})$;
length of latus rectum 3

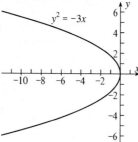

41. the parabola $x^2 = 16y$

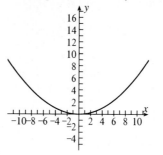

43. the left half of the parabola $x^2 = 16y$, excluding the origin

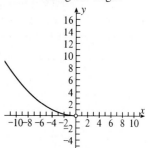

5. vertex $\left(-1 - \frac{48}{\pi}, -8\right)$;
axis $y = -8$;
focus $\left(-1 - \frac{48}{\pi} + \frac{\pi}{4}, -8\right)$;
directrix $x = -1 - \frac{48}{\pi} - \frac{\pi}{4}$;
endpoints of latus rectum
$\left(-1 - \frac{48}{\pi} + \frac{\pi}{4}, -8 - \frac{\pi}{2}\right)$,
$\left(-1 - \frac{48}{\pi} + \frac{\pi}{4}, -8 + \frac{\pi}{2}\right)$;
length of latus rectum π

49. (a)

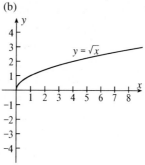

(b)

(c)

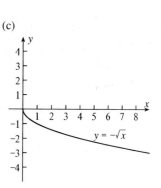

7. vertex $(1, 0)$; axis $x = 1$;
focus $(1, -\frac{1}{3})$; directrix $y = \frac{1}{3}$;
endpoints of latus rectum
$(\frac{1}{3}, -\frac{1}{3})$, $(\frac{5}{3}, -\frac{1}{3})$;
length of latus rectum $\frac{4}{3}$

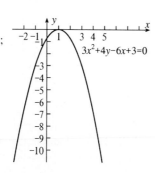

51. $\frac{9}{4}$ feet or 27 inches **53.** 3 cm
55. (a) $\frac{5}{2}$ yards (b) 40 yards

Section 8.2

1. center $(0, 0)$; $a = 8$; $b = 3$;
$c = \sqrt{55}$; foci $(\sqrt{55}, 0)$ and
$(-\sqrt{55}, 0)$; vertices $(8, 0)$
and $(-8, 0)$; endpoints of
minor axis $(0, 3)$ and
$(0, -3)$; $e = \frac{\sqrt{55}}{8}$; directrices
$x = \pm\frac{64}{\sqrt{55}}$; length of each
latus rectum is $\frac{9}{4}$

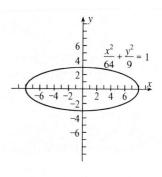

13. $y^2 = -8x$ **15.** $x^2 + 6x + 48y - 87 = 0$ **17.** $x^2 = -8y$
19. $y^2 + \frac{2}{3}y - 6x + \frac{28}{9} = 0$ **21.** $y^2 - y - 10x - \frac{337}{12} = 0$
23. $x^2 + 16y - 64 = 0$ **25.** $3y^2 - 12y - 16x + 28 = 0$
27. $7x^2 - 70x + 12y - 127 = 0$
29. $y^2 - 2y - 8x + 17 = 0$ or $x^2 - 4x + 4y = 0$
31. $y = \frac{7}{6}x^2 - \frac{3}{2}x - \frac{2}{3}$ **37.** $y = -2x - 19$

Answers to Exercises

3. center $(0, 0)$; $a = \sqrt{30}$; $b = 5$; $c = \sqrt{5}$; foci $(0, \sqrt{5})$ and $(0, -\sqrt{5})$; vertices $(0, \sqrt{30})$ and $(0, -\sqrt{30})$; endpoints of minor axis $(5, 0)$ and $(-5, 0)$; $e = \frac{1}{\sqrt{6}}$; directrices $y = \pm 6\sqrt{5}$; length of each latus rectum is $\frac{50}{\sqrt{30}}$

5. center $(-3, 4)$; $a = 9$; $b = 8$; $c = \sqrt{17}$; foci $(-3, 4 + \sqrt{17})$ and $(-3, 4 - \sqrt{17})$; vertices $(-3, 13)$ and $(-3, -5)$; endpoints of minor axis $(5, 4)$ and $(-11, 4)$; $e = \frac{\sqrt{17}}{9}$; directrices $y = 4 \pm \frac{81}{\sqrt{17}}$; length of each latus rectum is $\frac{128}{9}$

11. center $(\frac{1}{2}, -\frac{3}{2})$; $a = 4$; $b = 2$; $c = 2\sqrt{3}$; foci $(\frac{1}{2}, -\frac{3}{2} + 2\sqrt{3})$ and $(\frac{1}{2}, -\frac{3}{2} - 2\sqrt{3})$; vertices $(\frac{1}{2}, \frac{5}{2})$ and $(\frac{1}{2}, -\frac{11}{2})$; endpoints of minor axis $(\frac{5}{2}, -\frac{3}{2})$ and $(-\frac{3}{2}, -\frac{3}{2})$; $e = \frac{\sqrt{3}}{2}$; directrices $y = -\frac{3}{2} \pm \frac{8}{\sqrt{3}}$; length of each latus rectum is 2

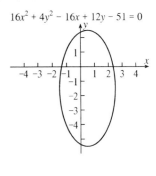

$16x^2 + 4y^2 - 16x + 12y - 51 = 0$

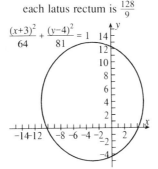

$30x^2 + 25y^2 = 750$

17. $\frac{(x-8)^2}{20} + \frac{(y-5)^2}{36} = 1$ **19.** $\frac{(x+3)^2}{202} + \frac{(y-9)^2}{81} = 1$

21. $\frac{(x-1)^2}{9} + \frac{y^2}{4} = 1$ **23.** $\frac{(x+2)^2}{5} + \frac{(y+6)^2}{9} = 1$

25. $\frac{(x-2)^2}{3\pi^2} + \frac{(y-\pi)^2}{4\pi^2} = 1$ **27.** $\frac{(x-2)^2}{16} + \frac{(y+3)^2}{4} = 1$

29. $\frac{(x-16)^2}{256} + \frac{(y-9)^2}{192} = 1$ **31.** $\frac{(x+4)^2}{44} + \frac{(y+2)^2}{144} = 1$

33. $\frac{(x-2)^2}{16} + \frac{(y+3)^2}{15} = 1$ **35.** $\frac{(x+\frac{13}{7})^2}{\frac{1296}{49}} + \frac{(y-9)^2}{\frac{81}{7}} = 1$

7. center $(4, -3)$; $a = 9\sqrt{2}$; $b = 8\sqrt{2}$; $c = \sqrt{34}$; foci $(4 + \sqrt{34}, -3)$ and $(4 - \sqrt{34}, -3)$; vertices $(4 + 9\sqrt{2}, -3)$ and $(4 - 9\sqrt{2}, -3)$; endpoints of minor axis $(4, -3 + 8\sqrt{2})$ and $(4, -3 - 8\sqrt{2})$; $e = \frac{\sqrt{17}}{9}$; directrices $x = 4 \pm 81\sqrt{\frac{2}{17}}$; length of each latus rectum is $\frac{128\sqrt{2}}{9}$

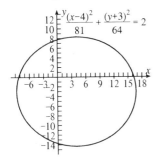

$\frac{(x-4)^2}{81} + \frac{(y+3)^2}{64} = 2$

9. center $(-1, 2)$; $a = 3$; $b = 1$; $c = 2\sqrt{2}$; foci $(-1 + 2\sqrt{2}, 2)$ and $(-1 - 2\sqrt{2}, 2)$; vertices $(2, 2)$ and $(-4, 2)$; endpoints of minor axis $(-1, 3)$ and $(-1, 1)$; $e = \frac{2\sqrt{2}}{3}$; directrices $x = -1 \pm \frac{9}{2\sqrt{2}}$; length of each latus rectum is $\frac{2}{3}$

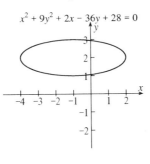

$x^2 + 9y^2 + 2x - 36y + 28 = 0$

43. (a) has limit infinity (b) has limit 1 (c) approaches the parallel lines $y = \pm b$

47. (a)

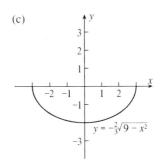

$y^2 = \frac{4}{9}(9 - x^2)$

(b) $y = \frac{2}{3}\sqrt{9 - x^2}$

(c) $y = -\frac{2}{3}\sqrt{9 - x^2}$

51. 90,436,000 miles

53. about major axis: egg; about minor axis (if $a \approx b$): Earth

Answers to Exercises

Section 8.3

1. center (0, 0); $a = 8$; $b = 3$; $c = \sqrt{73}$; foci ($\sqrt{73}$, 0) and ($-\sqrt{73}$, 0); vertices (8, 0) and (−8, 0); endpoints of conjugate axis (0, 3) and (0, −3); asymptotes $y = \pm\frac{3}{8}x$; $e = \frac{\sqrt{73}}{8}$; directrices $x = \pm\frac{64}{\sqrt{73}}$

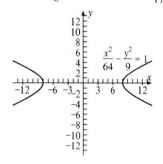

3. center (0, 0); $a = 5$; $b = \sqrt{30}$; $c = \sqrt{55}$; foci (0, $\sqrt{55}$) and (0, $-\sqrt{55}$); vertices (0, 5) and (0, −5); endpoints of conjugate axis ($\sqrt{30}$, 0) and ($-\sqrt{30}$, 0); asymptotes $y = \pm\frac{5}{\sqrt{30}}x$; $e = \sqrt{\frac{11}{5}}$; directrices $y = \pm 5\sqrt{\frac{5}{11}}$

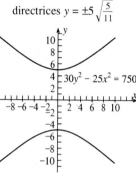

9. center (−2, 1); $a = 6$; $b = 3$; $c = \sqrt{45}$; foci (−2, 1 + $\sqrt{45}$) and (−2, 1 − $\sqrt{45}$); vertices (−2, 7) and (−2, −5); endpoints of conjugate axis (1, 1) and (−5, 1); asymptotes $y = 1 \pm 2(x + 2)$; $e = \frac{\sqrt{45}}{6}$; directrices $y = -2 \pm \frac{36}{\sqrt{45}} = -2 \pm \frac{12}{\sqrt{5}}$

11. center ($\frac{1}{2}$, −2); $a = \frac{3}{2}$; $b = 2$; $c = \frac{5}{2}$; foci (3, −2) and (−2, −2); vertices (2, −2) and (−1, −2); endpoints of conjugate axis ($\frac{1}{2}$, 0) and ($\frac{1}{2}$, −4); asymptotes $y = -2 \pm \frac{4}{3}(x - \frac{1}{2})$; $e = \frac{5}{3}$; directrices $x = \frac{7}{5}$ and $x = -\frac{2}{5}$

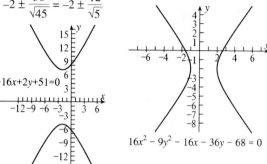

5. center (3, −4); $a = 9$; $b = 8$; $c = \sqrt{145}$; foci (3, −4 + $\sqrt{145}$) and (3, −4 − $\sqrt{145}$); vertices (3, −13) and (3, 5); endpoints of conjugate axis (11, −4) and (−5, −4); asymptotes $y = -4 \pm \frac{9}{8}(x - 3)$; $e = \sqrt{\frac{145}{9}}$; directrices $y = -4 \pm \frac{81}{\sqrt{145}}$

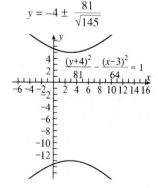

7. center (4, −3); $a = 8\sqrt{2}$; $b = 9\sqrt{2}$; $c = \sqrt{290}$; foci (4 + $\sqrt{290}$, −3) and (4 − $\sqrt{290}$, −3); vertices (4 + 8$\sqrt{2}$, −3) and (4 − 8$\sqrt{2}$, −3); endpoints of conjugate axis (4, −3 + 9$\sqrt{2}$) and (4, −3 − 9$\sqrt{2}$); asymptotes $y = -3 \pm \frac{9}{8}(x - 4)$; $e = \frac{\sqrt{145}}{8}$; directrices $x = 4 \pm 64\sqrt{\frac{2}{145}}$

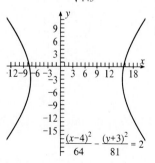

15. $4\sqrt{8}$ **17.** $\frac{8}{3}$ **19.** $(y-5)^2 - \frac{(x-8)^2}{15} = 1$

21. $\frac{(x+3)^2}{21} - \frac{(y-9)^2}{100} = 1$ **23.** $\frac{(x-11)^2}{\frac{81}{13}} - \frac{(y-9)^2}{\frac{36}{13}} = 1$

25. $\frac{(y-5)^2}{16} - \frac{(x-8)^2}{\frac{64}{9}} = 1$ **27.** $\frac{(x+1)^2}{9} - \frac{y^2}{16} = 1$

29. $\frac{(y-3)^2}{36} - \frac{(x+4)^2}{28} = 1$ **31.** $\frac{(y-\pi)^2}{\frac{4\pi^2}{9}} - \frac{(x-2)^2}{\frac{5\pi^2}{9}} = 1$

33. $\frac{(x-2)^2}{16} - \frac{(y+3)^2}{4} = 1$ **35.** $\frac{(x-16)^2}{16} - \frac{(y-9)^2}{48} = 1$

37. $\frac{(y+2)^2}{64} - \frac{(x+4)^2}{36} = 1$ **39.** $\frac{x^2}{16} - \frac{y^2}{4} = 1$

41. $\frac{(x-2)^2}{324} - \frac{(y+3)^2}{\frac{1377}{16}} = 1$ **43.** $7x^2 - 9y^2 - 124x + 162y - 365 = 0$

53. (a)

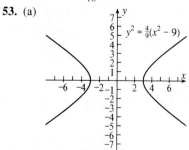

Answers to Exercises

(b)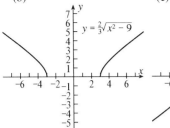

(c)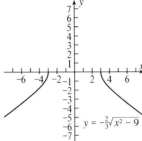

(d, e)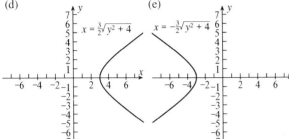

57. approximately (338.1165, 761.8835)

Section 8.4

1.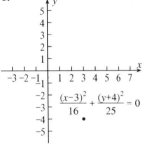

3.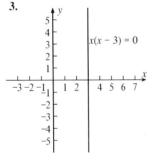

5.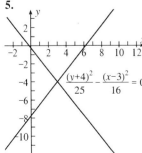

7.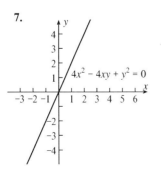

9. The graph is the empty set.
11. discriminant = 0; parabola **13.** discriminant = –24; ellipse
15. discriminant = 1; hyperbola
17. discriminant = 0; parabola
19. $\left(\dfrac{9-9\sqrt{3}}{4}\right)x'^2 + \left(\dfrac{3+9\sqrt{3}}{4}\right)y'^2 + \left(\dfrac{9-3\sqrt{3}}{2}\right)x'y' + 4 = 0$
21. $3x'^2 + y'^2 - 2\sqrt{3}x'y' - 4x' + 4\sqrt{3}y' - \pi = 0$
23. $\left(3 + \dfrac{3\sqrt{3}}{4}\right)x'^2 + \left(1 - \dfrac{3\sqrt{3}}{4}\right)y'^2 + \left(\dfrac{3}{2} - 2\sqrt{3}\right)x'y' - (4\sqrt{3}+3)x' + (4 - 3\sqrt{3})y' = 0$
25. $2x'y' - \sqrt{2} = 0$

27. $\alpha = \dfrac{\pi}{4}$; hyperbola; vertices (1, 1) and (–1, –1); foci $(\sqrt{2}, \sqrt{2})$ and $(-\sqrt{2}, -\sqrt{2})$

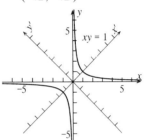

29. $\alpha = \arcsin\dfrac{1}{\sqrt{5}} \approx 0.4636$ ($\approx 26.5651°$); parabola; vertex (2.49, 1.295); focus (2.45, 1.275)

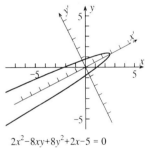

$2x^2 - 8xy + 8y^2 + 2x - 5 = 0$

31. $\alpha = \dfrac{\pi}{4}$; ellipse; vertices $(\sqrt{7}, \sqrt{7})$ and $(-\sqrt{7}, -\sqrt{7})$; foci $\left(\sqrt{\dfrac{70}{11}}, \sqrt{\dfrac{70}{11}}\right)$ and $\left(-\sqrt{\dfrac{70}{11}}, -\sqrt{\dfrac{70}{11}}\right)$

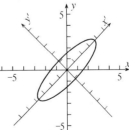

$3x^2 - 5xy + 3y^2 - 7 = 0$

33. $\alpha = \dfrac{\pi}{4}$; pair of parallel lines

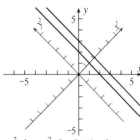

$x^2 + 2xy + y^2 - 5x - 5y + 6 = 0$

43. $xy = 0$ **45.** $9x^2 + xy - y^2 = 0$

Answers to Exercises

Section 8.5 (Review Exercises)

1. parabola; vertex (0, 0); axis $y = 0$; focus (4, 0); directrix $x = -4$; length of latus rectum is 16

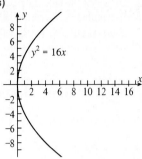

3. hyperbola; center (3, −5); $a = 2$; $b = 5$; $c = \sqrt{29}$; foci $(3, -5 + \sqrt{29})$ and $(3, -5 - \sqrt{29})$; vertices (3, −3) and (3, −7); $e = \dfrac{\sqrt{29}}{2}$; directrices $y = -5 \pm \dfrac{4}{\sqrt{29}}$; length of each latus rectum is 25; endpoints of conjugate axis (8, −5) and (−2, −5); asymptotes $y = -5 \pm \dfrac{2}{5}(x - 3)$

5. ellipse; center (−3, 1); $a = 3$; $b = 2$; $c = \sqrt{5}$; foci $(-3 + \sqrt{5}, 1)$ and $(-3 - \sqrt{5}, 1)$; vertices (0, 1) and (−6, 1); $e = \dfrac{\sqrt{5}}{3}$; directrices $x = -3 \pm \dfrac{9}{\sqrt{5}}$; length of each latus rectum is $\dfrac{8}{3}$; endpoints of minor axis (−3, 3) and (−3, −1)

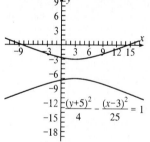

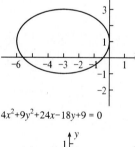

$4x^2 + 9y^2 + 24x - 18y + 9 = 0$

7. parabola; vertex (0, 0); axis $x = 0$; focus $(0, -\tfrac{2}{3})$; directrix $y = \tfrac{2}{3}$; length of latus rectum is $\tfrac{8}{3}$

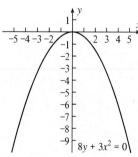

$8y + 3x^2 = 0$

9. $(x - 3)^2 = 40(y + 2)$ 11. $y^2 = 4(x + 1)$
13. $x^2 + 2x + 8y - 7 = 0$ 15. $\dfrac{(x+6)^2}{25} + \dfrac{(y+8)^2}{16} = 1$
17. $\dfrac{(x+3)^2}{9} + \dfrac{(y+8)^2}{4} = 1$ 19. ellipse; $\dfrac{(x-2)^2}{24} + \dfrac{(y-9)^2}{49} = 1$

21. $\dfrac{(x-2)^2}{63} + \dfrac{(y+2)^2}{144} = 1$ 23. $(x-5)^2 - \dfrac{(y-9)^2}{8} = 1$

25. $\dfrac{(x-13)^2}{25} - \dfrac{(y-6)^2}{\tfrac{100}{9}} = 1$ 27. $\dfrac{(y-12)^2}{36} - \dfrac{(x-8)^2}{28} = 1$

29. hyperbola; $41x^2 - 54xy - 31y^2 - 160y - 160 = 0$
31. (b), (c), and (d)

33.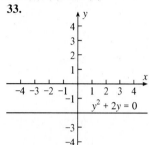

$y^2 + 2y = 0$

35.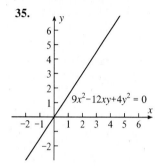

$9x^2 - 12xy + 4y^2 = 0$

37.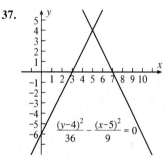

$\dfrac{(y-4)^2}{36} - \dfrac{(x-5)^2}{9} = 0$

39. discriminant = 12; hyperbola   41. discriminant = 0; parabola

43. $\dfrac{x'^2 - y'^2}{2} - 7 = 0$

45. $\alpha = \dfrac{\pi}{6}$; hyperbola; vertices $\left(\dfrac{\sqrt{105}}{10}, \dfrac{\sqrt{35}}{10}\right)$ and $\left(-\dfrac{\sqrt{105}}{10}, -\dfrac{\sqrt{35}}{10}\right)$; foci $\left(\dfrac{\sqrt{70}}{5}, \dfrac{\sqrt{210}}{15}\right)$ and $\left(-\dfrac{\sqrt{70}}{5}, -\dfrac{\sqrt{210}}{15}\right)$

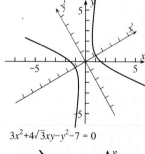

$3x^2 + 4\sqrt{3}xy - y^2 - 7 = 0$

47. $\alpha = \dfrac{\pi}{4}$; parabola; vertex $(-\tfrac{7}{4}, \tfrac{25}{12})$; focus $(-\tfrac{11}{6}, \tfrac{13}{6})$

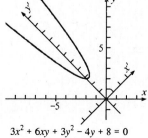

$3x^2 + 6xy + 3y^2 - 4y + 8 = 0$

Answers to Exercises

49. $\alpha = \frac{\pi}{8}$; hyperbola;
vertices are approximately
$(2.3784, 0.9852)$ and
$(-2.3784, -0.9852)$;
foci are approximately
$(6.2151, 2.5744)$ and
$(-6.2151, -2.5744)$

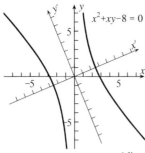

51. the parabola $y^2 = -2x$

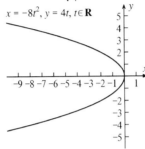

53. 4 inches **55.** $(2 - 2\sqrt{2})x^2 + 3y^2 - 6y - 8 + 8\sqrt{2} = 0$
57. $6.84[x - 0.878655]^2 - 0.035[y - 6]^2 = 6.84(0.878655)^2 - 0.035(36) - 4 \approx 0.0207167$. This simplifies approximately
to $\frac{(x - 0.878655)^2}{a^2} - \frac{(y - 6)^2}{b^2} = 1$, which is a hyperbola
with $a^2 \approx 0.0030288$ and $b^2 \approx 0.5919064$. Then
$c^2 = a^2 + b^2 \approx 0.5949352$. Thus $c \approx \sqrt{0.5949352} \approx 0.7713204$. Also $a \approx \sqrt{0.0030288} \approx 0.0550345$. Hence
$e = \frac{c}{a} \approx \frac{0.7713204}{0.0550345} \approx 14.015207 \approx 14.0152$.

Section 9.1

1. (a) and (d)
3. It's true because the two systems have the same (empty) solution set.
5. (a) and (d)
9. (a) $\begin{cases} \sqrt{2}x - 5z = 0 \\ 4x - \frac{y}{2} + z = 6 \\ -6y + 4z = \frac{3}{2} \end{cases}$ (b) $\begin{cases} 4x - \frac{y}{2} + z = 6 \\ \sqrt{2}x - 5z = 0 \\ -3y + 2z = \frac{3}{4} \end{cases}$
(c) $\begin{cases} \sqrt{2}x - 5z = 0 \\ 4x - \frac{y}{2} + z = 6 \\ -16x - y - 2z = -\frac{93}{4} \end{cases}$

11. (a), (b), and (c): in each case, the "effect" is to recover the original matrix
15. $x = 1, y = -\frac{1}{2}, z = 2$ **17.** The solution set is empty.
19. $(x, y, z) = \left(\frac{5 - 9t}{10}, \frac{7t}{5}, t\right)$
21. $(x_1, x_2, x_3) = (17 - 3t, 9 - 2t, t)$

23. $(x, y, z) = \left(\frac{60 + 5s - 6t}{3}, s, t\right)$ **25.** The solution set is empty.
27. The solution set is empty. **29.** The solution set is empty.
31. $x_1 = -1, x_2 = 3, x_3 = 0$ **33.** $x_1 = 4, x_2 = -5, x_3 = 2$
35. $(x, y, z) = (2 + 3s - 4t, s, t)$ **37.** $(x_1, x_2, x_3) = (20 + t, t, -16)$
39. 38 kg of Mixture A, 42 kg of Mixture B, and 120 kg of Mixture C
41. $x^2 + y^2 - 3x - 5y - 4 = 0$
43. 36 hours for Louise, 30 hours for Lewis, and 40 hours for Lois
45. (a) none (b) $k = 0$ (c) $k \neq 0$
47. (a) none (b) none (c) any number
49. (a) $k = 1$ (b) $k \neq 1$ (c) none

Section 9.2

1. (a) 2×3 (b) 3×3 (c) 4×3 (d) 2×1
(e) 1×2 (f) 1×1
3. $m = n = 1$
5. (a) $\begin{bmatrix} -12 & 9 \\ 15 & 9 \\ & 2 \end{bmatrix}$ (b) $[-8, 4, 0, 20]$ (c) $\begin{bmatrix} -1 \\ -2 \\ 0 \\ -\pi \end{bmatrix}$ (d) $\begin{bmatrix} 0 & 0 & 0 \\ 0 & 0 & 0 \end{bmatrix}$

9. (a), (c), (d), and (f) **11.** $\begin{bmatrix} 10 & 13 & 4 & 3 \\ -7 & -5 & -6 & 1 \end{bmatrix}$ **13.** $\begin{bmatrix} 0 & 0 & 0 \\ 0 & 0 & 0 \\ 0 & 0 & 0 \end{bmatrix}$

15. (a) $(AB)C = \begin{bmatrix} 0 \\ 13 \end{bmatrix} [6 \; 1 \; 5 \; 3] = \begin{bmatrix} 0 & 0 & 0 & 0 \\ 78 & 13 & 65 & 39 \end{bmatrix}$

$A(BC) = \begin{bmatrix} 4 & -3 & 0 \\ 1 & 0 & 2 \end{bmatrix} \begin{bmatrix} 18 & 3 & 15 & 9 \\ 24 & 4 & 20 & 12 \\ 30 & 5 & 25 & 15 \end{bmatrix} = \begin{bmatrix} 0 & 0 & 0 & 0 \\ 78 & 13 & 65 & 39 \end{bmatrix}$

(b) $(AB)C = \begin{bmatrix} 0 & 0 \\ 0 & 0 \end{bmatrix} \begin{bmatrix} -2 & -7 & 0 & 4 \\ 0 & 1 & 3 & 0 \end{bmatrix} = \begin{bmatrix} 0 & 0 & 0 & 0 \\ 0 & 0 & 0 & 0 \end{bmatrix}$

$A(BC) = \begin{bmatrix} 6 & 9 \\ 4 & 6 \end{bmatrix} \begin{bmatrix} -12 & -51 & -27 & 24 \\ 8 & 34 & 18 & -16 \end{bmatrix} = \begin{bmatrix} 0 & 0 & 0 & 0 \\ 0 & 0 & 0 & 0 \end{bmatrix}$

19. yes **21.** no
25. $A = \begin{bmatrix} -3 & 4 & 0 \\ \frac{1}{2} & -1 & 6 \end{bmatrix}, X = \begin{bmatrix} x \\ y \\ z \end{bmatrix}, B = \begin{bmatrix} 2 \\ 0 \end{bmatrix}$

The coefficient matrix is A.

The augmented matrix is $\begin{bmatrix} -3 & 4 & 0 & 2 \\ \frac{1}{2} & -1 & 6 & 0 \end{bmatrix}$.

27. (b), (c), (f), and (g)

29. (a) $\begin{bmatrix} -3 & 12 & 8 & -9 \\ -1 & 0 & 3 & -5 \\ -1 & 4 & 0 & -2 \end{bmatrix}$ (b) $\begin{bmatrix} -3 & 12 & 8 & -9 \\ 2 & -8 & 0 & 4 \\ -1 & 0 & 3 & -5 \end{bmatrix}$

(c) $\begin{bmatrix} 1 & -4 & 8 & -1 \\ -1 & 0 & 3 & -5 \\ 2 & -8 & 0 & 4 \end{bmatrix}$

31. $\begin{bmatrix} 1 & 0 & -1 \\ 0 & 1 & 2 \end{bmatrix}$ **33.** $\begin{bmatrix} 1 & 0 \\ 0 & 1 \end{bmatrix}$ **35.** $x = \frac{33}{4}, y = 27, z = -3$

Answers to Exercises

37. $(x, y, z) = (-2, s, t)$ **39.** The solution set is empty.

41. $\begin{bmatrix} \frac{5}{57} & \frac{5}{57} & -\frac{2}{19} \\ -\frac{53}{57} & \frac{4}{57} & -\frac{13}{19} \\ \frac{2}{57} & \frac{2}{57} & \frac{3}{19} \end{bmatrix}$

45. (a) $\begin{bmatrix} 7 & 5 & -7 \\ 19 & -6 & 6 \end{bmatrix}$ (b) $\begin{bmatrix} 9 & -5 & -1 \\ 13 & -6 & -2 \end{bmatrix}$ (c) $\begin{bmatrix} -5 & 5 & -1 \\ -5 & 3 & 3 \end{bmatrix}$ (d) $\begin{bmatrix} 26 & -10 & -6 \\ 42 & -18 & -2 \end{bmatrix}$

Section 9.3

1. (a) -2 (b) -15 (c) π (d) 0
3. (a) $M_{12} = -15$; $M_{24} = -20$ (b) $M_{12} = 0$; $M_{24} = 4$
5. (a) 84 (b) 55 (c) 11 **7.** (a) $\sqrt{2}\pi e$ (b) $-2\pi e$
13. (a) 72 (b) -50 (c) 86

21. (a) $\begin{bmatrix} -2 & \sqrt{2} \\ \pi & 0 \end{bmatrix}$ (b) $\begin{bmatrix} -1 \\ 4 \\ 0 \\ 7 \end{bmatrix}$ (c) $\begin{bmatrix} e & \pi \\ 4 & -1 \\ \frac{1}{2} & 0 \end{bmatrix}$ (d) $\begin{bmatrix} -1 & 2 & e \\ 0 & 1 & 4 \end{bmatrix}$

23. (a) $\begin{bmatrix} -1 & -1 & 1 \\ 0 & -9 & 6 \\ 3 & 18 & -12 \end{bmatrix}$ (b) $\begin{bmatrix} 0 & -4 \\ -8 & 2 \end{bmatrix}$

25. $\mathrm{adj}(A) = \begin{bmatrix} 0 & 24 & -\frac{1}{15} \\ 0 & 0 & \frac{1}{20} \\ 30 & -60 & \frac{1}{6} \end{bmatrix}$ and $A \cdot \mathrm{adj}(A) = \begin{bmatrix} 6 & 0 & 0 \\ 0 & 6 & 0 \\ 0 & 0 & 6 \end{bmatrix} =$
$6I = |A|I = \mathrm{adj}(A) \cdot A$

27. $\begin{bmatrix} -\frac{2}{19} & \frac{3}{19} \\ -\frac{7}{19} & \frac{20}{19} \end{bmatrix}$

29. $x = -35$, $y = \frac{131}{5}$, $z = 55$ **39.** $x = 9$, $y = \frac{8}{3}$, $z = 2$
41. $x_1 = 2 - \sqrt{2}$, $x_2 = 2 - 2\sqrt{2}$ **43.** The solution set is empty.
45. $x = \frac{1}{2}$, $y = \frac{5}{3}$ **47.** $(x, y, z) = (-1, z + 5, z)$
49. $(x_1, x_2, x_3, x_4) = \left(\frac{x_3 - 3x_4 + 32}{7}, \frac{2x_3 + x_4 - 27}{7}, x_3, x_4 \right)$

Section 9.4

1. (a) yes (b) no (c) no (d) yes

3.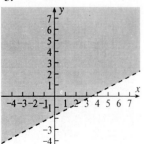

5.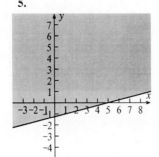

7.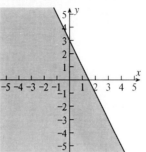

9.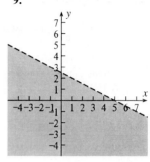

11.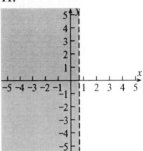

13.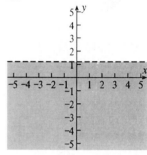

15. (a) open (b) closed (c) closed (d) open
17. Graph is unbounded; the only vertex is $(0, 4)$.

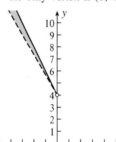

19. Graph is bounded; vertices are $(-\frac{13}{4}, -\frac{5}{2})$, $(-1, 2)$, and $(-\frac{1}{4}, \frac{1}{2})$.

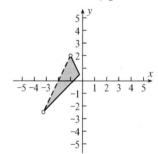

21. Graph is empty (hence bounded, with no vertex).

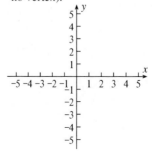

Answers to Exercises

23. Graph is bounded; vertices are $\left(\frac{1}{2}, -\frac{\sqrt{99}}{2}\right)$ and $\left(\frac{1}{2}, \frac{\sqrt{99}}{2}\right)$.

25. Graph is unbounded; vertices are $\left(\frac{3}{4}, \frac{3}{2}\right)$ and $(3, 0)$.

27.

29. maximum = 50; minimum = $-\frac{110}{3}$
31. maximum = 80; minimum = 60
33. 100 of Brand A and 450 of Brand B
35. 17 sticks and 17 pucks
37. $\frac{8000}{377} \approx 21.22$ acres of corn and $\frac{18,220}{377} \approx 48.33$ acres of wheat, with $\frac{11,480}{377} \approx 30.45$ acres fallow
39. 12.4 ounces of Brand A and no ounces of Brand B

Section 9.5 (Review Exercises)

1. (b), (c), and (e)
3. (c) is dependent and (b) has a unique particular solution.
7. $x = -\frac{419}{162}, y = \frac{34}{81}, z = -\frac{11}{81}$ **9.** The solution set is empty.
11. $(x, y) = \left(\frac{t-6}{2}, t\right)$
13. weaker tube at 28 ml/min, stronger tube at 112 ml/min, drain at 80 ml/min
15.

	Inconsistent	Dependent	Consistent with Unique Particular Solution
(a)	$k = 0$	none	$k \neq 0$
(b)	$k = 2$	$k \neq 2$	none

17. (a) 3×2 (b) 3×3 (c) 2×1 (d) 1×1
19. (a) $\begin{bmatrix} 2 & -4 \\ \frac{2}{3} & 0 \end{bmatrix}$ (b) $[0 \ -3 \ 12 \ 7]$ (c) 2 (d) 0

21. $\begin{bmatrix} 0 & 2 & -1 & 0 \\ 3 & 1 & -14 & 0 \end{bmatrix}$ **23.** $\begin{bmatrix} 0 & 0 \\ 0 & 0 \end{bmatrix}$ **27.** $x = \frac{14}{3}, y = \frac{2}{3}, z = \frac{7}{3}$

29. (a) (i) $\begin{bmatrix} -1 & 0 & 2 \\ 0 & 4\sqrt{2} & -4 \\ -3 & \pi & 5 \\ 0 & 1 & 0 \end{bmatrix}$ (ii) $\begin{bmatrix} -3 & \pi & 5 \\ 0 & \sqrt{2} & -1 \\ -1 & 0 & 2 \\ 0 & 1 & 0 \end{bmatrix}$ (iii) $\begin{bmatrix} -1 & 0 & 2 \\ -3 & \sqrt{2} & 5 \\ -3 & \pi & 5 \\ 0 & 1 & 0 \end{bmatrix}$

(b) (i) $\begin{bmatrix} \frac{1}{12} & 0 & \frac{1}{6} & 1 \\ 3 & -4 & -2 & 0 \\ 0 & 2 & -4 & 5 \end{bmatrix}$ (ii) $\begin{bmatrix} 0 & 2 & -4 & 5 \\ \frac{3}{4} & -1 & -\frac{1}{2} & 0 \\ \frac{1}{12} & 0 & \frac{1}{6} & 1 \end{bmatrix}$

(iii) $\begin{bmatrix} \frac{1}{12} & 0 & \frac{1}{6} & 1 \\ 1 & -1 & 0 & 3 \\ 0 & 2 & -4 & 5 \end{bmatrix}$

31. $\begin{bmatrix} 1 & 0 \\ 0 & 1 \\ 0 & 0 \end{bmatrix}$ **33.** $(x, y) = \left(\frac{28 - 63t}{2}, t\right)$

35. The solution set is empty. **37.** (a) no (b) $\begin{bmatrix} \frac{1}{3} & \frac{7}{2} \\ -\frac{1}{14} & \frac{3}{4} \end{bmatrix}$

39. (a) π (b) 13 (c) $-6\sqrt{2}$ (d) $-6\sqrt{2}$ (e) 0 (f) -79
41. (a) 0 (b) 0 (c) -3 (d) -2

43. (a) $\begin{bmatrix} 6 & -5 & 4 \\ 9 & 8 & 7 \end{bmatrix}$ (b) $\begin{bmatrix} \frac{1}{2} & -5 & 4 \\ 0 & 1 & 6 \\ -1 & 3 & -\frac{1}{2} \end{bmatrix}$ (c) $[\pi \ \ 0 \ \ -4 \ \ i]$

(d) $\begin{bmatrix} 0 & \sqrt{3} \\ 1 & 0 \\ -4 & 6 \end{bmatrix}$

45. (a) $\begin{bmatrix} 2 & 1 & -8 \\ 0 & -5 & 40 \\ 8 & 4 & -42 \end{bmatrix}$ (b) $\begin{bmatrix} -27 & -\frac{4}{5} \\ -20 & -\frac{2}{3} \end{bmatrix}$

(c) $\begin{bmatrix} -24 & -\frac{4}{5} \\ -20 & -\frac{2}{3} \end{bmatrix}$ (d) $\begin{bmatrix} 0 & 0 & 0 \\ 0 & 0 & 0 \\ 0 & 0 & 0 \end{bmatrix}$

47. $x = \frac{5}{53}, y = -\frac{46}{53}, z = \frac{69}{53}$ **49.** $x = -1, y = -3, z = 5$
51. $(x_1, x_2, x_3, x_4) = (3, 2 + x_3 - 5x_4, x_3, x_4)$
55. 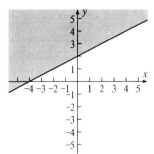 **57.**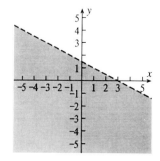

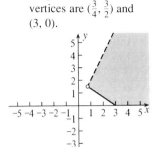

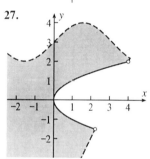

Answers to Exercises

59.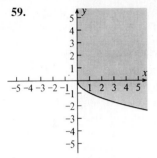

61. Graph is bounded; vertices are (0, 0), (0, 10), and (2, 8).

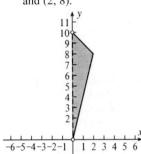

63. Graph is unbounded; vertices are (–1, 0) and (0, 1).

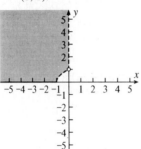

65. maximum = 56; minimum = 8
67. 150 of Brand A and 400 of Brand B

Section 10.1

1.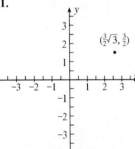

3.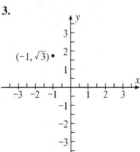

5.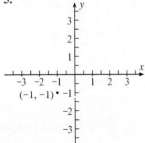

7.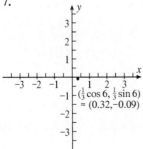

9.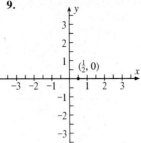

11.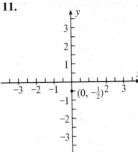

13.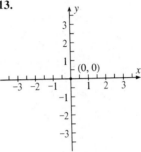

17. $\left(4, \dfrac{\pi}{3}\right)$ and $\left(-4, \dfrac{4\pi}{3}\right)$ **19.** $\left(4\sqrt{2}, \dfrac{5\pi}{4}\right)$ and $\left(-4\sqrt{2}, \dfrac{\pi}{4}\right)$

21. $(\sqrt{13}, 5.3003916)$ and $(-\sqrt{13}, 2.158798931)$

23. $(\sqrt{27}, 1.846439126)$ and $(-\sqrt{27}, 4.988031780)$

41. circle, center at the pole, radius 2

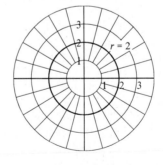

43. line, passing through the pole, slope = $\dfrac{1}{\sqrt{3}}$

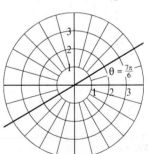

Answers to Exercises

45. circle, center at the pole, radius 3

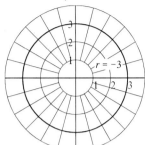

47. circle, center at (0, 4), radius 4

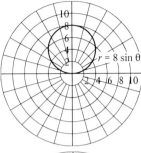

49. circle, center at $(\frac{1}{4}, 0)$, radius $\frac{1}{4}$

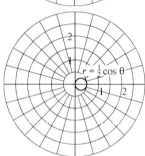

51. $r^2 = 25$ **53.** $r^2 - 18r \cos \theta = 0$ (or $r = 18 \cos \theta$)
55. $r = 2\sqrt{2} \sin \theta$ **57.** $r^2 + 4r \cos\left(\theta - \frac{5\pi}{3}\right) = 32$
59. radius $\frac{1}{2}$, center (1, 1) **61.** radius 5, center (–3, 4)
63. $r \cos \theta = \sqrt{3}$ **65.** $r \sin \theta = -\sqrt{3}$

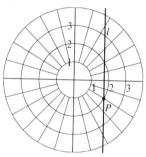

67. $r \cos\left(\theta - \frac{2\pi}{3}\right) = 5$

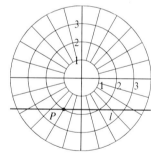

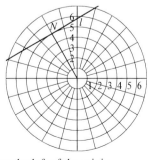

69. $x = -2$, vertical line, 2 units to the left of the origin

71. $\sqrt{2}x - 3y = 4$, line with slope $\frac{\sqrt{2}}{3}$ and y-intercept $(0, -\frac{4}{3})$
73. $x^2 + y^2 - 2x = 10$, circle, center at (1, 0), radius $\sqrt{11}$
75. $x = 2$, vertical line, 2 units to the right of the origin
77. The polar graph of $r(\theta) = \cos n\theta$ has $2n$ loops if n is even, and n loops if n is odd.

Section 10.2

1. cardioid

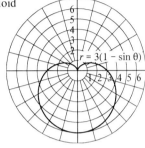

3. a 5-leaved rose, one petal is swept out by θ in $\left[\frac{\pi}{10}, \frac{3\pi}{10}\right]$

5. an 8-leaved rose, one petal is swept out by θ in $\left[0, \frac{\pi}{4}\right]$

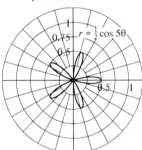

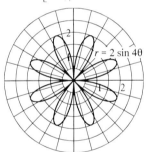

7. limaçon with two loops, the smaller loop is swept out by θ in $\left[\frac{5\pi}{3}, \frac{7\pi}{3}\right]$

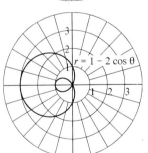

Answers to Exercises

9. limaçon

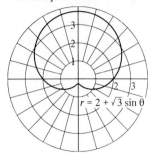

11. cardioid

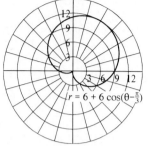

21.

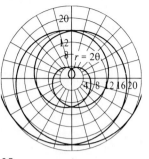

23.

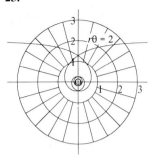

13. ellipse, one focus at the pole, corresponding directrix $x = 2$, eccentricity $e = \frac{3}{4}$

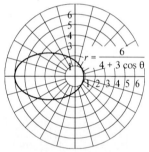

15. hyperbola, one focus at the pole, corresponding directrix $x = -3$, eccentricity $e = 4$

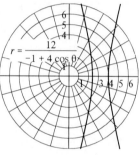

25.

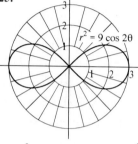

27. $r^2 \cos\theta \sin\theta = 4$ **29.** $r^2 = 9$ or $r = 3$ or $r = -3$
31. $x^2 - y^2 = 1$: an "east-west" hyperbola, center at origin, with $a = b = 1$
33. $y^2 = -4(x - 1)$: a parabola opening to the left, with vertex $(1, 0)$ and $p = 1$
35. symmetric with respect to the polar axis, $\frac{\pi}{2}$-line, and the pole
37. symmetric about the $\frac{\pi}{2}$-line
41. all points
43. $r = \dfrac{5\sqrt{2}}{1 + \sqrt{2}\cos\theta}$ **45.** $r = \dfrac{3}{2 - \sin\theta}$
47. $r = \dfrac{1}{4 + 4\sin\theta}$ **49.** $r = \dfrac{16}{5 + 3\sin\theta}$

17. parabola, focus at the pole, directrix $y = \frac{3}{2}$, eccentricity $e = 1$; exclude the points $(\frac{3}{2}, 0)$ and $(-\frac{3}{2}, 0)$ from the graph

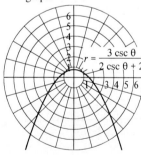

19. parabola, focus at the pole, directrix $r\sin\left(\theta - \dfrac{\pi}{4}\right) = -\sqrt{2}$, eccentricity $e = 1$

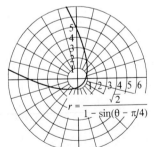

Answers to Exercises

Section 10.3

1. $(0, 0)$ and $(\frac{3}{2}, \frac{3}{2})$ are the points of intersection.

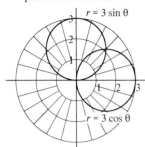

3. $(0, 0)$ is the point of intersection.

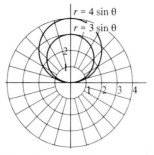

5. no points of intersection

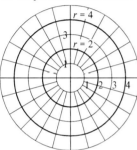

7. $(0, 0)$ and $(1, 1)$ are the points of intersection.

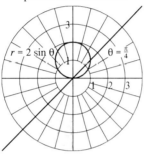

9. $(0, 0)$ is the point of intersection.

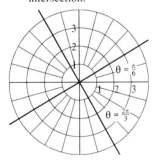

11. $(4, 0)$, $(0, 4)$, $(-4, 0)$, and $(0, -4)$ are the points of intersection.

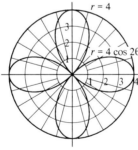

13. $(0, 3)$ and approximately $(2.8531696, 0.9270510)$, $(-2.8531696, 0.9270510)$, $(-1.7633558, -2.4270510)$, and $(1.7633558, -2.4270510)$ are the points of intersection.

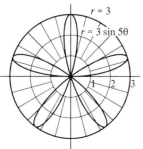

15. The approximate points of intersection are
$(1.6730326, 0.4482877)$,
$(-1.6730326, 0.4482877)$,
$(-1.6730326, -0.4482877)$,
$(1.6730326, -0.4482877)$,
$(0.4482877, 1.6730326)$,
$(-0.4482877, 1.6730326)$,
$(-0.4482877, -1.6730326)$,
and
$(0.4482877, -1.6730326)$.

17. $(0, 0)$

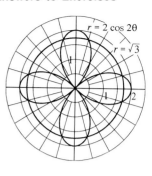

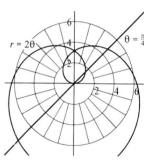

19. $(0, 0)$ and infinitely many intersection points corresponding to pairs of polar coordinates having forms such as
$\left(\frac{\pi}{2} + k4\pi, \frac{\pi}{4} + k2\pi\right)$ and
$\left(\frac{5\pi}{2} + k4\pi, \frac{5\pi}{4} + k2\pi\right)$ for
k a nonnegative integer

21. $(1, 0)$, $(-1, 0)$,
$\left(\frac{\sqrt{5}}{3}, \frac{-2}{3}\right)$, $\left(\frac{-\sqrt{5}}{3}, \frac{-2}{3}\right)$

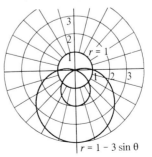

23. $(4, 0)$

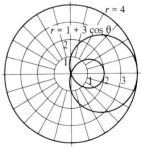

Answers to Exercises

25. $(0, -2)$, $(0.7336674, -1.3944487)$, and $(-0.7336674, -1.3944487)$

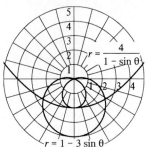

27. $(0, 0)$, $\left(\dfrac{\sqrt{3}}{4}, \dfrac{1}{4}\right)$, and $\left(\dfrac{-\sqrt{3}}{4}, \dfrac{1}{4}\right)$

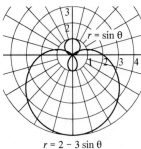

29. $(0, 0)$ and $(1, \sqrt{3})$

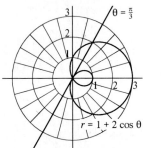

31. $(0, 0)$, $(0, 1)$, and $(0, -1)$

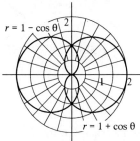

33. $(0, 0)$, $\left(\dfrac{1+\sqrt{2}}{2}, \dfrac{1+\sqrt{2}}{2}\right)$, and $\left(\dfrac{1-\sqrt{2}}{2}, \dfrac{1-\sqrt{2}}{2}\right)$

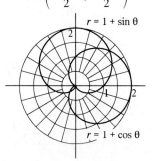

35. $(2, 0)$ and $(-2, 0)$

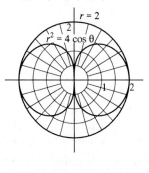

37. $(0, 0)$

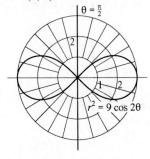

39. $(3, 0)$ and $(-3, 0)$

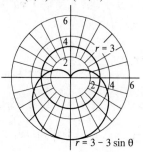

41. $(0, 0)$

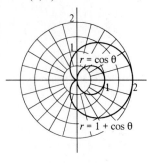

43. $(0, 0)$, $(2 - \sqrt{3}, 2\sqrt{3} - 3)$, and $(2 + \sqrt{3}, 2\sqrt{3} + 3)$

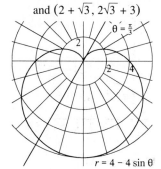

45. approximately $(0.4844562, 0.123702)$ and $(-0.4844562, 0.123702)$

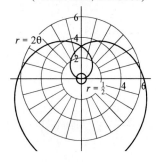

47. 3

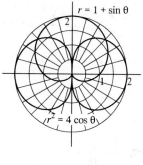

Section 10.4

1. $x = t$, $y = t^3 - t + 7$

3. $x = (2 + 3\sin\theta)\cos\theta$, $y = (2 + 3\sin\theta)\sin\theta$

5.

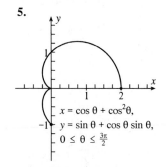

$x = \cos\theta + \cos^2\theta$, $y = \sin\theta + \cos\theta\sin\theta$, $0 \le \theta \le \dfrac{3\pi}{2}$

7. $y = \dfrac{x^2}{9} + 1$

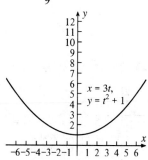

$x = 3t$, $y = t^2 + 1$

Answers to Exercises

9. $x = y^2 - 2y$
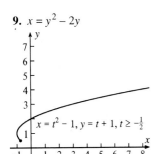

11. $y = 4x$
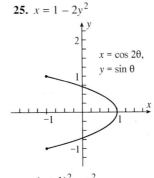

25. $x = 1 - 2y^2$

27. $y = 2$
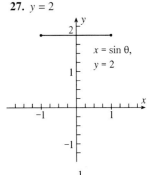

13. $x = (y-1)^4$
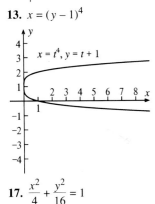

15. $y = x^2$
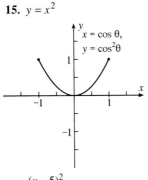

29. $\dfrac{(x+1)^2}{9} - \dfrac{y^2}{4} = 1$
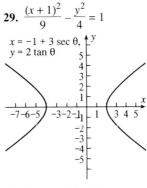

31. $x = 1 + \dfrac{1}{y-1}$
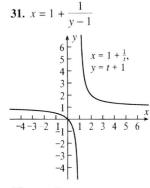

17. $\dfrac{x^2}{4} + \dfrac{y^2}{16} = 1$
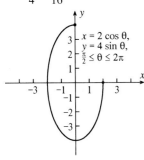

19. $\dfrac{(x-5)^2}{9} + (y-1)^2 = 1$
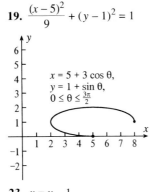

33. $2x - y - 11 = 0$ **35.** $y = -2x$
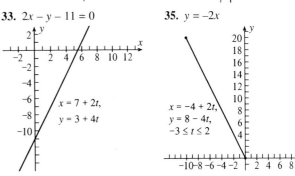

21. $(x-5)^2 + (y+3)^2 = 25$
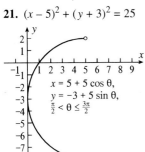

23. $x = y - 1$
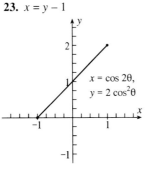

37. $x = \pi \cos \theta + 3,\ y = \pi \sin \theta - 2,\ I = [0, 2\pi)$
39. $x = \sqrt{2} \cos \theta,\ y = 2 \sin \theta,\ I = [0, 2\pi)$
41. $x = -2 + \sqrt{2} \sec \theta,\ y = 3 + \sqrt{3} \tan \theta,\ I = \left(-\dfrac{\pi}{2}, \dfrac{\pi}{2}\right) \cup \left(\dfrac{\pi}{2}, \dfrac{3\pi}{2}\right)$
43. $x = 3 - \dfrac{8}{7}t,\ y = 2 - \dfrac{4}{7}t^2,\ I = (-\infty, \infty)$
45. $x = 3t + 5,\ y = -8t + 2,\ I = (-\infty, \infty)$
47. $x = 5,\ y = t,\ I = (-\infty, \infty)$ **49.** $x = 4t,\ y = 3t + 2,\ I = (-\infty, \infty)$

Section 10.5

1. 46   **3.** 65   **5.** $-\dfrac{609}{125}$
27. (a) $\$100(1.07)^n$
    (b) $r = [100(1.07)^{n+1}] \div [100(1.07)^n] = 1.07$
29. (a) $e^k = 1.15$
31. (a) arithmetic, $d = 2$ (b) neither (c) geometric, $r = \dfrac{3}{4}$
    (d) arithmetic, $d = 0$ (e) arithmetic, $d = -\dfrac{3}{2}$ (f) neither
35. 3741   **37.** 414   **49.** (a) 70 (b) 120 (c) 12 600

Answers to Exercises

Section 10.6

1. 1, 8, 28, 56, 70, 56, 28, 8, 1
3. $a^8 + 8a^7b + 28a^6b^2 + 56a^5b^3 + 70a^4b^4 + 56a^3b^5 + 28a^2b^6 + 8ab^7 + b^8$
5. 220 7. $x^4 + 8x^3 + 24x^2 + 32x + 16$
9. $128x^7 - 2240x^6 + 16{,}800x^5 - 70{,}000x^4 + 175{,}000x^3 - 262{,}500x^2 + 218{,}750x - 78{,}125$
11. $\pi^5 x^5 - 15\pi^4 x^4 + 90\pi^3 x^3 - 270\pi^2 x^2 + 405\pi x - 243$
13. $x^{15} - \dfrac{12x^{10}}{\sqrt{y}} + \dfrac{48x^5}{y} - \dfrac{64}{y\sqrt{y}}$ 15. $3x^2h + 3xh^2 + h^3$
17. 120 19. 840 21. 364 23. 153 25. 28 27. $1365a^{11}b^4$
29. $-\dfrac{541{,}282{,}500}{x^6}$ 31. $(2{,}555{,}190)2^{86}x^4$ 33. $-(23{,}400)3^{24}x^{24}$
35. $147{,}420x^{24}y^4$ 37. $\binom{22}{16}(x^{2/3}y)^6(-y^2 x^{-1/4})^{16} = 74{,}613y^{38}$
39. (a) $\cos^9\theta - 36\cos^7\theta\sin^2\theta + 126\cos^5\theta\sin^4\theta - 84\cos^3\theta\sin^6\theta + 9\cos\theta\sin^8\theta + i(9\cos^8\theta\sin\theta - 84\cos^6\theta\sin^3\theta + 126\cos^4\theta\sin^5\theta - 36\cos^2\theta\sin^7\theta + \sin^9\theta)$
41. (a) $1 + 0.1 + 0.0045 + 0.000\,120 = 1.104\,620$
45. (a) 12 (b) $2m$ (c) $3n$ (d) mn 47. $2^4 = 16$
49. $\dfrac{50!}{(50-4)!} = 50 \cdot 49 \cdot 48 \cdot 47 = 5{,}527{,}200$
51. $\binom{50}{5} = 2{,}118{,}760$
53. (a) 30 (b) $\binom{n}{r}\binom{m}{s} = \dfrac{n!m!}{r!(n-r)!s!(m-s)!}$
55. (a) $\binom{50}{6} = 15{,}890{,}700$ (b) $\binom{49}{5} = 1{,}906{,}884$
(c) $\binom{49}{6} = 13{,}983{,}816$
(d) Each set of six finalists satisfies exactly one of the following two conditions: (i) it contains Dobbs-Peterson; (ii) it does not contain Dobbs-Peterson. So, by determining how many sets satisfy (i) and adding this to the number of sets satisfying (ii), you have counted (exactly once each) all the members in the collection of sets of finalists.

Section 10.7 (Review Exercises)

1.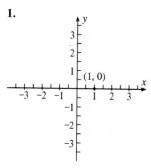

3. $\left(8, \dfrac{2\pi}{3}\right)$ and $\left(-8, \dfrac{5\pi}{3}\right)$
5. symmetric with respect to the polar axis
9. circle, center at the pole, radius 3

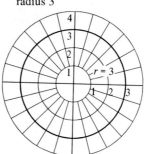

11. line, passing through the pole, slope $= -\sqrt{3}$

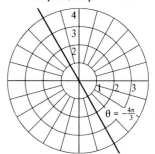

13. circle, center at (3, 0), radius 3

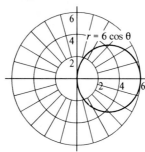

15. line, passing through $(\sqrt{2}, \sqrt{2})$, slope $= -1$

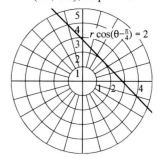

17. cardioid

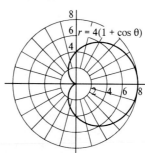

19. 4-leaved rose

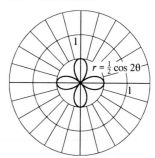

21. limaçon with two loops

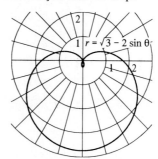

Answers to Exercises

23. hyperbola, one focus at the pole, corresponding directrix $y = \frac{2}{3}$, eccentricity $e = \frac{3}{2}$

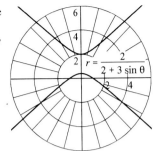

25. hyperbolic spiral

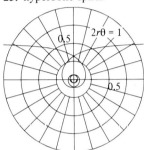

27.

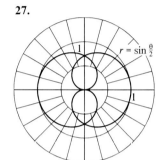

29. $4r^2 \cos^2 \theta + 9r^2 \sin^2 \theta = 36$
31. $r^2 - 2\sqrt{3}r \cos \theta + 2r \sin \theta - 1 = 0$ or
$r^2 + 4r \cos\left(\theta - \frac{5\pi}{6}\right) - 1 = 0$
33. $r = \dfrac{24}{5 - 3 \cos \theta}$
35. $(0, 0)$ and $(-1, 1)$ **37.** $(2, 0)$ and $(-2, 0)$

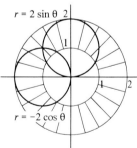

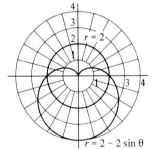

39. There are no points of intersection.

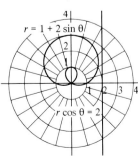

41. $x = 2t^3 + \frac{1}{3}t^2 + \sqrt{2}t - \pi,\ y = t$
43. $x + 2 = \frac{3}{25}y^2$ **45.** $y = 2x^2$

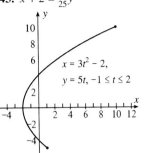

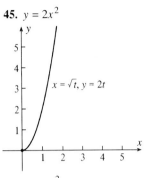

47. $x = y + 2$ **49.** $y = 2x^2 - 1$

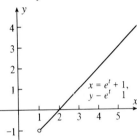

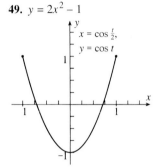

51. $x = \sqrt{3} \cos \theta + e,\ y = \sqrt{3} \sin \theta + \pi,\ I = [0, 2\pi)$
53. $x = 3 \sec \theta + 4,\ y = 2 \tan \theta - 5,\ I = \left(-\frac{\pi}{2}, \frac{\pi}{2}\right) \cup \left(\frac{\pi}{2}, \frac{3\pi}{2}\right)$
55. $x = t - 3,\ y = -3t + 2,\ I = (-\infty, \infty)$ **57.** $\frac{7}{4}$ **59.** $1 - (-2)^n$
67. (a) geometric, $r = \frac{1}{2}$ (b) neither (c) arithmetic, $d = -2$
  (d) arithmetic, $d = 5$ (e) geometric, $r = -\frac{3}{2}$ (f) neither
69. (a) 56 (b) 1640   **71.** $\frac{8}{729}$
73. (a) 1, 7, 21, 35, 35, 21, 7, 1
  (b) $a^7 + 7a^6b + 21a^5b^2 + 35a^4b^3 + 35a^3b^4 + 21a^2b^5 + 7ab^6 + b^7$
  (c) $a^7 + 7a^6b + \dfrac{7 \cdot 6}{2}a^5b^2 + \dfrac{7 \cdot 6 \cdot 5}{2 \cdot 3}a^4b^3 + \dfrac{7 \cdot 6 \cdot 5 \cdot 4}{2 \cdot 3 \cdot 4}a^3b^4$
  $+ \dfrac{7 \cdot 6 \cdot 5 \cdot 4 \cdot 3}{2 \cdot 3 \cdot 4 \cdot 5}a^2b^5 + \dfrac{7 \cdot 6 \cdot 5 \cdot 4 \cdot 3 \cdot 2}{2 \cdot 3 \cdot 4 \cdot 5 \cdot 6}ab^6 + b^7$
75. $x^5 + 10x^4 + 40x^3 + 80x^2 + 80x + 32$
77. $4x^3 + 6x^2h + 4xh^2 + h^3$
79. (a) $1 + 0.10 + 0.0040 = 1.1040$ (b) $1.104\,080\,803$
  (c) four

Index

Absolute value
 of complex numbers, 70, 74 (ex. 21–23), 364–365
 equation, 17–18
 function, 83
 inequalities, 27–28
 properties of, 4, 27
 of real numbers, 4
Addition
 of complex numbers, 67
 of functions, 120
 identities, 330–336, 358
 of matrices, 521
 of ordinates, 347–351
 of real numbers, 2
 of vectors, 404–405
Addition property of inequalities, 22
Additive identity of a matrix, 511
Additive inverse of a matrix, 521, 522 (ex. 46)
Adj(A), 528
Adjoint of a matrix, 528
Algebra
 Fundamental Theorem of, 163–164
 of matrices, 508
Ambiguous case, 389, 391–395, 396 (ex. 33)
Amplitude, 284–286, 295, 300 (ex. 14), 429
Angle, 248
 coterminal, 249
 of depression, 383–386
 of elevation, 266 (ex. 35–39), 282 (ex. 24, 26, 27), 383–386
 of inclination, 266 (ex. 28)
 initial side of, 248
 inscribed in semicircle, 396 (ex. 35), 424–425
 negative, 248
 positive, 248
 quadrantal, 249, 275
 reference, 276
 sides of, 248
 standard position of, 249
 terminal side of, 248
 between two vectors, 418–419
 vertex, 248
Angular speed, 255–256, 429
 relationship to linear speed, 256
Angular velocity (*see* Angular speed)
Antilogarithms, 223–224

Applications
 air pressure, 233, 235 (ex. 17–18)
 continuous compounding, 217
 decibel, 225 (ex. 72), 247 (ex. 24)
 doubling time, 234 (ex. 5), 247 (ex. 20–21)
 electrical, 233–234, 432–433
 ellipse, 462
 exponential functions, 214–217
 functions, 87
 harmonic motion, 433–435
 Hooke's Law, 31 (ex. 66)
 hyperbola, 473–474
 inverse square law of universal gravitation, 119 (ex. 38–39)
 linear functions, 116–117
 linear relations, 45–46, 54 (ex. 61–70)
 linear speed, 255–256
 linear system, 503, 547–550
 LORAN C, 474
 music, 431–432
 navigation, 385, 387 (ex. 26–28), 410–411
 Newton's Law of Cooling, 233, 235 (ex. 15–16), 247 (ex. 23)
 Newton's Second Law of Motion, 462
 parabola, 450–452
 population growth, 118 (ex. 7), 233, 234–235 (ex. 4–7), 247 (ex. 21)
 quadratic functions, 116–118
 radian measure, 254–256
 Richter, 225 (ex. 71)
 sinusoidal, 293–294 (ex. 43, 44), 428–435
 trigonometry, 381–440
 vectors, 410–414, 422–426
 whispering gallery, 462, 464 (ex. 53)
 work, 422–423
Arccos, 305
Archimedian spiral, 582 (ex. 21, 22)
Arcsin, 303
Arctan, 306
Area
 of circle, 402 (ex. 34)
 of parallelogram, 537 (ex. 37, 38)
 of regular *n*-gon, 402 (ex. 33)
 of sector, 254
 of triangle, 65 (ex. 60), 399, 401 (ex. 24–30), 438 (ex. 37), 537 (ex. 35–36)

Argument of a complex number, 365
Arithmetic sequence, 607 (ex. 31–42)
Associative property
 generalized, 3
 of matrix addition, 521
 of matrix multiplication, 511
 of real number addition, 3
 of real number multiplication, 3
 of vector addition, 404
Asymptote, 188–195
 definition, 188
 of exponential function, 209, 214
 horizontal, 190–192
 of hyperbola, 197 (ex. 57), 467–469, 483
 oblique, 188
 of rational function, 194
 slant, 188, 192–193
 of trigonometric graph, 299
 vertical, 188–190
Augmented matrix, 513
Average rate of change, 94
Axis
 conjugate, 466
 of ellipse, 454
 of hyperbola, 465
 imaginary, 364
 as line of symmetry, 98
 major, 454
 minor, 454
 of parabola, 112, 442
 real, 364
 rotation of, 480–486, 581–582
 semiconjugate, 466
 semimajor, 456
 semiminor, 456
 semitransverse, 466
 transverse, 465, 466

Back substitution, 495, 496–499
Base, 7
 of exponential function, 209–210
 of logarithmic function, 219
Bearing, 385, 387 (ex. 26–28), 410–411
Between, 3
Binomial Theorem, 377, 611–616
 compact statement of, 613
 *m*th term in, 613
 proofs of, 615–616, 618 (ex. 44), 619 (ex. 54)
 statement of, 611

Bound vector, 404
Boundary, 539
Bounded graph, 544
Bounds for real roots, 178–179
Branches of a hyperbola, 465
Butterfly, 577–578

$_nC_r$, 614
Calculators
   converting between degrees and radians, 251–254
   and e^x, 216
   and exponential functions, 210
   and factorials, 612, 614
   graphing, 87–92, 95–96 (ex. 80–81), 158–161, 162 (ex. 44), 172–173, 242–245, 246 (ex. 53), 271–272, 304, 354–356, 390, 444–446, 541–545, 562, 568–571, 612, 614
   and inverse trigonometric functions, 304, 308
   and limits of sequences, 608–609 (ex. 52, 54, 56, 58)
   and logarithms, 221, 224
   and nested multiplication, 148
   and trigonometric functions, 279–280, 390
Cardioid, 574
Cartesian coordinate system, 6–7
Cartesian coordinates, 6–7
   converting to polar coordinates, 560–562
Cartesian equation for a parameterized curve, 590
Cauchy-Schwarz inequality, 428 (ex. 53, 54)
Center
   of circle, 54
   of ellipse, 454
   of hyperbola, 465
Central rectangle, 468
Change of base relationship for logarithms, 221
Circle
   circumscribed, 396 (ex. 35), 438 (ex. 33, 34)
   definition, 54
   degenerate, 56
   inscribed, 402 (ex. 35), 438 (ex. 35, 36)
   intersections with, 59–61, 63–64
   parametric equation of, 273 (ex. 25–30), 593
   polar equation of, 564–566
   standard equation of, 55
   unit, 267
Circular functions, 259, 267–272
Cis θ, 365, 370 (ex. 24, 25, 27)
Closed half-plane, 539
Closure property
   of real number addition, 3
   of real number multiplication, 3
Coefficient matrix, 512

of kinetic friction, 266 (ex. 34)
Coefficients
   of polynomial, 8
   of polynomial function, 136
Cofactor, 524
   expansion method, 525, 535 (ex. 9)
Cofunction identities, 263
Cofunctions, 260
Collinearity, 37–38
Column vector, 509
Combinatorics, 618–619 (ex. 45–56)
Common logarithms, 220
Commutative law
   for dot product, 417
   of matrix addition, 521
   of real number addition, 3
   of real number multiplication, 3
   of vector addition, 404
Completing a square, 5
Complex number, 65–74
   absolute value, 70, 364
   argument, 365
   basic operations, 67
   conjugate, 68–70
   definition, 67
   dictionaries with planar vectors and some matrices, 416 (ex. 52), 428 (ex. 52), 522 (ex. 50), 537 (ex. 33)
   modulus, 365
   nth root of, 374
   polar form of, 363–369
   powers of, 371–374
   roots of, 374–377
   trigonometric form of, 363–369
Complex plane, 364
Complex polynomial, 136
Components, vector, 406–410
Composite function, 123, 129–130 (ex. 35, 36, 39)
Composition of functions, 123
Compound interest formula, 215
Computers
   and inverse trigonometric functions, 309
   trigonometric values on, 280
Conditional equation, 5, 311
Conic section, 441, 477–479, 578–582
Conjugate
   axis, 466
   of complex number, 68–70
   properties of, 69–70
   Root Theorem, 168–169
Consistent equation, 5
Consistent system, 493
Constant function, 85, 150
Constant polynomial, 8
Constraints, 546
Continuity, 149 (ex. 35), 150–151
Continuous compounding, 217
Continuous function, 150
Conversion
   Cartesian equation to parametric equation, 592–593
   degrees to radians, 251–254

parametric equation to Cartesian equation, 590–592
polar coordinates to rectangular coordinates, 559
radians to degrees, 251–254
rectangular coordinates to polar coordinates, 560–562
xy to $x'y'$ coordinates, 480, 488 (ex. 48)
$x'y'$ to xy coordinates, 480
Convex set, 546
Coordinate of a point, 2
Coordinates
   Cartesian, 6–7
   polar, 558–571
   rectangular, 6–7
Cost, 117
Coterminal angles, 249
Counterexample, 109 (ex. 61), 265 (ex. 19), 291 (ex. 4(d)), 311, 328, 330 (ex. 44–48, 55), 337 (ex. 43), 583 (ex. 39)
Cramer's Rule, 530–535, 538 (ex. 51)
Crucial values of θ, 573
Curtate cycloid, 595 (ex. 51)
Cycloid, 594, 595 (ex. 51)

d'Alembert, 163
Damped cosine, 352
Damped simple harmonic motion, 433–435
Damping factor, 433
De Moivre's Theorem, 371–374
   proof of, 603
Decay, exponential, 232–235
Decibel scale, 225 (ex. 72)
Decreasing direction, 2
Decreasing function, 85–87
Degenerate circle, 56
Degenerate conic sections, 442, 477–479, 487 (ex. 46), 538 (ex. 52)
Degenerate ellipse, 477
Degenerate hyperbola, 478–479
Degenerate parabola, 477–478
Degenerate segment, 99
Degree, 249
   converting to radians, 251–254
   of polynomial, 8, 135–136, 143 (ex. 43, 44)
Denominator, rationalizing, 14
Dependent system, 49, 493
Dependent variable, 79
Depressed equation, 164
Depression, angle of, 383–386
Descartes' Rule of Signs, 176, 184 (ex. 11)
Det A (see Determinants)
Detached coefficients, method of, 145
Determinants, 522–530
   elementary operations on, 525–527, 535–536 (ex. 10–12, 17)
   evaluating, 523, 525
   of order n, 522
   role in solving linear systems, 530–535
Difference identities, 358–361
Difference of vectors, 405

Index

Directed line segment, 403
Direction
  in aerial navigation, 385
  of real number line, 2
  of vector, 407
Directrix
  ellipse, 460
  hyperbola, 471
  parabola, 442
Discontinuous function, 150
Discriminant, 5, 62, 72, 163, 479, 487
  (ex. 41–42)
Disjoint sets, 1
Displacement vector, 403, 422
Distance
  between point and line, 46–47, 53
    (ex. 40), 423–424
  between two points, 4, 32–33
  formula, 32–34
Distributive property
  for dot product, 417
  generalized, 3
  of real number addition, 3
  of real number multiplication, 3
Dividend, 10, 139
Division
  algorithm, 139, 143 (ex. 45), 604–605
  of complex numbers, 67, 367–369
  of functions, 120
  of polynomials, 10–11
  of real numbers, 2
  synthetic, 143–147
Divisor, 10, 139
Domain, 79, 111, 120, 123, 125, 142
  (ex. 9), 185, 209, 214, 219, 241, 299, 307
Dot product, 416–417
  definition, 417
  properties of, 417
Double-angle identities, 338–342
Double-napped right-circular cone, 441, 477
Double root, 71
Double subscripts, 492, 507
Doubling time, 234 (ex. 5–6), 247
  (ex. 20–21)

e, 215–216, 616–617
  on calculators, 216
Eccentric angle, 593
Eccentricity
  ellipse, 460, 461, 472
  hyperbola, 471, 472
  parabola, 472, 582
Echelon form, 495–496, 498, 506–507
  (ex. 50)
Effect of elementary operations on a determinant, 526
Element
  of matrix, 507
  of set, 1
Elementary operations
  on determinant, 525–527

on linear system, 499–504, 505
  (ex. 11–12)
Elementary row operations on a matrix, 516
Elevation, angle of, 266 (ex. 35, 37, 38), 282 (ex. 24, 26, 27)
Eliminating parameter, 590–592
Elimination, Gauss-Jordan, 515–519
Ellipse, 454–462
  applications, 462
  center, 454
  degenerate, 477
  directrix, 454
  east-west, 455
  focal radius, 454
  focus, 454
  horizontal, 455
  latus rectum, 454, 463 (ex. 16)
  major axis, 454
  minor axis, 454
  north-south, 455
  parametric equations of, 463 (ex. 39, 40), 593
  semimajor axis, 456, 462–463 (ex. 14)
  semiminor axis, 456
  standard form equation of, 454–460
  vertex, 454
  vertical, 455
Ellipsoid of revolution, 462, 464 (ex. 55)
Empty set, 1
Endpoint of a ray, 248
Entry
  of cofactor, 524
  of matrix, 507
  of minor, 523–524
Epicycloid, 595 (ex. 52)
Equal functions, 82
Equal matrices, 508
Equal sets, 1
Equal vectors, 404
Equality of complex numbers, 67
Equation
  absolute value, 17–18
  conditional, 5, 311
  consistent, 5
  depressed, 164
  exponential, 236–238
  general linear, 45–46
  inconsistent, 5
  linear in n variables, 492
  linear in one variable, 5, 70, 163
  logarithmic, 238–240
  parametric (*see* Parametric equation)
  particular solution of, 5
  polar, 562–568, 573–582
  quadratic, 5, 163
  radical, 16–17
  root of, 5
  solution of, 5, 311
  trigonometric, 311–316
Equations
  of circles, 55, 370 (ex. 23), 506
    (ex. 41–42), 537 (ex. 34), 563–566, 593

  of ellipses, 454–462, 593
  of hyperbolas, 465–473, 595
    (ex. 41, 42)
  of lines (*see* Lines, equations of)
  of parabolas, 57, 443–444, 446–450, 595
    (ex. 43, 44)
Equilateral hyperbola, 471
Equilibrium, 411, 430, 433
Equivalent equations, 5
Equivalent inequalities, 22, 539
Equivalent systems, 496
Even function, 100
Expansion of $(a+b)^n$, 609
Exponent, 7
Exponential decay, 218 (ex. 39), 232–234, 247 (ex. 18–19)
Exponential equation, 236–238
Exponential function, 208–217, 230–234
  asymptote, 209, 214
  base of, 208–209
  on calculators, 210
  definition, 208–209
  domain, 209
  general characteristics of, 214
  graph of, 209–214
  and logarithmic function, 219
  range, 209
Exponential growth, 218 (ex. 33, 34, 37, 38, 40), 225 (ex. 73–76), 230–232, 246 (ex. 46–47), 247 (ex. 19–22)
Exponential inequalities, 240–241
Exponents, 7–8
  properties of, 7

Factor Theorem, 141
Factorial, 612
Factors of polynomials, 8–10, 141
Feasible point, 546
Feasible region, 546
Focal chord
  ellipse, 454
  hyperbola, 465
  parabola, 444
Focal radius
  ellipse, 454
  hyperbola, 465
Foci (*see* Focus)
Focus
  ellipse, 454
  hyperbola, 465
  parabola, 442
Force vector, 403
Fractional expressions, 11–13, 14
  rationalizing, 14
Fractional inequalities, 26–27, 29–30, 196
  (ex. 51–54)
Free vector, 404
Frequency, 429
Function
  absolute value, 83
  arccos, 305
  arccot, 307
  arccsc, 307

Index

arcsec, 307
arcsin, 303
arctan, 306
circular, 259, 267–272
composite, 123
constant, 85, 150
continuous, 150
decreasing, 85–87
defined by more than one equation, 83–84
definition, 79
difference, 120
discontinuous, 150
domain, 79
even, 100
exponential, 208–217, 230–234
greatest integer, 83
identity, 126
increasing, 85–87
inverse cosecant, 308
inverse cosine, 305–306
inverse cotangent, 308
inverse secant, 308
inverse sine, 302–305
inverse tangent, 306–307
inverse trigonometric, 302–309
linear, 94 (ex. 72), 110, 130 (ex. 41), 150
logarithmic, 218–224
natural domain, 79
objective, 546, 553 (ex. 43–44)
odd, 102
one-to-one, 128
periodic, 283
polynomial, 135
power, 136, 137
product, 120
quadratic, 111, 150
quotient, 120
range, 79
rational, 185–195
root of, 82
smooth, 151
sum, 120
trigonometric, 258–264
value at a point, 82
zero of, 82
Functional notation, 81–83
Functions
  addition of, 120
  composition of, 123
  division of, 120
  equal, 82
  inverse, 124–128
  multiplication of, 120
  operations on, 122–123
  products of, 351–354
  subtraction of, 120
  sum of periodic, 350
Fundamental Counting Principle, 618–619 (ex. 46–48)
Fundamental Theorem of Algebra, 163–164

Gauss, 163

Gauss-Jordan elimination, 515–519
General linear equation, 45–46
General quadratic equation, 476
Generalized associative property, 3
Generalized distributive property, 3
Geometric sequence, 601
Geometric triangle inequality, 34, 402 (ex. 37, 38, 40), 428 (ex. 55)
Graph
  of algebraic statement, 31
  of cardioid, 574
  of combinations of sine and cosine functions, 347–357
  of constant function, 150
  of cosine function, 270–272, 284
  of exponential functions, 209–214
  of inequalities, 22, 539
  of inverse cosecant, 307
  of inverse cosine, 305–306
  of inverse cotangent, 307
  of inverse function, 127
  of inverse relations, 126–128
  of inverse secant, 307
  of inverse sine, 303–305
  of inverse tangent, 306–307
  of limaçon, 576
  of linear functions, 94 (ex. 72), 110, 150
  of polar equations, 562–568, 573–582, 584–587
  of polynomial functions, 149–161, 167
  of products of functions, 351–354
  of quadratic functions, 111–116, 150
  of rational functions, 185–195
  of real numbers, 2
  of reciprocal trigonometric functions, 297–299
  of rose, 575
  of sine function, 282–284
  of tangent function, 294–297
  of zero polynomial, 150
Graphing calculator activity, 87–92, 95–97 (ex. 80–81), 158–161, 162–163 (ex. 44), 172–173, 242–245, 246 (ex. 53), 271–272, 346–347 (ex. 58), 354–356, 541–545, 568–571, 572 (ex. 77)
Greater than, 3
Greatest integer function, 83
Growth, linear, 45–46 (see also Exponential growth)
Guidelines
  for finding asymptotes of rational functions, 194
  for finding points of intersection of polar graphs, 587
  for graphing conic sections (rotated), 482
  for graphing polar equations, 573
  for graphing polynomials, 155
  for graphing rational functions, 194
  in a mathematical induction proof, 598
  for using the Law of Cosines, 397
  for using the Law of Sines, 389

Half-angle identities, 342–345
Half-life, 232, 235 (ex. 11)
Half-plane, 539
Harmonic motion, 428–435
Hero's formula (see Heron's formula)
Heron's formula, 65 (ex. 60), 399, 402 (ex. 37)
Hooke's law, 31 (ex. 66)
Horizontal asymptote, 190–192
Horizontal line
  equation, 44
  slope, 36, 40 (ex. 34)
  test, 127–128
Horizontal shrinking, 106–107, 211, 286–287
Horizontal stretching, 106–107, 211, 286–287
Horizontal translation, 103, 287–288
Hyperbola, 465–474
  applications, 473–474
  asymptotes, 197 (ex. 57), 467–469, 483
  branches, 465, 466, 474
  center, 465
  conjugate axis, 466
  degenerate, 478–479
  directrix, 471–472
  east-west, 466
  equilateral, 471
  focal chord, 465
  focal radius, 465
  focus, 465
  horizontal, 466
  latus rectum, 474
  north-south, 466
  parametric equation of, 475 (ex. 47), 595 (ex. 41, 42)
  rectangular, 471
  semiconjugate axis, 466
  semitransverse axis, 466
  standard form equations of, 465–473
  transverse axis, 465, 466
  vertex, 465, 475–476 (ex. 55)
  vertical, 466
Hyperbolic spiral, 583 (ex. 23)

i, 406
Identities, 5, 323
  addition, 330–336, 358
  cofunction, 263, 265 (ex. 27)
  double-angle, 338–342
  half-angle, 342–345
  product-to-sum, 358–359
  proving, 259–260, 324–329
  Pythagorean, 264, 324
  quotient, 259, 260, 324
  ratio, 259, 260, 324
  reciprocal, 259, 324
  subtraction, 330–336, 358, 402 (ex. 36)
  sum and difference, 358
  sum-to-product, 359–362, 363 (ex. 47–50)

Index

trigonometric, 323–329
Identity function, 126
Identity matrix, 511
Identity property
  of real number addition, 3
  of real number multiplication, 3
Imaginary axis, 364
Imaginary number, 66
Imaginary part of a complex number, 67
Imaginary unit, 66
Inclination, angle of, 266 (ex. 28)
Inconsistent equation, 5
Inconsistent system, 49–51, 493
Increasing direction, 2
Increasing function, 85–87
Independent variable, 79
Index
  of radical, 13
  of summation, 596
Induction
  basis, 598
  step, 598
Inequalities
  absolute value, 27–28
  equivalent, 22, 539
  exponential, 240–241
  fractional, 26–27, 29–30, 196 (ex. 51–54)
  greater than, 3
  greater than or equal to, 3
  involving products and quotients, 24–27
  less than, 3
  less than or equal to, 3
  linear, 22–24, 539–545
  logarithmic, 241–242
  polynomial, 179–183
  properties of, 21–22
  radical, 28–30
  rational, 196 (ex. 51–54), 197 (ex. 59)
  solving linear, 22–24, 539–541
  strict, 3, 539
  systems of, 541–545
  trigonometric, 316–318, 341–342
Inequality
  Cauchy-Schwartz, 428 (ex. 53, 54)
  geometric triangle, 34, 402 (ex. 37, 38, 40), 428 (ex. 55)
  graph of, 22, 539
  particular solution of, 22, 539
  solution of, 22, 539
  triangle, 4, 33–34
Infinite series, 609 (ex. 60)
Initial point of a vector, 403
Initial side of an angle, 248
Inner product, 417, 509
Integers
  negative, 1
  nonnegative, 1
  positive, 1
Interest, 214
Intermediate Value Theorem for Polynomials, 151
Intersection
  of circles, 63

of hyperbolas, 474, 476 (ex. 60)
of lines and circles, 59–61
of parabola with lines and circles, 63–64
of sets, 1
Interval
  closed, 3
  half-closed, 3
  half-open, 3
  open, 3
Invariant
  under rotation, 487 (ex. 35, 36)
  under translation, 487 (ex. 40)
Inverse
  cosecant, 308
  cosine, 305–306
  cotangent, 308
  elements, 3
  function, 124–128
  of matrix, 511–512
  of nonconstant linear function, 130 (ex. 41)
  relation, 124, 126–128, 131 (ex. 66)
  secant, 308
  sine, 302–305
  tangent, 306–307
  trigonometric functions, 302–309
Invertible matrix, 511
Irrational numbers, 2
Irreducible polynomial, 10
Irreducible quadratic, 180

j, 406
Jacobian, 522
Joule, James Prescott, 422

Kepler's Laws, 462
Key point, 25, 27, 179, 240, 316, 540

Latera recta (see Latus rectum)
Latus rectum
  ellipse, 454, 460
  hyperbola, 465
  parabola, 444, 452 (ex. 9)
Law of Cosines, 397–400
  proof of, 397
  when to use, 397, 398
Law of Sines, 387–395
  and the ambiguous case, 389, 391–395, 396 (ex. 33)
  cosine version of, 396 (ex. 31, 32)
  proof of, 388–389
  in terms of circumscribed circle, 396 (ex. 35)
  when to use, 389, 398
Laws of exponents, 7
LCM, 11
Leading coefficient, 8, 136, 152
Leading Term Test, 152–153
  proof of, 162 (ex. 26)
Least common multiple (LCM), 11
Lemniscate, 583 (ex. 25, 26)
Length of a line segment, 4

Limaçon, 576, 583 (ex. 50)
Limit
  of function, 153–157, 188–190, 192, 197 (ex. 57), 207 (ex. 53), 209, 216, 222, 281 (ex. 16–19), 300 (ex. 15, 16), 337 (ex. 45), 357 (ex. 22, 26–31), 402 (ex. 34), 434, 453 (ex. 46), 463 (ex. 42–44), 467, 475 (ex. 48, 50), 553 (ex. 43), 583 (ex. 23, 50), 617
  of sequence, 215–216, 218 (ex. 42, 43), 608 (ex. 51–60), 616
Limiting tendency (see Limit)
Line (see also Lines, equations of)
  of reflection, 98
  of symmetry, 97–98
Line segment
  directed, 403
  length of, 4, 33
Line symmetry, 97–98
Linear equations
  general, 45–46
  involving complex numbers, 70–71
  in n variables, 492
  in one variable, 5, 163
  point-slope form, 42
  slope-intercept form, 43
  system of (see Linear system)
Linear Factor Theorem, 165, 175 (ex. 53)
Linear function, 94 (ex. 72), 110, 130 (ex. 41), 150
Linear general equation, 45–46
Linear inequalities, 22–24, 539–545
  equivalent, 539
  solving, 22–24
  strict, 539
  systems of, 541–545
Linear programming, 538, 545–550
Linear relationship, 54
Linear speed, 255
  relationship to angular speed, 256
Linear system, 47–52, 491–504
  consistent, 48, 493
  dependent, 49, 493, 505 (ex. 7)
  echelon form, 496
  elementary operation on, 499–504, 505 (ex. 8)
  equivalent, 496
  general solution, 494
  inconsistent, 48, 51, 493, 538 (ex. 50)
  solving, 48
  solving by determinants, 530–535
  solving by Gauss-Jordan elimination, 512–518
  solving by substitution method, 49–52
Linear velocity (see Speed, linear)
Lines, equations of
  general form, 45–46
  horizontal, 44
  parametric, 593
  point-slope form, 42
  polar, 566–568
  slope-intercept form, 43

Index

in terms of angle of inclination, 266 (ex. 29)
in terms of normal vector, 427 (ex. 43)
using determinant, 536 (ex. 18–20)
vertical, 42
Logarithmic equation, 238–240
Logarithmic functions, 218–224
    definition, 219
    general characteristics of, 222
Logarithmic inequality, 241–242
Logarithmic spiral, 583 (ex. 24)
Logarithms
    on calculators, 221, 224
    change of base relationship, 221
    common, 220
    natural, 220
    properties of, 225–229
    using different bases, 221
LORAN C, 474
Lower bound, 178
Lowest terms, 11, 13

Magnitude of a vector, 403, 407
Main diagonal entries, 507
Major axis, 454
Mathematical induction, principal of, 598
Matrices (*see also* Matrix)
    algebra of, 508–512, 521, 555 (ex. 24–25)
    equal, 508
Matrix, 507
    addition, 521, 522 (ex. 46)
    adjoint, 528
    augmented, 513
    coefficient, 512
    element, 507
    elementary row operation on, 516
    entry, 507
    identity, 511
    inverse, 511–512, 518, 520 (ex. 17, 23), 529
    invertible, 511, 518, 521 (ex. 43), 530, 537 (ex. 30–32), 557 (ex. 53)
    $m \times n$, 507
    main diagonal entry, 507
    multiplication, 510
    nonsingular, 511
    partitioned, 518
    rank, 532
    reduced row-echelon form, 514–515
    scalar multiple, 508
    size, 507
    square of order n, 507
    submatrix of, 532
    subtraction, 521
    transpose, 527, 528, 536 (ex. 22), 556 (ex. 44)
    upper triangular, 526, 535 (ex. 8)
    Vandermonde, 536 (ex. 15)
Method
    of detached coefficients, 145

of partial fraction decomposition, 198
of repeated multiplication, 9
Midpoint formula, 34–36
    generalization of, 40 (ex. 32)
Minor axis, 454
Minor of entry a_{ij}, 523–524
Minutes, 249
Miscellaneous simplifications, 15
Modulus of a complex number, 365
Mollweide, 400, 402 (ex. 39–40)
Monomial, 8
Motion
    damped simple harmonic, 428–435
    simple harmonic, 433–435
Multiple roots, 164, 167
Multiplication
    of complex numbers, 67, 367–369
    of functions, 120
    of matrices, 510
    method of repeated, 9
    nested, 147–148
    of polynomials, 8
    properties of inequalities, 22
    of real numbers, 2
    scalar, 405
Multiplicative identity of a matrix, 511
Multiplicative inverse of a matrix, 512
Multiplicity of a root, 167, 175 (ex. 53)

n factorial, 612
Natural domain, 79, 111
Natural logarithms, 220
Nautical mile, 257 (ex. 21)
Navigation, 385, 387 (ex. 26–28), 410–411
Negative angle, 248
Negative direction, 2
Negative integers, 1
Negative multiplication property of inequalities, 22
Negative number, 2
Negative reciprocals result, 38–39, 41 (ex. 58), 338 (ex. 46), 427 (ex. 48)
Nested multiplication, 147–148, 604
Newton's Law of Cooling, 233, 235 (ex. 15–16), 247 (ex. 23)
Newton's Second Law of Motion, 462
Nonlinear systems
    of equations, 59–64
    of inequalities, 545
Nonnegative integers, 1
Nonsingular matrix, 511
Notation
    arc function, 311 (ex. 34)
    $_nC_r$, 614, 619 (ex. 50)
    $n!$, 612
    $\left(\dfrac{n}{r}\right)$, 614
    product, 608 (ex. 49, 50)
    summation, 596–598

nth roots
    of complex number, 374, 378 (ex. 28)
    of unity, 376, 378 (ex. 33)
Numerator, rationalizing, 14

Objective function, 546
Oblique asymptote, 188
Odd function, 102, 137
One-to-one correspondence, 2
One-to-one function, 128
Open half-plane, 539
Operations
    elementary, 499–504
    elementary row, 516
Opposite of a vector, 405
Ordinate, 7
Ordinates, addition of, 347–351
Origin, 2, 6
    as pole, 559
    symmetry about, 102
Orthogonal vectors, 420

Pair of polar coordinates, 559
Parabola, 57–59, 63–64, 112, 116, 150, 442–452
    axis, 112, 442
    basic types, 57
    degenerate, 477–478
    directrix, 442, 452 (ex. 10)
    focal chord, 444
    focus, 442
    intersection with lines and circles, 63–64
    latus rectum, 444
    parametric equation of, 453 (ex. 39–44)
    standard form equations of, 443–444, 446–450
    vertex, 57, 112, 116, 442, 452 (ex. 11), 453 (ex. 54)
Paraboloid of revolution, 451
Parallel vectors, 419
Parallelogram law of vector addition, 404
Parameter, 495, 588
Parameterized curve, 588
Parametric equation, 553 (ex. 42), 588–594
    of circle, 593
    converting to Cartesian equation, 590–592
    of curtate cycloid, 595 (ex. 51)
    of cycloid, 594
    deriving, 592–594
    of ellipse, 593
    of epicycloid, 595 (ex. 52)
    of hyperbola, 595 (ex. 41, 42)
    of line, 593
    of parabola, 595 (ex. 43, 44)
Partial fraction decomposition, 198
Partial fractions, 197–205
Particular solution
    of equation, 5
    of inequality, 22, 539
    of system, 48, 493
Pascal's Triangle, 609–610, 614–615, 619 (ex. 56)

Index

Period, 283, 286–287, 299, 430
Periodic functions, 283, 291 (ex. 3, 5), 292 (ex. 20), 357 (ex. 22)
   sum of, 350
Perpendicular vectors, 420
Phase constant, 429
Phase shift, 287–288, 292 (ex. 23), 295
Point-slope form for equation of a line, 42
Point symmetry, 99
Polar axis, 560
   symmetry about, 562
Polar coordinates, 558–571
   converting to rectangular coordinates, 559
Polar equation, 562–568, 573–582
   of Archimedian spiral, 582 (ex. 21–22)
   of butterfly curves, 577–578
   of cardioids, 574, 576
   of circles, 563–566
   of conic sections, 578–582
   graph of, 562–568
   guidelines for graphing, 573
   of hyperbolic spiral, 583 (ex. 23)
   of lemniscates, 583 (ex. 25, 26)
   of limaçons, 576, 583 (ex. 50)
   of lines, 566–568
   of logarithmic spiral, 583 (ex. 24)
   of roses, 574–577
   of spirals, 582 (ex. 21) , 583 (ex. 23, 24)
Polar form of a complex number, 363–369
Polar graphs
   definition, 562
   guidelines for, 573
   of inequality, 583 (ex. 41)
   intersection of, 584–587
Pole, 559
   symmetry about, 562
Polynomial, 8
   complex, 136
   constant, 8
   degree of, 8, 135–136
   inequalities, 179–183
   irreducible, 10
   nested multiplication format, 148
   real, 136, 185 (ex. 40)
   root of, 138
   zero, 8, 136
   zero of, 138
Polynomial function, 135ff
   definition, 135–136
   graphing of, 149–161, 167
Polynomials, 8–11
   division of, 8–10
   products of, 8–9
Population growth, 118 (ex. 7), 233, 234–235 (ex. 4–7), 247 (ex. 21)
Positive angle, 248
Positive direction, 2
Positive integers, 1
Positive multiplication property of inequalities, 22
Power function, 136

Powers
   of complex numbers, 371–374
   of i, 68
   Principle of, 16
Principal, 214
Principal argument of a complex number, 365
Principal nth root of b, 13
Principal square root of a negative real number, 66
Principle
   of mathematical induction, 598–605
   of powers, 16
Product
   of complex numbers, 368
   dot, 416–418
   of functions, 351–354
   identities, 358–361
   inner, 417, 509
   of matrices, 510
   notation, 608 (ex. 49, 50)
   of polynomials, 8–9
   scalar, 417
   -to-sum identities, 358–359
Profit, 117
Proj$_\mathbf{u}$ $\mathbf{v}$, 420–422
Projection of vectors, 420–422
Properties
   of absolute value, 4, 27
   of conjugates of complex numbers, 70
   of dot product, 417
   of exponents, 7
   of inequalities, 21–22
   of inverse functions, 125
   of logarithms, 225–229
   of radical inequalities, 28
   of radicals, 13–14
   of real numbers, 2–3
Proving identities, 324–329
Pure imaginary number, 66
Pythagorean identities, 264, 324

Quadrantal angle, 249, 275
Quadrants, 7
Quadratic equation, 5, 163
   involving complex numbers, 71–74
   roots of, 6, 163
Quadratic formula, 5, 163
Quadratic function, 111, 150
   graph of, 111–116, 150
Quasiperiod, 433
Quotient, 10, 139, 368
Quotient identities, 259, 260, 324

Radian, 250–256
   converting to degrees, 251–254
Radical equation, 16
Radical inequalities, 28–30
   properties for solving, 28
Radical sign, 13
Radicals, 13–14

   properties of, 13–14
Radicand, 13
Radius, 54
Radius vector, 387 (ex. 30)
Range, 79, 111, 115–116, 125, 209, 214, 219, 299, 307
Rank of a matrix, 532
Ratio identities, 259, 260, 324
Rational expression, 11
Rational function, 185, 196 (ex. 17, 18)
   guidelines for graphing, 194
Rational inequalities, 196 (ex. 51–54), 197 (ex. 59)
Rational numbers, 1
Rational Root Test, 170
Rationalizing
   denominator, 14
   fractional expressions, 14
   numerator, 14
Ray, 248
Real axis, 364
Real number line, 2
Real number system, 2
Real numbers, 2
Real part of a complex number, 67
Real polynomial, 136
Reciprocal identities, 259, 324
Rectangular coordinates, 6–7
   converting to polar coordinates, 560–562
Rectangular equation for a parameterized curve, 590
Rectangular hyperbola, 471
Reduced row-echelon form of a matrix, 514–515
Reference angle, 276
Region, feasible, 546
Relation, 78
   inverse, 124
Remainder, 10, 139
Remainder Theorem, 141
Repeated multiplication, method of, 9
Resultant vector, 405, 411
Revenue, 117
Richter scale, 225 (ex. 71)
Right triangle applications, 381–386
Roots
   of complex numbers, 374–377
   double, 71
   of functions, 82
   multiple, 164, 167
   of multiplicity k, 167
   of polynomials, 138
   of quadratic equations, 6, 163
   of real polynomials, 168
   of unity, 376
Roses, 575, 576, 577
Rotation of axes, 480–486, 581–582
Row vector, 508

Satisfy an equation, 5, 311
Scalar, 403

Index

multiple, 405, 508
multiplication, 405, 409
product, 417
quantity, 403
Scales
  decibel, 225 (ex. 72)
  Richter, 225 (ex. 71)
Second, 249
Sector, area of a, 254
Semiconjugate axis, 466
Semimajor axis, 456
Semiminor axis, 456
Semiperimeter, 399
Semitransverse axis, 466
Sequence
  arithmetic, 607 (ex. 31–42)
  geometric, 601–602, 606–607 (ex. 27–31, 39, 42)
  limit of (see Limit of sequence)
Set, 1
  convex, 546
  element, 1
  empty, 1
  solution, 5
Sets
  disjoint, 1
  equal, 1
  intersection of, 1
  union of, 1
Sides of an angle, 248
Sign chart method, 24–30, 179–183
  with absolute value inequalities, 27–28
  with fractional inequalities, 26–27, 29–30
  with logarithmic inequalities, 242
  with polynomial inequalities, 179–183
  with product inequalities, 24–26
  with radical inequalities, 28–30
  with trigonometric inequalities, 342
Signs
  Descartes' Rule of, 176
  variation of, 175
Simple harmonic motion, 428–435
  damped, 433–435
Simple trigonometric equation, 311
Simplification, miscellaneous, 15
Simplified rational expression, 186
Simultaneous particular solution, 493
Simultaneous solution, 493
Sine wave, 283
Sinusoidal, 283
Size of a matrix, 507
Slant asymptote, 188, 192
Slope, 36–40
  definition, 36
  of horizontal lines, 36, 40 (ex. 34)
  negative reciprocals result, 38–39
  of nonvertical parallel lines, 38
  of perpendicular lines, 38
  in terms of angle of inclination, 266 (ex. 28)
  of vertical lines, 36
Slope-intercept form for equation of line, 43
Smoothness, 151
Solution
  of equation, 5, 311
  of inequality, 22, 539
  simultaneous, 493
  simultaneous particular, 493
  by substitution, 49
Solution set, 5
  of inequality, 22
  of system, 48, 493
Solving
  absolute value equations, 17–18
  absolute value inequalities, 27–28
  conditional equations, 311
  equations, 5, 311
  exponential equations, 236–238
  exponential inequalities, 240–242
  linear inequalities, 22–24, 539–541
  logarithmic equations, 238–239
  logarithmic inequalities, 240–242
  polynomial inequalities, 179–183
  radical equations, 16–17
  radical inequalities, 28–30
  systems of linear equations, 48–52, 492–504, 512–518, 530–535
  systems of linear inequalities, 541–550
  trigonometric equations, 311–316
  trigonometric inequalities, 316–318
Special points, 347
Speed
  angular, 255
  linear, 255
Spiral
  Archimedean, 582 (ex. 21, 22)
  hyperbolic, 583 (ex. 23)
  logarithmic, 583 (ex. 24)
Square matrix, 507
Standard equation
  of circle, 55
  of ellipse, 454–462
  of hyperbola, 465–473
  of parabola, 443–444, 446–450
Standard position of an angle, 249
Submatrix, 532
Subset, 1
Substitution method
  to detect an inconsistent system, 51
  for solving a linear system of equations, 49–52
Subtraction
  of complex numbers, 67
  of functions, 120
  identities, 330–336, 358, 402 (ex. 36)
  of matrices, 521
  of real numbers, 2
  of vectors, 405, 409
Sum
  identities, 358–361
  -to-product identities, 359–362
  of vectors, 404–405
Summation
  formulas, 600
  notation, 596–598
Superposition, 431
Symmetry
  line, 97–98
  about origin, 102, 107 (ex. 12)
  about $\pi/2$-line, 562, 583 (ex. 40)
  point, 99, 109 (ex. 62–63)
  about pole, 562
  with respect to line $y = x$, 109 (ex. 61)
  with respect to polar axis, 562, 571 (ex. 31)
  with respect to x-axis, 98, 101
  with respect to y-axis, 98, 100–101
Synthetic division, 143–147
System
  of linear equations (see Linear system)
  of linear inequalities, 541–545
  of nonlinear equations, 59
  of nonlinear inequalities, 545

Tangent lines
  to circle, 61–62, 64 (ex. 28), 65 (ex. 38)
  to ellipse, 463 (ex. 37–38)
  to hyperbola, 475 (ex. 45–46)
  to parabola, 452 (ex. 35–36)
Terminal point of a vector, 403
Terminal side of an angle, 248
Test
  Horizontal Line, 127–128
  Leading Term, 152–153, 162 (ex. 26)
  lower bound, 178
  Rational Root, 170
  upper bound, 178
  Vertical Line, 84–85
Thales' Theorem, 425
Theorem
  Binomial, 611–616
  change of base for logarithms, 221
  Conjugate Root, 168
  De Moivre's, 372–374, 603
  Factor, 141
  Fundamental, of Algebra, 163–164
  horizontal translation, 103
  Intermediate Value, for polynomials, 151
  Linear Factor, 165, 175 (ex. 53)
  linear programming, 547, 553 (ex. 42)
  Mollweide's, 400
  Remainder, 141
  Thales', 425
Transitive property for inequalities, 22
Translations
  horizontal, 103, 287–288
  vertical, 104, 288
Transpose of a matrix, 527
Transverse axis, 465, 466
Triangle, Pascal's (see Pascal's Triangle)
Triangle inequality, 4, 33–34
Triangles
  oblique, 387, 397
  right, 381–390
Trichotomy property of real numbers, 4

Index

Trigonometric equation, 311–316
Trigonometric form of a complex number, 363–369
Trigonometric functions, 258–264
   graphing of, 282–299, 347–357
   and right triangles, 261–264
   values of, 273–280
Trigonometric identities, 323–329
Trigonometric inequalities, 316–318
Trigonometric inverse functions, 302–309
Trigonometric substitution, 330 (ex. 40–43)
Trigonometric values, 269–270, 273–280
Turning point, 151

Unbounded graph, 544
Union of sets, 1
Unit circle, 267
Unit vector, 404, 406, 408
Unity, roots of, 376
Universal gravitation, inverse square law, 119 (ex. 38–39)
Upper bound, 178
Upper triangular matrix, 526

Vandermonde determinant, 536 (ex. 15–16)
Variation of sign, 175
Vector, 403
   bound, 404
   column, 509
   components, 406–410
   direction of, 407
   displacement, 403, 422
   force, 404
   free, 404
   initial point of, 403
   magnitude of, 403, 407
   multiplication of by scalar, 405, 409, 415 (ex. 49–50), 508
   opposite of a, 405
   quantity, 403
   row, 508
   terminal point of, 403
   unit, 404, 406, 408
   velocity, 403
   zero, 404
Vectors
   addition of, 404–405, 409, 521–522 (ex. 44)
   angle between, 418–419
   difference of, 405, 409
   dot product of, 416–418, 509
   equal, 404
   inner product of, 417
   orthogonal, 420
   parallel, 419
   perpendicular, 420
   projection of, 420–422
   scalar product of, 417
   subtraction of, 409
Velocity (*see* Speed)
Velocity vector, 403
Vertex
   of angle, 248
   of convex set, 546
   of ellipse, 454
   of hyperbola, 465
   of parabola, 57, 112, 116, 442
Vertical asymptote, 188
Vertical line
   equation, 42
   slope, 36
   test, 84–85
Vertical shrinking, 105–107, 284–285
Vertical stretching, 105–107, 284–285
Vertical translation, 104, 288, 295
Vertices (*see* Vertex)

Whispering gallery, 462, 464 (ex. 53)
Work, 422–423

x-axis, 6
   symmetry about, 101
x-coordinate, 7

y-axis, 7
   symmetry about, 101
y-coordinate, 7

Zero
   of a function, 82
   of a polynomial, 138
   polynomial, 8, 136, 150
   product property, 3
   vector, 404

QUESTIONS:
1. p67 part (v)

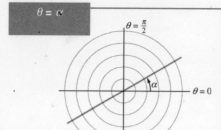
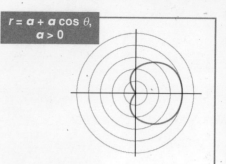

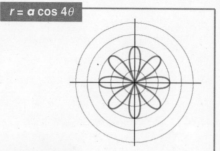
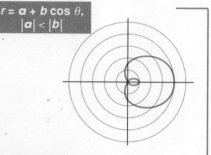
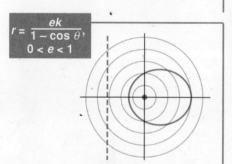
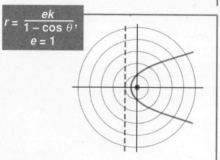
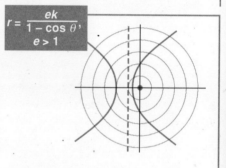

Basic Operations on Complex Numbers

1. $a + bi = c + di$ if and only if $a = c$ and $b = d$
2. $(a + bi) + (c + di) = (a + c) + (b + d)i$
3. $(a + bi) - (c + di) = (a - c) + (b - d)i$
4. $(a + bi)(c + di) = (ac - bd) + (ad + bc)i$
5. $\dfrac{a+bi}{c+di} = \dfrac{ac+bd}{c^2+d^2} + \dfrac{-ad+bc}{c^2+d^2}i$, if $c + di \neq 0$
6. $\overline{a+bi} = a - bi$
7. $|z| = |a + bi| = \sqrt{a^2 + b^2}$

Summation Formulas

$\sum_{i=1}^{n} i = \dfrac{n(n+1)}{2}$

$\sum_{i=1}^{n} i^2 = \dfrac{n(n+1)(2n+1)}{6}$

$\sum_{i=1}^{n} i^3 = \left[\dfrac{n(n+1)}{2}\right]^2$

Sequences

Arithmetic Sequence:
$a_i = a_1 + (i-1)d;$
$a_1 + a_2 + a_3 + \cdots + a_n = na_1 + \dfrac{(n-1)nd}{2}$

Geometric Sequence:
$a_i = a_1 r^{i-1};$
$a_1 + a_2 + a_3 + \cdots + a_n = \dfrac{a_1(1-r^n)}{1-r}$, if $r \neq 1$

Binomial Theorem

$(a+b)^n = \sum_{r=0}^{n} \dfrac{n!}{(n-r)!r!} a^{n-r} b^r$

Absolute Value

$|a| = \begin{cases} a & \text{if } a \geq 0 \\ -a & \text{if } a < 0 \end{cases}$

$-|a| \leq a \leq |a|$

$|ab| = |a||b|$

$|a+b| \leq |a| + |b|$

$|x| < a \Leftrightarrow -a < x < a$

$|x| > a \Leftrightarrow$ either $x > a$ or $x < -a$

Factoring Methods

1. $ax + ay = a(x + y)$
2. $x^2 \pm 2xy + y^2 = (x \pm y)^2$
3. $x^2 - y^2 = (x - y)(x + y)$
4. $x^3 + y^3 = (x + y)(x^2 - xy + y^2)$
5. $x^3 - y^3 = (x - y)(x^2 + xy + y^2)$

Completing the Square

$x^2 + kx = \left(x + \dfrac{k}{2}\right)^2 - \left(\dfrac{k}{2}\right)^2$

Quadratic Formula

$ax^2 + bx + c = 0,\ a \neq 0 \Rightarrow$

$x = r_1,\ r_2 = \dfrac{-b \pm \sqrt{b^2 - 4ac}}{2a},$

and $ax^2 + bx + c = a(x - r_1)(x - r_2)$

Composition of Functions

$(f \circ g)(x) = f(g(x))$